Annual Session of Scientific Papers IMT ORADEA 2018

AA003925

MATEC Web of Conferences Volume 184 (2018)

Baile Felix, Romania
31 May - 1 June 2018

Editors:

G. Grebenisan **A. V. Pele**

ISBN: 978-1-5108-6858-8

Printed from e-media with permission by:

Curran Associates, Inc.
57 Morehouse Lane
Red Hook, NY 12571

Some format issues inherent in the e-media version may also appear in this print version.

This work is licensed under a Creative Commons Attribution license:
http://creativecommons.org/licenses/by/2.0/
You are free to:
Share – copy and redistribute the material in any medium or format.
Adapt – remix, transform, and build upon the material for any purpose, even commercial.
The licensor cannot revoke these freedoms as long as you follow the license terms.
Under the following terms:
You must give appropriate credit, provide a link to the license, and indicate if changes were made. You may do so in any reasonable manner, but not in any way that suggests the licensor endorses you or your use. The copyright is retained by the corresponding authors.

Printed by Curran Associates, Inc. (2018)

For additional information, please contact EDP Sciences – Web of Conferences at the address below.

EDP Sciences – Web of Conferences
17, Avenue du Hoggar
Parc d'Activité de Courtabœuf
BP 112
F-91944 Les Ulis Cedex A
France

Phone: +33 (0) 1 69 18 75 75
Fax: +33 (0) 1 69 28 84 91

contact-edps@webofconferences.org

Additional copies of this publication are available from:

Curran Associates, Inc.
57 Morehouse Lane
Red Hook, NY 12571 USA
Phone: 845-758-0400
Fax: 845-758-2633
Email: curran@proceedings.com
Web: www.proceedings.com

TABLE OF CONTENTS

Statement of Peer review

In submitting conference proceedings to MATEC Web of Conferences, I certify to the Publisher that I adhere to the **Policy on Publishing Integrity** of the journal in order to safeguard good scientific practice in publishing.

1. All articles have been subjected to peer review administered by the proceedings editors.
2. Reviews have been conducted by expert referees, who have been requested to provide unbiased and constructive comments aimed, whenever possible, at improving the work.
3. Proceedings editors have taken all reasonable steps to ensure the quality of the materials they publish and their decision to accept or reject a paper for publication has been based only on the merits of the work and the relevance to the journal.

Title, date and place of the conference

ANNUAL SESSION OF SCIENTIFIC PAPERS IMT ORADEA 2018,
May 31st-June 1st, Felix SPA

Proceedings editor(s):

GAVRIL GREBENIŞAN

ALEXANDRU VIOREL PELE

Date and editor's signature

https://doi.org/10.1051/matecconf/201818401001

Optimizing the mechanical device of a mono-axial sun tracking mechanism

Cătălin Alexandru*

Transilvania University of Braşov, Product Design, Mechatronics and Environment Department, 500036 Braşov, Romania

Abstract. The paper deals with the dynamic optimization of the mechanical device for a mono-axial sun tracking system, with application for photovoltaic and solar-thermal systems, which convert the solar radiation in electrical or respectively thermal energy. The actuating of the mono-axial tracking mechanism is carried out by using a linear actuator, the optimization study intending to determine the optimal configuration of the mechanism with the aim to minimize the motor force developed by actuator in order to perform the imposed step-by-step tracking algorithm (motion law). The design variables in the optimal design process are represented by the global coordinates of the connection points of the linear actuator to the adjacent parts (the solar panel and the fixed support). The optimization study is conducted in virtual prototyping environment.

1 Introduction

In order to increase the efficiency of the solar panels (photovoltaic and solar thermal panels/collectors), a commonly used solution consists of equipping them with tracking systems (also called solar trackers), which allow the panel/collector to follow the trajectory of the sun on the celestial vault, so as to capture a greater amount of incident solar radiation [1-4]. By using such systems, the energetic efficiency of the solar panels/collectors can be significantly increased, in correlation with the specific geographic and climatic conditions [5-7].

The active tracking mechanisms of the solar panels are mechatronic systems, which integrate mechanical and electronic components. Generally, the optimization of the dynamic behavior of the tracking systems addresses two aspects (corresponding to the two main components of the mechatronic product):

- the geometric optimization of the mechanical structure (device), in order to minimize the driving forces developed by the actuators for performing the required motion (tracking) laws;
- the optimization of the control system in order to reduce the transient period, and minimize the tracking errors.

The both optimization ways aim, in fact, to reduce the energy consumption necessary to achieve the orientation, with a beneficial effect on the energy efficiency of PV tracking system. For the present work, the study is focused on the first optimization way (optimal design of the mechanical device). The dynamic model of the solar tracker was designed in mechatronic concept, by using MBS (Multi-Body Systems) and DFC (Design for Control) software solutions (namely, ADAMS and EASY5 of MSC.Software).

2 Setting the optimization problem

The use of a virtual prototyping platform offers the possibility to easily optimize the behavior of the mechanical & mechatronic systems by quickly testing a multitude of design variants (in terms of location, orientation and characteristics of the specific elements, mass distribution and so on), at reduced costs compared to the optimization process based on building & testing of physical (experimental) prototypes [8-10].

The optimization of the solar panel tracking systems from the point of view of the geometrical configuration of the mechanical device is based on the following stages: parameterizing the MBS virtual model; modeling the design variables, constraints and objectives; carrying out parametric design studies to identify the main design variables (with significant influence on the dynamic behavior of the tracking system, which is expressed by the design objectives); optimizing the virtual model by identifying the optimal values of the main design variables that assure, by case, the minimization or maximization of the design objective functions.

In this study, the goal of the dynamic optimization of the solar panel tracking systems is to minimize the motor force developed by the driving actuator, which in fact leads to a reduction in energy consumption (demand) to achieve the orientation. The case study is developed for a mono-axial tracking system, with one degree of freedom (DOF), corresponding to the diurnal movement of the solar panel, while the elevation is maintained at a fixed position. The solar panel is operated by means of a linear actuator disposed between the panel frame and the fixed support, the step-by-step motion law being modeled so as to achieve a minimum number of actuations (for energy efficiency considerations).

The virtual model of the mechanical device of the mono-axial solar tracker, which was developed by using the MBS software solution ADAMS, is shown in figure 1.

* Corresponding author: calex@unitbv.ro

© The Authors, published by EDP Sciences. This is an open access article distributed under the terms of the Creative Commons Attribution License 4.0 (http://creativecommons.org/licenses/by/4.0/).

On the other hand, the control system diagram block (figure 2) was developed by using a specific DFC (Design for Control) software solution (EASY5), the detailed description of the control device following to be presented in a future paper.

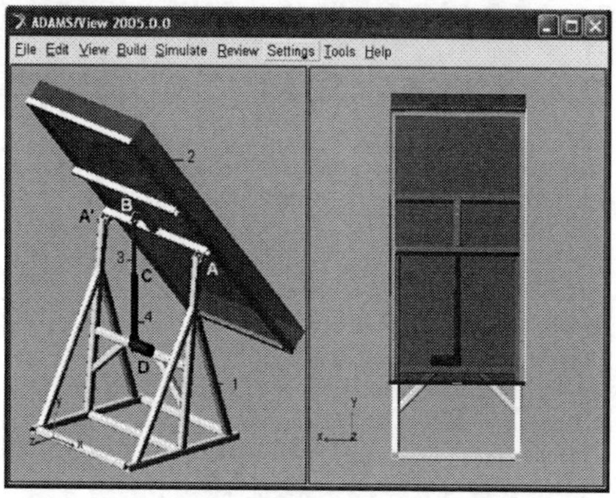

Fig. 1. The MBS model of the tracking mechanism.

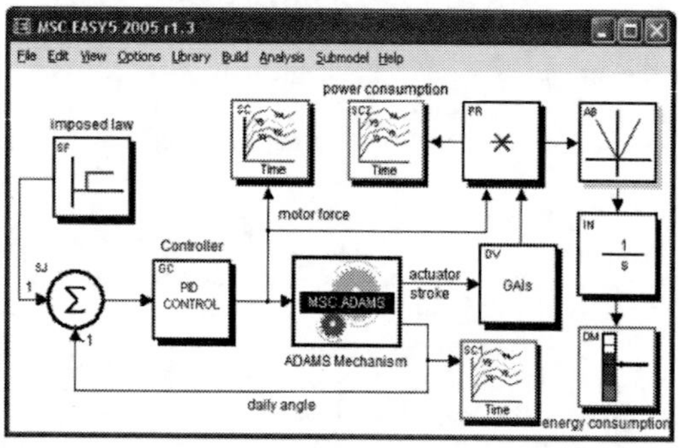

Fig. 2. The DFC model of the control system.

The controlled parameter (which is input in the control system model) is represented by the diurnal angle of the solar panel, while the motor force developed by the linear actuator defines the input in the mechanical device model. The communication between the mechanical & control components is managed through the ADAMS/Controls interface.

For the present paper, the numerical simulations are made considering the data specific to the Braşov geographical area (latitude - 45.6333 North, longitude - 25.5833 East), on the day of the summer solstice (21 of June, sunshine local time - 5.466, sunset local time - 21.183). The step-by-step motion (tracking) program was designed in accordance with the algorithm presented in [5], as follows: the angular domain for the diurnal movement - $\beta^* \in [64°, -64°]$, where $\beta^*=0°$ corresponds to the solar noon position; the actuating timing (in local time): 9.319 ($\Delta\beta^*=14°$), 10.319 ($\Delta\beta^*=14°$), 11.239 ($\Delta\beta^*=12°$), 12.089 ($\Delta\beta^*=12°$), 12.919 ($\Delta\beta^*=12°$), 13.719 ($\Delta\beta^*=12°$), 14.549 ($\Delta\beta^*=12°$), 15.399 ($\Delta\beta^*=12°$), 16.319 ($\Delta\beta^*=14°$), 17.319 ($\Delta\beta^*=14°$); return to the initial position (local time): 21.183.

For these input data, the time history variations of the daily angle, which is measured in the revolute joint A-A' between panel and support, and the corresponding angular velocity at which the diurnal movement is performed are shown in figure 3. By the inverse kinematics of the tracking mechanism, the correlation between the position angle of the solar panel and the linear stroke in the driving actuator is established.

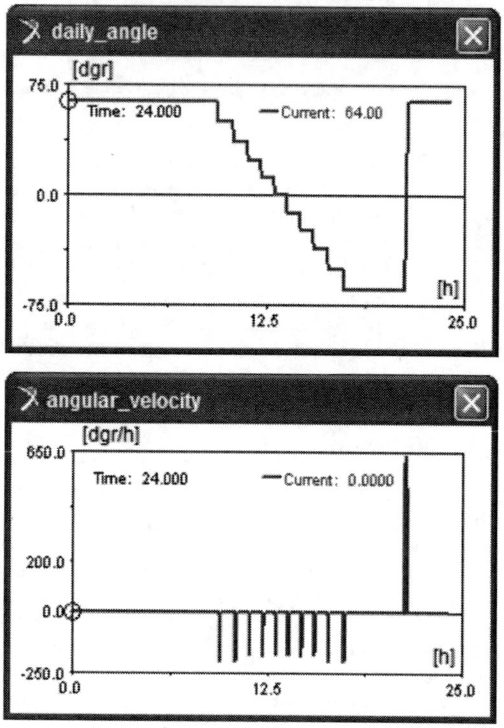

Fig. 3. The step-by-step tracking law.

The parameterization of the virtual model of the tracking system shown in figure 1 is based on the design points that define the structural scheme of the system, in this case the locations of the joints between bodies. For the optimization study, the connection points of the linear actuator to the adjacent elements (piston to panel frame - B, and cylinder to support - D) have been considered. The locations of the revolute joints between the panel frame and the fixed support (A - A ') are considered to be established on constructive criteria.

In this way, 6 design variables, which will control the model during the optimization, have been obtained, namely the global Cartesian coordinates (in the global reference frame XYZ) of the design points B and D: DV_1 → X_B, DV_2 → Y_B, DV_3 → Z_B, DV_4 → X_D, DV_5 → Y_D, and DV_6 → Z_D. Each design variable is defined by an initial value, and a variation range relative to the initial value, which was set in view of maintaining the system within acceptable constructive limits.

In order to maintain the relative location and orientation in the translational joint between the two components of the actuator (cylinder and piston), the two joint coordinate systems have been parameterized using the following expressions: LOC_RELATIVE_TO ({0, 0, 0}, POINT_C) - the function sets the location of the coordinate systems that define the translational joint relative to the position of the design point C; ORI_ALONG_AXIS (POINT_D, POINT_B, "Z") - the function sets the Z-axis orientation of the coordinate systems, representing the translational motion axis, in the direction defined by the design points D and B.

On the other hand, a design constraint has been defined to control the value of the pressure angle measured between the force vector generated by the actuator and the linear velocity vector on the solar panel at point B. Theoretically, the tracking mechanism self-locks when the pressure angle reaches 90°, but practically, due to the friction, the self-locking occurs at values less than the theoretical boundary. For safety reasons, the maximum acceptable value of the pressure angle was set at 65°. In ADAMS, the sign of the measure defining a constraint has a special significance. For example, if a constraint is positive, then it is violated, or if it is negative, the constraint has not been violated. In these terms, the design constraint was defined by a measure that subtracts the maximum accepted value from the current value of the pressure angle.

3 Results and conclusions

The parametric design studies have been carried out to determine the sensitivity of the objective function (motor force) to the modification of the six design variables. Such a study is performed for each variable, by swapping it in the specific variation domain (value range), considering the other variables with the initial (standard) value. As result of these studies, the main design variables (with significant influence on the design objective) have been identified, namely DV_2, DV_3 and DV_6. The other design variables (DV_1, DV_4 and DV_5) are secondary variables (with small influence on the design objective), which will be neglected in the optimization process (they will be considered with the initial/standard values).

In the initial design, before optimization, the dynamic behaviour of the tracking system is defined by the results shown in figure 4 (the time history variations of the motor force, power & energy consumption, and pressure angle). The optimization of the mechanical device was performed on the basis of the main variables, by seeking the optimal values of these variables in the considered value ranges (by simultaneously changing all the main design variables DV_2, DV_3 and DV_6). The optimal design, after optimization, is defined by the results shown in figure 5.

The optimization leads to a significant reduction in power/energy consumption for performing the tracking, with minor changes in the geometric configuration of the mechanism. At the same time, the pressure angle remains within the required limits.

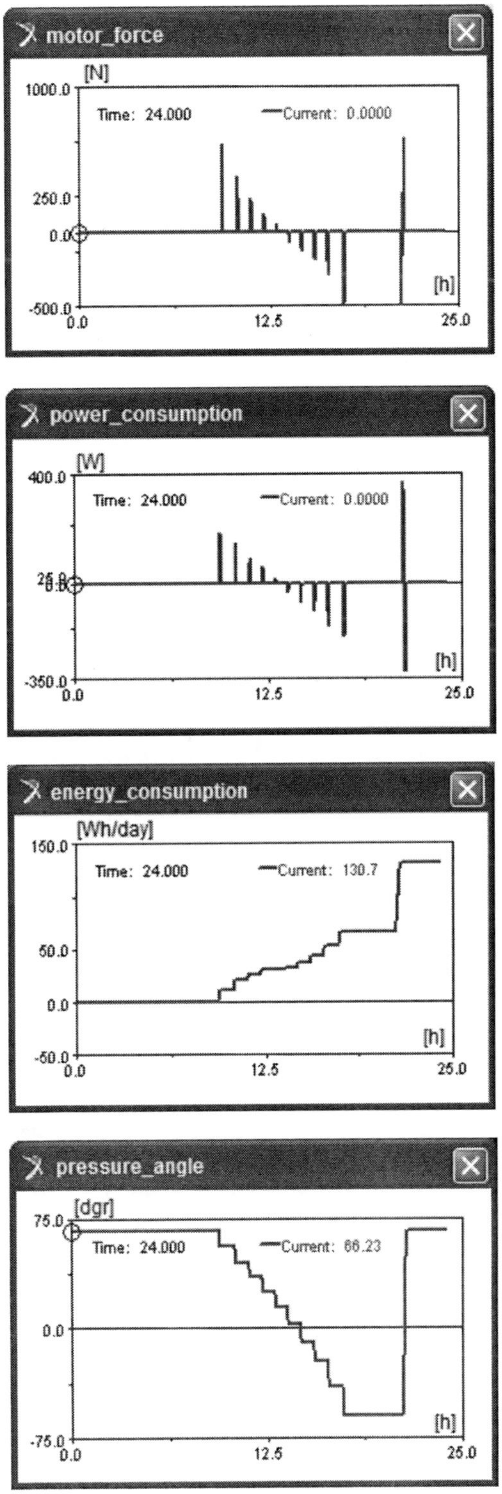

Fig. 4. Results specific to the initial design of the mechanism (before optimization).

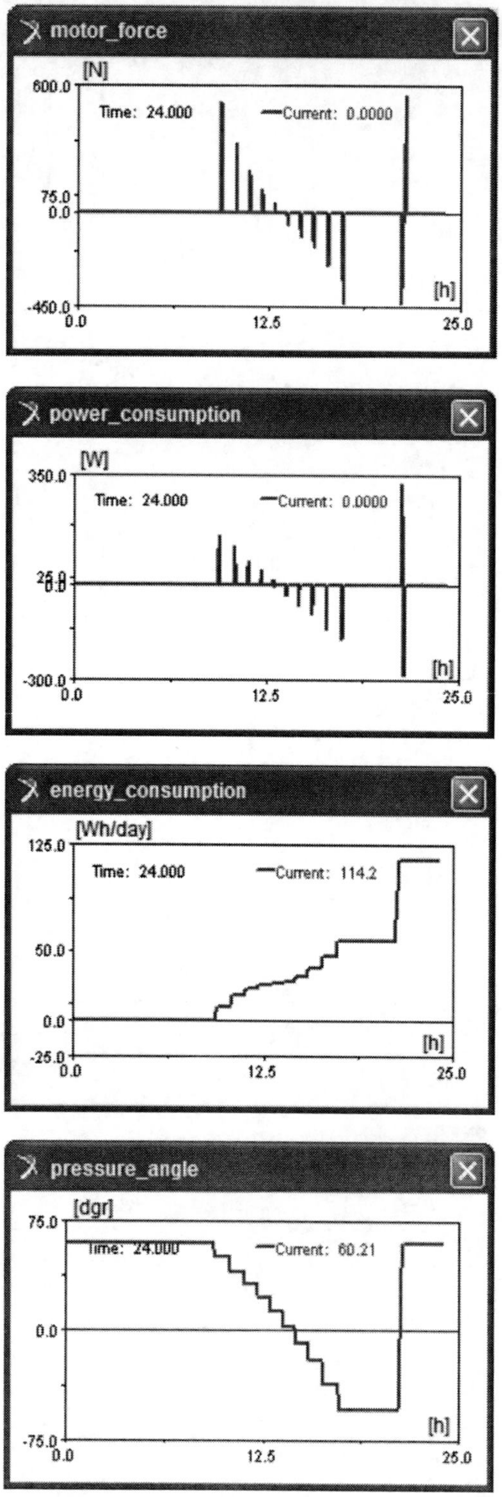

Fig. 5. Results specific to the optimal design of the mechanism (after optimization).

The application presented in this work is a suggestive example on the facilities offered by the virtual prototyping technique in the optimal design process of the mechatronic tracking systems of the solar panels. The virtual prototype of the solar tracker allows the assessment of various tracking strategies (in terms of angular motion domain, number of actuation steps, actuating timing) in order to increase the energy efficiency and reliability of the sun tracking mechanism (by maximizing the solar radiation input, minimizing the power/energy consumption to achieve orientation, minimizing the operating time of the driving actuators).

References

1. C. Alexandru, M. Comşiţ. *The energy balance of the photovoltaic tracking systems using virtual prototyping platform*. 5th IEEE International Conference on the European Electricity Market EEM, pp. 253-258 (2008)
2. C. Alexandru. *The design and optimization of a photovoltaic tracking mechanism*. 2nd IEEE International Conference on Power Engineering, Energy and Electrical Drives POWERENG, pp. 436-441 (2009)
3. D.A. Flores-Hernández, S. Palomino-Resendiz, N. Lozada-Castilloc, A. Luviano-Juárez. *Mechatronic design and implementation of a two axes sun tracking photovoltaic system driven by a robotic sensor*. Mechatronics, **47**, pp. 148-159 (2017)
4. F.M. Hoffmann, R.F. Molz, J.V. Kothe, E.O.B. Nara, L. Pablo, C. Tedesco. *Monthly profile analysis based on a two-axis solar tracker proposal for photovoltaic panels*. Renewable Energy, **115**, pp. 750-759 (2018)
5. C. Alexandru, I.N. Tatu. *Optimal design of the solar tracker used for a photovoltaic string*. Journal of Renewable and Sustainable Energy, **5**, pp. 1-16 (2013)
6. S. Seme, B. Stumberger, M. Hadziselimovic. *A novel prediction algorithm for solar angles using second derivative of the energy for photovoltaic sun tracking purposes*. Solar Energy, **137**, pp. 201-211 (2016)
7. S.A. Sharaf Eldin, M.S. Abd-Elhady, H.A. Kandil. *Feasibility of solar tracking systems for PV panels in hot and cold regions*. Renewable Energy, **85**, pp. 228-233 (2016)
8. C. Alexandru, C. Pozna, C. *Dynamic modeling and control of the windshield wiper mechanisms*. WSEAS Transactions on Systems, **8**, pp. 825-834 (2009)
9. C. Alexandru, P. Alexandru. *The virtual prototype of a mechatronic suspension system with active force control*. WSEAS Transactions on Systems, **9**, pp. 927-936 (2010)
10. D. Tarniţă, M. Catană, D.N. Tarniţă. *Contributions on the modeling and simulation of the human knee joint with applications to the robotic structures*. Mechanisms and Machine Science, **20**, pp. 283-297 (2014)

MATEC Web of Conferences 184, 01002 (2018) https://doi.org/10.1051/matecconf/201818401002
Annual Session of Scientific Papers IMT ORADEA 2018

Optimizing the control system of a single-axis sun tracking mechanism

Cătălin Alexandru*

Transilvania University of Brașov, Product Design, Mechatronics and Environment Department, 500036 Brașov, Romania

Abstract. The work shows the optimization of the control system for the single-axis solar tracker that equips a solar panel, with the aim to increase the energetic efficiency of the system by maximizing the quantity of incident solar radiation that is captured - absorbed by the panel. The single-axis solar tracker is driven by a linear actuator, the optimization study intending to determine the optimal configuration (in terms of tuning factors) of the controller, which is a PID (Proportional- Integral-Derivative) device, in order to accurately achieve the motion (tracking) law imposed on the solar panel. The solar tracker was approached as a mechatronic system, the mechanical device (developed in ADAMS - Automatic Dynamic Analysis of Mechanical Systems) and the control system (developed in EASY5 - Engineering Analysis System) being integrated at the level of virtual prototype, in the concurrent engineering concept.

1 Introduction

Virtual prototyping is a modern modeling, simulation and optimization technique, which is based on conception of complex virtual models, close to the real ones [1-4]. This technique is being used successfully in domains established as being suitable for testing in virtual environment (e.g. automotive, aerospace, railways), but also in new interdisciplinary areas, such as the renewable energy systems (RES). Among the RES applications, the design and testing of the mechatronic tracking systems used for the solar panels/collectors is appropriate to be addressed through the use of specific virtual prototyping tools [5-11].

The main issue to be considered when designing solar tracking systems is the energetic and economic efficiency, and in this area virtual prototyping offers a great support. The optimization of the solar tracking systems can be focused in two directions, corresponding to the two main components of a mechatronic system: the mechanical device, and the actuating & control system. The optimization of the mechanical device intends to minimize the geometric configuration of the tracking mechanism so that to minimize the motor forces/torques generated by the actuating sources to allow the panel to follow as accurately the trajectory of the sun on the celestial vault, thus minimizing the energy/power consumption for performing the tracking. On the other hand, the optimization of the control system intends to minimize the tracking errors (i.e. the differences between the imposed and current trajectories of the solar panel), thus maximizing the amount of incident solar radiation captured by the solar panel.

The present work is focused on the optimization of the control system for a single-axis solar tracker, whose dynamic virtual model was developed in mechatronic concept by using a virtual prototyping platform that integrates three types of software solutions CATIA - to create the 3D-solid model of the mechanism, ADAMS - to design and simulate the multi-body system (MBS) model, EASY5 - to design the control system block diagram. The specific data have been integrated at the level of MBS model conceived in ADAMS, which was also used for the dynamic optimization of the solar tracking mechanism.

2 Setting the optimization problem

In ADAMS, there are the following steps for performing the optimization of a mechanical or mechatronic systems: parameterizing the virtual model; modeling the design variables, constraints and objectives; carrying out parametric design studies to identify the main design variables (with significant influence on the dynamic behavior of the tracking system, which is expressed by the design objectives); optimizing the virtual model by identifying the optimal values of the main design variables that assure, by case, the minimization or maximization of the design objective functions.

In the present work, the aim of the optimization consists in the minimization of the tracking errors, as stated in the first section of the paper. The case study is developed for a single-axis tracking mechanism, with one degree of freedom (DOF), corresponding to the daily motion of the solar panel, while the elevation is maintained at a fixed optimal position, corresponding to the local geographic area. The actuating source of the tracking mechanism is a linear actuator, whose disposing was the subject of optimization from the point of view of the mechanical device [12].

* Corresponding author: calex@unitbv.ro

© The Authors, published by EDP Sciences. This is an open access article distributed under the terms of the Creative Commons Attribution License 4.0 (http://creativecommons.org/licenses/by/4.0/).

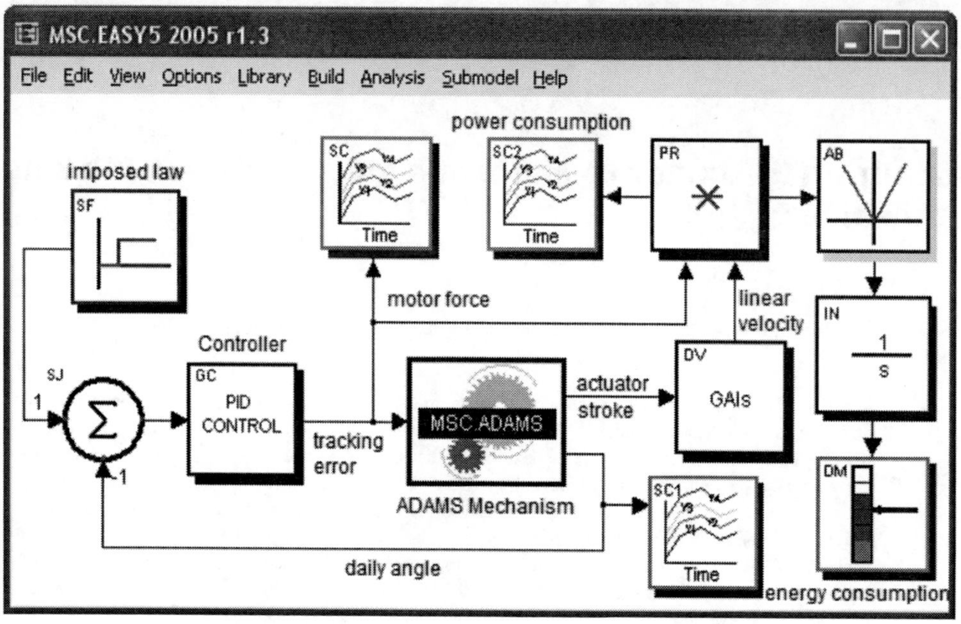

Fig. 1. The control system block diagram of the single-axis tracking mechanism (EASY5).

The control system block diagram of the single-axis tracking mechanism, which was developed by using the DFC software solution EASY5, is shown in figure 1. The control system includes the ADAMS block interface (ADAMS mechanism), which represents the link to the MBS model of the solar tracker (shown in figure 2), whose detailed description was presented in [12].

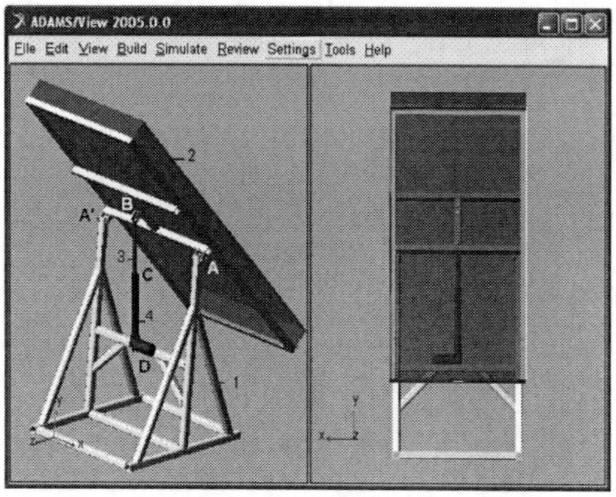

Fig. 2. The MBS model of the tracking mechanism (ADAMS).

The DFC block diagram shown in figure 1 is based on a single-loop control, the controlled parameter (which is input in the control system model) being represented by the daily angle of the solar panel, while the motor force developed by the linear actuator defines the input in the MBS (mechanical device) model. The daily angle is measured in the revolute joint A-A' between the solar panel and the fixed support, relative to the horizontal axis (X in figure 2).

Excepting the ADAMS interface block, the following specific components (blocks) have been used in order to conceive the virtual dynamic model of the control system [13]:

- SF - step function generator, which generates a step signal that is used to model the imposed tracking law (i.e. the imposed daily angle of the solar panel);
- SJ - summing junction, which is used to compare the imposed daily angle (the output from the SF block) with the current/measured angle provided by the MBS model of the tracking mechanism, the result of the subtraction being the tracking error that will be minimized by the controller;
- GC - general controller, which models a PID (Proportional-Integral-Derivative) control element; the input in this block is the tracking error, while the output is represented by the motor force that will be applied as actuating element in the MBS mechanical model of the tracking mechanism;
- DV - derivative block, which is used to calculate the derivative of the input, in this case the actuator stroke (i.e. the relative displacement between the two components of the linear actuator - piston & cylinder) that is provided by the MBS mechanical model, the output being the linear velocity in actuator;
- PR - product block, which calculates the product of two inputs, in this case the linear velocity in actuator and the motor force generated by the driving source, thus resulting the power consumption for performing the tracking (motion) law;
- AB - absolute block, which calculates the absolute value of the input parameter;
- IN - integrator block, which integrates the input signal with respect to time to produce the output signal, in

* Corresponding author: calex@unitbv.ro

this case the energy consumption for performing the tracking;

- SC - strip chart, which is used to plot component outputs while the interactive simulation is running;
- DM - digital meter, which is used to digitally display the value of a component output.

It must be specified that for a complete step-by-step tracking law (formed by more tracking steps), the SF block can be replaced by a tabular function block (T1 in EASY5), which allow to model the imposed variation of the daily angle as a function of time.

The PID controller is a control loop feedback device that assures important benefits, such as [14]: easy practical implementation, low cost, ability to solve various targets related to stationary error or overshoot, enable multi-loop control. Figure 3 shows the general scheme of the PID controller, the transfer function having the following form [13]:

$$S\_Out\_GC = ERI + GKP \cdot (REF\_GC - FBS \cdot S\_Feedback) \quad (1)$$

where:

$$FBS = \frac{ERV + S\_Feedback \cdot TC1 \cdot GKF}{TC2}$$
$$\frac{d(ERV)}{dt} = GKF \cdot S\_Feedback - FBS \quad (2)$$
$$\frac{d(ERI)}{dt} = GKI \cdot (REF\_GC - FBS)$$

with the following notations: REF - controller input (the tracking error); S_Out - controller output (the motor force generated by the linear actuator, which is transmitted to the MBS mechanical model of the solar tracker); GKP - proportional control gain; GKI - integration control gain; TC1 - derivative action time constant (which is used as a lag time constant to calculate an approximate derivative from the error signal); S_Feedback - controller feedback; TC2 - feedback damping time constant (which is used in the feedback line to help prevent an implicit loop); ERI - integrated error signal; ERV - intermediate output; s - Laplace transform.

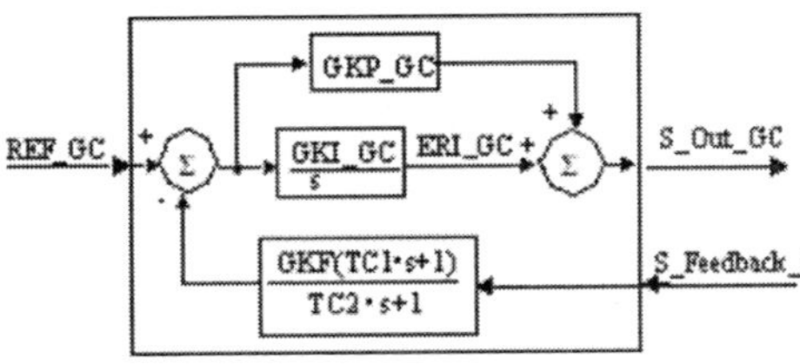

Fig. 3. The general model of the PID controller [13].

By neglecting the feedback line of the controller (which is integrated in the control system block diagram shown in figure 1), a simplified form of the transfer function of the PID controller is obtained, as follows:

$$\frac{S\_Out\_GC}{REF\_GC} = GKP \cdot \left[1 + \frac{1}{GKI \cdot s} + TC1 \cdot s\right] \quad (3)$$

The optimal design of the control system is defined by the following data: independent design variables - the tuning factors of the PID controller (GKP, GKI, TC1); design objective - the tracking error (the difference between the imposed and current/measured daily angles of the solar panel); monitored value of the design objective - the root mean square (RMS) during simulation; optimization purpose - the minimization of the design objective.

In the next stage, the control system model has been transfered from EASY5 to ADAMS, with the view to access the parametric optimization tools integrated in ADAMS (the same technique was used in [12] for the optimization of the mechanical device of the mono-axial tracking mechanism). In EASY5, the transfer is managed through the ADAMS External System Library (ESL) interface, the so generated ESL file being then imported in ADAMS as a general state equation [15]. In this way, the virtual prototype (including MBS & DFC components) is ready for the dynamic optimization. Each design variable (GKP, GKI, TC1) is defined by an initial value, and a variation domain / range limits [minimum value, maximum value].

Thus, a mono-objective optimization problem is configured with the aim to seek the optimal values of the three tuning factors of the PID controller in order to achieve the objective pursued by minimizing the tracking error (in terms of root mean square during simulation), thus increasing the energetic efficiency of the tracking system (by maximizing the rate of incident radiation that is captured by the solar panel).

3 Results and conclusions

For this research, the numerical computations (simulations) have been carried out by considering the data specific to the Brașov geographical area (latitude - 45.6333 North, longitude - 25.5833 East), on the day of the spring echinoctium (March, 21), the tracking law being configured in line with those presented in [9]. The angular domain for the daily motion is $\beta^* \in [60°, -60°]$, where the daily angle β^* is null when the panel reaches the solar noon position, with 8 equal motion steps of angular size $\Delta\beta^*=15°$. In these terms, the time-history variation of the daily angle of the solar panel (i.e. the step-by-step tracking law) is shown in figure 4.

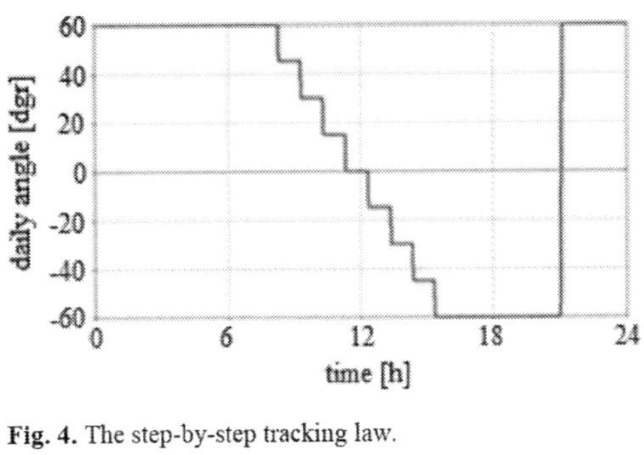

Fig. 4. The step-by-step tracking law.

https://doi.org/10.1051/matecconf/201818401002

The OPTDES-GRG generalized reduced gradient algorithm, which is provided with ADAMS/View, was used to find the optimal design of the PID control element. The solution is obtained by combining the values of the design variables within the specific variation domains, the optimal design corresponding to the set of values that assure the minimal root mean square during simulation of the tracking errors (namely, GKP=2200, GKI=1530, TC1=1500), the time-history variation of the tracking error shown in figure 5 proving the performance of the so obtained optimal solar tracker. Thus, the optimization of the control system (controller) minimizes the tracking errors, with a positive effect on the energetic efficiency of the solar tracker, by maximizing the rate of incoming solar radiation.

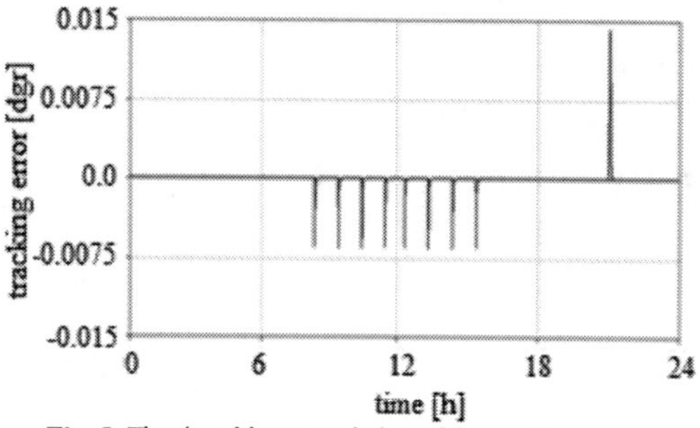

Fig. 5. The time-history variation of the tracking error.

As result of the dynamic simulation, two representative graphical simulation frames are shown in figure 6, corresponding to the extreme positions of the system (β*= ±60°), when the solar panel is oriented facing East (a), and respectively West (b).

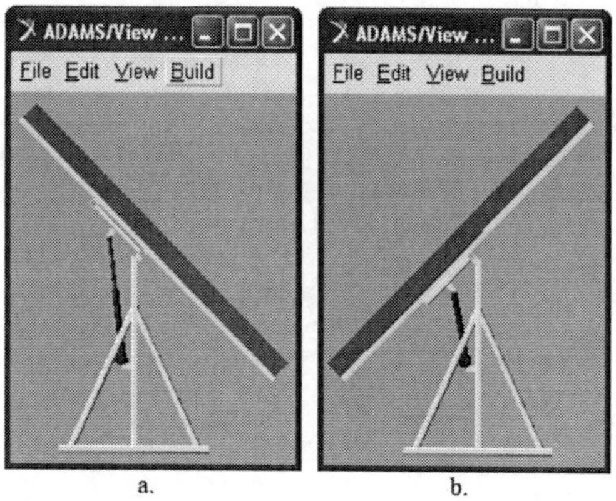

a. b.

Fig. 6. Graphical simulation frames.

Concluding, the study developed in this paper reveals the important benefits assured by the virtual prototyping in the solar trackers optimal design. The testing in virtual environment allows the comparative evaluation of a large variety of tracking strategies with the aim to improve the energy balance of the tracking mechanisms to much reduced costs compared to traditional experimental testing based on physical prototypes.

References

1. C. Alexandru, C. Pozna, C. *Dynamic modeling and control of the windshield wiper mechanisms*. WSEAS Transactions on Systems, **8**, pp. 825-834 (2009)
2. C. Alexandru, P. Alexandru. *The virtual prototype of a mechatronic suspension system with active force control*. WSEAS Transactions on Systems, **9**, pp. 927-936 (2010)
3. D. Tarniţă, D. Marghitu. *Analysis of a hand arm system*. Robotics and Computer-Integrated Manufacturing, **29**, pp. 493-501 (2013)
4. D. Tarniţă, M. Catană, D.N. Tarniţă. *Contributions on the modeling and simulation of the human knee joint with applications to the robotic structures*. Mechanisms and Machine Science, **20**, pp. 283-297 (2014)
5. C. Alexandru, M. Comşiţ. *The energy balance of the photovoltaic tracking systems using virtual prototyping platform*. 5th IEEE International Conference on the European Electricity Market EEM, pp. 253-258 (2008)
6. C. Alexandru. *The design and optimization of a photovoltaic tracking mechanism*. 2nd IEEE International Conference on Power Engineering, Energy and Electrical Drives POWERENG, pp. 436-441 (2009)
7. D.A. Flores-Hernández, S. Palomino-Resendiz, N. Lozada-Castilloc, A. Luviano-Juárez. *Mechatronic design and implementation of a two axes sun tracking photovoltaic system driven by a robotic sensor*. Mechatronics, **47**, pp. 148-159 (2017)
8. F.M. Hoffmann, R.F. Molz, J.V. Kothe, E.O.B. Nara, L. Pablo, C. Tedesco. *Monthly profile analysis based on a two-axis solar tracker proposal for photovoltaic panels*. Renewable Energy, **115**, pp. 750-759 (2018)
9. C. Alexandru, I.N. Tatu. *Optimal design of the solar tracker used for a photovoltaic string*. Journal of Renewable and Sustainable Energy, **5**, pp. 1-16 (2013)
10. S. Seme, B. Stumberger, M. Hadziselimovic. *A novel prediction algorithm for solar angles using second derivative of the energy for photovoltaic sun tracking purposes*. Solar Energy, **137**, pp. 201-211 (2016)
11. S.A. Sharaf Eldin, M.S. Abd-Elhady, H.A. Kandil. *Feasibility of solar tracking systems for PV panels in hot and cold regions*. Renewable Energy, **85**, pp. 228-233 (2016)
12. C. Alexandru. *Optimizing the mechanical device of a mono-axial sun tracking mechanism*. Submitted to IMT 2018
13. *** *EASY5 library user guides*. MSC Software, 2005
14. M.M. Sabir, T. Ali. *Optimal PID controller design through swarm intelligence algorithms for sun tracking system*. Applied Mathematics and Computation, **274**, pp. 690-699(2016)
15. *** *Getting started using ADAMS/Controls*. MSC Software, 2005

Dynamical analysis of a 2-degrees of freedom spatial pendulum

Stelian Alaci[1,*], *Florina-Carmen* Ciornei[1], *Sorinel-Toderas* Siretean[1], *Mariana-Catalina* Ciornei[2] and *Gabriel Andrei* Ţibu[1]

[1]Stefan cel Mare University, Suceava, 13 University Str., 720229, Romania
[2]Carol Davila University of Medicine and Pharmacy, Dpt. Clinic 2, 8 Bd. Eroii Sanitari, 050474, Bucharest, Romania

Abstract. A spatial pendulum with the vertical immobile axis and horizontal mobile axis is studied and the differential equations of motion are obtained applying the method of Lagrange equations. The equations of motion were obtained for the general case; the only simplifying hypothesis consists in neglecting the principal moments of inertia about the axes normal to the oscillation axes. The system of nonlinear differential equations was numerically integrated. The correctness of the obtained solutions was corroborated to the dynamical simulation of the motion via dynamical analysis software. The perfect concordance between the two solutions proves the rightness of the equations obtained.

1 Introduction

Oscillatory motions occur frequently in natural world [1]. In mechanical engineering, they are met as vibrations when the elastic elements are associated with elements characterized by inertial properties (mass and moment of inertia). Another class of oscillatory motions are the oscillations themselves performed by bodies with one or several degrees of freedom actuated by a field (gravity, electrostatic etc.). The pendulum with one degree of freedom was in the attention of the scientist from old times [2]. An important application is the pendulum watch that allowed for precise time measurement. The attention for the oscillatory motions of the pendula has not lost the importance even in nowadays. One of the modern theories concerning the evolution of dynamical systems is related to the integration of differential equations that describe their behaviour. Thus, for the integration of a differential equation of an actual case, the initial conditions must be stipulated. The attempt of modelling more complex phenomena lead to the conclusion that it is possible that a minor variation of initial conditions may produce significant changes in the evolution of the system [3]. This property of a system to essentially modify its evolution for the smallest variations of initial conditions is known as *chaos* and the mathematical domain that studies the phenomenon is named the chaos theory [4].

2 Obtaining the equations of motion of the 2 DOF spatial pendulum

Looking for simple models where the chaos occurrence is possible it was recognized that the plane double pendulum is one of the simplest dynamical systems where chaos may happen. A plane pendulum with two degrees of freedom is made of two joined rigid bodies, one of the bodies has an additional immobile joint and the two axes are parallel. For studying the dynamical behaviour of the system, the equations of motion must be firstly established. The equations of motion have generally a complex form. Particular cases are considered for obtaining analytical solutions, though. Many of the scientific references consider the double mathematical pendulum or, at most, the double physical pendulum made of two identical homogenous rods, as in Fig. 1.

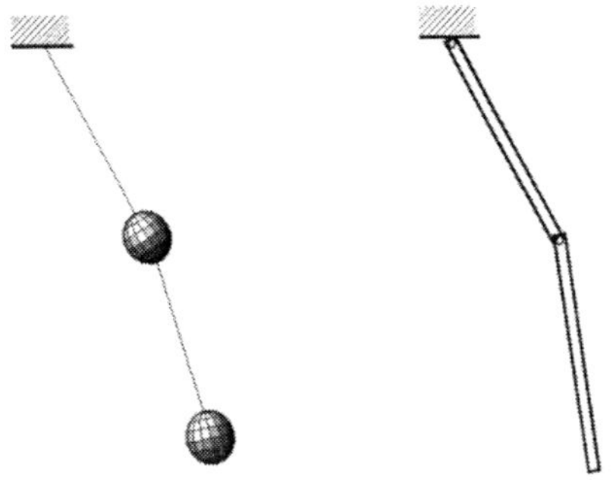

Fig. 1. Particular types of plane pendula with two DOF

The paper presents the analysis of the dynamical behaviour of a spatial pendulum with two DOF. The spatial pendulum with two degrees of freedom is presented in Fig. 2. It consists in two homogenous rods with negligible cross section, joined via a rotation pair. One of the rigid bodies is also jointed to the ground using a joint with vertical axis. Three coordinate systems are considered for the study of the motion: an immobile system $Ox_0y_0z_0$ and two Cartesian mobile frames with the origins O_1 and O_2 positioned in the centers of the

* Corresponding author: alaci@fim.usv.ro

© The Authors, published by EDP Sciences. This is an open access article distributed under the terms of the Creative Commons Attribution License 4.0 (http://creativecommons.org/licenses/by/4.0/).

immobile and mobile joint, respectively. The axes of the rotation pairs are the O_1z_1 and O_2z_2 axes. The axes O_1x_1 and O_2x_2 will pass through the centers of mass G_1 and G_2 of the two rigid bodies. The O_1y_1, O_2y_2 axes complete the coordinate systems in such a manner that they are right oriented. There are known: the masses of the bodies M_1 and M_2; the central moments of inertia J_{G_1} and J_{G_2} with respect to axes parallel to O_1z_1 and O_2z_2; the positions of the centers of mass précised by the distances ξ_1 and ξ_2. The external force acting upon the pendulum are the weights of the two rigid bodies, $m_1\boldsymbol{g}$ and $m_2\boldsymbol{g}$. The position of the pendulum is completely specified by the angles θ and φ that characterize the motions from the two rotation pairs.

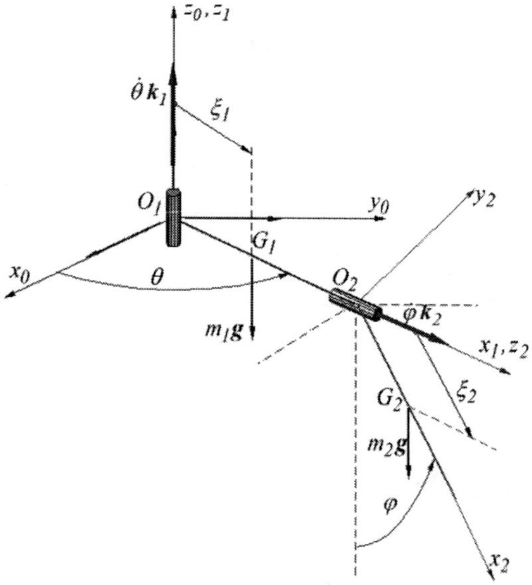

Fig. 2. Spatial pendulum with two DOF

The motion of the pendulum is characterized using the method of Lagrange equations [5] under the assumption that the friction is negligible in the two rotation pairs:

$$\frac{d}{dt}\frac{\partial E}{\partial \dot{q}_k} - \frac{\partial E}{\partial q_k} = \sum_i \boldsymbol{F}_i \cdot \frac{\partial \boldsymbol{r}_i}{\partial q_k} \quad (1)$$

where E is the kinetical energy of the system, q_k are generalized coordinates and $\boldsymbol{F}_i$ are the external forces. The next relation is applied for the calculus of kinetical energy:

$$E = \frac{1}{2}\omega_1^T \boldsymbol{J}_{G_1}\omega_1 + \frac{1}{2}M_1 v_{G_1}^2 + \frac{1}{2}\omega_2^T \boldsymbol{J}_{G_2}\omega_2 + \frac{1}{2}M_2 v_{G_1}^2 \quad (2)$$

In order to find the velocities of the centers of mass of the two elements, the position vectors of the centers of mass $\boldsymbol{r}_{G_1}$ and $\boldsymbol{r}_{G_2}$ are required. For the rigid linked to the ground:

$$\boldsymbol{r}_{G_1} = [\xi_1 \cos\theta \;\; \xi_1 \sin\theta \;\; 0]^T \quad (3)$$

and the velocity of the center of mass G_1 is:

$$\boldsymbol{v}_{G_1} = \dot{\theta}[-\xi_1 \sin\theta \;\; \xi_1 \cos\theta \;\; 0] \quad (4)$$

The angular velocity of the rod joined to the ground is:

$$\omega_1 = \dot{\theta}[0 \;\; 0 \;\; 1]^T \quad (5)$$

The inertia matrixes of the rods with respect to the principal central axes have the general form:

$$\boldsymbol{J}_{G_{1,2}} = \begin{bmatrix} J_{1,2x} & 0 & 0 \\ 0 & J_{1,2y} & 0 \\ 0 & 0 & J_{1,2z} \end{bmatrix} \quad (6)$$

Assuming that the dimensions of the cross-section of the rods are negligible compared to their lengths, the following approximation is justified:

$$J_{x_{1,2}} = J_{y_{1,2}} \cong 0 \quad (7)$$

The coordinate transformation relation [6-7] from the system 2 to the ground system is applied for finding the position vector of the center of mass G_2 in the ground frame:

$$\boldsymbol{r}_{G_2} = \boldsymbol{Z}(\theta,0)\boldsymbol{Y}(\frac{\pi}{2},0)\boldsymbol{Z}(\varphi,L_1)[\xi_2 \;\; 0 \;\; 0 \;\; 1]^T \quad (8)$$

Where, according to [7]:

$$\boldsymbol{Z}(\gamma,c) = \begin{bmatrix} \cos\gamma & -\sin\gamma & 0 & 0 \\ \sin\gamma & \cos\gamma & 0 & 0 \\ 0 & 0 & 1 & c \\ 0 & 0 & 0 & 1 \end{bmatrix} \quad (9)$$

and

$$\boldsymbol{Y}(\beta,b) = \begin{bmatrix} \cos\beta & 0 & \sin\beta & 0 \\ 0 & 1 & 0 & b \\ -\sin\beta & 0 & \cos\beta & 0 \\ 0 & 0 & 0 & 1 \end{bmatrix} \quad (10)$$

are the roto-translation operators about the axes Oz and Oy respectively and L_1 is the length of the first rod:

* Corresponding author: alaci@fim.usv.ro

$$\boldsymbol{r}_{G_2} = \begin{bmatrix} -\xi_2 \sin\theta \sin\varphi + L_1 \cos\theta \\ \xi_2 \cos\theta \sin\varphi + L_1 \sin\theta \\ -\xi_2 \cos\varphi \end{bmatrix} \quad (11)$$

The derivative with respect to time of the relation 11 provides the velocity:

$$\boldsymbol{v}_{G_2} = \begin{bmatrix} -\xi_2\dot{\theta}\cos\theta\sin\varphi - \xi_2\dot{\varphi}\sin\theta\cos\varphi - L_1\dot{\theta}\sin\theta \\ -\xi_2\dot{\theta}\sin\theta\sin\varphi + \xi_2\dot{\varphi}\cos\theta\cos\varphi + L_1\dot{\theta}\cos\theta \\ \xi_2\dot{\varphi}\sin\varphi \end{bmatrix} \quad (12)$$

The angular velocity of the rod with respect to the system in which the matrix $\boldsymbol{J}_{G_2}$ is presented is required for calculating the rotation kinetical energy of the element 2. The angular velocity of the bar 2 is:

$$\omega_2 = \dot{\theta}\boldsymbol{k}_1 + \dot{\varphi}\boldsymbol{k}_2 = \begin{bmatrix} -\dot{\theta}\cos\varphi \\ \dot{\theta}\sin\varphi \\ \dot{\varphi} \end{bmatrix} \quad (13)$$

and it is employed in finding the kinetical energy of the pendulum:

$$E_c = \frac{1}{2}\left[J_{O_1} + M_2(L_1^2 + \xi_2^2 \sin^2\varphi)\right]\dot{\theta}^2 + \frac{1}{2}J_{O_2}\dot{\varphi}^2 + M_2\xi_2\dot{\theta}\dot{\varphi}L_1\cos\varphi \quad (14)$$

where, according to Steiner theorem, [8] there were used the notations:

$$J_{O_{1,2}} = J_{G_{1,2}} + M_{1,2}\xi_{1,2}^2, \quad (15)$$

for the axial moments of inertia of the rigid bodies with respect to axes normal to the plane of motion, passing through the points O_1 and O_2 respectively. For the left member of the equation 1, to the angles θ and φ from relations 3 and 11, the increase $\delta\theta$ and $\delta\varphi$ are considered and then there are calculated the following increases:

$$\begin{aligned} \delta\boldsymbol{r}_{G_1} &= \boldsymbol{r}_1(\theta + \delta\theta) - \boldsymbol{r}_1(\theta) \\ \delta\boldsymbol{r}_{G_2} &= \boldsymbol{r}_2(\theta + \delta\theta, \varphi + \delta\varphi) - \boldsymbol{r}_1(\theta, \varphi) \end{aligned} \quad (16)$$

Afterwards, the virtual work is calculated using the relation:

$$\delta L = M_1\boldsymbol{g}\delta\boldsymbol{r}_{G1} + M_2\boldsymbol{g}\delta\boldsymbol{r}_{G_2} \quad (17)$$

From relation 17 it is obtained, after the calculus is made:

$$\delta L = -M_2\xi_2 g\,\delta\varphi \sin\varphi \quad (18)$$

The above relations can be employed in establishing the Lagrange equations for the motion of the system, under the following form:

$$\begin{cases} (M_2\xi_2^2\sin^2\varphi + J_{O_1} + M_2L_1^2)\ddot{\theta} + M_2L_1\xi_2\ddot{\varphi}\cos\varphi - M_2L_1\xi_2\dot{\varphi}^2\sin\varphi + M_2\xi_2^2\dot{\theta}\dot{\varphi}\sin 2\varphi = 0 \\ M_2L_1\xi_2\ddot{\theta}\cos\varphi + J_{O_2}\ddot{\varphi} - M_2\xi_2^2\dot{\theta}^2\sin\varphi\cos\varphi + M_2\xi_2 g\sin\varphi = 0 \end{cases} \quad (19)$$

$$\begin{cases} \ddot{\theta} = \begin{vmatrix} -M_2\xi_2L_1\dot{\varphi}^2\sin\varphi + M_2\xi_2^2\dot{\varphi}\dot{\theta}\sin(2\varphi) & M_2L_1\xi_2\cos\varphi \\ -M_2\xi_2^2\dot{\theta}^2\sin\varphi\cos\varphi + M_2\xi_2 g\sin\varphi & J_{O_2} \end{vmatrix} \Big/ \begin{vmatrix} J_{O_1} + M_2L_1^2 + M_2\xi_2^2\sin^2\varphi & M_2L_1\xi_2\cos\varphi \\ M_2L_1\xi_2\cos\varphi & J_{O_2} \end{vmatrix} \\ \ddot{\varphi} = \begin{vmatrix} J_{O_1} + M_2L_1^2 + M_2\xi_2^2\sin^2\varphi & -M_2L_1\xi_2\dot{\varphi}^2 + M_2\xi_2^2\dot{\theta}\dot{\varphi}\sin 2\varphi \\ M_2L_1\xi_2\cos\varphi & -M_2\xi_2^2\dot{\theta}^2\sin\varphi\cos\varphi + M_2\xi_2 g\sin\varphi \end{vmatrix} \Big/ \begin{vmatrix} J_{O_1} + M_2L_1^2 + M_2\xi_2^2\sin^2\varphi & M_2L_1\xi_2\cos\varphi \\ M_2L_1\xi_2\cos\varphi & J_{O_2} \end{vmatrix} \end{cases} \quad (20)$$

3 Comparison between analytical and numerical solutions

The form of the system of differential equations 19 reveal that it is unlikely to find an analytical solution and a numerical method should be used in obtaining the solution [9]. Given that the system is resolvable with respect to the highest order derivative with respect to the variables θ and φ, (relations 20) the Runge-Kutta 4 methodology [10-11] can be applied in solving it. In order to verify the correctness of the equations of motion 20, a pendulum made of two identical rods, the rod with mobile joint being launched from an angle $\varphi = \pi/3$ was considered. The pendulum was modelled both with

* Corresponding author: alaci@fim.usv.ro

the equations 20 and with the software dedicated to dynamical analysis, MSC.ADAMS, as seen in Fig. 3. In Figs. 4 and 5 there are presented comparatively the angular velocities from the two joints of the pendulum, obtained with the simulation software and by integrating the system of differential equations 20.

Fig. 3. The pendulum modelled using MSC.ADAMS

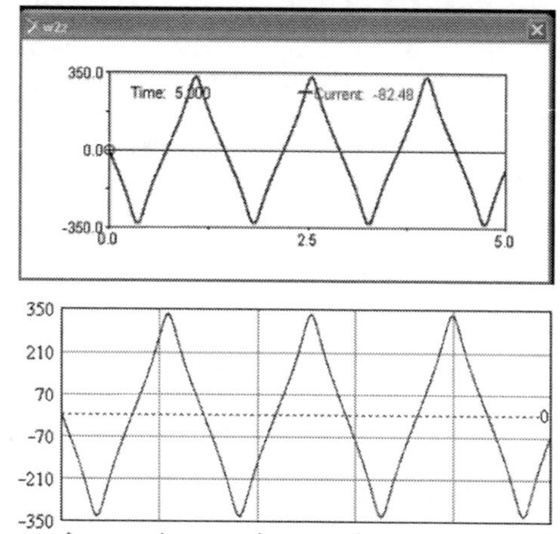

Fig. 4. The angular velocity from the immobile joint of the pendulum

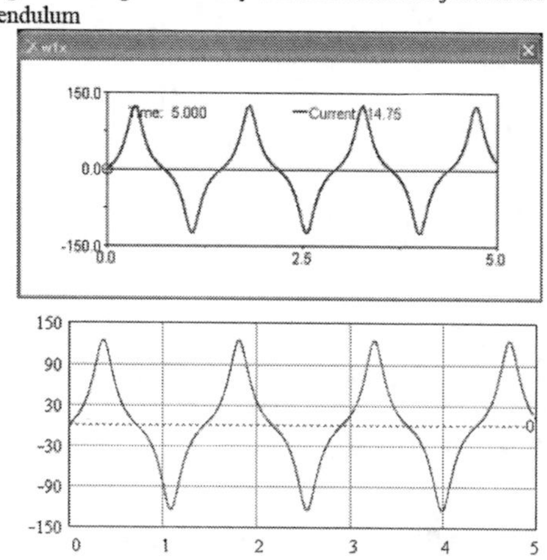

Fig. 5. The angular velocity from the mobile joint

From both plots one can observe the perfect similarity between the results given by the software and the solutions obtained from the equations of motion.

4 Conclusions

The paper presents the equations of motion obtained for the spatial pendulum with vertical immobile axis and horizontal mobile axis. The Lagrange's equations of motion were applied for establishing the equations of motion. The equations of motion are obtained for the general situation except for the simplifying assumption that, for each pendulum, the central axial moment of inertia with respect to the axis perpendicular to the axis of oscillations is zero. The nonlinear system of equations was numerically integrated for an actual case. The same pendulum was modelled using dedicated software to dynamical analysis. The perfect concordance between the solutions of the two methods confirms the correctness of the equations of motion obtained in the present work.

This work was supported by a grant of the Romanian Ministry of Research and Innovation, CCCDI – UEFISCDI, project number PN-III-P1-1.2-PCCDI-2017-0404 / 31PCCDI/2018, within PNCDI III.

References

1. L. Meirovitch, *Elements of Vibration Analysis* (2nd ed.) (McGraw-Hill Science / Engineering / Math. 1986)
2. H. Alan Lloyd, *Some Outstanding Clocks Over 700 Years 1250 – 1950* (Leonard Hill 1958)
3. E. N. Lorenz, J. Atmospheric Sciences, ***20*** (2) 130–141(1963).
4. S. H. Strogatz, *Nonlinear Dynamics and Chaos with Applications to Physiscs, Biology, Chemistry and Enginering* (CRC Press 2000)
5. D. M. Ardema, *Analytical Dynamics: Theory and Applications* (Kluwer Academic 2005)
6. J. M. McCarthy, *Introduction in Theoretical Kinematics* (MIT Press 1990)
7. I. Fischer, *Dual-Number Methods in Kinematics, Statics and Dynamics* (CRC Press 1995)
8. D. M. Ardema, *Newton-Euler Dynamics* (2005 Springer)
9. D. W. Jordan, P. Smith, *Nonlinear differential equation. Problems and Solutions. A Sourcebook for Scientists and Engineers. Oxford Texts in Applied and Engineering Mathematics* (Book 11) (Oxford University Pres 2007)
10. R. Hamming, *Numerical Methods for Scientists and Engineers* (Dover Publications 1987)

* Corresponding author: alaci@fim.usv.ro

https://doi.org/10.1051/matecconf/201818401003

11. C. W. Gear, *Numerical Initial Value Problems in Ordinary Differential Equations* (Englewood Cliffs, NJ Prentice-Hall 1971)

[*] Corresponding author: alaci@fim.usv.ro

Continuous and discrete models in the mechanics of deformable solid bodies

Géza Lámer

University of Debrecen, Faculty of Engineering, Department of Engineering Management and Enterprise, 4028 Debrecen, Ótemető u. 2-4., Hungary

Abstract. The study provides an overview of modelling possibilities for the mechanical behaviour of media. The discrete, continuous or differential geometric as well as the discrete nature and continuous description grid continuum model in particular are highlighted. We point out that the differential geometric model is based on the concept of continuity and interprets a continuous medium model. We reveal that the grid continuum model is based on the application of numerical method and interprets a discrete medium model.

1 Introduction

The system of discretely located elements, classical continuum and generalised continuum can be used in literature to describe the mechanical behaviour of deformable solid bodies. The discrete system can be characterised as describe the system in both physics space *and* phase space as discrete domain functions. The classical continuum can be characterised by interpreting it as a continuous geometric locus in the physical space, applying continuous functions in the phase space to describe its condition. The generalised continuums can be described by modelling a discrete geometric locus retaining the internal structure of the matter in the physical space (e.g. with „micro-continuums" sitting on the grid points of some kind of grid system), we characterise the condition of the internal structure matter with continuous functions for the phase space description.

The continuity of the three models differs in physical and phase space. We examine the three models according to this distinction in this study.

2 Characterisation of discrete modelling in physical space

2.1 Description of the discrete model

There are the bodies, degrees of kinematic freedom of bodies, methods of interaction existing among bodies, dynamic degrees of freedom, matter equations, the conditions of equilibrium, and the relationships describing the equilibrium (movement), the initial positions of the bodies and the initial values of dynamic effects in the initial positions, those dynamic, occasionally kinematic conditions, which exist when we seek the equilibrium of the system and its movement.

The system can be examined limited to mass points and rigid body sitting in grid points (it can be demonstrated that its deformability does not play any role), solution for the system can be examined without actually applying equilibrium (movement) equations. It is enough to limit ourselves to what degree of kinematic freedom does an element sitting on the grid points possesses and what equilibrium (movement) equations can be applied.

Hereinafter, the equation system of the discrete mechanic system is formally described. Provided there are n numbers of elements. The location of elements (reference points) is marked with $\mathbf{r}_i$, the position of elements (basis vectors of tangential space) is marked with ψ_i. Elements can have displacement $\mathbf{u}_i$ and rotation φ_i, ($i = 1,2,3 \ldots n$) degrees of freedom. The case of two degrees of freedom is examined separately and jointly as well.

Case of displacement: The internal force between the ith and j^{th} element, $\mathbf{F}_{ij}$ and $\mathbf{F}_{ji} = -\mathbf{F}_{ij}$, depends on the distance of the two elements. The distance of the two elements is marked with $\mathbf{r}_{ij} = \mathbf{r}_i - \mathbf{r}_j$ in the initial state. This vector changes to value $\mathbf{R}_{ij}$ with the shift of the two elements, where $\mathbf{R}_{ij} = \mathbf{r}_{ij} + \mathbf{u}_i - \mathbf{u}_j$. Accordingly, the following forces act upon the ith element:

$$\mathbf{F}_{ij} = \mathbf{F}_{ij}(c_{ij}, \mathbf{r}_{ij}, \mathbf{u}_i, \mathbf{u}_j), \quad (i = 1,2,3\ldots n, j = 1,2,3\ldots m_i; m_i \le n). \tag{1}$$

Here, the $m_i \le n$ inequality refers to the fact that there is not always a relationship between all elements in fact, different elements can have a relationship with a different number of elements. The fact that the magnitude of force between the ith and the jth elements can depend on various parameters was marked with c_{ij} in the context regarding $\mathbf{F}_{ij}$ force. The equilibrium of the system for every i^{th} element is given by force equations:

$$\sum\nolimits_{j=1}^{m_i} \mathbf{F}_{ij}(c_{ij}, \mathbf{r}_{ij}, \mathbf{u}_i, \mathbf{u}_j) + \mathbf{P}_i = \mathbf{0}, \quad (i = 1,2,3 \ldots n). \tag{2}$$

* Corresponding author: glamer@eng.unideb.hu

© The Authors, published by EDP Sciences. This is an open access article distributed under the terms of the Creative Commons Attribution License 4.0 (http://creativecommons.org/licenses/by/4.0/).

The $\mathbf{P}_i$ in the equation is the „external" force acting upon the ith element.

It can be seen from the equation system (we are in a three-dimensional space), that $3n$ equations are available for determining $3n$ variables. Starting from the known $\mathbf{r}_i$ positions of the individual elements in the task, the $\mathbf{R}_i = \mathbf{r}_i + \mathbf{u}_i$ ($i = 1,2,3 \dots n$) is the new position of elements that can be determined under the effect of external forces: $\mathbf{P}_i$ ($i = 1,2,3 \dots n$).

Case of rotation: The internal torque between the ith and j^{th} elements, $\mathbf{M}_{ij}$ and $\mathbf{M}_{ji} = -\mathbf{M}_{ij}$, depends on the relative position of the two elements. The relative position of the two elements is marked with $\psi_{ij} = \psi_i - \psi_j$ in the initial state. This vector changes to Ψ_{ij} value with the rotation of the two elements, where $\Psi_{ij} = \psi_{ij} + \varphi_i - \varphi_j$. Accordingly, the following torques act upon the ith element:

$$\mathbf{M}_{ij} = \mathbf{M}_{ij}(d_{ij}, \psi_{ij}, \varphi_i, \varphi_j), \quad (i=1,2,3 \dots n, j=1,2,3 \dots m_i; m_i \le n) . \tag{3}$$

See the annotation regarding the $m_i \le n$ inequality above. The fact that the magnitude of torque between the ith and j^{th} elements can depend on various parameters was marked with d_{ij} in the context regarding the $\mathbf{M}_{ij}$ torque. The equilibrium of the system for every i^{th} element is given by torque equations:

$$\sum_{j=1}^{m_i} \mathbf{M}_{ij}(d_{ij}, \psi_{ij}, \varphi_i, \varphi_j) + \mathbf{M}_i = \mathbf{0}, \quad (i = 1,2,3 \dots n). \tag{4}$$

The $\mathbf{M}_i$ in the equation is the „external" torque acting upon the ith element.

It can be seen from the equation system that $3n$ equations are available for determining $3n$ variables. Starting from the known ψ_i position of the individual elements in the task, the $\Psi_i = \psi_i + \varphi_i$ ($i = 1,2,3 \dots n$) is the new position of elements that can be determined under the effect of external torques: $\mathbf{M}_i$ ($i = 1,2,3 \dots n$).

Case of displacement and rotation: The internal force between the ith and j^{th} elements, internal torque, depends on the distance and relative position of the two elements, see above. Accordingly, the following forces and torques act upon the ith element:

$$\mathbf{F}_{ij} = \mathbf{F}_{ij}(c_{ij}, \mathbf{r}_{ij}, \mathbf{u}_i, \mathbf{u}_j, d_{ij}, \psi_{ij}, \varphi_i, \varphi_j), \tag{5}$$

$$\mathbf{M}_{ij} = \mathbf{M}_{ij}(c_{ij}, \mathbf{r}_{ij}, \mathbf{u}_i, \mathbf{u}_j, d_{ij}, \psi_{ij}, \varphi_i, \varphi_j), \quad (i=1,2,3 \dots n, j=1,2,3 \dots m_i; m_i \le n) . \tag{6}$$

See the annotations regarding the $m_i \le n$ inequality and c_{ij} and d_{ij} above. The equilibrium of the system for every i^{th} element is given by force and torque equations:

$$\sum_{j=1}^{m_i} \mathbf{F}_{ij}(c_{ij}, \mathbf{r}_{ij}, \mathbf{u}_i, \mathbf{u}_j, d_{ij}, \psi_{ij}, \varphi_i, \varphi_j) + \mathbf{P}_i = \mathbf{0}, \tag{7}$$

$$\sum_{j=1}^{m_i} \mathbf{M}_{ij}(c_{ij}, \mathbf{r}_{ij}, \mathbf{u}_i, \mathbf{u}_j, d_{ij}, \psi_{ij}, \varphi_i, \varphi_j) + \mathbf{M}_i = \mathbf{0}, \tag{8}$$

where ($i = 1,2,3 \dots n$).

The $\mathbf{P}_i$ in the equation is the „external" force acting upon the ith element, while the $\mathbf{M}_i$ is the „external" torque acting upon the ith element.

It can be seen from the equation system that $2\times3n$ equations are available for determining $2\times3n$ variables. Starting from the known $\mathbf{r}_i$ position and ψ_i direction of the individual elements in the task, the new position $\mathbf{R}_i = \mathbf{r}_i + \mathbf{u}_i$, and its new direction $\Psi_i = \psi_i + \varphi_i$ of elements ($i = 1,2,3 \dots n$) can be determined under the effect of $\mathbf{P}_i$ external forces and $\mathbf{M}_i$ external torques ($i = 1,2,3 \dots n$).

2.5 Possibilities and limitations of the discrete model

The discrete method provides an opportunity to examine the mechanical condition of a system consisting of a finite number of mass points or rigid bodies with assumption that the mass point has three displacement, and the rigid body has three displacement *and* three rotation degrees of freedom, the known position vectors and the direction of the individual rigid bodies enables the determination of force between two-two mass points and the determination of force *and* torque between two-two rigid bodies. These internal forces depend on the displacement of mass points, and the internal forces *and* torques depend on the displacement *and* rotation of rigid bodies. The equations of equilibrium (movement) of mass point and rigid body provide the same number of equations as the number of kinematic variables. The solution of equilibrium equations provides the locations of mass points and the location *and* direction of rigid bodies depending on external forces, or external forces *and* torques.

The system provides a solution regarding the discrete points of space. That means that the theory for unknown quantities (as functions) whether they are displacements, rotations, forces or torques can only determine values in the discrete points of the physical space.

3 Continuous in physical space model

3.1 Description of continuous in physical space model

A body is given as a continuous model, along with its location in space, position, shape, characteristics of material behaviour, support conditions of the body, effects on the body, kinematic and dynamic values in the initial position of the body.

We identify a deformable body in the continuous model with its region occupied in the Euclidean space. Let there be given the $\{q^i\}$ coordinate system. The position vector is marked with $\mathbf{r}$ before the deformation, and with an $\mathbf{R}$ after the deformation.

Firstly, overview the quantities describing the geometry; see the relationships themselves e.g. [1,2].

- Basis vectors in the $\mathbf{r}$ position vector point.
- Components of the metric tensor.
- Partial differentials of basis vectors.
- Affine coefficients (Christoffel symbols).
- Christoffel–Riemann curvature tensor for expressing the curvature of the Riemann space.

The following geometric terms can be expressed with the abovementioned quantities:

- scalar products of vectors **a** and **b**,
- angles between vectors **a** and **b** (cosine),
- regular arc length between two points in space,
- absolute differentials,
- parallel shift along the curve.

The basis of a continuous model is the continuity itself. This enables the application of Taylor Series, which is applied in practice in the following form:

$$F(x+\Delta x, y, z) = F(x, y, z) + \{\partial F(x, y, z) / \partial x\} \Delta x. \quad (9)$$

The consequence of this is that differential relationships are established among the unknown quantities.

According to the continuous model, the position vector of all points of the deformable rigid body changes under external effect. In accordance with this, the differential geometric description of space can be used for the kinematic characterisation of the body. The interpreted terms are the following [1,2].

The difference of position vectors is the displacement vector: $\mathbf{u} = \mathbf{R} - \mathbf{r}$, this also describes the changes of the coordinate system at the same time. The components of the metric tensor in the deformed state are as the scalar product of basis vectors. The difference of the metric tensors can be relating to the deformation. The difference of the two tensors is the measurement tensor of the deformation: $\boldsymbol{\gamma} = \mathbf{G} - \mathbf{g}$. The relative stretch of a direction vector is described by the main diagonal elements, the change in the angle between two direction vectors is described by the off-diagonal elements of the measurement tensor [2], when the relative strains (stretches *and* angel changes) are small [3-5]. This leads to the tensor of small strains: $\boldsymbol{\varepsilon} = (\mathbf{G} - \mathbf{g}) / 2$. The change of affine connexion coefficient, $\lambda_{ij}^{k} = \Gamma_{ij}^{k} - \Gamma_{ij}^{k}$, the relationships required for the equilibrium equations of the deformed state. The change of the curvature tensor provides the compatibility equations (if we make it equal to zero). The expression of different geometric objects with the components of the gradient tensor of displacement vector is required for establishing the theory. Here, we only provide a relationship for one geometric object regarding the measurement tensor of the strain: $\boldsymbol{\gamma} = \nabla\mathbf{u} + (\nabla\mathbf{u})^{*} + (\nabla\mathbf{u})(\nabla\mathbf{u})^{*}$. The other relationship can also be expressed with the gradient tensor of the displacement, see e.g. [1-5].

By describing deformation with displacement two further characterisations of changes in space can be provided. The rigid body rotation of the neighbourhood of a points (the tangential basis vectors to the point) is given by orthogonal component of the polar decomposition of the gradient tensor of the deformed position vector. The other is the rotation of a direction. This is provided by the whole gradient tensor of the deformed position vector.

The displacement of the points of the continuum clearly determines the rotation of neighbourhood of all the points. This cannot be an independent quantity. Consequently, neither the rotation of a direction, nor the rotation of the neighbourhood of a point can be interpreted as an independent kinematic variable.

The kinematic characterisation of the continuum can be summarised as follows. The primary kinematic variable of the continuum is the displacement field, while the secondary kinematic variable is the symmetrical strain tensor. The displacement of the continuum per points clearly provides the relative stretch and rotation of every unit vector, the angle change per point between two-two unit vectors and the rigid-body like rotation of the neighbourhood of every point (that is to say the tangential basis vectors belonging to the point).

Note: We have 3 + 6 kinematic variables, with six strain-displacement relationships (the kinematic equations). The kinematic description of the continuum indefinite.

We wish to interpret internal forces in the continuous model in physical space as continuous. As a point has no dimension, no weight and only one concentrated force can be assigned to it in the physical space, the Newtonian model has to be modified so that we assign dynamic quantity to a small but finite volume rather than to a point. Avoiding details, let there be $\boldsymbol{\sigma}_{\mathbf{n}}(x,y,z)$ vector quantity on an surface with **n** outward normal, which if we sum (integrate) on finite size surface figure, then ultimately we obtain concentrated force (force vector):

$$\int_{\Delta A} \boldsymbol{\sigma}_{\mathbf{n}}(x, y, z) dA = \mathbf{F}_{\mathbf{n},\Delta A}(x, y, z). \quad (10)$$

The $\boldsymbol{\sigma}_{\mathbf{n}}(x,y,z)$ quantity introduced this way is called stress vector [2]. The concentrated torque on a small but finite size surface figure can be interpreted similarly [2]:

$$\int_{\Delta A} \mathbf{r} \times \boldsymbol{\sigma}_{\mathbf{n}}(x, y, z) y dA = \mathbf{M}_{\mathbf{n},\Delta A}(x, y, z). \quad (11)$$

The equilibrium of elemental tetrahedron can be examined with the help of forces and torques interpreted on a plane figure. It can be verified with force equations that stresses form a second order tensor quantity, and it can be verified with torque equations that the stress tensor is symmetrical [2].

The equilibrium of the elemental cuboid can be examined with the help of forces and torques interpreted on a surface figure. The general form of force equations is

$$\frac{\partial \sigma_{xi}}{\partial x} + \frac{\partial \sigma_{yi}}{\partial y} + \frac{\partial \sigma_{zi}}{\partial z} = 0, \qquad (i = x, y, z), \quad (12)$$

which can be extended with body force and the inertia parameter [2], the torque equations are met due to the symmetry of the stress tensor.

Note: We have 6 dynamic variables, with three relationships (the equilibrium equations) among them. The dynamic description of the continuum is indefinite.

There are relationships and material equations between quantities characterising the kinematic and dynamic side and provide exactly as many equations as the number of variables in the system.

3.2 Possibilities and limitations of the continuous in physical space model

The continuous model is based on many, primarily topological and metric assumptions (see [6,7]).

The model can describe the change in shape of the deformable solid body (locally the relative stretches and also locally the angle change between two curves), and the internal forces (stresses) distributed in the body.

The model cannot express the rotation of a point, the force acting upon a point, and the torque acting upon a point as a point has no dimensions and surfaces act upon each other rather than points. At the same time the model is suitable for expressing the rotation of a direction vector and the neighbourhood of a point as well as assigning force and torque to a small but finite size surface figure with any outward normal vector in the point.

The model only describes the change kinematically correctly in case of fixed atomic-molecular order or strained (unchanged topology) particles. Therefore e.g.: in case of atomic structure the strain of the crystalline grid is described accurately by the continuous model. It cannot describe the rearrangement of the atomic order accurately in case of plastic flow as the rearrangement is accompanied by the damage of the topologic order. Similarly, the model does not reveal the rearrangement of atoms in fluids, the collisions of atoms-molecules in gases and their changeovers either.

The model can only describe the collective movement of molecules. Therefore the model is suitable for describing acoustic vibrations not for the optical branch. It is capable of describing laminar flow in fluids and gases with adequate accuracy but no the cross diffusion.

The model is not suitable for determining force between atoms-molecules and particles as it can only interpret defined force and torque in the integrated sense assigned to the surface figures of a small but finite size.

The model does not reveal kinematic and dynamic parameters related to dimension of elements (atoms, molecule, particle) considered to be finite. Therefore, the values of forces between atoms and molecules constituting the matter cannot be determined in the model nor the values of rotation of particles and the torques occurring during rotation.

4 Discrete in the physical space, continuous in the phase space model

4.1 Description of the discrete in the physical space, continuous in the phase space model

The matter is still considered to be the system of finite dimension elements in the physical space. (The data listed in the discrete model are known for the task; see the first paragraph of point 2.1.) We presume about this model that it is periodical and fixed. Because it is fixed, the behaviour of the medium can be characterised with the fact that the elements constituting the medium do not change places, they only move around their rest position at most. We presume in the study that in the grid points a mass point, a rigid body or a deformable solid body are located.

We assigned *one* state function to all elements collectively and not to each and every element in the phase space. We consider displacement as the example. Instead of the $\mathbf{u}(P(x_i, y_j, z_k))$ displacement vectors assigned to the various $P(x_i, y_j, z_k)$ points, we consider *one*, the displacement vector field $\mathbf{u}(x, y, z)$. Consequently, instead of the $\mathbf{u}(P(x_i, y_j, z_k))$ „exact" solution, we consider the „approximate" solution $\mathbf{u}(x, y, z)$. The „transition" is briefly marked:

$$\mathbf{u}(P(x_i, y_j, z_k)) \Rightarrow \mathbf{u}(x, y, z). \tag{13}$$

In this case we have not yet made the theoretical description continuous in phase space. The mechanical relationships interpreted in discrete points are still interpreted in discrete points, but it is not written in the $\mathbf{u}(P(x_i, y_j, z_k))$ form, but in the $\mathbf{u}(x, y, z)|_{P(x_i, y_j, z_k)}$ form. (There are the same numbers of variables in both forms.)

The next step of continuous model construction in the phase space is to find such continuous equations, which describe the mechanical state of the discrete system of the continuous kinematic $\mathbf{u}(x,y,z)$ and dynamic analogue with this variables, as well as the material relations. It has to be noticed that we approach the „transition" analogue under (13) with one of the steps applied in the numerical method, the selection of approaching functions: the unknown functions are approached by the linear combination of known basis (functions), and we set up algebraic equations for determination of unknown real numbers (namely coefficients of the known basis functions). The next step of model construction in the phase space has to be analogue with the other step of numerical method: the distance of the exact and approximate solution has to be determined and this distance has to be minimised or made orthogonal for any (complete) vector of the space of the applied basis function. The numerical method applies to the approximate solution of the *known* equation. The task during the establishment of the continuous model in the phase space is to set up (continuous) state equation related to the state functions interpreted as continuous one in the phase space. As we do not have a continuous operator in the phase space, we cannot interpret the distance of the exact and approximate solution for the numerical method in an analogue way. Consequently, it is not sufficient to consider the state functions as continuous when creating a continuous in the phase space model, but the related state equations also have to be „selected" continuous. The selection of the equation cannot be arbitrary as it is to be applied to describe a given mechanical system. The only possibility is to apply analogy. Avoiding details, in order to create a continuous in the phase space model, the Lagrange-function of the discrete system has to be made continuous as the example in (13):

$$\begin{aligned} &L(\mathbf{u}(x_i, y_j, z_k), \mathbf{F}(x_i, y_j, z_k), \ldots, P(x_i, y_j, z_k)) \Rightarrow \\ &\Rightarrow L(\mathbf{u}(x, y, z), \boldsymbol{\sigma}(x, y, z), \ldots, P(x, y, z)). \end{aligned} \tag{14}$$

https://doi.org/10.1051/matecconf/201818401004

Not the difference of the function values taken up in the discrete points but its continuous counterpart has to be included in the continuous Lagrange-function; applying the Taylor Series, for displacement:

$$\mathbf{u}(P(x_i, y_j, z_k)) - \mathbf{u}(P(x_{i-1}, y_j, z_k)) \Rightarrow \partial\mathbf{u}(x, y, z) / \partial x. \quad (15)$$

Interpreting displacement ($\mathbf{u}$), rotation ($\boldsymbol{\varphi}$) and the strain modus ($\boldsymbol{\varepsilon}$) as primary kinematic variable, the secondary kinematic variables are interpreted with the relationships below:

$$\boldsymbol{\varepsilon}_u(x, y, z) \Rightarrow \nabla\mathbf{u}(x, y, z), \quad (16)$$

$$\boldsymbol{\varepsilon}_\varphi(x, y, z) \Rightarrow \nabla\boldsymbol{\varphi}(x, y, z), \quad (17)$$

$$\boldsymbol{\varepsilon}_\varepsilon(x, y, z) \Rightarrow \nabla\boldsymbol{\varepsilon}(x, y, z). \quad (18)$$

Formally, to the analogy (13) of the discrete-continuous „transition", a $\mathbf{F}(P(x_i, y_j, z_k)) \Rightarrow \boldsymbol{\sigma}(x, y, z)$ type discrete-continuous „transition" should be also applied in the case of internal forces. In practice, we consider the partial differential quotients according to kinematic variables of the Lagrange-function as (generalised) dynamic variables. For these, in the case of the $\mathbf{u}$, $\boldsymbol{\varphi}$ and $\boldsymbol{\varepsilon}$ variables

$$\boldsymbol{\sigma}_u(x, y, z) \Rightarrow \partial L(\mathbf{u}(x, y, z), \boldsymbol{\varepsilon}_u(x, y, z), \ldots, \mathbf{r}) / \partial\boldsymbol{\varepsilon}_u, \quad (19)$$

$$\boldsymbol{\sigma}_\varphi(x, y, z) \Rightarrow \partial L(\boldsymbol{\varphi}(x, y, z), \boldsymbol{\varepsilon}_\varphi(x, y, z), \ldots, \mathbf{r}) / \partial\boldsymbol{\varepsilon}_\varphi, \quad (20)$$

$$\boldsymbol{\sigma}_\varepsilon(x, y, z) \Rightarrow \partial L(\boldsymbol{\varepsilon}(x, y, z), \boldsymbol{\varepsilon}_\varepsilon(x, y, z), \ldots, \mathbf{r}) / \partial\boldsymbol{\varepsilon}_\varepsilon \quad (21)$$

relationships can be set up. The generalised Hooke's law for the generalised continuum can be set up in the form:

$$\begin{bmatrix} \boldsymbol{\sigma}_u \\ \boldsymbol{\sigma}_\varphi \\ \boldsymbol{\sigma}_\varepsilon \end{bmatrix} = \begin{bmatrix} \mathbf{C}_{uu} & \mathbf{C}_{u\varphi} & \mathbf{C}_{u\varepsilon} \\ \mathbf{C}_{\varphi u} & \mathbf{C}_{\varphi\varphi} & \mathbf{C}_{\varphi\varepsilon} \\ \mathbf{C}_{\varepsilon u} & \mathbf{C}_{\varepsilon\varphi} & \mathbf{C}_{\varepsilon\varepsilon} \end{bmatrix} \cdot \begin{bmatrix} \boldsymbol{\varepsilon}_u \\ \boldsymbol{\varepsilon}_\varphi \\ \boldsymbol{\varepsilon}_\varepsilon \end{bmatrix}. \quad (22)$$

Both in (19-21), and (22) we set out from the fact that the material equations, namely relations between the generalised strains and the generalised internal forces, are known.

The state equations (the description limited to equilibrium), in accordance with the method of the model creation, are provided with by determining the maxima of the L functional describing the state. Formally, the following three equations are set up for variables $\mathbf{u}$, $\boldsymbol{\varphi}$ and $\boldsymbol{\varepsilon}$:

$$\begin{aligned} &\partial L(\mathbf{u}(x, y, z), \boldsymbol{\varepsilon}_u(x, y, z), \ldots, P(x, y, z)) / \partial\boldsymbol{\varepsilon}_u = \\ &= \partial L(\mathbf{u}(x, y, z), \boldsymbol{\varepsilon}_u(x, y, z), \ldots, P(x, y, z)) / \partial\mathbf{u}, \end{aligned} \quad (23)$$

$$\begin{aligned} &\partial L(\boldsymbol{\varphi}(x, y, z), \boldsymbol{\varepsilon}_\varphi(x, y, z), \ldots, P(x, y, z)) / \partial\boldsymbol{\varepsilon}_\varphi = \\ &= \partial L(\boldsymbol{\varphi}(x, y, z), \boldsymbol{\varepsilon}_\varphi(x, y, z), \ldots, P(x, y, z)) / \partial\boldsymbol{\varphi}, \end{aligned} \quad (24)$$

$$\begin{aligned} &\partial L(\boldsymbol{\varepsilon}(x, y, z), \boldsymbol{\varepsilon}_\varepsilon(x, y, z), \ldots, P(x, y, z)) / \partial\boldsymbol{\varepsilon}_\varepsilon = \\ &= \partial L(\boldsymbol{\varepsilon}(x, y, z), \boldsymbol{\varepsilon}_\varepsilon(x, y, z), \ldots, P(x, y, z)) / \partial\boldsymbol{\varepsilon}. \end{aligned} \quad (25)$$

Finally, the functional – the Lagrange-function of the system –, which describes the state of the system also has to be given. The continuous Lagrange-function in the phase space can only be set up based on analogy. Regularly, we set out from the fact that kinetic energy is proportional to the square of velocity, while elastic energy is proportional to the square of the strain. Taking this into consideration, *based on analogy*, the Lagrange-function can be constructed. The state equations of the system for variables $\mathbf{u}$, $\boldsymbol{\varphi}$ and $\boldsymbol{\varepsilon}$ can be set up in the following operator form:

$$I_{uu}\ddot{\mathbf{u}}_u - \nabla(\mathbf{C}_{uu}\nabla\mathbf{u}) - \nabla(\mathbf{C}_{u\varphi}\nabla\boldsymbol{\varphi}) - \nabla(\mathbf{C}_{u\varepsilon}\nabla\boldsymbol{\varepsilon}_u) = \mathbf{Q}_u. \quad (26)$$

$$I_{\varphi\varphi}\ddot{\boldsymbol{\varphi}}_\varphi - \nabla(\mathbf{C}_{\varphi u}\nabla\mathbf{u}) - \nabla(\mathbf{C}_{\varphi\varphi}\nabla\boldsymbol{\varphi}) - \nabla(\mathbf{C}_{\varphi\varepsilon}\nabla\boldsymbol{\varepsilon}_\varphi) = \mathbf{Q}_\varphi. \quad (27)$$

$$I_{\varepsilon\varepsilon}\ddot{\boldsymbol{\varepsilon}}_\varepsilon - \nabla(\mathbf{C}_{\varepsilon u}\nabla\mathbf{u}) - \nabla(\mathbf{C}_{\varepsilon\varphi}\nabla\boldsymbol{\varphi}) - \nabla(\mathbf{C}_{\varepsilon\varepsilon}\nabla\boldsymbol{\varepsilon}_\varepsilon) = \mathbf{Q}_\varepsilon. \quad (28)$$

The I_{ij} quantities are inertias of the „spread" matter in the generalised continuum versus the displacement, rotation and strain. The analogy is perfect with the motion equations of the classic continuum because the formulas were compiled according to its formalism. Such conditions from mathematical aspects exist, which ensure the correctness of the targeted boundary values ([8]).

4.2 Characterisation of discrete in the physical space, continuous in the phase space model

The continuous in the phase space model is constructed based on analogy. The continuous, secondary kinematic variables, the continuous Lagrange-function of the system and the continuous internal forces are interpreted based on analogy. Following this, we set up the equations describing the state of the system based on the known mathematical algorithms. A peculiarity of the model is that no such experiments exist, which would enable direct measurement of material constants. While in the case of classic continuum, from the differential geometric description it can be concluded such experimental arrangement (uniaxial pull and shear), which connects the six strains of an elemental cube to stress assigned to three surfaces of the elemental cube can be interpreted, in the case of the generalised continuum no such geometrical shape exists, where the dynamic variables could be clearly assigned to its displacements. In case of generalised continuums not simply single-parameter experiments are required for the verification of the model. Firstly, the introduced kinematic and dynamic variables have to be interpreted, secondly, such experimental arrangements have to be prepared, where these phenomena can be detected, measured *and* the experiments have to be carried out, thirdly, the theoretical tasks regarding the targeted experimental arrangement have to be set *and* have to be solved. Finally, the experimental results have to be compared with the theoretically determined results. As many experimental states have to be interpreted as the number of material constants exist in the theory. If these material constants are determined in accordance with the experiments, that is to say the parameters in the theory are fitted to the measured results then the theory can be used for the task to be solved.

The model describes the state of points located on grid points; the values of continuous functions taken up on grid points have mathematical and mechanical senses.

The model can be created preferably when the mechanical relationship among elements located on grid points is known.

5 Summary

We overviewed three possibilities of modelling the mechanical behaviour of media. The *first possibility*, is that the motion of each and every mass particle constituting the medium described by an individual equation. The functions describing the state of the medium obtain a value in the location of mass particles characterised by it. The system is discrete from this aspect. This possibility can be restricted to the concurrent examination of a few thousand or ten thousand mass particles; it is not suitable for the examination of millions of particles or at atomic level for the concurrent examination of the scale of 10^{23} gas molecules present in one cubic decimetre. As a second possibility, the application of some continuous functions has to be mentioned instead of the discrete domain functions. This in the first approach means that the state of the system described by a few functions interpreted in the continuous domain of some continuum consisting of many points including independent points instead of functions interpreted in a finite number of independent points. Two such spaces can be identified. One is the physical space, where the elements of the medium move, while the other is the phase space, where the state of the elements of the medium are mathematically characterised. Therefore, *the second possibility* is to consider a region of the Euclidean space (which has continuum cardinality) instead of the examined finite number of particles, and we extend the mechanical relationships related to the particles to this domain. In this model construction, we *„spread"* all mass present in points by default in one continuous region of the Euclidean space modelling the physical space, we have no mass particles but a medium with continuous distribution exists, along with this the functions characterising the system are continuous functions interpreted on a continuum. This model creation is based on differential geometry. This method leads to the term of classic continuum and as such only one exists. The *third possibility* is to keep the finite number of many points with their independent, unique size, characteristics as models, at the same time we embed them into the finite region of the Euclidean space with the help of their reference points, then we interpret the continuous functions describing the system on this range and extend relationships related to the discrete mechanical system for these continuous domain functions. In this model construction, we „spread" the functions describing the states in the phase space by default: continuous domain (con tinuous) functions are included instead of discrete domain functions. We create the continuous domain Lagrange-function of the system and determine the state indicators and state equations for continuous functions. This model creation applies steps of the numerical method (selection of basic functions, creation of error principle). This method is suitable for describing fixed, periodical structure rigid bodies and leads to the generalised continuums. The generalised continuum, contrary to its name, is not continuous but a discrete system with typical discrete states (e.g.: optical vibration branch of atoms), which do not exist in the classic continuum. Many generalised continuums exist, these form a hierarchy.

References

[1] Love, A.E.H. *A Treatise on the Mathematical Theory of Elasticity*. Fourth ed. Cambridge, At the University Press (1927)

[2] Lur'e, A.I.: *Theory of Elasticity*. In Russian. Nauka, M. (1970)

[3] Lámer G.: Notes on the Theory of Large Displacement with Small Strain = *Periodica Politechnica* **29** (1-2), pp. 53-65 (1985)

[4] Lámer G.: *Mathematical Foundations of the Theories of Perfect Elastic Shells and Rods Underdoing Large Displacement with Small Strains*. PhD Dissertation. (*A kis alakváltozások mellett nagy elmozdulásokat végző tökéletesen rugalmas héjak és rudak elméleteinek matematikai alapjai*) Budapest (1990)

[5] Lámer G.: On the Kinematics of the Continuum Underdoing Large Displacement with Small Strains. (Kis alakváltozások mellett nagy elmozdulást végző kontinuum kinematikájáról) = *Építés-, Építészettudomány* **XXIII** (1-2), pp. 35-59 (1992-93)

[6] Lámer G.: Opportunity and Limit of Applications of the Topological Tools in the Mechanical Models for the Media. (Topológiai eszközök alkalmazásának lehetőségei és korlátai a közegek mechanikai modellezésében). Proceedings of *XII. MAMEK* (Miskolc, 2015. aug. 25-27.) Ed.: Baksa A. – Bertóti E. – Szirbik S. paper 211. p. 13 (2015)

[7] Lámer G.: Opportunity and Limit of Applications of the Metrical Tools in the Mechanical Models for the Media. (Metrikus eszközök alkalmazásának lehetőségei és korlátai a közegek mechanikai modellezésében. Proceedings *XII. MAMEK* (Miskolc, 2015. aug. hó 25-27.) Ed.: Baksa A. – Bertóti E. – Szirbik S. paper 326. p. 13 (2015)

[8] Kunin, I.A.: *Theory of Elastic Media with Microstructure*. In Russian. Nauka, M. (1975)

Vibroacoustics and psychoacoustics investigation of biodegradable composites for musical instruments

Jorge Segura Alcaraz[1], *José María Gadea* Borrell[1], *Ernesto Juliá* Sanchis[1*], *José Enrique Crespo* Amorósl[1], *Isaac Montava* Belda[1], *Sofía Rebagliato* Torregrosa[1]

[1]Universitat Politècnica de València, Campus d'Alcoi, Spain

Abstract. Due to the extinction of some wood species used to manufacture guitars, there is a trend to search new alternative materials with the same acoustic performance. There is a wide range of tonewoods: ebony, rosewood, yellow and red cedar, mongoy and sapelli, among others. These woods have high resonance qualities and the requirements are both acoustical and mechanical. In this work, alternative composite materials have been manufactured to compare dynamic and psychoacoustic parameters with the commonly used woods. This comparison will allow to conclude the possibility of using these materials as substitutive of tonewoods.

1 Introduction

Deforestation in some parts of the planet is becoming a threat for the use of some woods in musical instruments, specially the tropical species. The current situation makes a need of substituting these materials for these applications. Some researchers focus their works on other types of wood [1] [2].
In this work, the vibroacoustic parameters of biodegradable materials, as substitutive of woods, are studied. Table 1 shows some common woods used for guitars:

Table 1. Commonly used woods.

Material	Density (g/cm³)
German Fir	0,325
Sitka Fir	0,274
Engelmann Fir	0,296
Red Cedar	0,283
Yellow Cedar	0,299
Mongoy	0,807
Sapelli	0,624

Other authors focus their investigation on alternative materials to wood ([3][4]).

Table 2 shows the characteristics of the three materials manufactured as alternative to woods.

Table 2. Alternative materials.

Material	Density (g/cm³)
Thick wool	1,1
Hemp	1,07
Jute	1,42

With these materials, the goals of the investigation are:

- To compare the results of the FFT.
- To compare the psychoacoustic parameters.

The results will allow to evaluate the possibilities of using these alternative materials.

1.1 Psychoacoustics parameters

Noise is defined as an audible sound which disturbs the silence or leads to annoyance. The evaluation of noise depends on the physical characteristics of the sound, on the psychoacoustical features of the human ear and on the psychological aspects of people [5]. The main psychoacoustic parameters are the following:

1.1.1 Loudness

The loudness is a subjective measurement of the content of energy of a perceived sound. It is related to the decibel as it ranges from strong to weak sounds.

1.1.2 Sharpness

Sharpness is the frequency content of a sound. A sound with higher frequencies is sharper.

1.1.3. Tonality

Tonality is a quality that allows to classify a sound from low to high frequencies.

* Corresponding author: erjusan@mes.upv.es

© The Authors, published by EDP Sciences. This is an open access article distributed under the terms of the Creative Commons Attribution License 4.0 (http://creativecommons.org/licenses/by/4.0/).

https://doi.org/10.1051/matecconf/201818401005

1.1.4. Roughness

Roughness is the subjective perception of the amplitude changes. It depends on the modulation of the frequency and on the sound level pressure.

2 Materials and Methods

2.1 Manufacturing

To manufacture the samples, the Vacuum Infusion Process (VIP) technique has been used. It is a method to manufacture composite materials reinforced with fibers. This process utilizes the atmospheric pressure to push the resin into the mold cavity. A polystyrene resin is used as cohesive material.

The process consists of impregnating a dry preformed fiber with a thermosetting resin of low viscosity, by a pressure difference between the cavity of the preform and the deposit of the resin, sealed with a vacuum bag. The pressure difference is made by means of a vacuum pump, which forces the resin to move gradually inside the cavity.

This process requires a control over the time of impregnation of the fibers, not being too fast (which would cause porosity) or too slow (causing a cure that prevents the advance of the resin in the preform). Figure 1 shows the manufacturing process of one of the samples.

Fig. 1. Vacuum Infusion Process (VIP).

Figure 2 shows the three test samples manufactured.

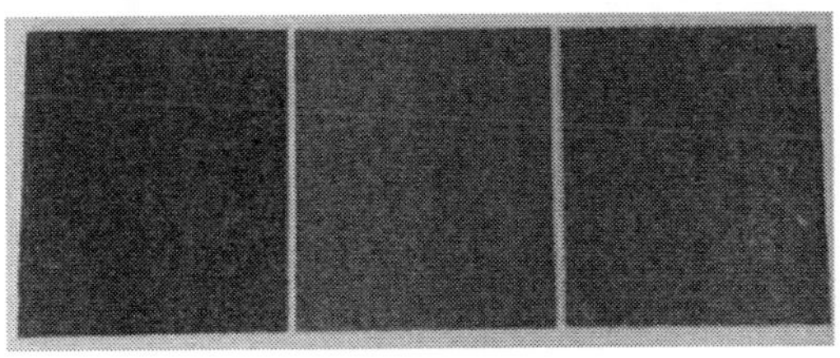

Fig. 2. The three test samples: thick wool, hemp and jute (left to right).

2.2 Experimental Tests

Two different tests have been carried out to determine the acoustic parameters required to compare the materials:

2.2.1 First experimental set-up.

Figure 3 shows the first experimental set-up for a dynamic test. The FFT is obtained by means of this test.
The working conditions of the shaker are: 5 Hz, 4 V and 0,33 A.

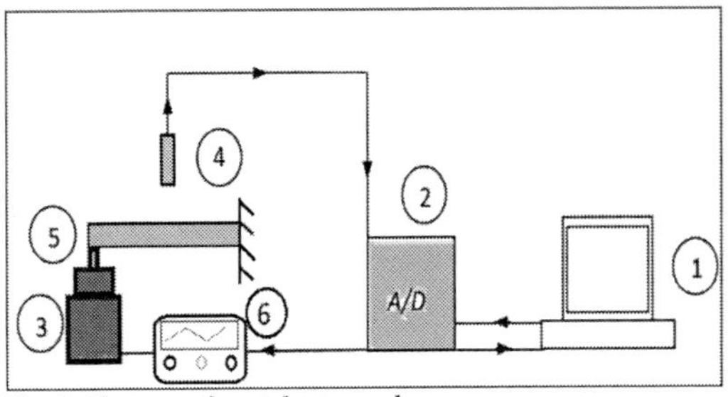

Fig. 3. First experimental set-up scheme.

The equipment consists in:

1) Computer,
2) Data acquisition card,
3) Shaker,
4) Microphone,
5) Test sample and
6) Function generator.

2.2.2 Second experimental set-up

Figure 4 shows the scheme of the experimental set-up to determine the psychoacoustic parameters.

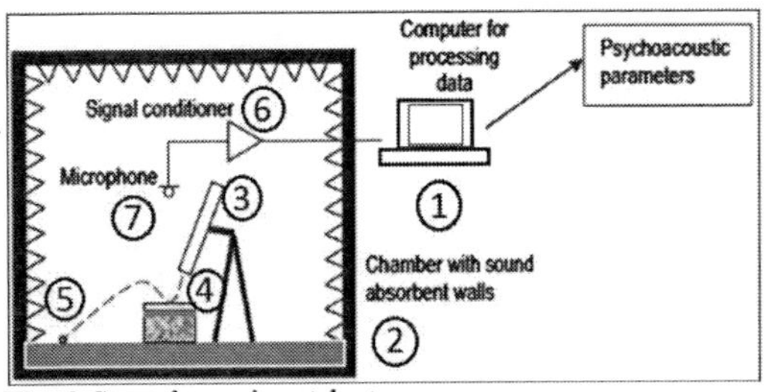

Fig. 4. Second experimental set-up.

Where:

1) Computer,
2) Anechoic chamber,
3) Tube to throw the impact steel ball,
4) Test sample,
5) Impact steel ball,
6) Signal conditioner and
7) Microphone.

https://doi.org/10.1051/matecconf/201818401005

3 Results

3.1 Dynamic results

To compare the behaviour of the materials, the FFT are represented in Figures 5 to 11. The frequency range is from 20Hz to 4200 Hz, which is inside the human hearing field. The reference is a piano which is from 27,5 Hz to 4186 Hz.

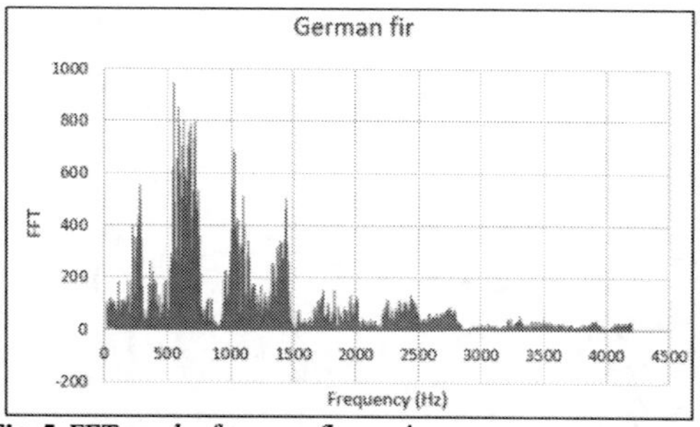

Fig. 5. FFT graph of german fir wood.

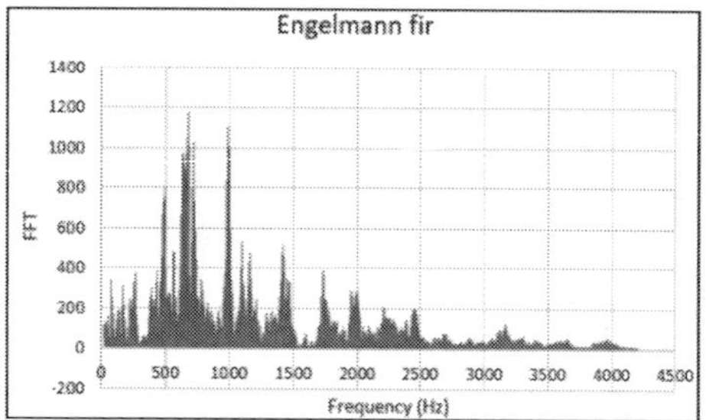

Fig. 6. FFT graph of Engelmann fir wood.

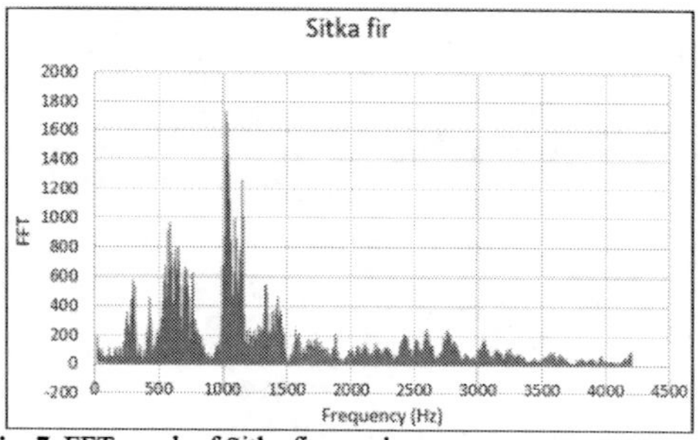

Fig. 7. FFT graph of Sitka fir wood.

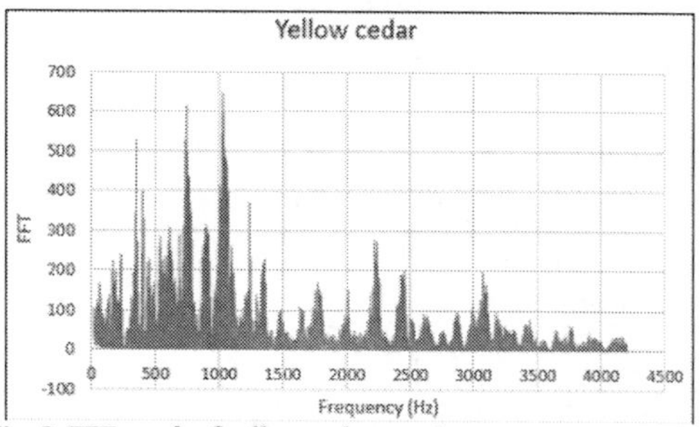

Fig. 8. FFT graph of yellow cedar wood.

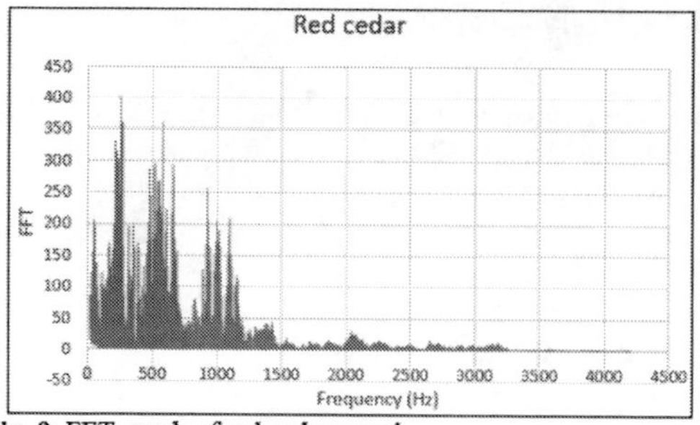

Fig. 9. FFT graph of red cedar wood.

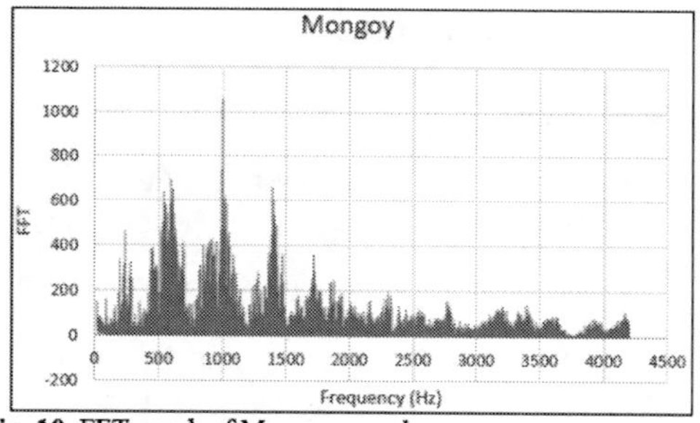

Fig. 10. FFT graph of Mongoy wood.

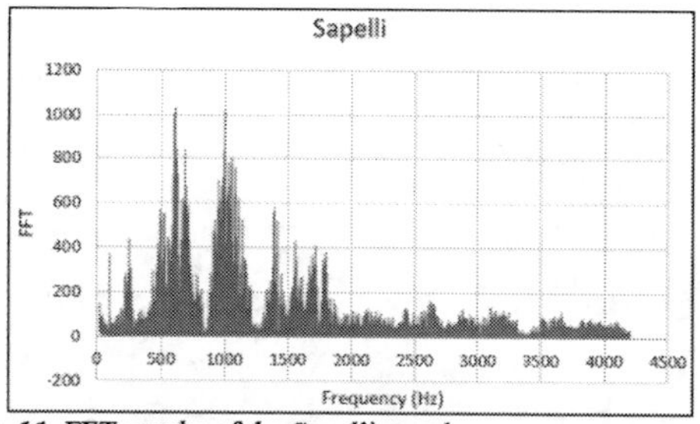

Fig. 11. FFT graphs of the Sapelli wood.

The same analysis is carried out for the alternative materials: thick wood, hemp and jute (Figures 12 to 14), so that the FFT graphs are compared to the new materials.

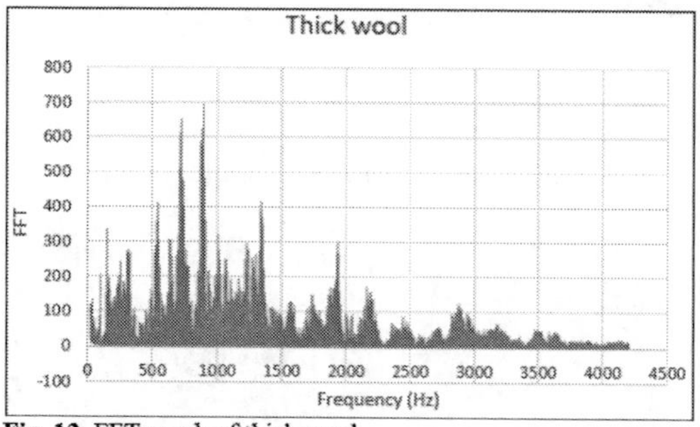

Fig. 12. FFT graph of thick wool.

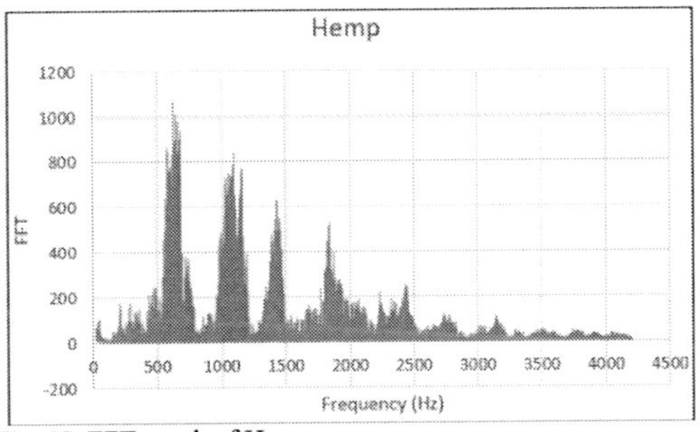

Fig. 13. FFT graph of Hemp.

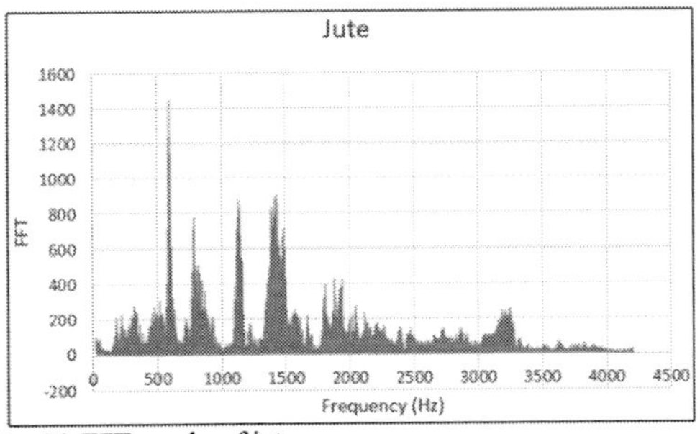

Fig. 14. FFT graphs of jute.

3.2 Psychoacoustic results

For the psychoacoustic parameters there are results from a survey in which people compare the subjective sensation. Other parameters are obtained with software dBFA.

Tables 3 and 4 show the psychoacoustic results for the common woods and for the new materials, respectively.

Table 3. Psychoacoustics results of common woods.

Psychoacoustics Criteria	German fir	Engelmann fir	Sitka fir	Yellow cedar	Red cedar	Mongoy	Sapelli
Loudness	36,6225	37,2675	33,665	33,5125	33,57	38,2	37,09
Sharpness	1,365	1,2875	1,2775	1,3175	1,2625	1,4025	1,415
Tonality	0,5925	0,685	0,6675	0,775	0,6475	0,3875	0,755
Roughness	1,115	1,12	1,01	1,1825	1,1925	1,14	1,21
Subjective sensation	0,19	0,215	0,2725	0,2425	0,2425	0,1475	0,185

Table 4. Psychoacoustics results of the new materials.

Psychoacoustics Criteria	Hemp	Wool	Jute
Loudness	34,765	29,515	35,49
Sharpness	1,72	1,41	1,59
Tonality	0,25	0,615	0,505
Roughness	1,525	1,45	1,115
Subjective sensation	0,075	0,19	0,155

4 Conclusions

It can be concluded that the thick wool material has a similar behaviour to Engelmann fir and yellow cedar woods. The most representative peaks are at 500Hz, 700Hz, 900Hz, 1300Hz, 1900Hz y 2900Hz.

For this reason, these materials present more representative peaks in the audible range and have harmonics wider and balanced.

After the study, the thick wool composite material would be applicable as an alternative to the commonly used woods in musical instruments.

References

[1] B. C. Bennett, "The Sound of Trees: Wood Selection in Guitars and Other Chordophones," *Econ. Bot.*, vol. 70, no. 1, pp. 49–63, 2016.

[2] U. G. K. Wegst, "Wood for sound," *Am. J. Bot.*, vol. 93, no. 10, pp. 1439–1448, 2006.

[3] J. A. Decker, "Production technology: Commercial composite-materials acoustic guitars," *Evol. Technol. Compet. Edge, Books 1 2*, vol. 42, p. 582–592\r1572, 1997.

[4] J. A. Torres and R. Torres-Martínez, "Evaluation of Guitars and Violins Made using Alternative Woods through Mobility Measurements," *Arch. Acoust.*, vol. 40, no. 3, pp. 351–358, 2015.

[5] K. Genuit and A. Fiebig, "Prediction of psychoacoustic parameters," *J. Acoust. Soc. Am.*, vol. 118, no. 3, pp. 1874–1874, 2005.

MATEC Web of Conferences **184**, 01006 (2018)
Annual Session of Scientific Papers IMT ORADEA 2018
https://doi.org/10.1051/matecconf/201818401006

Aspects regarding the experimental research of the stressors of the human pregnancy in case of road events

Alexandru Thierheimer[1,*], *Cătălin* Alexandru[1], and *Walter* Thierheimer[2]

[1]Transilvania University of Braşov, Product Design, Mechatronics and Environment Department, Braşov, Romania
[2]Transilvania University of Braşov, Food and Tourism Management and Engineering Department, Braşov, Romania

Abstract. In the present paper, it is presented a research facility whose purpose is to analyze the stresses that a pregnant woman's abdomen is feeling during the kicking of the baby. It is able to measure, store and simultaneously process 16 parameters of the human body's movement independently, but also in relation to the movement of the vehicle. The partial information that appears after the processing has finished on a numeric computer, a PC, will allow us to assume the way in which these occupants will behave in all the phases of a car accident: before, during, when coming out of it and after the collision, as well as to design new constructive measures to ensure that there is a non-lethal space within the occupant-interior elements-vehicle frame.

1 Introduction

The information given by the measurement and data purchasing technical systems after the experimental research was performed make the processing of the said data easy and lead to a series of practical considerations such as: choosing the correct measurement method, adapting and connecting the machines that make up the said technical system, ensuring the conditions for normal functioning by isolating them from exterior factors, the presence of a simple user interface.

The safety of the occupant who is driving the car depends on a significant number of geometric-constructive factors and the ones present in the driver's seat and their technical state, respectively. Among these factors, it can be observed the geometry of the char, the spatial configuration of the board, the passive safety systems, the elastic characteristics of the material used on the bottom and upper part of the chair, the pressurizing characteristics of the airbag systems, etc. At the same time, the safety of the occupant is influenced mainly by the type and the manner in which the collision takes place, by the type of safety belt that exists in the car, by the correlation between the position of the driver with the location where the main force of the collision was applied, as well as the rigidity of the vehicle.

One can say that the safety of the vehicles in general and the safety of the occupants to be more exact are very complex issues that determine the position and the state of the vehicle, as well as the movements of its occupants. Because of this, these issues are given special attention. This field does not focus on theoretical and experimental researches, but the purpose is to identify adequate constructive solutions that would lead to an improvement of the safety of the passengers when they come in contact with sharp bodies or interior elements that can cause damage.

2 Research methodology

Starting from these considerations, the present work presents a few aspects regarding experimental research focused on the safety of the driver when the driver is a pregnant woman. For this, it was developed a technically complex measuring system that registers and stores the movement parameters of the victim present in a vehicle, as well as those of the vehicle when it collides with another obstacle.

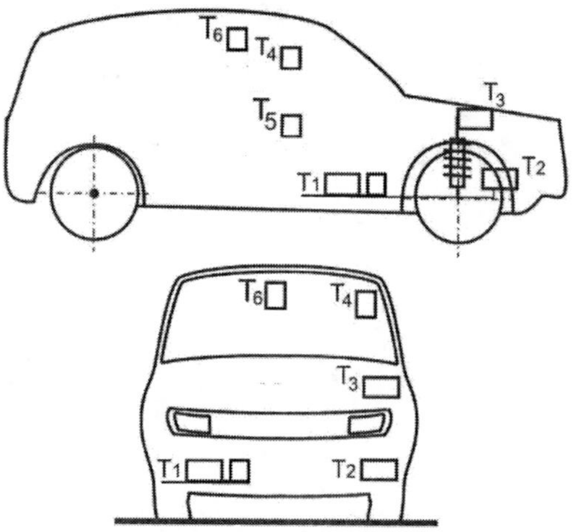

Fig. 1. The sketch to place the sensors

* Corresponding author: th.alex@yahoo.com

© The Authors, published by EDP Sciences. This is an open access article distributed under the terms of the Creative Commons Attribution License 4.0 (http://creativecommons.org/licenses/by/4.0/).

The main scheme of this technical system is illustrated in Figure 1 and it is made up of the following measurement and registering machines: T_1 DSD PicDAQ - Dr. Steffan Datentechnik: Data Acquisition System, measurement and storing unit that is specialized in acquiring data in digital form, which contains a triaxial accelerometer [1]. One can notice that the information received using the DSD PicDAQ, Xsens MTi-Inertial Measurement Unit are saved and archived on an Secure Digital SD memory card of 8GB in order to process and interpret them later on with the help of a PC computer that contains special Program for the evaluation of vehicle dynamics data PocketDAQ Analyzer, and Next VIEW 2.5 respectively. T_2 is a potentiometric transducer used to measure the angle the wheel is turning to; T_3 is a transducer used to measure the internal wing accelerations [2]; T_4 is a platform Xsens MTi, gyrometer, accelerometer, triaxial magnetometer at the same level with the head; T_5 platform Xsens MTi, triaxial gyrometer, accelerometer and magnetometer at the same level with the pelvis within the iliac-sacral bone complex; T_1 is the PocketDAQ triaxial accelerometer mounted in the center of the vehicle.

Considering the components of the data measuring and registering machine, it can be stated that the main measures that were taken into account when researching the stresses that the human body is subjected to, in this case that of a pregnant woman, during a collision focus on the general movement of the driver of the vehicle in relation to the interior elements of the cabin, as well as those of the mechanical stresses of the victim that are created when she comes into contact with these interior elements or with those that end up penetrating her body (her own components or foreign bodies) in the interior of the vehicle and which are strictly related to the bruises of the victim [3]. This can help us follow the kinematic and dynamic behavior of the body during the collision.

In order to perform the experimental research of the passive safety in general and of the degree of protection offered to the occupants of the vehicle, it was obtained a program that focuses on large-scale attempts [4]. This program was meant to store information regarding the mechanism that produces the bruises on the human body, the lower torso area during the events when there is a full frontal collision, a partial centric one or one that took place at a specific angle. All of these tests were performed with different speeds both for the situations when the systems aimed to restrict the movement (airbag, safety belt) worked, as well as for the situations when they were not being used.

The oscilograms that was obtained after processing the information are presented in Figures 2, 3 and 4 respectively, showing the graphic illustration of the variation curves of the measured units that have been measured and filtered with the help of Cebîsev polynomials in the case of NextView, which are fitting for a frontal collision [5]. In Figure 2, it can be seen the monitoring curves of the dynamic behavior of the vehicle such as longitudinal, transversal and vertical acceleration, in this case the rolling, pitching and rotation movements.

In Figure 3, it is shown the curves that illustrate the behavior of the technical system that recreates the embryo-uterus-abdomen area. In the upper part of the central region of the figure, in black background you will notice an animation of the movement of the embryo/baby in real time [6].

The dynamic behavior of the vehicle that pushes, the one who will collide is shown in Figure 4. For this,

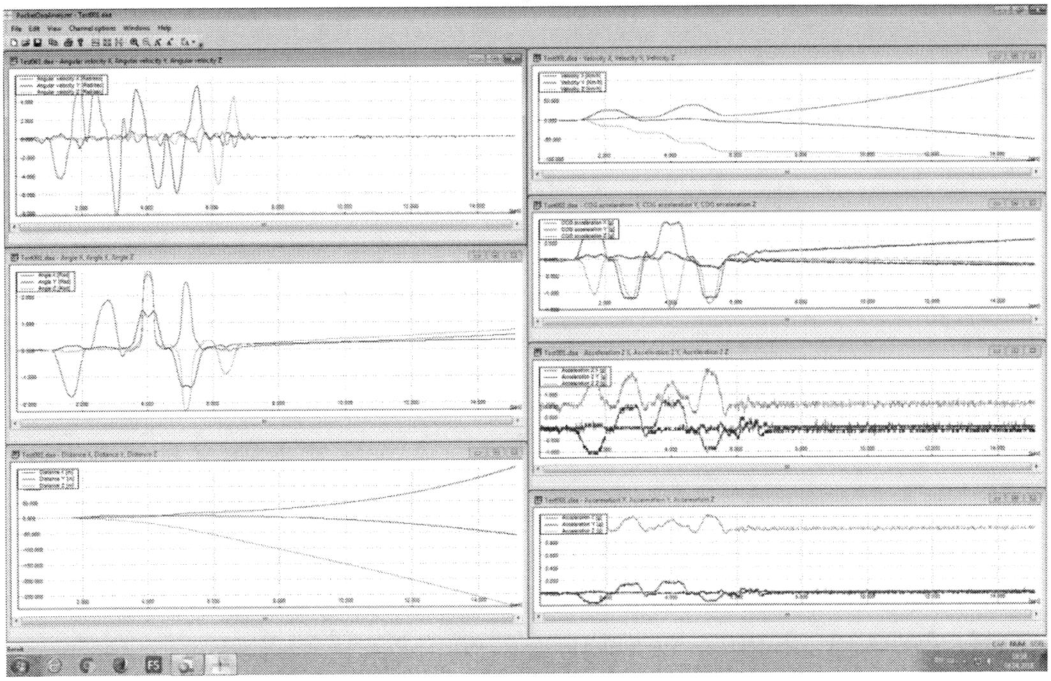

Fig. 2. The interface of the specialized program used for analysis is PocketDAQ Analyzer

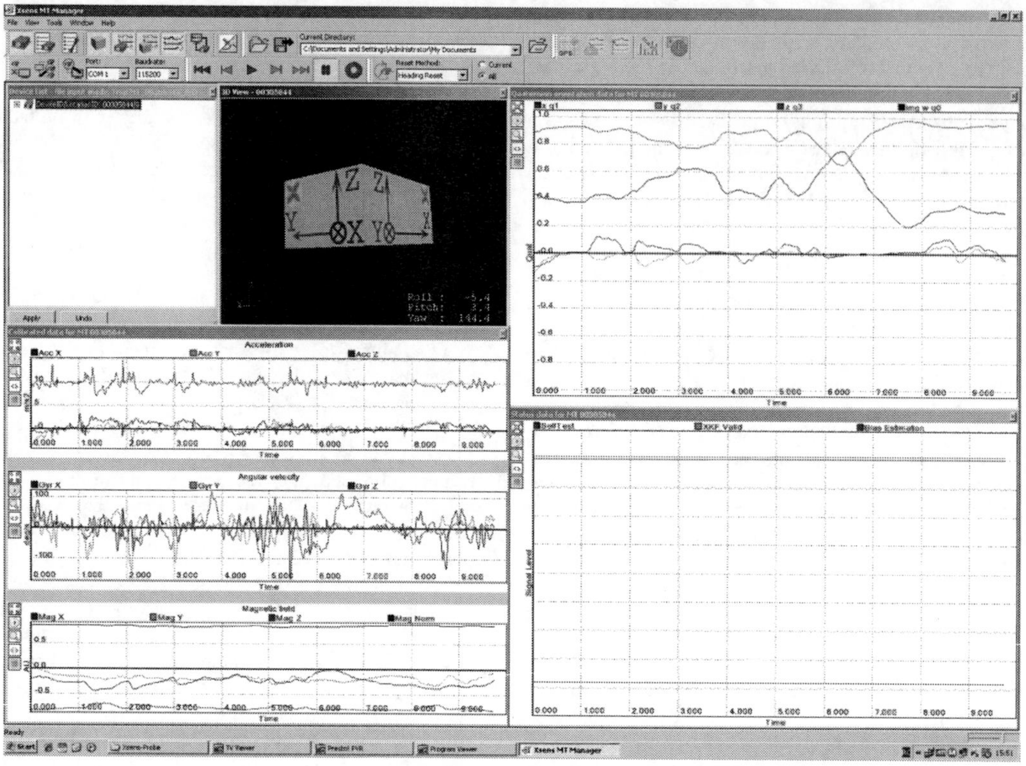

Fig. 3. The specialized analysis interface of the program is Xsens MTi

following hardware was used: a PC Intel(R) Core™ i5 3.6 GHz, RAM 16.0 GB and the programs that were used have a user license [7].

The curves presented here have been obtained after a frontal collision with a partial cover with a speed of 10 m/s it can be noticed, if you will look closely, the behavior of the vehicles during such a collision. As you can see from the diagrams, the collision has the same characteristics of a prolonged collision, which is a transient dynamic motion regime a dynamic and because of this, it can be seen a significant acceleration of the three axis of the vehicle.

There were a few experimental try-outs performed for different means of protection of the driver when it is a pregnant woman in order to study the stressors applied on a pregnancy of up to 8 weeks [8].

The theoretical and experimental research that are currently being undertaken have one purpose, to come up with new constructive solutions of the technical system used to investigate the mechanical stressors applied on an existing human pregnancy and to provide us with a thorough and global analysis of the results obtained.

3 Conclusions

In order to validate the data that was obtained from the measurements performed in laboratory conditions, as well as those obtained in real time conditions, the measurements were analyzed by comparing them with each other. Each of the two investigating tactics require a certain set of steps to be followed which will provide solutions to a few certain problems such as: the manner in which the bruises are created; analyzing the data; isolating the factors that do not have a direct connection with the manner in which the bruises appeared on the body; checking the documents, the norms and the current laws in force that govern the active and passive safety and the protection given to all the people who participate in the road traffic. Each step of the investigations leads to feedback which implies not only redoing some analyses in case some uncertain or contradictory results are obtained, but rather coming up with a series of new hypothesis that can lead to abnormal functioning conditions, which will then be compared with the real manners of bruising or obtaining some extra information regarding the areas of the body close to the area that is studies (when this is necessary).

It is obvious that there is no perfect method to obtain test samples in laboratory conditions that would match the real events when the abdominal stresses could occur (stresses on the pregnancy). The advantages and the disadvantages of each type of experimental tryout must be compared while considering the area of the body that the experiments are focused on.

The main purpose of these experiments is to offer guidelines that would allow the experimental collisions to be performed correctly, while also following the manner in which the tests are performed in general and underlining each equipment in order to detect and

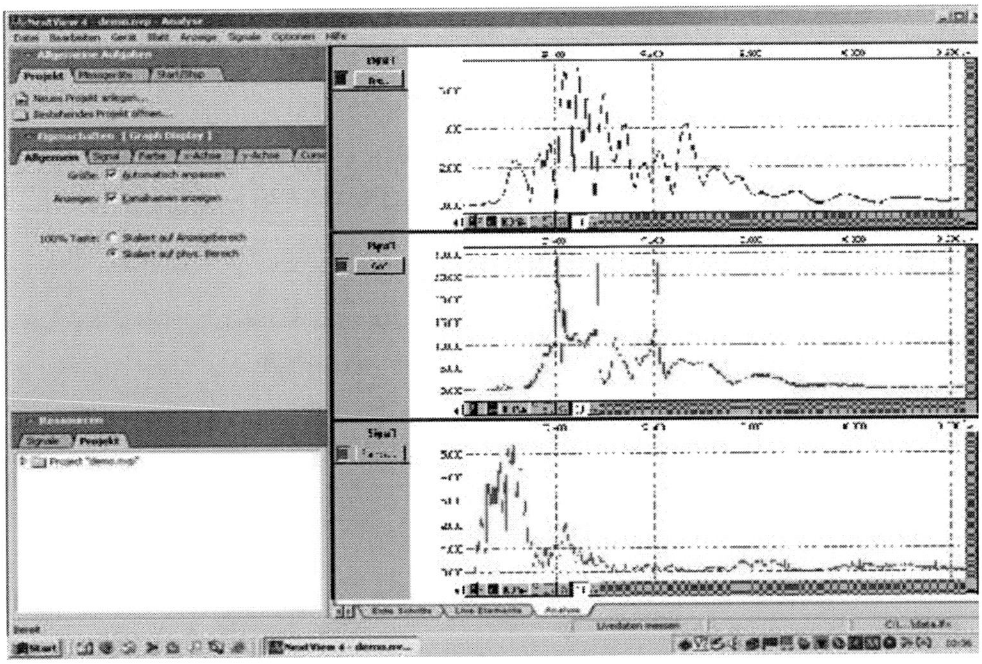

Fig. 4. The interface of the specialized program used for analysis is Next VIEW 4 (2.5)

quantify the type and the size of the stressor acting on the abdomen and on the pregnancy in particular. In these conditions, two main elements will be taken into account that occur when the experimental tests are performed: checking the samples obtained from the experiments by ensuring a structural approach to them and producing some stressing forces on the specially designed stand that would characterize the type of stressors that occur in real collisions. Thus, it is intended to develop the technical systems used for the tests that would allow us to place them in the spaces where they would be most useful.

In this work, it was analyzed only the stressors that act on the victim after a single type and manner of collision, i.e. the frontal collision with a full cover.

References

1. T.M. Best, J.H. McElhaney, W.E. Garett, B.S. Myers, *Axial strain measurements in skeletal muscle at various strain rates, Journal of Biomechanical Engineering JoBE,* **11,** 7, 262-265, (1995)
2. F. Harold, *Mathematical Methods for Accidents Reconstruction*, CRC Press Taylor & Francis Group, USA, (2010)
3. Alexandru. C., *Modeling the connections from the mechanical systems through multi body szstems softwares (Modelarea legăturilor din sistemele mecanice prin utilizarea de softuri multibody)*, 8th International Conference, Targu Jiu, May 24-26, 2002;
4. W. Thierheimer, *Technical Systems from the industry of food and agriculture: from agriculture and food industry: initiation and theoretical foundations (Sisteme tehnice din agricultură şi industrie alimentară: iniţiere şi fundamente teoretice)*, "Transilvania" University of Brasov, Romania, (2001);
5. R. Thomas, *MC-32 System-Bedienungsanleitung, BMC Dr. Schetter*, Puchheim, Munchen-Berlin, Germany, (2003);
6. J.S. Wilson, *Sensor Technology Handbook*, British Library Cataloguing-in-Publication Data Elsevier, USA, (2005);
7. G. Jr. McGwin, S.R. Russell, R.L. Rux, et al. *Beliefs, and practices concerning seat belt use during pregnancy*. J Trauma **56**(3):670-675, (2004);
8. Barsan-Anelli, A., Samaha, R.R., Prasad, P., Kommineni, V amalakkannan, S., *Development and Validation of a 2009 Toyota Venza MADYMO Frontal Occupant Model*, The National Crash Analysis Center - NCAC 2013-W-003, iulie 2013.

MATEC Web of Conferences 184, 01007 (2018) https://doi.org/10.1051/matecconf/201818401007
Annual Session of Scientific Papers IMT ORADEA 2018

Study regarding the influence of airbag deployment time on the occupant injury level during a frontal vehicle collision

Alexandru Ionut Radu[1,*], *Corneliu* Cofaru[1], *Bogdan* Tolea[2] and *Dragoș* Dima[1]

[1]Department of Automotive and Transport, Faculty of Mechanical Engineering, 500091 University "TRANSILVANIA" of Brasov, Romania
[2]Department of Mechanical Engineering and Automotive, 410610 University of Oradea, Romania

Abstract. The aim of this paper was to analyse the influence of airbag deployment delay upon the head of the occupant in the case of frontal collision using simulations in PC Crash and MADYMO dummy as the occupant. The study will also take into account the pretension delay of the seat-belt which is activated along with the airbag. Frontal airbags on both the passenger and the driver were analysed including the occupant kinematics during the collision. Also, to validate the simulation, a comparison was done with a real crash test. We predict that by increasing the delay of deployment, the head acceleration will increase due to the fact the head travels close to the instrument panel/steering wheel, and the force of the airbag will generate a significant acceleration upon the head. To better assess the potential injury of the occupant, the head injury criteria (HIC) will be calculated and correlated with the Abbreviated injury scale (AIS) code.

1 Introduction

The increasing number of vehicles during the last decades in the urban area, as well as the intensification of congestions in intersections has led to considerable road-accident recordings [1]. Road traffic accident research has been conducted over the years in order to study of the dynamic and kinematic behaviour of the occupant in the impact phase as well as the injury assessment [2]. Since the introductions of airbags, lots of lives have been saved in motor vehicles crashes [3]. The airbag provides a good protection for all kinds of crashes (side and frontal) that could seriously injure or kill the occupants of the vehicle [4]. Huere has shown in a study that airbags can reduce head injuries by up to 82% for the 56-65 km / h range. Using the airbag, it was demonstrated that in 85% of cases, head injuries were classified as minor injuries [5].

Airbags have been demonstrated to reduce car accident mortality by 19%, and reduce morbidity from facial fractures and chest injuries. In purely frontal crashes, airbags reduce the fatality risk by 34% [6].

Airbags inflates when an electrical current provided by the airbag control unit is distributed to the detonator. The ignition starts a chemical reaction that produces nitrogen gas which rapidly inflates the nylon fabric bag. After the airbag has been fully inflated, the nitrogen gas is released through small vent holes at the side when the occupant's head hits and compresses the airbag. The small holes are constructed specifically in a size and spaced out to reduce the volume of the airbag at a different rate, depending on the type of vehicle. This gas is released alongside with dust particles that are from the material used to lubricate the airbag (typically talcum powder and corn-starch) [7]. There are multiple types of airbags (front, sides, knee and curtains) [8].

The head injury can be assessed by a parameter called HIC (Head Injury Criteria), defined by a mathematical formula. The HIC criterion is a way of assessing the risk of a cranial trauma injury as a result of an accident. The HIC criterion is the maximum standard value of the integral of the head acceleration. Depending on the interval for which it is calculated, the HIC criterion is:
- Unlimited - HIC;
- 36 ms maximum - HIC36;
- 15 ms maximum - HIC15.

In order to predict the injury level of the occupant, the HIC formula takes into account the head acceleration on the occupant's head. The HIC formula is presented below [9], [10]

$$HIC36 = \max_{t1,t2}[(t_2 - t_1)(\frac{1}{t_2 - t_2}\int_{t1}^{t2} a(t)dt)^{2.5} \quad (1)$$

Where: t2 – final time value (s), t1 – initial time value (s), a(t) – head acceleration resultant (g)

The acceleration value, according to the impulse duration, shows a maximum limit of 40 g for the head. In the case of pedestrians the situation is more dangerous, at the impact with the ground, much higher head accelerations (120 - 200 g) can cause severe injuries. The head injury assessment is done using the HIC (Head Injury Criteria) criteria in a time interval of 36 milliseconds for the occupant, respectively 15 ms for the pedestrian [11], [12].

* Corresponding author: alexandru.radu@unitbv.ro

© The Authors, published by EDP Sciences. This is an open access article distributed under the terms of the Creative Commons Attribution License 4.0 (http://creativecommons.org/licenses/by/4.0/).

The injury probability chart shown in Fig. 1. represents the correlation of the AIS scale code with the HIC36 value as discovered by Alberto [13].

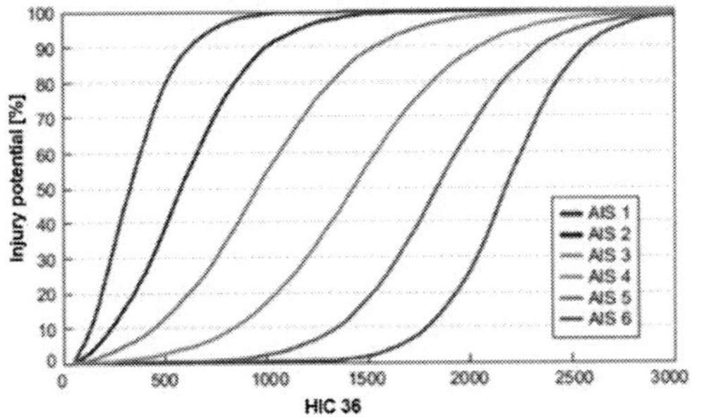

Fig. 1. Injury potential according to HIC 36 value in correlation with the AIS scale code.

The graph presents the evolution of the probability curves for AIS injury based on HIC36 value. Body tolerance according to the HIC criterion is 1000 and corresponds to a probability of 15-20% of the AIS-4 + injury. This increases to 60% for HIC = 1500. The probability of 100% (death for the occupant) indicate an AIS-6 and occurs at HIC36 above 2500.

Sances researched the effects of airbag delay on the neck [14]. He observed that the peak tension force and extension moment in neck are well below the neck injury critical intercept values of 3200 N and extension moment of 60 Nm proposed discovered by NHTSA. In contrast, in the case of late airbag deployment, the peak tension and extension moment are well above the neck injury tolerance. This suggests the high probability of serious neck injury with late deployment airbag.

An analysis by Frieder and Kumar using accident simulations with a virtual occupant model revealed that HIC36 values of the virtual dummy increased with the increase in airbag deployment delay. It was observed a percentage increase of the HIC36 value by up to 600% in the event of an airbag deployment delay of 100 ms. If the delay was 150 ms, the gas force inside the airbag "throws" the occupant's head in the rear and it comes in contact with the head restraint causing injury by increased acceleration values of the head [15]. The study at hand follows up this analysis and test new variants including passenger injury risk a seatbelt pretensioning system.

2 Occupant Model

The occupant is represented by a multibody dummy, male 50th percent, MADYMO model integrated in PC Crash, used to simulate the kinematics of the occupant [16]. Dynamic simulation were conducted before in order to analyse the occupant behaviour using the LS-DYNA model [17]. By using mathematical models of the human body, the movement of the body elements can be imprinted extensively [18].

Using this model, airbag deployment and kinematics can be analysed for both the driver and the passenger along with the seatbelt pretension system. For the driver, the main characteristics of the airbag are presented in Fig. 2.

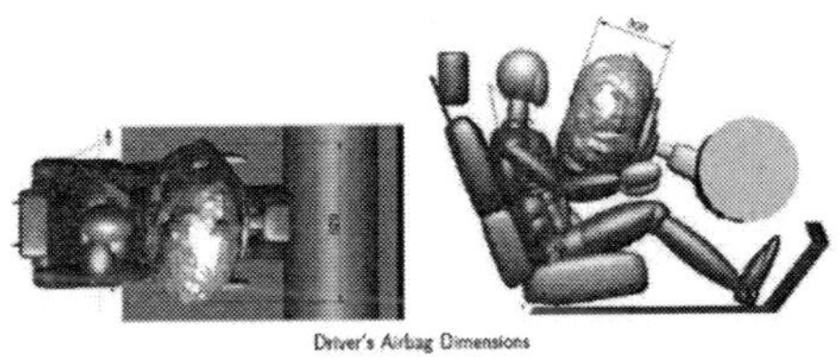

Fig. 2. Driver airbag technical specifications [16].

The inflation dimensions of the airbag are 300 mm in length and 620 mm in width. The total duration of inflation is 25 ms, during this time, the massflow reaches 1.8 kg/s in the first 10 ms, then it drops to 0 in the next 40 ms. The mass of the airbag expands from 0 to 30 g in 50 ms.

In Fig. 3, the technical specifications of the passenger airbag are presented in the similar way as the driver. It is to mention that on the passenger side, the airbag is much larger.

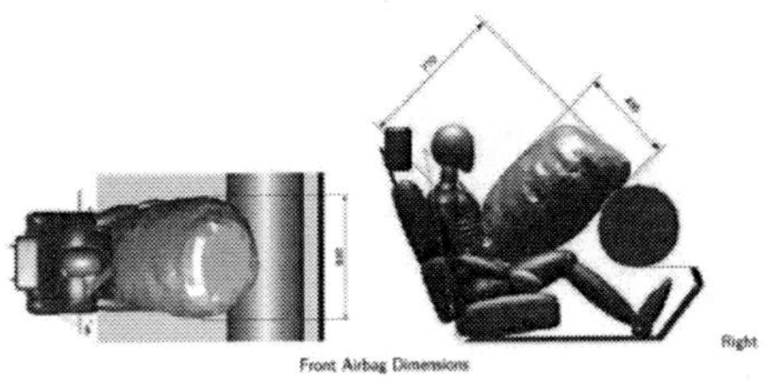

Fig. 3. Passenger airbag technical specifications [16].

The inflation dimensions of the passenger airbag are 450 mm in length, 610 mm in width and 710 mm in height. During inflation, the massflow goes from 0 to 3.3 kg/s in the first 20 ms, then it drops to almost 0 in the next 80 ms. The mass of the airbag expands from 0 to 100 g in 100 ms. It can be observed that the passenger airbag is 2 time larger that the driver side. The technical characteristics of the airbags are fixed and cannot be modified in the program, however, there are some key parameters that can be adjusted, such as the ignition time of the airbag and seatbelt pretention.

Using this screen, the ignition time could be set to a specific value, making this study possible. Also, the airbag and seatbelt could be either set as active, or one or both set as inactive. There are other settings can set be se such as the seatbelt configuration (only shoulder belt or lap belt) and if the current crash simulation is a side impact.

3 Methods Used

The method involves running simulations in PC-Crash using the MADYMO model where the ignition time of the airbag is modified in order to analyse the delay time and the occupant kinematics. Also, to validate the simulation, a frontal crash test will be conducted where the airbag and seatbelt are used. The crash test involves a stationary

 https://doi.org/10.1051/matecconf/201818401007

vehicle (VEH 1) where an occupant dummy is positioned, equipped with an accelerometer to measure the head acceleration and a striking vehicle (VEH 2) accelerated at 35 km/h.

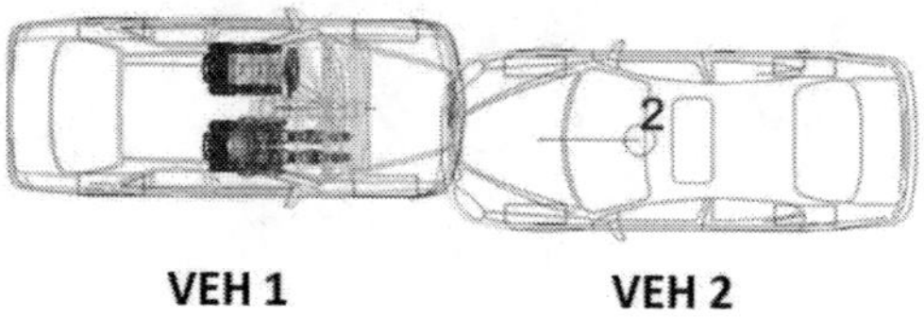

Fig. 4. Crash test configuration.

The crash test was filmed with a high speed camera in order to analyse the occupant kinematic. In Fig. 5. the kinematic analysis is presented. Also the comparison with the MADYMO simulation was done.

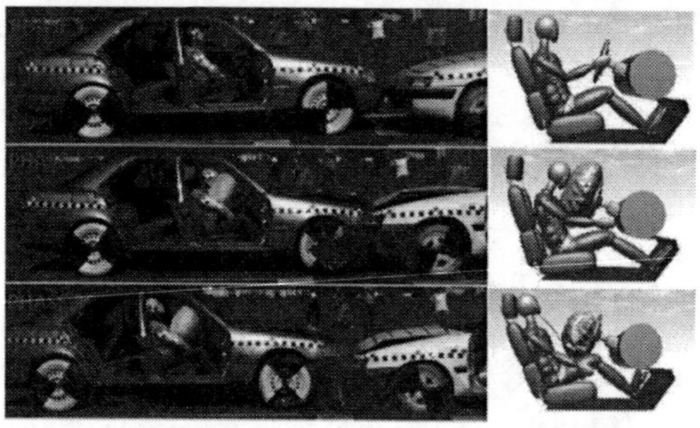

Fig. 5. Occupant kinematic comparison between the crash test and MADYMO model.

The crash was divided into 3 phases, the first phase consists the pre-impact phase when the vehicles make first contact. Phase 2 is the impact phase when the deformation of the vehicles take place, the energy is transferred and also the airbag and seatbelt pretention are deployed and limit the movement of the occupant. In phase 3, the vehicles detach and occupant's head rebound back and hits the headrest. It can be observed a good kinematic similarity between the crash test dummy and the MADYMO occupant.

4 Results

In Fig. 6. the vehicle velocities are presented during the collision. It can be seen that vehicle VEH 1 was standing still while VEH 2 was accelerated to the impact velocity of 34 km/h.

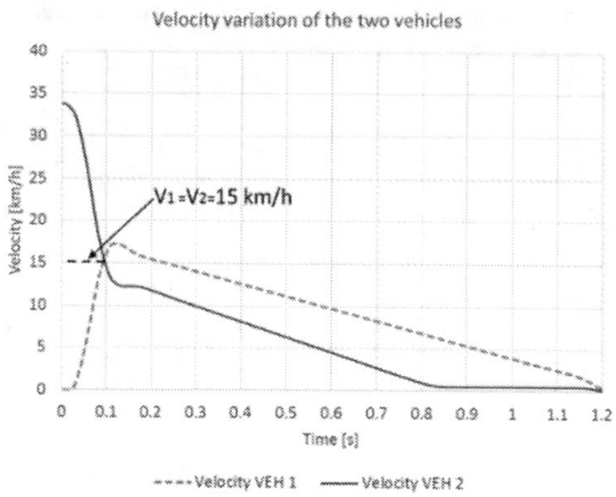

Fig. 6. Vehicles velocity variation during the collision.

The velocity variation shows that at 15 km/h, both vehicles had the same velocity, afterwards one vehicle slightly accelerates and the other tends to come to a stop. VEH 1 reached a velocity of about 17 km/h during the first 100 ms of the impact then coming to a complete stop at the time of 1.2 seconds. VEH 2 registered a rapid decrease of velocity in the first 100 ms of impact from 34 to 13 km/h, afterward coming to a complete stop at 0.9 seconds. In Fig. 7., the head acceleration comparison of the occupants (test and simulation) is presented.

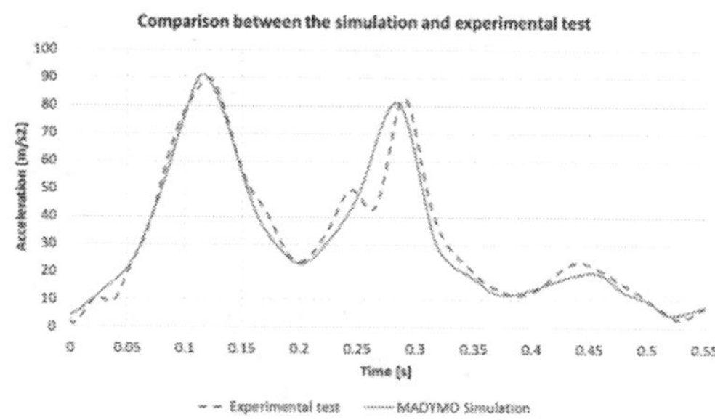

Fig. 7. Head acceleration comparison for the occupants.

The results of the simulation showed that the variation in the head acceleration resulted in two peaks, the first being the contact between the head and the airbag, and the second one between head and head restraint. The data obtained from the comparative analysis between the simulation and the experimental test indicated a high degree of similarity between them, with a 9 % error.

Following the validation of the virtual occupant's behaviour with the actual collision test, the effects of the delayed airbags on the head were analysed. In order to assess the severity of injuries suffered by motor vehicle occupants due to delayed deployment of the airbag, it was necessary to carry out 7 simulations. In all these situations the occupant did not have the seat belt, but only the airbag. No passive safety system has been used in one of the simulations. In Fig. 8 the occupant movement is presented for the 7 times the airbag trigger time has varied at the time of 100 ms of the collision. The analysis was performed up to the 100 ms delay value. The acceleration reference value (90 m/s2) corresponds to the 0 ms time.

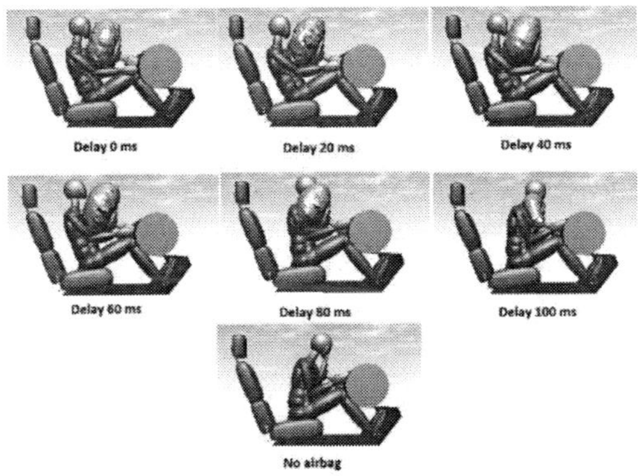

Fig. 8. Occupant kinematics at 100 ms of the collision when the airbag trigger time was modified.

It can be observed that the lack of passive safety systems results in collision of the occupant's torso with the steering wheel. The 0 ms delay is the ideal situation where the airbag is fully expanded at the moment of contact between the occupant and the airbag. Minor differences occur with delays of 20 and 40 ms, but after 60 ms, it can be seen that the position of the occupant approaches the situation where there are no passive safety systems. At 100 ms the occupant comes into contact with the steering wheel and the airbag deployment produces a head impact force. A further analysis of the occupant movement was done by observing the consequences at the time of 100 and 200 ms presented in Fig. 9.

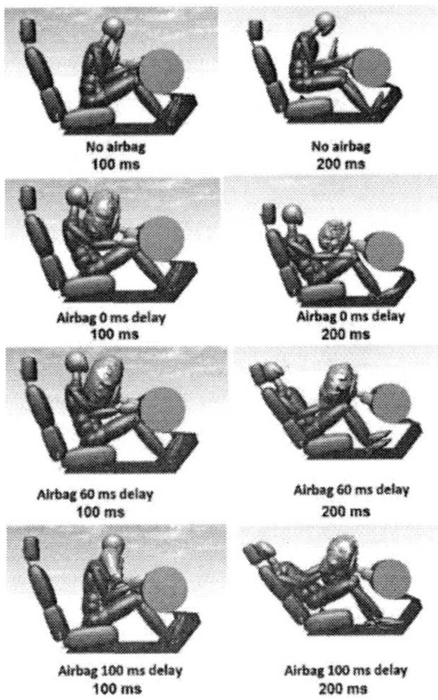

Fig. 9. Occupant kinematics at 100 ms and 200 ms of the collision when the airbag trigger time varied.

It can be seen that at 200 ms, for the trigger times of 60 and 100 ms, the force of the expanding airbag is so great, that it repels the occupant to the back of the seat generating and extra acceleration for the head. Following the simulations, a comparative analysis of the head acceleration was performed (Fig. 10).

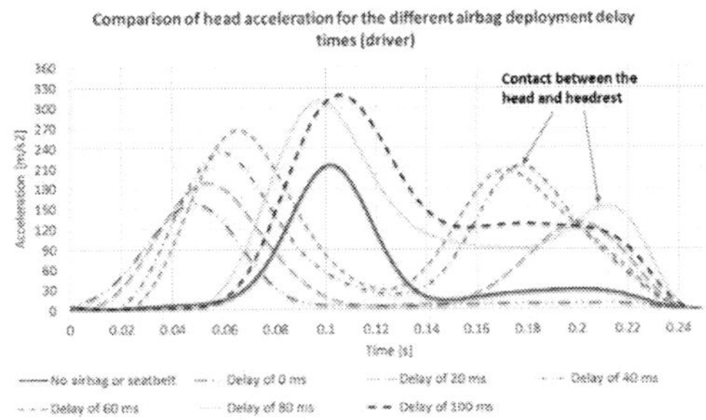

Fig. 10. Comparative head acceleration for the different airbag delay times.

From the diagram analysis, it was concluded that after the airbag trigger delay time of 80 ms, the acceleration value at the head level doubles due to the expansion force of the airbag that is transmitted to the occupant's head. In the unfavourable situation where the belt and the airbag are missing, there was an increase in acceleration value of 38%. To calculate the risk of injury of the occupant, the head injury criteria (HIC) was calculated.

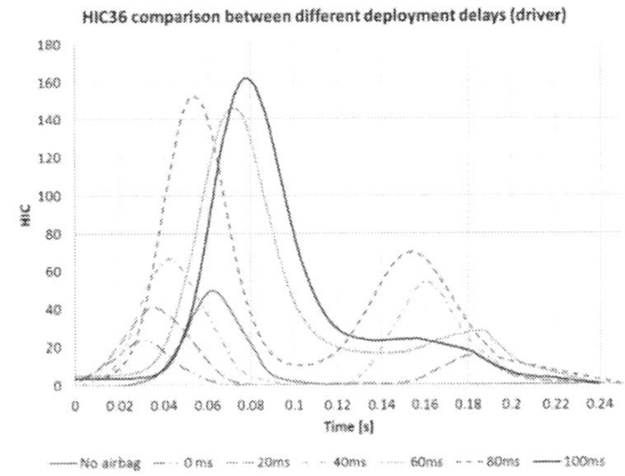

Fig. 11. HIC values for the different airbag delay times.

Head injury criteria (HIC) values have been used to express the occupant's injury potential by correlating them with the AIS severity scale. Since the maximum HIC values were below the threshold of 300, the injury potential is considered to be low (AIS-1 - minor injuries). So Fig. 12 presents the likelihood of occurrence of lesions for the related study.

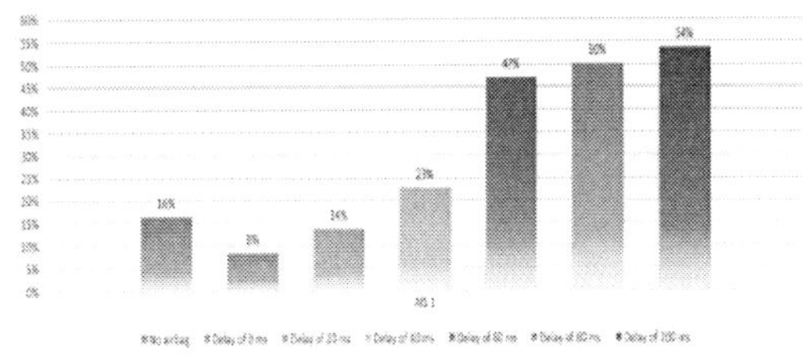

https://doi.org/10.1051/matecconf/201818401007

Fig. 12. The rate of occurrence of AIS-1 type lesions depending on the airbag deployment times.

The data revealed that the delay in triggering the airbag can induce the increase in head acceleration by 46%, directly affecting the risk of injury. The explanation is due to the fact that the distance between the head and the steering wheel is reduced if the airbag is delayed by 80-100 ms, and at the moment of triggering, all the force generated by the expansion of the volume of gas inside the airbag is transmitted directly to the occupant's skull. The proper operation of the airbag (0 ms delay) requires that at the moment of contact between the head and the airbag, the airbag has reached the maximum expanded volume.

This study also contains analysis for the passenger airbag trigger delay. Using the same procedures as above, 3 simulations were performed (airbag and seatbelt with no delay, only the airbag with no delay and only the airbag with the delay of 40 ms). In Fig. 13 the occupant kinematics are presented at the collisions time of 100 and 200 ms for the situations mentioned.

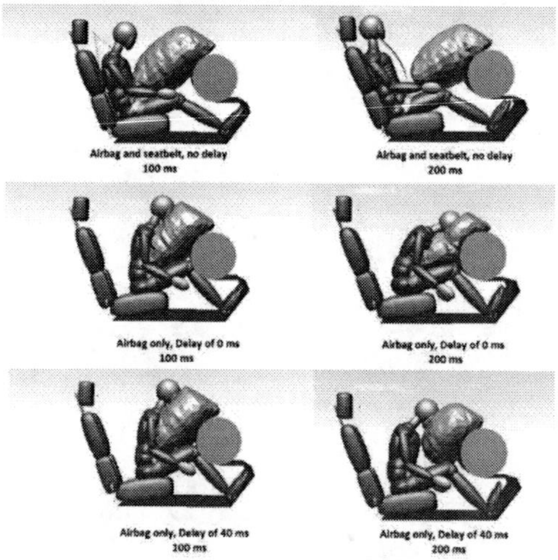

Fig. 13. Passenger occupant kinematics at 100 ms and 200 ms of the collision when the airbag trigger time varied.

It can be seen that when the airbag and seatbelt are used the occupant is well protected. When only the airbag is set with no delay, the occupant impact the airbag when it's completely inflated. With the delay of 40 ms, at 200 ms time, the airbag force hits the occupant, but has lesser effect than the driver due to the fact that the distance between the head and the instrument panel is greater than in the case of the driver, the steering wheel is closer to the head. In Fig. 14 the head acceleration comparison is presented.

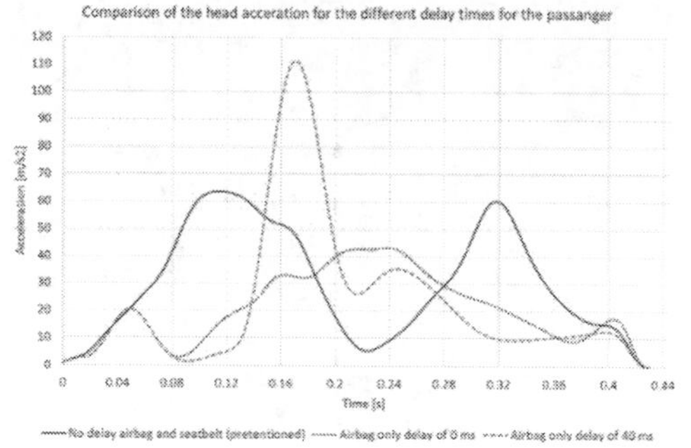

Fig. 14. Comparative head acceleration for the different airbag delay times (passenger).

With the seatbelt and airbag, the head acceleration is pretty low, with the value of 45 m/s2. When only the airbag is present, with no delay, the acceleration increases by 44% and when the trigger time is delayed with 40 ms, the acceleration increases even further by 144%. To evaluate the injury potential in the similar way for the driver, HIC was calculated and presented in Fig. 15.

The HIC values are very low in this case, the maximum value being 11 when the airbag was delayed. In Fig. 19 the injury potential is presented in regards to the AIS scale.

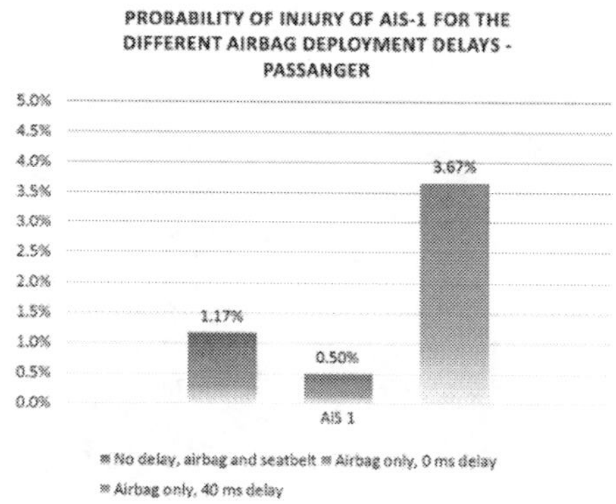

Fig. 15. The rate of occurrence of AIS-1 type lesions depending on the airbag deployment times (passenger).

Based on the AIS scale, the injury potential is very low (AIS-1) with the procent of 0.5% when the seatbelt and airbag are present and increases a bit, to 3.67% when the delay of the airbag was 40 ms. These results show that airbag delay times does increase the injury potential for both the driver and passenger, but the passenger has a lesser chance of injury.

Further simulations were carried out to study the delay in triggering the seat belt pretensioner. Using the same impact configuration as in the previous analysis, it was intended to use only the seatbelt and varying the lag time when triggering it. The values for the delay times were: 0 ms, 10 ms, 20 ms and 30 ms. In Fig. 16 the occupant's movement is presented for the related study.

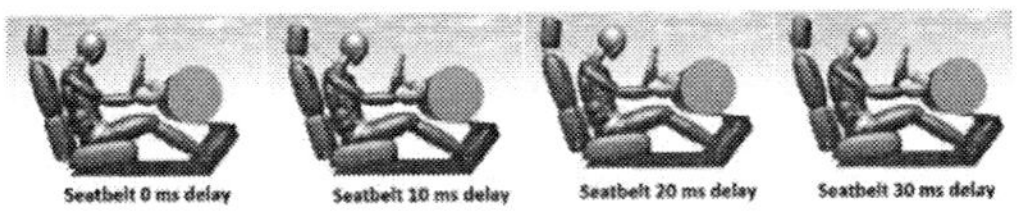

Fig. 16. The kinematics of the occupant according to the delay times when the pretensioner is triggered.

From the point of view of the occupant's behavior, small differences were found in all situations except for the 20 and 30 ms times where the movement was identical. The comparative analysis of head acceleration is shown in Fig. 17.

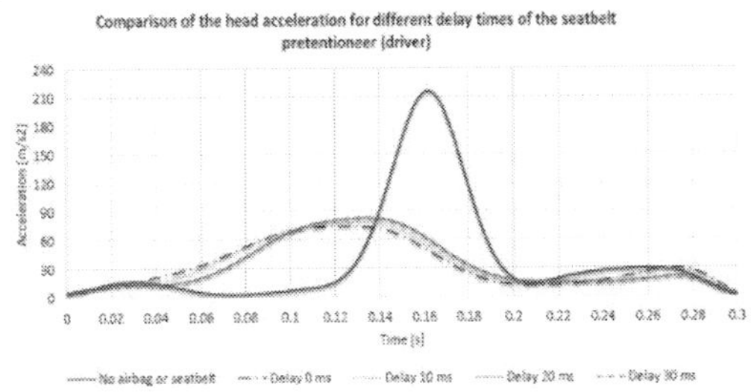

Fig. 17. Head acceleration comparison for different delay times of the seatbelt pretentioner.

From the analysis of the obtained data, it was noted the existence of small value differences for the 4 analysed situations, thus concluding that the delay of the pretensioning system does not have a massive impact on the degree of injury of the occupant.

5 Conclusions

Following the study regarding the influence of airbag deployment delay on the occupant injury potential, the following conclusions were drawn.

It has been found that by increasing the delay of the airbag deployment time in the event of a frontal impact, the probability of injury to the occupant's head increases by up to 46%. Reducing the distance between the occupant's head and the dashboard / steering wheel when the airbag ignites would result in a force expansion of the gas that is transmitted to the occupant's head generating an extra acceleration and also throws back the occupant increasing the injury potential due to the impact between the head and headrest. Thus, an increase in injury probability of 8% was observed in the 0 ms delay of the airbag deployment, while a 100 ms delay resulted in a 54% increase in the head acceleration value. So the role of the airbag is reversed, it no longer has the role of cushioning the collision, but to generate injuries.

In the case of the passenger, similar results were obtained, although a lower overall injury risk was obtained. In this case, the increase of acceleration was of only 3.67% witch is considered negligible.

The results of the seatbelt pretensioning delay study showed that there were no major differences in head acceleration values due to the fact that the seat belt secures the occupant well in the seat regardless of the pretensioning system.

References

1. O. Oţăt, ACTA Univ. Cib. **1;67(1)**:115-20, (2015)
2. O. Oţăt, N. Dumitru, V. Oţăt, In AMM, **823**, 187-192, (2016)
3. https://www.nhtsa.gov/equipment/airbags, (2017)
4. W.J. McFeely Jr, D.I. Bojrab, K.G. Davis and D.F. Hegyi, OHNS, **121(4)**: 367-73, (1999)
5. J.F. Huère, J.Y. Foret-Bruno and G. Faverjon, In *Proceedings: International Technical Conference on the Enhanced Safety of Vehicles* **2001**, 6, (2001)
6. G.R Toganel, and A.O. Soica, In *IOP Conference Series: Materials Science and Engineering,* **252(1)** 012020, (2017)
7. J. Kendall, IRSA, (2014)
8. http://www.lexomans.com/manguide-454.html, accessed, (2016)
9. B. McHenry, Head Injury Criterion and the ATB, (2004)
10. R. Eppinger, E. Sun, F. Bandak, M. Haffner, N. Khaewpong and M. Maltese, NHTSA, (1999)
11. B. Kang, S.W. Moon, J. Lim, B.H. Choi, In *Proceedings of the Institution of Mechanical Engineers, Part D: Journal of Automobile Engineering*. ; **229(7)**: 866-74, (2015)
12. M. Kramer, AUTOMOBILTECHN. Z ; **74(10),** (1972)
13. http://www.biomeccanicaforense.com/criteri-di-valutazione-del-trauma-cranico.html, (2017)
14. A. Sances, S. Kumaresan, W. Broadhead and K. Weiss, In *Proceedings of the 2003 Summer Bioengineering Conference*, 615-616 (2003)
15. R. Frieder, and S. Kumar, In *ASME 2007 Summer Bioengineering Conference*, 615-616, (2007)
16. S. Datentechnik., DSD, (2015)
17. O. Oţăt, S.C. Castravete and V. Oţăt, In *Proceedings of the European Automotive Congress EAEC-ESFA 2015*, 723-73, (2016)
18. D. Trusca, B. Benea, A. Soica and S. Tarulescu, In *2nd WSEAS Intrenational Conference on Multivariate Analysis and its Application in Science and Engineering MAASE09*, (2009)

An improved model for the damped impact of composite materials applicable to wind turbine blades

Florina-Carmen Ciornei[1,*] *Stelian* Alaci[1], *Sorinel-Toderas* Siretean[1] and *Mariana-Catalina* Ciornei[2]

[1]Stefan cel Mare University, Suceava, 13 University Str., 720229, Romania
[2]Carol Davila University of Medicine and Pharmacy, Dpt. Clinic 2, 8 Bd. Eroii Sanitari, 050474, Bucharest, Romania

Abstract. Composite materials are used in manufacturing a vast game of products, from usual objects, medical devices, aerospatial parts to military equipment. One of the features considered when a composite material is designed as target aims the energy dissipation with respect to the projectile. The paper presents an improved alternative of a model which is broadly used in the impact of projectile and composite target. The initial model has the weakness that does not present the hysteresis loop. The improvement of the model consists in including a nonlinear damper placed between the projectile and the target, the damping force being simultaneously proportional to the impact velocity and the elastic force. The new model has a hysteresis loop closed in the origin. The proposed model is applied for concrete situations and the results are presented in graphical manner.

1 Introduction

Composite materials are widely used in developing the products met in everyday life [1] but also in specialized domains, like medical devices [2-5], military equipment or wind electric systems [6]. Theoretically they can be designed for achieving preferred characteristics but as Abrate highlights [7-10] the domain of application is narrowed by the impact behavior. The numerous papers dedicated lately to this matter prove the increasing attention produced by this subject.

A number of papers study the impact at low velocities and the impact strength. Abrate [11] underlines that a first step in understanding the effect consist in the development of a model capable to allow the prediction of time evolution of the impact force and of the reaction of the structure. The analysis of impact behavior emphasizes that any impact model is situated between two limit situations: the first case, when the entire energy of the projectile is transferred to the target, and small deformations are produced that finally lead to crack micro initiators; the second case, when the composite, due to its properties, returns the entire impact energy to the projectile. A typical case for the situations mentioned above are the rotor blades of the wind turbines made of composite materials, Mishnaevsky [12]. A recent paper due to Gaudern [13] proofs the destructive effect produced by the erosion upon the rotor blades of the wind turbine. The statement is demonstrated with probes obtained in wind tunnels. To be remarked that the leading edge is the most affected region of the rotor blade. As known, the wind turbine farms must be placed in regions where minimum impact upon the environment is produced. Onshore, it is expected to view wind turbines working quite remote from popular areas, even in agricultural lands or in desert regions. A characteristic of these regions is the presence of dust particles that will be swept up into motion by the large air masses with velocities up to $13m/sec$, [14] and thus becoming actual micro-projectiles on the surface of the turbine. The mentioned facts are good reasons for the conception of models capable to estimate the impact force and lastly to allow for optimum design of rotor blades of wind turbines.

2 The Abrate spring-mass models for the impact of composite structures

Abrate makes a comprehensive review [11] concerning the representations that study the projectile–composite material impact and identifies several models.

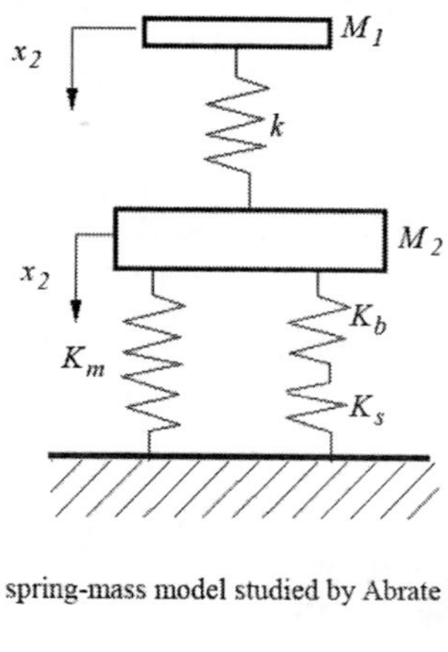

Fig. 1. The spring-mass model studied by Abrate [11]

* Corresponding author: florina@fim.usv.ro

© The Authors, published by EDP Sciences. This is an open access article distributed under the terms of the Creative Commons Attribution License 4.0 (http://creativecommons.org/licenses/by/4.0/).

Among these the spring-mass models are present. These models have the advantage of being simple and offering precise solutions for certain impact types, frequently met during the tests where small projectiles are used.

A two degree of freedom model is presented in [11] and consists in a deformable structure (2) collided by a projectile of small dimensions (1), as presented in Fig. 1.

The system is modeled using two masses and three springs. The equations describing the motion of the system are:

$$\begin{cases} M_1\ddot{x}_1 + P = 0 \\ M_2\ddot{x}_2 + K_{bs}x_2 + K_m x_2^3 - P = 0 \end{cases} \quad (1)$$

where M_1 is the mass of the projectile, K_{bs} is the elastic constant of the structure with linear behaviour, K_m is the elastic constant corresponding the the nonlinear behaviour of the structure. In equation 1 P is the impact force between the structure and the projectile, expressed as a strongly nonlinear dependency on the deformation $x_2 - x_1$.

Next it is assumed that the impact force P presents a Hertzian form. Thus, for the case of centric impact of two spheres, the impact force has the expression:

$$P = K(x_2 - x_1)^{\frac{3}{2}} \quad (2)$$

$$K = \frac{4}{3\pi}\left(\frac{R_1R_2}{R_1 + R_2}\right)^{\frac{1}{2}} \quad (3)$$

where $R_{1,2}$ are the radii of curvature of the contacting bodies and η is the contact stiffness, defined according to Johnson [15] as:

$$\eta = \frac{1 - v_1^2}{E_1} + \frac{1 - v_2^2}{E_2} \quad (4)$$

In the above relation, v_{12} and $E_{1,2}$ are the Poisson coefficients and the Young moduli, respectively, of the two materials. Next it is considered the impact between a ball-projectile and plane membrane-target. For a complete clamped isotropous circular membrane, the coefficients K_m and K_b are expressed by the relations:

$$K_b = \frac{4\pi Eh}{3(1 - v^2)a^2} \quad (5a)$$

$$K_m = \frac{(353 - 191v)\pi Eh}{648(1 - v^2)a^2} \quad (5b)$$

where h is the thickness and a is the radius of the membrane, [16]. To be mentioned that the effective mass of the membrane taking part to the oscillation motion is a quarter of the total mass of it. The equations 1 considered together form a system of nonlinear differential equations ,[17] and solving it requires applying a numerical procedure, [18]. The initial conditions are necessary in order to integrate the system and in the present case they are:

$$\begin{aligned} x_1 &= 0 \\ x_2 &= 0 \\ \dot{x}_1 &= v \\ \dot{x}_2 &= 0 \end{aligned} \quad (6)$$

The next concrete situation was considered for integrating the system:

$$\begin{aligned} a &= 0.1m, \\ h &= 0.005, \\ R_1 &= 0.02m, \\ R_2 &= 2 \cdot 10^3 m \end{aligned} \quad (7)$$

Both the ball and the membrane are made of steel and have the elastic modulus $E_{1,2} = 2.1 \cdot 10^{11} Pa$. The Runge-Kutta IV method, [18] was applied for integrating the system of equations 1.

The absolute displacements of the projectile x_1, of the center of the membrane x_2 and the relative displacement of the projectile with respect to the membrane are presented in Fig. 2.

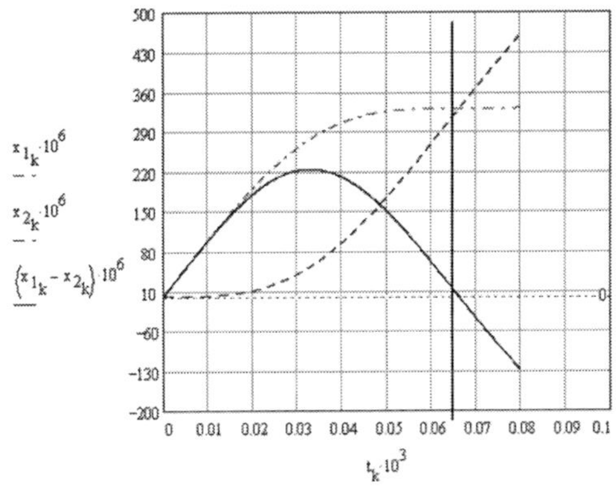

Fig. 2. Absolute deformations of the ball x_1, of the membrane x_2, and the relative deformation ($x_1 - x_2$)

From Fig. 2 it is noticed that at a certain instant, the deformation of the membrane equals the deformation of the ball.

The deformations from Fig. 2 are represented for a longer period in Fig. 3. It can be remarked that both the projectile and the membrane get an oscillatory motion as a result of at least two harmonics and therefore, the relative displacement of the projectile with respect to the target is a quasi sinusoidal oscillation.

The plot force-relative deformation of the projectile presented in Fig. 4 is of particular attention.

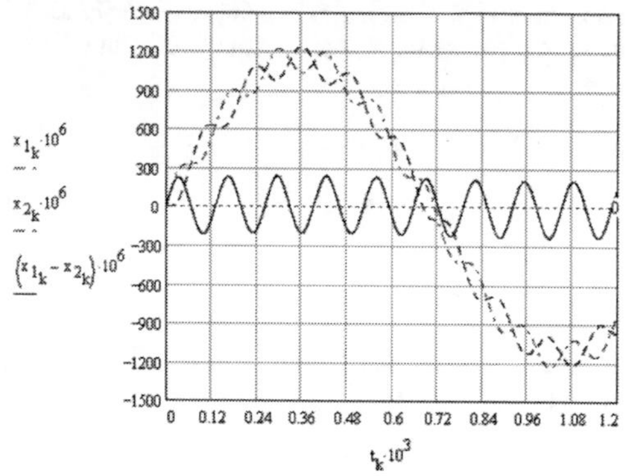

Fig. 3. Absolute deformations (of the ball x_1, of the membrane x_2) and relative deformation ($x_1 - x_2$) for longer time

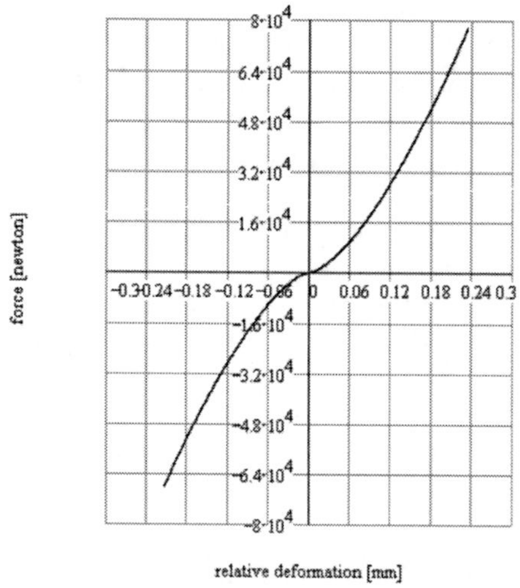

Fig. 4. Force-relative deformation curve for the projectile

From Fig. 4 it can be concluded that the system of differential equation 1 describe the behavior of an un-damped system since for a damped action, a closed force-deformation curve entrapping an area equal to the work of damping should occur.

The fact that the model of a composite material exhibits un-damped behaviour is a major deficiency of the representation.

3 Proposal of improved model

In order to improve the model proposed by Abrate, a viscous dashpot (damper) is attached to the model, connected in parallel to the spring between the projectile and the membrane. As it can be observed form Fig. 5, the damping constant was considered variable.

This requirement was applied by Lankarani and Nikravesh [19], and Flores et al. [20]; for modeling an impact phenomenon with viscous damping, they establish the model based on the remark made by Hunt and Crossley [21], who showed that when the viscous damping depends only on the relative velocity the hysteresis loop is open in origin, this meaning that at the end of the impact, the two bodies attract to each other instead of rejecting.

Based on the observations made by Dubowsky and Freudenstein [22-23], Hunt and Crossley [21] showed that for obtaining a loop closed in origin it is necessary that the damping force should be proportional both to relative velocity and to the contact force exerted when no damping is present.

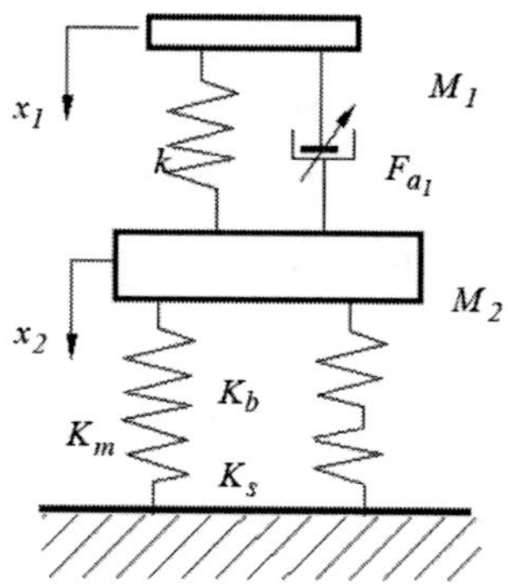

Fig. 5. Proposed model with damper

Founded on this remark, a modified equation of motion, that considers the damping occurring in the ball-composite contact, is proposed under the following form:

$$\begin{cases} \ddot{x}_1 = \left[-\dfrac{K}{M_1} - \dfrac{\kappa}{M_1}(\dot{x}_1 - \dot{x}_2)\right] |x_1 - x_2|^{\alpha} \, sgn(\dot{x}_1 - \dot{x}_2) \\ \ddot{x}_2 = \left[-\dfrac{K_b}{M_1} x_2 - \dfrac{\kappa}{M_2} x_2^3\right] + \\ \quad + \left[\dfrac{K}{M_1} + \dfrac{\kappa}{M_1}(\dot{x}_1 - \dot{x}_2)\right] |x_1 - x_2|^{\alpha} \, sgn(\dot{x}_1 - \dot{x}_2) \end{cases} \tag{8}$$

In the relations 8, the parameter κ is the characteristic of damping and $\alpha = 3/2$ is the exponent typical to the Hertzian contact [24-27].

The system of differential equations 8 was integrated numerically for the parameters 7 and for the same initial conditions 6, for two values of the κ constant, that is: $\kappa = 10^9$, $\kappa = 3 \cdot 10^9$ (units $N \cdot s / m^{1+\alpha}$).

For thes3e two values of the κ constant, there are presented comparatively:

- the absolute displacements of the projectile and target and the relative displacement, in Fig. 6;
- the impact forces versus time in Fig. 7;
- the hysteresis loops in Fig. 8.

https://doi.org/10.1051/matecconf/201818401008

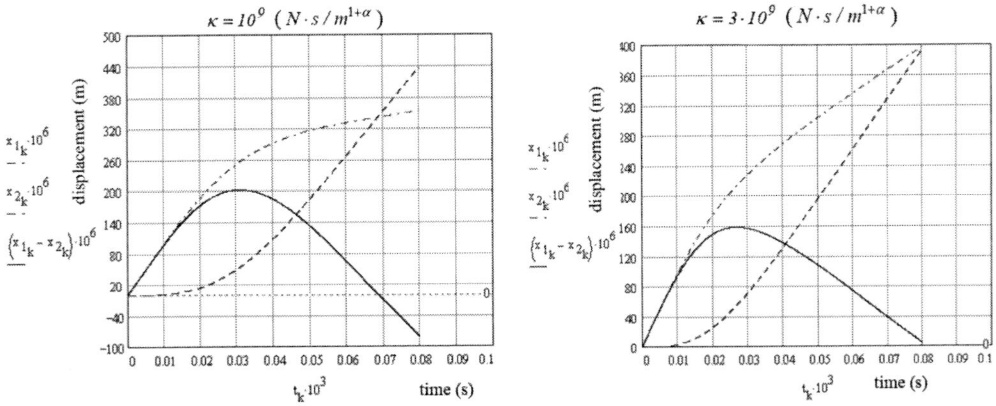

Fig. 6. Displacements of the projectile and target

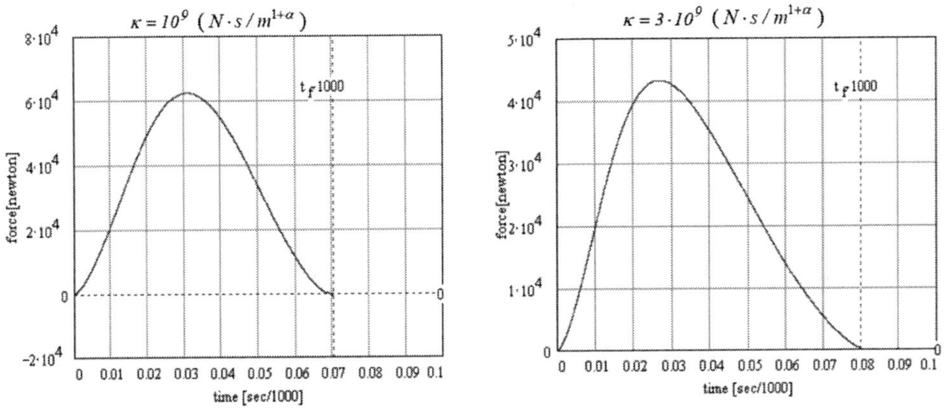

Fig. 7. Impact force variation versus time

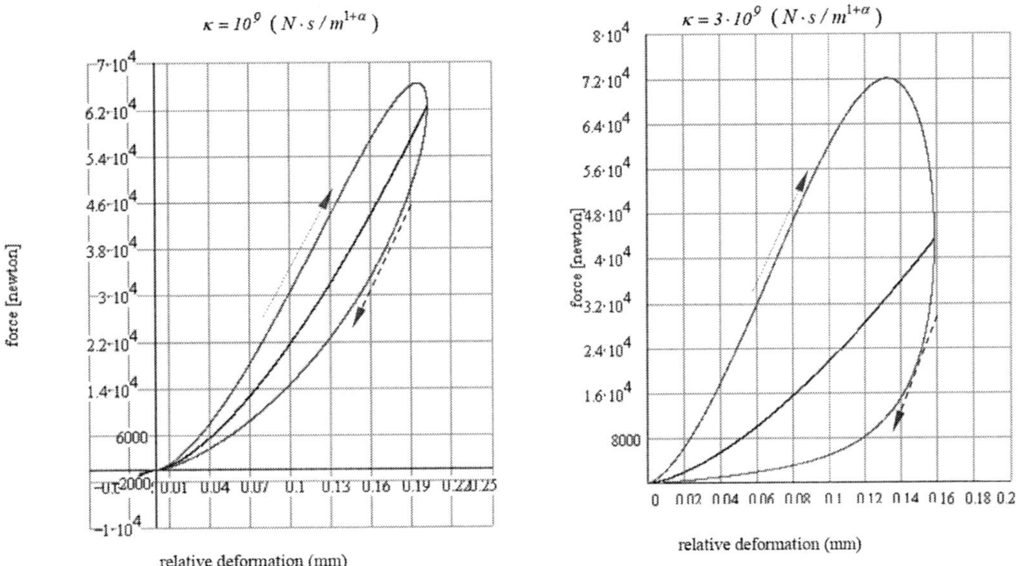

Fig. 8. Hysteresis loops

* Corresponding author: florina@fim.usv.ro

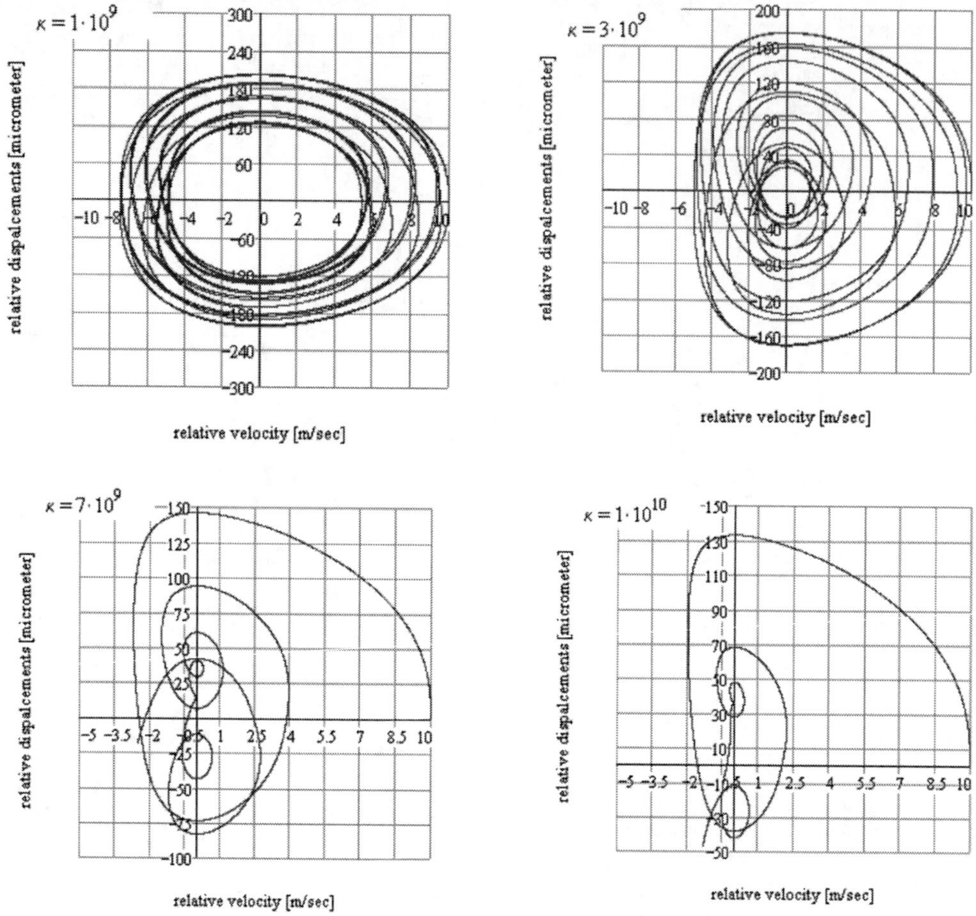

Fig. 9. Phase diagram for different damping values

A very useful instrument in dynamical analysis of comportment of a system is the phase diagram. In Fig. 9, for the model from Fig. 5, the phase diagrams corresponding for four value of damping are shown.

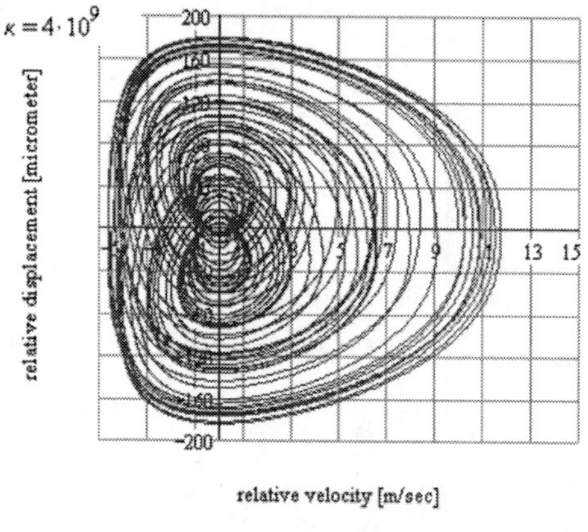

Fig. 10. Phase diagram for critical damping

It can be observed that for lower damping there is only an asymptotic point, while for higher damping values there are two asymptotic points. That means that there is a critical value of damping which splits the two aspects.

In Fig. 10 it is presented the phase diagram for critical damping of the system from Fig. 5.

A number of important conclusions result from Figs. 6-8:

1. The proposed model demonstrates damping with the hysteresis loop closing in origin.
2. The damping work is more important in the approaching phase than during the detaching phase.
3. The effect of increasing the damping is a slight increase of the time when the displacements of the projectile and target are equal.
4. The area of the hysteresis loop increases significantly when the damping augments.
5. It is observed that the maximum impact force is not in phase with the minimum relative displacement. When the maximum impact force is attained, the minimum relative displacement is delayed with respect to this moment.

5. Exaggerate augmentation of damping may lead to the increase of maximum impact force.
6. There is an optimum value of damping to which a minimum value of the maximum impact force corresponds.
7. In the phase diagram, the characteristic point describes a selfintersecting path, occurring the possibility that, for higher damping, two asymptotic points exist as indication of bifurcation phenomenon.

Conclusions

The present paper aims to provide a model for the impact between the sand or dust particles and the surface of the rotor blade of a wind turbine made of composite material. The analysis starts from a model for the impact of composites, broadly applied, and it is highlighted the advantage of this model - the fact that the impact force presents a nonlinear variation, but also the weak point - namely the lack of energy dissipation, phenomenon viewed as a fundamental requirement for composite materials. An improved variant of the existent model is proposed in the present work consisting in introducing a nonlinear viscous damper between the projectile and the target made of composite material. Based on recent papers, the damping force is assumed simultaneously proportional to the velocity and to the elastic force. In this manner, the hysteresis loop closed in origin is ensured. The equation of the model is integrated for a few particular cases and the results are presented in graphical form.

This work was supported by a grant of the Romanian Ministry of Research and Innovation, CCCDI – UEFISCDI, project number PN-III-P1-1.2-PCCDI-2017-0404 / 31PCCDI/2018 , within PNCDI III".

References

1. B. Attaf ed., *Advances in Composite Materials - Ecodesign and Analysis* (InTech Rijeka Croatia 2011)
2. E. Salernitano, C. Migliaresi, J. Appl. Biomater. Biomech. **1**(1) 3-18 (2003)
3. Arif Iftekhar, *Standard Handbook of Biomedical Engineering and Design 12.1, chapter 12, Biomedical Composites,* Downloaded from Digital Engineering Library @ McGraw-Hill (www.digitalengineeringlibrary.com), (The McGraw-Hill Companies 2004)
4. T. Kondo, *Piezoelectric and Acoustic Materials for Transducer Applications* Safari A., Akdoğan E.K. (eds) (Springer, Boston, MA 2008)
5. A. C. Baptista, I. Ferreira, J.P. Borges, J. Comp. Biodegr. Polymers, **1**, 56-65 (2013)
6. L. Mishnaevsky, Comput. Mech. 50, 195 (2012)
7. S. Abrate, Appl. Mech. Rev. **44** (4), 155 (1991)
8. S. Abrate, Appl. Mech. Rev. **47**(11), 517 (1994)
9. S. Abrate, Appl. Mech. Rev. **50** (2), 69, (1997)
10. S. Abrate, *Impact on composite structures* (Cambridge University Press 1998)
11. S. Abrate, Composite Structures, **51**, 129 (2001)
12. L. Mishanevsky, K. Branner, H. Nørgaard Petersen, J. Beauson, M. McGugan, B. F. Sørensen, Materials, **10**, 1285 (2017)
13. N. Gauder, J. Phys. Conf. Ser., **524** 012031 (2014)
14. D. M. Helgren, J. M. Prospero, J. of Climate and Applied Meteorology **26**, 9, 1147 (1987)
15. K. L. Johnson, Proc. of the Instit. of Mech. Eng. **196**, 363 (1982)
16. R. Olsson, AIAA J, **30**(6), 1587 (1992)
17. W. Jordan, P. Smith, *Nonlinear differential equation. Problems and Solutions. A Sourcebook for Scientists and Engineers. Oxford Texts in Applied and Engineering Mathematics* (Book 11) (Oxford University Pres 2007)
18. W. Gear, *Numerical Initial Value Problems in Ordinary Differential Equations* (Englewood Cliffs, NJ Prentice-Hall 1971)
19. H. M. Lankarani, P. E. Nikravesh, J. Mech. Design, **112** 369 (1990)
20. P. Flores, J. Ambrósio, J. C. Pimenta Claro, H. M. Lankarani, *Kinematics and Dynamics of Multibody Systems with Imperfect Joints Models and Case Studies* (Springer 2011)
21. K. H. Hunt, F. R. E. Crossley, J. Appl. Mech. **7**, 440 (1975)
22. S. Dubowsky, F. Freudenstein, 1971, J. of Eng. Ind., **B 93**(1), 305 (1971)
23. S. Dubowsky, F. Freudenstein, 1971, J. of Eng. Ind., **B 93**(1), 310 (1971)
24. W. Goldsmith, *Impact, The Theory and Physical Behaviour of Colliding Solids* (Edward Arnold Ltd, London, England 1960).
25. S. Alaci, F. C. Ciornei, D. Amarandei, C. Filote, D. Cerlinca, S. Spinu, S.: Ann. Oradea Univ. Fasc. Manag. Techn. Eng., **X (XX)**, 1.2102, (2011)
26. S. Alaci, F. C. Ciornei, D. Amarandei, C. Filote, L. Irimescu, S. Spinu, Ann. Oradea Univ. Fasc. Manag. Techn. Eng., **X (XX)**, 2.2221 (2011)
27. S. Alaci, D. A. Cerlincă, F. C. Ciornei, C. Filote, G. Frunză, J. of Physics: Conf. Series **585**, 1, 2015.

MATEC Web of Conferences **184**, 01009 (2018)
Annual Session of Scientific Papers IMT ORADEA 2018
https://doi.org/10.1051/matecconf/201818401009

Considerations on finding the rolling and spinning friction coefficients

Florina-Carmen Ciornei[1,*] *Stelian* Alaci[1], *Sorinel-Toderas* Siretean[1], *Mariana-Catalina* Ciornei[2], *Ioan-Bogdan* Dragoi[1] and *Vlăduț-Constantin* Lazăr[1]

[1]Stefan cel Mare University, Suceava, 13 University Str., 720229, Romania
[2]Carol Davila University of Medicine and Pharmacy, Dpt. Clinic 2, 8 Bd. Eroii Sanitari, 050474, Bucharest, Romania

Abstract. The complex motion from a point contact has as consequence the occurrence of a friction torsor within both spinning and rolling friction exist. Additionally, recent researches showed that the two friction moments, spinning and rolling torques, are not proportional to the normal force. Here, the power law dependency is accepted and therefore, besides the coefficients of spinning and rolling friction, the exponents from the relations friction torque-normal force should be first determined. The paper proposes as method for finding the four parameters the use of the inclined plane principle. The acceleration of a revolution body in descending motion on the inclined plane is found for four different values of the tilting angle and a system of four equations is obtained. The detailed procedure of finding the solutions of the system is presented in the present paper.

1 Introduction

The Hertzian type contacts are frequently met in machine building in diverse systems such as: cam mechanisms, gear transmissions, elasto-hydrodynamic variators etc. These mechanical contacts are characterized by the existence between the two bodies of a sliding motion described by the sliding velocity v_{al}, contained in the common tangent plane (P) and a rotation motion, the last described by the relative angular velocity ω with a component about the common normal - the spinning angular velocity ω_s, and another component contained in the tangent plane – the rolling angular velocity ω_r. The presence of the relative motion in the nonconforming contact generates a friction torsor, composed by: the friction force T collinear with the relative velocity v_{al} but of opposite sense; the spinning friction torque M_s, collinear and of opposite sense to the angular spinning velocity and the rolling friction torque, M_r, [1].

The precise evaluation of the components of the friction torsor is a fundamental task in tribological research since the values of these parameters together to the relative motions are directly in control of the energy losses from a mechanical transmission, losses revealed by the value of the mechanical efficiency of the transmission, [2].

Concerning the velocity and the sliding friction T, the theory is well established both for lubricated contacts by the theory of elasto-hydrodynamic lubrication and for dry friction contacts, by the theory based on the Amonton-Coulomb laws, [3]; as for spinning and rolling friction torques, the theory is an open topic and is the subject of numerous scientific works and monographs.

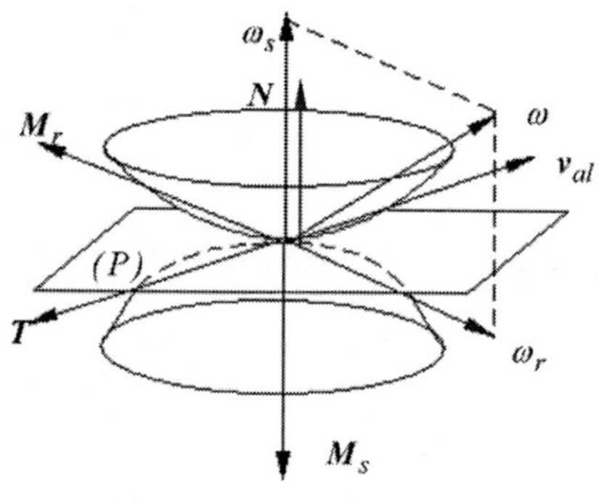

Fig. 1. The relative motions and the elements of the friction torsor in a Hertzian contact

The dynamical treatises accept, in a first approximation, analogous to the case of sliding friction, that both the spinning and rolling friction torque are proportional to the normal force from the contact N, the proportionality factors being the coefficient of spinning friction s_s and the coefficient of rolling friction s_r respectively:

$$M_s = s_s N \quad (1)$$

$$M_r = s_r N \quad (2)$$

[*] Corresponding author: florina@fim.usv.ro

© The Authors, published by EDP Sciences. This is an open access article distributed under the terms of the Creative Commons Attribution License 4.0 (http://creativecommons.org/licenses/by/4.0/).

Recent papers based on elements from the theory of elasticity conclude that equally the spinning friction torque Popov [4] and the rolling friction torque Chjerepanov [5] present a dependency on the normal force of power law type, specifically:

$$M_s = s_s N^{\alpha_s} \tag{3}$$

and

$$M_r = s_r N^{\alpha_r} \tag{4}$$

Under this power-law dependency the tribological and energetically study is obviously more complicated given that not only the coefficients s_s and s_r must be found, but also the exponents α_s and α_r from the relations 3 and 4. In a series of recent works, [6-10], there were presented several methods and the devices used in the estimation of either the coefficient of spinning friction or of the coefficient of rolling friction.

As principle, these techniques employ oscillatory motions between the contacting surfaces or continuous motions. From the schemes based on monotonous velocity motions, one of the most employed is the inclined plane method. But one of the major drawbacks of the method resides in the extremely small value of the slope of the plane. This method gives precise results when the ratio s_r / r (where r is the rolling radius) is comparable to the inclination angle γ of the plane.

$$s_r / r \cong \gamma[rad] \tag{5}$$

When the condition 5 is disobeyed, the results are comparable to the measuring errors. Assuming non-linear dependence between the friction torques and the normal force, the relations 3 and 4, the principle of superposition from linear elasticity cannot be applied and therefore, there are required experimental techniques which allow for simultaneous finding of the characteristic friction parameters of spinning and rolling.

2 The principle of the proposed method

The need of synchronized estimation of spinning and rolling friction torques conducted to the method and device [11] consisting in an inclined conduct (groove) with respect to horizontal plane. A bearing ball set on the top of the V-groove, contacting both walls, is let into to free motion and the time necessary for running an imposed length along the channel permits finding the acceleration and implicitly gives information about the spinning friction and rolling friction between the wall and the ball.

Accepting linear dependency between the friction torques and the normal force, two experiments carried out for two slopes of the groove are required. The method was provisionally neglected, due to the impossibility of precise controlling of the tilting angle.

The present paper analyses the possibility of finding the characteristic parameters of spinning friction and rolling friction under the assumption of nonlinear dependency on the normal force.

The principle of the method is presented in Fig. 2. The operating principle of the test rig is relatively simple. A bearing ball of R radius that has a cylindrical ring attached to the equatorial zone is set on top of two cylindrical parallel rods. The parallelism between the axes of the rods is accurately controlled with two prismatic parts having each two holes of diameters equal to the ones of the rods. The holes from the two parts are machined simultaneously in order to have precisely the same distance between axes. The horizontality of the plane containing the rods can be controlled with accuracy by placing a ball in contact with the two rods. When the balls are immobile, the plane of their axes can be considered horizontal. A cord winded over the ring passes over a pulley of negligible mass.

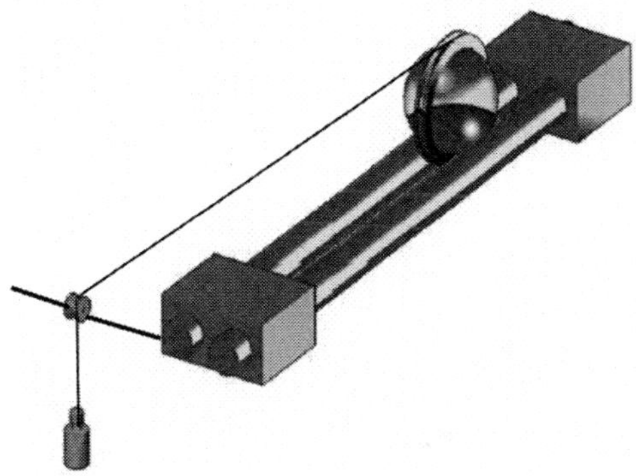

Fig. 2. The principle of the experimental device

For the dynamical analysis, the scheme from Fig. 3 is considered. A body with the mass m is hanging at the end of the cord and has the function of actuating the ball into motion. The ball is contacting the rods in two points, C' and C''. Assuming that there is no slip in these points, the straight line $C'C''$ is the instantaneous axis of rotation and the angular velocity is parallel to it.

In each of the two contact points occur: the friction force, T', T'' parallel to the Ox axis; the moment of spinning friction M'_s, M''_s directed along the normal on the contact point and the moment of rolling friction M'_r, M''_r normal to the M'_s, M''_s, respectively. Applying the hypothesis of dry friction, all the components of the friction torsor from the two contact points are constant and thus it is expected that the motion of the ball is uniformly accelerated.

The theorem of motion of center of mass and the moment of momentum theorem are applied with respect to the center of mass in order to find the law of motion of the ball.

The center of mass moves along the horizontal direction and the abscissa x_G describes its position. The ball also has a rotation motion characterized by the angle φ. It is considered that initially:

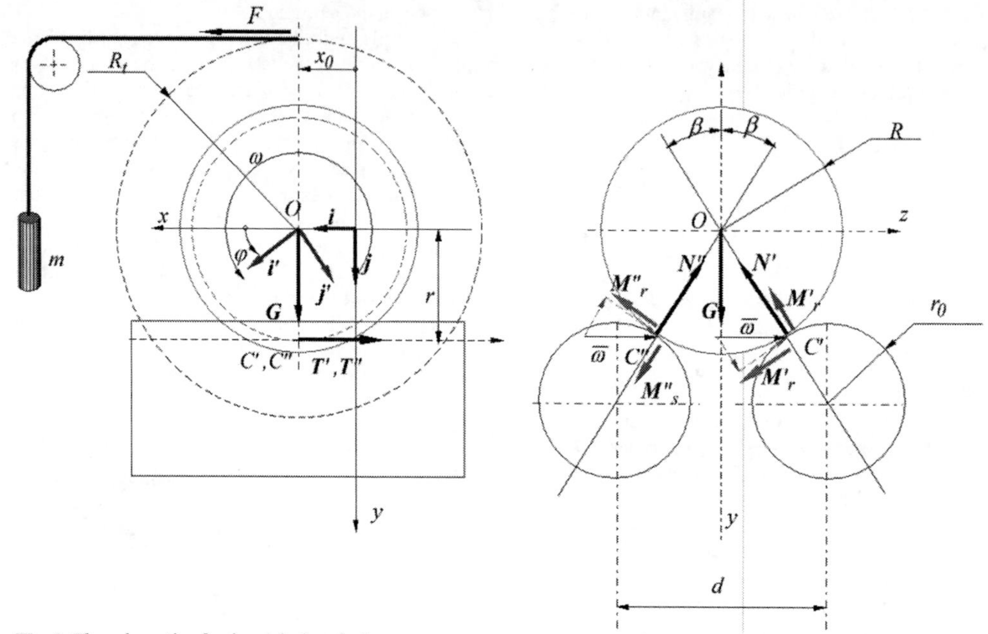

Fig. 3. The schematics for dynamical analysis

$$x = 0, \varphi = 0 \tag{6}$$

The pure rolling condition is expressed as:

$$x_G = r\varphi \tag{7}$$

where r is the distance from the center of the ball to the line of the contact points $C'C''$. Thus, one of the parameters φ or x_G is sufficient to characterize the position of the ball. The equation of motion of the center of the ball has the following form:

$$M\ddot{\boldsymbol{r}}_G = \boldsymbol{G} + \boldsymbol{N}' + \boldsymbol{N}'' + \boldsymbol{T}' + \boldsymbol{T}'' + \boldsymbol{F} \tag{8}$$

where $\boldsymbol{F}$ is the tension from the rope, equal in modulus with:

$$F = mg \tag{9}$$

Due to symmetry, the following equalities are valid:

$$N' = N'', T' = T'' \tag{10}$$

and are considered in the projections of the equation 8 on the axes. It results the following system:

$$Mr\varepsilon = mg - 2T \tag{11}$$

$$0 = 2N\cos\beta - Mg$$

And the following expressions result:

$$T = \frac{1}{2}mg - \frac{1}{2}Mr\varepsilon \tag{12}$$

$$N = \frac{1}{2}Mg\cos\beta$$

Based on Fig. 1, the angle can be expressed as

$$\beta = asin\frac{d}{2(R + r_0)} \tag{13}$$

where r_0 is the radius of the rods. The moment of momentum theorem, expressed as:

$$J\varepsilon\boldsymbol{k} = \boldsymbol{r}_{C'} \times (\boldsymbol{N}' + \boldsymbol{T}') + \boldsymbol{r}_{C''} \times (\boldsymbol{N}' + \boldsymbol{T}'') + mgR_t\boldsymbol{k} + \boldsymbol{M}'_s + \boldsymbol{M}''_s + \boldsymbol{M}'_r + \boldsymbol{M}''_r \tag{14}$$

where R_t is the radius of winding of the ring, has projection only on Oz axis. Based on the scheme from Fig. 2, the following scalar equation is obtained:

$$\varepsilon = \frac{mgR_t + 2rT\cos\beta - 2M_t\sin\beta - 2M_r\sin\beta}{J_G} \tag{15}$$

In the equation 15 there are introduced the proposed dependencies 3 and 4 for the two friction torques and the relation 12 for the modulus of the friction force; the final form of the equation of motion of the ball is obtained:

$$\varepsilon = \frac{\left[\begin{array}{l} mgR\cos^2\beta - 2s_s\left(\dfrac{Mg}{2\cos\beta}\right)^{\alpha_s}\sin\beta \\ -2s_r\left(\dfrac{Mg}{2\cos\beta}\right)^{\alpha_r}\cos\beta + mgR_t \end{array}\right]}{J_G + MR^2\cos\beta^3} \tag{16}$$

The relation 16 can now be regarded as a constraining relationship linking the parameters $s_r, s_s, \alpha_r, \alpha_s$:

$$\varepsilon(J_G+MR^2\cos\beta^3)+2s_s\left(\frac{Mg}{2\cos\beta}\right)^{\alpha_s}\sin\beta$$
$$+2s_r\left(\frac{Mg}{2\cos\beta}\right)^{\alpha_r}\cos\beta-mgR\cos^2\beta=0 \qquad (17)$$

3 Discussions concerning the possibilities of finding the tribological parameters

As affirmed previously, the experimental tests aim to find the parameters s_s, , α_s and α_r and these will permit establishing the spinning friction torque and rolling friction torque, M_s and M_r, respectively. It results that four tests with different values of the parameters involved in equation 17 are necessary. As it can be noticed, the constructive parameters of the device that can be varied are the angle β and the mass m of the actuating weight. The design of the device allows for accurate adjustment by using prismatic parts for supporting the rods with different distances between the guide-holes. Four equations having the following form are obtained:

$$\varepsilon_k(J_G+MR^2\cos{\beta_k}^3)+2s_s\left(\frac{Mg}{2\cos\beta}\right)^{\alpha_s}\sin\beta_k$$
$$+2s_r\left(\frac{Mg}{2\cos\beta_k}\right)^{\alpha_r}\cos\beta_k-m_kgR\cos^2\beta_k=0, \qquad (18)$$
$$k=1\div4$$

The system 18 is system of transcendental equations and for solving it the use of a numerical method (Newton-Raphson) is required. The calculus can be simplified if it considered the notice that the equations 18 are linear in s_s and s_r. Therefore, from the last two relations, s_s and s_r are expressed as functions of α_s and α_r.

$$s_s=s_s(\alpha_r,\alpha_s) \qquad (19)$$
$$s_r=s_r(\alpha_s,\alpha_r)$$

Due to lack of space the explicit form of the equation 19 is not presented. The equations 19 are introduced in the first two equations 18 and it results a system of transcendental equations in α_s, α_r having the form:

$$H_k(\alpha_s,\alpha_r)=\varepsilon_k(J_G+MR^2\cos{\beta_k}^3)-m_kgR\cos^2\beta_k$$
$$2s_s(\alpha_s,\alpha_r)\left(\frac{Mg}{2\cos\beta}\right)^{\alpha_s}\sin\beta_k$$
$$+2s_r(\alpha_s,\alpha_r)\left(\frac{Mg}{2\cos\beta_k}\right)^{\alpha_r}\cos\beta_k=0\quad k=1,2$$
$$(20)$$

The numerical solution of the system 20 written in the simplified manner:

$$\begin{cases}H_1(\alpha_s,\alpha_r)=0\\H_2(\alpha_s,\alpha_r)=0\end{cases} \qquad (21)$$

was strongly dependant on the precision of experimental finding of the angular acceleration ε_k.

To illustrate this affirmation, the following example is given. For the parameters of the rotor $R=0.03\,m$, $R_t=0.033\,m$, $M=0.882kg$ and $J_G=3.493kg\cdot m^2$, the desired parameters were assumed known $\alpha_s=1.5$, $\alpha_r=1.3$ $s_r=2\cdot10^{-4}$ $s_s=5\cdot10^{-4}$ and additionally the values of the mass $m_1=0.07$ $m_2=0.05$ $m_3=0.03$ $m_4=0.05$ (kg) and the angles $\beta_1=0.29rad$, $\beta_2=0.443$, $\beta_3=0.608$, $\beta_4=0.796$ (rad) and the angular acceleration for descending motion were found.

$$\varepsilon_1=35.123;\quad \varepsilon_2=22.765;\quad \varepsilon_3=7.51;\quad \varepsilon_4=15.254. \qquad (22)$$

Now, the values of the accelerations ε_k and masses m_k are introduced in the equations 20 with the unknowns α_s and α_r and the solution obviously must be:

$$\alpha_s=1.5,\ \alpha_r=1.3 \qquad (23)$$

$$\begin{bmatrix}\varepsilon_1\\\varepsilon_2\\\varepsilon_3\\\varepsilon_4\end{bmatrix}=\begin{bmatrix}35.088\\22.765\\7.51\\15.254\end{bmatrix}$$

x := 1.5   y := 1.5

Given

H1(x, y) = 0   H2(x, y) = 0

$$\text{find}(x,y)=\begin{pmatrix}1.497\\1.34\end{pmatrix}$$

Considering that the angular acceleration ε_1 was found with a *0.2%* error, it is obtained (from the Mathcad program written for the study) that:

$$\begin{bmatrix}\varepsilon_1\\\varepsilon_2\\\varepsilon_3\\\varepsilon_4\end{bmatrix}=\begin{bmatrix}35.159\\22.765\\7.51\\15.254\end{bmatrix}$$

x := 1.5   y := 1.5

Given

H1(x, y) = 0   H2(x, y) = 0

find(x, y) =

did not find solution

To clarify this behaviour, the variations of the two function from system 21 were represented for values of α_s and α_r in the range $(1,2)$. The two functions are represented in Fig. 4 by the level curves.

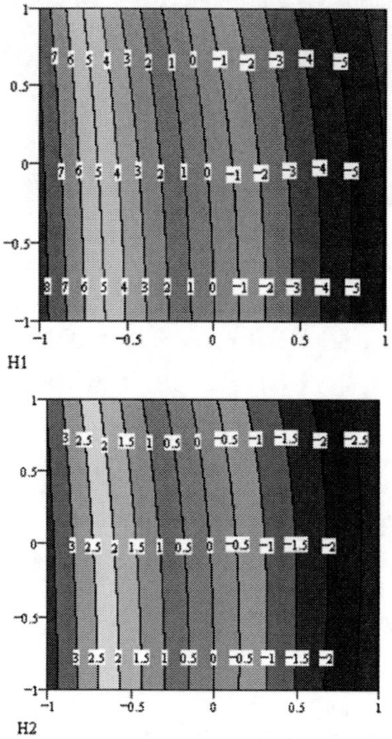

Fig. 4. The functions $H_1; H_2$ plotted by level curves

As it can be noticed form Fig. 4, the curves $H_1 = 0$ and $H_2 = 0$ are very closed to each other and it is difficult to identify the point of intersection.

The following function is defined:

$$H = \sqrt{H_1^2 + H_2^2} \qquad (24)$$

If the system 21 has a stable solution the plot of the function H should present distinct minimum points.

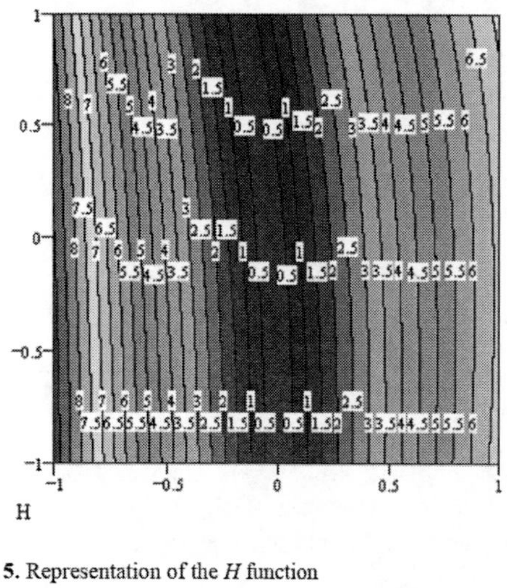

Fig. 5. Representation of the H function

As it can be observed from Fig. 5, the region where the H presents values close to zero is very wide on one direction and this fact justifies the instability of the solution.

Next, the above methodology applied to a problem with a stable solution is presented. More exactly, it is considered a problem where the functions H_1 and H_2 have simple expression, so the solution of the system 21 can be obtain analytically.

The problem of finding the intersection between a circle and an ellipse is considered. The equation of the circle is:

$$H_1(x,y) = x^2 + y^2 - r^2 = 0 \qquad (25)$$

where r is the radius of the circle. The equation of the ellipse is:

$$H_2(x,y) = \frac{x^2}{a^2} + \frac{y^2}{b^2} - 1 = 0\,, \qquad (26)$$

where a,b are the ellipse semi-axes. The graphical solution of the problem is presented in Fig. 6.

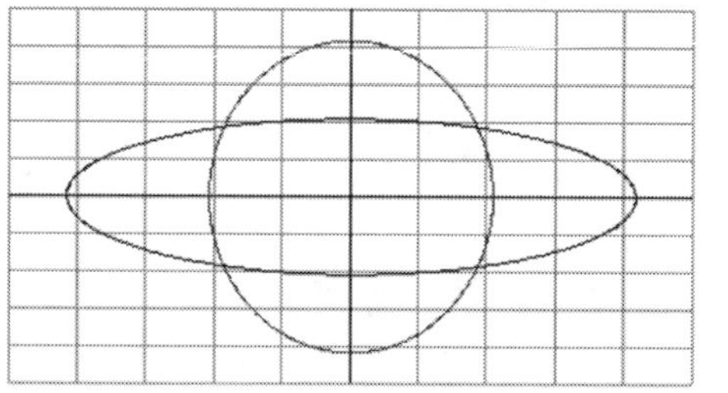

Fig. 6. The graphical solution of the problem

Next, in Figs. 7 and 8 the contour plots of the functions $H1$ and $H2$ (relations 25 and 26, respectively) are traced.

The contour lines of function H are plotted in Fig. 9. Due to the existence of stable solutions of the problem, the domains from xy plane where the solutions should be looked for are well defined

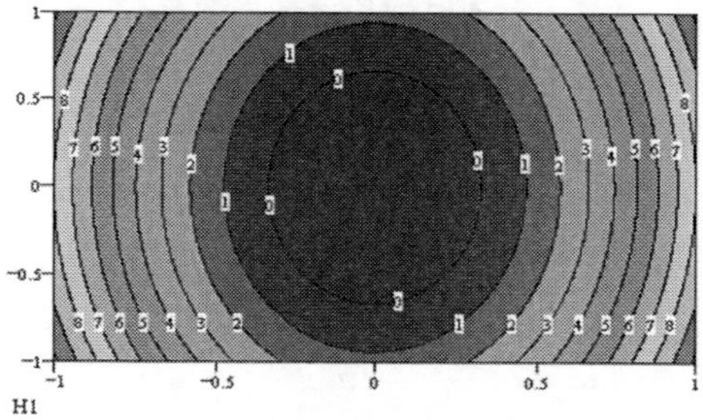

Fig.7. Contour plot of H1 function

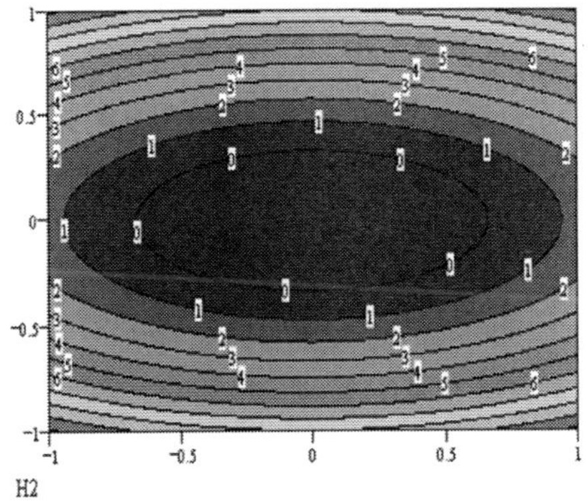

Fig.8. Contour plot of H2 function

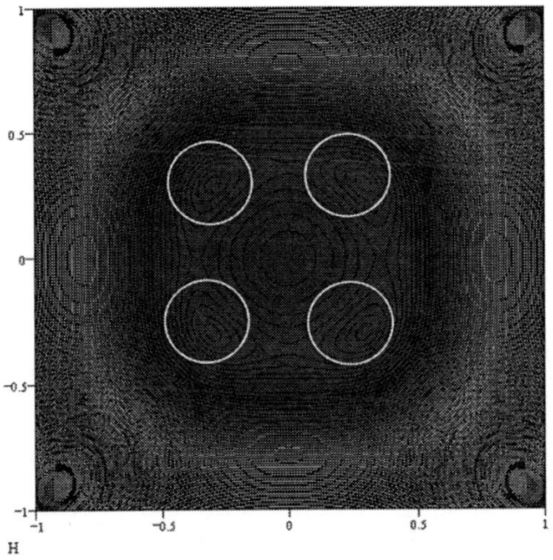

Fig. 9. The contour plots of H function and the domains of the solutions

4 Conclusions

The paper shows that when a bearing ball is contacting two identical metallic parallel rods, two contact points occur and here there are present both spinning and rolling friction torques. It is assumed that each of the two moments depends on the normal force according to a power law relation and four parameters are necessary for characterizing the two moments: the coefficients and the exponents from the two relations moment-normal force.

The method proposed in order to establish these parameters is finding the acceleration of the ball for downward motion for four different inclination angles of the plane of the two rods. Consequently, a system of four equations with four unknowns is obtained. The system is linear with respect to the coefficients of spinning and rolling friction allows for reducing the system to a linear one with two unknowns, the two exponents.

For an actual case, when the characteristics of the system and the values of the four unknowns were assumed identified, it is proved that the system is extremely instable. Practically, an insignificant variation of the known parameters produces a substantial variation of the solution.

This work was supported by a grant of the Romanian Ministry of Research and Innovation, CCCDI – UEFISCDI, project number PN-III-P1-1.2-PCCDI-2017-0404 / 31PCCDI / 2018 , within PNCDI III".

References

1. K. L. Johnson, *Contact Mechanics* (Cambridge University Press 1985)
2. F. Schmelz, H.-C. Seherr-Thoss, E. Aucktor, *Universal Joints and Driveshafts, Analysis, Design, Applications* (Springer 2006)
3. N. Alves, M. T. Peixinho, P. Silva, P. Flores, H. M. Lankarani, Mech. and Mach. Theory, **85**, 172 (2015)
4. V. L. Popov, *Contact Mechanics and Friction. Physical Principles and Applications* (Springer 2017)
5. G. P. Cherepanov, J. Appl. Mech. Techn. Physics **55** (1) 182 (2014)
6. S. Alaci, F. C. Ciornei, D. Amarandei, L. Irimescu, I. C. Romanu, M. A. Rotar, IOP Conf. Ser. - Mater. Sci. Eng. **174** 012008 (2017)
7. M. C. Ciornei, S. Alaci, F. C. Ciornei., I. C. Romanu, IOP Conf. Ser. - Mater. Sci. Eng, **227**, 012027 (2017)
8. S. Alaci, L. Irimescu, M. A. Popescu, D. Fodorcan, D. Amarandei, F. C. Ciornei, MATEC Web of Conf., **112,** 07002 IManE&E (2017)
9. F. C. Ciornei, S. Alaci, R. I. Bedrule, C. M. Drelciuc, Robotica & Management **18,** 2, 13 (2013)
10. S. Alaci, F. C. Ciornei, A. Ciogole, M. C. Ciornei, IOP Conf. Series: Materials Science and Engineering, **200,** 012005 (2017)
11. S. Alaci, E. Diaconescu, F. C. Ciornei, L. Irimescu, D. Cerlinca, Proceedings Rotrib 10, 1 (2010)

Graphical-analytical analysis of the mechanism with rotating cam and flat-face follower

Stelian Alaci[1,*], *Florina-Carmen* Ciornei[1], *Sorinel-Toderas* Siretean[1] and *Mariana-Catalina* Ciornei[2]

[1]Stefan cel Mare University, Suceava, 13 University Str., 720229, Romania
[2]Carol Davila University of Medicine and Pharmacy, Dpt. Clinic 2, 8 Bd. Eroii Sanitari, 050474, Bucharest, Romania

Abstract. The graphical-analytical methods are mentioned as being straightforward and expedite. The paper presents a graphical-analytical method for the kinematical analysis of a mechanism with rotating cam and oscillating flat-face follower. The manner supposes considering the contact point as a separate element of the mechanism and using concurrently the replacing mechanism with lower pairs and the actual mechanism, the relative motions from the contact point of the cam and contact point of the follower, respectively, are established. The methodology is exemplified for an actual case.

1 Introduction

The cam mechanisms are identified taking into account the motion of the cam, the motion of the follower and the type of contact between the cam and the follower [1]. Considering the last criterion, the followers are of knife-edge type, when a single contact point exists with the cam, and from flat-face category, when the follower is tangent permanently to the profile of the cam. The kinematical study mechanisms can be made applying graphical, graphical-analytical or analytical procedures. The graphical-analytical methodologies have as foremost advantages the straightforward mathematical apparatus and quickness of solving. The main drawback of the method consists in the fact that it can be applied for a specified position of the cam. This inconvenience can be eliminated by the analytical methods. The disadvantage of the analytical methods resides in the complex mathematical apparatus, [2-3].

2 General considerations upon cam mechanisms

The graphical-analytical kinematics of a cam mechanism requires as principle, establishing the analytical expressions for the velocity/acceleration of a point of the mechanism (for the actual case, the contact point) in two manners, and afterwards solving graphically these equations by means of the velocities and accelerations polygons. In the case of the mechanisms with flat face follower, these systems are undetermined systems (as it will be proved later). An artifice is presented in [4] for overcoming this drawback, that is using a fictive follower with the knife edge placed in the cam-follower contact point. The fictive follower creates two mechanisms, the first one with the cam and the second one with the follower. The constraint imposed to the fictive follower, to have the same motion in both mechanisms, conducts to the absolute motion of the follower and the motion from the higher pair. Pelecudi [5] remarks that there are several equivalent mechanisms for the cam mechanism and proposes for solving the problem the selection of a certain mechanism. A method for the kinematical graphical-analytical analysis was propose and applied in a recent paper [6] for a mechanism with cam and curved follower with non-zero curvature. In the paper it is noticed that the method cannot be applied for mechanisms with rotating cam and flat face follower, Fig. 1, since the two mechanisms are not structurally equivalent. For this reason, a method for kinematical analysis of the mechanism with flat face follower is presented next with emphasis upon the causes impeding the direct establishing of the motion of the follower and of the relative cam-follower motion.

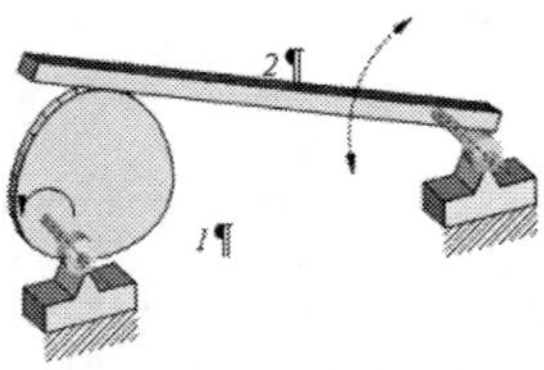

Fig. 1. Mechanism with rotating cam and oscillating flat face follower

To this end, the mechanisms with oscillating flat-face follower from Fig. 2 is considered, where the cam is made from a disc with R radius mounted eccentrically at distance δ with respect to the centre. The equivalent mechanism with plane pairs is also shown in Fig. 2. The vector contour from Fig. 3 is considered and the vector equation of closed vector contour is:

$$\boldsymbol{L_1} + \boldsymbol{L_2} + \boldsymbol{L_3} + \boldsymbol{L_0} = \boldsymbol{0} \quad (1)$$

* Corresponding author: alaci@fim.usv.ro

© The Authors, published by EDP Sciences. This is an open access article distributed under the terms of the Creative Commons Attribution License 4.0 (http://creativecommons.org/licenses/by/4.0/).

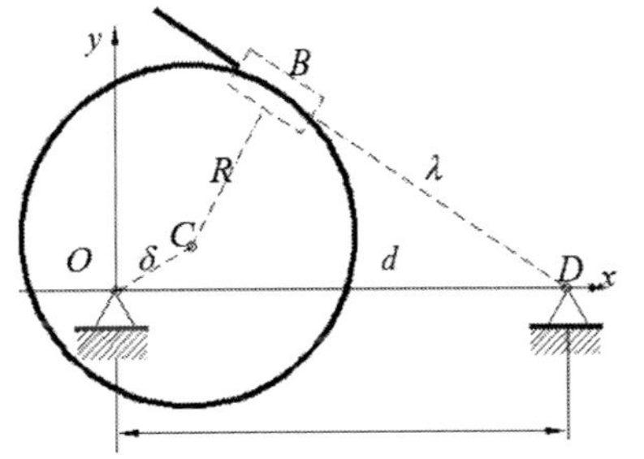

Fig. 2. Circular eccentric with oscillating flat-face follower

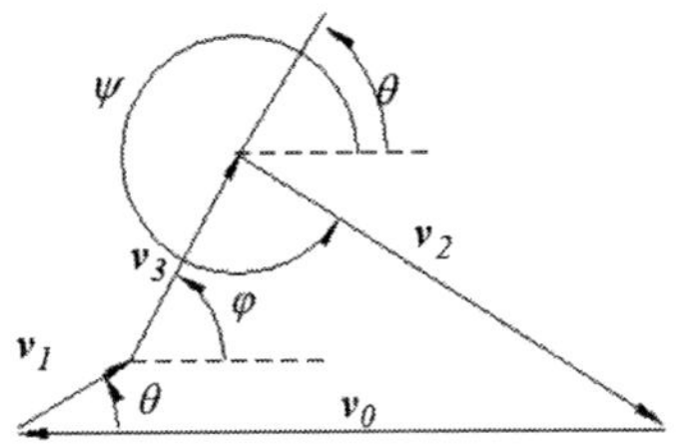

Fig. 3. The equivalent mechanism of the cam mechanism

and has the equations of projections:

$$\begin{cases}\delta \cos\theta + R\cos\varphi + \lambda\cos\psi - d = 0 \\ \delta \sin\theta + R\sin\varphi + \lambda\sin\psi = 0\end{cases} \tag{2}$$

with φ, ψ and ψ as unknowns. Adding the equation:

$$\psi = \varphi + 3\pi/2 \tag{3}$$

the system becomes compatible and has the solutions:

$$\varphi_{1,2} = 2\,atan\frac{d\sin\theta \pm \sqrt{\delta^2 - R^2 - 2\delta d\cos\delta + \delta^2}}{\delta\cos R - d - R} \tag{4}$$

In Fig. 4 there are presented the cam and the follower occupying the two positions and afterwards, using the relations:

$$\begin{aligned} x_{C_{1,2}} &= \delta\cos\theta + R\cos\varphi_{1,2} \\ y_{C_{1,2}} &= \delta\sin\varphi + R\sin\varphi_{1,2}\end{aligned} \tag{5}$$

the two trajectories of the contact points from the two mechanisms were traces. There were also represented two knife-edge followers having the same joint D with the ground having the knife-edges in the two contact points. From this representation one can notice the essential difference between the two mechanisms. While for the knife-edge follower the contact point is immobile on the follower, the distance λ being constant, and the absolute trajectory is represented by the circles $\Gamma_{1,2}$ with the centers in the joint D, for the case of the mechanisms with plat-face follower, the contact point is mobile both on the cam and on the follower. As well, the trajectories of the contact points are two unknown curves and therefore, for a given position of the cam, the directions of the absolute velocities $\boldsymbol{v}_{C_{12}}$ of the contact points are not known, compared to the case of the knife-edge follower, when the velocity of the contact point is always normal to the direction of the follower.

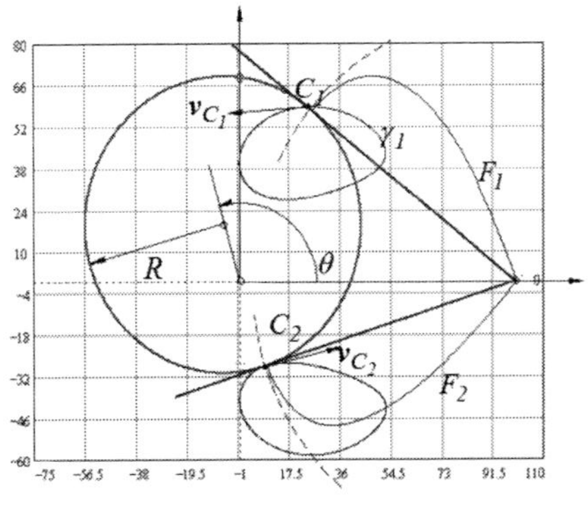

Fig. 4. The cam and the follower occupying the two positions corresponding to the solutions

From here it results the conclusion that three unknowns occur in the system of plan vector equations that expresses the absolute velocity of the contact point: the absolute motion of the follower and the relative motions of the contact point with respect to the cam and to the follower.

3 The kinematical graphical-analytical analysis of the mechanism

The motion of a point B_2 on a mobile plane curve is considered and it is assumed that at the moment of the analysis the point B_2 overlies on the B_1 point placed on the curve. The following relations are valid between the velocities of the two points:

$$\boldsymbol{v}_{B_2} = \underline{\underline{\boldsymbol{v}_{B_1}}} + \underset{\|tt}{\underline{\boldsymbol{v}_{B_2B_1}}} \tag{6}$$

and between the accelerations:

$$\boldsymbol{a}_{B_2} = \underline{\underline{\boldsymbol{a}_{B_1}}} + \underset{\|nn}{\underline{\boldsymbol{a}^{tn}_{B_2B_1}}} + \underset{\|nn}{\underline{\boldsymbol{a}^{c}_{B_2B_1}}} + \underset{\|tt}{\underline{\boldsymbol{a}^{tn}_{B_2B_1}}} \tag{7}$$

The corresponding equation in the polygon of velocities:

$$\overline{p_v b_2} = \overline{p_v b_1} + \overline{b_2 b_1} \tag{8}$$

and in the polygon of accelerations, respectively:

$$\overline{p_a b'_2} = \overline{p_a b'_1} + \overline{b'_1 n_{B_2B_1}} + \overline{n_{B_2B_1} k_{B_2B_1}} + \overline{k_{B_2B_1} b'_2} \tag{9}$$

In the above relations, nn and tt are the notations for the directions parallel to the normal and tangent in the contact point, respectively.

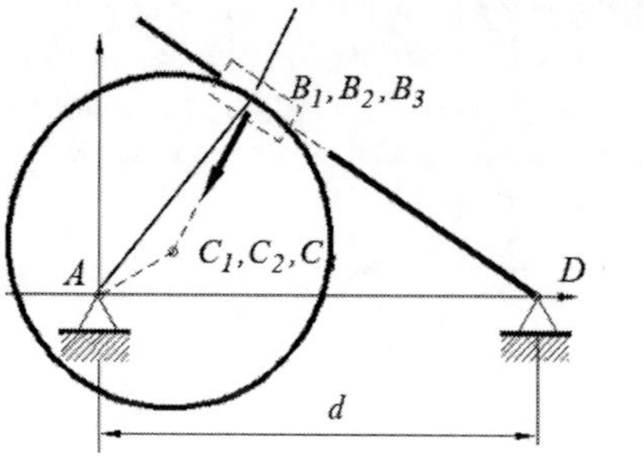

Fig. 5. The actual mechanism and the replacing mechanism

The normal transport acceleration $a^{tn}_{B_2B_1}$ is defined by the relation:

$$a^{tn}_{B_2B_1} = -(v^2_{B_2B_1} / \rho)v \tag{10}$$

where ρ is the curvature radius and n is the versor of the normal. The relative Coriolis acceleration is:

$$a^{c}_{B_2B_1} = 2\omega_1 \times v_{B_2B_1} \tag{11}$$

where ω_1 is the angular velocity of the plane containing the curve. The relative acceleration is $a^{r}_{B_2B_1}$, directed about the tangent in the contact point. Both the actual mechanism and the replacing mechanism are presented in Fig. 5. The element 3 replaces the higher pair and has at the ends a rotation joint placed in the centre of curvature of the cam and a prismatic joint in the contact point. In the centre of curvature of the cam and in the contact point there are the points C_1, C_2, C_3 and B_1, B_2, B_3 placed on the cam, on the follower and on the replacing element, respectively. Applying the relation 5 it is obtained, for the points B_3 and B_1:

$$\begin{aligned} v_{B_3} &= \underline{\underline{v_{B_1}}} + \underline{v_{B_3B_1}} \\ v_{B_3} &= \underline{\underline{v_{B_2}}} + \underline{v_{B_3B_2}} \end{aligned} \tag{12}$$

and for the points B_3 and B_2, respectively:

$$\begin{aligned} a_{B_3} &= \underline{\underline{a_{B_1}}} + a^{tn}_{B_3B_1} + a^{c}_{B_3B_1} + a^{tn}_{B_3B_1} \\ a_{B_3} &= \underline{\underline{a_{B_2}}} + a^{tn}_{B_3B_2} + a^{c}_{B_3B_2} + a^{tn}_{B_3B_2} \end{aligned} \tag{13}$$

Concerning the velocities, the directions of the relative velocities $v_{B_3B_1}, v_{B_3B_2}$ are unknown but this can be overcome by subtracting the equations 12, member by member and writing the results under the form:

$$v_{B_2} = v_{B_1} + v_{B_2B_1} \tag{14}$$

In the equation obtained above, v_{B_1} is completely known and the other two vectors have known directions, $v_{B_2} \perp DB$ and $v_{B_2B_1} \parallel tt(BD)$; therefore the equation allows for finding the velocity of the point B_2 and implicitly the angular velocity of the follower and the relative velocity from the higher pair. The challenge of applying the same methodology for the analysis of the accelerations reaches the barrier that there are necessary the relative velocities of the contact point with respect to the cam and to the follower; these velocities must be known for finding the magnitudes of the accelerations $a^{tn}_{B_3B_1}, a^{c}_{B_3B_1}$ and $a^{c}_{B_3B_2}$; the acceleration $a^{tn}_{B_3B_2}$ is zero since the curvature radius of the follower is $\rho_2 \equiv \infty$. To surmount this difficulty it is recalled that due to the prismatic joint from point B the next equality is always valid:

$$v_{B_3B_2} = v_{C_3C_2} \tag{15}$$

In the velocity polygon, the relation 15 is written as:

$$\overline{b_2b_3} = \overline{c_2c_3} \tag{16}$$

Moreover, it is taken into consideration that due to the revolution joint from C, $v_{C_1} \equiv v_{C_3}$ and the points c_1 and c_3 are always overlapped in the velocity polygon. The position of the point c_1 is found by applying the theorem of similar triangles between the known ABC and ab_1c_1. Now, the point c_2 can be found by translation of the point c_3 with the vector b_3b_2. For establishing the accelerations distribution, the equation 4 is applied twice, for the points B_3 and B_1, and for B_3 and B_2 respectively:

$$\begin{cases} \bar{a}_{B_3} = \underline{\underline{\bar{a}^{n}_{B_1A}}} + \underline{\underline{\bar{a}^{t}_{B_1A}}} + \underline{\underline{\bar{a}^{c}_{B_3B_1}}} + \underline{\underline{\bar{a}^{tn}_{B_3B_1}}} + \underset{\parallel tt}{\underline{\bar{a}^{t}_{B_3B_1}}} \\ \bar{a}_{B_3} = \underset{\perp DB}{\underline{\bar{a}^{t}_{B_2D}}} + \underline{\underline{\bar{a}^{n}_{B_2D}}} + \underline{\underline{\bar{a}^{c}_{B_3B_2}}} + \underline{\underline{\bar{a}^{tn}_{B_3B_2}}} + \underset{\parallel tt}{\underline{\bar{a}^{t}_{BB_2}}} \end{cases} \tag{17}$$

The right members of the equations 17 are equalled and introducing the notations:

$$\bar{a}^{c}_{B_2B_1} = \underline{\underline{\bar{a}^{c}_{B_3B_1} - \bar{a}^{c}_{B_3B_2}}} \tag{18}$$

$$\bar{a}^{tn}_{B_2B_1} = \underline{\underline{\bar{a}^{tn}_{B_3B_1}}} \tag{19}$$

it results the following relation:

$$\bar{a}^{t}_{B_3B_1} - \bar{a}^{t}_{B_3B_2} = (\bar{a}^{t}_{B_3} - \bar{a}^{t}_{B_1}) + (\bar{a}^{t}_{B} - \bar{a}^{t}_{B_2}) = \bar{a}^{t}_{B_2B_1} \tag{20}$$

formally rewritten as:

$$\underline{\underline{\overline{a}^n_{B_2D}}} + \underset{\perp DB}{\overline{a}^t_{B_2D}} = \underline{\underline{\overline{a}^n_{B_1A}}} + \underline{\underline{\overline{a}^t_{B_1A}}} + \underline{\underline{\overline{a}^c_{B_2B_1}}} + \underline{\underline{\overline{a}^{tv}_{B_2B_1}}} + \underset{\| tt}{\overline{a}^\tau_{B_2B_1}} \tag{21}$$

The above equation is formally identical to the equation of accelerations corresponding to the knife-edge follower with the remark that $\overline{a}^c_{B_2B_1}$, $\overline{a}^{n\tau}_{B_2B_1}$ are just notations and have different relations compared to the ones known for the two terms. The consequent equation from the polygon of accelerations is:

$$\overline{p_a n_{B_2D}} + \overline{n_{B_2D} b'_2} = \overline{p_a n_{B_1A}} + \overline{n_{B_1A} b'_1} + \overline{b'_1 k_{B_2B_1}} + \overline{n_{B_2B_1} b'_2} \tag{22}$$

The solving of the equation in the accelerations' polygon is presented in Fig. 6.

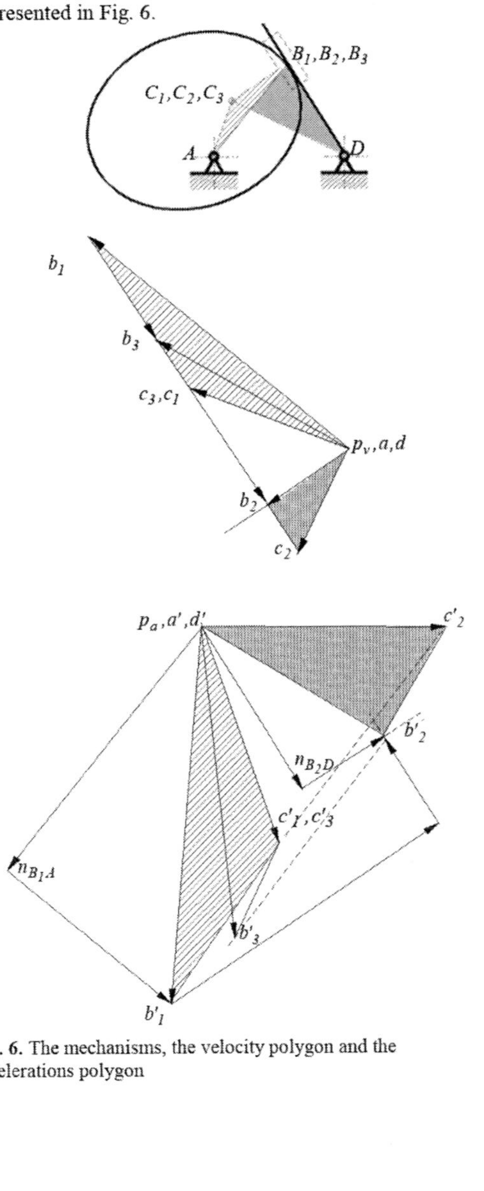

Fig. 6. The mechanisms, the velocity polygon and the accelerations polygon

The acceleration $a^t_{B_2D}$ is found and implicitly the angular acceleration $\overline{\varepsilon}_2$ together to the tangential component of relative acceleration $a^t_{B_2B_1}$ .With the points b'_1 and b'_2 known in this polygon, the points c'_1 and c'_2 are found in a manner similar to the one applied for finding the points c_1 and c_2 from the velocities polygon. In the polygon of accelerations there are constructed the triangles $a'b'_1c'_1 \sim ABC_1$ and $d'b'_2c'_2 \sim DBC_2$ respectively. Finally, the point b'_3 is found from the condition similar to the one from velocities, namely b' divides $c'_1c'_2$ in the same ratio as the point B divides the segment C_1C_2 :

$$\overline{b'_2 b'_3} = \overline{c'_2 c'_3} \tag{23}$$

At this moment, the analysis is complete.

4 Conclusions

The present work proposes a graphical-analytical method for the kinematics analysis for a plane mechanism with rotating cam and oscillating flat-face follower. The principle of the method consists in considering simultaneously both the actual mechanism and the equivalent mechanisms with lower pairs. In this manner, the motion from the cam-follower joint can be decomposed into two motions of the contact point, a first motion of the contact point with respect to the profile of the cam and a second motion between the contact point and the profile of the follower. These two motions are necessary for establishing a number of components of the relative accelerations of the contact point-cam and of the contact point-follower.

This work was supported by a grant of the Romanian Ministry of Research and Innovation, CCCDI – UEFISCDI, project number PN-III-P1-1.2-PCCDI-2017-0404 / 31PCCDI / 2018 , within PNCDI III".

References

1. J. Uicker, G. Pennock, J. Shigley, *Theory of Machines and Mechanisms* (Oxford University Press 2010)
2. J. Angeles, C.S. Lopez-Cajun, *Optimization of Cam Mechanisms* (Kluwer Academic Publishers, Dordrecht 1991)
3. H. A. Rothbart, *Cam Design Handbook* (McGraw Hill 2004)
4. S. Alaci, F. C. Ciornei, C. F, Bull. IP Iasi, **LII (LVI)** 6X (2006)
5. C. Pelecudi,D. Maroş, V. Merticaru, N. Pandrea, I. Simionescu, *Mechanisms,* (EDP Bucureşti 1985)
6. S. Alaci, F.C. Ciornei, F. Buium, I.C. Romanu, O.T. Rusu, ACME 2018 (to be published)

MATEC Web of Conferences **184**, 01011 (2018) https://doi.org/10.1051/matecconf/201818401011
Annual Session of Scientific Papers IMT ORADEA 2018

Analysis of the Evolution of the Number of Fatal Road Accidents in Romania

Lucian Tarnu[1], and *Cristian* Deac[1,*]

[1]Lucian Blaga University of Sibiu, Romania, Department of Industrial Engineering and Management, Emil Cioran str., 4, Sibiu, Romania

Abstract. Fatal car accidents are a major issue of concern for both the police, the local and regional administrations and all those involved in providing road safety. The current paper presents an analysis of the number of fatal car accidents on Romanian roads, in the context of the Road Safety Programme of the European Commission, emphasising the efforts being made in this regard, the areas where road safety is still lacking and recommending some measures for improving the situation.

1 Introduction

Nowadays, road traffic has reached in some areas impressive proportions. Unfortunately, the increase of cars driving on roads and motorways has led in many countries and regions also to an increase in the number of car accidents. Fatal car accidents are especially tragic and costly, both in terms of the material losses and in terms of human lives, representing a major problem of the human society.

According to [1] and [2], respectively, at world level, while in 2002 road traffic injuries were the second-most-frequent cause of death for the age groups 5-14 years and 15-29 years and the third-most frequent cause of death for the age group 30-44 years, in 2012 road traffic injuries were the main cause of death for the age group 15-29 years and it is projected to be ranked 7th among all age groups in 2030.

In the European Union, a first attempt to address this major problem was the setting of a goal to reduce road fatalities on all roads of the Union from 2001 to 2010 by 50%. However, the measures adopted in correlation with this goal were insufficient, so that the achieved reduction level was of only 35%. In 2011 more than 31,500 people lost their lives in the European Union due to car accidents (figure 1 [3]).

Consequently, in 2011, the European Commission's White Paper on Transport and the Action Programme on Road Safety have called for a 50% reduction of the road traffic fatalities in the European Union from 2011 to 2020, aiming towards a rate of zero fatalities from road accidents in 2050 [4].

As can be seen from figure 1, however, there are once more growing discrepancies between the set goal and the achieved reduction in road fatalities.

For its part, Romania has adopted this goal, aiming to reduce road fatalities by half over the period 2011-2020.

In the current paper, the authors analyse the situation of fatal car accidents in Romania at national and county level, compared to the goal set as mentioned above, indicating the main causes identified for these accidents and suggesting a series of measures that would need to be implemented, both at national, regional and local level.

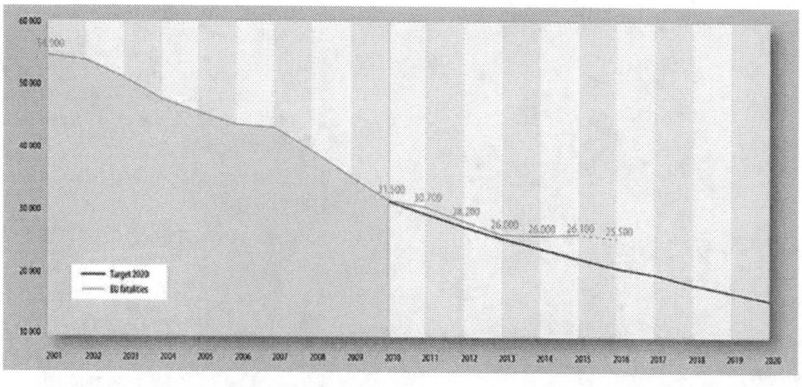

Fig. 1. The number of road fatalities in the EU between 2001 and 2016 compared to the goal set for 2020 [4]

* corresponding author: cristian.deac@ulbsibiu.ro

© The Authors, published by EDP Sciences. This is an open access article distributed under the terms of the Creative Commons Attribution License 4.0 (http://creativecommons.org/licenses/by/4.0/).

2 Road Accidents – Considerations and Causes

Since their invention at the end of the 19th century, cars have gradually transformed and improved human life, they imposed an accelerated rhythm, shortened travel distances and durations and became one of the symbols of modern civilisation [5].

Across the European Union, the number of passenger cars grew by 4.5% over the last five years, from 241 to 252 million [6].

In Romania, cars first appeared very soon after this means of transportation was invented, more precisely around the year 1890. Currently, there are around 5.7 million cars registered in Romania and more than 6.6 million persons have driver's licenses [5].

Unfortunately, car traffic is sometimes accompanied by accidents, so that cars are sometimes called "lethal weapons". Road accidents are a scourge of the civilised world, that leads each year to important material damage, to minor or major injuries or even the death of participants in the road traffic: car drivers, pedestrians, passengers, bicycle riders, motorcycle riders etc [5]. Statistics at world level show that each year, 1.5 million people die in car accidents, while other 35 million receive severe or light injuries [2].

The consequences of fatal road accidents are even more tragic when taking into account the fact that most victims are young people. An accident-produced death is not only a family tragedy, but a significant loss for the whole society, given the efforts undertaken for the education, professional training and training for life of the young generations [5].

Since many people use road vehicles to commute to the workplace or even to carry out their daily jobs, road traffic accidents can be regarded as work accidents, so risks related to road safety can be determined with a methodology similar to that used for determining the risks at the workplace [7].

Road safety issues have become a major point of interest for many governments and administrative institutions.

Road safety policies have to be devised considering the citizens as major action factors and have to encourage people to take on the responsibility for their own safety and for the safety of the other participants in the road traffic situation.

In the European Union, the goal is to achieve an integrated approach towards the various policies. This means that the road safety policy needs to be closely connected to the policies from areas such as energy, environment, workforce market, education, youth, health, research, development and innovation, justice, insurances, trade etc. The policies in these areas need to take into account the goals of the road safety policy and vice-versa.

The road safety policy of the European Union aims to increase the road safety level so as to allow the safe and ecological movement of citizens on all roads of the member states. It promotes the equality of all road traffic participants, seeking to increase the road safety level for the most vulnerable among them [8, 9].

The European Commission considers following three actions as having paramount importance in the area of road safety [9, 10]:

- Implementing a structured and coherent cooperation framework based on the best practices in the member countries, as a necessary condition for efficiently implementing the orientations for the European road safety policy 2011-2020;
- Elaborating a strategy regarding injuries and granting first aid, in order to answer the more and more acute need to reduce the number of road accidents leading to injuries.
- Increasing the safety of vulnerable participants in the road traffic, especially with regard to motorcycle riders, in whose case statistics are worrying

The concept of integrated management in the area of road safety and of traffic incidents prevention is used successfully in the member states of the European Union, aiming at the coordination and cooperation of all specialised institutions and authorities, in order to facilitate the achievement of goals set in the area of road safety.

As with all policies of the European Union, the road safety policy too needed to be assumed and implemented by all member states. In Romania, the National Strategy for Road Safety 2016-2020 [11] was drafted and approved as a document containing coherent and unitary long-term policies in the area of road safety, following the unsatisfactory results of the previously valid policies in this area. The strategy presents the main directions for the policy of Romania with regard to road safety, as well as the means for transposing and improving it. This strategy has already been implemented by the institutions and factors involved in the area of road safety, both at national and at local level.

The key actors involved in road safety aspects in Romania are presented in table 1 [12].

Table 1. Key actors in road safety in Romania [12]

Key functions	Key actors
1. - Formulation of national RS strategy - Setting targets - Development of the RS programme	- The Sectoral Operational Programme "Transport" - Romanian Road Authority - Ministry of Transports and Infrastructure - Directorate for Road Transport
2. Monitoring of RS development in the country	- Romanian Road Authority
3. Improvements in road infrastructure	- Romanian Road Authority
4. Vehicle improvement	- Romanian Automotive Register
5. Improvement in road user education	- Romanian Road Authority - Road Traffic Assistance Assoc.
6. Publicity campaigns	- Traffic Police

 https://doi.org/10.1051/matecconf/201818401011

7. Enforcement of road traffic laws	- The Traffic Police (under Ministry of Administration and Interior)

A road accident in itself does not represent a legal offence, but its causes can be generally traced back to an infringement of the legal provisions related to road safety. The main causes of road accidents in Romania have been identified as follows [5, 8, 9]:

- Excessive speed;
- Not granting the proper passage priority to pedestrians or other vehicles;
- Illegal crossing of the road by pedestrians;
- Carelessness of bicycle and motorcycle riders;
- Insufficient road safety education among the road traffic participants;
- Driving without a valid driver's license;
- Driving under the influence of alcohol Driving
- Driving in a state of advanced tiredness.

In order to improve the situation with regard to the number of fatal car accidents in Romania, there have been taken several measures aimed at fighting the mentioned causes, which led to some progress.

3 Analysis of the road fatalities situation at national and county level in Romania

In Romania, the Inspectorate General of the Romanian Police is the authority mandated to collect and manage the data referring to road accidents. It can then transmit the data to the National Statistics Institute and to the interested international institutions.

For their analysis, the authors resorted to data retrieved from the National Registry of Car Accidents [13].

In 2010, statistics indicated the fact that Romania was ranked last in the European Union with regard to road deaths (figure 2, [14]).

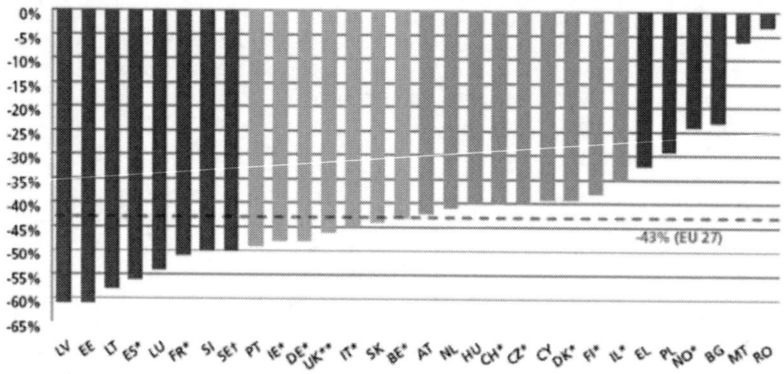

Fig. 2. Status of achievement of the goal of 50% reduction of road fatalities in 2010 [13].

Since then, especially in the context of the new goalsetting for the period 2011-2020, there has been some progress in Romania towards reaching the European goal of a 50% reduction in road fatalities.

According to data from the National Registry of Car Accidents [13], the number of fatalities due to car accidents has varied between 2010 and 2017 as shown by the chart presented in figure 2.

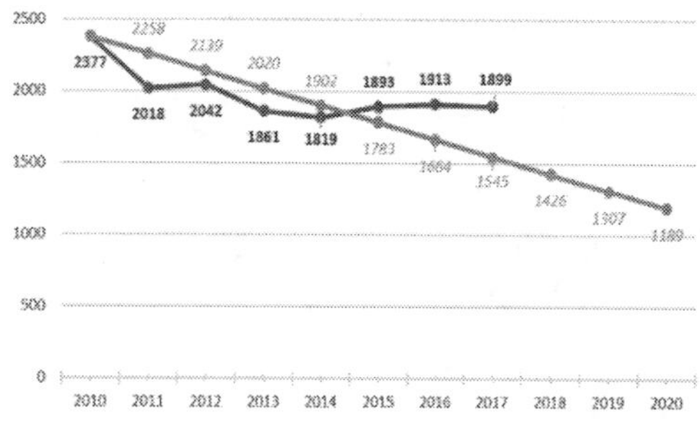

Fig. 3. Number of actual fatalities vs. the goal set for the period 2010-2020 (data from [13])

As shown, despite an initial encouraging reduction in road deaths, the numbers have remained by and large in the same range of values, which should be worrying, given the relatively short time left until the end of the target period (2020).

At county level, the average number of road fatalities per county in the period 2010-2017 was 372. However, there can be noticed a very heterogeneous situation – some counties have registered significant improvements in road safety, visible in a steady reduction of road fatalities, while in others the number of road fatalities has remained roughly at the same level since 2010 and in others the number of road fatalities has even increased.

In 19 counties, the number of road fatalities was above the national average, as indicated in figure 3.

In some of these counties there could be noticed a stagnation or even an increase of the number of road fatalities over the last few years. Figures 5 and 6 present the numbers found in the National Registry of Car Accidents related to the counties of Dolj and Timis.

It should be also mentioned that, while the number of accidents with fatalities on all Romanian motorways represents only a small percentage of the total number of deaths, the number of deaths increased in 2017 to almost twice the number recorded in 2010 (34 versus 18 deaths).

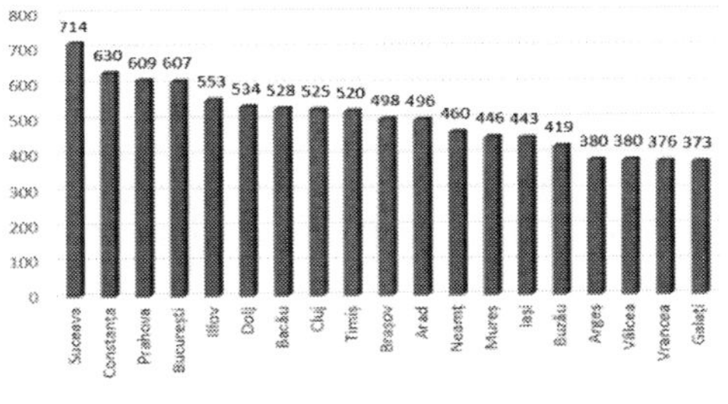

Fig. 4. Counties that recorded a number of fatalities above the national average between 2010 and 2017

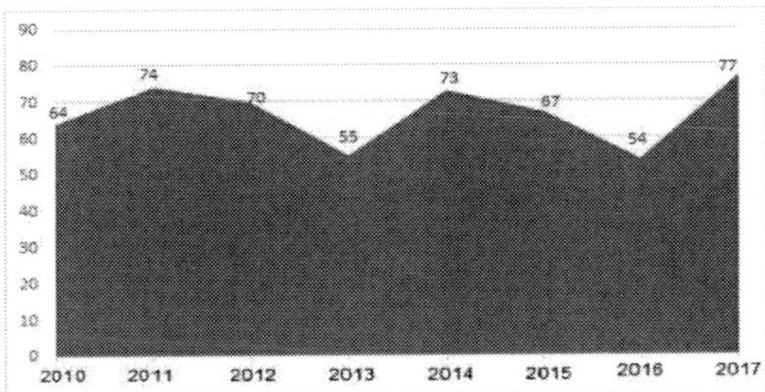

Fig. 5. Evolution of the number of fatal road accidents in the county of Dolj between 2010 and 2017

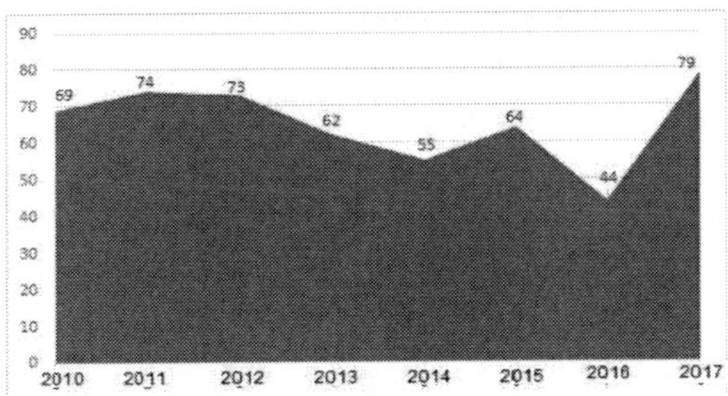

Fig. 6. Evolution of the number of fatal road accidents in the county of Timis between 2010 and 2017

In order to achieve the desired reduction in the number of road deaths, there should be taken some measures, both by the Police and by the other participants in road traffic. Some of these measures could be:

- Periodic checking of the technical status of the vehicles and of the legality of the transportation activities carried out;
- Firm attitude of the policemen towards any infringements of the law on the roads.
- Dissemination of information on road education issues and on road safety campaigns;
- Meetings between road traffic participants on road safety topics
- Improvement of the cooperation between the various actors involved in road works for a better direction of traffic in those areas;
- Implementation of an intelligent road traffic monitoring system
- Improvement of the attitude of passenger car drivers towards pedestrians, public transportation drivers, bicycle and motorcycle riders etc.

While the current values might seem to move the goal for 2020 out of reach, there is still some time left and by implementing the mentioned measures, it is still possible to achieve a reduction of the number of road deaths by 50% by 2020 in Romania and in the European Union.

4 Conclusions

Road safety is an issue of interest for all citizens, each of them being able to contribute to increasing the safety of traffic on public roads.

As the current paper has shown, both at European level and in Romania, there is much to do if the ambitious goal of reducing road fatalities by 50% in 2020 compared to 2011 is to be achieved. Still, if thoroughly implementing the measures indicated in the paper, the targeted reduction would be still feasible.

In future, the authors intend to extend their researches by applying a questionnaire on road safety to the various categories of participants in the road traffic, especially in the identified problem areas.

References

1. M. Peden, *World report on road traffic injury prevention* (World Health Organization, 2004)
2. World Health Organization, *Global status report on road safety 2015* (World Health Organization, 2015)
3. European Commission, *White Paper on Transport* (European Commission, 2011)
4. European Commission, *Road Safety: Statistics accidents data* (European Commission, 2018)
5. L.I. Tarnu, *Analysis, investigation and reconstruction of road accidents* (Translated from Romanian), (Editura Universul Juridic, Bucuresti, 2012)
6. ACEA, *Vehicles in use 2017* (European Automobile Manufacturers Association, 2017)
7. A. Gligor, Bull. Univ. Ploiesti, **LXVI**, 4 (2015)
8. L.I. Tarnu, *Investigation and reconstruction of road accidents* (Translated from Romanian), (Editura Universitatii "Lucian Blaga" din Sibiu, Sibiu, 2009)
9. Inspectorate-General of the Romanian Police, *Road Safety Bulletin. Annual Report 2016* (Romanian Police, 2017)
10. L.I. Tarnu, *Aspects of the systematisation and signalling of the road traffic in Romania* (Translated from Romanian), (Editura Universitatii „Lucian Blaga" din Sibiu, 2016)
11. Government of Romania, *Decision 755/2016 on the Approval of the National Strategy for Road Safety 2016-2020* (Official Monitor of Romania, 2016)
12. European Commission, *Road Safety Country Overview: Romania* (European Commission, 2016)
13. Inspectorate-General of the Romanian Police, *National Registry of Road Accidents* (Romanian Police, 2017)
14. ETSC, *4th PIN Report: Road Safety Target in Sight: Making up for Lost Time* (ETSC, 2010).

MATEC Web of Conferences 184, 01012 (2018) https://doi.org/10.1051/matecconf/201818401012
Annual Session of Scientific Papers IMT ORADEA 2018

Considerations on the Right of Way as Cause of Road Accidents in Romania

Lucian Tarnu[1], and *Cristian* Deac[1,*]

[1]Lucian Blaga University of Sibiu, Romania, Department of Industrial Engineering and Management, Emil Cioran str., 4, Sibiu, Romania

Abstract. While cars and road traffic are an essential component of modern human society, the accidents that may occur in traffic are one of the largest sources of deaths and injuries. A first important step towards reducing these road accidents is the identification of their main causes. In the the current paper, the authors analyse the main causes of road accidents in Romania and then discuss the conditions surrounding the accidents that had as cause the failure to properly grant right of way, taking into account both the national, regional (county-level) and local statistics.

1 Introduction

Road transportation is one of the most common expressions of modern civilisation. On the one hand, cars make it possible to move between various locations in a much shorter time than it was possible 200 years ago and on the other hand they can represent powerful status symbols.

However, the popularity of this means of transportation has led in many places to very busy roads. Moreover, the increase in the number of cars driving on roads and motorways has led in many countries and regions also to an increase in the number of car accidents. Fatal car accidents are especially tragic and costly, both in terms of the material losses and in terms of human lives, representing a major problem of the human society.

Statistics at world level [1] show that in 2012 road traffic injuries were the main cause of death for the age group 15-29 years (Fig. 1 [1]) and it is projected to be ranked 7th among all age groups in 2030. This situation is worse than that recorded for 2002 [2], when it was reported that road traffic injuries were the second-most-frequent cause of death for the age groups 5-14 years and 15-29 years and the third-most frequent cause of death for the age group 30-44 years. On the other hand, the number of road deaths seems to have remained at roughly the same level from 2007 to 2013 – 1.25 million deaths worldwide [1].

The European Union is the safest world region in this regard, mainly due to ambitious goals set in this regard by the European Commission. While the target set in 2001 for 2010 – a 50% reduction of road traffic fatalities – has not been fully reached (the reduction was of only 35%), these targets were renewed in the White Paper on Transport and the Action Programme on Road Safety for the period from 2011 to 2020, even aiming towards a rate of zero road traffic fatalities in 2050. [3]

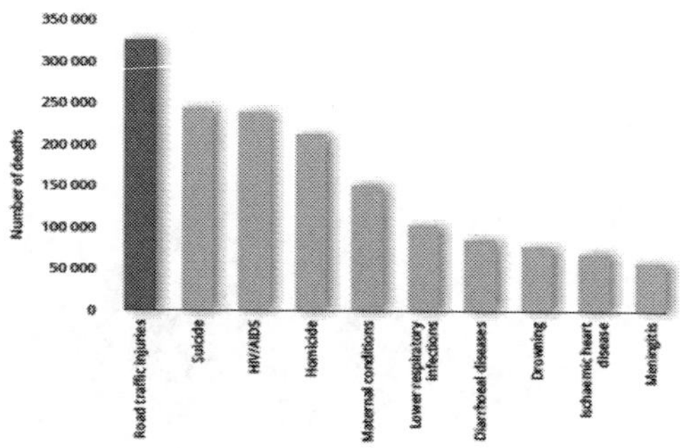

Fig. 1. The main causes of death for people in the age group 15-29 years, at world level, in 2012 [1]

One of the main conditions for achieving these ambitious goals is to precisely identify the causes that can lead to road accidents and especially to road fatalities and to develop corresponding measures for combatting them.

In the current paper, the authors analyse the main causes of road accidents in Romania and especially the failure to properly yield the right of way as one of the main causes, suggesting a series of measures that need to be implemented, both at national, regional and local level in order to help reduce the number of accidents that have this cause.

2 Road Accidents – Situation and Main Causes at World Level and in Romania

Since their invention at the end of the 19th century, cars have gradually transformed and improved human life, they imposed an accelerated rhythm, shortened travel distances and durations and became one of the symbols of modern civilisation. [4]

* corresponding author: cristian.deac@ulbsibiu.ro

© The Authors, published by EDP Sciences. This is an open access article distributed under the terms of the Creative Commons Attribution License 4.0 (http://creativecommons.org/licenses/by/4.0/).

Across the European Union, the number of passenger cars grew by 4.5% over the last five years, from 241 to 252 million [5]

In Romania, cars first appeared very soon after this means of transportation was invented, more precisely around the year 1890. Currently, there are around 5.7 million cars registered in Romania and more than 6.6 million persons have driver's licenses. [4]

Unfortunately, car traffic is sometimes accompanied by accidents that lead each year to important material damage, to minor or major injuries or even the death of participants in the road traffic: car drivers, pedestrians, passengers, bicycle riders, motorcycle riders etc. [4] Statistics at world level show that each year, 1.2-1.5 million people die in car accidents, most of whom are young people, while other 35 million people receive severe or light injuries. [1] An accident-caused fatality is a family tragedy, but represents also a significant loss for the whole society, given the efforts undertaken for the education, professional training and training for life of the young generations. [4]

Because road vehicles are a favourite way for people to commute to and from the workplace or to carry out their daily jobs, road traffic accidents can be regarded as work accidents, so risks related to road safety can be determined with a methodology similar to that used for determining the risks at the workplace. [5]

In recent years, road safety has become a major topic for many governments and administrative institutions, including at the level of the European Union. The road safety policy of the European Union aims to increase the road safety level so as to allow the safe and ecological movement of citizens on all roads of the member states. It promotes the equality of all road traffic participants and for this seeks to increase the road safety level for the most vulnerable among them. [3, 6, 7]

The European Commission considers following three actions as having paramount importance in the area of road safety [3, 8]:

- Implementing a structured and coherent cooperation framework based on the best practices in the member countries, as a necessary condition for efficiently implementing the orientations for the European road safety policy 2011-2020;
- Elaborating a strategy regarding injuries and granting first aid, in order to answer the more and more acute need to reduce the number of road accidents leading to injuries;
- Increasing the safety of vulnerable participants in the road traffic, especially with regard to motorcycle riders, in whose case statistics are worrying.

A study published by the World Health Organisation in 2015 [1] has found that the main risk factors and causes for road accidents are the following ones:

- Excessive and/or inadequate speed;
- Risky/reckless behaviour by pedestrians and cyclists;
- Poor road engineering works;
- Poor state of the cars/vehicles.
- Inexperience, thrill-seeking and low alcohol tolerance of (especially young) drivers and riders;

One way to systematise the main causes and to aid in elaborating more efficient measures to combat them is the so-called Haddon matrix (Fig. 2 [9]), used either in the simplified form presented or in expanded forms as base for proactive initiatives.

		Factors		
		Human factors	Vehicles and equipment factors	Environmental factors
Phases	Pre-crash	Alcohol and drugs	Defective brakes	Darkness, rain, fog, snow, ice
	Crash	No safety belt worn	No airbag	Tree too close to road
	Post-crash	No or inadequate first aid	Fire due to fuel leakage	Slow response of rescue services

Fig. 2. Haddon matrix [9]

The main causes of road accidents in Romania include [6, 10, 11]:

- Offenses by pedestrians;
- Inadequate speed;
- Failure to yield right of way to vehicles;
- Failure to yield right of way to pedestrians;
- Failure to maintain an adequate distance between vehicles;
- Offenses by bicycle riders;
- Reckless driving;
- Irregular overtaking manoeuvres;
- Driving under the influence of alcohol;
- Inadequate manoeuvres to switch lanes.

Between 2005 and 2012, in Romania there were reported 252633 road accidents, the causes of which were distributed as indicated in Fig. 3 (data from [12]).

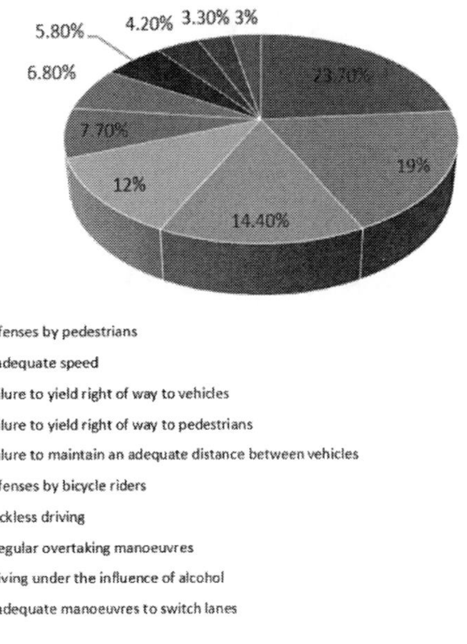

Fig. 3. Weight of the main causes of road accidents in Romania between 2005 and 2016 [12]

 https://doi.org/10.1051/matecconf/201818401012

3 Analysis of number of accidents caused by the failure to yield right of way to other vehicles in Romania

In Romania, the Inspectorate General of the Romanian Police is the authority mandated to collect and manage the data referring to road accidents. It can then transmit the data to the National Statistics Institute and to the interested international institutions.

For their analysis, the authors resorted to data retrieved from the National Registry of Car Accidents [12].

As can be seen from Fig. 3, the failure to yield right of way to vehicles is one of the most important causes for road accidents in Romania.

The number of accidents and the number of victims stemming from the failure to properly yield right of way to vehicles in Romania, in the years 2007-2016, is shown in Fig.s 2 and 3, respectively. In all, between 2007 and 2016 there were reported 32252 accidents that led to 46739 victims (fatalities, severe injuries or light injuries).

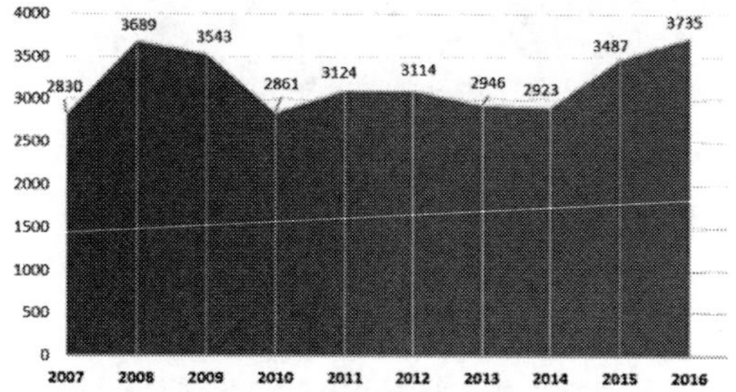

Fig. 4. Evolution of the number of road accidents caused by the failure to yield right of way to vehicles, between 2007 and 2016, in Romania.

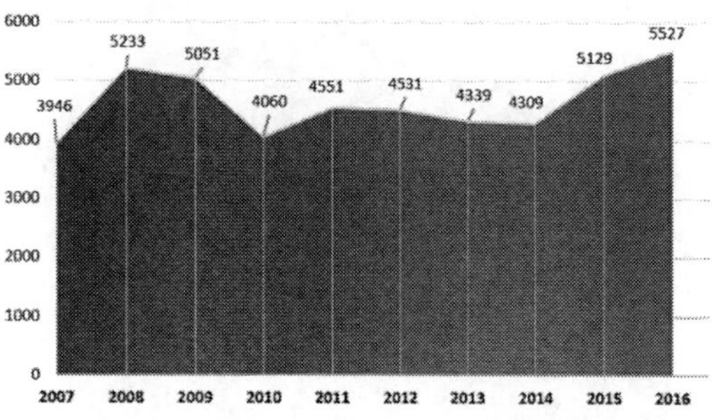

Fig. 5. Evolution of the number of victims of road accidents having as cause the failure to yield right of way to vehicles, between 2007 and 2016, in Romania.

Table 1 presents the evolution of the number of fatalities, of severe injuries and of light injuries in the period between 2007 and 2016.

From these Fig.s it can be seen that, while the numbers have fluctuated from year to year, overall the number of accidents caused by the failure to yield right of way to other vehicles has increased. However, the increase is largely due to the increase in the number of light injuries, while the number of fatalities has decreased.

Table 1. Evolution of the number of fatalities, severe injuries and light injuries in accidents having as main cause the failure to yield right of way.

Year	Number of fatalities	Number of severe injuries	Number of light injuries
2007	152	666	3128
2008	132	905	4196
2009	129	874	4048
2010	95	753	3212
2011	108	780	3663
2012	87	841	3603
2013	61	724	3554
2014	69	658	3582
2015	83	910	4199
2016	66	663	4798
Total	982	7774	37983

A county-level analysis reveals an average number of 768 accidents caused by failure to yield right of way per county from 2007 to 2016. However, this average is misleading, since the county-level municipality of Bucharest recorded a much higher number of accidents than any other county (7424). Furthermore, as can be seen from Fig. 1, several other counties – Constanta, Timis, Ilfov, Prahova, Cluj - reported figures above the average, while others – Vaslui, Covasna, Salaj, Tulcea, Bistrita-Nasaud - remained well below this average.

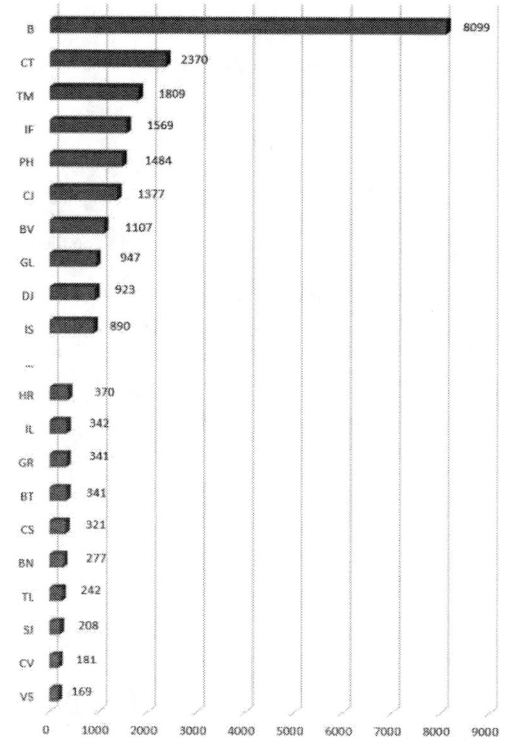

Fig. 6. Number of accidents caused by the failure to yield right of way to vehicles, by county, between 2005 and 2017

The vast majority (91%) of the total number of accidents took place within the limits of settlements and only a small part can be ascribed to the higher speeds allowed outside of settlements.

Also, 75% of the accidents were lateral (sideways) collisions, 15% were frontal collisions, 6% were due to tailgating and 4% were other types of collisions, such as chain collisions (Fig. 7).

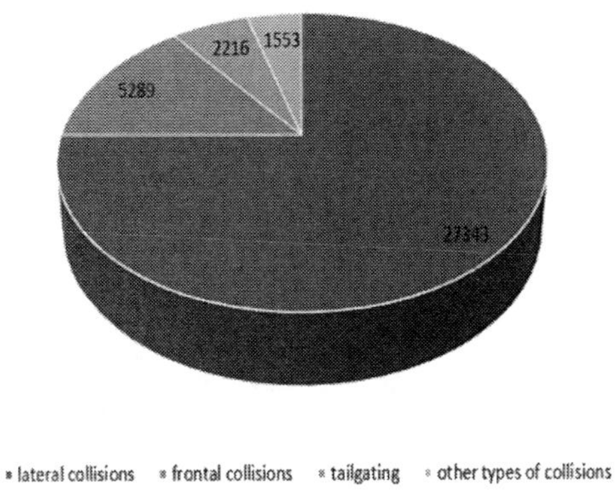

Fig. 7. Number of accidents caused by the failure to yield right of way by vehicles grouped by the type of collision.

4 Measures for the reduction of the number of accidents caused by the failure to yield right of way to other vehicles in Romania

As a conclusion of the analysis presented in the previous paragraph and in agreement with the speciality literature in this area, following measures can be considered in order to reduce the number of accidents that have this cause:

- Dissemination of information on road education issues and on road safety campaigns;
- Firm attitude of the policemen towards any infringements of the law on the roads;
- Periodic checking of the technical status of the vehicles and of the legality of the transportation activities carried out;
- Improvement of the cooperation between the various actors involved in road works for a better direction of traffic in those areas and for the provision of a larger area of visibility around crossroads;
- Periodic checking and improvement of the traffic signs and other signalling devices (placing them in more visible positions, increasing their reflectivity at night etc.), especially with regard to the right of way signs and in the areas with potentially dangerous crossroads or curves;
- Installation of more video surveillance equipment and eventually the implementation of an intelligent road traffic monitoring system;
- Meetings between road traffic participants on road safety topics;
- Improvement of the attitude of passenger car drivers towards pedestrians, public transportation drivers, bicycle and motorcycle riders etc.

5 Conclusions

Road safety is an issue of interest for all citizens, each of them being able to contribute to increasing the safety of traffic on public roads.

As the current paper has shown, the failure to yield right of way to vehicles is an important cause of road accidents, at world level and in Romania, even if the numbers vary significantly between the various areas and counties.

A thorough implementation of the measures suggested in the paper would certainly lead to a significant reduction in the number of accidents with this cause.

In future, the authors intend to extend their researches also to other causes of road accidents and by applying a questionnaire on road safety to the various categories of participants in the road traffic, especially in the identified problem areas.

References

1. World Health Organization, *Global status report on road safety 2015* (World Health Organization, 2015)
2. M. Peden, *World report on road traffic injury prevention* (World Health Organization, Geneva, 2004)
3. European Commission, *White Paper on Transport* (European Commission, 2011)
4. L.I. Tarnu, *Analysis, investigation and reconstruction of road accidents* (Translated from Romanian) (Publishing House Universul Juridic, Bucharest, 2012)
5. A. Gligor, Bull. Univ. Ploiesti, **LXVI**, 4 (2015)
6. L.I. Tarnu, *Investigation and reconstruction of road accidents* (Translated from Romanian) (Publishing House of the "Lucian Blaga" University of Sibiu, Sibiu, 2009)
7. ETSC, *4th PIN Report: Road Safety Target in Sight: Making up for Lost Time* (ETSC, 2010)
8. L.I. Tarnu, *Aspects of the systematisation and signaling of road traffic in Romania* (Translated from Romanian) () (Publishing House of the "Lucian Blaga" University of Sibiu, Sibiu, 2016)
9. DEKRA, *Road Safety Report 2017 – Best Practices* (DEKRA, 2017)
10. Inspectorate-General of the Romanian Police, *Road Safety Bulletin. Annual Report 2016* (Romanian Police, 2017)
11. Government of Romania, *Decision 755/2016 on the Approval of the National Strategy for Road Safety 2016-2020* (Official Monitor of Romania, 2016)
12. Inspectorate-General of the Romanian Police, *National Registry of Road Accidents* (Romanian Police, 2017)

Research on combustion process of gasoline homogenous charge compression ignition engine

Corneliu Cofaru[1], *Mihaela Virginia* Popescu[1]

[1]Transilvania University of Brasov, 29, Eroilor blvd. 50036, Romania

Abstract. The paper presents the research designed to develop a HCCI (Homogenous Charge Compression Ignition) engine starting from a spark ignition engine platform. The chosen test engine was a single cylinder, four strokes provided with a carburettor. The results of experimental research data obtained on this version were used as a baseline for the next phase of the research. In order to obtain the HCCI configuration, the engine was modified, as follows: the compression ratio was increased from 9.7 to 11.5 to ensure that the air – fuel mixture auto- ignite and to improve the engine efficiency; the carburettor was replaced by a direct fuel injection system in order to control precisely the fuel mass per cycle taking into account the measured intake air-mass; the valves shape were modified to provide a safety engine operation by ensuring the provision of sufficient clearance beetween the valve and the piston; the exchange gas system was changed from fixed timing to variable valve timing to have the possibilities of modification of quantities of trapped burnt gases. The cylinder processes were simulated on virtual model. The experimental research works were focused on determining the parameters which control the combustion timing of HCCI engine to obtain the best energetic and ecologic parameters.

1 Introduction

Transport plays an important role in today's economy and society and has a large impact on economic growth and employment. It is generally agreed that transport sector is one of the major contributors with pollutants and GHG of environment.

Reducing exhaust emissions and increasing the fuel economy of internal combustion engines are of global importance. To meet the demand of economy, energy conservation, less environmental harmful exhaust emissions, especially carcinogenic nitrogen oxides (NOx) and responsibility for a greenhouse effect carbon dioxide (CO2) and higher thermal efficiency present generation engines must be characterized by: low fuel consumption; high efficiency; reliability; low price and low cost of usage. To achieve these tasks regarding diminishing fuel consumption, pollutants and CO_2 emissions the implementation of the concept of homogenous charge compression ignition engine (HCCI) is required.

The combustion technology of HCCI combines the main advantages of the gasoline engines (the homogeneous charge) with the main advantages of the diesel engines (higher compression ratios leading to a higher efficiency) determining the potential of reducing the emissions while increasing fuel economy.

Analysing the main strength and weak points of a conventional spark ignition engine and compression ignition engines, it is possible to summarize the advantages and disadvantages of HCCI engine.

Firstly, the spark ignition engines provided with a three-way catalyst can be considered as being very clean engines, problem appears during on part-load conditions. Low efficiency at part-load results from pumping losses due to specific spark ignition (SI) engine load control. SI engine is controlled by the changing of the air flow rate, adapting the fuel flow, in order to obtain the desired air fuel ratio. The SI engines operate with stoichiometric mixture ($\lambda \approx 1$) which produces a large amount of CO_2, this is responsible for the greenhouse effect. In other terms, low efficiency means high fuel consumption especially at part loads and high emissions of CO_2. Worthy mentioning the fact that in urban traffic the engines mainly run at part loads, resulting a very low level of overall efficiency on urban normal driving conditions

Secondly, the compression ignition engines are widely used in small displacement passenger cars, vans, trucks, buses, and locomotives, as well as large high power marine and power-generation applications. Load control is achieved by changing the amount of fuel injected in engine cylinders; the air flow rate is almost the same for all load conditions for naturally aspirated CI engines or, in the case of turbocharged engines, the air mass flow increases with engine load. Compression ignition (CI) engine in opposition to SI engine has no pumping losses and, thereby, has higher part-load efficiency. Naturally aspirated or supercharged engines operation in combination with high compression ratio caused that the part-load efficiency is higher for a CI than for a SI engine. The CI engines are the important source of particulate emissions, 0.2% – 0.5% from the fuel mass injected in the

* Corresponding author: ccornel@unitbv.ro

© The Authors, published by EDP Sciences. This is an open access article distributed under the terms of the Creative Commons Attribution License 4.0 (http://creativecommons.org/licenses/by/4.0/).

cylinder is emitted as small (~ 0.1μm diameter) particles which consist primarily of soot with some additional absorbed hydrocarbons. Particulates are mainly formed in the rich region of combustion area and NOx - in the hot temperature burned gas regions. Latest technologies used on CI engines for exhaust gas after treatment are very complex systems and expensive.

The HCCI could give 50% improvement in engine efficiency at part load compared to spark ignition engines and 30% compared to compression ignition engines, zero particulate matters and very low NOx emissions (< 10 ppm – compared to >500 ppm for diesel engines), as a result of using a diluted air-fuel mixture. In HCCI engine, the soot formation due to the use of homogeneous charge is not of a main concern.

The HCCI engines having a simplified construction, do not require:

- sophisticated shapes of inlet duct (cylinder head) to generate specified turbulence in combustion chamber;
- high pressure injection system;
- very complex aftertreatment exhaust gas systems.

The HCCI engine can be regarded as a hybrid of both types (SI and CI) of engines with attractive properties as: higher specific heat ratio of very lean mixture; higher cycle efficiency caused by using higher compression ratio; closer to ideal cycle due to the faster combustion.

The HCCI engines can operate on gasoline, diesel fuel, and most of alternative fuels.

2 Combustion process in HCCI engine

The combustion process in HCCI engine is different from conventional engine as SI and CI engines because of lacks of flame propagation. Ignition of homogenous air-fuel mixture occurs simultaneously at whole volume of combustion chamber and burns at the same time.

The chalange of HCCI engine is to control the start of combution timing and the combustion parameters as comubustion heat release rate and combustion temperatures.

The gasoline HCCI combustion is achieved by controlling the levels of the cylinder charge temperatures, pressures, and mixtures composition through the following parameters: the EGR rate, air-fuel ratio, compression ratio, intake temperature, inlet manifold pressure, the injection timing, fuel properties, the air-fuel homogenization, the heat transfer to the engine.

Because the entire mass of air fuel of combustion chamber burns simultaneously, the combustion process is caracterized by high values of heat-released rate, cylinder's pressures and a rapid increased of cylinder's temperatures, these parameters can be controlled by air-fuel mixture dilution with burnt gases.

In case of gasoline HCCI engine, the management of burnt gases for HCCI combustion can follow two strategies:

Residual gas trapping, known as negative valve overlap strategy (NVO). The great amount of residual gas is trapped in cylinder when the exhaust valve is closed early, during the exhaust stroke. In order to keep the trapped burnt gases in cylinder and to avoid the back flow of these gases into inlet manifold, the intake valve is opened later, after TDC at the beginning of the inlet stroke.

This strategy involves the recirculation of burnt gases after they left the engine cylinder, and the valve timing permit the positive valve overlap. The burnt gases recirculation can be realized using the so-called internal gas recirculation (IEGR) method. The method consists in sucking the exhaust gases existing in the exhaust system by reopening for a short period of time the exhaust valve during the intake stroke or by closing of the exhaust valve later during the inlet stroke.

With increasing amounts of exhaustgas recirculation (EGR), the combustion process becomes slower, resulting in lower peak pressure and lower rate of heat release and therefore longer combustion rates. Both the combustion and gross indicated efficiencies increased with increasing EGR.flows.

Other strategies applied to control the combustion of HCCI engines is the fuel injection process by injector location, fuel-spray patterns and by varying injection timing and duration.

In the gasoline HCCI engine development, three injection patterns may be implemented, as follows:

a) injection during negative overlap duration to determine fuel reformation;
b) injection during to intake stroke to form homogenous air-fuel mixture;
c) Injection during compression stroke to form stratified air-fuel mixture.

Research regarding combustion process shows that varying the charge in homogeneity, a method for controlling the combustion duration in HCCI engines can be used.

For combustion process Vibe function is used to aproximate the rate of heat released rate during the combustion process:

$$\frac{dx}{d\alpha} = a \cdot (m+1) \cdot y^m \cdot e^{-a \cdot y^{(m+1)}} \quad (1)$$

$$dx = \frac{dQ}{Q} \quad (2)$$

$$y = \frac{\alpha - \alpha_0}{\Delta\alpha_c} \quad (3)$$

Where:
Q - the total heat released,
α – crankshaft position,
α_0 - crankshaft position at the begining of combustion process,
$\Delta\alpha_c$ – Combustion process duration,
m - Shape parameter,
a - the Vibe parameter (6.9 for a complete combustion).

To obtain the mass-fraction-burned equation (2.1) has to be integrated:

$$x = \int \frac{dx}{d\alpha} \cdot d\alpha = 1 - e^{-a \cdot y^{(m+1)}} \tag{4}$$

Mass-fraction burned for each engine's cycle represents a normalized quatity on a scale from 0 to 1 (0 until the start of combustion, 1 after the complete fuel burned during the combustion process).

Mass-fraction burned is used to appreciate the release of chemical energy as function of crankshaft position.

To obtain a more accurate approximation of the heat release a combination of two or more Vibe functions has to be used. For the gasoline HCCI engines, where the heat is released in a single stage, the best results were obtained by using a combination of three Vibe functions:

$$\frac{dx}{d\alpha} = pv_1 \cdot a \cdot (m_1 + 1) \cdot y^{m_1} \cdot e^{-a \cdot y^{(m_1+1)}} + pv_2 \cdot a \cdot (m_2 + 1) \cdot y^{m_2} \cdot e^{-a \cdot y^{(m_2+1)}} + pv_3 \cdot a \cdot (m_3 + 1) \cdot y^{m_3} \cdot e^{-a \cdot y^{(m_3+1)}} \tag{5}$$

(2.5)

and for mass-fraction burned:

$$x = \frac{\left(1 - e^{-a \cdot y^{(m_1+1)}}\right) + \left(1 - e^{-a \cdot y^{(m_2+1)}}\right) + \left(1 - e^{-a \cdot y^{(m_3+1)}}\right)}{pv_1 + pv_2 + pv_3} \tag{6}$$

Where *pv1*, *pv2* and *pv3* are the shares of the corresponding Vibe functions.

3 Experimental setup

To collect the experimental data in this research a single cylinder four-stroke naturally aspirated engine was used. The test engine having the specifications presented in Table 1 was connected to a SIEMENS 1PH6167 4NG00 dynamometer, in order to obtain the engine torque and power. The HCCI test engine was realized starting from a SI engine equipped with a carburettor as a platform. In order to obtain a HCCI operation cycle some modifications were implemented to the engine. The compression ratio was increased from 9.7:1 to 11.5:1, the fuelling system was changed, replacing the carburettor by direct injection system provided with an electromagnetic actuating injector. The injector was fitted by replacing the spark plug.

Table 1. Engine's specifications

Engine type	SI	HCCI	[-]
No. of cylinders	1	1	[-]
Fuel	Gasoline	Gasoline	[-]
Displacement	652	652	[cm^3]
Compression Ratio	9.7:1	11.5:1	[-]
Bore	100	100	[mm]
Stroke	83	83	[mm]
Con-rod length	149	149	[mm]
Valve position	DOHC	DOHC	[-]

To modify the quality of cylinder's mixture in order to control the combustion process, the engine was provided with a variable gas distribution system using a camshaft phaser. The original camshaft was replaced by a camshaft with new redesigned cams' profile. These cams provide a short valves opening period and a lower valve lifts to offer a safety clearance (valve-piston) when the valve timings are changed during the research work. The standardized equipment was used for tests.[1,2,3,4,5,6,7].

During the tests, one could register data regarding: ambient, intake and exhaust pressures, injection and cylinder pressure, oil pressure, ambient, cylinder, oil, water temperatures, torque, the valves timing and pollutant emissions (HC, CO, NOx).

The inlet air has the possibility to be conditioned in conformity with research requirements to obtain the control of combustion timing. The inlet and exhaust pressure was measured through KISTLER 6041A and KISTLER 4075A10 transducers. The cylinder's pressure data were obtained with a KISTLER 6041A piezoelectric transducer. The information regarding crankshaft's position was obtained by using a transducer type ROD 426A.

The fuel consumption was measured by using a fuel-mass flow meter and the air-flow was measured by using a laminar anemometer.

The engine test incorporated a full electronic and manual control.

The equipment used for pollutant measurements were NDIR Pierburg for O_2/N_2 and CO, CO_2, CLD Pierburg for NO/NO_x, HFID AVL for HC.

The "BremoXP" soft-ware was used for thermodynamic analysis of combustion process.

4 Experimental results and discussions

The main problem of gasoline HCCI engine is the control of combustion timing.

The auto-ignition process is a result of various complex chemical, physical and energetic processes that take place during the auto-ignition process. It was observed that before the main ignition of the mixture, a first-stage smaller-scale reaction, called "cool flame" was also present. By simulating the cylinder's processes, it was found that the pressure required to initiate the first-stage reaction was a linear function of the compression pressure at TDC, while depending on the fuel, the required compression pressure to initiate autoignition decreased with increasing fuel concentration. To increase the compression pressure to improve auto-ignition properties, the engine's compression ratio was rised from 9.7:1 to 11.5:1.

To control the start of combustion timing (SOC) and to moderate the pressure rise rate during the combustion process internal exhaust gas recirculation (IEGR) was used, and met the both requirements, to rise the mixture temperature and to realize the mixture's dilution, exploiting the burnt gases effects: heating, dilution, heat capacity and chemical activity. The amount of trapped residual gases (TRG) in the cylinder can be changed, by changing the valves timing of the engine which determine modifications of temperature, pressure, and composition of the cylinder mixture at inlet valve closing (IVC), achieving a stable HCCI combustion. The negative valve overlap strategy (NVO) was adopted for this experimental research. The amount of residual gas was trapped in cylinder when the exhaust valve is closed early, during the exhaust stroke. In order to keep the trapped burnt gases in cylinder and to avoid the back flow of these gases into inlet manifold, the intake valve is opened later, after TDC at the beginning of the inlet stroke.

The test was realized for exhaust vale closing (EVC) values: 639[CA] *(81[CA BTDC])*, 644[CA] *(76[CA BTDC])*, 646[CA] *(74[CA BTDC])* and 649[CA] *(71[CA BTDC])*, (Fig.1.) and (Fig.2.).

The main goal of the experiments was to investigate the possibility of controlling combustion phasing and combustion duration using various internal exhaust gas recirculation (IEGR) fractions.

By increasing the advance of closing of the exhaust valve timing from 71[CA BTDC] 649 [CA] to 81[CA BTDC] 639 [CA], the peak of heat release rate rises by 31 [%]. These increases determine a vigorous combustion process due to the rise quantity of trapped burnt gases.

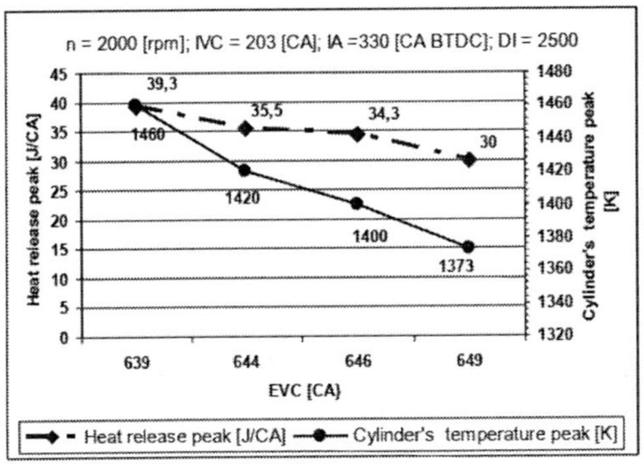

Fig.1. Influence of the exhaust valve closing timing on heat release rate peak and cylinder's temperature peak

Cylinder's temperature peak increases from 1373 [K] to 1460 [K] (Fig.2.) representing 6.34 [%] for the analyzed interval, despite of the rise of the air-fuel mixture dilution with burnt gases, then, the NOx emission increases from 17.86 [ppm] to 25.72 [ppm] representing 44 [%] (Fig.2.) due to the influence of cylinder's temperature.

Diminishing the quantity of burnt gases in the cylinder's mixture determinates the moving of combustion in the expansion stroke, the angle of heat release rate peak moves from 363 to 367 [CA] and the angle of peak of cylinder's temperature change from 371 [CA] to 377 [CA].

The influence of the intake valve's closing was realized for 204, 210 and 215 [CA], maintaining constant the following parameters: EVC at 649 [CA], the injection advances at 330 [CA], the injection duration ID=2550 [µs] and the engine speed n = 2000 [rpm], in these cases, the quatities of burnt gases and fuel remainig constant in the cylinder.

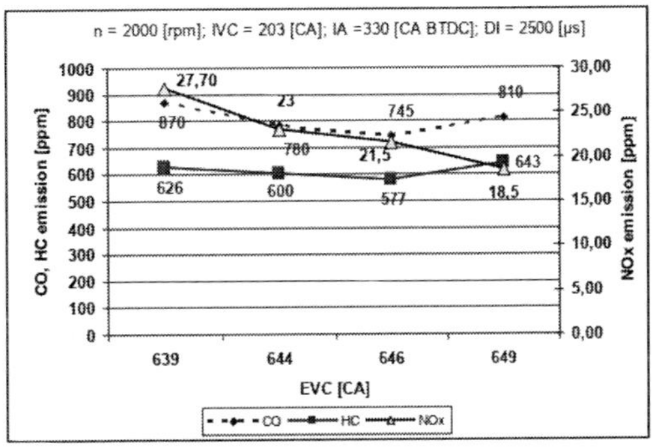

Fig.2. Influence of the exhaust valve closing timing on CO, HC and NOx emission.

By changing of values of IVC, the intake valve's closing timing determines the modification of the fresh-air content in the cylinder's mixture, in fact, the quality of mixtures are changed. Increasing the values of retarding the inlet valve's closing from 204 [CA] to 210 and 215 [CA], the data of experimental research show that the peak of heat release rate decreases from 29,8 [J/CA] at IVC = 204 [CA] to 25,1[J/CA] at IVC = 210 [CA] (- 15,8%) and to 23,8 [J/CA] at IVC = 215 [CA] (- 20,1%), (Fig.3.).

The peaks of heat release rate are placed at the same crank angle. The peak of cylinder's temperature decreases from: 1377 [K] to 1347 [K] and, 1320 [K] for established IVC values. The pollutant emissions follow the values of cylinder's temperatures, HC increases from 650 [ppm] to 681 [ppm] and 775 [ppm], CO increases from 800 [ppm] to 1000 [ppm] and 1445 [ppm], and NOx decreases from 17, 5 [ppm] to 15, 2 [ppm] and 15, 6 [ppm], (Fig.4.).

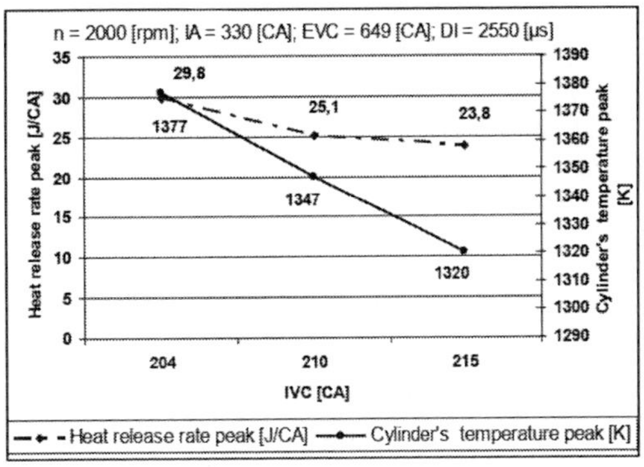

Fig.3. Influence of the inlet valve closing timing on heat release rate peak and cylinder's temperature peak

Fuel injection into the cylinder at different stages of the engine cycle allows HCCI combustion timing to be advanced by improving mixture ignitability or retarded by increasing fuel stratification, creating the possibility of expanding the low- and high-load operating limits.

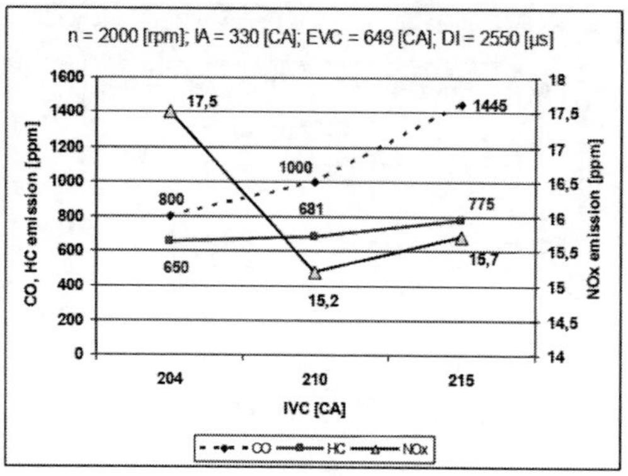

Fig.4. Influence of the exhaust valve closing timing on CO, HC and NOx emissions

Direct injection can be a good way to control HCCI combustion, but it depends heavily on the type of fuel and the timing of the direct injection.

For this research, the early direct injection of fuel in the intake stroke produced near identical results to a premixed charge. The values of injection advances were, as follow: 325 [CA]; 335[CA]; 345[CA]; 350[CA], or 35 [CA ATDC] 25 [CA ATDC] 15 [CA ATDC] 10 [CA ATDC] of the engine cycle. These starts of injection point were placed during the intake stroke after the exhaust valve had been closed, but before the intake valve had been opened.

In order to keep constant the quantity and quality of air-fuel burnt gases mixture during injection process, the following parameters were established at constant values: engine speed n =2000 [rpm], intake valve's closing timing IVC = 203 [CA], exhaust valve's closing timing EVC = 644 [CA], injection duration ID = 2500 [µs].

By incresing the fuel injection advance from 325{CA] to 350 [CA] the fuel was injected in hotter burnt gases, the jet dropplets evaporate and mixes with the burnt gases which have a remainig quntity of oxigen, hydrocabon molecules had same reforming chemical reactions which improve the ignition and combustion of cylinder mixture. After the opening of the intake valve, the burnt gases and the evaporated fuel will mix with the air, and this mixture will be homogenized during the intake and compression strokes.

For this reason, the peak of heat's release rate increases from 25.5 [J/CA] to 31.3 [J/CA] representing 22.75[%], (Fig.6.). The improvement of combustion process detemine an increased peak of cylinder's temperatures from 1327 [K] to 1373 [K], this representing 3.7%. The cylinder's temperature is an important parameter of NOx formation, (Fig.5.)

By increasing the injection advance, the NOx emission rises from 8.6 [ppm] at IA=325 [CA] to 14.4 [ppm] at IA=350 [CA] representing 67.44 [%], HC and CO emissions increases with decreasing o fuel injection advances (Fig.6.).

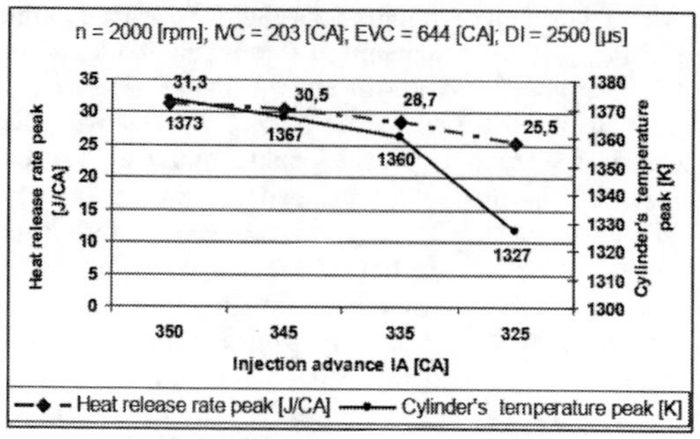

Fig.5. Influence injection advance on heat release rate peak and cylinder's temperature peak

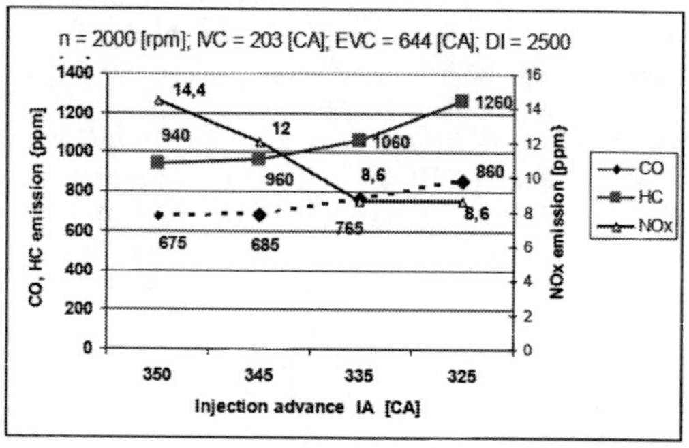

Fig.6. Influence of the injection advance on CO, HC and NOx emissions

The influence of engine's speed on the combustion processes was realized for 1900, 2000, 2100, 2200 and 2300 rpm, keeping constant the same parameters which maintain constant content of the cylinder.

On the one hand, the engine's speeds have an important influence on the heat exhange process which decreases at the increasing of engine's speed; on the other hand, the homogenization process of cylinder's content is intensified. By increasing of engine's speed, the combustion process is improved, it is represented by an increased heat release peak from 23,5 [J/CA] to 31,0[J/CA], representing +21,6%, (Fig.7.)

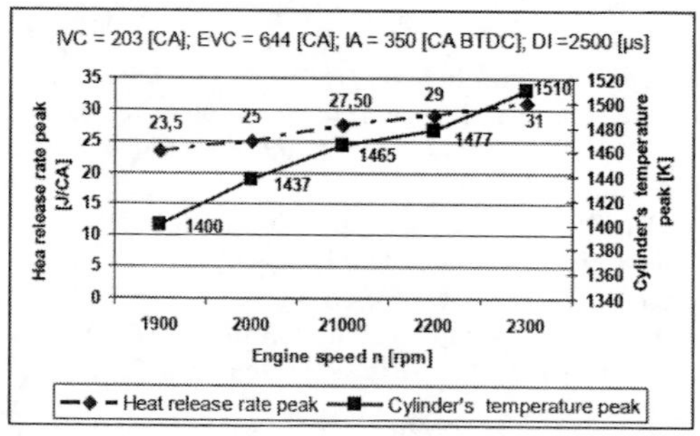

Fig .7. Influence of engine's speed on heat release rate peak and cylinder's temperature peak

https://doi.org/10.1051/matecconf/201818401013

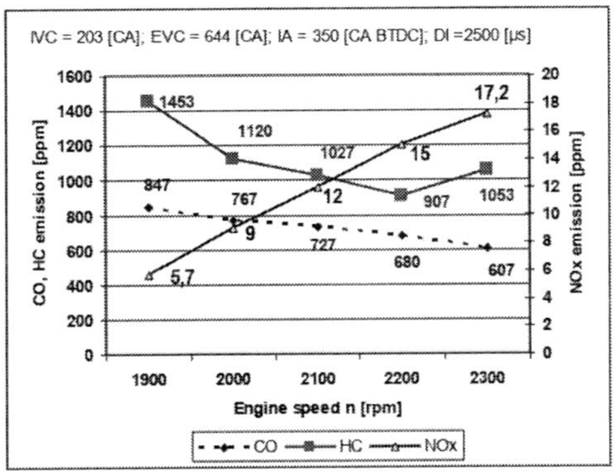

Fig.8. Influence of the engine's speed on CO, HC and NOx emission

The influence of the fuel quantity injected (engine's load) on cylinder's processes was realised by modyfing the fuel injection duration, the used values being 2400 [μs], 2500 [μs], 2600 [μs] and 2700 [μs], keeping constant same parameters, as presented in Fig.10.

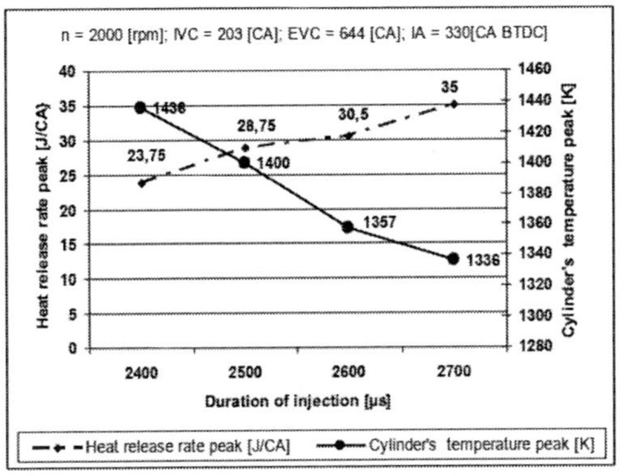

Fig.9. Influence of injection duration (engine's load) on heat release rate peak and cylinder's temperature peak

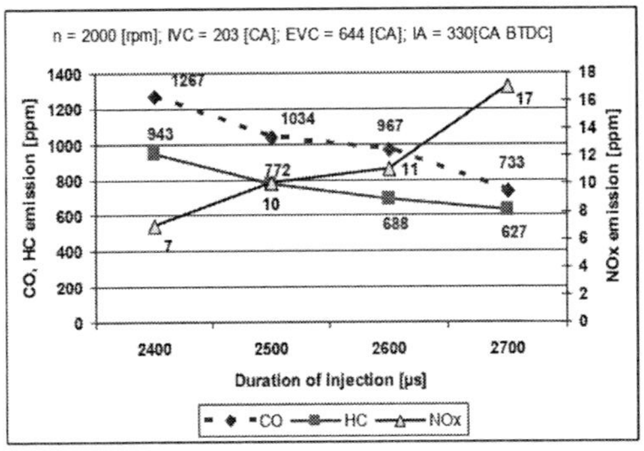

Fig.10. Influence of injection duration (engine's load) on CO, HC and NOx emission

5 Conclusions

The amount of trapped burnt gases has a great influence on the combustion process.The amount of residual gas trapped in the cylinder was changed by modifying the exhaust valve's closing timing earlier during the exhaust stroke. During the experimental research, the negative overlap strategy to control the engine's combustion processes by residual gas trapped in cylinder was adopted.

The intake valve's timings were chosen in order to keep the trapped burnt gases in the cylinder and to avoid the back flow of these gases into inlet manifold; the intake valve is opened later, after the TDC at the beginning of the inlet stroke. The closing of the intake valve was adopted after the TDC to maintain the volumetric efficiency in an acceptable range. For this reason, the intake and exhaust valves timing needs to be tuned and depends on the engine's load and speed.

The obtaining of homogenous mixture in HCCI needs to select carefully the advance of injection (start of injection). Within the experimental tests, the point (SOI) was placed during the intake stroke after the exhaust valve had been closed, but before the intake valve had been opened. However, injection close to TDC improved the combustion efficiency of very low fuel load mixtures

The influence of the operational parameters as: engine speed and engine load on HCCI combustion were tested.

References

1. C. Cofaru, Research on cylinder processes of gasoline homogenous charge compression ignition (HCCI) engine *CAR 2017 , IOP Conf. Ser.: Mater. Sci. Eng.* **252** 012070 (2017)
2. C.Cofaru, Research on Gasoline Homogenous Charge Compression Ignition (HCCI) Engine *CONAT 2016,* pp 357-370 (2016)
3. R. Cosgarea, M. Aleonte, C. Cofaru. The influence of the valve lift strategies on the combustion characteristics of a homogeneous charge compression ignition engine model. *International Journal of Mechanics* Issue 3, Vol. **5**, pp 191-201 (2011)
4. R. Cosgarea, M. Aleonte, C. Cofaru, The Influence of the Internal Exhaust Gas Recirculation (EGR) On the Pumping Losses. In: *Bulletin of the Transylvania University of Brasov* • Vol. **4** (53) No.1 – 2011 Series I, ISSN 2065-2119 (Print), ISSN 2065-2127, pp. 1-7 (2011)
5. R. Cosgarea, C, Cofaru , M. Aleonte, G. Sandu , An approach for modelling the homogeneous charge compression ignition (hcci) combustion using the vibe functions. *CAR 2011* Universitatea din Pitesti , ISSN 1453-1100 (2011)
6. R.Cosgarea, Research on using of advanced combustion systems by using homogeneous mixtures/partial homogeneous mixtures in compression-ignition engines *PhD Thesis,* Universitatea Transilvania din Brasov (2011)
7. H. Zhao HCCI and CAI engines for the automotive industry, Cambridge edition: *Woodhead Publishing Limited.* (2007)

The possibility of extending work time of tribocouple welding improved stainless steel with improved carbon steel

Vlatko Marušić[1], *Predrag Dašić*[2], *Ivan Opačak*[1*]

[1]Mechanical Engineering Faculty in Slavonski Brod, Trg Ivane Brlic-Mazuranic 2, Croatia
[2]SaTCIP Publisher LTD., Tržni centar Pijaca 106a, Serbia

Abstract. In the conditions of tribo-corrosion wear, extending of life time of parts can be achieved by using stainless steel, which is hardened to sufficiently high hardness. In the tribopair bolt/shell/slice of the bucket elevator transporter conveyor machine, the previously improved martensitic stainless steel for bolts is hardened at ≈ 45 HRC and welded with the improved high yield carbon steel for bolts. As the additional material, an electrode based on Cr-Ni-Mo (18/8/6) is used. The structure and hardness of welded samples are tested. On the tensile tester, resistance of the welded joint is tested with a simulated experiment. Dimensional control of wear of elements of tribopair was performed after six months in service. The analysis of test results showed that with regulated additional material and by applying the recommended welding parameters, the properties of the weld, which provide the necessary reliability of the joint in the conditions of tensile load, can be achieved, with a significant increase in the wear resistance of bolt.

1 Introduction

In case of complex mechanisms wear is esspecially evident need for choosing proper type of materials and as well technological and construction parameters [1, 2]. In terms of tribo-corrosion wear basic elements of tribocouple material should be resistant to aggressive media, sufficiently high hardness with the satisfaction of the conditions of tribological compatibility [3]. For research of specifics that appear in tribo-corrosion wear is selected tribocouple in which damage of the elements is caused by corrosion due to abrasive and adhesion wear. A striking example of such wear is a tribosystem of bucket elevators for transporting of oil cake [4]. In order to determine the possibilities of extending the life to tribo-corrosion exposed parts recording condition must be done. Then choose your own variant materials and procedures to protect against wear. At the laboratory samples perform metallographic examination and hardness measurements. Samples-parts install in the drive, and comparing them with the original parts of elevator monitor wear resistance under real conditions. Final control of metallographic examinations traces of wear and intensity of wear should provide insight into the possibility of extending the life of tribosystem elements.

2 Wear of original parts of elevators

Visual check of the original parts of elevators is performed: internal and external slices, internal and external link shoes, shell and bolt and buckets that are bolted on link shoes (every third segment of the chain). According to the original documentation: breaking force of elevator chain is about 2500 N and declared service life of a chain is two years. Figure 1 shows the parts of the original elevator chain after about six months of work.

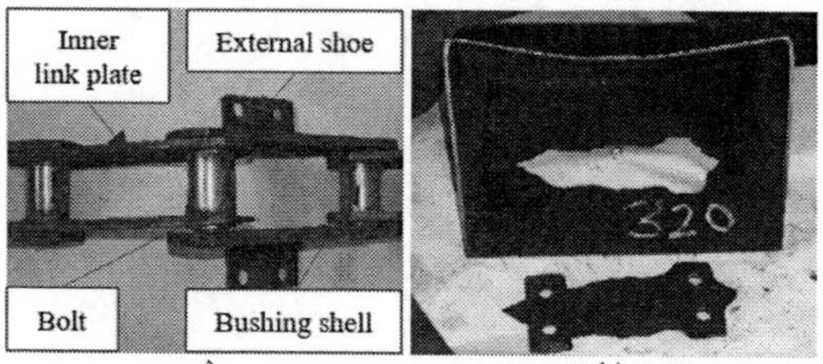

a) b)

Fig. 1. Appearance a damaged chain (a); buckets severed of link shoes (b).

Dimensional control measured: 1. wear of a bolt is ≈ 3 mm, only the surface of bolt in contact with capsule is worn; 2. only the surface of bolt in contact with capsule is worn; 3. wear of a shell on the outside is ≈ 2,2 mm and on the inside to ≈ 1 mm, with emergence of upsetting compund shell/slice and increasing the gap at about 0,8 mm by an upsetted compound. Chain pitch is 160 mm. Wear of a tribocouple, which causes a change in chain pitch (slice, shell and bolt) is about 8 mm, and there is 6,25 tribocouples on meter of a chain. Conclusion can be that the chain extension, as a result of wear parts, can reach more than 50 mm per meter, resulting in overall chain elongation of 1,600 mm in total length chain of 32 m. Wear so big reflects to chain pitch, and this sometimes results in ripping buckets of link shoes.

*Corresponding author: iopacak@sfsb.hr

© The Authors, published by EDP Sciences. This is an open access article distributed under the terms of the Creative Commons Attribution License 4.0 (http://creativecommons.org/licenses/by/4.0/).

2.1 Laboratory tests on the damaged original parts of chain

Chemical analysis showed that the slice and link shoes, are made of improved steel, C45E. The measured surface hardness of slices and link shoes, mean is about 210 HB. The thickness of these parts of the chain before the installation was 8 mm. Dimensional control was found that the corrosion caused local reducing of thickness on max. 6,8 mm. Metallographic search were carried out on a sample taken from a damaged external slice, Figure 2. In slice is pressed bolt Ø30 mm which is additonally welded to the outer side on the slice.

Fig. 2. Original damaged slice.

Chemical analysis showed that the bolts (Ø30 x 83 mm) and shells (Ø42/Ø30 x 65 mm) made of austenitic corrosion steel X5CrNiMo17-12-2 [5]. From damaged bolt Ø30 mm, shown in Figure 3.a, sample was cutted and control of hardness is performed. The cross-section of the test sample of damaged bolt is shown on Figure 3.b.

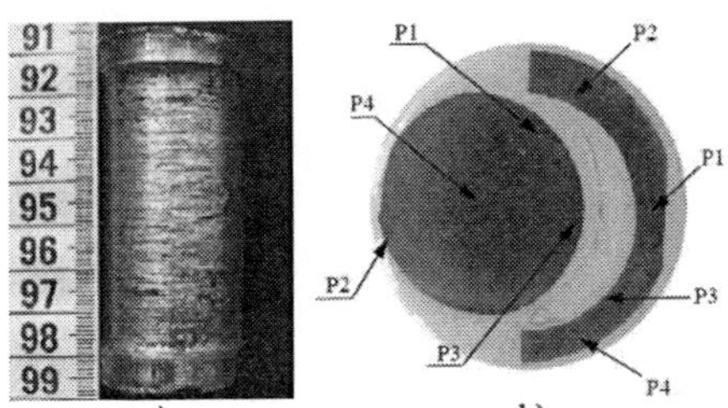

a) b)

Fig. 3. The appearance of the original bolts in place „worn place" (a); macro recording cross-sectional sample with marked positions of measuring the hardness of the bolt and the bushing shell (b).

Test of shell is carried out on spot of damaging in towing part with the bolt, Figure 4.a, and in the outer side in contact with the sprocket, Figure 4.b.

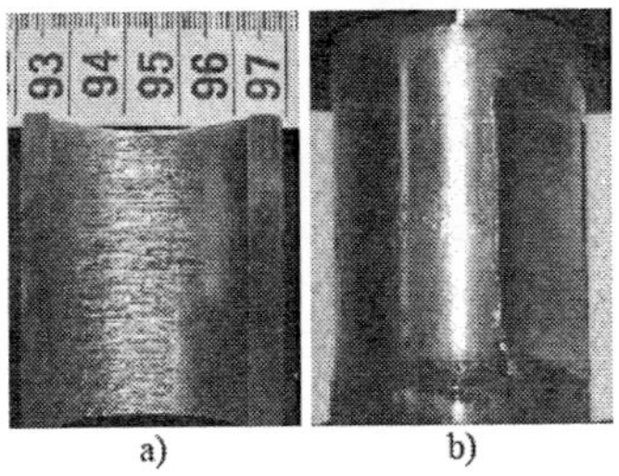

a) b)

Fig. 4. The appearance of the original bolt on spot „worn place" (a); macro recording of cross-sectional sample with marked positions of measuring the hardness of the bolt and shell (b).

At the macro recording of cross-section of damaged shells, positons of hardness HV1 measuring are marked, Figure 3.b. Hardness values measured HV1 from surface to core of sample taken from damaged bolt and shell are showed in Table 1.

Table 1. Result of hardness measurements HV1 on sample of damaged bolt and shell.

Distance from edge, mm	Hardness, HV1			
	Bolt		Shell	
	position S1	position S2	position C1	position C2
0,1	294	269	269	348
0,25	283	276	269	330
0,5	239	269	251	357
1	193	251	229	219
1,5	182	234	214	189
2	168	214	210	193
3	171	182	205	185

3 The results of own variant materials

Analyze of recording results of condition are: 1. slices and link shoes are made from shell for improving, and the cause of their damage is corrosion due to the presence of fat acids in the cake; 2. shells and bolts are made of austenitic stainless steel and were damaged by abrasion particles of crust, but also by adhesion in the area of mutual contact; 3. because of a corrosion impact of agressive media on link plates and link shoes, damaging of their connecting spots with pressed bushing shells has been observed. A final result of all wear processes is increasing of a chain step and „upon contact" of the sprocket. This could cause the appearance of inadmissible stress and appear of breakout of connect spots with bushing shell.

3.1 The choice of own variant materials for wear resistance research

Based on the results of the recording situation, conclusion was that the most critical element in tribosystem are bolt and shell. The dominant mechanism of wear is abrasion with particles of grind and adhesion in area of mutual contact, and shell contact with sprocket. Another critical spot is contact of shell and bolt with slice link shoes. For creating samples/bolts on which will be performed experiments are selected martensitic stainless steel in an improved condition X20Cr13 [5]. These bolts are then induction hardened. It was measured that the surface hardness of the bolt is in the range of 55 ÷ 57 HRC. Test bolts are welded to slice with MMA procedure. As additional material is used electrode based on Cr-Ni-Mn (18/8/6), classified by EN 1600 [6]. Electrode diameter is Ø3,25 mm, instesity of current is $I \approx 80$ A. Material of slice is C45E. Samples/bolts are incorporated in new original elevator chain at the beggining of processing sunflowers.

 https://doi.org/10.1051/matecconf/201818401014

In this chain shell are made from steel for cementation 16MnCr5. Effective layer depth of bushing shell was ≈ 0,8 mm from inside and outside. Measured surface hardnesses of shell are in range from 55 to 58 HRC.

3.2 Test results of own variant samples/parts

In order to verify welded compound bolt/slice were tested on the testing machine with static tensile tests. The load is added untill the termination of welded compound, and broken places served for metallographic examination and hardness control. The appearance of the welded compound slice/bolt after testing is showed on Figure 5.a. The cross-section of tribosystem shell/bolt/slice where will be performed metallographic examination and hardness measurement is shown in Figure 5.b.

a) b)

Fig. 5. External slice after breaking bolt (a); the cross-section shell/bolt/slice (b).

Characteristic traces of bolt wear are shown in Figure 6.a and 6.b.

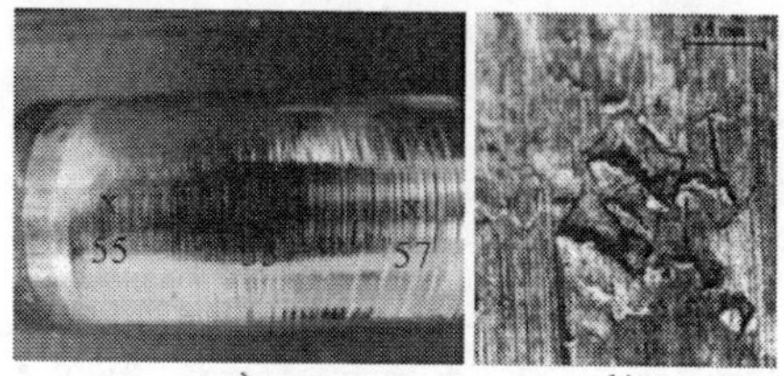

a) b)

Fig. 6. Characteristic wear traces of bolts, macro (a); micro (b).

Diagram in Figure 7.a shows the flow of hardness HV1 from the edge to the core of the bolts on the unexhausted part (curve a) and on worn parts (curve b). Diagram in Figure 7.b shows the flow of hardness HV1 from the edge to the core of the shell after use.

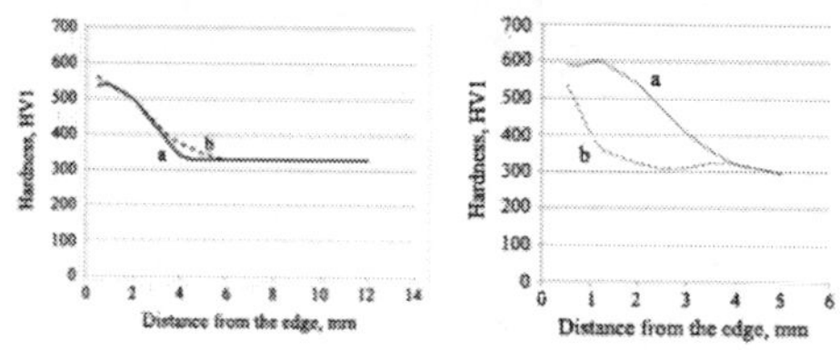

a) b)

Fig. 7. The flow of the hardness HV1 from the edge to the core after use of the bolts (a) and of the shell (b).

The edge zone of bolt, hardness ≈ 600 HV1 (unexpended part)	Hardness of clear weld are from 184 to 193 HV10	Hardnesses range from weld to HAZ on link plate side is 212 to 253 HV1

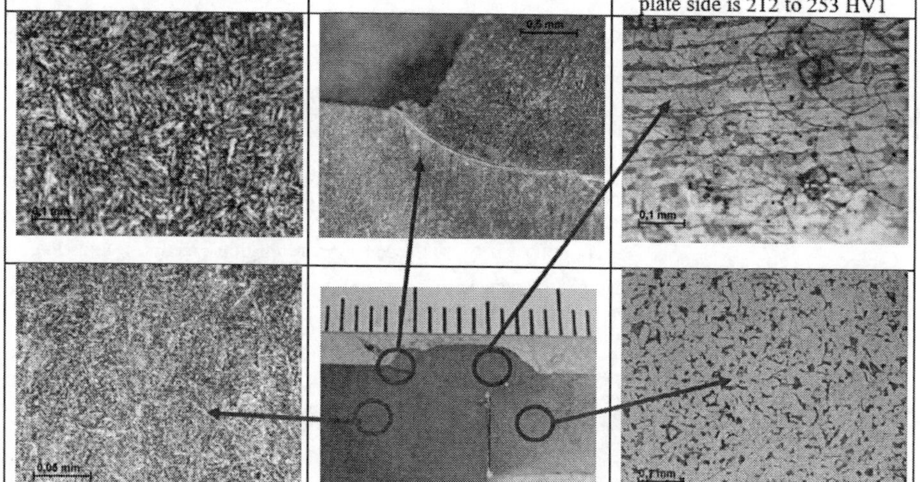

Adler		Nital
Hardness core of bolt is 330 HV1		Hardness core of slice is from 150 to 165 HV1

Fig. 8. Microstructure and hardness of bolt and slice.

Microstructure of bolt, slice, weld and HAZ are showed in Figure 8.

On test sample of bolts with floating benchmarks is conducted dimensional control.

Based on measurements on ten positions, conclusion was that on the most spended part diamater is 29,7mm, which is the higher for ≈ 0,3 mm compared to the built measure of bolt. Traces of wear in the towing part of the stud characterize radial furrows caused by abrasive action of tiny particles of sunflower shell. These particles, during the work came in contact with bolt/shell, and considering the hardness of micro abrasives ≈ 6 Mohsa [7] „furrows" are interrupted at the point of encountering on harder carbides are logical look of traces of wear. Figure 9 shows the cross-section of the welded joint shell/slice and values of measured hardness HV1.

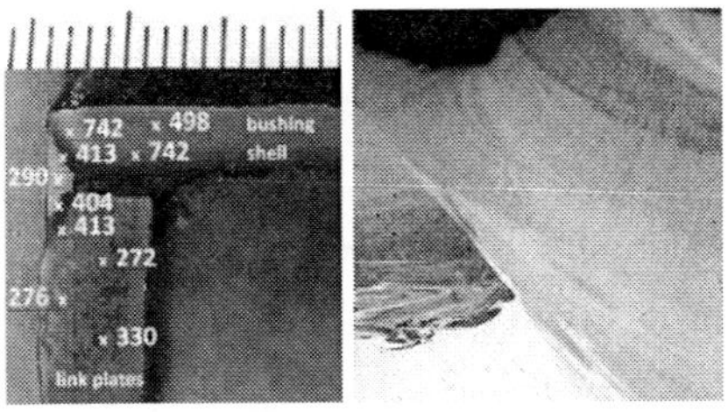

Fig. 9. Cross-section of welded compound link plate/bushing shell and values of measured hardness HV1.

Figure 10 shows results of dimensional shell control. There is a intesive wear in contact area with the bolt, wall thickness is reduced from 6 on ≈ 3 mm.

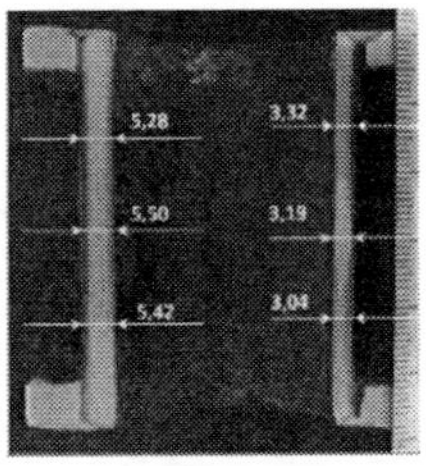

Fig. 10. Results of dimensional control of bushing shell.

Figure 11.a shows characteristic traces of wear of external surface of shell, Figure 11.b shows traces of shell wear from the inside in contact with bolt.

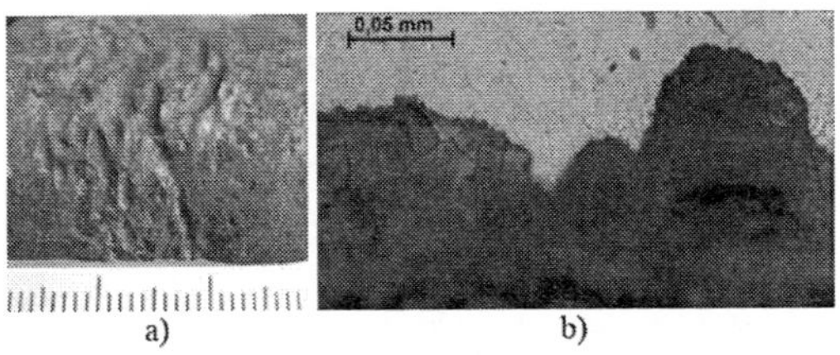

a) b)

Fig. 11. The typical wear traces of external carburized surface of bushing shell, the outer surface – macro (a); inner surface (b).

4 Analyze of results and conclusion

Based on the results of laboratory tests and comparisons of wear of original parts of chain elevator with variant bolts and shells, conclusion is: 1. the dominant wear mechanism of bolts made of martensitic corrosion-resistant steel is abrasion with tiny particles of husks, same as carburized shells in contact with bolt; 2. carburuzed shells are damaged by abrasion and corrosion due action of acid media which has pH value ≈ 5.2; 3. wear traces of external shell carburized surface with obvious local damage, suggest that dominant mechanism is abrasion; 4. maximum measured force at break of weld joint bolt/slice is higher than the budgetary breaking force of chain. The analysis results, search of structure and the measured hardness of triboelements show that the application of this approach to solving complex problems of wear elements of straight-chain transporter can achieve significant prolongation of life. Additionally, perhaps the most important, welding shells and bolt to appropriate slice avoide the risk of fracture. this is for all structures including tribosystem most important thing, because intensity of wear based on analyze of structure, hardness, traces of wear and middle media can predict. These results make conditions for possible continuation of research which will, except to increase the wear resistance of triboelements, could go in the direction of more detailed analysis of the impact of additional material for welding shells and bolts with slices so the choice of other parameters for welding process.

5 References

1. K. Grilec, S. Jakovljević, G. Marić: *Tribology in Mechanical Engineering*, Faculty of Mechanical Engineering and Naval Architecture, Zagreb, 2015.
2. S. Perić, B. Nedić, D. Trifković, M. Varuna: An Experimental Study of the Tribological Characteristics of Engine and Gear Transmission Oils, Strojniški vestnik **59** (2013) 7-8, 443-450.
3. H. Czichos: *Tribology: a systems approach to the science and technology of friction, lubrication and wear*, Federal Institute for Testing Materials, Berlin, 2009.
4. V. Marušić, B. Kovačević, A. Milinović: *Possibility of application of stainless steel in fabrication of parts of oil-cake chain elevator*, Faculty of Technology - Institute of Field and Vegetable Crops, Novi Sad, 2011., 115-121.
5. EN 10088-2-2005: Stainless steels. Technical delivery conditions for sheet/plate and strip of corrosion resisting steels for general purposes.
6. BS EN 1600-1997: Welding consumables. Covered electrodes for manual metal arc welding of stainless and heat resisting steels. Classification.
7. B. Nedić, Ž. Rosandić, D. Jovanović, M. Čupović, V. Marušić: Influence of Tribological Processes of the Mechanisms of Free Motion on Operational Characteristic of the Impulsive Friction Variators, Strojarstvo **54** (2012) 1, 49-58.

MATEC Web of Conferences **184**, 01015 (2018) https://doi.org/10.1051/matecconf/201818401015
Annual Session of Scientific Papers IMT ORADEA 2018

Selection criteria of optimal characteristic material and technologies for precision processing of basic working surface of human hip-joint implant

Raul Turmanidze[1,*], *Predrag* Dašić[2], and *Giorgi* Popkhadze[1]

[1]Georgian Technical University, Department of Production Technologies of Mechanical Engineering,Kostava St. 77, Tbilisi 0175,Georgia
[2] Faculty for Strategic and Operational Management (FSOM), Str. Jurija Gagarina 149, 11070 Novi Beograd,

Abstract. In view of the fact that the endo-prosthesis heads of human hip-joint are operated in extreme conditions, in respect of load, the selection of corresponding material and also increase of precision and quality of machining of spherical surfaces is rather topical task. In the submitted work are reviewed the problems connected with definition of the influence degree of orientation of the sapphire crystal on its workability during diamond grinding with a butt of the ring and elaboration of the perspective, original scheme of formation of the incomplete spherical surface, particularly, of the sapphire head of endo-prosthesis of the human hip-joint.

1. INTRODUCTION

Endo-prosthetics is effective and often the only method of the function restoration of the human joint. It is established that every year in the world are made about a million operations of exchange of the human hip-joint.

The endo-prosthesis heads of the human hip-joint from the point of view of the character and value of load are operated in extreme conditions. The contemporary joint endo-prostheses consist of acetabular (cup) and hip component (leg) and the head from metals or ceramic materials on their base. At present, there is an acute problem of creation of wear resistant inert materials for Implant- ology.

Therefore, in each specific case the selection of the necessary material with corresponding physical-mechanical characteristics and also increasing of precision and quality of the most significant part of endo-prosthesis – spherical surfaces is rather actual task the acuteness of which intensively grows in recent years. The number of used endo-prostheses is some tens of millions of pieces a year and the statistics shows that unfortunately this number increases every year.

The production of implants from bio-ceramic materials is rather profitable direction. Up to now, a powerful industry of fabrication of implants, tools and accompanying materials has been created and the western market of this product is evaluated in 2,5-3 billion Dollars a year. The basic developing and producing countries of implants are USA, Japan, Germany, France, Great Britain, Russia, Italy, South Korea, countries of Asia region and other countries.

It is conditioned by the fact that if earlier the necessity of similar operations was caused by the age factor of the man or traumatology fractures in recent two decades abruptly increased the number of patients at young age of 30-40 years of both men and women without any injuries and fractures. In opinions of physicians, the principal reasons of it are non-active way of life of youth, composition of contemporary artificial food products and metabolic disorder. All the above-mentioned reasons determine the number of used endo-prostheses in some tens of millions of pieces a year.

The medical practice proves that the repeated prosthetics of the human hip joint is connected with big problems. In many cases, the implementation of such operations becomes practically impossible. Therefore, the durability of the endo-prosthesis of the human hip joint to the end of the patient's life especially at young age has especially significant meaning.

The contemporary elaborations of the endo-prosthesis structure are directed to exchange of chirulen by ceramic material that leads to the change of the endo-prosthesis structure, development of new ceramic materials with the improved physical and mechanical characteristics, development of processing of internal and external spherical surfaces, development of diamond tools providing the high quality of the machined surface, definition of the optimal wear resistance of the pair of materials for manufacture of the endo-prosthesis etc.

For nowadays in the world practice these heads are manufactured from various alloys, composite materials and ceramics, which mainly are isotropic materials. Therefore, the data of the above-mentioned works do not give the necessary information on machining of

* Raul Turmanidze: inform@gtu.ge

© The Authors, published by EDP Sciences. This is an open access article distributed under the terms of the Creative Commons Attribution License 4.0 (http://creativecommons.org/licenses/by/4.0/).

anisotropic materials, particularly artificial crystal of sapphire.

Together with indisputable positive qualities, the contemporary structures of endo-prostheses from metal, ceramics polymers have certain defects. Among them there is an insufficient biological inertness and redundant wear of components of the friction pair that lead to short duration of work of the artificial hip-joint. Besides that as the result of wear of the material in the friction pair of endo-prosthesis hinge toxic and oncological danger products of the dissociation in various organs and tissues accumulate that in 30-40% leads to hard complications and requires complex disabling interferences [1, 2].

The circle of materials that satisfy the criterion of biological compatibility is rather limited. Because of it and requirements of the resource of articles in condition of action of alternating loads, corrosion active environment titanium and its alloys as the material for fabrication of endo-prostheses have the advantage over Co-Cr-Mo alloys.

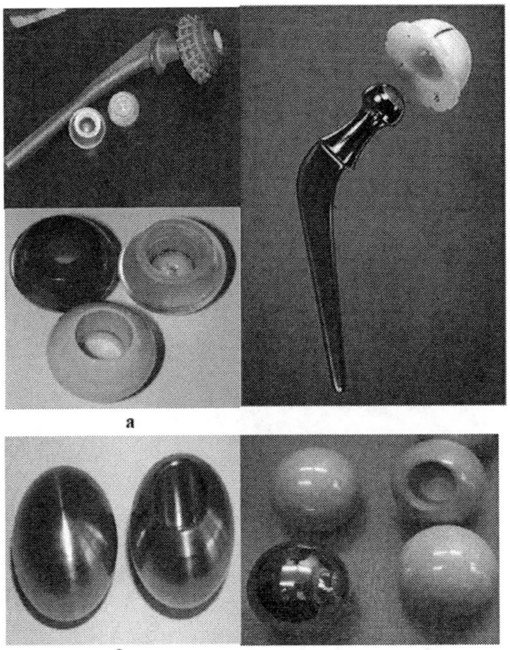

Fig. 1. Hip heads from sapphire a) zirconium ceramics b) biologically pure titanium alloy c) stainless steel (SS)

2. Grind ability of monocrystalline sapphire, experimental results and discussions

The wear of heads from zirconium ceramics is 7-9 nm a year while titanium 105 nm a year. Coming out of this the fabrication of implants from bio-ceramic materials is a perspective direction of development of the science intensive technologies of creation of materials and their machining. Most biocompatible with the human organism, wear resistant and durable material for fabrication of the above-mentioned article is the artificial mono-crystal of sapphire.

For nowadays, the circle of materials for fabrication of pairs of bearing surfaces of endo-prostheses of joints with minimally possible number of products of wear has been determined. Such pairs of bearing surfaces with excellent characteristics of wear resistance are the friction pairs of ceramics-ceramics, metal-metal and polyethylene with a high degree of cross-section links in combination with ceramics or metal. Together with that, the basic unresolved problem of the last decade was the development of bearing surfaces that could endure much higher loads with young and movable patients. The surfaces that are being investigated at present in laboratory conditions due to their hopeful characteristics of wear are ceramic matrix (82% of aluminum oxide, 17% of zirconium dioxide, 0,3% chrome oxide), zirconium dioxide and ceramics in pair with cobalt-chrome alloy.

Together with that sapphire representing the mono-crystal of aluminum oxide as a material for bearing surfaces possesses unique inertness including electrolyte passiveness, perhaps the best of the well-known materials, biocompatibility, corrosion resistance and hardness. Resistance of sapphire towards any acids and alkali is rather higher than that of metals and even poly-crystalline oxide of aluminum. Probably because of it, sapphire does not change the immune status of the patient. If metals and poly-crystalline materials used for the bearing surfaces have different speed of wear of the micro-sites it leads to the coefficient increase of the pair friction and wear increase then this effect with sapphire is absent.

Tribological investigations of the friction pairs of materials were conducted by the scheme of rotational friction with the contact geometry of "ball on disk" type. The flat disk was manufactured from the investigated material and the ball from the material of counter body (Fig. 2).

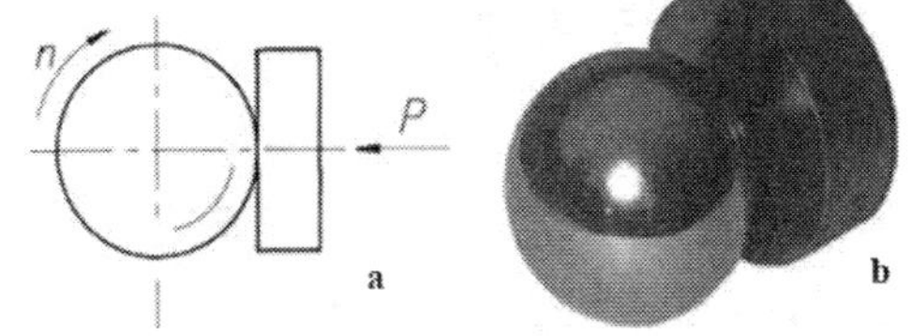

Fig. 2. Schematic arrangement of rotational friction – a, and example of counter bodies – b.

The obtained dependencies of the friction force of pairs: sapphire-sapphire, sapphire-ruby from the time of tests (number of cycles) has mainly non-monotonous abrupt character.(fig. 3-4) From them one can mark out three stages of wear, namely: running-in wear – creation of the work roughness and necessary supporting surface on frictional surfaces. At the moment of beginning of work the contact of bodies takes place in the point, accordingly the specific load is sufficiently high that leads to the abrupt growth of the friction force and because of wear of the surface material.

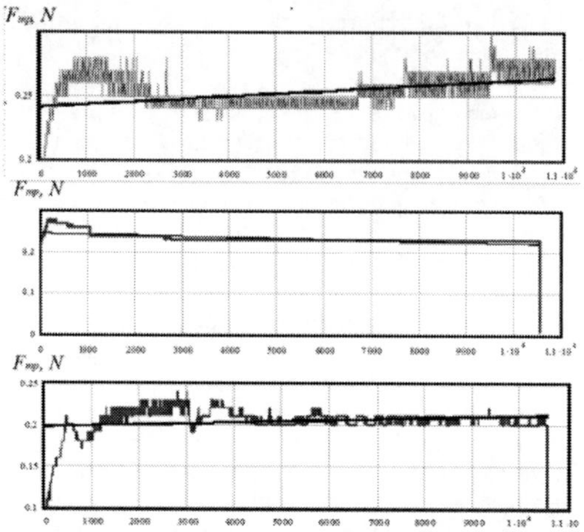

Fig. 3. Dependence of the friction force of the sapphire-sapphire couple from the time of tests and its approximation by straight: flat 1018 – a, flat 1080 – b, flat 0001 – c.

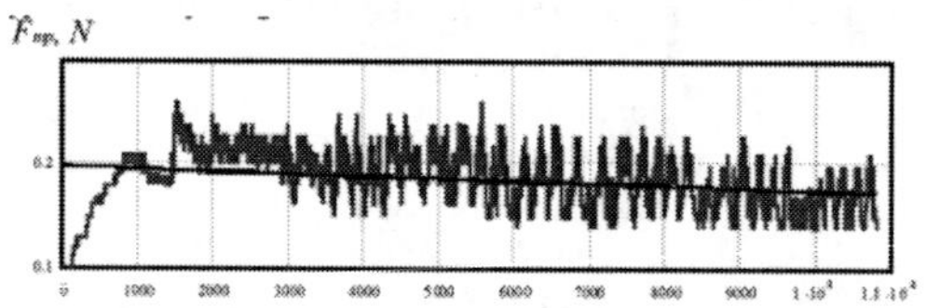

Fig. 4. Dependence of the friction force of the sapphire-ruby couple from the time of tests and its approximation by straight.

Normal wear is after achievement of the surface wear meaning, at which the optimal meanings of characteristics of supporting surface are reached, the stable process of friction of bodies has place with gradually decrease force of friction.

Catastrophic wear is in the process of friction of counter bodies of the wear products which fill the pockets on the friction surface and because of the weak (drop) feeding of the Ringer solution into the area of friction the wear products accumulate occupying all the vacant place in the pockets and as a result of this take part in the process of wear of the investigated surfaces as a free abrasive.

The pointed out above stages of wear can have different length of time or generally to be absent in the wear process.

From the dependencies follows that the best results in relation of the friction force to the pressing force shows the friction pair of sapphire-sapphire with orientation of the crystalline lattice on flatness 0001, linear wear of the friction pair of sapphire-sapphire with the orientation of the crystalline lattice on the flatness 0001 and sapphire-ruby.

Titanium and alloys on its basis are widely used in medicine as implants and other articles. From the point of view of biocompatibility for the implants, working for a long time in the live organism the use of titanium is preferable that unlike alloys does not contain alloying additives harmful for the live organism. However, titanium in its usual state has low mechanical properties in comparison with its alloys. This problem was solved by means of formation of Nano - and composition structure in technically pure titanium.

The use of the friction pairs of sapphire/titanium consisting of non-toxic materials would help us to solve the pointed out problem. However, for nowadays it is considered that from titanium and its alloys is impossible to manufacture a friction pair because of their high inclination to a contact snatch and therefore increased wear during friction. This property makes dangerous the use of titanium in the friction pairs.

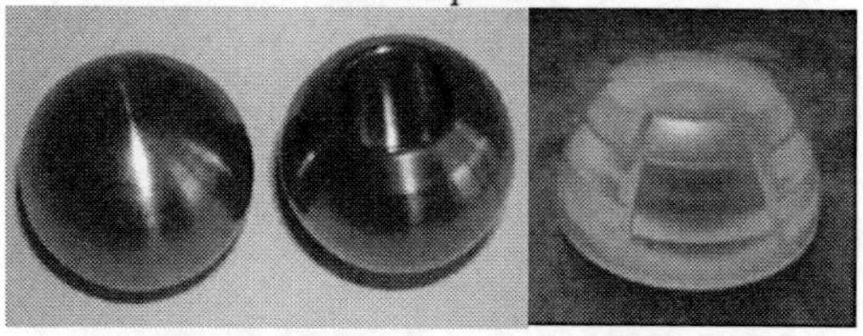

Fig. 5. Experimental specimens of titanium head and sapphire cup for endo-prosthesis of the hip-joint.

Thus, the work objective is the creation of a new bearing connection of endo-prosthesis of the hip-joint having the improved quality at the expense of use of hardened sapphire and biologically inert technically pure titanium with modification of the surface layer of IPD and subsequent nitrating as materials improving tribological properties of the connection.

a

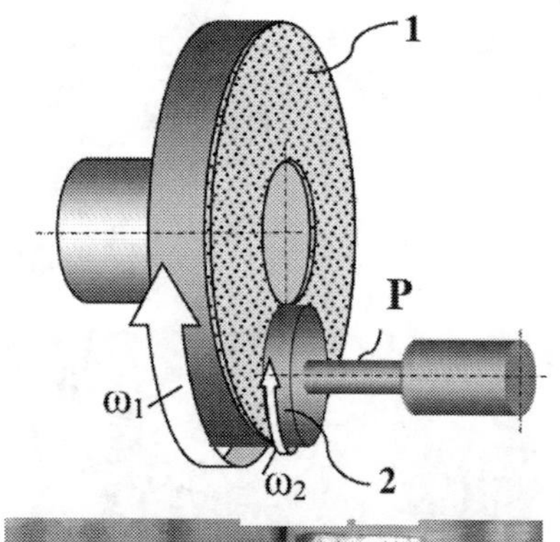

b

Fig. 6. a) Schematic arrangement of LPG: 1- Grinding ring, 2- Cassette with parts, b) Laboratory setting for LPG.

For conduct of investigations of machinability of the sapphire crystal, the method of Low Temperature Precision Grinding (LPG) has been chosen developed at the Department of Mechanical Engineering of the Georgian Technical University as a version of progressive methods of diamond grinding of hard and brittle non-metallic materials.

In Fig.6a the machined parts set on the cassette in separators or by other method of mounting, for example, gluing carry out rotational movement with angular speed of ω_2 and the grinding ring with speed of ω_1 in the same direction. In the cutting area, the pressing is carried out by force P of the machined surfaces of parts to the work surface of the grinding ring.The experimental investigations were conducted on the laboratory setting equipped with the special precision head (Fig. 6b).

By the data analysis of complex, all-round experimental investigations conducted by us one can make the following conclusion:

By other equal conditions of machining, the most intractable is orientation 0001. For all the tested diamond rings, the ration of meanings of the linear removal of material q is within q_{0001}/q_{1010}= 0,25…0, 5, at that, q_{1012}/q_{1010} = 0,75…1.

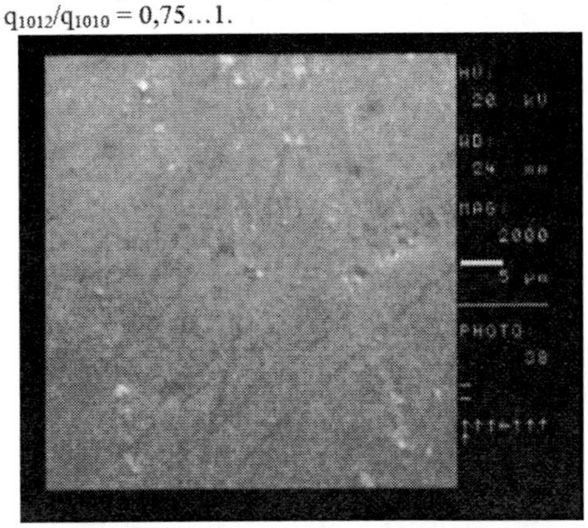

a

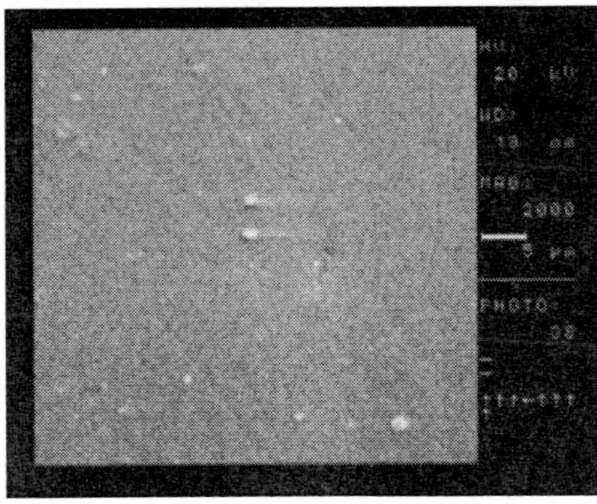

b

Fig. 7. Micro-photographs of surfaces of experimental specimens of sapphire machined by the LPG method: a - Orientation 1010, b - 1012. Diamond ring–ACM 14/10, bunch organic special, 50%. Cutting modes: V=1m/sec, P=750 kPa.

The productivity of machining grows in the range of cutting speed of V=1…6 m/sec while at the further increase of the cutting speed to 12m/sec it remain constant.With increase of the pressing force P in all the investigated range the productivity of machining grows, however, in the interval of P=1000…1500 kPa the productivity growth considerably decreases. From characteristics of the diamond tool on productivity by the prevailing way influence granularity and bunch of the diamond tool. The concentration influence is insignificant. With the increase of the grain size within d_3=14/10…28/20 the productivity grows 1.5…2.5 times. The maximum productivity of machining is provided by the tool on ceramic bunch ({0001}– 130mkm/min; {1010} –300mkm/min; {1012} –250mkm/min), then on metallic ({0001} –50 mkm/min; {1010} –200mkm/min; {1012} –170mkm/min) and organic ({0001} – 30mkm/min; {1010}– 120mkm/min; {1012} – 110mkm/min). The tool on the ceramic bunch works in the mode of self-grinding [3, 4].

From characteristics of the diamond, tool the granularity and material of the tool bunch influence in prevailing way on the surface quality. With the grain increase in the investigated range the height of unevenness R_z grows within 1…1.5 classes and the depth of the violated layer H 1.5…2 times. In other equal conditions of machining on orientation, (0001) higher quality of the surface is reached than on the rest two. The difference is in 1…1.5 classes of roughness. By this indication, the best results are shown by the diamond rings on organic bunch. So, for example, on the diamond rings on bunches BC-11 and organic special are obtained the following results: R_z=0,25 mkm; t_{P03}=35...45%; H=2...5 mkm. At that the meaning of parameter Rz is in order below than parameter t_{P03}– 1,5 times higher of parameter H- 3…5 fewer than the meanings of corresponding parameters that are obtained on the diamond rings on ceramic and metallic bunches.

The influence character of the process factors on output parameters for the chosen orientations of the sapphire crystal ((0001), (1010), (1012)) is constant.

By study of morphology of the machined surface the possibility of cutting the sapphire material by plastic deformation of the removed layer at low cutting speeds V=1…3 m/sec with the least depth of the violated sub-relief layer has been proved (Fig. 7).

3. Kinematics of forming the spherical heads of endo-prosthesis

It is known that machining of vitreous materials, particularly, the sapphire crystal by means of plastic deformation of the removed layer instead of fragile destruction – dispersion, the pledge of acquisition of the machined surface practically without hereditary defect – without sub-relief layer. Value H appeared to be the least namely on these specimens of sapphire. The obtained meaningful result requires individual investigations the conduct of which is being planned.

The development of a new or updating of the existing technological process of machining of the sapphire head sets the actual task of creation of new highly effective schemes of formation. The optimization criteria of technological operations such as productivity, indices of the surface quality and precision of machining determine the place of new schemes-methods of formation in technological process taking into account their advantages.

The schematic arrangement of the spherical head of endo-prosthesis is shown in Fig. 8. The partial sphere is determined by its radius and angle β of spherical segment. There are several ways of machining of the spherical head. The closest to the LPG process in kinematics is the scheme of grinding of the partial sphere with use of the end grinding ring. The machined part rotates with angular speed of ω_2 and ω_3 around axes 2 and 3 subsequently (Fig. 9). This kinematics forming the incomplete surface of the sphere is taken as the basis of development and possible realization of more effective schemes of grinding of spherical heads of endo-prosthesis taking into account the kinematics and other positive features of the LPG method [5, 6].

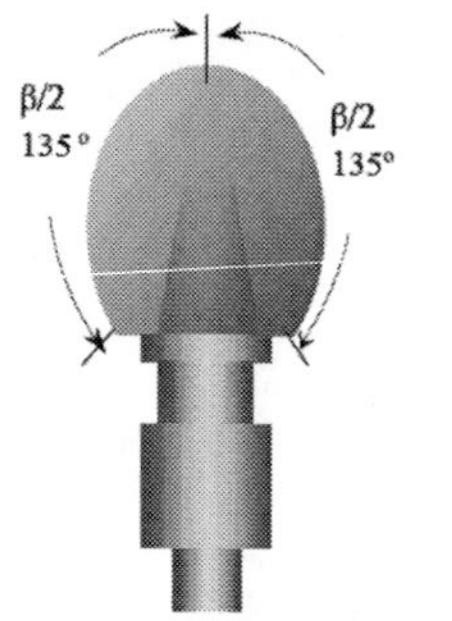

Fig. 8. Endo-prosthesis heads: β – angle of spherical segment.

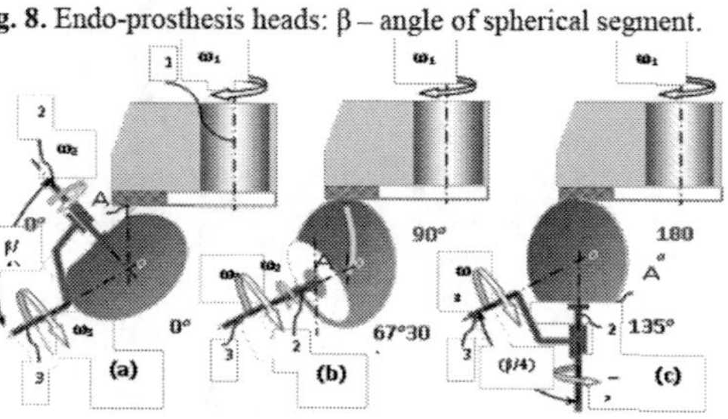

Fig. 9. Schematic arrangement of grinding of the partial sphere with use of the grinding ring.

The end grinding ring rotates with the angular speed of ω_1 removing the tightness for machining from sphere with the required speed of cutting and rate of feeding. For full machining, the spherical head must rotate simultaneously around axes 2 and 3 with the angular speed of ω_2 и ω_3. Axes 2 and 3 intersect in the centre of sphere in the point *O*. As it is seen in Fig.4, the angle between axes, 3 and 4 must be equal to β/4 but for to machine the whole surface of partial sphere angular speeds ω_2 and ω_3 must cinematically correlate as the both of them specify the values of components of the feeding rate. The linear speed on the ring/ball juncture and correlation of angular speeds ω_2/ω_3 will determine the surface texture of the half-complete or complete head of endo-prosthesis. The detailed analysis of kinematic connections between rotational speeds and other relative parameters of the grinding process are out of the Report and this will be discussed in the following publications. One of the formation schemes of the incomplete spherical surface with the forming grinding ring is shown in Fig. 10.

The partial spherical head rests on the forming grinding ring consisting of internal and external cones with abrasive layers (Fig. 10 a, b) and rotates around axes 2 and 3 with angular speeds of ω_2 and ω_3 in the same manner as it was described above in Fig.8. The machined part by means of a strong spring P is pressed to the forming ring and is continuously fed to the required depth in direction of the ring. The applied pressing is equally distributed on the end of the ring work i.e. in points A and B (Fig. 8 b). In Fig. 8a are shown eight subsequent positions of the spherical head in one cycle. For clarification all, the positions are equally distributed along the ring although such subsequent positions are in the same place M. This scheme enables also carry out machining in "machining stations" around the ring. The main defect of this scheme is a complexity of the grinding ring and difference of speeds in points A and B. The latter can be compensated in the same way as in the LPG process.

Comparatively simple schemes of formation of the spherical heads of endo-prosthesis with the use of commercially available shapes of the ring are shown below in Fig.9 a, b, c and d. The schemes presented in Fig.9 a, b and c differs only by the shape of used rings. The kinematics of formation of the partial spherical head is the same as the first scheme presented in Fig. 7 and 8.

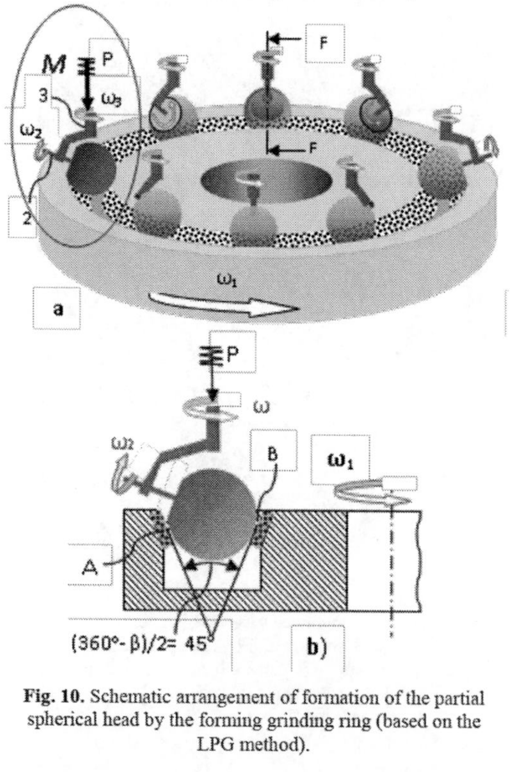

Fig. 10. Schematic arrangement of formation of the partial spherical head by the forming grinding ring (based on the LPG method).

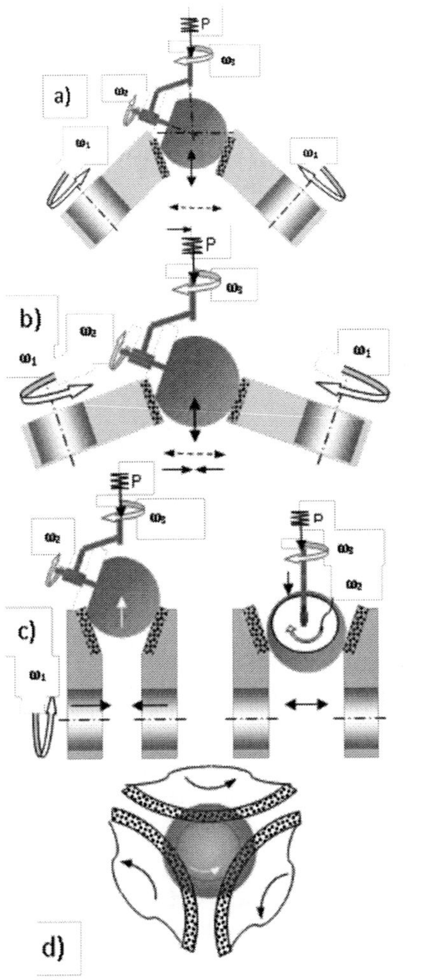

Fig. 11. Possible versions of grinding of the endo-prosthesis during the use of commercially available usual rings.

4. Conclusion

Scale of influence of single-crystal sapphire of the crystallographic flat orientation and the grinding conditions in the mode of removal of material, surface finishing and the state of sub-surface layer studied during low-temperature precise grinding have revealed that workability of single-crystal sapphire considerably depends on the crystallographic flat orientation. The relative values of removal of the material against the specimen with crystallographic flat orientation (1010) are within $q_{0001}/q_{\underline{1010}}$=0.25…0.5, and $q_{1012}/q_{\underline{1010}}$=0.75…1. In other equal conditions of the LPG process, the high quality of the ground surface is achieved for crystal orientation (1010). The difference with two other orientations of the crystallographic flatness is within 1 …1.5 class of roughness.

By the studies of morphology of the ground surface was proved the grinding possibility by single-crystal sapphire in compliant mode, i.e. removal of the deformed layer by cutting without cracks at the low speed of cutting. In such conditions of machining was detected the least depth of the faulty sub-surface layer.

The structural schemes of formation of the partial spherical heads of endo-prosthesis for human hip-joint need a further analysis and optimization with the aim of designing of advanced technological processes and the prototype of grinding machines.

There are proposed two variants of the kinematic scheme of a special grinding machine for obtaining incomplete spherical surfaces of high accuracy, of which the first option is based on the use of a diamond grinding wheel with a complex shaped profile of the working part, and the second involves the use of grinding wheels with simple profiles, but the machine itself must have a software control system of high accuracy and reliability. The choice of the customer will depend on the required productivity of the machine and the level of qualification of the maintenance staff.
A great application of modern nanomaterials in this area is also expected [7].

References

1. B.I. Batiashvili, D.S. Butskhrikidze, G.A. Mamulashvili, R.S. Turmanidze, K. Kromp, B. Mills, O. Mgaloblishvili, Technological Possibilities of Low Temperature Precision Grinding Process when Machining Hard and Brittle Materials. Fractography of advanced ceramics, International Conference, Stará Lesná, High Tatras, May 2001.
2. R.S. Turmanidze, D.S. Butskhrikidze, K. Kromp, B. Mills, "Low temperature precision grinding of hard and brittle materials". Problems of Mechanics and Physico-Chemistry of the Process of Abrasive Machining, Kiev, pp. 490-499 (2002).
3. O.A. Rozenberg, S.V. Sokhan, V.V. Vozny, A.G. Mamalis, J. Gavlik, D.-J. Kim, Trends and Development in the Manufacturing of Hip Joints: An Overview. Int. J.Adv. Manuf. Technology, **27**, pp. 537-542 (2006).
4. J.R. Lieberman, Two Alternative Bearings for Total Hip Arthroplasty: More Data Are Needed. J Am Acad Orthop Surg, pp. 61-62 (2009).
5. E. Dobrovinskaya, L. Litvinov, V. Pishchik, Mono-crystals of corundum – K: Science, 256 pp. (1994).
6. R.S. Turmanidze, Implants of the human hip joints and peculiarities of their manufacture with the high precision and quality of machining of the work surfaces. Proceedings of the 13th International Conference on Tools (ICT-2012), 27-28 March 2012. University of Miskolc, Hungary. (2012).
7. P. Dašić, D.Lj. Mirjanić, Basic properties and applications of graphene. Proceedings of the Conference on Contemporary Materials (CM-2017), 9-10 November 2017. Banja Luka: Academy of Sciences and Arts of the Republic of Srpska, 2018, pp. 557-568. (In Serbian).

Thermal degradation and stability of wood particle composites deployed as decorative components

Dana Luca Motoc[1*], *Santiago* Ferrándiz Bou[2], and *Adrian Petru* Pop[3]

[1]Transilvania University of Brasov, Automotive and Transport Department, 500107 Bd. Eroilor, Romania
[2]Polytechnic University of Valencia, Mechanical and Material Engineering Department, 03801 Plaza Ferrandiz-Carbonell, Spain
[3]University of Oradea, Mechatronics Department, 410087 Str. Universitatii, Romania

Abstract. The contribution aims to design, deliver and debate on thermal degradation and thermal stability of several wood/PP composite materials. The wood polymer-based composites (n. WPCs) were manufactured through injection moulding by deploying various wood species under 10% and 40% weight fraction within the thermoplastic matrix. Thermal degradation of WPC specimens revealed similarities in characteristics, small discrepancies in the degradation temperatures but higher contents in the char formation, between 10% to 35% with wood content increase.

1 Introduction

Wood particles reinforced plastic composites (WPCs) timely captured the headlines of publishing literature proven their extensive studies since, firstly, issued. A closer reading reveals the systemic approaches on the combination of reduced manufacturing cost, overall material performance, environmental friendly and recyclable attributes [1-4]. These are imperative with composite industry and other key stakeholders focused to fulfil customer demands, face an increased competition and drive economic development and innovation.

Furthermore, low or high-density polyethylene (HDPE) and polypropylene (PP) appear to prevail within the broad range of common thermoplastics used for WPCs manufacturing irrespective the incompatibility issues shown at wood reinforcements/polymer interface. Solutions were seeking in terms of various coupling agents, individual or combined chemical treatments of wood reinforcements prior to their embodiment to improve interfacial adhesion, homogeneity, processability and decrease water absorption.

A relatively large database of research papers is covering the aforementioned subject, generally, while the works of Danyady (2007, 2010), Matini Behzad (2012) or Csizmadia (2013) and their co-authors can be traced in relation to the constitutive of herein paper, particularly [5-8]. Their findings shown that irrespective of the technique used on interface adhesion properties of interest resulted dissimilarly. Moreover, they argued on targeted properties invariability with the coupling agent amount and need for constitutive optimization in order to manage the overall materials' performance.

In addition, interfacial inconsistencies were shown to highly impact the mechanical, thermal stability, weatherability or flammability performances. Tangible evidence on prior listed can be identified with the works of Srubar et al. (2012), Zhang et al. (2012), Liqing et al. (2013) or Soares (2013) and co-authors on opposite with use of additives, flame retardant materials or other surface conditioning techniques [9-11]. Admittedly, the resulted quantitative values favour on the addressed properties' enhancement.

Influence of wood species and content while engineering the overall material properties were issues tackled by researchers and accounted as primary influencing factors. Thus, works of Fabiyi or Nourbakhsh and their co-authors (2010), Bouza et al. (2011), Jeamtrakull et al. (2012) and Thumm (2013) can be traced as contribution approaching the subject [12-16].

Further literature insights enable tracks in conjunction with mechanical performances and thermal stability of composite materials out of thermoplastic to balance the manufacturing costs, environmental issues and material properties.

The major drawbacks impeding the widespread use of WPC in structural applications can be related to the limited information on their thermal stability properties. Contributions of Yang (2005), Chaharmahali (2010), Cui (2010) and Singh (2010) or Pengfei (2011) and co-authors it can be traced in relation to previous [17-21].

Their findings' emphasis a slightly decrease on both thermal stability and expansion properties as the lignocellulosic content increased, with or without use of additives to improve compatibility of composite components, irrespective of the experimental settings (e.g. controlled atmosphere, heating rate, etc.). Moreover, with respect to the mechanical performances, increased values were reported in comparison with an unmodified system used as reference.

Hence, in the line with above, the objective of the present work focused on the effect of wood constituents

* Corresponding author: danaluca@unitbv.ro

© The Authors, published by EDP Sciences. This is an open access article distributed under the terms of the Creative Commons Attribution License 4.0 (http://creativecommons.org/licenses/by/4.0/).

and content on thermal degradation and thermal stability of several wood/PP composite materials.

2 Experimental research

2.1 Material selection and specimen preparation

Wood species such as beech (lat. *fagus sylvatica*), oak (lat. *quercus robur*), sapele (lat. *E. cylindricum*), pine (lat. *pinum*) and a mixture of previous as agglomerate (n. wt.% of individual constitutive were not tackled as an issue for further debates) were reinforced in different weight fraction within the thermoplastic matrix, 10 wt.% and 40 wt.%, respectively.

Prior injection moulding into test specimens by aid of a Meteor 270/75 injection machine (Mateu& Sole®, Spain) at 185 °C, all wood particles were subjected to a moisture removal process within a controlled temperature chamber at 85 °C for 24 h.

2.2 Material testing procedure

2.2.1 Thermogravimetric analysis (TGA)

Experimental research on thermal degradation of wood particle reinforced composites was performed using a Netzsch TG/DTG thermogravimetric analyser called STA 449 F3 Jupiter.

The measurements were carried out according to the general procedures outlined by DIN ISO 11358-1:2014-10 standard followed by deployment of ASTM E1131-08:2014 for compositional analysis particular to thermogravimetric technique.

The TGA system enables detection of specimens' mass variation of 1 μg as a function of temperature. Wood plastic composite samples were subjected to pyrolysis in nitrogen environment (20 ml/min flow) to a maximum temperature of 650 °C at a linear heating rate of 10 K/min inside to an alumina sample holder.

Isothermals lasting of 5 min followed by dynamic cooling step at identical rate of 10 °C /min were discarded from data processing proven their deployment only for measuring completeness.

For every WPC specimen the weight change (%) and degradation speed (%/min) curves revealing the main peaks were managed by Netzsch Proteus Thermal Analyzer software. Supplementary, thermal degradation of PP specimen was also recorded to help the comparison purpose.

2.2.2 Dilatometry (DIL)

Thermal expansion was monitored using a Netzsch DIL 402 PC pushrod dilatometer. Combined DIN 51045-1:2005-08 and ASTM E228:1995 standard basic rules were deployed to assess the coefficients of thermal expansion (i.e. CTE) of wood plastic composite samples.

Thermal strain fields were recorded in static air atmosphere to a maximum temperature of 200 °C at a linear heating rate of 1 °C /min by simple laying the wood composites on two fused silica sample holders. By aid of a thermocouple, records on samples' dimensional changes as function of temperature were enabled.

3 Results and discussions

The thermal decomposition characteristics of WPC specimens herein were examined by means of TG/DTG. The thermogravimetric parameters retrieved for all WPC composite samples are summarized in Table 1.

In Fig. 1 and Fig. 2 were plotted the typical curves for mass loss (%) while in Fig. 3 and Fig. 4 plots of degradation speed (%/min) for individual specimens in accordance with their wt.% can be identified.

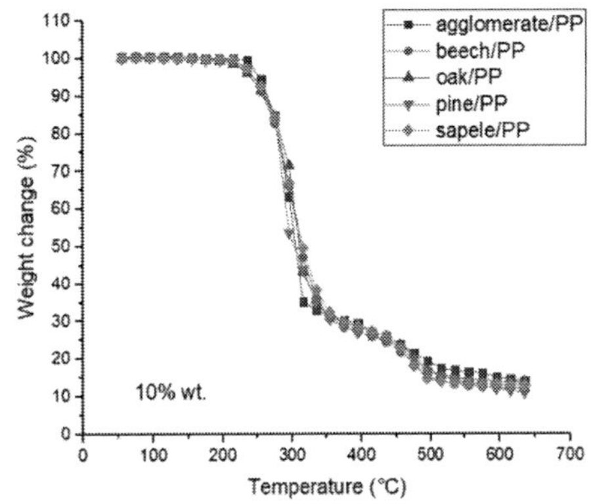

Fig. 1. Weight change over temperature range – 10 wt.% WPC

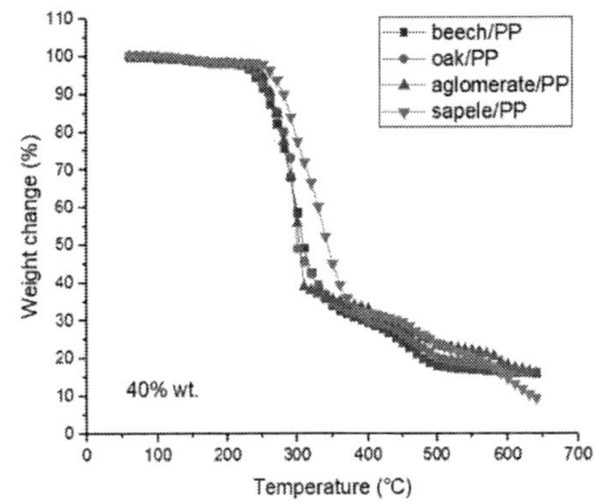

Fig. 2. Weight change over temperature range – 40 wt.% WPC

As it can be seen form Fig. 1 and Fig. 2 retrieved on WPCs specimens subjected to pyrolysis, two regions can be clearly identified. First, one can observe a specimens' slight weight loss occurring within 50 °C up to 200 °C that can be assigned mainly to thermal decomposition of PP and further to water evaporation and residues from manufacturing procedure (e.g. lubricants, residues from former deployment, etc.).

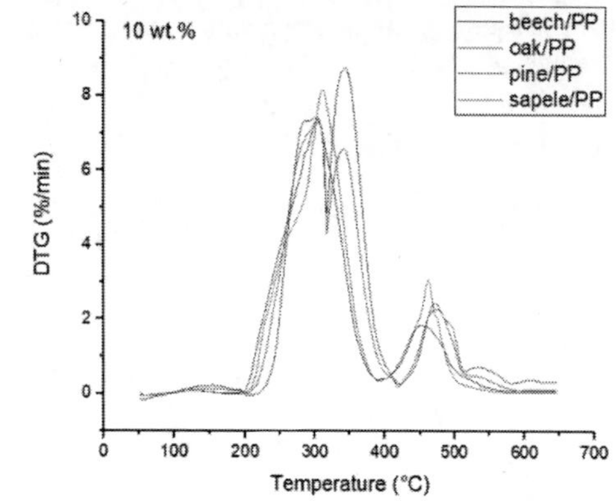

Fig. 3. Representative DTG curves for 10 wt.% wood/PP

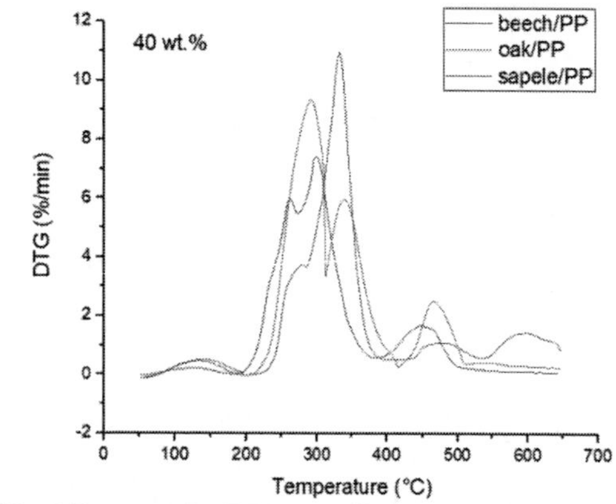

Fig. 4. Representative DTG curves for 40 wt.% wood/PP

Next, the second region, associated with temperature variation from 200 °C up to 650 °C, corresponds to the active pyrolysis due to the thermal degradation of wood particles embedded within. With respect to the latter, three principal components namely hemicellulose, cellulose and lignin are sought to degrade within above temperature range as it can be seen in Fig. 5 for a 10% wt. sapele/PP sample.

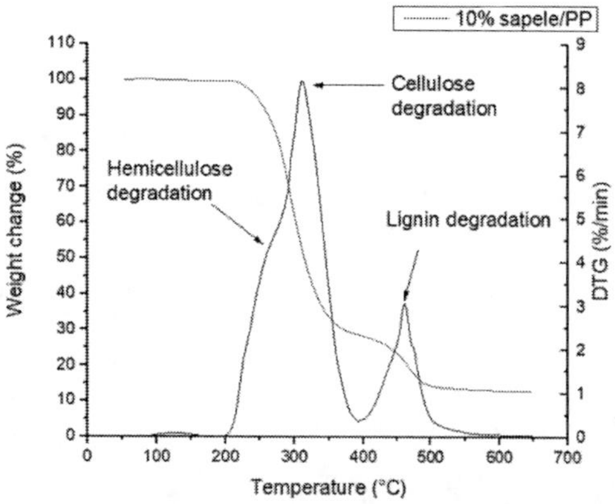

Fig. 5. TG/DTG curves for the 10 wt.% sapele/PP specimens

The first shoulder highlights the hemicellulose degradation from 200 °C up to 275 °C, while cellulose degradation follows next up to 400 °C.

Furthermore, the principal decomposition behaviour is not changed from wood species from other accounting the same weight fraction but one can see a shift in the weight loss with increasing wood content, especially for sapele/PP specimens (see Fig. 2).

It is well acknowledged that wood consist of different complicated high molecular compounds, namely cellulose (33-50% of absolute dry mass), lignin (20-30%) and hemicellulose (14-27%). Both cellulose and hemicellulose are natural polymers while lignin is a natural aromatic polymer that accounts for shape and size changes resistance of wood cells [22]. Decomposition of above components happen after the moisture evaporation, beyond 200 °C.

In relation with the derivative thermogravimetric curves (DTG) in figures 3 and 4, respectively, the curves main peaks differ with respect to their magnitude while switching from 10 wt.% to 40 wt.% reinforcements in all wood species under debate.

Lower particles content within the PP matrix seem to provide enhanced thermal stability during the cellulose degradation compared with an increased reinforcement content.

Table 1. TGA and DTA results of WPC samples

Sample	$T_{-5\%}$ °C	$T_{-50\%}$ °C	T_{max} °C	DTA_{peak} °C	Res. mass %
10 wt.%					
Agg./PP	253.3	302.9	426.9	389.1	13.62
Beech/PP	241.5	311.1	454.8	455.2	11.76
Oak/PP	240.8	301.8	455.4	456.7	13.28
Pine/PP	246.7	302.0	455.2	448.9	10.32
Sapele/PP	245.9	314.2	455.8	461.0	12.78
40 wt.%					
Agg./PP	247.6	303.6	427.0	385.8	15.09
Beech/PP	237.6	309.2	451.2	453.4	16.00
Oak/PP	230.5	295.1	456.5	449.5	15.57
Pine/PP	234.1	265.0	449.1	318.9	10.5
Sapele/PP	237.8	315.6	446.2	410.3	6.39

Supplementary, as it can be seen in the correspondingly graphs, a higher content of wood particles it leads to a degradation speed curve widening.

The last stage of the thermal decomposition is char combustion, that differ from wood species to another. Curves analysis reveal that char combustion occurs in the temperature range 450 °C – 550 °C. Accounting the decomposition processes holding under air, this process is considered to be an exothermic process.

Supplementary, is it well known that temperature and particle size of reinforcement influence the morphology of the char. It is beyond the purpose of herein article to debate on above, however it should be mentioned that the heating rate imposed to the measurements input was selected to surpass the impediment due to the particle size influence occurring at low heating rates.

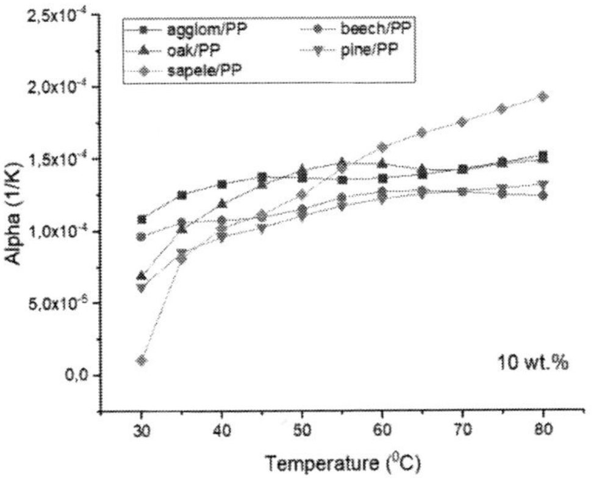

Fig. 6. Linear coefficient of thermal expansion with temperature range for 10 wt.% wood particles

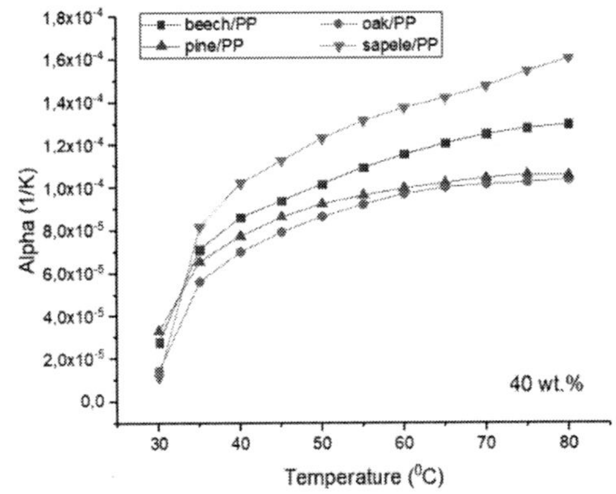

Fig. 7. Linear coefficient of thermal expansion with temperature range for 40 wt.% wood particles

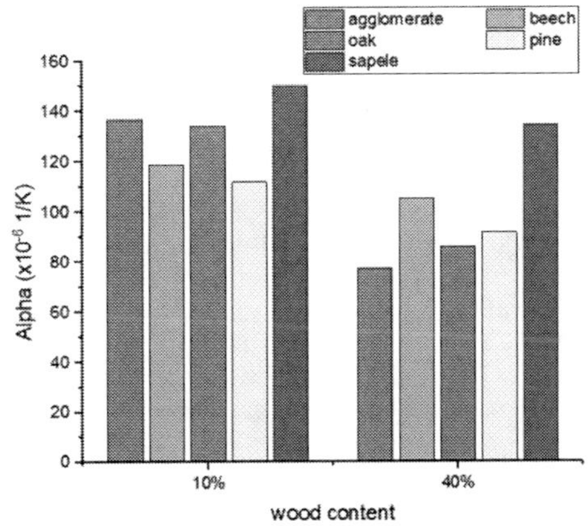

Fig. 8. Overall coefficient of thermal expansion with 10 wt.% and 40 wt.% wood particle content

With respect to residual mass on the wood particle composites herein, from the above table it can be sized relatively small differences from one species to another, or with increasing wood content.

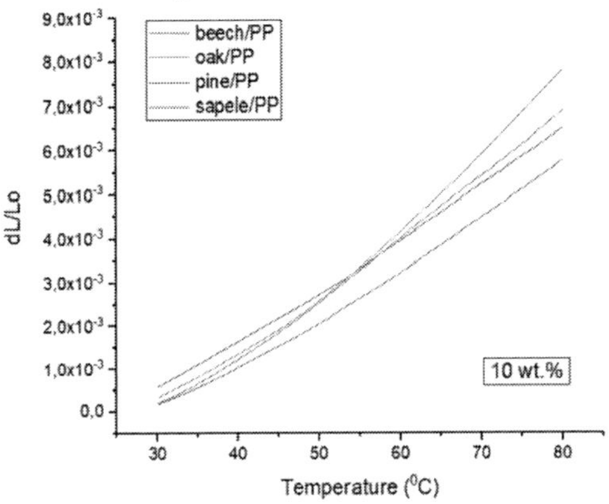

Fig. 9 Thermal stress field developed within 10 wt. % wood particles samples

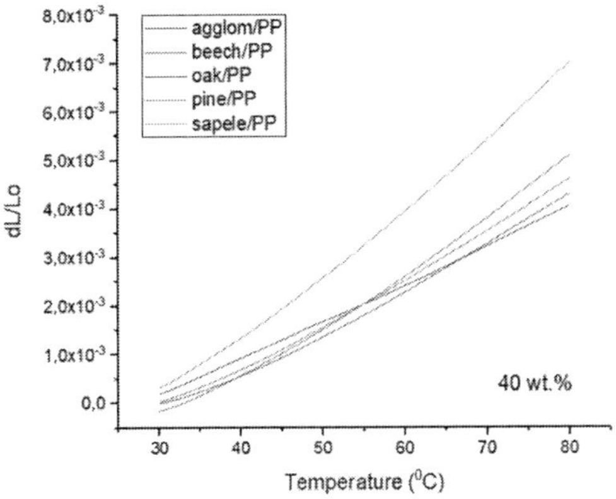

Fig. 10 Thermal stress field developed within 40 wt. % wood particles samples

In addition, a shift to lower temperatures occur in relation to the second peak that corresponds to the lignin degradation, both on TG curves and their first derivative (i.e. degradation speed).

The temperatures $T_{-5\%}$ and $T_{-50\%}$ of 5% and 50% weight loss, respectively, along with the temperature of maximum decomposition rate (T_{max}) were listed within Table 1 for all wood/PP composites under discussion.

A closer look into the values provided on individual wood particle content category, reveal differences that cannot be accounted to exhibit an increasing or decreasing trend, but different. These differences can be regraded to the decomposition of cellulose backbone, that is particular to the wood species accounted with the study.

Thermal stability in WPCs specimens can be monitored by heating up within a controlled atmosphere the samples. As expected, the wood particle content as well as woos species have a great contribution on the evolution over selected temperature range of the linear coefficient of thermal expansion (n. CTE) on individual samples. These can be sized from graphical representations from Fig. 6 to Fig. 8, on different wood particles content.

Thus, samples with 10 wt.% wood particle content within their PP matrix, exhibit CTE values experiencing approximatively a linear variation over the temperature range. The PP expansivity is dominant in the overall CTE variation with temperature.

Moreover, irrespective the wood species, the CTE values are very close in case of agglomerate, oak and sapele composites and slightly lower on the other samples.

Next, an increase in the wood content is impeding high expansivity as encountered in the 10 wt.% WPCs. Pine, oak and agglomerate reinforced thermoplastics are experiencing the highest decrease in their overall CTE.

Overall thermal stress field developed within the herein wood particle reinforced PP were plotted on Fig. 9 and Fig. 10. The values were collected from the second thermal cycle runs applied upon the wood particle reinforced PP plastic.

Data selection from second or higher thermal conditioning of thermoplastic composites herein can be regraded to manufacturing and environmental issues such as wax, chemical and other volatiles removal as well as moisture evaporation from samples' surface.

A closer look to the aforementioned graphs reveals similarities irrespective of the wood species considered, especially in case of lower particle content. Furthermore, there it can be seen a linear increase in the thermal stress field developed within the composite specimens with temperature increase.

Little discrepancies can be found with wood species reinforced thermoplastics having a 40 wt.% weight fraction that revealed smaller thermal stress values due to the higher particle content impeding large specimen dimension expansion, and thus high thermal stress development.

Additionally, in both cases it can be seen that thermal stress fields are crossing around 55 °C with exception of pine reinforced PP composites. This behaviour was identified within herein wood reinforced thermoplastics and can be sought to hold particularly to the specimens under scrutiny.

4 Perspectives

Accounting the huge interest of automotive players on WPCs development, the subject remains under scrutiny.

Wood plastic composites are known both as sustainable and biodegradable materials and redoubtable competitor for green polymer composites out of natural fibres reinforced within organic/inorganic polymer matrices.

In the near future, it is expected to identify automotive components out of WPCs or in combination with these, such as: headliners, sunroof sliders, front and rear door linens, as well as decorative appliances.

Supplementary research is needed on WPCs material properties, including mechanical, acoustics, thermo-physical or dynamic mechanical accounting for various wood content and/or species.

A niche can be easily identified in the direction of WPCs out of recycled components, especially in terms of thermoplastic polymer matrix. Nonetheless, despite the high number of articles approaching the natural constitutive pyrolysis and thermal stability, both reinforcements and matrices related, there are numerous potential subject to be approached.

In addition, specimen degradation under extreme environmental conditions remains a hot subject to be tackled accounting its importance on automotive components life time cycle.

From a manufacturing perspective, automotive producers were adapting their production lines and processes to adapt to natural/green reinforcements or combinations of these.

Novel architectures emerged naturally such is the one developed within Draexlmeier Group based on a combination between kenaf fibres and PP mat mixed into a cross-layering process [23]. The innovation moved further on displaying a translucent decorative film onto the heated combination by lamination process. The outcome panels were proven to comply with the sustainable criteria imposed for mass production in addition to the demands concerning the lightweight design and safety, surface feel and usability.

Lightweight strategies to achieve potential weight saving up to 50% and beyond for increased number of automotive components may consider the herein composites as candidates for their structures especially if economy to cost come as an imposed strategy of development.

Among the weight reducing strategies adopted by OEMs to minimize weight in automobiles can be mentioned the deployment of light weight materials, revision of manufacturing phases/processes, optimization of existing designs, irrelevant content/features removing and parts/systems resizing.

5 Conclusions

Thermal degradation and stability of several wood species reinforced in polypropylene matrix under two different content, 10 wt.% and 40 wt.% respectively, was monitored and debated.

Thermal degradation of WPCs considered reveal similarities in characteristics irrespective of their wood particle content. Little discrepancies on the degradation temperatures (T_{max}) with increase in the wood content were accounted.

Char formation is greatly influenced by the wood content, increasing along with it from 10% to 35% or higher. Thus, final decomposition temperature should be set to a value higher than 650 °C.

With respect to the WPCs thermal stability, an increase on the wood content lead to small samples expansivities, more pronounced with agglomerate, oak and pine specimens, from 20% up to 50% and more. The higher the content of wood reinforcement content, the enhanced thermal stability of the specimens. This property is particular important during sun exposure of wood particle reinforced thermoplastic polymer composites.

With respect to thermal stresses developed within composite specimens, a linear increase with temperature was identified to hold irrespective of wood species. Little discrepancies can be regarded with respect of the overall thermal stress field with increase of the wood particle content, from 10 wt.% up to 40 wt.%.

An optimization procedure to identify the wood content that enables enhanced thermal stability and diminished char formation can be approached inherently as a subsequent research step.

Other thermoplastic polymers can be further accounted as matrix material in addition of considering the wood particle reinforcement as a component within a hybrid combination with other materials prone to be deployed as engineering materials.

References

1. A. Ashori, Bioresource Technology **99** (11), 4661-4667 (2008).
2. Qiang Yuan, Donglyang Wu, J. Gotama and S. Bateman, Journal of Thermoplastic Composite Materials **21** (3), 195-208 (2008).
3. L. Sobczak, R. W. Lang and A. Haider, Composites Science and Technology **72** (5), 550-557 (2012).
4. M. Valente, F. Sarasini, F. Marra, J. Tirillò and G. Pulci, Composites Part A: Applied Science and Manufacturing **42** (6), 649-657 (2011).
5. L. Dányádi, T. Janecska, Z. Szabó, G. Nagy, J. Móczó and B. Pukánszky, Composites Science and Technology **67** (13), 2838-2846 (2007).
6. L. Dányádi, J. Móczó and B. Pukánszky, Composites Part A: Applied Science and Manufacturing **41** (2), 199-206 (2010).
7. H. Matini Behzad, A. Ashori, A. Tarmian and M. Tajvidi, Construction and Building Materials **35** (0), 246-250 (2012).
8. R. Csizmadia, G. Faludi, K. Renner, J. Móczó and B. Pukánszky, Composites Part A: Applied Science and Manufacturing **53** (0), 46-53 (2013).
9. W. V. Srubar Iii, S. Pilla, Z. C. Wright, C. A. Ryan, J. P. Greene, C. W. Frank and S. L. Billington, Composites Science and Technology **72** (6), 708-715 (2012).

10. L. Wei, A. G. McDonald, C. Freitag and J. J. Morrell, Polymer Degradation and Stability **98** (7), 1348-1361 (2013).
11. F. A. Soares and S. M. B. Nachtigall, Polymer Testing **32** (4), 640-646 (2013).
12. J. S. Fabiyi and A. G. McDonald, Composites Part A: Applied Science and Manufacturing **41** (10), 1434-1440 (2010).
13. A. Nourbakhsh, A. Karegarfard, A. Ashori and A. Nourbakhsh, Journal of Thermoplastic Composite Materials **23** (2), 169-174 (2010).
14. R. Bouza, C. Marco, M. Naffakh, L. Barral and G. Ellis, Composites Part A: Applied Science and Manufacturing **42** (8), 935-949 (2011).
15. S. Jeamtrakull, A. Kositchaiyong, T. Markpin, V. Rosarpitak and N. Sombatsompop, Composites Part B: Engineering **43** (7), 2721-2729 (2012).
16. A. Thumm and A. R. Dickson, Composites Part A: Applied Science and Manufacturing **46** (0), 45-52 (2013).
17. S. Singh, A. K. Mohanty and M. Misra, Composites Part A: Applied Science and Manufacturing **41** (2), 304-312 (2010).
18. H. S. Yang, M. P. Wolcott, H. S. Kim and H. J. Kim, J Therm Anal Calorim **82** (1), 157-160 (2005).
19. M. Chaharmahali, J. Mirbagheri, M. Tajvidi, S. K. Najafi and Y. Mirbagheri, Journal of Reinforced Plastics and Composites **29** (2), 310-319 (2010).
20. Pengfei Niu, Baoying Liu, Xiaoming Wei, Xiaojun Wang and Jie Yang, Journal of Reinforced Plastics and Composites **30** (1), 36-44 (2011).
21. Y. H. Cui, J. Tao, B. Noruziaan, M. Cheung and S. Lee, Journal of Reinforced Plastics and Composites **29** (2), 278-289 (2010).
22. P. Maryandyshev, A. Chernov, V. Lyobov, G. Trouve, A. Brillard, J. Brilhac, Journal of Thermal Analysis and Calorimetry **122**, 963-973 (2014).
23. I. Schmiedel, G. Barfuss, T. Nickel and L. Pfeufer, ATZ worldwide **116**, 20-23 (2014).

Contribution to experimental research of alternate fuel fumigation in single cylinder research diesel engine

Doru Baldean[1,*], *Dan* Moldovanu[1], and *Adela-Ioana* Borzan[1]

[1]Tehnical University of Cluj-Napoca, Automotive and Transportation Department, 400641 Muncii Boulevard, Romania

Abstract. The advanced engineering scientific paper encompass the problem of supplying fuel in engine's combustion chamber in order to outline the influence of operating regime and fumigation method upon injection and fuel consumption. It develops a theoretical and experimental approach of the system components from Lombardini compression ignited engine test bench in relation with consumption and fumigation parameters. There was monitored engine speed, dynamic behaviour at some distinct operating loads. The opportunity for experimental research of the fumigation method in compression ignited engine resides in the state of the art technology accessible today for alternate fuel supply inside engine through this process. The results are analysed and discussed in detail. The fuel fumigation method is applicable to the presented engine and leads to some positive behaviour in operation.

1 Introduction

This paper develops a theoretical and applied research protocol which is a part of the complex phenomenological mapping of a Single Cylinder Research Diesel Engine (SCRDE), which has as result even an innovation patent with the fumigation features.

In the present age, when the expectations concerning the fuel supply systems for internal combustion engines are higher and strictly evaluated and the operation is electronically controlled it is both necessary and important to reconsider and adjust the existing knowledge and methods for air-fuel mixture formation in order to improve engine performances at economy and pollution levels.

The research paper shows a set of operating theories from the bottom of this system, a study of the fuel supply system from the internal combustion engine and a set of results from the experimental research concerning the dynamic behaviour in various operating conditions during the laboratory tests made with a hydraulic brake.

2 State of the art in the field

2.1. Review of the scientific literature

The fumigation problem is being studied and treated with serious concern by the scientific and engineering community for a time now. Bio-fuels and fumigation are two major directions that are closely considered by a specific part of the scientific research literature [1-3].

Bio-fuel has come under consideration due to the effect of fossil oil crisis. Bio-fuels are acting as a renewable replacement of petroleum fuels due to some environmental and economic benefits. Bio-fuel can be produced from different kinds of raw materials. Researchers have seen that absolute utilization of bio-fuel is not appreciable as it will affect the food chain but the blend of bio-fuel with conventional fuel could precisely reduce its use and become beneficial to greenhouse effect. It has been inferred that in the hot and cold environment bio-fuel is not fully convenient to replace fossil fuel. In the controlled environment with modified combustion equipment, biodiesel can be used as an alternate fuel 5.

Cheng et al. [1] and Zhang et al. [4,5] have described the fumigation methanol effect on different emissions of exhaust gases and particulate matter (PM) [4-11].

Intensive investigation on different alcohols, mainly ethanol and methanol and combination of them with diesel fuel is used for reducing the emissions of nitric oxides (NO_x) and PM [1]. The diesel fuel and alcohols are generally employed together either in the blended mode or even in the fumigation mode [5-9], although there are less investigation being conducted on different approaches like dual fuel injection system. In comparison to the blended mode, the fumigation method is definitely seems more flexible despite extra fuel injection system is required. First it allows the quantity of alcohol to vary at the time of injection to meet the actual demand. Second, since the alcohol is not premixed while using as diesel fuel, an emulsion additive is added to ensure proper mixing of the alcohol [4-11].

2.2 Our activities related to the subject

The research team has a constant preoccupation for developing studies and research programs for optimizing internal combustion engines and for smoothing the

* Corresponding author: doru.baldean@auto.utcluj.ro

© The Authors, published by EDP Sciences. This is an open access article distributed under the terms of the Creative Commons Attribution License 4.0 (http://creativecommons.org/licenses/by/4.0/).

 https://doi.org/10.1051/matecconf/201818401017

transition toward a cleaner environment and less pollutant automotive field.

Last year, the effects of an experimental bioethanol fumigation application using an experimental ultrasound device on performance and emissions of a single cylinder Diesel engine have been experimentally investigated. Engine performance and pollutant emissions variations were considered for three different types of fuels (biodiesel, biodiesel-bioethanol blend, and biodiesel-fumigated bioethanol). Reductions in brake specific fuel consumption and NO_x pollutant emissions are correlated with the use of ultrasonic fumigation of bioethanol fuel, comparative to use of biodiesel-bioethanol blend [5, 6].

Considering the fuel consumption as Diesel engine's main performance parameter, the proposed bioethanol's fumigation method, offers the possibility to use more efficient renewable biofuels (bioethanol), with immediate effects on environmental protection [5-7].

2.3 Continuation of the study at the present moment

In the present stage there is a patent proposal in progress on a method for multi-fuel supply in the internal combustion engines with aerosols produced through ultrasonic waves applied to biofuels based on alcohols and a research concerning fumigation in the air-fuel supply system from compression ignited engine. Also the phases of development for the internal research project at Technical University from Cluj-Napoca are treated with full responsibility in order to achieve the proposed objectives.

3 Mathematical models

To study all the engine's parameters concerning the consumption, optimal operation, output power, torque and break specific consumption it is necessary to highlight the mathematical models which are at the basis of each calculation after experimental measurements.

Hourly fuel consumption is determined with the mathematical relation 8:

$$C_h = 3.6\frac{V \cdot \rho}{\tau} = 3.6\frac{m}{\tau}, \quad (1)$$

in which: V is the injected fuel volume in a given time [ml]; τ – the time in which the fuel volume V is injected, measured [s]; ρ – fuel density, in [kg/h]; m – the fuel mass in [g].

Brake specific fuel consumption is determined with the mathematical relation [8]:

$$c_e = \frac{C_h}{P_e}, \quad (2)$$

in which: C_h is the hourly fuel consumption [kg/h]; P_e – the brake mean effective power [kW].

4 Materials and method

In order to develop the research work and to measure all the envisaged parameters the engineering staff and technicians prepared a test bench with an internal combustion engine from tractor company Lombardini, that operates on the fundamental cycle of four stroke compression ignition, with a single cylinder and direct injection, being air cooled. Beside the single cylinder research engine there were also used some devices to measure the temperature, the engine speed, the output torque and power.

Figure 1 represents the schematic principle of ultrasonic fumigation process and technology [6].

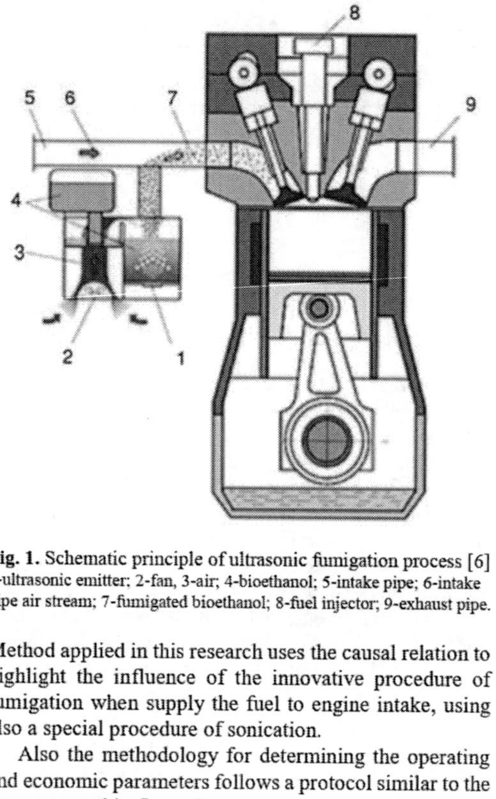

Fig. 1. Schematic principle of ultrasonic fumigation process [6] 1-ultrasonic emitter; 2-fan, 3-air; 4-bioethanol; 5-intake pipe; 6-intake pipe air stream; 7-fumigated bioethanol; 8-fuel injector; 9-exhaust pipe.

Method applied in this research uses the causal relation to highlight the influence of the innovative procedure of fumigation when supply the fuel to engine intake, using also a special procedure of sonication.

Also the methodology for determining the operating and economic parameters follows a protocol similar to the one presented in figure 2.

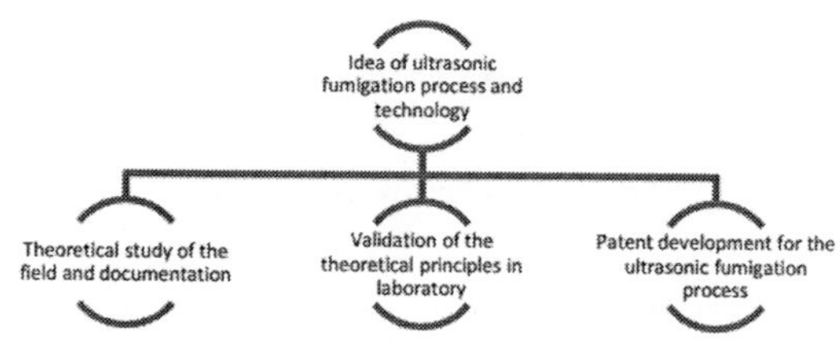

Fig. 2. Simplified methodological protocol followed in the research

5 Analyze of the experimental results

5.1. Test bench configuration

Figure 3 presents the engine test bed and the equipment linked to the bench in order to measure and determine the fuel consumption parameters.

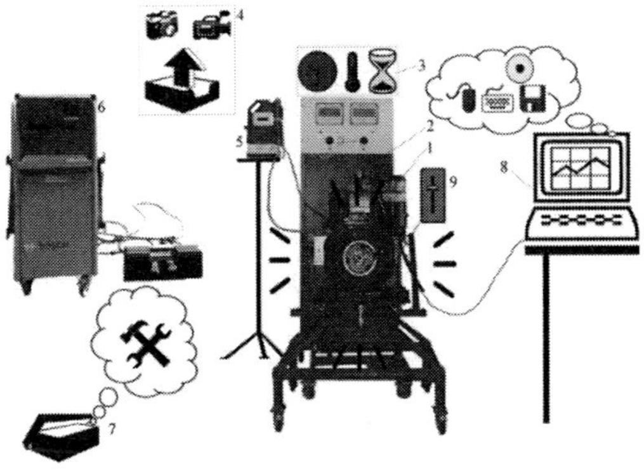

Fig. 3. Experimental test bed [3]
1-single cylinder engine; 2-hydraulic brake, 3-measuring and control devices; 4-recording equipment; 5-mass measuring equipment; 6-gas analyzer; 7-tools; 8-data acquisition system; 9-engine load adjustment lever.

Monitoring the parameters and tracing the engine operation diagrams during the tests were realized with a specialized data acquisition system, which is made of auxiliary devices and also optical units for electronic data storage. Single cylinder engine is mounted on a test bed mechanically coupled with a hydraulic brake and electronically linked with data acquisition/storage system. Also the engine exhaust pipe is connected with the gas sensor of the pollutant analyzer.

5.2. Torque and pressure measurement

In order to obtain reliable torque measurements (Table 1) the engine's crankshaft was break down by a hydraulic brake and the indicated values were recorded from a corresponding display panel (Fig. 4). The level of engine speed (Table 2) was measured with sensors put on the engine and upon hydraulic brake. In relation with the engine speeds and combustion chamber gas pressure data, the processing system has represented the indicated diagram (pressure vs. volume), thus being possible a series of complex analyzes concerning energetic performances during the operation in different fuel supply (injection, fumigation etc.) conditions.

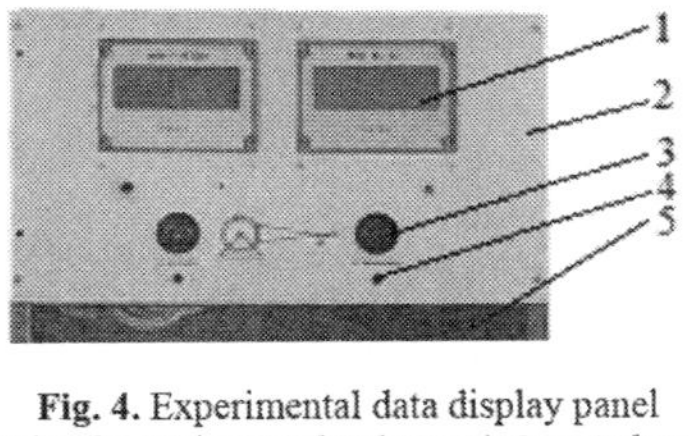

Fig. 4. Experimental data display panel
1-digital display (i.e. engine speed and torque); 2-central panel, 3-analog display; 4-conecting socket; 5-hard frame support.

Table 1. Values for Engine Torque in the case of two fuels tested at full load (100%)

Symbol	Quantity	Experimental values, [N]
M100.100%.M	Engine torque	1.655
	determined in the	5.020
	case of injecting	10.137
	in cylinder	15.077
	conventional Euro	20.089
	5 diesel fuel	25.472
B20.100%.M	Engine torque in	1.672
	the case of 20%	5.036
	biodiesel volume	10.112
	mixture with	15.050
	conventional	20.090
	diesel fuel	25.378

Table 2. Values for Engine Speed during testing

Symbol	Quantity	Experimental values, [rot/min]
M100.100%.n	Engine speed	2986
	determined in the	2911
	case of injecting	2835
	in cylinder	2830
	conventional Euro	2700
	5 diesel fuel	2510
B20.100%.n	Engine speed in	2978
	the case of 20%	2889
	biodiesel volume	2828
	mixture with	2825
	conventional	2698
	diesel fuel	2500

5.3. Engine load results displayed on chart

The practical determinations were presented on graphic charts in order to facilitate their observation, to describe their trends and to study engine's behavior (Fig. 5). The engine's load coefficient presents high levels when torque output riches the upper scale indicated by the manufacturer.

At low torques the engine loads are similar, but at the maximum points the load differs between the biodiesel and diesel fuel delivered to the combustion chamber.

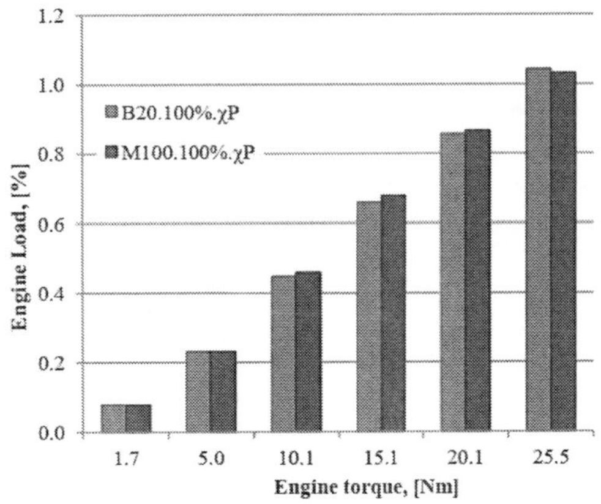

Fig. 5. Variation of engine load coefficient (χP) versus torque

5.4. Fuel consumption determination

Engine's (hourly and specific) fuel consumptions determined on test bench in laboratory conditions are represented in comparative graphics (Fig. 6 and Fig. 7).

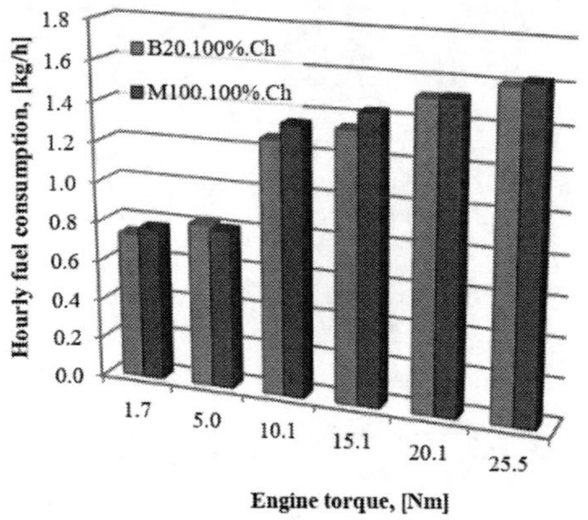

Fig. 6. Variation of hourly fuel consumption versus torque

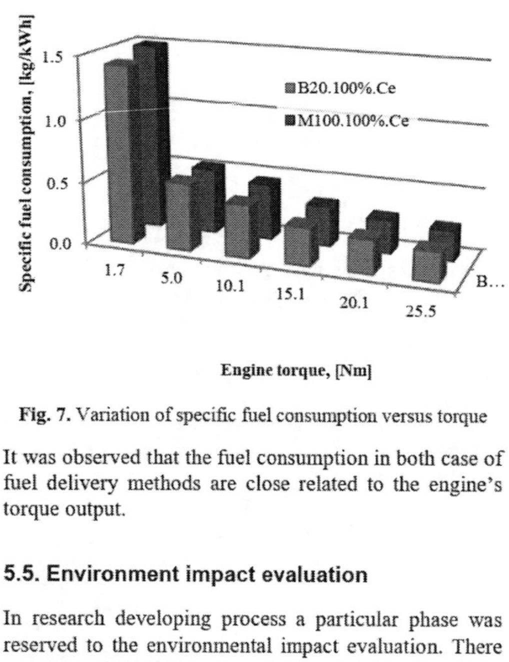

Fig. 7. Variation of specific fuel consumption versus torque

It was observed that the fuel consumption in both case of fuel delivery methods are close related to the engine's torque output.

5.5. Environment impact evaluation

In research developing process a particular phase was reserved to the environmental impact evaluation. There were recorded all the relevant data related to the smoke opacity (Fig. 8) and nitric oxides (Fig. 9).

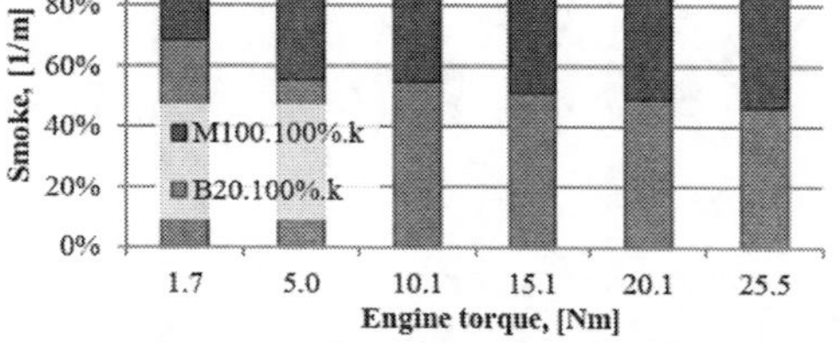

Fig. 8. Variation of exhaust smoke coefficient (k) versus torque

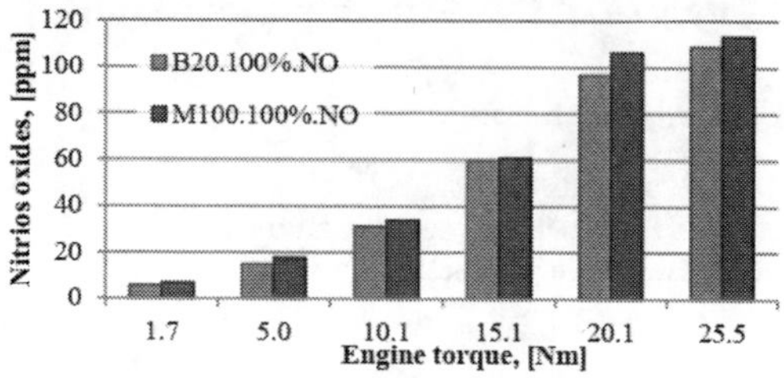

Fig. 9. Variation of nitric oxides (NO_x) versus torque

5 Conclusion

Engine's (hourly and specific) fuel consumptions determined on test bench in laboratory conditions are represented in comparative graphics (Fig. 6 and Fig. 7).

The paper was developed in the period of internal research project at Technical University of Cluj-Napoca, with the reference code number UTCN 11/1.2/2015

References

1. A.K. Agarwal, "Biofuels (alcohols and biodiesel) applications as fuels for internal combustion engines." Progress in Energy and Combustion Science, 33, pp. 233-271, 2007.
2. C. Cheng, C. Cheung, T. Chan, S. Lee, C. Yao, "Experimental investigation on the performance, gaseous and particulate emissions of a methanol fumigated diesel engine," Science of the Total Environment, 389, pp. 115-124, 2008.
3. D. Baldean, "Studies and researches concerning modeling the combustion process in compression ignited engine in biodiesel use conditions for improving engine's performances" Doctoral thesis, UTPress, 2011.
4. D. Baldean, et al "Experimental research regarding the possibility of biofuel fumigation supply method on a single cylinder compression ignited engine." in CONAT 2016 International Congress of Automotive and Transport Engineering, ISBN 978-3-319-45447-4, DOI 10.1007/978-3-319-45447-4, S.I P. 2017
5. E. Sadeghinezhada, S.N. Kazia, Foad Sadeghinejadb, A. Badarudina, Mohammad Mehralic, Rad Sadria, Mohammad Reza Safaeia, "A comprehensive literature review of bio-fuel performance in internal combustion engine and relevant costs involvement," Ren. and Sustain. En. Rev., 30, pp. 29-44, 2014.
6. F. Mariasiu, N.V. Burnete, D. Moldovanu, B.O. Varga, C. Iclodean, L. Kocsis, "Effects of bioethanol ultrasonic generated aerosols application on diesel engine performances." Thermal Science, Vol. 19, No. 6, pp. 1931-1941, 2015.
7. M. Abu-Qudais, O. Haddad, M. Qudaisat, "The effect of alcohol fumigation on diesel engine performance and emissions." Energy Conversion and Management, 41, pp. 389-399, 2000.
8. N. Bataga, "Thermal engines. Practical works" (Motoare termice. Lucrari practice) I.P.C-N., pp. 57-60, 1988.
9. N.C. Surawski, B. Miljevic, B.A. Roberts, R.L. Modini, R. Situ, R.J. Brown, et al. "Particle emissions, volatility, and toxicity from an ethanol fumigated compression ignition engine." Environmental Science & Technology, 44, pp. 229-235, 2009.
10. Z. Zhang, C. Cheung, T. Chan, C. Yao, "Emission reduction from diesel engine using fumigation methanol and diesel oxidation catalyst" Science of the Total Envir. 2009, 407, 4497-4505.
11. Z.H. Zhang, K.S. Tsang, C.S. Cheung, T.L. Chan, C.D. Yao Effect of fumigation methanol and ethanol on the gaseous and particulate emissions of a direct- injection diesel engine. Atmospheric Environment 2011, 45, 2001-2008.

Contributions on experimental research of fuel consumption and management system in K9K892 Diesel engine from Dacia-Renault

Doru Baldean[1,*], and *Adela-Ioana* Borzan[1]

[1]Tehnical University of Cluj-Napoca, Automotive and Transportation Department, 400641 Muncii Boulevard, Romania

Abstract. The present paper develops an experimental study that highlights some aspects of fuel consumption in engine's cylinders in order to outline the influence of engine's management system and operating conditions upon economy and fuel consumption, with corresponding effect on sustainability and adequate socio-economic development. The present work makes a theoretical and applied enquiry in the system features from Euro 5 diesel engine management in relation with fuel consumption and different driving scenarios. There were closely monitored engine temperatures, driving stiles and the values displayed ON-BOARD-DIAGNOSIS screens. The importance and opportunity for experimental inquiry of the fuel consumption and economy problems in compression ignited engine resides in the state of the art equipment and managing systems available today for monitoring all the engine's activities and for making possible to outline the economical operating regime, in order to reduce undesired losses. The experimental data are analyzed in detail.

1 Introduction

This study is continuing a theoretical and experimental research program which is a part of the internal research enquiry started a while ago at Technical University from Cluj-Napoca, being for the present moment a particular phase of a particular project.

Nowadays, pollutant gas reduction and fuel economy is less an alternative or secondary priority but more like one of the important aspects to be considered with all rigorousness. We have to prioritize a lot in this direction. Besides dreaming to eco-friendly engines, which will consume and pollute less (even if this is possible at all), there have to be taken into consideration meanwhile also the operating engines, and the eco-performance level, which were not designed taking into account all the aspects mentioned by strictest criteria in the field [1].

Today, the expectations concerning management and fuel supply systems for internal combustion engines are at high level and delicately considered. The engine's operation cycle is electronically controlled by a complex management system which reduces fuel and thermal loses. Thus it is important and quite necessary to reorganize and redefine the classical knowledge and analyzing methods for engine's operation in order to improve eco-performances and reduce pollution [2-6].

The research work outlines a set of engine operating features from the electronic management system, and a specific study of the fuel supply system from Dacia Logan Euro 5 diesel engine, and also a data package from the experimental measurements concerning the eco-level in various operating conditions during the field tests developed on a real motor vehicle.

2 State of the art

2.1. Study of the scientific literature

For a diesel engine, fuel injection pressure (FIP) and injection timings are very important parameters, which influence the engine performance, emissions, and combustion. Other operational and injection parameters affecting engine performance are rate of injection, injection pattern, number of injections [1], coolant temperature, working stability level etc. [2], [3], [4], [5].

Great concerns are growing up on the environmental impact of fossil fuel and poor air quality in urban area due to traffic-related air pollution. In recent years particular attention was paid mainly to the toxic emissions (HC-hydrocarbons, CO-carbon oxide, PM-particulate matter and NOx-nitrous oxides) of diesel engines for their negative effects on human health and photochemical smog. Nowadays, the overall concern about the global warming problem has changed the social interest of the pollutant emissions so that an increasing attention is building up also for CO2 emissions, one of the major greenhouse gases. A significant improvement can be obtained with the increase of the use of "clean" and renewable fuels. Some of them may offer the possibility to overcome the emission regulations without significant changes to engine layout and fuel economy. As it is known, great attention is given to the fatty-acid methyl-

* Corresponding author: doru.baldean@auto.utcluj.ro

© The Authors, published by EDP Sciences. This is an open access article distributed under the terms of the Creative Commons Attribution License 4.0 (http://creativecommons.org/licenses/by/4.0/).

esters (generally named with its acronym FAME or more generally as "biodiesel") [2].

The Dieselgate scandal has already raised questions in Europe about the accuracy of emissions testing, putting other automakers on the defensive position. Environmental groups have long complained that existing tests understate the amount of pollution that cars generate under real driving conditions [6].

2.2 Our related research of the problem

Our research staff promotes a constant evolution for knowledge and scientific programs for optimizing automotive engines and for a healthy environment and less polluted society.

Reductions in brake specific fuel consumption and pollutant emissions are correlated with the use of bio-fuel and ultrasonic technology, comparative to use of known fuel blends. Considering the fuel consumption as Diesel engine's main performance parameter, the proposed fuel supply method, offers the possibility to use more efficient renewable biofuels (bioethanol), with immediate effects on environmental protection [5].

A similar research, as an overview of the problem, showed that the fuel supply system influence on the engine economy may be affected by injection characteristics and also by each auxiliary system proper performance which may be monitored with the state of the art technologies which are available today [3].

3 Mathematical apparatus

Outlining all the important parameters related to the fuel consumption, vehicle speed, engine power, and pollutant generation is significant in order to understand the mathematics which stands behind the results.

Fuel consumption is determined with the mathematical relation [7]:

$$C_h = 3.6\frac{I_V \cdot \rho}{t} = 3.6\frac{I_m}{t} \quad (1)$$

in which: I_V is the fuel injected volume, in [ml]; t – measuring time interval, in [s]; ρ – fuel density, in [kg/h]; I_m – the fuel mass in [mg].

Effective power output is calculated with the mathematical expression [8]:

$$P_e = \frac{p_e \cdot n \cdot i \cdot V_s}{30000 \cdot \tau}, \quad (2)$$

in which: pe is the brake mean effective pressure; n – engine speed; i – number of cylinders; Vs – cylinder-unit volume; τ – number of engine strokes.

Effective fuel specific consumption is calculated with the relation [7]:

$$c_e = \frac{C_h}{P_e} = 3.6\frac{I_m}{P_e \cdot t}, \quad (3)$$

in which: Ch is the hourly fuel consumption [kg/h]; Pe – effective power output [kW].

4 Methodology

Methodology for testing the eco-performance and fuel consumption parameters follows a protocol similar to the one presented in figure 1.

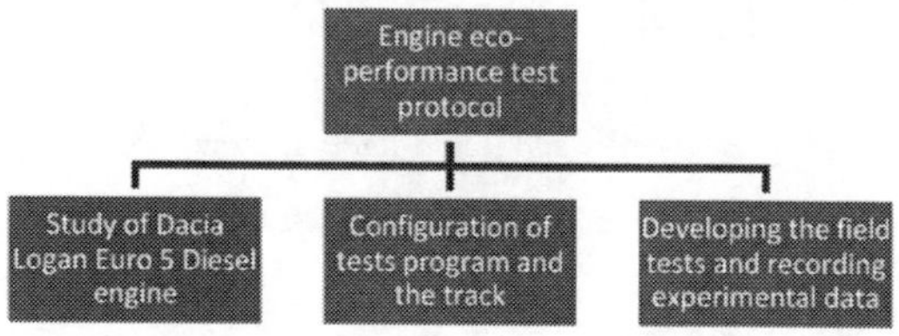

Fig. 1. Main protocol applied in the field tests

Testing method relies on the connection of digital devices to the engine's management system in order to collect and store the experimental data. Useful apparatus for the development of experimental tests were the Bosch KTS OFF-BOARD-DIAGNOSIS system with a user-friendly graphic interface and a motor vehicle (Dacia Logan with K9K892 Euro 5 engine).

The link through the CAN-BUS electronic communication interface has the objective of allowing a research staff member to configure the engine-scanning protocol and to define all parameters in which the management system of the engine works and acquire the available experimental data during operating time in field tests (Figure 2).

For enquiring the field measurements are used ESI-tronic applications and KTS Bosch 650 equipment. At their operating electronic base these programs have complex mathematical models. Main apps are using the analog values measured by the sensors, converted then to digital code. With the technology progress experimental protocols became very complex and the restrictions which were in the former systems are now diminished.

This work presents an innovative procedure for field testing of the engine management system in real operating scenarios by using state of the art equipment and latest communication strategies in order to monitor and record the engine eco-behavior with the electronic control unit (ECU).

This innovative method of computer assisted testing (CAT) allows the researcher to evaluate the phenomenology of actual values variation in real operation conditions in order to correlate the important parameters with signs and symptoms. Thus car and especially the engine's semiotics may be improved and thoroughly defined.

It took some years in training to bring more packages of data in order to improve this new method of study by testing and to implement it while the vehicle is in motion and all the parameters are oscillating due to different working conditions.

Cumulating operational data from engine in various regimes and scenarios it improves the whole "picture" of entire phenomenology, which will be never exhaustive and absolutely completed, just better defined.

Fig. 2. Experimental set for ECU-KTS communication
1-experiment site map; 2-track, 3-start-point; 4-end-point; 5-tested vehicle; 5a-ON-BOARD display with total number of km; 5b-consumed fuel quantity during tests; 5c-mileage; 5d-fuel quantity in the vehicle's tank; 6-analog-digital converter; 7-KTS-Bosch 650; 8-graphic interface; 8a-engine module link; 8b-airbag; 8c-Anti-lock Braking System; 8d-fault details.

5 Analyze of the experimental data

5.1. Test protocol configuration

Recording the measurements and tracking the engine eco-performance during the field tests were realized with a specialized electronic data management system, which consists of portable devices and also data storage units for post-processing. Euro 5 compression ignited engine is installed on a Dacia Logan vehicle linked with a electronic management and scanning system. Also the fuel tank is monitored with the ON-BOARD available default program.

Figure 3 shows the ECU available systems on the tested vehicle at the time of test protocol configuration.

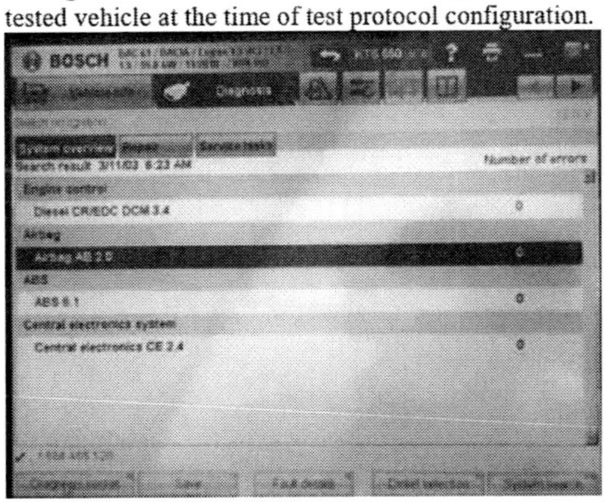

Fig. 3. ECU available systems

5.2. Electronic Diesel Common-Rail control module

In order to acquire reliable engine operational data the electronic diesel control (EDC) module is activated (Figure 4) which allows us to identify the management system.

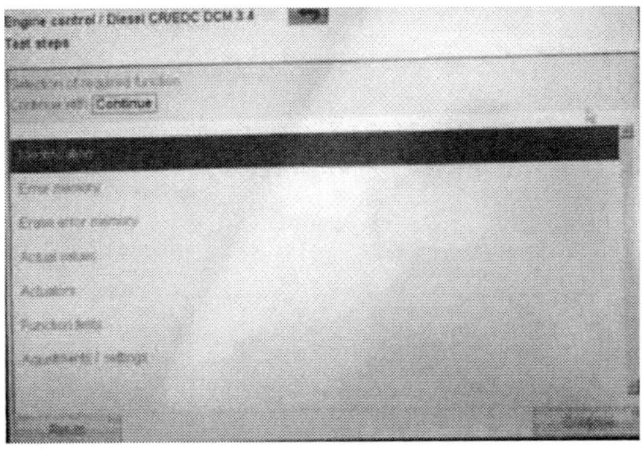

Fig. 4. Engine control module for identification

After the engine identification procedure (Figure 4) we proceed to actuators scanning (Figure 5) and actual values monitoring (Figure 6).

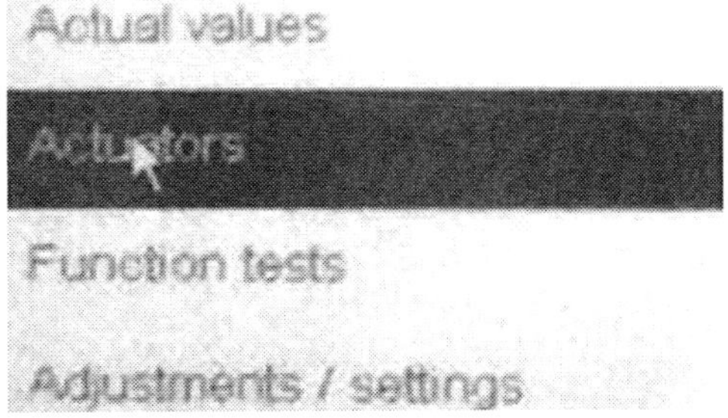

Fig. 5. Digital scanning of the engine's actuators

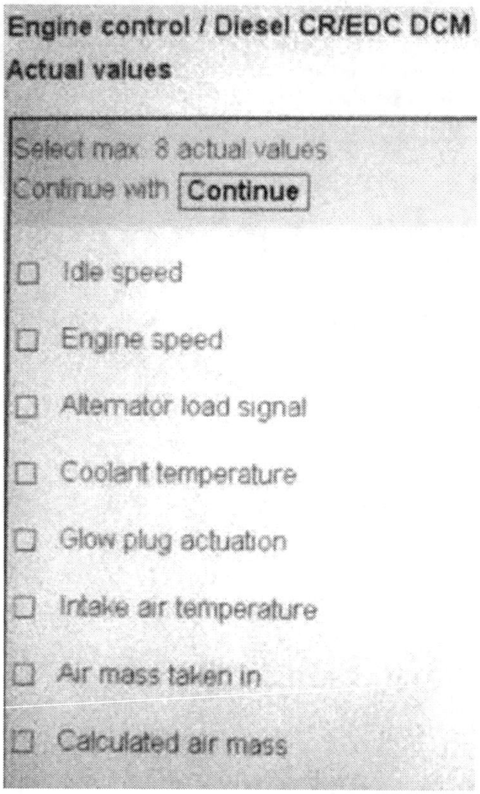

Fig. 6. Selecting the monitored and recorded actual values

Fuel consumption was monitored ON-BOARD in different operation conditions (Table 1).

Table 1. Values for fuel consumption in urban driving conditions

Symbol	Quantity	Experimental values
C_1	Fuel consumption (FC) when defensive driving strategy is applied	
C_2	FC in defensive driving, but engine is under 60°C coolant temperature	
C_3	Rush hour defensive driving, engine is 60÷80°C coolant temperature	
C_4	Air-conditioning on, engine is 60÷80°C coolant temperature	
C_5	Not defensive driving, engine is 60÷80°C coolant temperature (headlights on,)	
C_6	Beginner style driving, engine is 60÷80°C coolant temperature (headlights on,)	

5.3. Actual values and accelerator sensors

The readings from electronic control module corresponding to the actual values were accessed progressively (Fig. 5-9). Also the battery voltage, power supply for sensors and glow control were closely monitored.

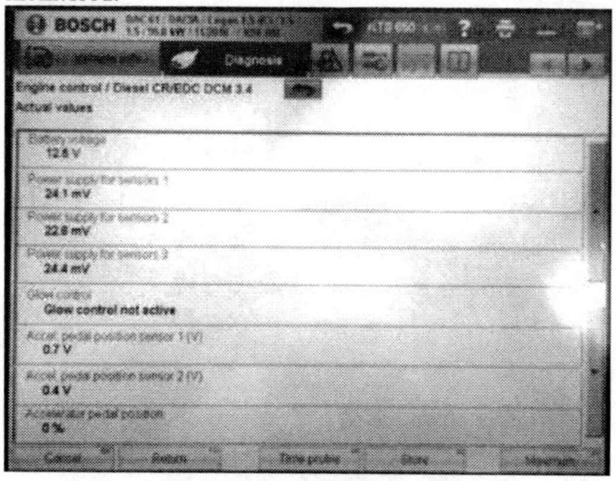

Fig. 7. Actual values for power voltage and accelerator pedal

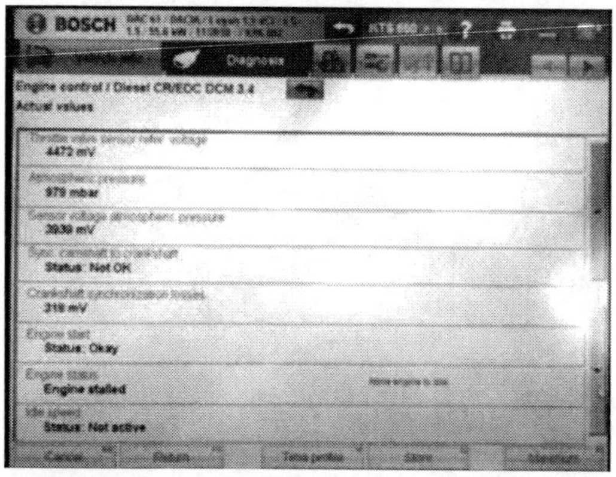

Fig. 8. Actual values for throttle valve and crankshaft sensors

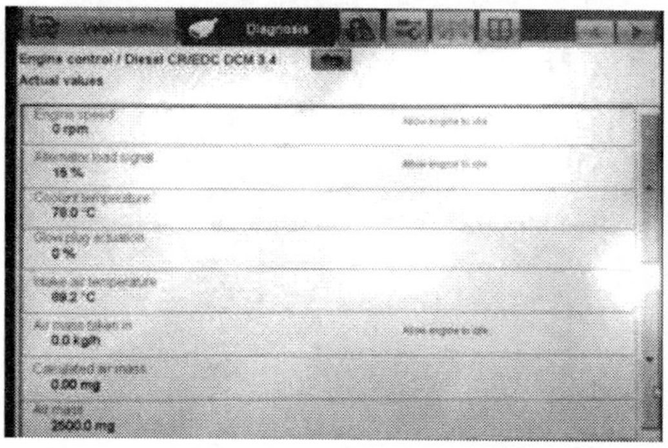

Fig. 9. Engine speed and air mass when engine stalled/stopped

It was observed that the fuel consumption and the fuel injected quantity in all cases of driving and operation conditions are close related to the engine's temperature and load. Also, the accelerator pedal position is essential for fuel injection management.

5.4. Fuel injection and exhaust gas recirculation

Fuel nominal pressure in the common rail system is at idle engine speeds 271 bar (Figure 10).

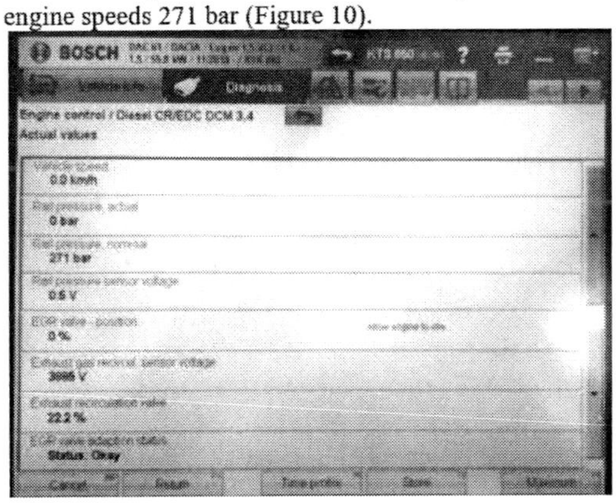

Fig. 10. Injection system and exhaust gas recirculation

Exhaust gas recirculation (EGR) deviation and cooler regeneration, besides fuel temperature and injected quantity are scanned with the ECU and KTS (Figure 11).

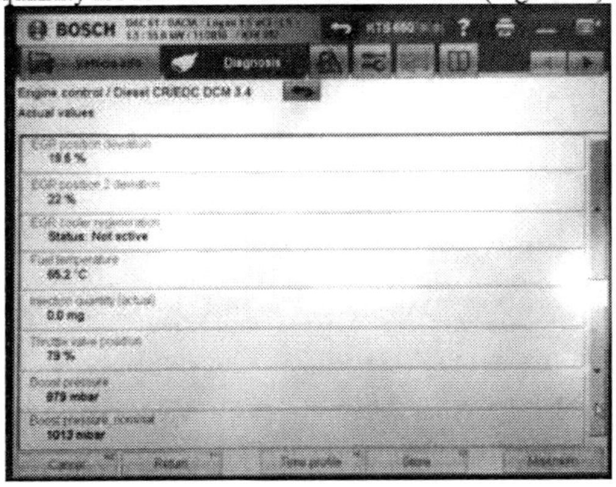

Fig. 11. Injection, EGR and air boost systems

Boost pressure sensor voltage for the supercharge system and particulate filter soot impact are monitored like shown in the (Figure 12).

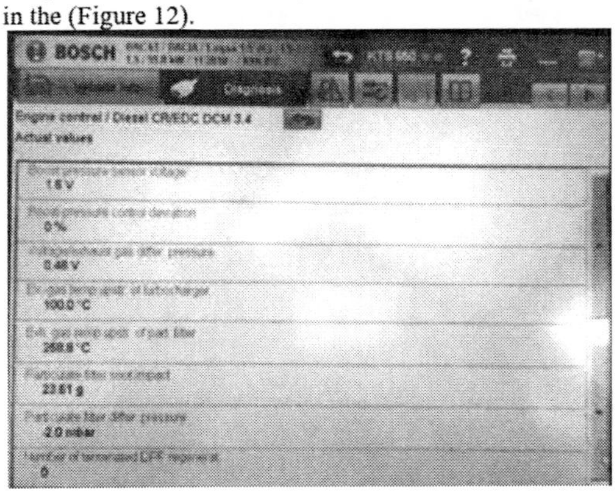

Fig. 12. Supercharger and particulate filter status

Diesel particulate filter (DPF) details like time since last regeneration and its duration, traction control and the possibility of air conditioning activation are also monitored (Figure 13).

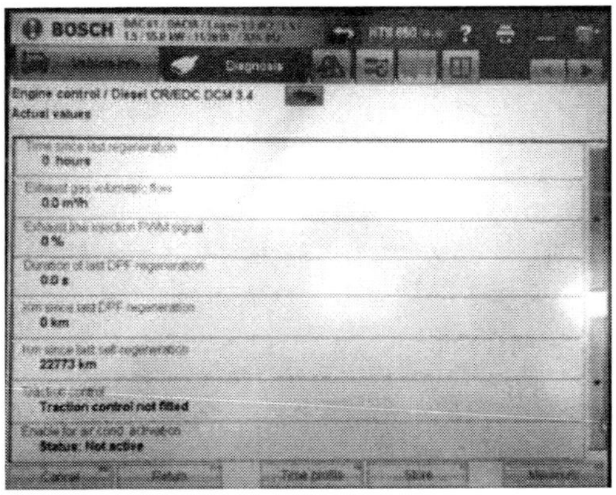

Fig. 13. Diesel particulate filter regeneration details

Cruise control if existing, vehicle speed, exhaust line and gear features were also monitored (Figure 14).

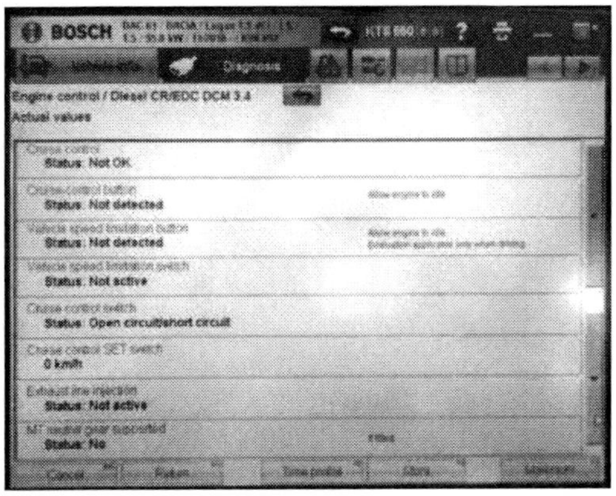

Fig. 14. Dynamic control monitoring

Transmission, reverse gear, vehicle speed, and auxiliary features were recorded (Figure 15).

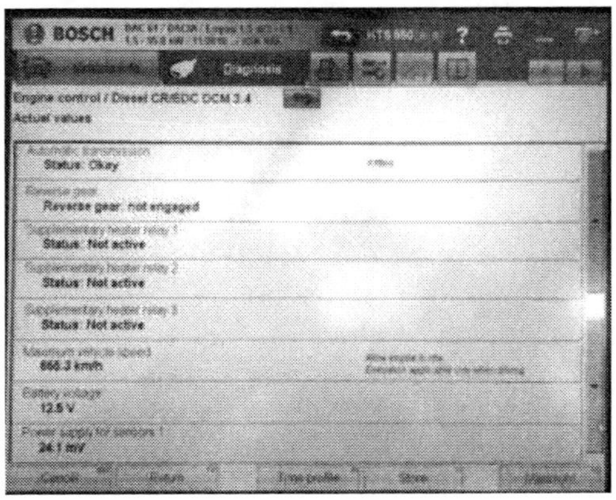

Fig. 15. Transmission and vehicle control monitoring

5.5. Common rail high-pressure test

In the experimental testing process there was also monitored and followed the high pressure test for the common-rail (CR) fuel pump (Figure 16).

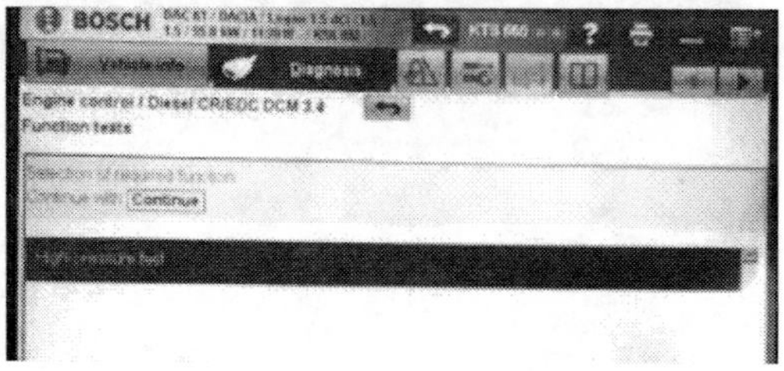

Fig. 16. Access of the high pressure test in CR system

Other adjustments and settings were made in injection quantity, crankshaft adaption, intake manifold, and exhaust systems (Figure 17).

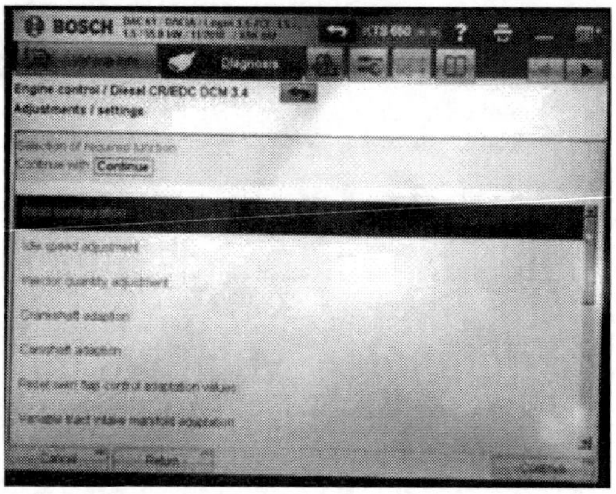

Fig. 17. Configurations, adjustments and settings in engine's electronic management system

ECU facilitate the EGR adaptation values, rail pressure sensor adjustment, fuel pump setting and to access other features in the management system (Figure 18).

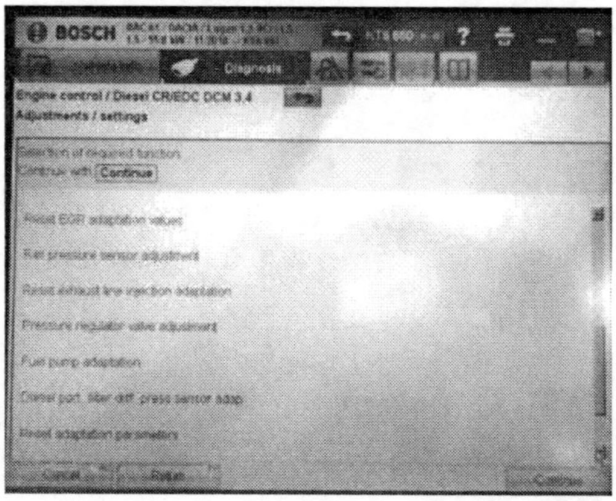

Fig. 18. Adjustments and settings in engine's electronic diesel management system

Vehicle identification number and diagnosis version are shown also (Figure 19).

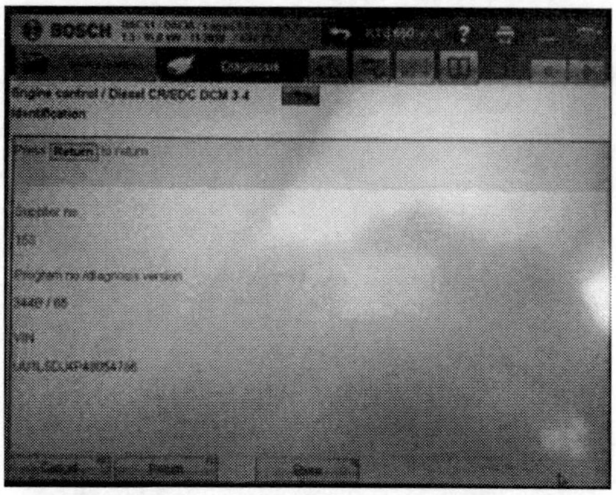

Fig. 19. Diagnosis version and vehicle identification

5 Conclusion

The paper presents an experimental enquiry in the field of fuel consumption and engine management system on real diesel engine during operation. Future papers will study alternative fumigation methods on classic systems.

References

1. A. K. Agarwala, D. K. Srivastavaa, A. Dhara, R. K. Mauryaa, P. C. Shuklab, A. P. Singha, "Effect of fuel injection timing and pressure on combustion, emissions and performance characteristics of a single cylinder diesel engine," Fuel, 111, pp. 374-383, 2013.
2. B. Carlo, C. Guido, S. Di Iorio, "Experimental analysis of alternative fuel impact on a new "torque-controlled" light-duty diesel engine for passenger cars," Fuel, 89(11), pp. 3278-3286, 2010.
3. D. Baldean, "Contributions to the research and testing of modern diesel engines for certification and homologation of performances and emissions." Vol. 45, Inginerie agrară şi transport auto, Chişinău, R. Moldova, pp. 146-150, 2015.
4. D. Baldean, "Studies and researches concerning modeling the combustion process in compression ignited engine in biodiesel use conditions for improving engine's performances" (Studii şi cercetări privind modelarea procesului de ardere în motorul cu aprindere prin comprimare în condiţiile utilizării biodieselului pentru îmbunătăţirea performanţelor motorului) Doctoral thesis, UTPress, 2011.
5. F. Mariasiu, N.V. Burnete, D. Moldovanu, B.O. Varga, C. Iclodean, L. Kocsis, "Effects of bioethanol ultrasonic generated aerosols application on diesel engine performances." Thermal Science, Vol. 19, No. 6, pp. 1931-1941, 2015.
6. J. Ewing, "Volkswagen C.E.O. Martin Winterkorn Resigns Amid Emissions Scandal." New York Times, International Business, SEPT. 23, 2015.
7. N. Bataga, "Motoare termice. Lucrari practice" I.P.C-N., pp. 57-60, 1988.
8. N. Bataga, N. et al: "Internal combustion engines" (Motoare cu ardere interna). E.D.P., pp. 15-17, 1995.
9. L.-V. Crişan-Lupa, et al: Research applied to exhaust gas after-treatment systems in 1.6 l ZSG 416 Ford engine. in Proceedings of the European Automotive Congress EAEC-ESFA 2015, ISBN 978-3-319-27276-4, DOI 10.1007/978-3-319-27276-4, retrieved 22.03.2018.

Numerical analysis of the injection law influence on NOx emissions for a compression ignited engine

Dan Moldovanu[1,*], *Florin* Mariașiu[1]

[1]Technical University of Cluj-Napoca, Cluj-Napoca, Romania

Abstract. Under the present conditions, where sustained efforts are being made by researchers and automotive industry developers to reduce toxic gases (and fitting them into existing pollution standards), research is required to define the possible directions to be followed for this purpose. One of these directions is also presented in this paper.The paper presents research done by using numerical analysis of the influence of the injection law on NOx emission on compression ignited engines. Instead of a normal injection law, a particular injection law was considered to reduce the effect of the cavity on the fuel flow inside the injector in order to achieve a more homogeneous fuel jet exit at the injector.It is noted that there are NOx emission reductions of up to 20% (obtained at 760 degrees CA) compared to the initial injection law, which confirms the need to continue studying the presented research topic. There are also differences in the formation and development of NOx emission zones within the combustion chamber (the volume and positioning of the contact surface towards the walls of the combustion chamber)..

1. Introduction

Considering the stringent European norms regarding emissions, there are many articles and dedicated journals on emission reduction.

Nehmer and Reitz have written a manuscript regarding the measurement of the effect of injection rate and split injections on diesel engine Soot and NOx emissions. The study on a heavy-duty diesel engine was made to understand how rate-shaped and split injection can affect both the soot and the NOx emissions. The provided results shown that split injections allowed peak pressures to be reduced by more than 45% and have a significant effect on the overall rate of pressure rise. The authors indicated that pulsed injection may provide a means to reduce particulate emissions and allow for reduced NOx from controlled pressure rise [1].Patterson et.al. also investigated a way of reducing the Soot and NOx emissions by modelling the effects of fuel injection characteristics, using the three-dimensional KIVA code. Given that the Soot and NOx emissions are very sensitive to factors that influence the chamber gas temperatures, the results show that the combustion is controlled by the details of the spray model. Furthermore, computations have showed that soot and NOx emissions increase at low injection pressures [2].

Kook et.al. investigated the influence of charge dilution and injection timing on low-temperature diesel combustion and emissions. They have emphasized that by lowering the flame temperature and by influencing the local charge equivalence ratio through intake charge dilution, this offers the potential of avoiding both soot and NOx formation. The conclusion they highlighted was that although the ignition delay increased with high dilution and early injection, the heat release analysis indicated that a large portion of the combustion and emissions formation processes was still dominated by the mixing-controlled phase [3].

Many papers were concentrated on reducing the soot and NOx emissions using multiple injections. Although each paper has approached the problem from a different point of view, all the authors have reached the conclusion that multiple injections indeed lead to a reduced quantity of emissions [4]-[7].

Payri et.al. investigated the influence of post-injection pattern on performance, soot and NOx emissions in a HD diesel engine. During the tests, EGR (exhaust gas recirculation) was introduced to reduce NOx emissions to a given value, and then different post-injection patterns were produced.

The final results obtained allowed the authors to discern the effects produced by the different post-injection patterns and to assess their potential for Soot emissions reduction, keeping NOx emissions and fuel consumption at the original values [8].

Many authors tend to go towards renewable energy, seeing how badly the emissions from presently used energy fuels have increased and how harmful they are. Others on the other hand chose to concentrate on making our present systems and fuels better.

The authors either choose experimental investigations to control the emissions, try to better understand the fuel properties and composition to lower the NOx emissions, or they even concentrate more in detail regarding the emissions from DME (Dimethyl Esther) combustion in

*Corresponding author: dan.moldovnau@auto.utcluj.ro

© The Authors, published by EDP Sciences. This is an open access article distributed under the terms of the Creative Commons Attribution License 4.0 (http://creativecommons.org/licenses/by/4.0/).

diesel engines and their implications on meeting future emission norms [9]-[12].

In order to further investigate the properties and influences of the injection law influence on NOx emissions for a compression ignited engine, the authors proposed a study using AVL FIRE and AVL FIRE ESE DIESEL (by AVL GmbH), presented in the following chapter.

2. Methodology

The methodology used for the current simulation starts with the correct input data. The simulation mesh and boundary were made using FIRE ESE DIESEL, where the general data were introduced (presented in Table 1), the piston was created from a template piston shape, after measuring the real piston section. The introduced data for the piston geometry parameters are presented in Fig. 1. After introducing the geometry of the piston, it is important to check that the actual compression ratio is still close to the imposed compression ratio.

Table 1. General data for the simulated engine.

Paramenter	Dimension
Bore	70 mm
Compression ratio	21
Crank radius	37.5 mm
Connecting rod length	80 mm
Piston offset	0

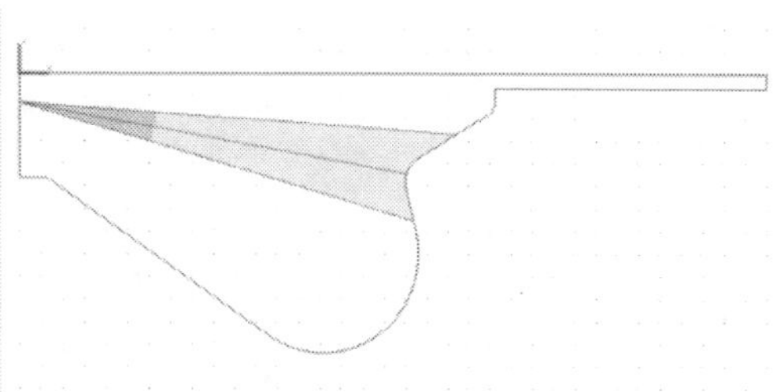

Fig. 1. Piston geometry parameters.

The next step was to generate the 3D meshes and check all the moving meshes for triangle faces that can diverge the simulation. The final mesh and all the selections are presented in Fig. 2.

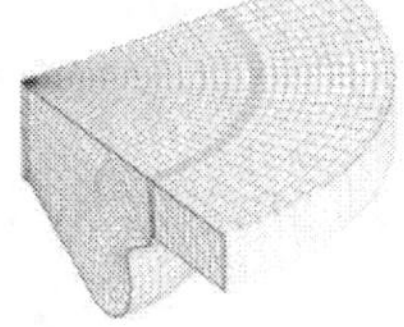

Fig. 2. Final mesh and selections

The important selections that were made, are: piston bowl, cylinder liner, cylinder head, and the symmetry selection.

After having the moving mesh, the Solver Steering File was addressed. Starting with Crank-angle (CA) run mode (from 540 to 800 deg CA), with module activation for species transport, combustion (the module that simulates the turbulent combustion of inhomogeneous mixtures of hydrocarbon fuel, air and exhaust gas.), spray (the module that simulates sprays based on the Discrete Droplet Method (DDM), which represents a spray as a statistical sample of discrete parcels.) and emissions.

The boundary conditions that were introduced are: Mesh movement for the piston selection and an imposed temperature of 575.15 K, wall boundary condition for the cylinder liner, with a temperature of 475.15 K, a simple symmetry boundary for the symmetry selection, and a wall boundary condition for the cylinder head, with a temperature of 550.15 K presented in Fig. 3.

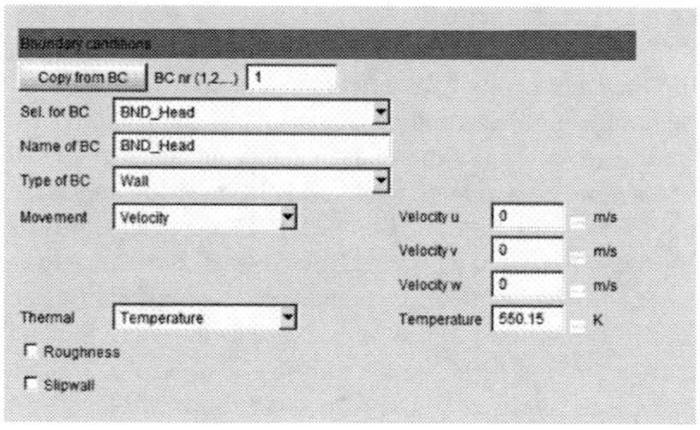

Fig. 3. Boundary conditions for the cylinder head

For the liquid properties, DIESEL was selected and a set temperature of 330.15 K.

The injected mass per cycle was 10 mg. The start of injection angle (SOI) was 719 deg CA, and the end of injection angle (EOI) was 734 deg CA (Fig. 4).

The injection law was changed, by modifying the injection rate, but the injected quantity was kept at 10 mg.

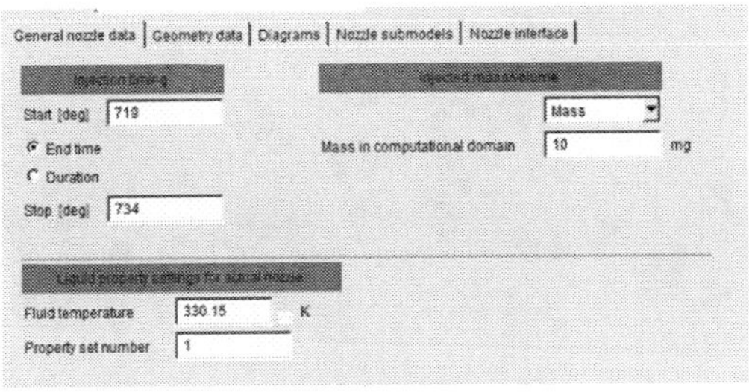

Fig. 4. Injection timing and quantity

The main idea of the paper, was to change the opening of the injector in steps, so that the same quantity of fuel is introduced, but with different flow velocities. The injection rates for the normal and step injection are presented in Fig. 5 and 6.

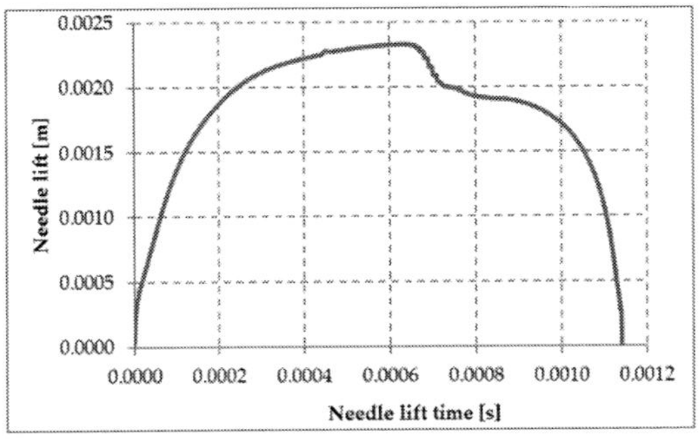

Fig.5. Standard injection rate

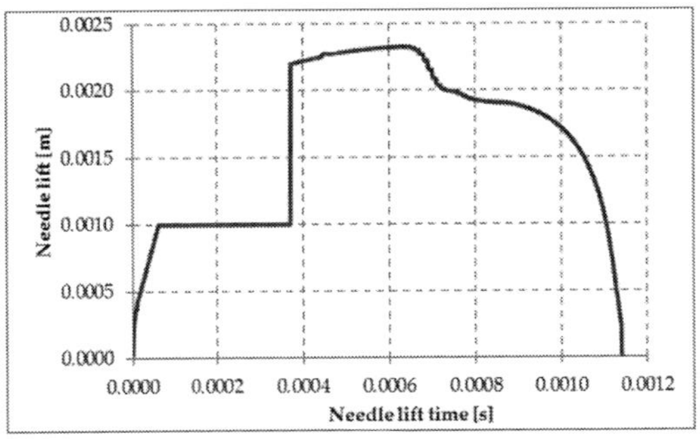

Fig. 6. Step injection rate

In AVL FIRE the k-zeta-f model was activated, with the aim to improve numerical stability by solving the transport equation for the velocity scale ratio $\zeta = \overline{\vartheta^2}/k$. The eddy-viscosity is obtained from (Eq. 1):

$$v_t = C_\mu \zeta \frac{k^2}{\varepsilon} \quad (1)$$

The turbulent time scale T and length scale L are given by (Eq.2 and 3):

$$T = \max(\min\left(\frac{k}{\varepsilon}, \frac{a}{\sqrt{6}C_\mu|S|\zeta}\right), C_T\left(\frac{v}{s}\right)^{\frac{1}{2}}) \quad (2)$$

$$T = C_L \max(\min\left(\frac{k^{\frac{3}{2}}}{s}, \frac{k^{\frac{1}{2}}}{\sqrt{6}C_\mu|S|\zeta}\right), C_\eta \frac{v^{\frac{3}{4}}}{s^{\frac{1}{4}}}) \quad (3)$$

, where C_μ=0.22, C_T=6, C_L=0.36 and C_η=85.

After the simulation was done for the standard injection rate, the step injection simulation was made and the results in NOx formation and

3. Results

To monitor the input of the simulation, the combustion equivalence ratio was monitored. The results are presented in Fig. 7, 9 and 11 for the standard injection at 720, 724 and 730 deg CA and in Fig. 8, 10 and 12 for step injection, at the same CA.

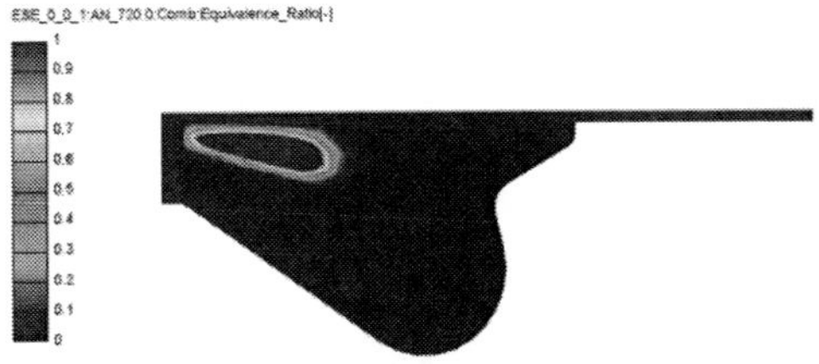

Fig. 7. Equivalence ratio for standard injection at 720 deg CA

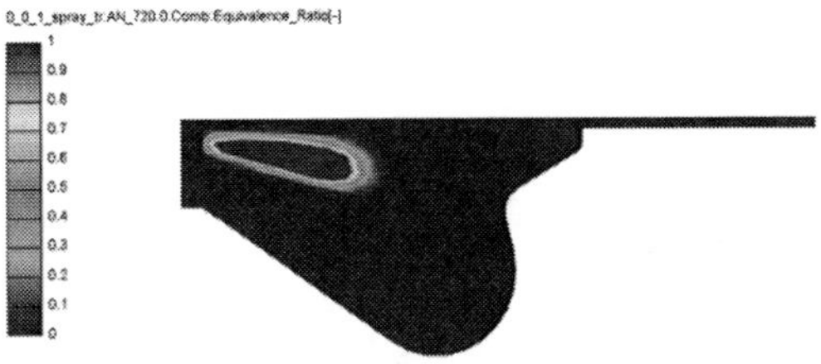

Fig. 8. Equivalence ratio for step injection at 720 deg CA

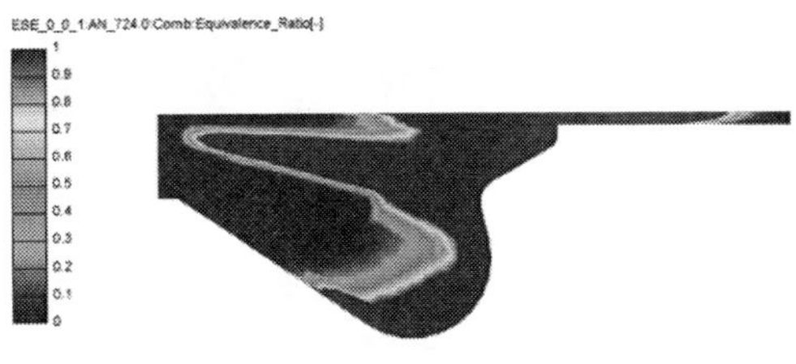

Fig. 9. Equivalence ratio for standard injection at 724 deg CA

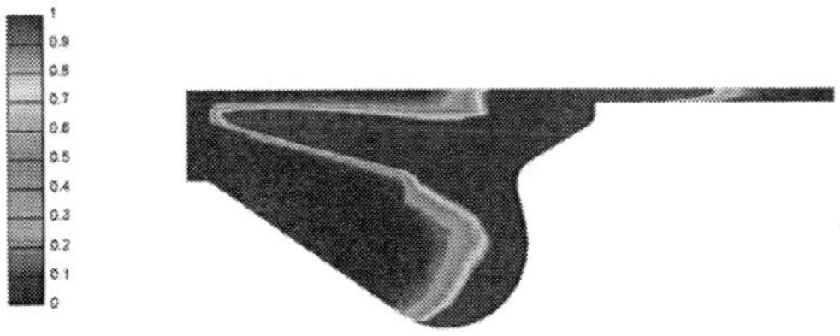

Fig. 10. Equivalence ratio for step injection at 724 deg CA

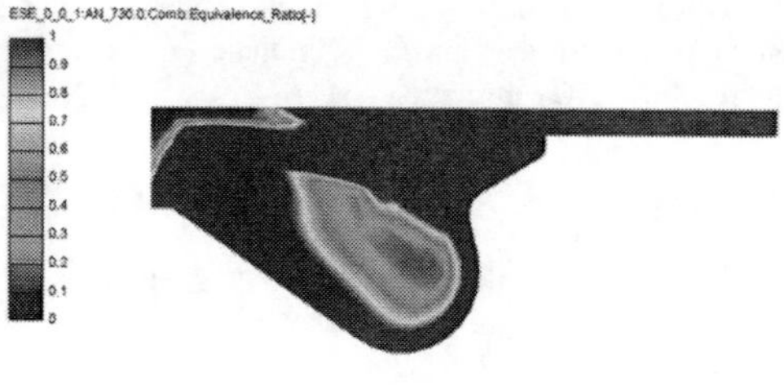

Fig. 10. Equivalence ratio for standard injection at 724 deg CA

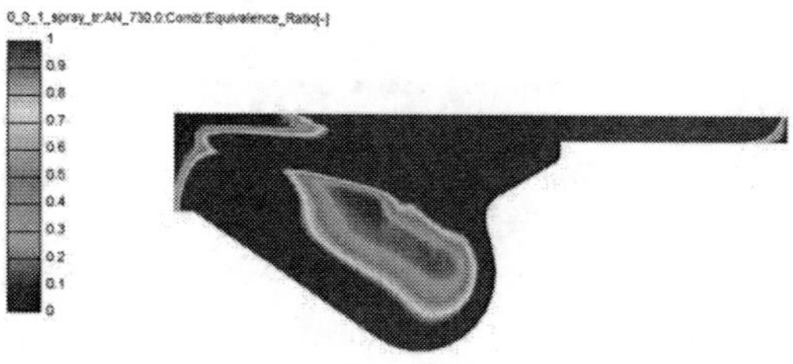

Fig. 11. Equivalence ratio for step injection at 724 deg CA

From Fig. 7-11 the observation is that the penetration of the standard injection is larger than the step injection, at the same timing, and because the quantity of the injected fuel is the same, the emissions will be different.

In order to monitor the NOx emissions, sections were made at 745, 760 and 780 deg CA respectively (presented in Fig. 12-17) and even though the maximal values were different, the scales were chosen the same so that colours mean the same NOx concentration in all sections.

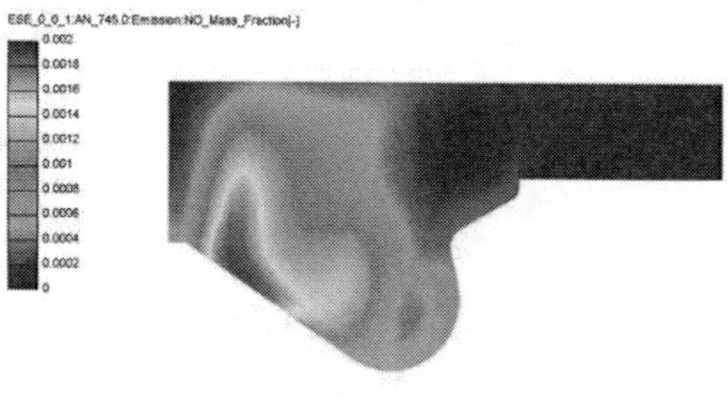

Fig. 12. NOx for standard injection at 745 deg CA

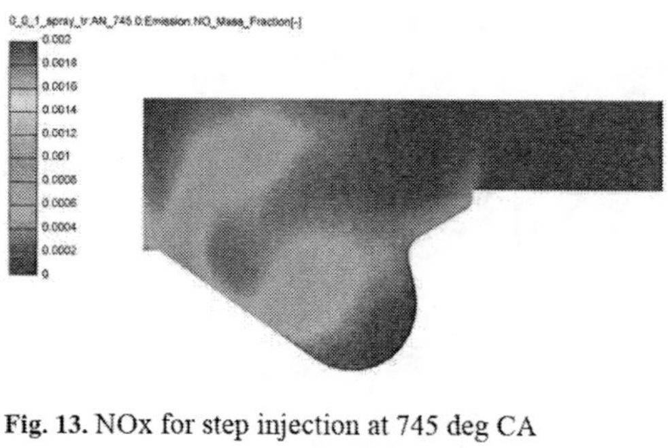

Fig. 13. NOx for step injection at 745 deg CA

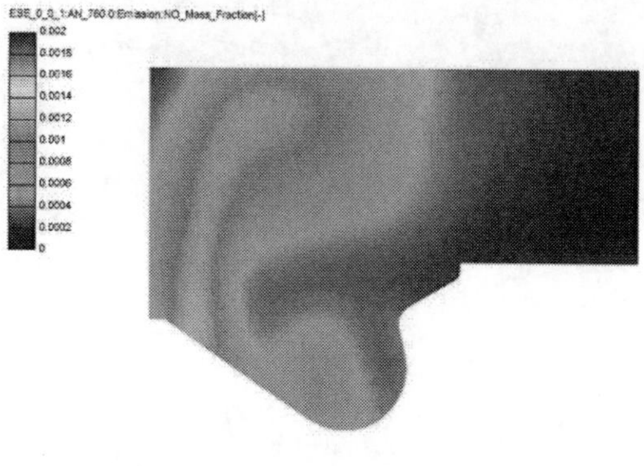

Fig. 14. NOx for standard injection at 760 deg CA

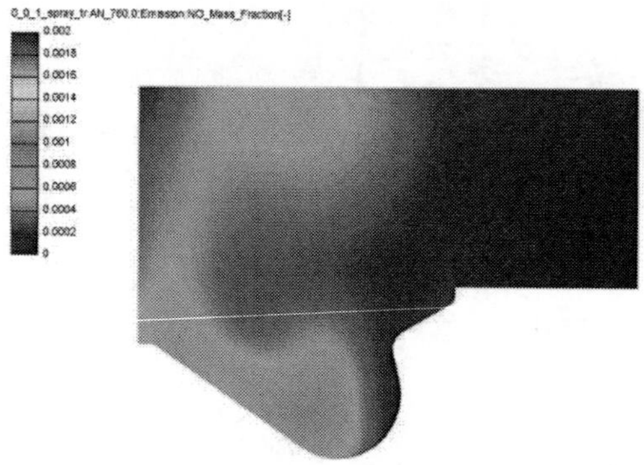

Fig. 15. NOx for step injection at 760 deg CA

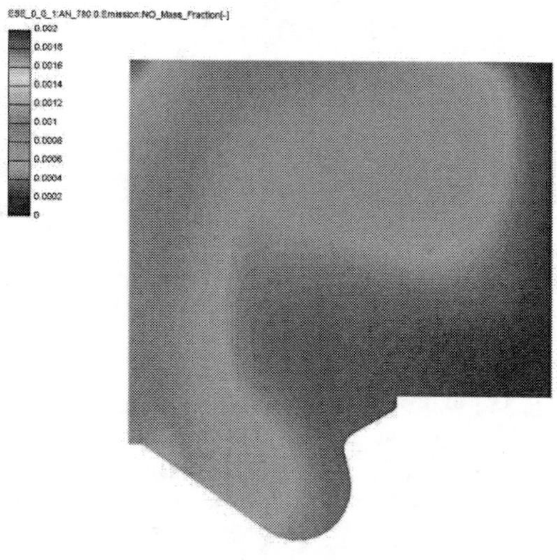

Fig. 16. NOx for standard injection at 780 deg CA

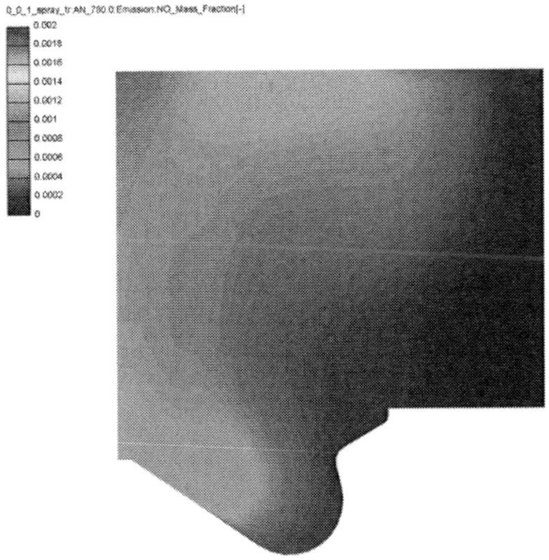

Fig. 17. NOx for step injection at 780 deg CA

The results for Max NOx on the entire mesh and Average NOx value area weighted for both standard injection and step injection were extracted for different CA and are presented in Tables 2 and 3 and are presented in Fig. 18 and 19.

Table 2. Max NOx on the entire mesh and Average NOx value area weighted for standard injection.

CA [deg]	730	745	760	780	800
Max NOx on entire mesh [-]	2.31E-03	1.99E-03	1.32E-03	1.15E-03	1.06E-03
Average NOx value area weighted [-]	7.54E-05	5.95E-04	5.36E-04	5.63E-04	5.07E-04

Table 3. Max NOx on the entire mesh and Average NOx value area weighted for step injection.

CA [deg]	730	745	760	780	800
Max NOx on entire mesh [-]	1.50E-03	1.75E-03	8.90E-04	6.90E-04	6.43E-04
Average NOx value area weighted [-]	1.40E-05	3.05E-04	3.25E-04	3.45E-04	3.40E-04

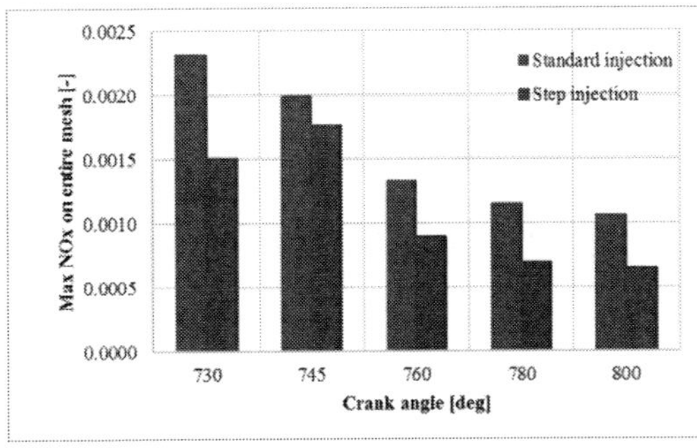

Fig. 18. Variation of Max NOx on the entire mesh

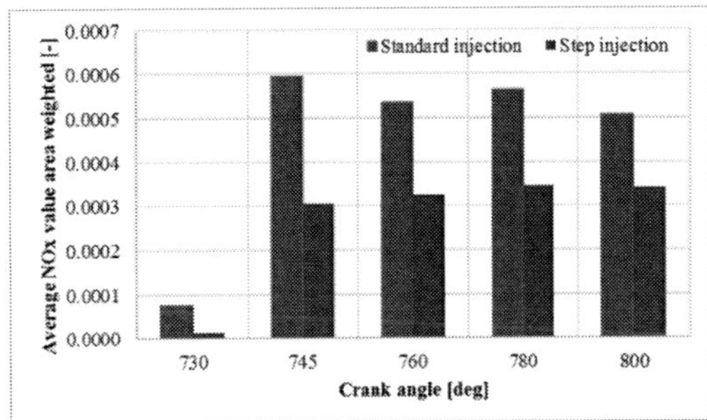

Fig. 19. Variation of Average NOx value area-weighted

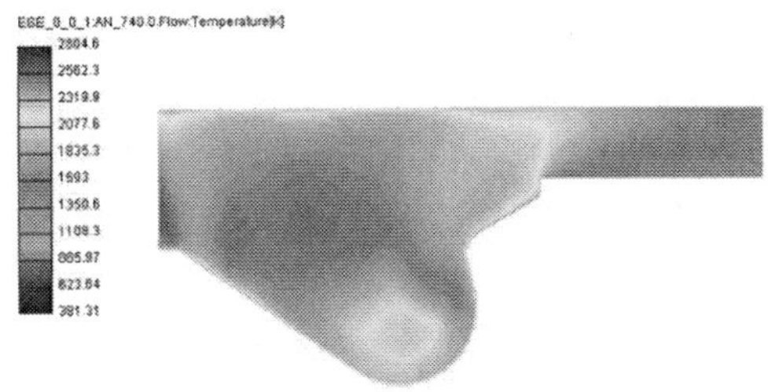

Fig. 20. Variation of the temperature inside the combustion chamber at 740 deg CA for standard injection

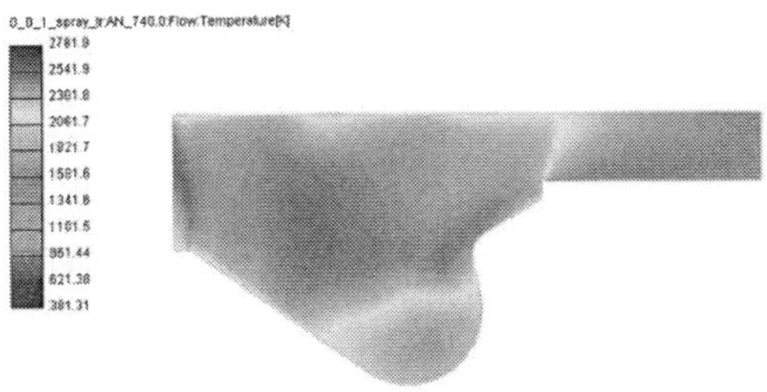

Fig. 21. Variation of the temperature inside the combustion chamber at 740 deg CA for step injection

4. Conclusions

By analyzing the results, it can be seen from Fig. 12-18 that the evolution of NOx formation is similar as shape and formation zone, but the values differ due to the different injection law.

Using the same scale and analysing the areas, it can be observed that for the standard injection the NOx emissions have a bigger area and value.

Graphs in Fig. 18 and 19 were made to underline the maximum NOx variation on the entire mesh and the Average area-weighted NOx value. For the standard injection the variation is clear, as the CA rises, the Max NOx concentration is lower for standard injection and also for the step injection (except 730 deg CA), mainly because the temperature inside the cylinder is lower as the cycle evolves. The average area-weighted NOx value is low at 730 deg CA, but at 745 deg CA the values jump to 0.0006 (standard injection) and 0.0003 (step injection), the lower value is due to the lower temperature inside the combustion chamber for the step injection (like seen in Fig. 20 and 21).

References

1. D.A. Nehmer, R.D. Reitz, “Measurement of the effect of injection rate and split injections on diesel engine soot and NOx emissions”. No. 940668. SAE Technical Paper, (1994).
2. M. A. Patterson, et al. “Modeling the effects of fuel injection characteristics on diesel engine soot and NOx emissions”. No. 940523. SAE Technical Paper, (1994).
3. S. Kook, C. Bae, P.C. Miles, D. Choi, L.M. Pickett,“The influence of charge dilution and injection timing on low-temperature diesel combustion and emissions”. No. 2005-01-3837. SAE Technical Paper, (2005).
4. D.T. Montgomery, and D.R. Rolf,“Six-mode cycle evaluation of the effect of EGR and multiple injections on particulate and NOx emissions from a DI diesel engine”. No. 960316. SAE Technical paper, (1996).
5. S.K. Chen, “Simultaneous reduction of NOx and particulate emissions by using multiple injections in a small diesel engine”. No. 2000-01-3084. SAE Technical Paper, (2000).
6. Z. Han, , et al. "Mechanism of soot and NOx emission reduction using multiple-injection in a diesel engine." SAE transactions **105**,837-852 (1996).
7. T.C. Tow, D.A. Pierpont, and D.R. Rolf,“Reducing particulate and NOx emissions by using multiple injections in a heavy duty DI diesel engine”. No. 940897. SAE Technical Paper, (1994).
8. F Payri, J Benajes, J.V. Pastor, S Molina, “Influence of the post-injection pattern on performance, soot and NOx emissions in a HD diesel engine”. No. 2002-01-0502. SAE Technical Paper, (2002).
9. D. Agarwal, S. Sinha, A.K. Agarwal, "Experimental investigation of control of NOx emissions in biodiesel-fueled compression ignition engine." Renewable energy **31.14**,2356-2369 (2006).
10. A.K. Agarwal, et al. "Effect of fuel injection timing and pressure on combustion, emissions and performance characteristics of a single cylinder diesel engine." Fuel **111**, 374-383 (2013).
11. G. Thomas, et al. "Emissions from DME combustion in diesel engines and their implications on meeting future emission norms: a review." Fuel processing technology **119**,286-304, (2014).
12. K. Varatharajan, and M. Cheralathan. "Influence of fuel properties and composition on NOx emissions from biodiesel powered diesel engines: A review." Renewable and Sustainable Energy Reviews **16.6**3702-3710 (2012).

Comparative analysis of automatic transmission and manual transmission behaviour on the worldwide harmonized light duty test cycle

Norbert Bagameri[1], *Bogdan* Varga[1], *Aron* Csato[1] and *Dan* Moldovanu[1,*]

[1]Technical University of Cluj-Napoca, Automotive and Transport Department, 400641 Cluj-Napoca, Romania

Abstract. The European Commission has been actively involved in the development of the World-wide harmonized Light duty Test Cycle (WLTC). The aim of that project was to develop a harmonized light duty test cycle, that represents the average driving characteristics around the world and to have a legislative world-wide-harmonized procedure put in place from 2017 and onwards for determining the level of CO_2 emissions. This work presents the results of the effect of automatic and manual transmissions as a drivetrain of a light duty vehicle on WLTC driving cycle. AVL Cruise software was used as simulation platform to analyse the CO_2 level and fuel consumption of dynamic vehicle model using different type of drivetrains. At first the simulation tool and the most influential parameters of the driving cycle procedure are described. In the second phase various components and modules for both dynamic models using automatic and manual transmission, as well as the respective input parameters, were defined. Analyses are based on a developed dynamic vehicle model in AVL Cruise. Finally, the simulation results were evaluated and presented for the two dynamic models according to a legislative driving cycle to provide the basis fuel consumption and exhaust gas emissions.

1 Introduction

Recently research and development trends in the automotive industry have been focused on developing vehicles with drivetrains that are strongly driven by steadily increasing demands on improvements in fuel consumption, emissions and performance. The transportation sector is currently responsible for over 20% of emission of greenhouse pollutants, emissions that directly affect the climatic conditions of the planet. Increased emissions in conjunction with achieving the maximum degree of natural regeneration of the environment lead to serious problems related to pollution [1, 2]. Since the early phase of the automotive development, the interaction between the environment and the traffic has been represented by an ever-increasing number of legal standards related to reducing the exhaust gas pollutants, hazardous substances and waste. The increasingly high traffic density, high demand of mobility and transportation adversely have a significant impact on the environment, causing the large societal problems such as a rapid depleting petroleum resources, an increasing air pollution and global warming [3].
The European Union automobile industry welcomes the shift to WLTC and has actively contributed to the development of this new test cycle.

2 Driving cycle procedure

The European Union has developed a new test, called the Worldwide Harmonised Light Vehicle Test Cycle (WLTC). The WLTC has been developed with driving profile obtained from in use driving data and a test procedure more realistic in terms of vehicle mass and resistance to progress. The New European Driving cycle (NEDC) significantly differs from WLTC characterized by a shorter duration and distance, longer idling and cruising time and lower speed and acceleration. Although WLTC driving cycle profile is more transient than NEDC, when these two cycles are tested under the same driving resistance. In WLTC are taking into account its minimum and maximum unladen mass, which is defined as the vehicle's standard weight without driver, fluid or any additional equipment, the maximum permissible weight, the difference in rolling resistance between different tire versions, as well as the difference in aerodynamic resistance expressed as the product of the drag coefficient and the frontal area between vehicle model with the best and worst aerodynamics. Because pollutant emissions from vehicles are caused by combustion of a fuel generally fossil fuels, increasing the efficiency and effectiveness of vehicles drivetrain is constantly the focus

* Corresponding author: dan.moldovanu@auto.utcluj.ro

© The Authors, published by EDP Sciences. This is an open access article distributed under the terms of the Creative Commons Attribution License 4.0 (http://creativecommons.org/licenses/by/4.0/).

of attention of researchers from the automotive industry [2]. The objective of WLTC was to provide a more robust test-basis and a procedure which is more representative of actual on-road vehicle operation. WLTP significantly differs from NEDC, its main differences affecting fuel consumption include the test cycle and gear-shifting sequence, vehicle mass definition, road load determination, chassis dynamometer preconditioning [3, 4].

3 Simulation software package

Modelling various natural or artificial processes and phenomena in engineering is achieved by applying and performance of complex mathematical models. Using the AVL Cruise software simulations can be developed, controlled and coordinated at the same time to achieve results and solutions to optimize and improve the performance of vehicle's technical system and are analysed in terms of performance and operating efficiency. In order to quantify and analyse the effect of the drivetrain in the emissions and fuel consumption on WLTC driving cycle simulations were performed on two different transmissions. Fig. 1 represents the WLTC driving cycle.

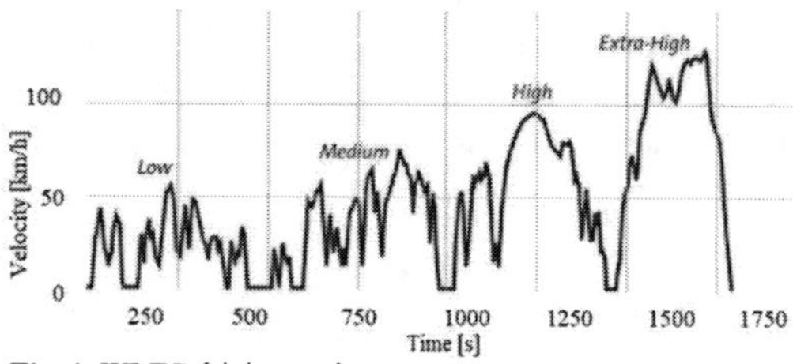

Fig. 1. WLTC driving cycle.

4 Model description

In the following paragraph provide a description of the drivetrains used for vehicles: with manual transmission and automatic transmission. The vehicle model includes the various sub-models. The vehicle model is a structure that has been developed to simulate accurately the two types of drivetrains. They are not different in the arrangement of components of the drivetrain. The developed model consists of components that exist in vehicles, such as the internal combustion engine, transmission, control and program of the transmission, brake system, steering system, wheels and the respective connections among them which can be either mechanical or electrical. Each component is described by several parameters and is configurable in such a way that it can represent a vehicle for this study. The main components of the simulation model with reference to Fig. 2, are the internal combustion engine and the attached mechanical consumer, manual/automatic transmission together with the controller that controls the operation of the gearbox such as the gearshift schedule, the differential together with the final drive, the brakes, the cockpit, the vehicle body, the electrical system which includes the battery and other modules essential for initiating the simulation. The Transmission Controller of Dual Clutch Transmission consists of shift state controller, which preselects the gear according to the current and desired vehicle speed, which can be characterized by gearshift schedule.

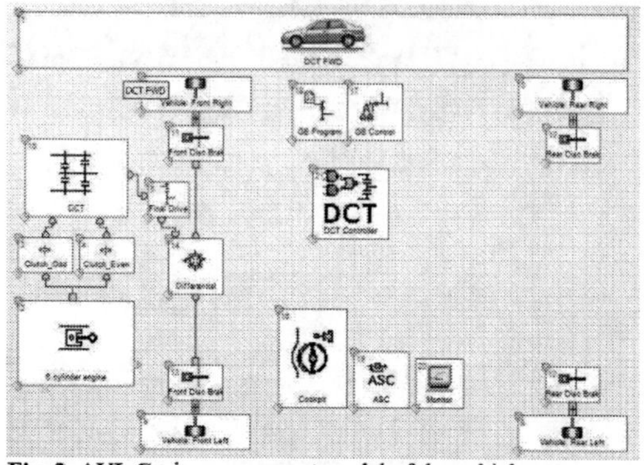

Fig. 2. AVL Cruise component model of the vehicle.

This simulation model was developed based on an existing vehicle using technical data provided by the manufacturer to simulate and to put in evidence the advantages and disadvantages of the drivetrain powered a front wheel drive vehicle equipped with manual transmission and with automatic dual clutch transmission. After every component of the vehicle model is set up the model is tested on the WLTC driving cycle. The basic parameters of the vehicle model are given in Table 1.

Table 1. DCT Vehicle model parameters

Parameters	Value
Vehicle mass	1490 [kg]
Engine displacement	2497 [cm^3]
Engine maximum power	160 [kW]
Engine maximum speed	6500 [rpm]
Engine Inertia	[kg/m^2]
Final drive ratio	3.27 [-]
Clutch 1 inertia	[kg/m^2]
Clutch 2 inertia	[kg/m^2]
Frontal area	1.88 [m^2]
Drag coefficient	0.32 [-]

Final drive is a gear step with fixed ratio. It can be used as a transmission step of the differential also called final drive unit. A drive torque will be transferred to a power take of torque of the transmission step by considering the transmission, the mass moments of inertia, and the moment of loss [4, 5]. The gear ratios of the transmissions are given in the Table 2.

Table 2. Gear ratio

Gear	Gear ratio
1st	4.325
2nd	2.689
3rd	1.746
4th	1.25
5th	0.942
6th	0.721
R	1.0

The driver should imitate the behaviour of a real driver as well as possible. Everything that would normally be controlled by a real driver, such as turning the steering wheel, stepping on the gas, brake and clutch pedal, shifting gears on the manual transmission is controlled by the virtual driver. As the driver controls the clutch pedal, it is necessary to express the clutch position as a function of time during clutch engagement phase [4].

5 Drivetrain of the vehicle

Two different transmission types were evaluated, one of the is the conventional manual transmission which has a limited number of gears, maximum up to six and the automatic transmission using two clutches, one for the odd gears and one for the even gears. In case of the manual transmission the entire synchronization process can be divided in nine stages Stage 1: free travel from neutral position, which has no torque transfer. This stage can also start from a disengagement from a previous gear called here as start of disengagement; Stage 2: beginning of the contact among strut and synchronization ring, initiating the indexing process or preselecting; Stage 3: indexing process concluded; Stage 4: contact of synchronization ring with the sleeve cone, initiating the synchronization phase; Stage 5: end of synchronization ring blocking and beginning of ring and gear turn due to cone friction; Stage 6: second free travel; Stage 7: tip contact of the sleeve against the gear clutch washer. In this stage is possible to detect the double bump phenomena; Stage 8: end of gear angular movement; Stage 9: full engagement [5, 6]. In the following in Fig. 3 the shifting schedule is presented for the upshift and downshift process.

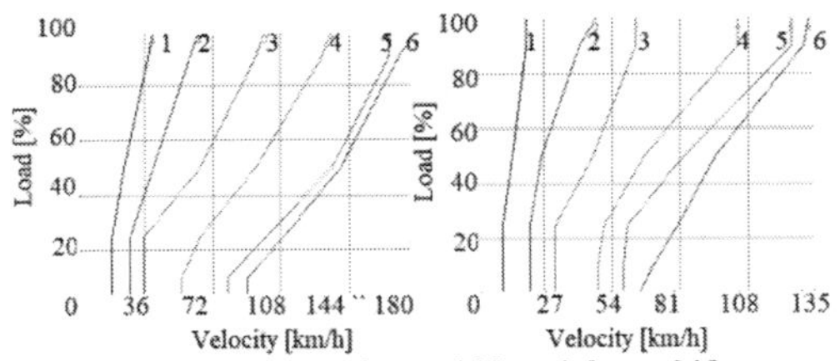

Fig. 3. Shifting strategy for upshift and downshift.

6 Simulation and results

Simulation were performed to investigate the vehicle fuel consumption and emission over time based driving cycle. The simulation is a method used to verify a model which represents the behaviour of a physical system. The maximum distance is 23266 meters and average speed of 46,5 km/h, maximum acceleration 5,66 m/s^2 and total time is 1800 seconds The fuel economy for dual clutch transmission is lower with 3.43 % than the manual transmission over the WLTC driving cycle due to the elimination of torque interruption that is associated with manual transmission. The gearbox program is the same for the manual and dual clutch transmission and the gear shifting time is 500 ms for manual transmission and 8ms for dual clutch transmission. Due to the different working conditions of the gearboxes the automatic transmission operates frequent gear changes, 22 more operations over the WLTC cycle than the manual transmission.

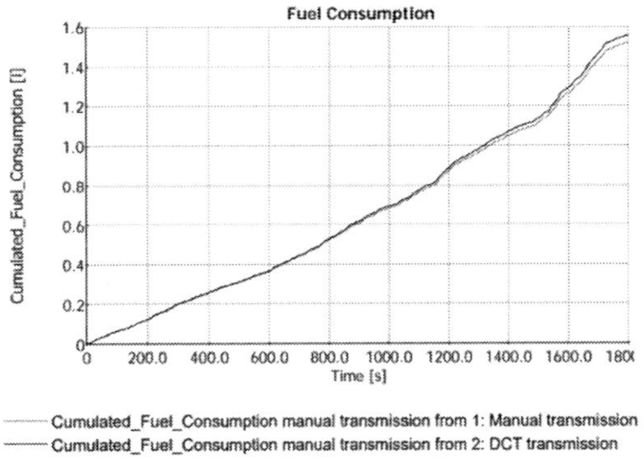

Fig. 4. Variation of the fuel consumption during WLTC driving cycle

In the simulation carried out with AVL Cruise WLTC regulation fixed points, the total CO emissions were found higher by 6 % higher for the case using automatic dual clutch transmission than using the manual transmission vehicle when compared to the WLTC CO emissions curve.

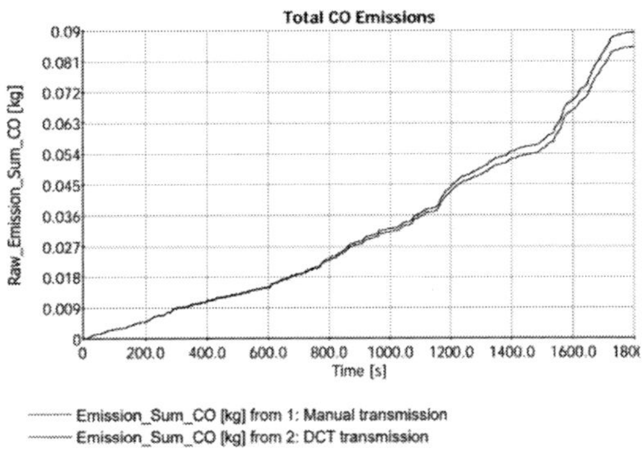

Fig. 5. Variaton of CO emissions over the WLTC driving cycle.

In the Fig. 6 it can be seen that the maximum value of the HC emissions is 0.015 [kg] and it is achieved using the manual transmission due to the higher load of the internal combustion engine and different gear shift strategy of the transmission.

 https://doi.org/10.1051/matecconf/201818401020

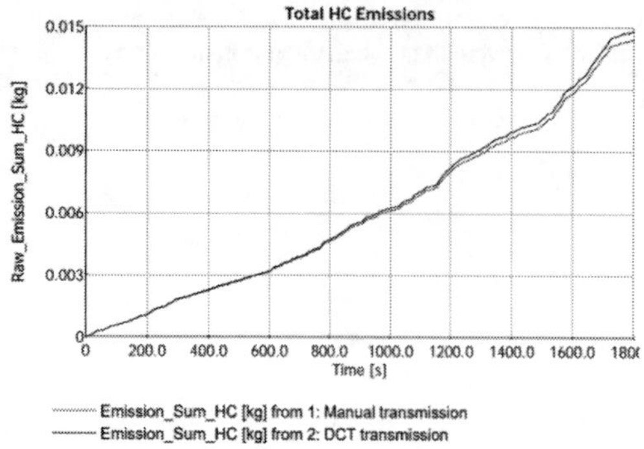

Fig. 6. Variaton of HC emissions over the WLTC driving cycle.

It is seen in the figure that especially there is a significant increase in the NO_x emission in the high speed area of the driving cycle due to the higher temperature into the combustion chamber and higher speed of the internal combustion engine using a lower gear ratio.

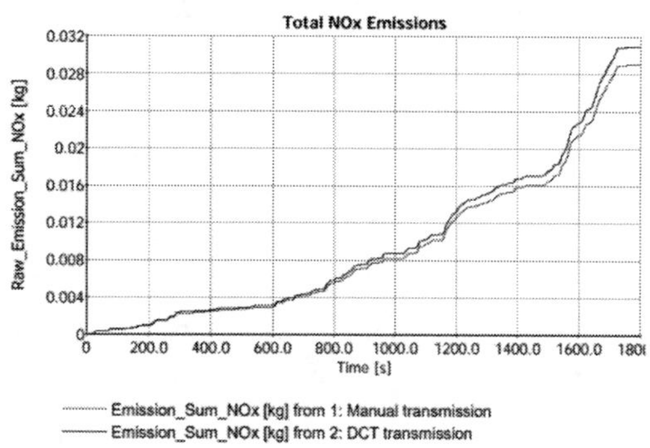

Fig. 7. Variaton of NO_X emissions over the WLTC driving cycle.

The Fig. 8 shows the necessary power of the engine to drive the vehicle during WLTC driving cycle. During low speed sector (0-589s), medium speed part (589-1022s), high speed sector (1022-1477s) and the extra high speed sector (1477-1800s) the average power requirements of vehicle is 24 Nm for the vehicle model with automatic dual clutch transmission and 27 Nm for the one with manual transmission.

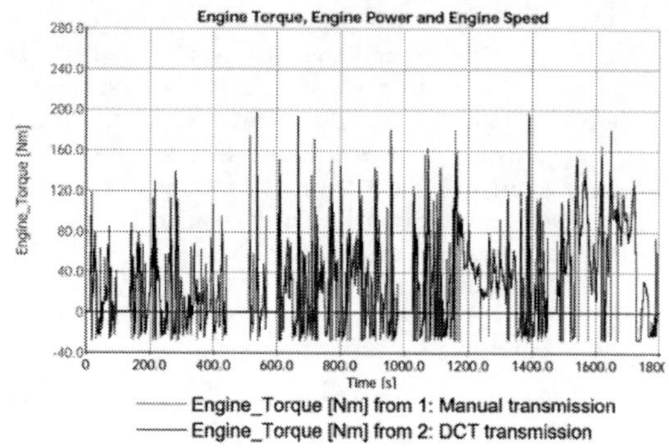

Fig. 8. Variaton of engine power necessary for propulsion the vehicel

7 Conclusion

Simulation software can significantly reduce the time required for developing new technologies and control strategies in the automotive field. The overall system engineering model is used to compare the two drivetrains topologies powered by the same engines during WLTC driving cycle. This investigation focuses on evaluation of the fuel consumption, emissions of the considered vehicle configuration. Finally simulation results demonstrates the differences of using two type of drivetrains for the vehicle and the importance of considering also the automatic dual clutch transmission in order to achieve lower fuel consumption and emissions.

Reference

1. B. Himadri, A. Simos, D. Samraj, "An objective analysis of drivability for two wheeler powertrain with control oriented dynamic model," IFAC-PapersOnLine 48-15, 292–299, 2015.
2. K. Sangmyeong, K. Jinsu, G. Sung, J. Lee, "Evaluation and development of improved braking model for a motor-assisted vehicle using MATLAB/simulink." Journal of Mechanical Science and Technology No.29, pp.2747-2754, 2015.
3. E. Luciano, T. Plinio, "Human Factors Analysis of Manual Gear shifting Performance in Passanger Vehicles," Procedia - Social and Behavioral Sciences, Vol.3, pp. 4350-4357, 2015.
4. X. Xiaohui "Shifting Control and Analysis of Dual Clutch Transmission of Automobile." International Journal of Control and Automation Vol.8, No.12, pp.241-252, 2015.
5. T. Stefanos, G. Fontaras, C. Biagio, Z. Samaras "A simulation-based methodology for quantifying European passenger car fleet CO_2 emissions." International Journal of Applied energy Vol.199, pp.447-465, 2017.
6. B. O. Varga, F. Mariaşiu, D. Moldovanu, C. Iclodean"Electric and Plug-in Hybrid Vehicles – Advanced Simulation Methodologies", Springer International Publishing Ed., ISBN: 978-3-319-18638-2, 2015.

Optimization of shifting strategy for a dual clutch transmission through mathematical model

Norbert Bagameri[1], *Bogdan* Varga[1] , *Aron* Csato[1] and *Dan* Moldovanu[1,*]

[1]Technical University of Cluj-Napoca, Automotive and Transport Department, 400641 Cluj-Napoca, Romania

Abstract. European Union legislation sets mandatory emission and fuel consumption reduction targets for new vehicles and therefore economy remains a key challenge for automotive engineers. Optimizing the transmission shift schedule for an automatic transmission with double clutch is an important part of improving fuel economy. Using simulation and optimization algorithms can ensure that a truly optimal shift schedule is implemented. The main objective of this paper focuses on study the performance of a vehicle with a dual clutch transmission (DCT), with special emphasis on optimizing the shifting schedule using optimization algorithms. In order to study and simulate the shifting strategy the dynamic model is developed based on an existing model in MATLAB Simulink software. In the first phase of the paper a detailed description of the DCT technology, its main functional components and dynamic characteristics are explored. In the second phase model development and simulation concerns the major part of this work, to find the optimal shift schedule for different engine speed and engine torque as it accelerates through the transmission gears by engaging the dual clutches. Finally, the simulation results were evaluated and presented which successfully demonstrate the performance improvement of a dual clutch transmission with an optimized shifting strategy.

1 Introduction

The function of a vehicle transmission is to adapt the traction available from the drive unit to suit the vehicle, the surface, the driver and the environment. It has a decisive effect on the reliability, fuel consumption, ease of use, road safety and transportation performance. The optimization of the transmission can significantly increase the performance of the vehicle efficiency and fuel economy [1]. One of the main challenges in the automotive industry is to improve mechanical efficiency of the drivetrain and in the same time to ensure high comfort level at the board of the vehicle. [2]. The Dual Clutch Transmission is a new type of automatic transmission device in the field of automotive automatic transmission. It realizes power transmission and shifting without power interruption through respective connection between the two-transmission shaft and two clutches, which improve better fuel economy. Shift control is the key technology of DCT vehicles, and an important index to measure the driving comfort of vehicles, therefore, to carry out targeted shift analysis and control of the DCT seems quite necessary [3, 4]. A dual clutch transmission is quite similar to the conventional automatic, the main difference being the double clutch structure compared to the single automatic clutch used in automatics. Automatics make the use of a torque converter to transfer engine torque from the engine to the transmission, DCTs on the other hand don't require torque converters [5].

2 Structure and principle of DCT

In the following paragraph, a short description of the dual clutch transmission is presented.. When shifting between gear ratios, the transmission control unit (TCU) should synchronize the engagement of the on-coming clutch and the disengagement of the off-going clutch, this process is called clutch-to-clutch shift control [6]. Under dynamic driving conditions, the transmission is required to shift in order to match the engine power with the changing drive power. Furthermore, a gearshift decision is expected to be consistent such that the vehicle can remain in the next gear for a period of time without deteriorating the acceleration capability. Therefore, an optimal conversion of the engine power plays a key role in improving the fuel economy and driveability. Similar with the traditional manual gearbox structure it includes cylindrical gear sets and synchroniser gear set. Fig. 1 shows the layout of the DCT mounted on a front wheel drive vehicle, with transversal internal combustion engine and gearbox. All the gears of the transmission have an identification number, which is used to specify the gear transmission ratios.

* Corresponding author: dan.moldovanu@auto.utcluj.ro

© The Authors, published by EDP Sciences. This is an open access article distributed under the terms of the Creative Commons Attribution License 4.0 (http://creativecommons.org/licenses/by/4.0/).

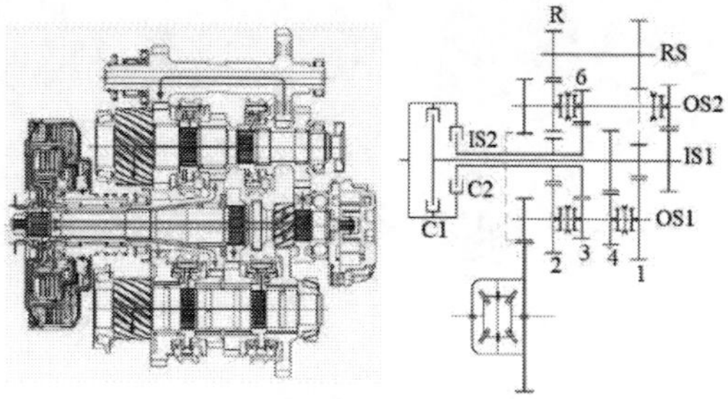

Fig. 1. Cross section and layout of the DCT [7].

The two clutches, that give the name to the transmission, can be single or multiple disk, dry or wet type, depending mainly on the amount of power they have to transfer and they are mounted between the engine flywheel and one of the two coaxial input shafts of the transmission, one for odd gears and the other for even gears. Gearshift and engagement of clutches are electronically managed by a control unit, which drives the mechatronic assembly integrated in the gearbox. The gear shift from first to second gear is automatically actuated by the transmission control unit (TCU) in a short time thanks to the concurrent disengagement of the first clutch and engagement of the second clutch, thus eliminating the dead times due to the traditional synchronising phase. Hence, this clutch-to-clutch shift allows not only to accomplish the gearshift, but also to have a continuous torque transfer to the wheels (power-shift) during the gear change process. Obviously, a similar procedure is repeated when the engagement of the second gear is completed: the system gets ready to the following gear shift inserting the synchroniser corresponding to the next gear [4], [7]. A distinctive characteristic of the DCT is represented by the fact that during normal operating conditions the dual clutch gearbox has two different gear ratios engaged at the same time: the TCU must avoid the concurrent lock of the two clutches in order to prevent damages to the gearbox [4].

3 Simulation Software

In order to analyse the dynamic characteristics of the vehicle and to study performance, a computer simulation model of the vehicle system, using Matlab Simulink software package has been developed. The software Simulink and Matlab are available from the Mathworks, which provides a graphical interface for building models as block diagrams. Matlab Simulink provides a platform to simulate the performance of the optimal shifting controller by running the transmission model in a certain driving cycle. In this model, the Transmission Controller block mainly has several inputs and outputs; the main inputs in this model are the vehicle speed, the internal combustion engine speed, the clutch slipping sensor signal, and the command speed, and the main outputs are the gear, pressure and torque demanded, which are used to drive the transmission.

With the definition of system notion and the development of a mathematical device to realize the connection between input parameters, output parameters and operating parameters of the system modeling process began to be increasingly used in all areas related to human activities. Expression of relations between input and output values of the systems is achieved by using mathematical functions. The simulation was performed using Matlab Simulink software package in order to obtain the optimal shifting schedule of the dual clutch transmission for fuel economy of the vehicle model.

4 Vehicle subsystems

The powertrain consists of functional components in the form of blocks such as vehicle dynamics, engine model, input and output shafts, transmission, transmission controller, wheels and tires. The characteristic curve of an internal combustion engine defines the torque supplied as function of engine speed and throttle valve opening. The transmission subsystem is the focal point of the powertrain model, this is dual clutch transmission with a six forward gear sets and one reverse gear set. The dual clutch transmission is utilize clutches to transfer torque at various speed ratios through the transmission. The synchronizers cause discontinuities in the passage from synchronization to engaged gear: the interaction between elements, initially due to friction is then realize by the contact of the dog-teeth. Stationary idle, transition to motion and interruption of the power flow are all made possible by the clutch. The clutch slips to compensate for the difference in the rotational speeds of engine and drivetrain when the vehicle is set in motion. When a change in operation conditions makes it necessary to change gears clutch disengages the engine from the drivetrain for the duration of procedure. The clutch is controlled by the driver via the cockpit. The maximum transferable torque is the maximum torque the clutch can transfer between the drive and the power take-off side [6].

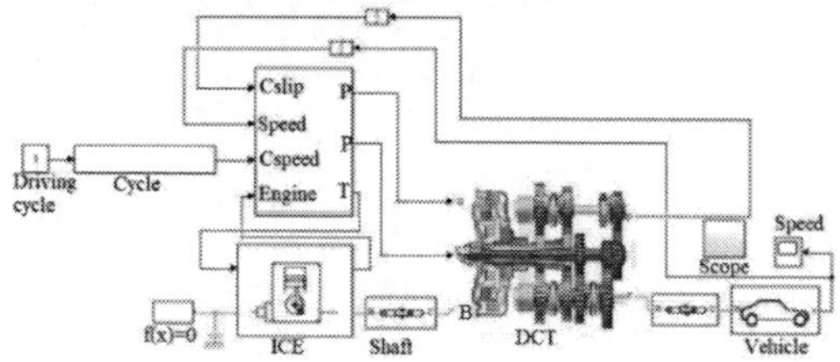

Fig. 2. Vehicle model

5 Shifting strategy

For conventional vehicles equipped with discrete ratio transmissions, e.g. automatic, dual clutch, automated (manual) transmissions, etc., a decision on selecting a different gear leads to a sudden change of the engine operating point due to a stepped change of the transmission ratio. With highly nonlinear characteristics of the engine fuel consumption and maximum torque, a change of the engine operating point due to a gearshift can result in a deficient fuel operation state, and/or a low driveability condition for the vehicle thereafter. Hence, an optimal gear shift strategy for a discrete ratio transmission plays an important role in achieving a high performance for the vehicle [8-9]. The shift process of dual clutch consists of two stages: the torque stage is the conversion process of torque, that is, the separation of the torque from the clutch converts to engaging to it; the inertial stage is the speed synchronization process, namely, the engine speed synchronizes with the target gear speed. When control the throttle to up shift the gear, first, entering the torque stage, and then enters the inertia stage, in the downshift process, it is the opposite. The second stage is the synchronous stage in this stage, the clutch is stable, the speeds of the master and the side driven plates are equal. When the transfer torque is less than the capacity of the clutch torque, the clutch torque is equal to the current transmission torque, while when the transfer torque is bigger than the capacity of the clutch torque, the clutch torque is equal to the maximum transfer torque of the clutch [8].

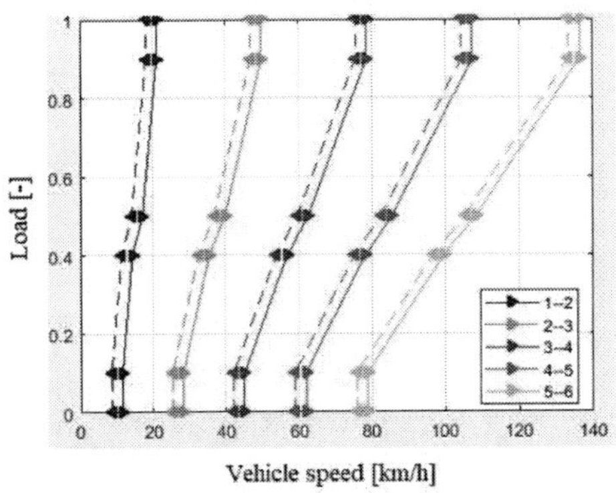

Fig. 3. Shifting strategy

In a real world application, the dual clutch transmission controller has several gear shifting strategy's for different driving conditions. According to the situation the TCU decides which strategy is using.

6 Simulation and results

In this chapter, an optimal control strategy for the gear shifting schedule of the automatic transmission is developed depending on the proper application of the decision-making algorithm for best fuel economy. The shifting schedule is optimized for FTP75 cycle, with helping of optimization solvers. At first parameters of the shifting schedule is optimized. Two parameters of the shifting schedule are identified the distance between the 0 load shifting lines and the distance between full load shifting lines. In the first and second optimization only the shifting schedule parameters are optimized with two different solvers. The third optimization solver was using 30 parameters to optimize the shifting schedule.

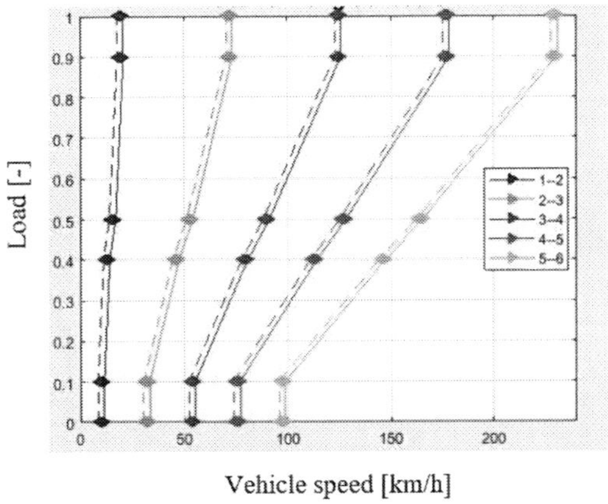

Fig. 4. First optimization shifting schedule

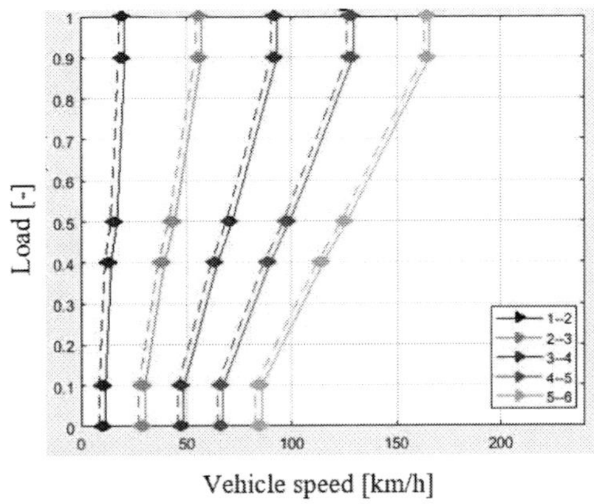

Fig. 5. Second optimization shifting schedule

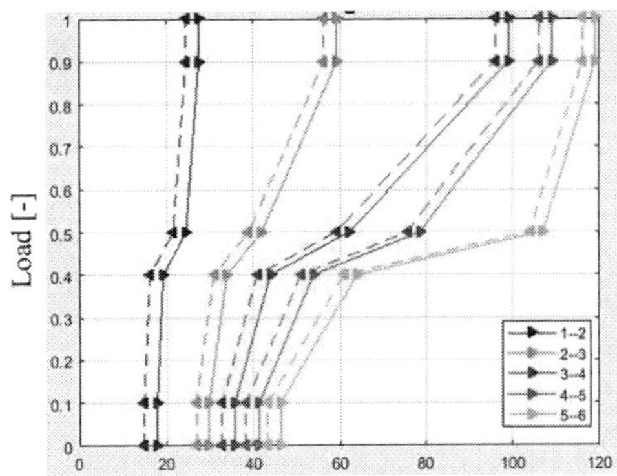

Fig. 6. Third optimization shifting schedule

After the integrated control analysis of the DCT shift and the formulation of control strategy, on the basis of MATLAB Simulink (Mathworks) simulation platform, simulation model and simulation analysis has been made on the shifting process for the six speed dual clutch transmission. An optimal gear shift strategy, allowing the engine to operate in the region closely to maximum torque of the engine in every gear because this offers minimal consumption.

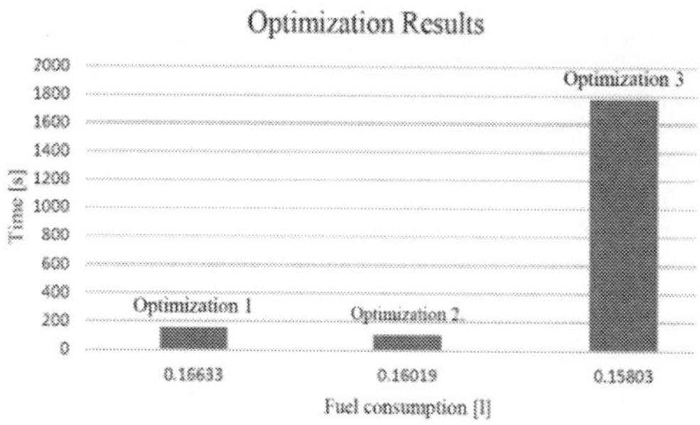

Fig. 7. Optimization results

The optimization process take several iteration, in the 3rd case is more than 10000 iteration to find the lowest fuel consumption. Once we have a complete model, we can run the simulation for any shift schedule that we want to test. Improved fuel economy is probably the most significant benefit of using a dual clutch transmission due to the fact that during gearshifts power delivery from the engine to the transmission is uninterrupted and therefore fuel efficiency increases dramatically.

7 Conclusion

The control strategy of the engine and clutch and the comprehensive shift control strategy are made in the DCT shifting process. Based on the Matlab Simulink software simulation platform, simulation model of vehicle shifting has been developed, and the simulation results validates affectivity of the shift model, and it also indicates the good control effect of the formulated shift control strategy. Modelling and simulation approach of technical and functional characteristics that influence the powertrain is widely presented in this paper. The potential for optimization the shifting strategy for increasing the overall efficiency of the vehicle strongly depends on the driving cycle that it is applied to, disturbances from traffic and driving behaviour. In Matlab Simulink model are simulated a powertrain and a drive train. For optimizing the shifting schedule the model is running on FTP 75 cycle and the solvers are minimizing the fuel consumption by trying different shifting schedules. Optimization process can be used for different engines that uses the same DCT transmission. These optimization solvers are based on finding minimum values so they are faster than brute force optimization. The fuel-optimal gearshift strategy, proposed on Fig. 6, yields an early and frequent upshift pattern for the transmission. It renders the engine, on average, to operate at a lower speed and high torque region, and hence improving the fuel economy. For facing the high requirements of the new emission regulations every optimization have to be considered, so a correct optimization for gear schedule could have a high impact on the emissions.

References

1. W. Sarawoot "Modeling and estimation for stepped automatic transmission with clutch-to-clutch shift technology" Dissertation Thesis, Ohio State University, 2004.
2. A. Faraz, "Modelling and simulation of a vehicle with a controlled dual clutch transmission" University of Bradford, 2015.
3. G. Enrico, A. Vigliani, M. Velardocchia, "Dynamic and kinematic model of dual clutch transmission," Mechanism and machine theory 794-805, 2011.
4. N. Chunsheng, L. Tongli, J. Zhang, "Gearshift control for dry dual-clutch transmissions" Issue 11, Vol.8, pp. 1177-1186, 2009.
5. X. Xiaohui "Shifting Control and Analysis of Dual Clutch Transmission of Automobile." International Journal of Control and Automation Vol.8, No.12, pp.241-252, 2015.
6. N. Dac, "Gear shift strategies for Automotive Trasnsmission" Eindhoven University of Technology, ISBN 978-90-386-3222-3, 2012.
7. N. Chunsheng, L. Tongli, J. Zhang, "Gearshift control for dry dual-clutch transmissions" Issue 11, Vol.8, pp. 1177-1186, 2009.
8. Y. Wang, "Optimal Gear Shifting Strategy for a Seven-speed Automatic Transmission Used on a Hydraulic Hybrid Vehicle" Dissertation Thesis, University of Toledo, 2012.
9. F. Meng, G. Tao, H. Chen, "Smooth shift control of an automatic transmission for heavy-duty vehicles," Journal Neurocomputing, 159, pp.197–206, 2015.

Influence of swirl and tumble motion inside the combustion chamber of a compression ignited engine on vertices formation

Dan Moldovanu[1,*], *Florin* Mariaşiu[1], *Norbert* Bagameri[1]

[1]Technical University of Cluj-Napoca, Cluj-Napoca, Romania

Abstract. In modern compression ignited engines, one of the key innovative directions is optimization of the burn process especially to reduce emissions. This optimization can be done by ensuring a better air-fuel mixture. To answer this problem, in the current paper, the authors analysed the vertices formation and velocity inside the combustion chamber in thirteen cases, by using numerical analysis (AVL FIRE Software), to generate pure tumble motion, pure swirl motion, no in-cylinder motion of the air, and other combinations on tumble and swirl motion. The cases that were chosen combine 0%, 50%, 70% and 100% tumble and swirl motion. The greatest velocity of the in-cylinder air motion was noted when a 100%tumble and 50% swirl motion was induced (77.2 m/s at 736 degrees CA), while the smallest velocity was noted at 100% swirl and 0% tumble (71.07 m/s at 736 degrees CA). Because of the different induced motions, the vertices that form inside the combustion chamber have a great influence on air-fuel mixture and implicitly on emissions.

1. Introduction

Current European laws restrict harmful emissions through several amendments, especially for Diesel engines. Therefore, a large number of papers investigate all possible directions for this cause.

Payri et.al. chose the direction of measuring tangential and radial velocity components of a Light Duty Vehicle (LDV) with accent on the flow patterns inside the 4-valve single cylinder diesel engine for various engine speeds, underlining swirl numbers and piston-bowl geometries, under similar conditions to those of a production engine. Their work was focused on the near-wall region of the axisymmetric combustion chambers because of the emissions formation. And their conclusion was that the piston bowl shape and intake configuration affect the flow patterns and the turbulence level [1].

The main in-cylinder motion inside a spark ignited engine is the tumble motion which affects the air-fuel mixture and therefore the emissions. Kim et.al. used a commercial code (STAR-CD) to investigate the mixture formation and combustion characteristics in a spray-guided GDI engine. To predict the breakup of the spray, the Reitz-Diwakar's breakup model was used for the ultra-lean mixture operating mode. The results shown that enhanced tumble flow can deteriorate the mixture distribution of the hollow cone spray, and therefore decreases the burning rate [2].

Yadollahi et.al. investigated the effect of combustion chamber geometry on injection and mixture preparation in a CNG direct injection SI engine. By changing the combustion chamber geometry, the air-fuel mixture changes. The results show a strong sensitivity to grid size across the nozzle diameter and the cylinder head forces a non-symmetrical flow field inside cylinder [3].

Hill and Zhang have written a paper regarding the rotating flow (swirl and tumble) in premixed spark-ignition engines and its effects on turbulence generation and flame propagation. Both motions, swirl and tumble, have had many research devoted to them, in recent years the increasing attention being paid to tumble. Because the two motions have different characteristics and they interact differently with the piston motion, is has been concluded that an optimum rotating flow field may create through combination of the two kinds of rotational motion. In their paper the authors have reviewed the principal features of the mean velocity and turbulence fields associated with these rotational motions, the mechanisms for decay, generation, transport and enhancement of in-cylinder turbulence [4].

Another paper regarding the two motions was the one written by Lee et.al, concerning the effects of tumble and swirl flows on flame propagation in a four-valve spark-ignition engine. Through many experiments the authors underlined that there is a correlation between the stronger tumble during induction and turbulence levels at the time of ignition, this leading to a faster flame development. Furthermore, the tumble and swirl mixture was found to be more effective than pure tumble in effecting rapid and stable combustion under lean mixture conditions [5].

Khalighi investigated the intake-generated swirl and tumble motions in a 4-valve engine with various intake configurations. The results of this study indicated that during the induction process the in-cylinder flow field is comprised of several large scale vertical motions (stable and repeatable) on different planes within the engine cylinder. Although no significant in-cylinder tumbling

* Corresponding author: dan.moldovnau@auto.utcluj.ro

© The Authors, published by EDP Sciences. This is an open access article distributed under the terms of the Creative Commons Attribution License 4.0 (http://creativecommons.org/licenses/by/4.0/).

 https://doi.org/10.1051/matecconf/201818401022

motion was observed in the engine with both intake valves operating, a well-defined tumbling flow structure was observed with shrouded intake valve configurations. The author also measured and evaluated the velocity fields, tumble and swirl ratios under transient conditions [6].

There were also many papers that have investigated the swirl and tumble motions in regards to the burner systems. For example, Syred and Beér have researched the combustion in swirling flows in order to underline that there is sufficient information to indicate that stratified or staged fuel or air entry may be used to minimize noise, hydrocarbon, and most importantly NOx emissions from swirl combustors [7]. Furthermore, there were papers who further researched the burner systems in regards to energy, combustion and flame and heat transfer and NOx emissions [8-11].

2. Methodology

The methodology used for the current simulation starts with the correct input data. The simulation mesh and boundary were made using FIRE ESE DIESEL, where the general data were, the piston was created from a template piston shape, after measuring the real piston section. After introducing the geometry of the piston, the actual compression ratio was checked.

The next step was to generate the 3D meshes and check all the moving meshes for triangle faces. The final mesh is presented in Fig. 1.

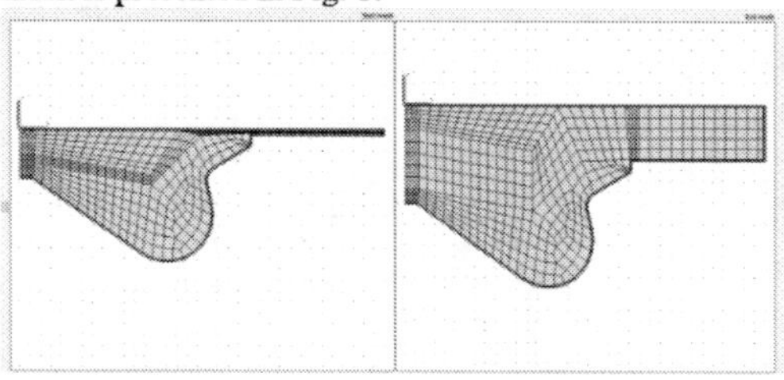

Fig. 1. Final mesh

The selections that were made are for:

✓ the piston bowl - mesh movement and an imposed temperature of 575.15 K,

✓ cylinder liner - wall boundary condition, with a temperature of 475.15 K

✓ cylinder head - wall boundary condition with a temperature of 550.15 K

✓ the symmetry selection - symmetry boundary chosen.

After having the moving mesh, the Solver Steering File was completed, starting with Crank-angle run mode (from 540 to 800 deg CA), with module activation for species transport, spray (the module that simulates sprays based on the Discrete Droplet Method (DDM), which represents a spray as a statistical sample of discrete parcels.) and emissions.

The main idea of the paper was to change the movement of the air inside of the cylinder before the start of injection so that the fuel is injected in different types of vertrices and different air-fuel mixtures form.

For the simulations, a pure swirl motion simulation was made and a pure tumble simulation was made, after that, the cases that were chosen combine 0%, 50%, 70% and 100% tumble and swirl motion.

Table 1. Simulated cases (T-Tumble, S-Swirl motions).

Case no.	1	2	3	4	5	6
Section	50% T 50% S	70%T 70%S	50%T 70%S	50%T 100% S	70%T 50%S	70%T 100% S

7	8	9	10	11	12	13
0%T 0%S	0%T 50% S	0%T 100% S	100% T 0%S	100% T 50%S	100% T 70%S	100% T 100% S

3. Results

To show the influences of the imposed motions, sections were made, monitoring the velocity and direction of the air at 716 deg CA (2 degrees before the start of injection) and at 736 deg CA (2 degrees after the end of the injection).

Since there are a lot of simulated cases, the results with the vertrices were presented for the most significant cases (Fig. 2 – 7).

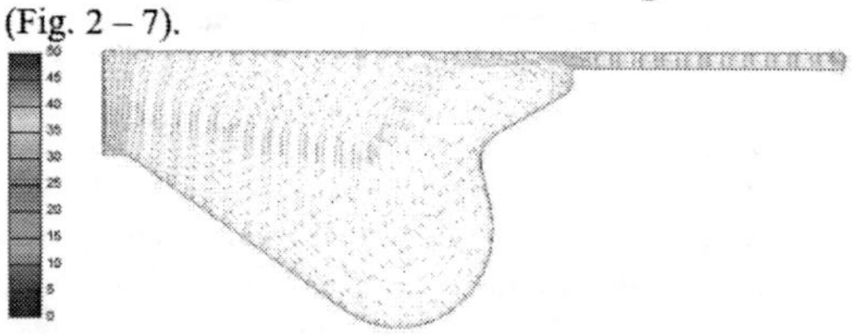

Fig. 2. Vertrices formation at 718 deg CA, case 10

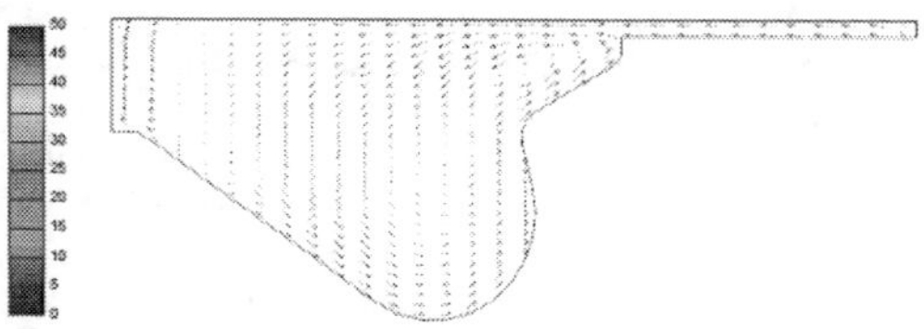

Fig. 3. Vertrices formation at 718 deg CA, case 11

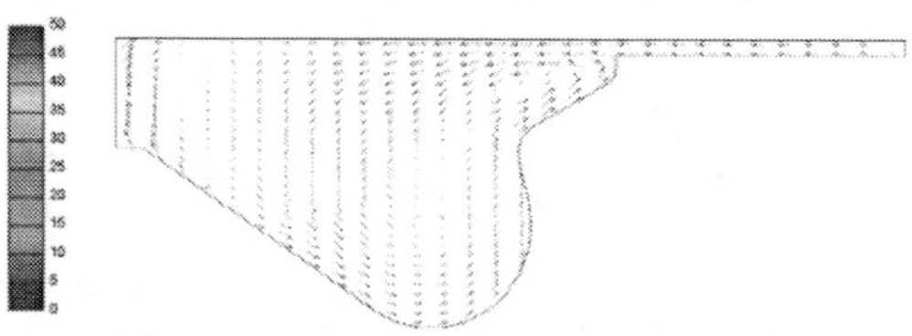

Fig. 4. Vertrices formation at 718 deg CA, case 5

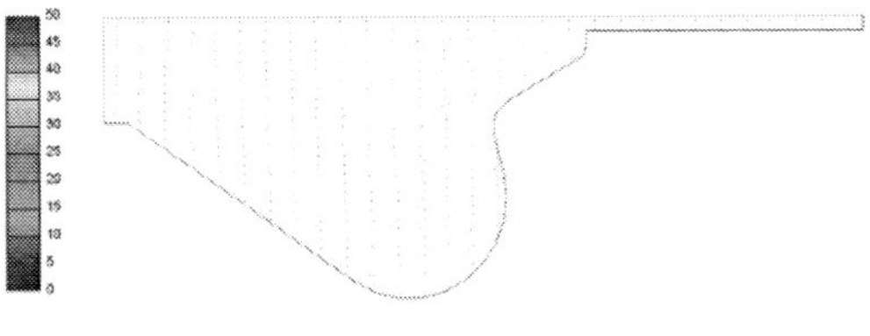

Fig. 5. Vertrices formation at 718 deg CA, case 7

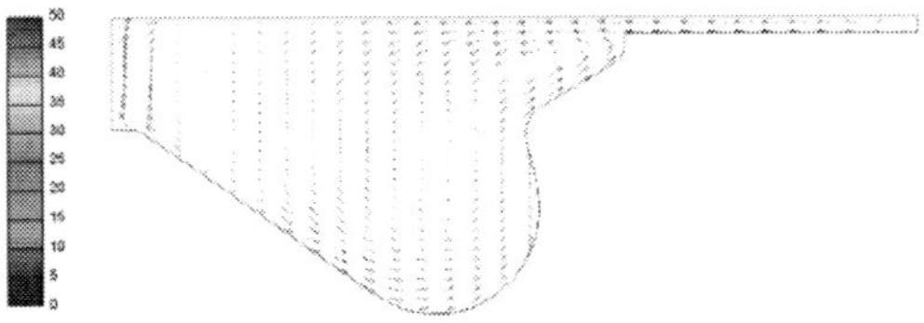

Fig. 6. Vertrices formation at 718 deg CA, case 2

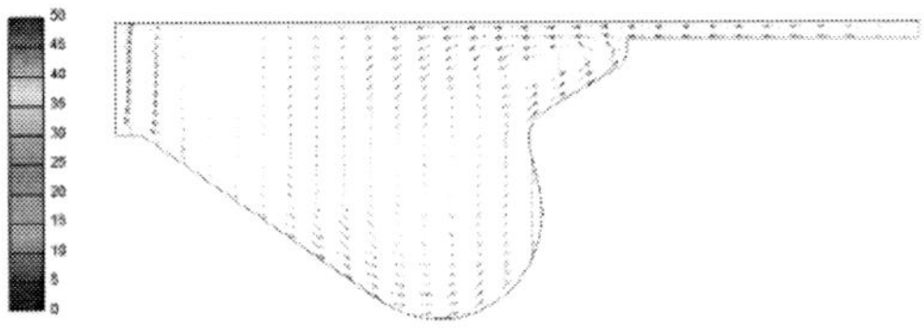

Fig. 7. Vertrices formation at 718 deg CA, case 13

For the same cases, sections were also presented at 736 deg CA (Fig. 8 – 12).

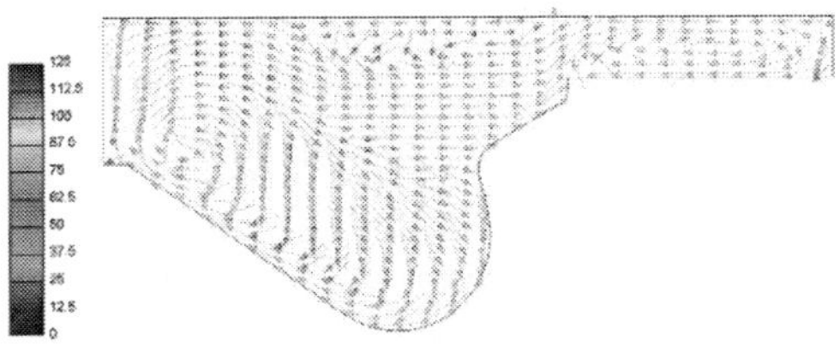

Fig. 8. Vertrices formation at 718 deg CA, case 10

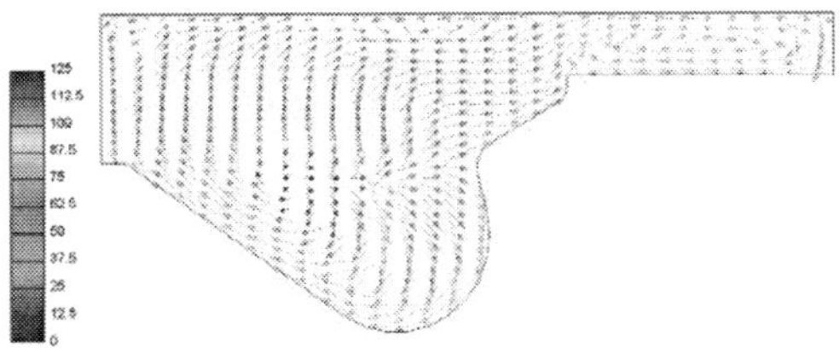

Fig. 9. Vertrices formation at 718 deg CA, case 5

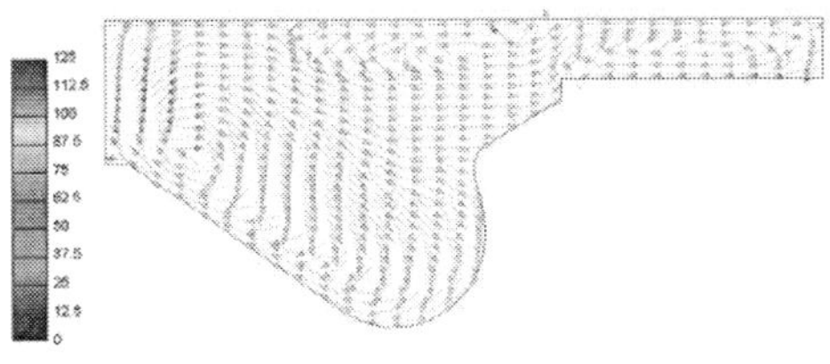

Fig. 10. Vertrices formation at 718 deg CA, case 7

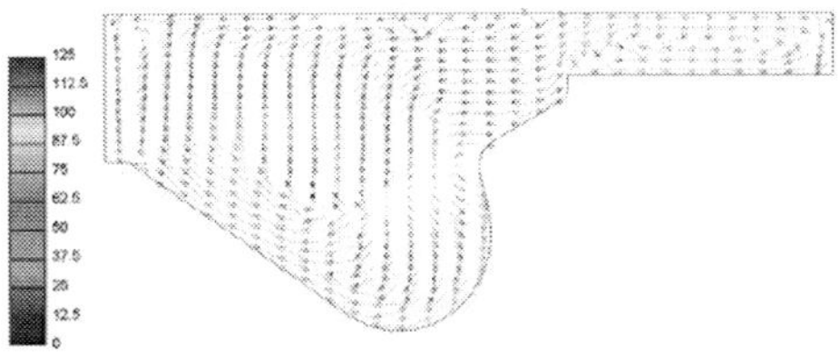

Fig. 11. Vertrices formation at 718 deg CA, case 2

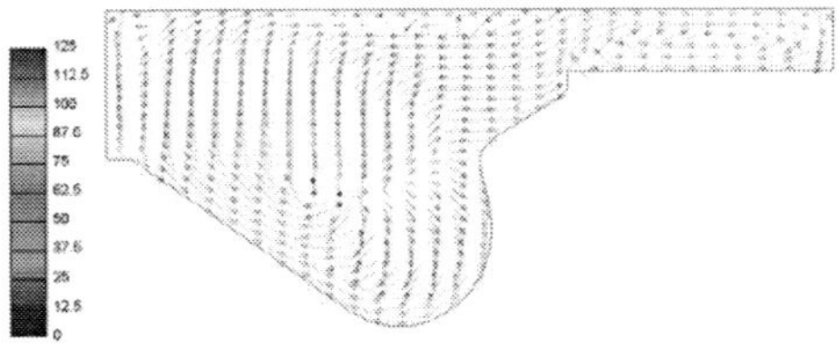

Fig. 12. Vertrices formation at 718 deg CA, case 13

After observing the location and orientation of the vertrices, the maximum velocities were extracted for all sections and inserted in Table 2.

Table 2. Velocity results for all simulated cases

Case no.	**1**	**2**	**3**	**4**	**5**	**6**
Velocity at 718 deg CA [m/s]	17,77	17,7	18,7	19,52	17,17	18,87
Velocity at 736 deg CA [m/s]	74,51	73,7	72,63	72,29	75,91	73,21

7	**8**	**9**	**10**	**11**	**12**	**13**
Velocity at 718 deg CA [m/s]	16,2	20,42	20,41	15,9	16,65	17,12
Velocity at 736 deg CA [m/s]	76,44	71,96	71,07	75,65	77,2	75,33

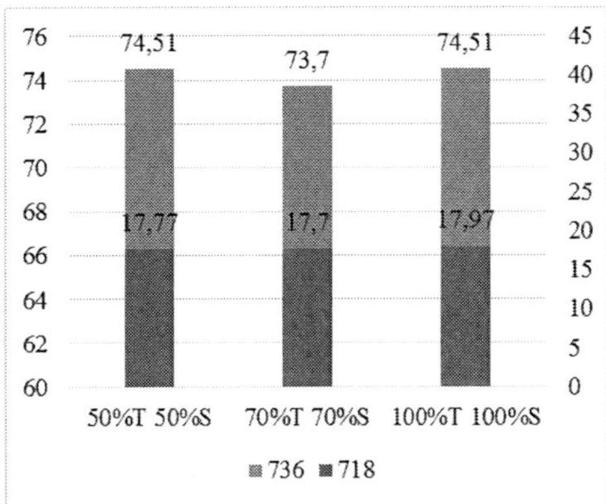

Fig. 13. Velocity [m/s] at 718 and 736 deg CA, cases 1, 2, 13.

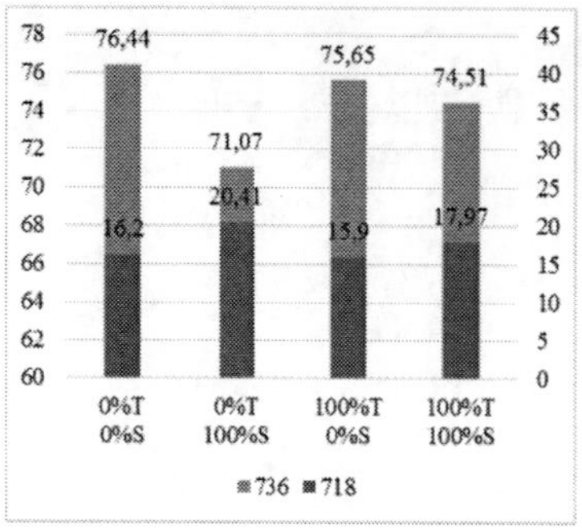

Fig. 14. Velocity [m/s] at 718 and 736 deg CA, cases 7, 9, 10, 13.

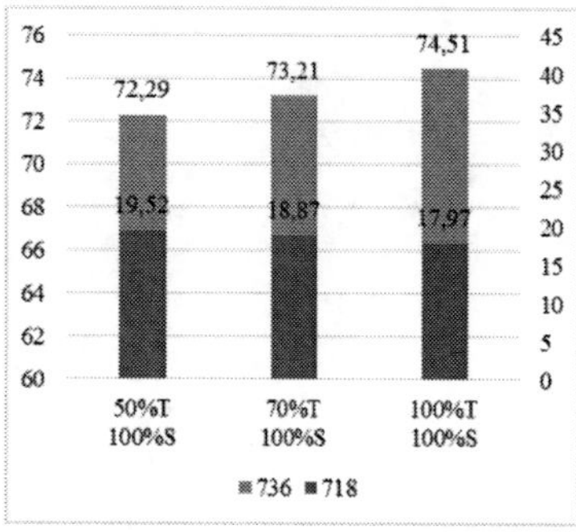

Fig. 15. Velocity [m/s] at 718 and 736 deg CA, cases 4, 6, 13.

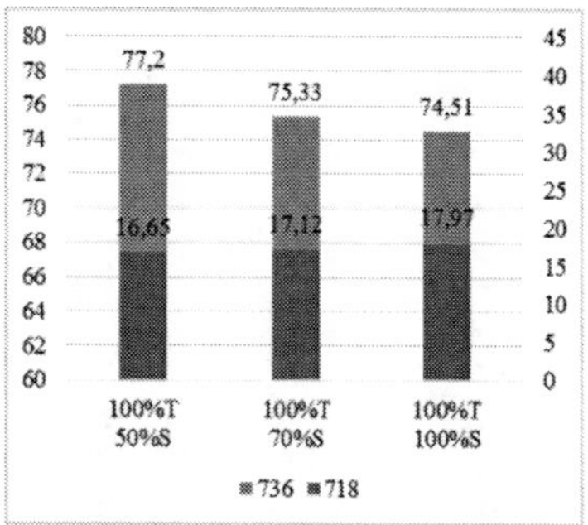

Fig. 16. Velocity [m/s] at 718 and 736 deg CA, cases 11, 12, 13.

4. Conclusions

Taking into account all the gathered data, the conclusion is that the intake process is very important due to the air vertrices formation inside the combustion chamber.
Results have shown multiple vertrices like in Fig. 8, but also vertrices formed only because of the injection of the fuel.

An interesting fact is underlined by Fig. 15 and 16: as the tumble coefficient rises, the maximum velocity inside the cylinder rises, but if the swirl increases, the velocity inside the cylinder drops by 3 m/s, even though the initial motion had similar velocities.

References

1. F., Payri, J. Desantes, J. Pastor, Experiments in Fluids **22** (1996):. https://doi.org/10.1007/s003480050029
2. S., Kim, Y., Kim, J., Lee, "Analysis of the In-Cylinder Flow, Mixture Formation and Combustion Processes in a Spray-Guided GDI Engine," SAE Technical Paper 2008-01-0142, (2008), https://doi.org/10.4271/2008-01-0142.
3. B., Yadollahi, and M. Boroomand. "The effect of combustion chamber geometry on injection and mixture preparation in a CNG direct injection SI engine." Fuel **107** 52-62 (2013).
4. P. G., Hill, and D. Zhang. "The effects of swirl and tumble on combustion in spark-ignition engines." Progress in energy and combustion science **20.5** 373-429 (1994).
5. K., Lee, B., Choongsik and K., Kernyong. "The effects of tumble and swirl flows on flame propagation in a four-valve SI engine." Applied thermal engineering **27.11-12** 2122-2130 (2007).
6. B., Khalighi, "Intake-generated swirl and tumble motions in a 4-valve engine with various intake configurations—flow visualization and particle tracking velocimetry." SAE transactions 354-374 (1990).
7. N., Syred, J.M., Beer. "Combustion in swirling flows: a review." *Combustion and flame* **23.2** 143-201 (1974).
8. J. C., Broda, et al. "An experimental study of combustion dynamics of a premixed swirl injector." *Symposium (International) on Combustion*. Vol. **27**. No. **2**. Elsevier, (1998).
9. P., Schmitt, et al. "Large-eddy simulation and experimental study of heat transfer, nitric oxide emissions and combustion instability in a swirled turbulent high-pressure burner." *Journal of Fluid Mechanics* **570** 17-46 (2007).
10. N., Syred, "A review of oscillation mechanisms and the role of the precessing vortex core (PVC) in swirl combustion systems." *Progress in Energy and Combustion Science* **32.2** 93-161 (2006).
11. L., Selle, et al. "Compressible large eddy simulation of turbulent combustion in complex geometry on unstructured meshes." *Combustion and Flame* **137.4** 489-505 (2004).

Determine of the wave equation in the task of electrical oscillations

Gordana V. Jelić[1,*], *Vladica* Stanojević[2], *Dragana* Radosavljević[1]

[1]Faculty of Technical Sciences in Kosovska Mitrovica, Serbia
[2]Faculty of Sciences and Mathematics in Kosovska Mitrovica, Serbia

Abstract: One of the basic equations of mathematical physics (for instance function of two independent variables) is the differential equation with partial derivatives of the second order (3). This equation is called the wave equation, and is provided when considering the process of transverse oscillations of wire, longitudinal oscillations of rod, electrical oscillations in a conductor, torsional vibration at waves, etc... The paper shows how to form the equation (3) which is the equation of motion of each point of wire with abscissa x in time t during its oscillation. It is also shown how to determine the equation (3) in the task of electrical oscillations in a conductor. Then equation (3) is determined, and this solution satisfies the boundary and initial conditions.

1 Formation of a wave equation

In mathematical physics, implied under a wire is a thin elastic thread. Let the wire of the length *l* at the initial moment match with O from O to *l*. Assume that the ends of the wire are fixed at points x=0 and x=*l*. If the wire is taken from its original position and then released, or without moving the wire, we provide a speed at its initial points, then the wire points will move - the wire starts to oscillate, flickers (oscillates). The task is to determine the shape of the wire at any time and to determine the law of motion of each point of the wire with the abscissa *x* at the moment *t*.

When the wire oscillates the deviations of the wire points from the initial position are small, on the Ox axis and are in one plane. The oscillations are determined by the function in *u(x,t)*, which gives the displacement value of the wire points from the abscissa x at the moment t (Fig.1).

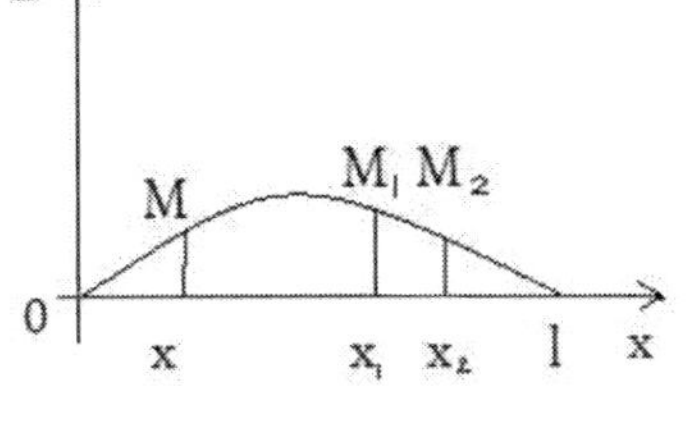

Fig. 1. Displacement value of the wire points

When we observe small wire deviations in the plane (x, u), we will assume that the length of the wire element is equal to its projection on the axis Ox, i.e. M_1M_2 =x_2-x_1. We also assume that the tension at all points of the wire is the same and is denoted by T.

Consider the wire element MM' (Fig. 2).

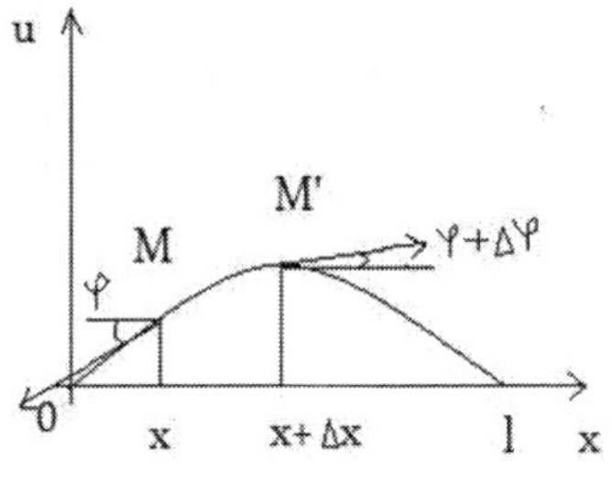

Fig. 2. Wire element MM'

At the ends of this element, at the tangents, the force T acts on the wire. Let the tangents form with Ox axes the angles φ and $\varphi+\Delta\varphi$. Then the projection on the axis of the Ou of the forces acting on the element MM', will be equal to $T\sin(\varphi+\Delta\varphi)-T\sin\varphi$.. When the angle φ is small, we can put that $tg\varphi \approx \sin\varphi$, so we have:

$$T\sin(\varphi+\Delta\varphi)-T\sin\varphi \approx Ttg(\varphi+\Delta\varphi)-Ttg\varphi =$$
$$=T\left[\frac{\partial u(x+\Delta x,t)}{\partial x}-\frac{\partial u(x,t)}{\partial x}\right]=T\frac{\partial^2 u(x+\theta\Delta x,t)}{\partial x^2}\Delta x \approx$$
$$\approx T\frac{\partial^2 u(x,t)}{\partial x^2}\Delta x, \qquad 0\angle\theta\angle 1 \qquad (1)$$

* Corresponding author: jelicgordana@ymail.com

© The Authors, published by EDP Sciences. This is an open access article distributed under the terms of the Creative Commons Attribution License 4.0 (http://creativecommons.org/licenses/by/4.0/).

Where Lagrange's theorem is applied in the expression in square brackets.

In order to get the equation of motion, it is necessary to equalize the external forces acting on the element with the force of inertia. Let ρ be the specific weight of the wire. Then the mass of the wire element will be $\rho\Delta x$. The element acceleration equals $\partial^2 u/\partial t^2$. According to D'Alembert's principle, we obtain:

$$\rho\Delta x\frac{\partial^2 u}{\partial t^2}=T\frac{\partial^2 u}{\partial x^2}\Delta x. \tag{2}$$

If we shorten this equation with Δx and put $T/\rho=a^2$, we get the following differential equation with partial derivatives of the second order [1-6]:

$$\frac{\partial^2 u}{\partial t^2}=a^2\frac{\partial^2 u}{\partial x^2} \tag{3}$$

It is a wave equation - the equation of oscillation of a wire. The equation (3) itself is insufficient to fully determine the motion of the wire. The required function in u(x, t) should satisfy even boundary conditions that determine what happens at the ends of the wire x=0 and x=1, and the initial conditions describing the position of the wire at the initial moment t=0.

Let, for example, as we have assumed, the ends of the wire in x=0 and x=l be immobile. Then for any t the equalities are exact:

$$u(0,t)=0,\ \ u(l,t)=0. \tag{4}$$

These equalities are boundary conditions in the task.

In the initial moment t=0 the wire has a certain shape, which we have given it. Let this form be determined by the function f(x). Therefore, there must be:

$$u(x,0)=u/_{t=0}=f(x). \tag{5}$$

Further, the speed needs to be given in the starting point at each point of the wire, which is determined by the function $\varphi(x)$. Accordingly:

$$\frac{\partial u}{\partial t}/_{\ t=0}=\varphi(x). \tag{6}$$

The conditions (5) and (6) are the initial conditions.

As indicated above, equation (3) is also obtained in the task of electrical oscillations in the conductor. This can be shown in the following way. The electrical current in the conductor is characterized by the size $i(x,t)$ and the tension $v(x,t)$, which depend on the coordinate of the conductor point h and on the time t. Observing the conductor element Δx, we can write that the decrease in the tension on the element Δx equals $v(x,t)-v(x+\Delta x,t)\approx-\frac{\partial v}{\partial x}\Delta x$. This decrease in tension is equal to the sum of resistance $iR\Delta x$ and induction $\frac{\partial i}{\partial t}L\Delta x$. Accordingly:

$$-\frac{\partial v}{\partial x}\Delta x=iR\Delta x+\frac{\partial i}{\partial t}L\Delta x, \tag{7}$$

where R and L are coefficients of resistance and induction on the unit of conductor length. The minus sign should be taken because the current flows in the opposite direction from growth of v. By shortening it with Δx, we get the equality:

$$\frac{\partial v}{\partial x}+iR+L\frac{\partial i}{\partial t}=0. \tag{8}$$

Furthermore, the difference in currents passing through the element Δx in time Δt is:

$$i(x,t)-i(x+\Delta x,t)\approx-\frac{\partial i}{\partial x}\Delta x\Delta t. \tag{9}$$

It is consumed on charging the element, it equals $C\Delta x(\partial v/\partial t)\Delta t$, and on the outflowing through the side surface of the conductor due to the imperfection of the insulation, equaling $Av\Delta x\Delta t$ (here A - is the coefficient of outflow). By equating these expressions and shortening with $\Delta x\Delta t$, we get:

$$\frac{\partial i}{\partial x}+C\frac{\partial v}{\partial t}+Av=0. \tag{10}$$

The equality (8) and (10) are called *telegraphic equations*.

Equations (8) and (10) can be obtained from the equation system, which contains only the required function $i(x,t)$, and the equations, which contains only the required function $v(x,t)$. Differential equality (10) by x; equation (8) is differentiated by t, multiplied by S and substract the two values. We obtain:

$$\frac{\partial^2 i}{\partial x^2}+A\frac{\partial v}{\partial x}-CR\frac{\partial i}{\partial t}-CL\frac{\partial^2 i}{\partial t^2}=0. \tag{11}$$

From (8) we express ∂v/∂x and replace it in (11). We get:

$$\frac{\partial^2 i}{\partial x^2}=CL\frac{\partial^2 i}{\partial t^2}+(CR+AL)\frac{\partial i}{\partial t}+ARi. \tag{12}$$

Analogously, we obtain the equation that determines $v(x,t)$.

Differential equality (8) by x; equation (10) is differentiated by t, multiplied by L and substract the two values. We obtain:

$$\frac{\partial^2 v}{\partial x^2}+R\frac{\partial i}{\partial x}-CL\frac{\partial^2 v}{\partial t^2}-AL\frac{\partial v}{\partial t}=0 \tag{13}$$

From the equation (10) we express ∂i/∂x and replace it in (13). We get:

$$\frac{\partial^2 v}{\partial x^2} = CL\frac{\partial^2 v}{\partial t^2} + (CR + AL)\frac{\partial v}{\partial t} + ARv. \quad (14)$$

If we can neglect the loss through insulation (A=0) and resistance (R=0), then the equations (12) and (9) become wave equations:

$$a^2\frac{\partial^2 i}{\partial x^2} = \frac{\partial^2 i}{\partial t^2}, \quad a^2\frac{\partial^2 v}{\partial x^2} = \frac{\partial^2 v}{\partial t^2}, \quad (15)$$

where: $a^2=1/CL$

2 Solving the oscillation (flickering) equation of a wire

A solution to the following equation needs to be found:

$$\frac{\partial^2 u}{\partial t^2} = a^2\frac{\partial^2 u}{\partial x^2} \quad (16)$$

satisfying the following conditions:

$$u(0,t) = 0, \quad u(l,t) = 0, \quad (17)$$

$$u(x,0) = f(x), \quad \frac{\partial u}{\partial t}/_{t=0} = \varphi(x). \quad (18)$$

We will seek the solution of the equation (16), which satisfies the boundary conditions (17) and (18) in the form of products of the two functions H(h) and T(t), the first of which depends on x, and the second on t:

$$u(x,t) = X(x)T(t). \quad (19)$$

By substitution in the relation (10), we obtain: $X(x)T''(t) =$. We divide the equation with a^2XT and we obtain:

$$\frac{T''}{a^2T} = \frac{X''}{X}. \quad (20)$$

and thus we have separated the variables.

On the left side of the equation is a function, which does not depend on x, and on the right the function which does not depend on t. The equation (20) is only possible in this case when the left and right sides do not depend on either x or t, i.e. they are equal to the constant. Let's denote it with $-\lambda$, where λ >0. Thus

$$\frac{T''}{a^2T} = \frac{X''}{X} = -\lambda. \quad (21)$$

From this equation we obtain two linear differential equations with constant coefficients:

$$X''+\lambda X = 0 \quad T''+a^2\lambda T = 0 \quad (22)$$

The general solutions of these equations are:

$$X(x) = A\cos\sqrt{\lambda}x + B\sin\sqrt{\lambda}x, \quad (23)$$

$$T(t) = C\cos a\sqrt{\lambda}t + D\sin a\sqrt{\lambda}t, \quad (24)$$

where *A, B, C, D* are the production constants.

By putting the expressions X(x) and T(t) into equality (19), we obtain:

$$u(x,t) = (A\cos\sqrt{\lambda}x + B\sin\sqrt{\lambda}x)(C\cos a\sqrt{\lambda}t + D\sin a\sqrt{\lambda}t).$$

$\sin a(\sqrt{\lambda}\, t)$ (25)

Let's find the constants A and B so that they satisfy the conditions (17) and (18). Since T(t) $\neq 0$ (otherwise it would be $u(x,t) \equiv 0$, which would be contrary to the set condition), then the function X(x) satisfies conditions (17) and (18), i.e. X(0)=0, X (*l*)= 0. Replacing the values x=0 and x=*l* in equality (23), on the basis of (17) we obtain:

$$0 = A\cdot 1 + B\cdot 0, \ 0 = A\cos\sqrt{\lambda}l + B\sin\sqrt{\lambda}l. \quad (26)$$

The first equality gives A=0. From the second equality follows:

$$B\sin\sqrt{\lambda}l = 0. \quad (27)$$

$B \neq 0$, because otherwise it would be $X \equiv 0$ and $u \equiv 0$, which contradicts the condition. Therefore, it must be $\sin\sqrt{\lambda}l = 0$, where there follows:

$$\sqrt{\lambda} = \frac{n\pi}{l}, \ (n=1, 2, 3, ...) \quad (28)$$

(we do not take the value n=0, because in this case it would be $X \equiv 0$ and $u \equiv 0$). We obtain:

$$X = B\sin\frac{n\pi}{l}x. \quad (29)$$

The values found λ are called *their own values* for the given boundary task. The corresponding functions X(x) are called *the own functions*.

If instead $-\lambda$ we take $+\lambda = k^2$, the equation (17) would have the form

$$X''-k^2X = 0. \quad (30)$$

The general solution of this equation is:

$$X = Ae^{kx} + Be^{-kx}. \quad (31)$$

A solution different from zero in this form cannot satisfy the boundary conditions (11) and (12).

Knowing $\sqrt{\lambda}$, using equality (18), we can write:

$$T(t) = C\cos\frac{an\pi}{l}t + D\sin\frac{an\pi}{l}t \quad (n=1, 2, ...).$$

(32)

For each n, therefore also for each λ , the expressions (29) and (32) are replaced in the equation (19) and we obtain a solution of the equation (3), which satisfies the boundary conditions (17). We designate this solution with $u_m(x,t)$:

$$u_n(x,t) = \sin\frac{n\pi}{l}x(C_n\cos\frac{an\pi}{l}t + D_n\sin\frac{an\pi}{l}t). \quad (33)$$

For each value n we can take different constants C and D and therefore we write C_n and D_n (the constant B is included in C_n and D_n). Since the equation (16) is linear, then the sum of solutions is also a solution, and hence the function represented in the order:

$$u(x,t) = \sum_{n=1}^{\infty}(C_n\cos\frac{an\pi}{l}t + D_n\sin\frac{an\pi}{l}t)\sin\frac{n\pi}{l}x, \quad (34)$$

is also the solution of the differential equation (16), which meets the boundary conditions (17).

The solution (34) needs to satisfy the initial conditions (18). Putting in (34) t=0 and taking into account (18), we obtain:

$$f(x) = \sum_{n=1}^{\infty}C_n\sin\frac{n\pi}{l}x. \quad (35)$$

If the function $f(x)$ is such that it can be explained in the Fourier series in the interval (0, l), then the condition (35) will be satisfied if we put:

$$C_n = \frac{2}{l}\int_o^l f(x)\sin\frac{n\pi}{l}xdx. \quad (36)$$

Differential equality (34) by t and put that t=0.

$$\frac{\partial u}{\partial t} = \sum_{n=1}^{\infty}(-C_n\frac{an\pi}{l}\sin\frac{an\pi}{l}t + D_n\frac{an\pi}{l}\cos\frac{an\pi}{l}t)\sin\frac{n\pi}{l}x,$$
$$t = 0. \quad (37)$$

We obtain:

$$\frac{\partial u}{\partial t} = \sum_{n=1}^{\infty}D_n\frac{an\pi}{l}\sin\frac{n\pi}{l}x. \quad (38)$$

From condition (27) we obtain the equality>

$$\varphi(x) = \sum_{n=1}^{\infty}D_n\frac{an\pi}{l}\sin\frac{n\pi}{l}x. \quad (39)$$

In this line Fourier's coefficient is>

$$D_n = \frac{2}{an\pi}\int_0^l \varphi(x)\sin\frac{n\pi}{l}xdx. \quad (40)$$

Different method for solving differential equations are shown in the papers [7-12].

3 Conclusion

The task of electrical oscillations in the conductor is solved. The solution is in the form of Fourier order and satisfies the initial conditions.

References

1. P. Blanchard, R. L. Devaney, and G. R. Hall, Differential equations. Thompson, 2006.
2. G. Jelić, Differential forms. Annals of the Oradea University - Fascicle of Management and Technological Engineering, **13**, 3 (2014), pp. 170-173. doi: 10.15660/AUOFMTE.2014-3.3052.
3. E. Kamke, A handbook on ordinary differential equations. Moscow: Nauka / Science, 1971. (In Russian).
4. I.G. Petrovskiy, Differential equations. Moscow: Nauka / Science, 1987. (In Russian).
5. N.S. Piskunov, Differential and integral calculus. Moscow: Nauka / Science, 1970. (In Russian).
6. D. Zwillinger, Handbook of differential equations. 3rd Edition. Boston: Academic Press, 1997.
7. G. Jelić, Asymptotic dynamics and spectral analysis for Schrödinger operator with weakly accelerated potential. Procedia Technology, **19** (2015), pp. 802-809. doi: 10.1016/j.protcy.2015.02.115.
8. Z. Damnjanović, D. Mančić, P. Dašić, D. Lazarević, and R. Pantović, Thermoelastic stress analysis based on infrared thermography. Technics Technologies Education Management (TTEM), **7**, 2 (2012), pp. 914-919.
9. M. Lekić, S. Cvejić, and P. Dašić, Iteration method for solving differential equations of second order oscillations. Technics Technologies Education Management (TTEM), **7**, 4 (2012), pp. 1751-1759.
10. M. Lekić, S. Cvejić, and P. Dašić, Oscillating two-amplitudinal solutions of the canonic differential equation of the second order. Metalurgia International, **18**, 7 (2013), pp. 133-137.
11. C.M. Pappalardo, and D. Guida, On the computational methods for solving the differential-algebraic equations of motion of multibody systems. Machines, **6**, 2 (2018), Article no. 20. doi: 10.3390/machines6020020.
12. C.M. Pappalardo, and D. Guida, On the Lagrange multipliers of the intrinsic constraint equations of rigid multibody mechanical systems. Archive of Applied Mechanics, **88** 3 (2018), pp. 419-451. doi: 10.1007/s00419-017-1317-y.

Snow fence load on a house roof with tiles

Cornel Cătălin GAVRILĂ[1] and *Mihai -Tiberiu* LATEŞ[2]

[1] Transilvania University of Brasov, Romania
[2] Transilvania University of Brasov, Romania

Abstract. Virtual modeling of mechanical structures, using computer performant software, is largely used in many fields. The paper presents some aspects regarding the loads on a snow fence mountig system, placed on a house roof with ceramic tiles. First, there are presented some general aspects of the snow fence types and mounting systems used. Then, there are Then there are presented the aspects about virtual modeling of these parts, using CATIA software. Also, there are presented some aspects about the Finite Element Analysis, respectively the results of this. In the paper final part, there are presented the conclusion.

1 Introduction

The virtual modelling, the detail design and also the FEM analysis of mechanical products, using computer software is a necessity and it saves time in product developing, reduces the number of physical prototypes and experiments, reduces the prices of product and also, increases the quality of product. In the paper there are presented some aspects regarding the virtual modelling and FEM analysis for a snow fence mounting system fixed on a house roof.

Generally, to prevent the snow fall from buildings, there are available many constructive solutions. Usually, the producers providing the adequate fixing kit for snow fence, but the solutions are related more to the new types of tiles used than the previously manufactured. For old roofs types, the kits provided are not matching very well as dimensions or are not manufactured as well. This mismatching lead usually to the early roof damages, snow falling, or water drip inside the construction. The ceramic tiles produced and used in Romania are standardized according to SR EN 1304 + A1: 2001 [1] (the former Romanian STAS 515 – Ceramic Tiles and Ridges).

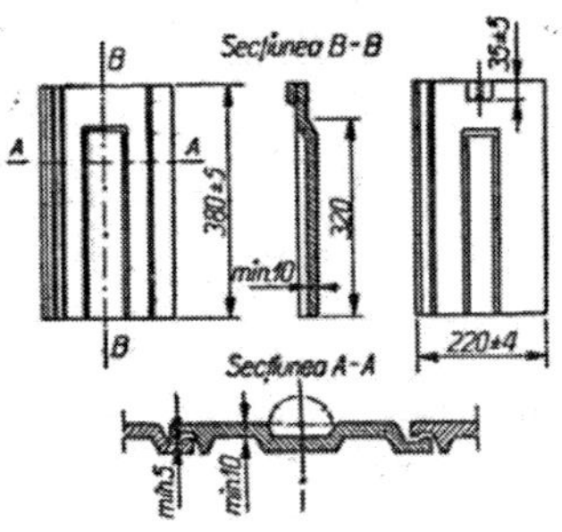

Fig. 1. The tile main dimensions.

According to the standard, for the big format tiles with drains and shafts with one lateral rib and one nose, the main dimensional characteristics are presented in Fig. 1. It is the most used tile type in Romania's last 60 years until today. This type of tiles can be arranged only in network (Fig. 2) on the roof structure made by thin wide woods placed on wood timbers. In this case, the wide woods distance is 33,5 mm [3].

Fig. 2. The tiles arrangement on roof

The most used constructive solution for snow fence mounting system consist in metal folded clips inserted between the tiles and fixed on roof wood structure, as is presented in Fig. 3, a, and b [2].

a.Snow fence folded clip

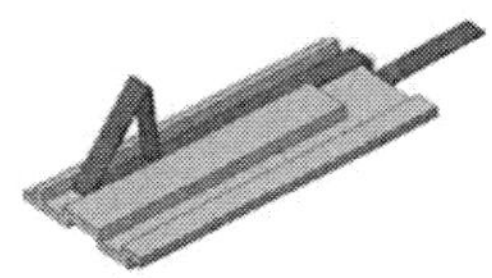

b. The clip placed on tile

Fig. 3. The snow fence clip and the tile

* Corresponding author: cgavrila@unitbv.ro

© The Authors, published by EDP Sciences. This is an open access article distributed under the terms of the Creative Commons Attribution License 4.0 (http://creativecommons.org/licenses/by/4.0/).

2 The Snow Load on Roof

According to "Technical regulation CR 1-1-3/2012 [3]", the snow load on roof depends by a lot of factors, as the thickness of the snow layer, roof shape and declination and, also, the snow distribution by wind and snow melting. When the snow layer slides on roof, the snow fence, considered as obstacle, is loaded with a force on length unit, given by relation

$$F_s = s\, b\, \sin\alpha_r\,, \qquad (1)$$

where

s – the characteristic snow load on roof, in kN/m^2

b – the horizontal distance between snow fence clips, in m

α_r – the roof declination, in degrees.

The characteristic snow load on roof can be calculated by relation

$$s = g_{IS}\, \mu_i\, s_k\,, \qquad (2)$$

where

g_{IS} – the exposing at snow load factor

μ_i – the shape factor of snow load on roof

s_k – the snow load on the ground, in kN/m^2.

The studied snow fence mounting system is considered placed on a building roof, with two sides, both declined at 30 degrees by horizontal reference. The building priority class is III. According to the mentioned technical regulations, the shape factor value is $\mu_i = 0.8$, for the considered case.

The snow load on the ground is s_k = 2 kN/m^2, a standard value given for Brasov location.

Because the snow fence clips are usually fixed on the roof wood timbers, the horizontal distance between them may vary between 0,8 and 1 m (fig. 4). An increased distance lead to premature bending for thin wide woods just under the tiles weight. Also, a shorter distance is not used.

a.Fixed on wood timber **b.** The horizontal distance

Fig. 4. The snow fence clip on the tiles

Taking account by relations (1) and (2), the resulted force along the roof declination, on a snow fence clip is 800 N, for considered horizontal distance value b =1 m. If the considered horizontal distance value is reduced to b =0,8 m, the force decrease to 640 N.

3 The Virtual Modelling

The virtual model is developed, using CATIA software, for the metallic clip and the ceramic tile, as the main parts involved in snow fence fixing operation on the roof [4, 5]. The wood roof structure, which serves to support the entire assembly, is considered as rigid.

For metallic snow fence clip, first step is to sketch the main section. The dimension are given by producer, but, to fit the clip in the roof structure, it has to be considered the difference angle between the declinations, previously introduced [6]. For the studied case, the difference angle between the declinations is α_d = 2 degrees. Then, using Part Design module, the virtual model is completed, as is presented in Fig. 3, a.

For the tile, first step is to sketch the main section, with drains and shafts and one lateral rib and one nose. Then, using Part Design module, the virtual model is completed, as is presented in Fig. 5.

a.The tile top view

b. The tile bottom view

Fig. 5. The tile virtual model

To assembly the components, the Assembly Design module is used. The assembled virtual model was already presented in Fig. 3, b, for the snow fence clip and tile, in Fig. 4, a, for clip fixed on the timber wood, and, respectively, in Fig. 4, b, for the entire roof ensemble.

Because of the tile shape, with drains and shafts, the metalic clip may be placed in few particular positions, as it is presented in Fig. 6, a - c.

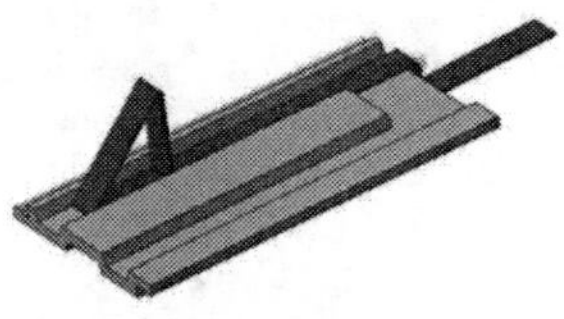

a.The left position

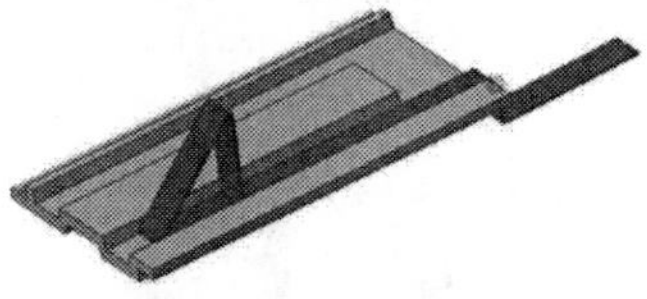

b. The right position

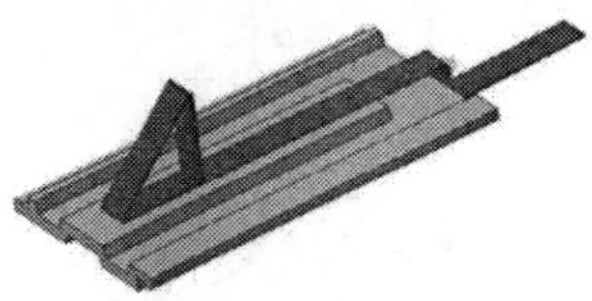

b. The middle position

Fig. 6. The snow fence clip particular positions

In the Fig. 6, a, the snow fence clip is placed in the tile left drain. In the Fig. 6, b, the snow fence clip is placed in the tile right drain. Also, in the Fig. 6, c, the clip is placed on the tile middle width sector.

4 The Finite Element Analysis

For analysis, the previous models for snow fence metalic clip and for the tile will be used. The material properties, as Young modulus, Poisson coefficient, Tensile Yield Strength, Ultimate Strength should be defined.

For snow fence metalic clip, the used material is steel [6, 7]. For the analysis, the loading force, F_s is applied along the roof declination corresponding to a tile declined at 30 degrees by horizontal reference. For simulation, the metalic clip is considered fixed on the surface asembled on timber wood and, respectively, on the bottom sector, where it loads the tile. To estimate the deformations, the metalic clip will be also considered with no tile above. The model needs to be meshed into finite elements, solids and deformable, type hexaedral and tetraedral, with integration node [7]. The metallic clip model, with defined force, is presented in Fig. 7.

Fig. 7. The snow fence metalic clip with defined force

For the tiles, the material is ceramic, with the main properties given by literature [8, 9]. The ceramic materials are fragile – under the load, are subjected to break before some elastic deformations. Also, the ceramic materials present the static fatigue – for an increased time of load application, the break resistance is decreasing. For analyse, the ceramic material for tiles is considered having fine texture: homogenous granulation, without material failures, as breaks or inclusions. For ceramic material, the limit stress is 15 MPa.

The tile model also needs to be meshed into finite elements, solids and deformable, type hexaedral and tetraedral, with integration node. The tile is declined at 28 degrees by horizontal reference, and the resultant force is applied on a contact surface with the bottom sector of the snow fence clip. The tile is considered fixed above the nose, on the line contact and on the bottom edges, which correspond to the placing on the tiles above. The tile model with defined force is presented in Fig. 8.

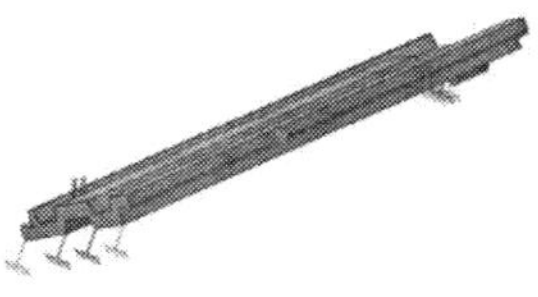

Fig. 8. The tile model with defined force

5 The Simulations and Results

The snow fence metalic clip simulation was made for both cases of load:

- F_s = 640 N, for horizontal distance b =0,8 m,
- F_s = 800 N, for horizontal distance b =1 m.

In Fig. 9 there are presented the snow fence metalic clip deformations and it can be observed that, the maximum deformation value appear on the top folded sector.

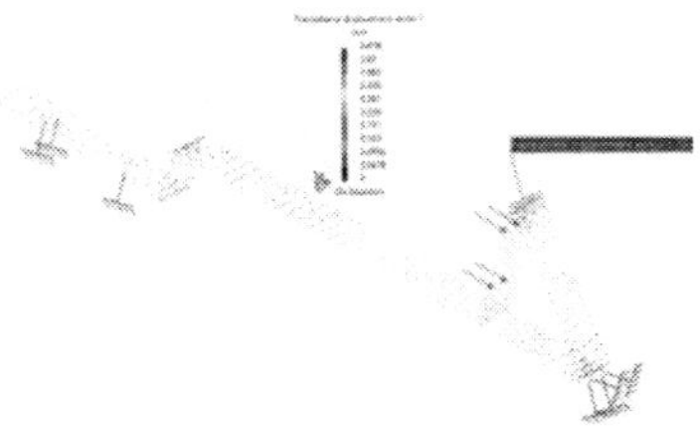

Fig. 9. The deformations on the snow fence metalic clip

In Fig. 10 is presented the stress on the metalic clip; the maximum value is 123 MPa, and appears on the bottom end, where the clip is folded.

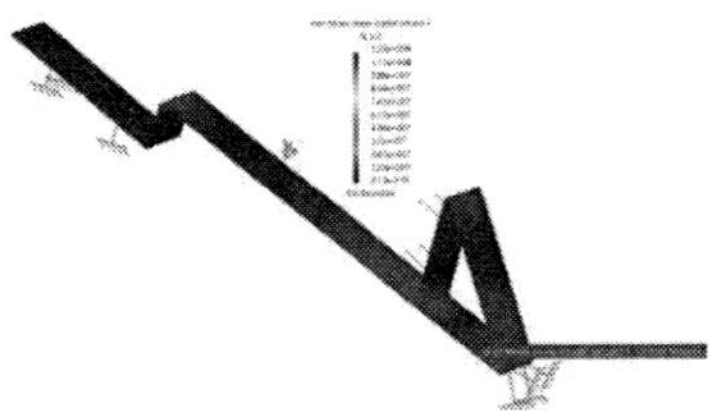

Fig. 10. The stress on the snow fence metalic clip

The snow fence metallic clip simulation results are presented in Table 1.

Table 1. The simulation results for snow fence metallic clip.

Horizontal distance, m	Load force, N	Maximum deformation, mm	Maximum stress, MPa
0,8	640	0,382	98,8
1	800	0,478	123

For the tile simulation, there are considered the each snow fence metallic clip positions on tile, presented in Fig. 6, a - c. The load on tile is applied distributed on the contact area with bottom end of the snow fence metalic clip. Because this contact area may vary, depending by the tile and clip folding radius and also by positioning on roof structure, there is considered the reduced contact area to 33x5 mm, which correspond to increased misalignments and increased snow fence clip deformations.

The tile simulation was also made for both cases of load force, previously presented.

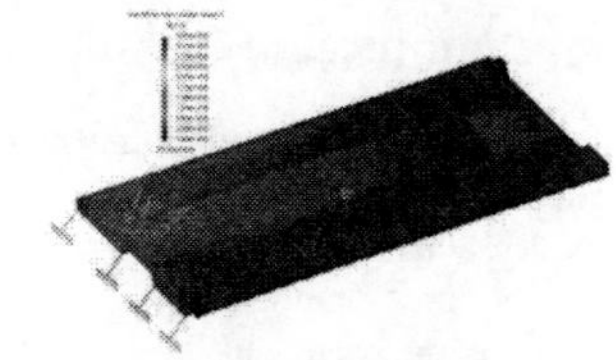

a. Stress values for the load applied on the left position

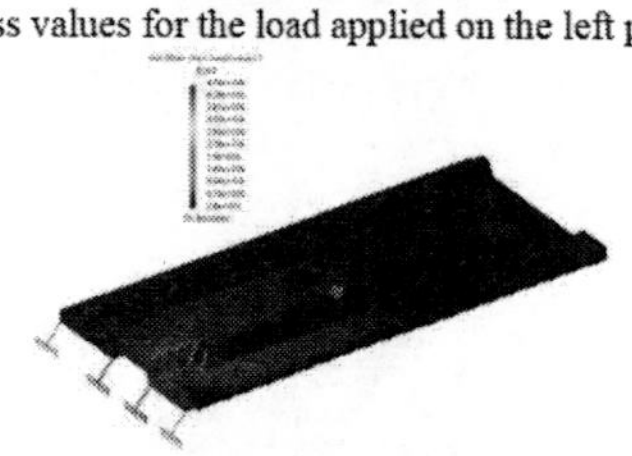

b. Stress values for the load applied on the right position

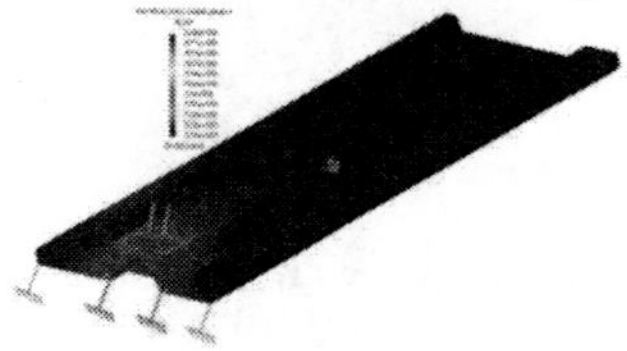

c. Stress values for the load applied on the middle position

Fig. 11. Stress values on tile

In Fig. 11, the obtained stress maximum values on tile is presented. The tile simulation results are presented in Table 2.

Table 2. The simulation results for tile.

The load position	Horizontal distance, m	Load force, N	Maximum stress, MPa
Left	0,8	640	2,76
	1	800	3,45
Right	0,8	640	3,8
	1	800	4,76
Center	0,8	640	4,19
	1	800	5,24

Conclusions

On real snow fence device assembly, there are situations when, because of surfaces misalignments, the contact surface between the studied parts may be different from the studied case.

For metallic clip, the deformations and stresses values are lower than the limit values recommended in literature for steel. But, the clip bending is possible, when the load accidentally increases.

For tiles, in all the studied cases, the maximum stress values are lower than the limit values recommended in literature for ceramic materials, so the tile resist to load. Comparative, from all three cases of tile load, the lowest values were obtained for the snow fence position as in Fig. 6, a; respectively the maximum values were obtained for the center position, as in Fig. 6, c.

Taking account that the ceramic materials are fragile and present the static fatigue – for an increased time of load application, the break resistance is decreasing in subjected locations.

To prevent the roof failures because of the snow load (the tile breaking or the metalic clip bending), the snow fence clips must be placed on the roof at prescribed distance and arrangement.

For other locations, the calculus and simulation is similar.

References

1. Romanian Standard, SR EN 1304 + A1 : 2001, *Ceramic Tiles and Ridges,* 2001.
2. https://www.accesoriipentruacoperis.ro/parazapezi/
3. Technical regulation CR 1-1-3/2012, *Design Code: Snow Action On Construction Evaluation (Cod de proiectare. Evaluarea actiunii zapezii asupra constructiilor),* 2012 (in romanian).
4. C. Alexandru, *Mechanical Articulated Systems Simulation on Computer (Simularea pe calculator a sistemelor mecanice articulate),* Ed. Lux Libris, Braşov, 2000 (in romanian).
5. I.G. Ghionea, *CATIA v5. Application in Mechanical Engineering (CATIA v5. Aplicaţii în ingineria mecanică),* Editura Bren, Bucureşti, 2009 (in romanian).
6. C. C. Gavrilă, R. Velicu, M. T. Lateş, *3D Modeling and FEM Analysis for Solar Panel Mounting System on a House Roof,* In: Annals of the Oradea University, Fascicle of Management and Technological Engineering, Volume XVI, (XXVI), Oradea 2017, pp. 1 – 4.
7. M. T. Lateş, R. Velicu, R. Papuc, *Multiscale modelling of chain-guide contact by using tests and FEM.* 11th World Congress on Computational Mechanics WCCM XI, Barcelona, Spain, 20-25 July, 2014, pp.1062-1069.
8. C. Bob, P. Velica, *Construction Materials (Materiale de construcţii),* Ed. Didactică şi Pedagogică, Bucureşti, 1978 (in romanian).
9. I. Teoreanu, N. Ciocea, A. Bărbulescu, N. Ciontea, *Ceramic and Refractory Material Technology (Tehnologia produselor ceramice şi refractare),* Vol. I, Ed. Tehnică, Bucureşti (in romanian).

MATEC Web of Conferences 184, 02002 (2018) https://doi.org/10.1051/matecconf/201818402002
Annual Session of Scientific Papers IMT ORADEA 2018

On the benchmarking of friction in timing chains

Lenard Jurj[1] and *Radu* Velicu[1,*]

[1]Transilvania University of Brasov, Product Design Mechatronics and Environment Department, Eroilor 29, 500069 Brasov, Romania

Abstract. This paper is dealing with the evaluation of friction in timing chains. Benchmarking friction is based on the experimentally determined friction losses in basic chain drives with transmission ratio i=1. This friction is not including friction losses in bearings. These results should be used, in the simplest manner, in order to evaluate friction in timing chains (i = 2). We are presenting the theoretical approach and experimental validation that have been developed for the timing chain drives, with transmission ratio i = 2. Friction losses are presented as resistant friction torques at the input shaft, depending on input rotational speed and tensioning force in the chain drive.

1 Introduction

This scientific research is developed in the frame of the 20/20/20 goals stipulated by the Directive for Energy Eficiency of the European Union (2012/27/UE). This imposes the goals of automotive industry for year 2020. Targets in the equivalent CO2 emissions and fuel economy have been set. American standards CAFE (Corporate Average Fuel Economy) are imposing a one third reduction of fuel consumption, compared to year 2008. European Commission set a goal for 2021 of average CO2 emissions of 95 grams CO2 / km for the whole fleet produced in 2020. This means an average fuel consumption of 4.1 l petrol/100 km or 3.6 l Diesel/100 km.

Benchmarking friction in timing chains is only a little step in the direction of reducing the losses in the timing system of the combustion engine. According to [1-4] 30–35% of the fuel energy consumed in automobiles are needed to overcome friction in the combustion engine systems.

Very few results on chain drive friction have been found in literature. Reference [5] presents a rig, and experimental results on the contribution of guides and chain separately (bearing friction included in chain friction) on friction losses in the timing chain drive of a combustion engine. The results are not showing any influence of speed, oil temperature or tensioning on the friction. The rig is consisted of a full engine equipped in order to measure separate friction in camshaft, chain and the two guides.

Benchmarking friction in timing chain is defined by: the equipment, the procedure of testing friction in a basic chain drive and the methodology for processing the experimental results. The output should be the data for evaluation of the friction in the timing chain depending on several variables like: rotational speed, tensioning, torque at the camshaft, parameters of the lubrication.

The research developed at Transilvania University of Brasov is using the rig presented in [6] with an image and the diagram presented in Figure 1.

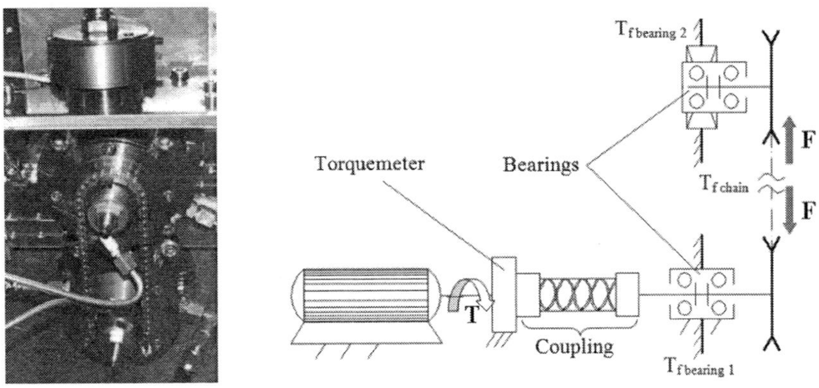

Fig. 1. Image and functional diagram of the chain friction rig

* Corresponding author: rvelicu@unitbv.ro

© The Authors, published by EDP Sciences. This is an open access article distributed under the terms of the Creative Commons Attribution License 4.0 (http://creativecommons.org/licenses/by/4.0/).

In the format presented in Figure 1, the rig measures the input torque T (global friction torque), for a basic chain drive with $i = 1$, identical chain sprockets with number of teeth $z_1 = z_2 = 23$, chain with 64 links, different rotation speed n in the range 500…5000 rot/min, different tensioning of the chain drive F and different temperature of the 5W30 oil for chain lubrication.

$$T = T_{chain+bearings} = T_{f\,chain} + T_{f\,bearing1} + T_{f\,bearing2} \quad (1)$$

The rig can be adapted in the format presented in [7], in order to measure separately the friction in the two bearings, results presented in [8, 9]

$$T = T_{bearings} = T_{f\,bearing1} + T_{f\,bearing2}. \quad (2)$$

Friction in chain results from extracting the measured friction torque in bearings from the measured global friction torque

$$T_{f\,chain} = T_{chain+bearings} - T_{bearings} \quad (3)$$

More details on the testing procedure are presented in [7].

2 Theoretical approach on chain friction

In the theoretical approaches of chain friction, the literature is generally considering only the friction in the cylindrical joints between links. The other possible friction between links and sprockets is considered negligible. The approaches presented in [10, 11] are calculating mechanical efficiency or power loss by friction, but not the resulted friction torque.

For the following theoretical approach, we consider a a chain transmission with two sprockets, tensioning with variable centre distance, transmission ratio i. Figure 2 presents the calculus diagram with a detail of a cylindrical joint with friction. A silent chain is presented but the results are valid also for bush and roller chains.

Friction losses appear in points 1' and 1" (entering and exiting the driving sprocket 1) and also in points 2' and 2" (entering and exiting the driven sprocket 2).

The friction forces in joints 1', 1", 2' and 2", F_{fy}, result as:

$$F_{fy} = \mu \cdot F_y \ , \quad (4)$$

where: μ is the friction coeficient between bolt and link, F_y is the force along the chain (F_1 – for points 1" and 2", F_2 – for points 1' and 2',

In all these points there is a relative rotation of the link on the bolt. If $\theta = 360°/z_1$ is the angular pitch of the driving sprocket 1, for a rotation with angle θ of the driving sprocket 1, in each point 1' and 1" we will have a relative rotation with angle θ of the link on the bolt. In each point 2' and 2" we will have a relative rotation with angle θ/i ($360°/z_2$ – angular pitch of sprocket 2) of the link on the bolt.

For a rotation with angle θ of the driving sprocket 1, the mechanical work L_{fy} of the friction forces, in joints 1', 1", 2' and 2" results as:

$$L_{f\,1'} = F_{f1'} \cdot \frac{d_b}{2} \cdot \theta = \mu \cdot F_2 \frac{d_b}{2} \cdot \theta \ , \text{ for joint1'} \quad (5)$$

$$L_{f\,1''} = F_{f1''} \cdot \frac{d_b}{2} \cdot \theta = \mu \cdot F_1 \frac{d_b}{2} \cdot \theta \ , \text{ for joint1''} \quad (6)$$

$$L_{f\,2'} = F_{f2'} \cdot \frac{d_b}{2} \cdot \frac{\theta}{i} = \mu \cdot F_2 \frac{d_b}{2} \cdot \frac{\theta}{i} \ , \text{ for joint 2'} \quad (7)$$

$$L_{f\,2''} = F_{f2''} \cdot \frac{d_b}{2} \cdot \frac{\theta}{i} = \mu \cdot F_1 \frac{d_b}{2} \cdot \frac{\theta}{i} \ , \text{ for joint 2''} \quad (8)$$

where d_b is the diameter of the bolt.

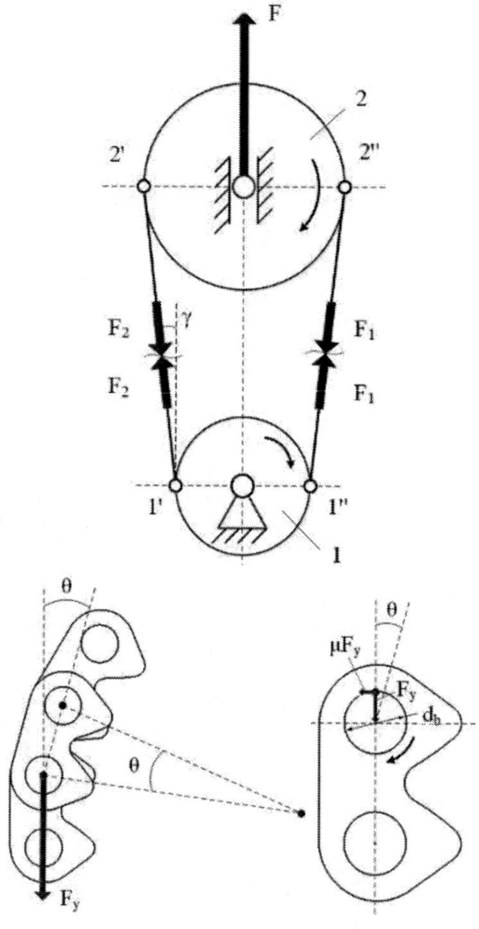

Fig. 2. Diagram of friction losses in chain drives

The friction torque reduced at the driving sprocket 1 results by equalizing the sum of mechanical works of friction forces with the mechanical work of the friction torque at the driving sprocket $T_f\,\theta$:

$$T_f = \frac{\sum L_{fy}}{\theta} = \frac{\mu(F_1+F_2)\frac{d_b}{2}\theta\left(1+\frac{1}{i}\right)}{\theta} =$$

$$= \mu(F_1 + F_2)\frac{d_b}{2}\left(1 + \frac{1}{i}\right) \quad (9)$$

Depending on the tensioning force, which can be aproximated as $F = (F_1+F_2)\cos\gamma$, the friction torque is:

$$T_f = \mu F \frac{d_b}{2}\left(1+\frac{1}{i}\right)\cos\gamma. \quad (9)$$

For $i = 1$, $\gamma = 0$

$$T_{f\,(i=1)} = \mu F d_b. \quad (10)$$

For $i = 2$,

$$T_{f\,(i=2)} = 0.75\mu F d_b\cos\gamma = 0.75T_{f\,(i=1)}\cos\gamma. \quad (11)$$

The power loss by friction is

$$P_f = T_f\omega = \mu F\omega\frac{d_b}{2}\left(1+\frac{1}{i}\right)\cos\gamma. \quad (12)$$

where ω is the angular speed of the driving sprocket.

As it can be noticed from equations (9) and (12), from the theoretical point of view, the friction torque and the power loss doesn't depend on number of teeth or chain pitch, but they are directly depending on friction coefficient, bolt diameter and initial tensioning.

3 Experimental results. Validation

Figure 3 presents the diagrams determining chain friction $T_{f\ chain}$ by extracting the measured friction torque in bearings $T_{bearings}$ from the measured global friction torque $T_{chain+bearings}$ (see eq. (3)). The values of all the torques are presented as percentage of the value of T_{fchain} resulted for n = 500 rot/min and F = 1 kN.

$$T_{\%} = {T}\Big/{T_{fchain\,(n=500\,rot/min,\ F=1kN)}}\ 100. \quad (13)$$

For a good comparison, the diagrams from Figure 3 have the same scale. The results presented in Figure 3 are from the testing of a basic silent chain drive (i=1) with a pitch of 6.35 mm [12], at a temperature of chain oil of 90 °C and a temperature of bearings oil of 50 °C, for two values of tensioning. The results show that:

- Friction torque in chain is depending mostly on the tensioning, while speed influence is irrelevant; This is partially validating the theory (see eq. (10));
- Friction in bearings has a very small influence from tensioning but an important influence from rotational speed.

For testing friction in the chain drive with i = 2 (like the case of timing chains) we used the same chains tested at i = 1 (23-23 teeth sprockets, 64 links), replacing one of the sprockets with a 46 teeth sprocket (see Figure 4).

Due to the fact that the upper bearings are working at half the speed of the diving shaft (lower bearings), we needed to measure the bearings friction also for input rotational speeds at half the values from initial testing, in the range 250...2500 rot/min. Because we only measure the sum of bearing friction of the two bearings at the same rotational speed $T_{bearings}$ (n), we calculated the bearings friction from situation with i = 2 as

$$T_{bearings\,(i=2)}(n) = T_{fbearing1}(n) + \frac{1}{i}T_{fbearing2}(n/2) =$$

$$\frac{1}{2}T_{bearings\,(i=1)}(n) + \frac{1}{4}T_{fbearings(i=1)}(n/2). \quad (14)$$

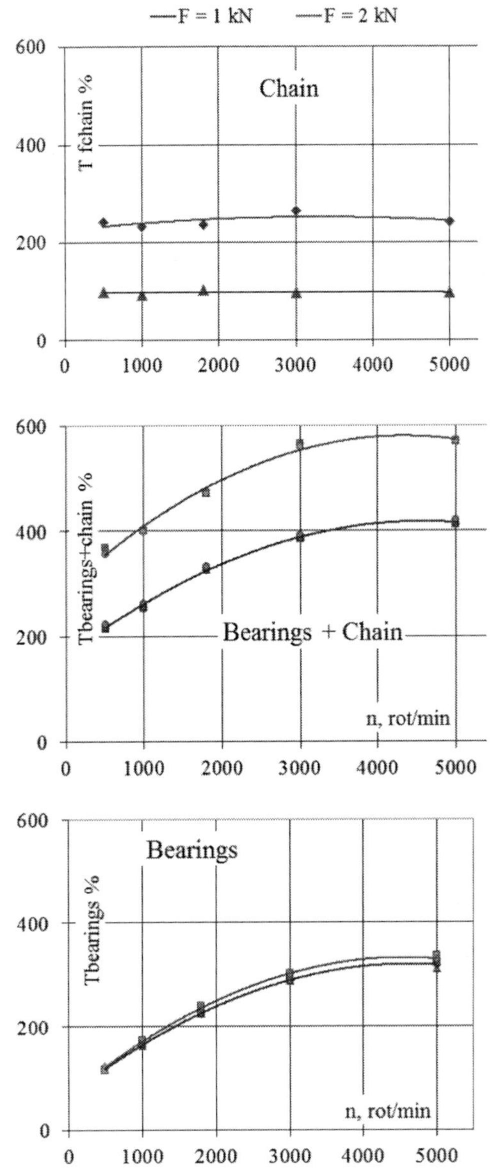

Fig. 3. Chain friction determined by subtracting bearings from global friction

Fig. 4. Testing chain drive with i = 2

https://doi.org/10.1051/matecconf/201818402002

Equation (14) is based on the following assumptions:

- Friction in the two bearings are equal;
- $T_{bearings\ (i=2)}$ is calculated as torque at the input (driving) shaft so the friction torque at upper bearing $T_{bearing2}$ must be reduced at the input shaft by multiplying with $1/i$.

Chain friction for i = 2 situation is calculated by subtracting the bearings friction $T_{bearings\ (i=2)}$ (eq. (14)) from the measured global friction $T_{chain+bearings\ (i=2)}$ Figure 5 presents the resulted friction torques measured for i = 2, for two type of chains [12] (a – bush chain with 7 mm pitch, 7B and b – silent chain with 6.35 mm pitch, 6S). Measurements have been made at a temperature of chain oil of 90 °C and a temperature of bearings oil of 50 °C, for two values of tensioning.

There are also presented, for comparison, the chain friction resulted for i = 1 situation and the theoretical calculus of $T_{f(i=2)}$ depending on measured $T_{f(i=1)}$ (see eq. (11)). For our calculus cos $\gamma = 0.9781$. For each diagram, the values of all the torques are presented as percentage of the value of $T_{f\,chain\ i=2}$ measured for n = 500 rot/min and F = 1 kN.

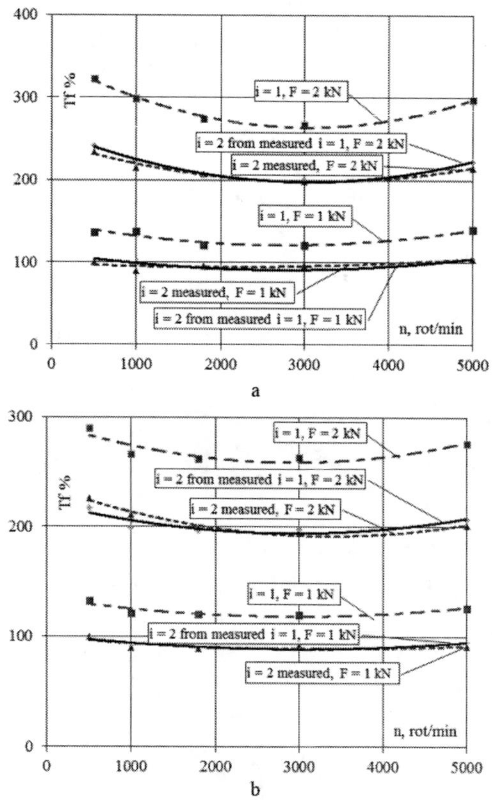

Fig. 5. Chain friction for $i = 2$

The comparison between the measured i = 2 situation and the resulted from i = 1 measurements shows very close values for all the range of rotational speed and tensioning. This is validating the calculated chain friction $T_{f(i=2)}$ depending on measured $T_{f(i=1)}$ (see eq. (11)).

4 Conclusion

A very simple equation is linking the friction measured on a basic transmission with the case of the timing chain (i = 2). The next research for benchmarking friction in chain drives should focus on the following directions:

- guide friction contribution to the global friction of the chain drive;
- optimizing the initial tensioning, this being the main parameter influencing chain friction.

References

1. S.C. Tung, M.L. McMillan, Tribol. Int. **37 (7)**, 517 (2004)
2. K. Holmberg, P. Andersson, A. Erdemir, Tribol. Int. **47 (14)**, 221 (2012)
3. M. Schwaderlapp, F. Koch, J. Dohmen, Seoul 2000 FISITA World Automotive Congress, 1 (2000)
4. C.M. Taylor, Wear **221**, 1 (1998)
5. T. Hyakutake, M. Inagaki, M. Matsuda, N. Hakamada, Y. Teramachi, JSAE Review **22**, 5 (2001)
6. A. Todi-Eftimie, R. Velicu, R. Saulescu, C. Jaliu, Adv. Mat. Res. **753-755**, 1110 (2013)
7. R. Velicu, M. Lates, Annals of the Oradea University, Fascicle of Management and Technological Engineering **XIV (XXIV)**, 59 (2015)
8. R. Velicu, S. Popa, Annals of the Oradea University, Fascicle of Management and Technological Engineering **XV (XXV)**, 5 (2016)
9. M. Lates, R. Velicu, R. Papuc, (2014) International J. Surf. Sc. and Eng., **8 (4)**, 345 (2014)
10. B. Horovitz, *Transmissions and variable transmissions with chains and belts (in Romanian)* (Ed. Tehnica, Bucuresti, 1971)
11. G. Niemann, *Machinenelemente. Band. II.* (Springer Verlag, Berlin, 1975)
12. Schaeffler Group Automotive *Schaeffler Chain Drive Systems Catalogue* (2009)

Geometry of silent chain - involute sprocket

Lenard Jurj[1], *Radu* Velicu[1,*] and *Radu* Săulescu[1]

[1]Transilvania University of Brasov, Product Design Mechatronics and Environment Department, Eroilor 29, 500069 Brasov, Romania

Abstract. Silent chains tend to have more and more usage in automotive industry. The study of these types of chains has a lot to consider due to the fact that silent chains are not standardized yet. The aim of this paper is to study and analyse the theoretical contact point between a silent chain's toothed plate and an involute sprocket. Varying different parameters of the chain and sprocket, there will be studied its effects on the theoretical contact point and the sprocket's profile geometry.

1 Introduction

Improving tribological performances of automotive power units can have significant benefits related to power increase, friction reduction, oil and fuel consumption reduction and not in the last place could increase also the reliability of engine components [1].

There are different approaches trying to reduce friction between automotive engine components, one of which is presented by [2], within there are considered different materials, surface treatments and surface coatings of engine timing chain transmissions. Chain transmissions due to their properties of high power and torque transmission, are used more often as component of internal combustion engine's timing system [3].

Within the group of chain transmissions used as timing chains, the silent chains besides the usual advantages of chain drives, has a reduced functioning noise, which tends to have a place between the necessary requirements of chain drives; a study that includes a few types of chain transmissions is presented by [4].

Improvements were done for the clearances compensation system of a timing chain with modelling, testing and geometric contact point optimization between the chain and the chain tensioning system for a bush chain in [5]. Geometric modelling of the contact point between the elements of a bush chain was established in [6].

A spring model for sprocket-chain load distribution was studied and then compared with traditional force-balance analysis in the work of [7]. Important facts were discussed about the pitch difference and its effect on the contact point that appears between the bushes of a bush chain with the chain sprocket [8]. These works contributed to a better understanding of the forces distribution and contact point position variation in case of bush and roller chains. Equations that describe the relation between the sprocket-silent chain link contacts were presented depending on the profile angle of the link's outer flange [9]. Influence of the centrifugal force and the number of teeth on the forces distribution that can be found in a silent chain transmission was analysed in [10] with the conclusion that the load is distributed not just on the first links but on several links. Another view on the force distribution in the case of a silent chain transmission was presented with the consideration of the friction force that appears between the silent chain's elements and the sprocket in [11]. The sprocket constructions with straight tooth profile are replaced with curved flanks constructions. In this paper, the contact point position will be analysed for a silent chain transmission with involute sprocket.

2 Theoretical approach

The dynamic behaviour of the silent chain transmission is seemingly complex, regarding the corresponding contacts between the silent chain elements with the sprocket's teeth. Importance was given in this paper to the study of the contact point between the silent chain's toothed plate that transmits power to the corresponding involute surface of a teeth.

Figure 1 presents the theoretical contact point geometry for the silent chain- involute profile sprocket.

The aim of this geometric modelling is to determine the contact point between the silent chain's toothed plate and the sprocket's teeth of the chain transmission, and to define some of the basic parameters of a sprocket, using the known dimensions of the silent chain. For this there are taken into account the following input geometric parameters of the chain transmission: the pitch of the chain (p), the number of teeth of the sprocket (z), the pitch pressure angle (α_0), the angle of the toothed plate's outer flange line to the center line (α_x), the distance between the pin hole centre and the plate's flank (d_x) and the distance between the theoretical contact point and a line traced perpendicular to the toothed plate's outer flange through the centre of the pin hole (d_z). The output should be the module of the sprocket m and its profile displacement x.

* Corresponding author: rvelicu@unitbv.ro

© The Authors, published by EDP Sciences. This is an open access article distributed under the terms of the Creative Commons Attribution License 4.0 (http://creativecommons.org/licenses/by/4.0/).

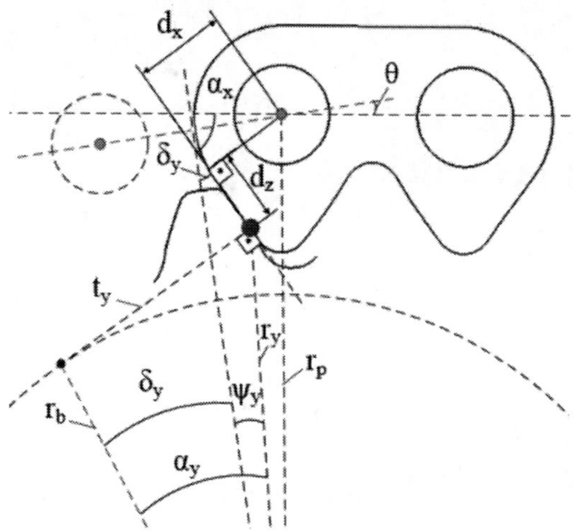

Fig. 1. Geometry of silent chain-sprocket contact.

The theoretical approach considers the definition of some basic parameters of the chain transmission:

- the angular pitch:

$$\Theta = \frac{360}{z} ; \tag{1}$$

- the primitive module of the sprocket:

$$m_p = \frac{r_p}{z} = \frac{p}{z \cdot \sin\left(\frac{\Theta}{2}\right)} . \tag{2}$$

The standardized module [12] of the sprocket m should be smaller than the primitive module, calculated with (2).

Pitch radius of the sprocket is calculated with:

$$r = \frac{m \cdot z}{2} , \tag{3}$$

the base circle radius is calculated with:

$$r_b = r \cdot \cos \alpha_0 , \tag{4}$$

the pitch radius of the silent chain transmission can be described as:

$$r_p = \frac{p}{\sin\left(\frac{\Theta}{2}\right)} . \tag{5}$$

The distance t_y results as

$$t_y = \sqrt{r_p^2 - (r_b + d_z)^2} - d_x . \tag{6}$$

The angle α_y can be calculated as

$$\alpha_y = \text{atan}\frac{t_y}{r_b} . \tag{7}$$

Using (5), (6) and (7) we obtain the radius of the circle where the contact point can be found:

$$r_y = \frac{r_b}{\cos \cdot \alpha_b} \tag{8}$$

The angle between the toothed plate outer flange and the line passing through the middle of the sprocket's tooth has the following formula based on Figure 1:

$$\delta_y = 90 - \alpha_x - \Theta \cdot \tag{9}$$

The angle between the line linking the theoretical contact point with the centre of the base circle and the line passing through the middle of the sprocket teeth:

$$\psi_y = \alpha_y - \delta_y \tag{10}$$

According to [12] we can write:

$$\psi_y = \frac{S}{d} + \text{inv}\alpha_0 + \text{inv}\alpha_y , \tag{11}$$

$$S = m\left(\frac{\pi}{2} + 2 \cdot x \cdot \tan \alpha_0\right) , \tag{12}$$

where: S is the sprocket tooth thickness at the pitch circle, and x is the tooth's profile displacement.

Using (11) and (12) we obtain the formula for the profile displacement:

$$x = \frac{z(\psi_y - \text{inv}\alpha_0 + \text{inv}\alpha_y) - \frac{\pi}{2}}{2 \cdot \tan \alpha_0} . \tag{13}$$

The analysis should be also developed with the goal of finding the position of the contact point on the flank of the chain plate, described by dimension d_z. In this case, the profile displacement x is also considered an input. By using (10), (11) and (12), the angle α_y is determined.

With t_y determined from (7), the position of the contact point defined by dimension d_z can be described by the generalized Pythagorean Theorem and has the following form:

$$d_z = \sqrt{r_p^2 - (t_y + d_x)^2} - r_b . \tag{14}$$

3 Numerical analysis – Results

The geometric model for the definition of the silent chain – involute sprocket contact has been applied for different values of the known input parameters. These parameters are presented in Table 1.

The values of the calculated parameters using the input parameters and the equations presented in the first subsection of this paper are presented in Table 2. The theoretical profile displacement x for given values of d_z and for different values of the sprocket's teeth z is show in Figure 2, while depending of the plate flank angle α_x is shown on Figure 3.

Table 1. Input parameters.

Parameters	Unit	Value
p	mm	8
d_x	mm	3.7
d_z	mm	2
α_x	deg	55
α_0	deg	20
z	-	38

The following conclusion can be drawn from the results presented in the diagrams from Figures 2 and 3:

- The needed profile displacement x is increasing with the number of teeth and with decrease of the distance dx on the chain plate; There could be issues of undercutting for small number of teeth and larger dimensions of the chain plate, dx.
- The needed profile displacement x is increasing with the plate flank angle α_x; As seen in Figure 3, the values of a valid plate flank angle are in a relatively narrow range, this is why this angle is very important to be set properly; If this angle is not in the range, involute sprockets may not be possible to use.

Figure 4 represents the evolution of the ratio between the contact point displacement dz and the chain pitch depending on the number of teeth, for profile displacement $x = 0$.

Figure 5 presents the ratio between the contact point displacement d_z and the chain pitch depending on the profile displacement $x = 0$.

The distance d_z is increasing with the number of teeth but it decreases with the distance dx and profile displacement x. It can prove to be very important if we impose a perfect laying of the chain plate on the teeth of the sprocket.

Table 2. Results.

Parameters	Unit	Value
Θ	deg	9.47368
m_p	mm	2.54938
m	mm	2.25
r	mm	42.75
r_b	mm	40.17185
t_y	mm	20.12854
α_y	deg	26.61365
r_y	mm	44.93257
δ_y	deg	25.52631
ψ_y	deg	1.08733
S_y	mm	1.70542
x	-	-0.03652

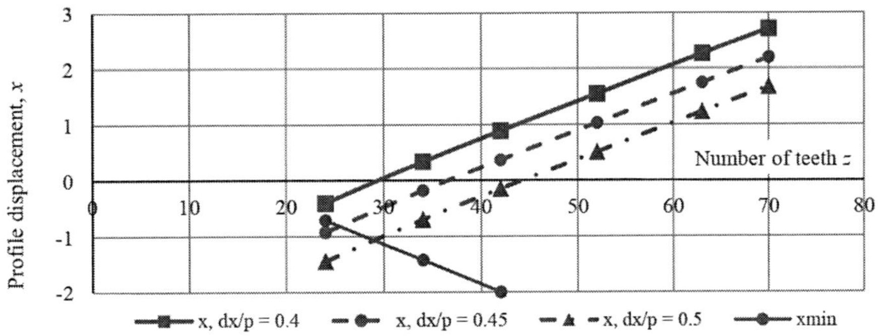

Fig. 2. Profile displacement depending on the number of teeth of sprocket.

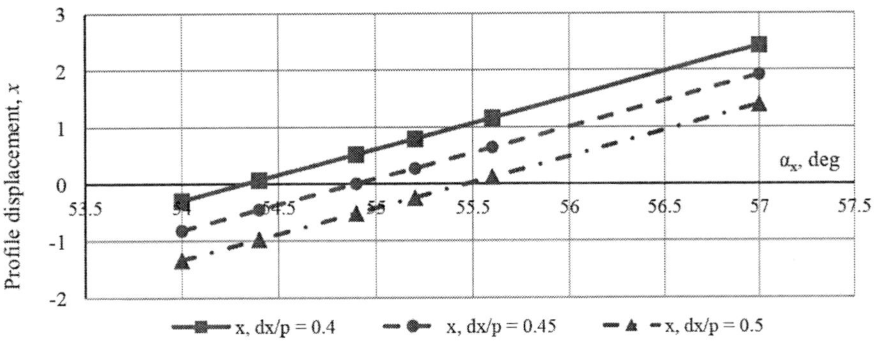

Fig. 3. Profile displacement depending on the angle of the flank of chain plate.

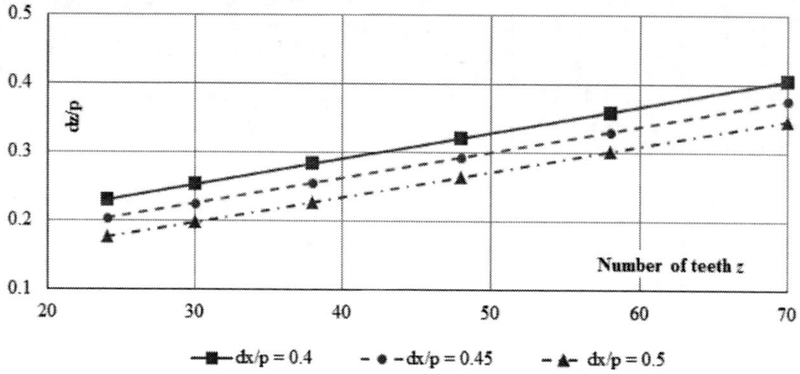

Fig. 4. Contact point displacement depending on number of teeth of the sprocket for $x = 0$.

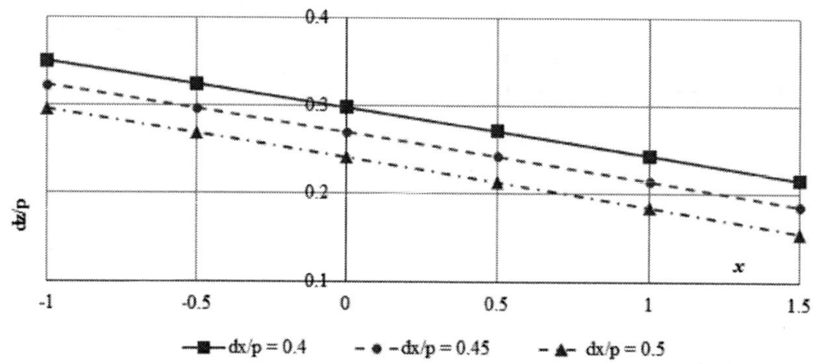

Fig. 5. Contact point displacement evolution over a given interval of the profile displacement of the sprocket.

3 Conclusion

The theoretical analysis starts with the presumption that during the functioning of the silent chain transmission the chain plate lays on the tooth without any sliding. This is not possible in the case of sprockets with straight tooth flank. This paper considers an involute sprocket. For this case, the perfect match between a chain and a sprockets leads to imposing a precise profile displacement. Even so, any lack of precision will create sliding between the tip of the plate and the toorth of the sprocket. Further proof of this can be found after a visual checking of the tooted links of the chain. They show on their outer flange marks of wear after use.
Parameters like tooth flange angle, profile displacement of the sprocket have direct influence on the contact point's position and it's sliding over the involute flank of the sprocket's teeth.

References

1. S.C. Tung, M.L. McMillan, *Tribo Int* **37** (7), 517 (2004).
2. T. Fink, B. Holger, MTZ, **72** (2011)
3. S. Belmer, T. Fink, MTZ (2006).
4. J.H. Pfalzgraf, *Experimentelle Reibleistungs-untersuchungen an Kettentrieben im Verbrennungsmotor*, *PhD Thesis,* Technical University Munchen (2011).
5. R. Papuc, *Theoretical and experimental research regarding tribological optimization of the clearances compensation system within timing chain transmissions, PhD Thesis*, Transilvania University of Brasov (2015)
6. A. Todi-Eftimie, *Theoretical and experimental study of friction losses in bush chain transmission PhD Thesis*, Transilvania University of Brasov (2015)
7. K. M. Marshek, Mech. and Mach. Theory **14**, 135 (1979)
8. R. Velicu, R. Saulescu, L. Jurj, Mat. Sc. and Eng. **147** (2016)
9. L. Jurj, R. Velicu, Annals of the Oradea University, Fascicle of Management and Technological Engineering **XVI (XXVI)**, 13 (2017)
10. L. Jurj, R. Velicu, SYROM 2017 (2017) (to be published)

11. L. Jurj, R. Saulescu, R. Velicu, Appl. Mech. and Mat. **880**, 21 (2017)
12. ISO 6336-1: 1996 (E). *Calculation of load capacity of spur and helical gears. Part 1: Basic principles, introduction and general influence factors (1966)*

The comparative study of the static friction coefficient between chain link and different types of polyamides

Radu Papuc[1,*], *Mihai-Tiberiu* Lates[2]

[1]Transilvania University of Brasov, Romania
[2]Transilvania University of Brasov, Romania

Abstract. This paper has as scope the comparative analysis of the experimental study of the static friction coefficients between the chain links and two guide types (segments cut from the tensioning guide of the transmission chain used in combustion engines). Basically, this study will be performed for the contact between steel and different types of polyamides PA46 and PA66, using two tribometers. In the first stage there will be presented the tribometers, devices and work methods followed by the tests results. The next stage presents with the help of finite element analysis the contact pressure variation and displacements distribution. At the end of the paper there are presented the conclusions of the tests performed on the two tribometers.

1 Introduction

It is known the fact that, if two bodies are put in contact, so that one exerts pressure on the other, results a tangent force on each of the bodies contact surface, called sliding friction force. It is also known, that this tangent force represents the action of a body onto the other body being equal and with opposite direction on each of the two contact surfaces G. W. Stachowiak, A. W. Batchelor [1].

Knowing the parameters of friction is an essential issue in tribological research, because friction always appears when there is a relative rotation between surfaces found in contact.

Friction is produced in the absence of lubricant (dry friction) or in the partial or total presence of the lubricant (limit friction, mixed, elastohydrodynamic, hydrodynamic, hydrostatic etc.) W. Shizhu, H. Ping [2].

The friction phenomenon is characterized by a multitude of parameters. Amongst them, one of the most important (if not the most important) is the friction coefficient F. C. Chiu, G. F. Kao [3].

The friction coefficients are of two types: static friction coefficient μ_s and kinetics friction coefficients (dynamic) μ_c B. Mouhmid, A. Imad, N. Benseddiq [4]

The static friction coefficients are the coefficients that appear at the limit between stop and movement, so at the start. Knowing the static friction coefficients represent a specific importance, especially in the friction joints where the starts and stops are frequent .

Because of the multiple factors that can interfere in the friction process, the calculation equations for determining the friction coefficients are complex and difficult to resolve. For this reason, it is very important that the friction coefficients are determined through experimental methods, but with a high precision J. Williams [5].

2 Describing the test and methodology of testing

For the experimental determination of the static friction coefficients between the chain links and guide segments, there will be used two types of tribometer, prismatic and UMT, presented in figures 1 and 2.

Fig. 1. Prismatic tribometer

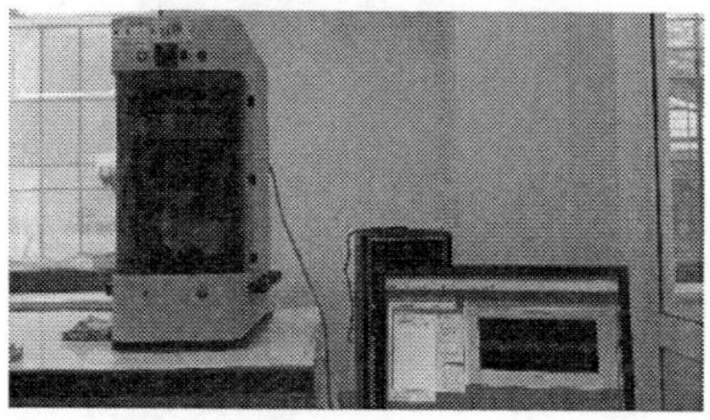

Fig. 2. Tribometer UMT

The prismatic tribometer has a simple functioning principle, being expressed with the relation 1 and detailed in figure 3:

$$Tg\alpha=\mu \qquad (1)$$

* Corresponding author: radu.papuc@unitbv.ro

© The Authors, published by EDP Sciences. This is an open access article distributed under the terms of the Creative Commons Attribution License 4.0 (http://creativecommons.org/licenses/by/4.0/).

https://doi.org/10.1051/matecconf/201818402004

Where α is the tilt angle of the mobile table and μ is the sliding friction coeffiecient.

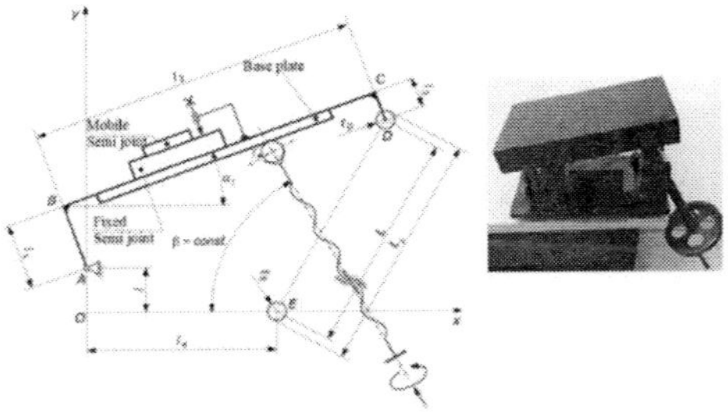

Fig. 3. Functioning principle of the prismatic tribometer

The other tribometer, UMT, works by being controlled through a software installed on a computer. In order to determine the static friction coefficients, on the UMT tribometer there will be connected two modules: reciprocating and rotary. These modules are shown in figures 4 and 5.

Fig. 4. Reciprocating module

Fig. 5. Rotary module

The reciprocating module, shown in figures 3 and 4, is suitable for performing tests on small samples of different pieces, and the motion is alternative rectilinear. The work frequency of the module is between 0.1 – 60 Hz, and the maximum load is F = 1000 N.
Also, the rotary module, presented in the same figure, is suitable for experimental determination in rotational motions. The torque value range 0.001 – 5000 rpm, and the maximum load force is F = 1000 N.

To evaluate the static friction coefficient between the tooth chain links and guide segments there were designed and built four testing devices C. C. Gavrilă [6]. In figure 6, there are presented the models used for tdetermining the friction coefficient, in the laboratory conditions from Transilvania University Research Institute. All models were designed with CATIA V5.

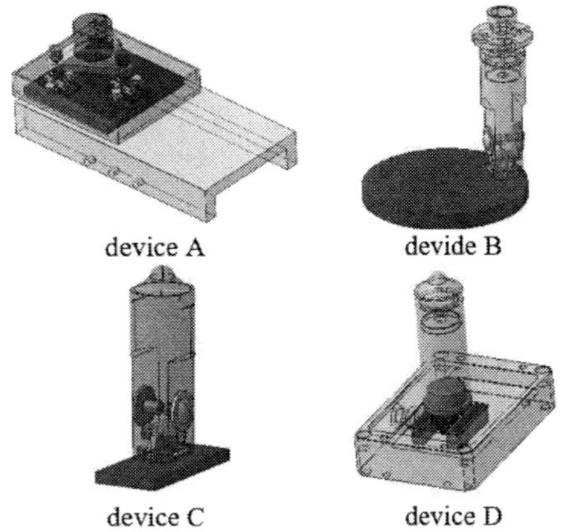

Fig. 6. Device used to determine the static friction coefficients

In the prismatic tribometer there were performed two study cases: case I - determining the static friction coefficient at contact between the tooth chain links and PA46 polyamide guide segments, in lubricated environment; case II - determining the static friction coefficient at contact between the tooth chain links and PA66 polyamide guide segments, in lubricated environment, presented in figure 7.

Fig. 7. Prismatic tribometer

Fig. 8. Reciprocating module

On the other testing tribometer, UMT, there were performed four study cases in order to determine the static friction coefficient at contact between tooth chain links and guide segments (made from polyamides, PA46 and PA66). These experimental tests were done on the Reciprocating, presented in figure 8 and Rotary modules. Thus the imposed condition of the tests were normal force F=5 N and lubricant: 15W-30 engine oil. The results of the experimental tests are presented in figures 9 and 10.

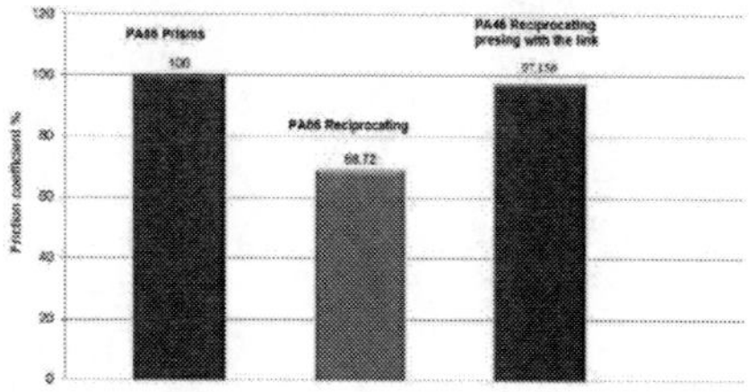

Fig. 9. Comparation prisms and reciprocating

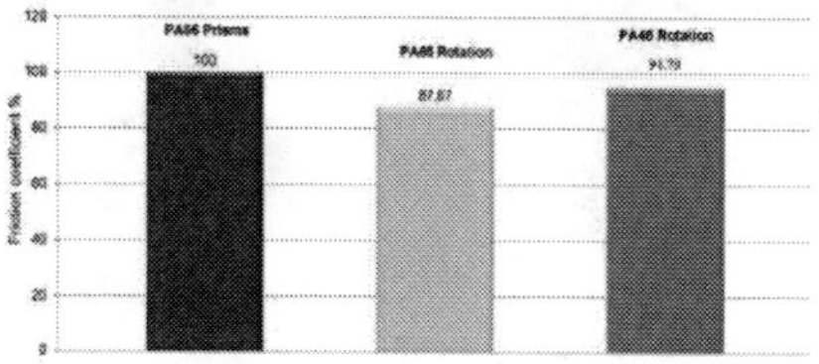

Fig. 10. Comparation prisms and rotary

It has to be mentioned that, in this paper, the values obtained for the reference measures as friction coefficients are presented as relative or percentage values, expressing dependence on parametes for which absolute values are used. This fact was imposed by the confidentiality conditions resulted from framing this research in the contract Chain Drive System Dynamic Friction.

3. The study of contact pressure

The finite elements method is a numerical method used at solving partial derivatives equations which model physical systems with an infinite number of freedom degrees. By using finite element analysis these equations are reduced to algebraic equations systems, meaning a discrete system with a finite number of freedom degrees.

In order to have a higher efficiency, in finite element analysis it is used a general structure concept and simpler than regular one. Usually, in finite element analysis by structure (of resistance) it is understood an assemble of bars, plates, layers and volumes (solids).

Finite elements method has applicability in various engineering domains (and not only), where there are found pysical phenomena described by partially derivatives equations. Amongst the main domains there are: structural analysis, fluids, magnetical and electrical analysis J. E. Akin [7].

The numerical evaluation of the previously studied concepts, here the determining the deformation state as well as the contact distribution between the interest elements (guide and bolt). Determining the tensioning state is not taken into consideration because the applied forces are hot big and the physical models are rigid enough so that they can withstand higher loads. For the numerical evaluation of the studied models it was used the ANSYS Workbench software M. T. Lateş, R. Velicu and R. Papuc [8].

The assembly model with finite elements is shown in figure 11, a and in figure 11, b is shown a detail on the bolt meshing.

On the elements found in contact there was made a finer meshing, meaning it had more layers of finite elements and nodes in order to have a better convergence of the results and contact from that area. After meshing there were obtained 16234 finite elements and 67696 modes.

For this model there are shown two calculation cases, as follows: case 1: guide material PA46, F=5 N and μ=0.28; case 2: guide material PA66, F=5 N and μ=0.28;

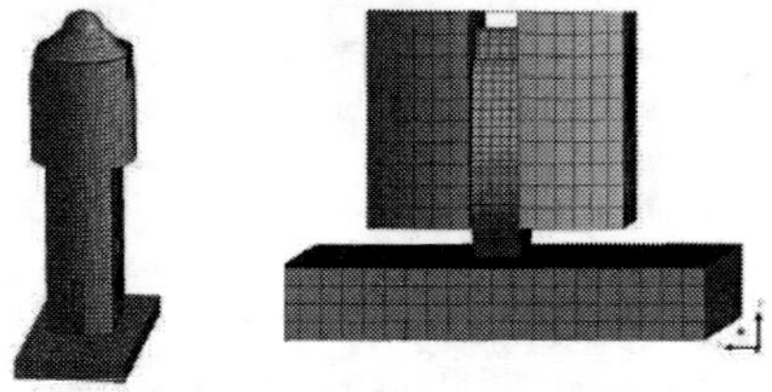

a-general view b-bold detailed view

Fig. 11. Finite elements model

The method of applying the force is presented in figure 12 being identical for both analyzed cases.

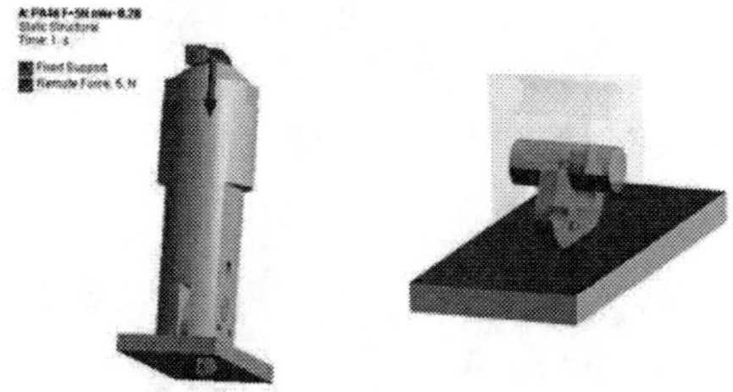

a-the method of applying the force b-defining contacts

Fig. 12. The method of applying the force and defining contacts in ANSYS

For load case 1, of this model, the obtained value of maximum displacement is 0.0005 mm, figure 13 a.

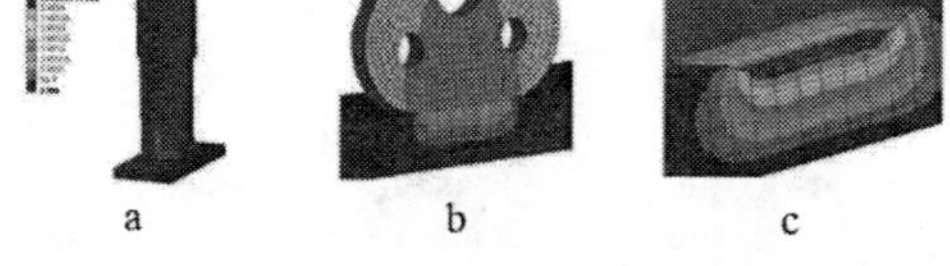

a b c

Fig. 13. The displacement distribution

If it is made a detailed value on the guide it is observed that the guide is deformed locally and only on the contact zone between the guide and bolt. In figure 13 b, it is shown the guide deformation at a scale 1:1, and in figure

13 c, it is shown the guide deformation at an enlarged scale by 3000 times, only for visualization. The distribution of the contact pressure between guide and bolt, in this case, it is shown in figure 14 and has an uniform distribution with higher values towards the interior of the bolt, the maximum value of the contact pressure being 0.54 MPa. The maximum contact pressure between bolt and sprocket is 4.57 MPa being distributes uniformly on the tangent line between the bolt and the sprocket generator.

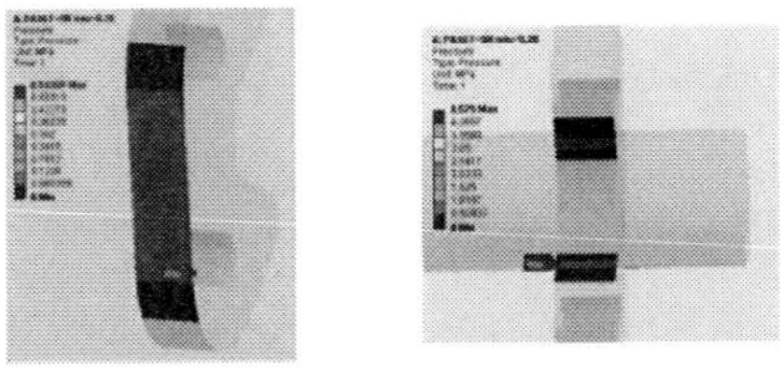

a-the contact pressure between guide and bolt
b-the contact pressure between stud and guidee

Fig. 14. The distribution of the contact pressure

For case 2, where the guide material is PA66, the maximum displacement is 0.00046 mm, figure 15. It is observed that the displacement value is lower than case 1, where it is applied the same loading force, from where it can be concluded that the PA66 material is more rigid than PA46, the friction coefficient not having any influence in this case.

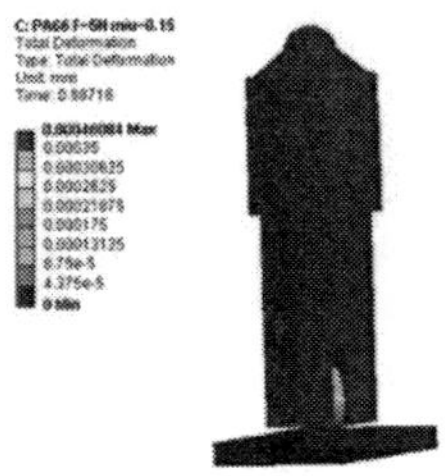

Fig. 15. Displacements distribution

The values of the contact pressures and their distribution on this model are identical with the ones obtained in loading case 1, but with material PA46. Thus, the maximum value of the contact pressure between guide and bolt is 0.54 MPa, and between stud and bolt is 4.57 MPa.

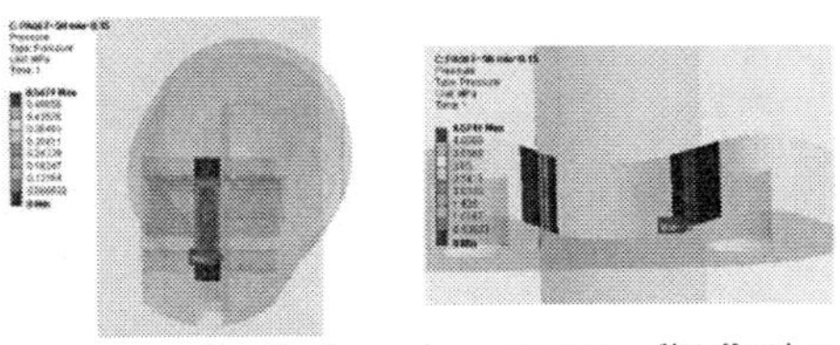

a-pressure distribution between guide and links
b- pressure distribution between bolt and links

Fig. 16. The distribution of the contact pressure

4. Conclusion

Based on the obtained results after the experimental tests and the comparative analysis of the two materials, it was found that PA66 polyamide has better characteristics in terms of friction reduction. These results show that from tribological point of view, the pair of materials analyzed have a specific behaviour different that the general know one characterized by the Stribeck curve.

On the basis of the results obtained after the study, using finite elements analysis it is highlighted that for the contact between the chain links and two guide segments there were obtained small values for the contact pressure due to high mechanical characteristics of the studied polyamide. It must be mentioned that these results are in accordance with the specific literature Van, J. Ruiten, R. Proost, and M. Meuwissen [9]. Because the transmitted force from the application point to the fastening point is done only throught the contacts between components, there were no significat equivalent tensions.

References

1. G. W. Stachowiak, A. W. Batchelor, *Engineering tribology* (Ed. Elsevier, 3rd ed., Burlington, 2005).
2. W. Shizhu, H. Ping, *Principles of Tribology* (Singapore, John Wiley & Sons Publishing House, 2012).
3. F. C. Chiu, G. F. Kao, Polyamide 46/multi-walled carbon nanotube nanocomposites with enhanced thermal, electrical and mechanical properties in Composites, Part A, nr. 43, 208 – 218, (2012).
4. B. Mouhmid, A. Imad, N. Benseddiq, D. Lecompte, An experimental analysis of fracture mechanics of short glass fibre reinforced polyamide 6.6 (SGFR-PA66) in Journal of Composite Science and Technology, nr. 69, 2521 – 2526, (2009).
5. J. Williams, *Engineering tribology* (Cambridge University Press, USA, 2011).
6. C. C. Gavrilă, 3D Modeling, FEM analysis and detail design for a testing device with spherical joint (In: Annals of the Oradea University, Fascicle of Management and Technological Engineering, vol. **XIII (XXIII)**, **ISSUE 1**, pp.39-42, 2014).
7. J. E. Akin, *Finite Element Analysis for Undergraduates* (Academic Press, Orlando, Florida, USA, 1996).
8. M. T. Lateş, R. Velicu, R. Papuc, Multiscale modelling of chain-guide contact by using tests and FEM. 11th World Congress on Computational Mechanics **WCCM XI**, Barcelona, Spain, 20-25 July, P.1062-1069, (2014).
9. Van, J. Ruiten, R. Proost, M. Meuwissen, How the choise of the polyamide type in timing chains tensioning systems affects the CO_2 emission and fuel economy of internal combustion engines in Presentation at **VDI Veetiltrieb un Zylinderkopf November 28th,** (2012).

https://doi.org/10.1051/matecconf/201818402005

Determining the optimal operating parameters of a wind system

Cornel Cătălin Gavrilă[1] and *Radu* Săulescu[2]

[1] Transilvania University of Brasov, Romania
[2] Transilvania University of Brasov, Romania

Abstract. The paper main objective is to determine the representative parameters for the geometrical modeling of the speed amplifiers used in the high-capacity wind systems, taking into account the assembling and also the neighboring conditions of the satellite wheels in the planetary units. The analysis is based on a case study taken from a representative producer in this field - a speed amplifier for which is known: wind rotor diameter, type of speed amplifier and electric generator. Starting from these data and from the power curve offered by each wind device producer, the paper presents the synthesis of the number of teeth and the efficiency of such an amplifier taking into account the electrical and wind parameters (wind velocity, density air, wind rotor diameter). Based on the electrical and wind parameters, the system's power factor is determined to determine the efficiency of the speed amplifier and the wind speed at which the system can work best.

1 Introduction

The main function of a speed multiplier is to transmit, with constant ratio, the mechanical power, converted from the wind, to the electric generator, for the production of electric power. This involves identifying a system that works best to achieve the functional correlation between the wind turbine and the electric generator.

Usually, the rotor blades rotate effectively at low speeds to maintain its peak velocity in the subsonic wavelength range. For example, large rotors are rotating at 20 ... 30 rpm, while generators have a nominal speed of 1500 or 3000 rpm; therefore, the amplification ratio of a speed amplifier is in the range 50 to 150 [1].

Most large-capacity wind turbine systems uses high-speed, single-input and output-output multipliers: a planetary unit and two outer cylindrical gears [2], two planetary units and a cylindrical outer gear [3] or three planetary units [4].

A similar variant to that proposed by Suzuki [3] is that proposed by Leimann [5]. The structural schemes of the both two solutions are shown in Fig.1, a, respectively b, and represents the constructive solution studied in the present paper.

2 The Speed Amplifier

In practice, most wind power manufacturer companies presents to potential customer some technical data, as rotor diameter (D_R) and wind speed rotor (n_R); the constructive type and the speed of the multiplier transmission and its multiplier ratio; the electrical power obtained by the electric generator and its efficiency, as well as the power curve of the wind system, without saying how efficient the system is, at what wind speed it works best or what should be the number of wheel teeth in the multiplier transmission, to allow the client to obtain an optimal transmission ratio on each gear that corresponds to the multiplied speed value provided by the prospectus.

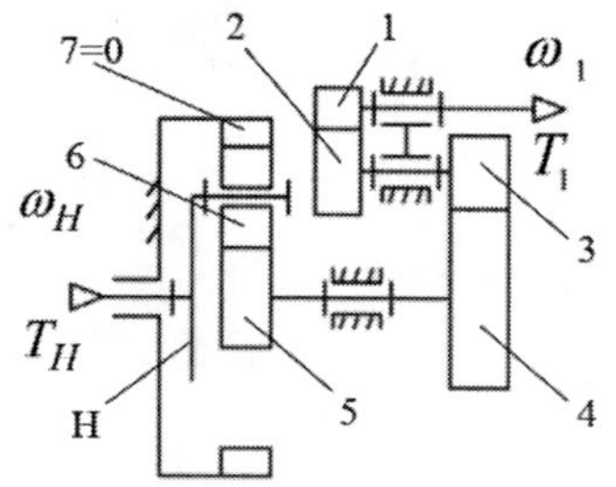

a. Structural scheme for Suzuki multiplier

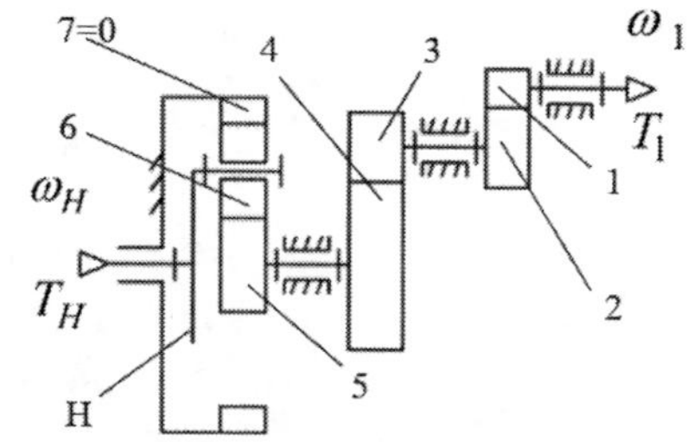

b. Structural scheme for Leimann multiplier

Fig.1. Structural schemes for speed multiplier [6]

According to the manufacturer recommendations [B] and structural schemes presented in Fig. 1, such systems consist in a planetary unit (composed by the satellite gear support carrier H, the sun gear 5, the satellite

* Corresponding author: rsaulescu@unitbv.ro

© The Authors, published by EDP Sciences. This is an open access article distributed under the terms of the Creative Commons Attribution License 4.0 (http://creativecommons.org/licenses/by/4.0/).

6 and also by the fixed crown gear 7) and two steps multiplier gears (3 and 4 respectively 1 and 2), connected in series.

The winding rotor is rigidly connected to the satellite gear support carrier H of the speed amplifier planetary unit, whose multiplied speed output is transmitted to the next multiplier step with gears and finally to the rotor of the electric generator. So, the multiplier transmission inputs are the rotation (ω_H) and torque (T_H), given by wind rotor. The outputs are the rotation (ω_1) and torque (T_1), to be transmitted to the electric generator rotor.

Starting from the structural scheme of the amplifier, it is proposed to determine the transmission ratios of the two gears and the planetary unit teeth numbers, so that the transmission has the multiplier ratio given by the manufacturer [7]. Based on this, is proposed the algorithm for determining the geometrical parameters of a speed multiplier based on wind parameters, electrical, multiplying ratio and wind power curve, as well as at what wind speed the system considered can work optimally.

3 Wind Power Efficiency

The electrical power of a wind turbine that works at a constant speed depends on the input speed of rotation. Thus, for a low rotation speed, the power reaches the maximum value at a low wind speed. A wind system that operates at high speed extracts a large amount of energy at high wind speeds, while functioning inefficiently [8, 9].

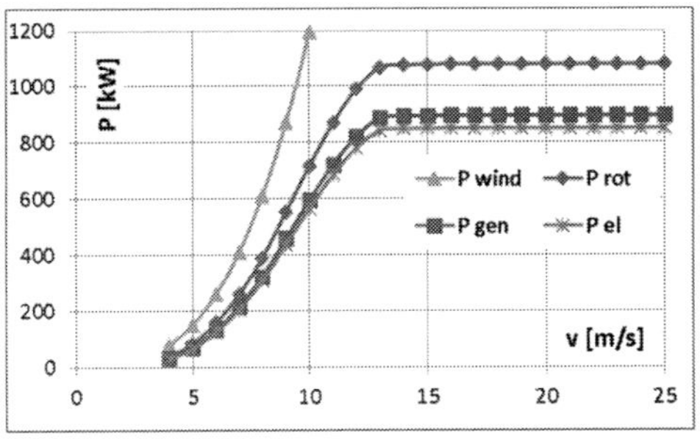

Fig.2. The power curves for the studied wind system

Starting from these, it is proposed to determine the power factor (the efficiency) of the system, based on the power curve, considered known (Fig. 2). The wind turbine power is given by relation

$$P_{el} = \frac{1}{2} \cdot \rho \cdot c_p \cdot A \cdot v^3 = P_{wind} \cdot c_p \ , \quad (1)$$

where P_{el} is the power of a wind turbine (for the studied case, the installed capacity, P_{el} = 850 kW), ρ is the air density, (the value ρ =1.2255 is considered at sea level at the temperature of 15°C), A is the swept area (the area covered by the studied rotor while rotating, A=2124 m^2) and also, v is the wind speed. From the relation (1) results the power factor for the studied wind system

$$c_p = \frac{P_{el}}{P_{wind}} = \frac{P_{el}}{\frac{1}{2} \cdot \rho \cdot A \cdot v^3} \ . \quad (2)$$

Taking accounting that the power factor (c_p) of the wind system depends by the efficiency of the winding rotor (η_{rot}), the speed multiplier efficiency (η_a) and also by the generator efficieny (η_{gen} = 0.95, value given by the manufacturer [7], adopting the theoretical yield of a fixed axle gear (η_{ang}), the efficiency of wind turbine is determined, as it is presented in Fig. 3, a, b).

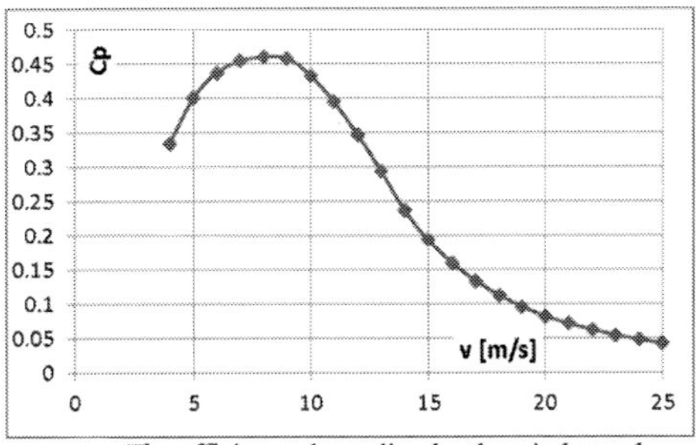

a. The efficiency depending by the wind speed

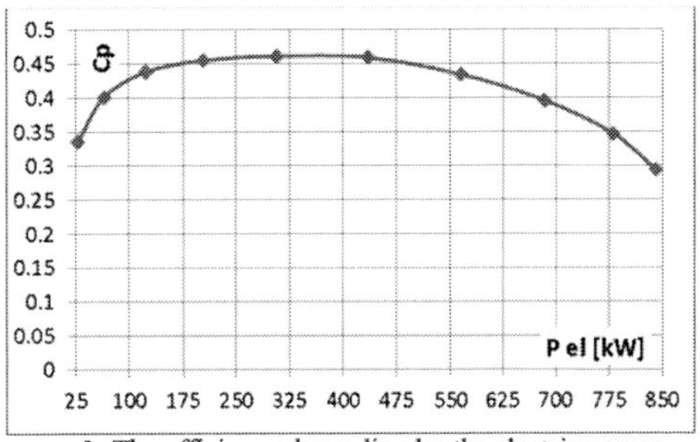

b. The efficiency depending by the electric power

Fig.3. The studied wind system efficiency

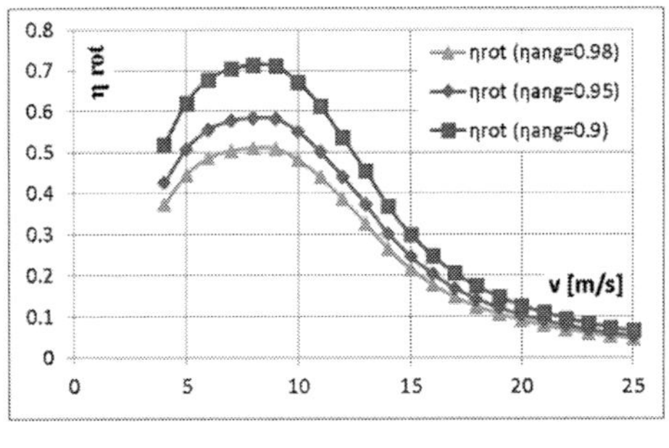

Fig. 4. The winding rotor efficiency, depending by various fixed axle gear efficiency values

The winding rotor efficiency is given by relation

$$\eta_{rot} = \frac{c_p}{\eta_{ang} \cdot \eta_{gen}} \ , \quad (3)$$

and can be determined, depending by the gear efficiency, as is presented in Fig. 4.

Once determined, it is compared with the theoretical values and then it is possible to perform the synthesis of the planetary speed multiplier for given efficiency for fixed axle gear.

Since the literature [1] recommends wind turbine efficiency 40 ... 60%, the value adopted for the efficiency of a fixed axle gear will be 95%, for which the average wind rotor yield is approx. 45%. The previous values are useful in the wind system power curves obtained for the wind power (P_{wind}), the wind power turbine (P_{rot}) and also for the mechanical power to the electric generator input (P_{gen}) (see also Fig. 2).

4 Statics and Kinematics

The static and kinematic modeling presented below is based on the block diagram of the speed multiplier presented in Fig. 6.

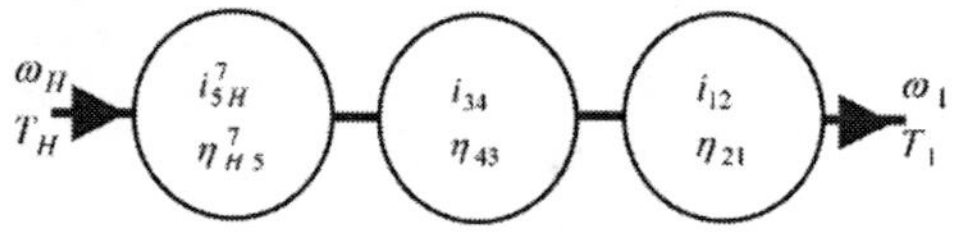

Fig.6. The speed amplifier block diagram

In the presented block diagram, the transmission ratio are written for each gear step of the speed amplifier, from the output to input, and led to the multiplier ratio.

According to Fig. 1, for given values the values for the planetary unit multiplying ratio (i_a=61.74) and the efficiency for fixed axle gear (η_{ang}=0.95), the main kinematic and static parameters are given by relations:

$$i_a = i_{1H} = i^7_{5H} \cdot i_{34} \cdot i_{12}, \tag{4}$$

$$\eta_a = \eta_{H1} = \eta^7_{H5} \cdot \eta_{43} \cdot \eta_{21}. \tag{5}$$

To perform the synthesis of the planetary gear teeth numbers of the speed multiplier, the standardized gear ratio [10] for the fixed axle gears are adopted (i_{12}= 3.15 respectively i_{34}= 3.55) and the transmission ratio of the planetary unit, i^7_{5H}, is determined by

$$i^7_{5H} = \frac{i_a}{i_{34} \cdot i_{12}} = \frac{61.74}{(-3.55)\cdot(-3.15)} = 5.52, \tag{6}$$

where:

$$i^7_{5H} = 1 - i^H_{57} = 1 - i_{0I} = 5.52. \tag{7}$$

By adopting the efficiency value for the, the efficiency values for the each multiplying fixed axle gear steps (3 and 4 respectively 1 and 2) as well as the internal efficiency for the planetary unit result as:

$$\eta_{43} = \eta_{21} = \eta_{ang} = 0.95, \tag{8}$$

$$\eta_{0I} = \eta^H_{57} = \eta_{56} \cdot \eta_{67} = 0.95^2 = 0.9025. \tag{9}$$

To determine the yield of the speed multiplier efficiency, the planetrary unit efficiency is determined by relation:

$$\eta^7_{H5} = \frac{-\omega_{57} \cdot T_5}{\omega_{H7} \cdot T_H} = -\frac{T_5/T_H}{i^7_{H5}} = \frac{(1 - i_{0I})}{1 - i_{0I} \cdot (\eta_{0I})^x}, \tag{10}$$

where $x = \mathrm{sgn}(\omega_{5H} \cdot T_5) = -1$.

Taking account by relations (5), and (7...10) for the multiplier efficiency the following value is obtained:

$$\eta_a = \frac{1 + 4.52}{1 + 4.52/0.9025} \cdot 0.95 \cdot 0.95 = 0.8291. \tag{11}$$

The previous results led to the multiplying functions for the analyzed speed amplifier:

$$\omega_1 = i_a \cdot \omega_H = 61.74 \cdot \omega_H, \tag{12}$$

$$T_1 = -i_a \cdot \eta_a \cdot T_H = -51.19 \cdot T_H. \tag{13}$$

So, the input rotation value is multiplied by 61.74 and the input torque value is reduced by 51.19.

5 Gears Teeth Numbers in Planetary Unit

A problem with this type of multiplier could be the load through the planetary unit, related with the wind turbine load, which is the most demanding; therefore a stage of constructive design is that of its geometric-kinematic synthesis. For this purpose, is considered the type of planetary unit presented in Fig. 1 and the internal kinematic ratio i_{0I} given by relation (7); also there are constraints given by the outer dimensions of the planetary unit, which led to the teeth number adopting for the crown wheel (z_7). With this data, it is proposed to determine the number of tooth numbers of the component wheels and the number of satellites (s).Then, the geometrical sizes of the gears can be determined and the mechanical power transmission through the speed amplifier established.

For study, the planetary unit can be considered as fixed axle gear, obtained by the motion inversion. Inside, there are needed the following constructive conditions: satellite parallel mounting, axis coincidence between the sun and crown gear, neighborhood condition between the satellites. The number of satellites (s) can be determined depending by their tooth numbers (where N is an integer), for which the neighboring condition [C] will be checked. The calculus example presented in Fig. 6 [10] led to a number of 4 satellites (s = 4). Based on the parameters from Fig. 6, the constructive scheme for the planetary unit is obtained (Fig. 7).

https://doi.org/10.1051/matecconf/201818402005

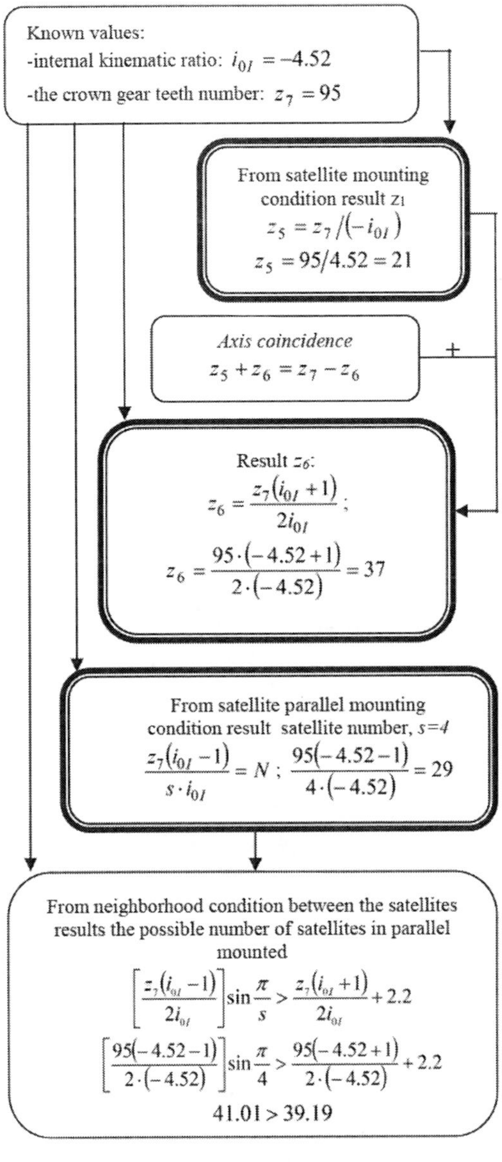

Fig.6. Teeth numbers synthesis algorithm

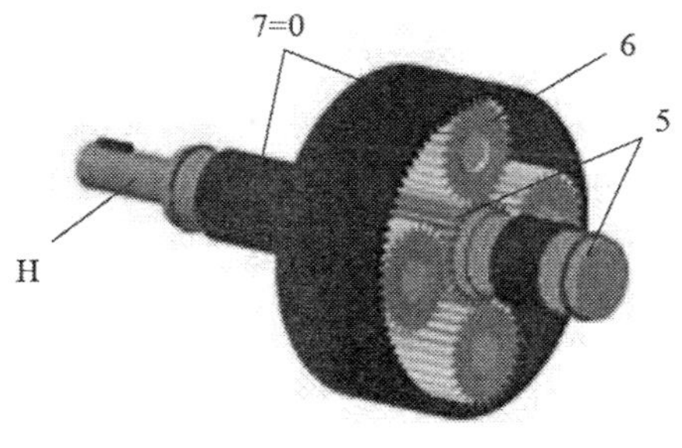

Fig.7. Planetary unit constructive variant

6 Conclusions

From the diagrams presented in Fig. 3, it can be observed that wind systems are more efficient before reaching the rated electrical power given by the generator.

Increasing wind speeds, and also the electricity production, increases the risks of optimal use of the wind system, so the additional loads on blades, wind speed amplifier respectively wind turbine. Taking into account the above, it can be said that the performance of a wind system depends mainly on the performance of the rotor, the speed amplifier and the electric generator.

The proposed synthesis algorithm led to the optimum number of wheel teeth inside the transmission to determine geometric parameters and loads transmitted by each gear, shaft and bearings; this being the subject of a future work. The analysis of the results highlights that the selection of certain combinations of transmission reports, which should be applied in a practical case, depends on the design requirements set, in a pre-feasibility study.

References

1. Hau E., *Wind Turbines: Fundamentals, Technologies, Application, Economics*, Springer, Germany, 2006.
2. Tong W., *Wind Power Generation and Wind Turbine Design*, WIT Press, 2010.
3. Suzuki K., a. o, *Planetary gear train with oiling system for supplying lubricant oil to bearing*, Patent no. US20130172144 A1, 2013.
4. Carriveau R., *Fundamental and Advanced Topics in Wind Power*, InTech, Rijeka, 2011.
5. Leimann D., *Planetary gear unit for a gearbox for a wind turbine*, Patent no. US20110077120 A1, 2011.
6. Vişa, I., a. o, *The Role of Mechanisms in Sustai-nable Energy Systems*. Ed. Universităţii Transilvania din Braşov, 2015 (in romanian).
7. Gamesa Catalogue, *Wind Turbine G52-850kW*, 2010
8. Burton T., Sharpe D., Jenkins N., Bossanyi E., *Wind Energy Handbook*, John Wiley & Sons, LTD, West Sussex, 2001.
9. Lateş, M. T., Velicu, R., *CFD analysis and theoretical modelling of multiblade small Savonius wind turbines*. Proceedings of The 4th International Conference on Sustainable Energy, 2014, Braşov, România.
10. Romanian Standard, no. 6055, *Gear ratio* (in romanian).
11. Săulescu, R., Jaliu, C., Climescu, O., Diaconescu, D. *On the use of 2 DOF planetary gears as "speed increaser" in small hydros and wind turbines*. Proceedings of the ASME 2011, IDETC/CIE 2011, Washington, DC, USA, CD Proceedings.

https://doi.org/10.1051/matecconf/201818402006

Modeling in ADAMS of a 6R industrial robot

Mariana Ratiu[1,*], *Alexandru* Rus[1], and *Monica Loredana* Balas[2]

[1]University of Oradea, Department of Mechanical Engineering and Automotive, 410087 Oradea, Romania
[2]Transilvania University of Brasov, Department of Product Design, Mechatronics and Environment, 500068 Brasov, Romania

Abstract. In this paper, we present the first steps in the process of the modeling in ADAMS MBS of MSC software of the mechanical system of an articulated robot, with six revolute joints. The geometric 3D CAD model of the robot, identical to the real model, in the PARASOLID format, is imported into ADAMS/View and then are presented the necessary steps for building the kinematic model of the robot. We conducted this work, in order to help us in our future research, which will consist of kinematic and dynamic analysis and optimization of the robot motion.

1 Introduction

Today, in the context of using of the robots in the most industries, due to the benefits related to the reduction of the costs and increasing productivity, there is, as stated by the [1], a special interest of the companies and researchers concerning the modeling and optimization of these mechatronic systems. In recent decades, alongside with the traditional methods of design and modern physical prototyping, the modeling and virtual prototyping are new trends in the industry and become more and more complex and powerful, thanks to the many advantages that they bring. As presented in [2], there are a large number of studies and multiple and complex software packages used for the modeling of the industrial robots. One of the most widely used Multi-Body System (MBS) software for the modeling, analysis, and simulation of the complex mechanical systems' motion, is ADAMS (Automatic Dynamic Analysis of Mechanical Systems) from MSC software, because of the benefits which it offers: performance, safety, and comfort, good visibility and good results in real-time, consistent with the results of the physical model. In this paper, we present the process of the modeling in ADAMS of the mechanical system of a 6DOF (degree of freedom) robot, with six revolute joints. This work is the first step in a longer research related to the modeling and optimization of the motion of this kind of robot, and that will be applied on a physical robot ABB IRB 1600-6/1.2, which exists in one of our laboratory, integrated into a flexible manufacturing cell.

2 Geometric and structural presentation of the robot ABB IRB 1600-6/1.2

The ABB IRB 1600-6/1.2 robot is a 6R robot (also called articulated robot, or revolute robot, or anthropomorphic robot) having the six revolute joints arranged as shown in Fig. 1. In accordance with the manufacturer's specifications [3], the main applications of this robot are: machine tending, material handling, arc welding, cutting, dispensing, assembly, palletizing and packaging, measuring, die casting, injection moulding.

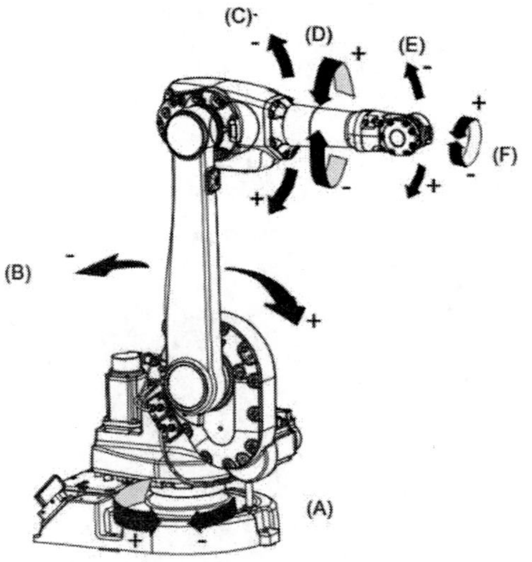

Fig. 1. Geometric model of the robot ABB IRB 1600-6/1.2.

The types of motion and the range of movement, according to [3], are presented in Table 1.

Table 1. Type of motion and range of movement.

Axis	Type of motion	Range of movement
A	Rotation motion	+180° la -180°
B	Arm motion	+110° la -63°
C	Arm motion	+55° la -235°
D	Rotation motion	+200° la -200°
E	Bend motion	+115° la -115°
F	Turn motion	+400° la -400°

* Corresponding author: mratiu@uoradea.ro

© The Authors, published by EDP Sciences. This is an open access article distributed under the terms of the Creative Commons Attribution License 4.0 (http://creativecommons.org/licenses/by/4.0/).

3 Modeling in ADAMS of the robot's mechanical system

The virtual prototyping process in ADAMS, respectively [4, 5], comprises the following steps:
- Building the model using: bodies, forces, contacts, joints, motion generators
- Testing the model using: measures, simulations, animations, plots
- Validating the model by: importing test data, superimposing test data
- Reviewing the model by adding: friction, flexible parts, forcing functions, control systems
- Iterating the model through variations using: parametrics, design variables
- Improving the model using: design of experiments, optimization
- Automating the modeling process using: custom menus, custom dialog boxes, macros.

As we have already mentioned, in this paper we will build the kinematic model of the robot, which contains the bodies (also called parts or kinematic elements) of the robot, connected through the revolute joints, and the geometric parameters specific to the mechanism, in fact, the locations of the joints; the inputs are made by using kinematic restrictions and motion generators, which controls the position or velocity of the elements [6]. Then, we will run a kinematic simulation, will animate simulation results and will plot results values from the simulation run.

In continuation, there are shown the necessary steps for building the kinematic model of the robot. There is the possibility to create the model of the robot directly in ADAMS/View or to design it in others virtual environments and then to transfer the 3D CAD model in ADAMS/View, as was done in [7, 8 and 9].

In this work, the geometric 3D CAD model of the robot, identical to the real model, in the PARASOLID format, is taken from the manufacturer's website and imported into ADAMS/View and the robot's bodies are identificated, as can be seen in Fig. 2.

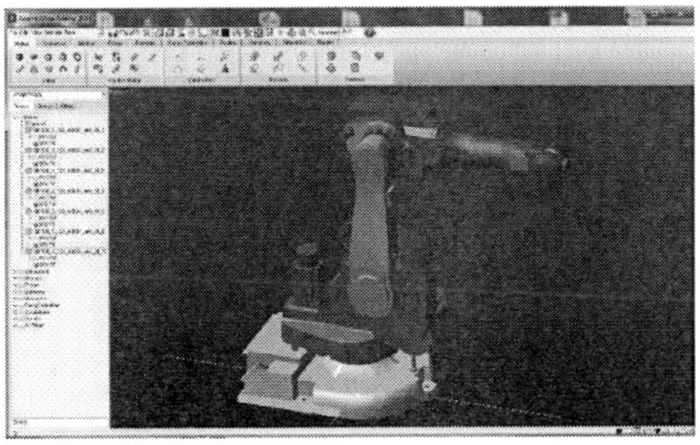

Fig. 2. Identification of the robot bodies in ADAMS/View

The necessary characteristic points (geometric parameters specific to the mechanism, in fact, the locations of the joints) are created, as can be seen in Fig. 3.

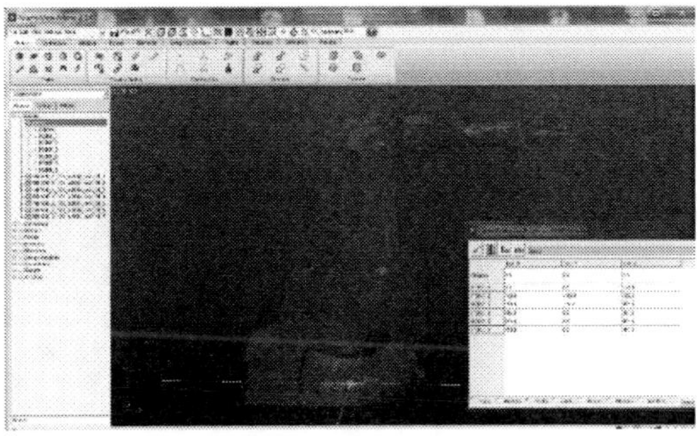

Fig. 3. Creating characteristic points

Then the kinematic elements are connected through six revolute joints, and the robot is fixed on the ground, like in Fig. 4.

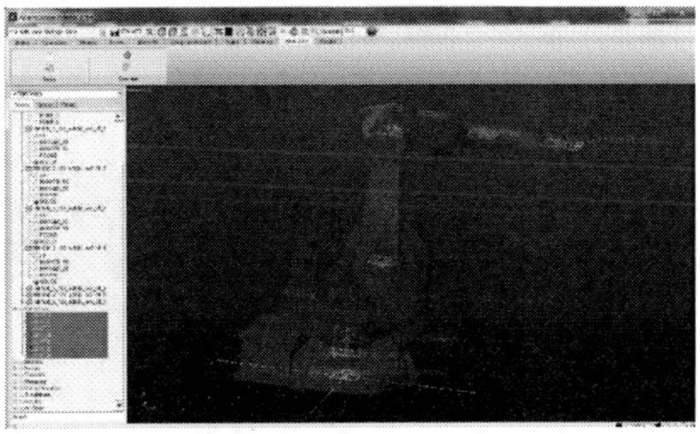

Fig. 4. Creating revolute joins

The material from which the robot is made is selected (aluminum), thereby obtaining the mass and inertia properties of the component elements (Fig. 5).

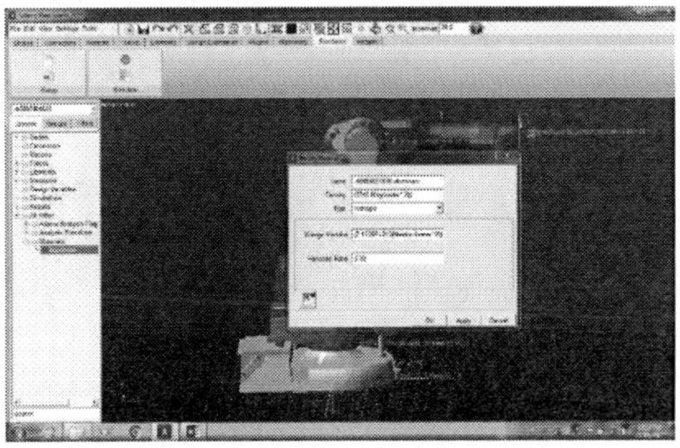

Fig. 5. Selection of the material

The motion generators of the six revolute joints are imposed, after a certain function, predefined in the program (30.0d * time), like in Fig. 6, and so, the MBS model of the robot, in ADAMS/View, is obtained.

It is simulated the motion of the robot, for 1 second and 100 steps and is obtained the position of the robot presented in Fig. 7.

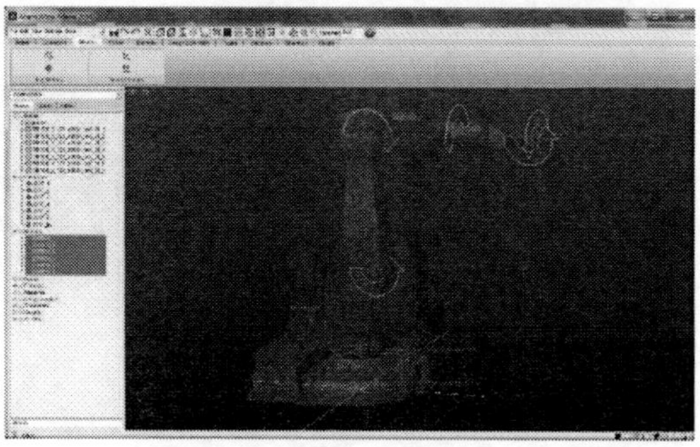

Fig. 6. Imposing the motion generators

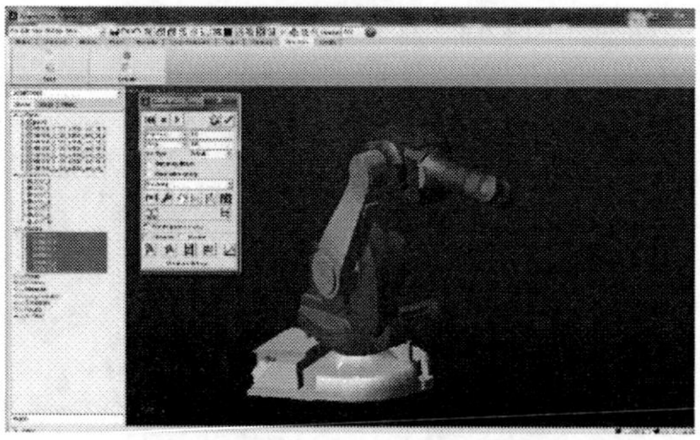

Fig. 7. Simulation of the MBS model of the robot in ADAMS

In order to be in line with those provided in Fig. 1, are changed the sense of rotation of some joints (Fig. 8) and is observed the changing of the robot position after a simulation like before (Fig. 9).

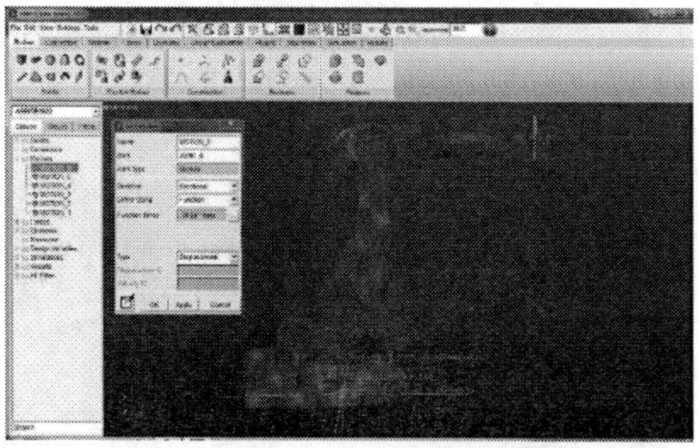

Fig. 8. Changing the sense of rotation for a joint

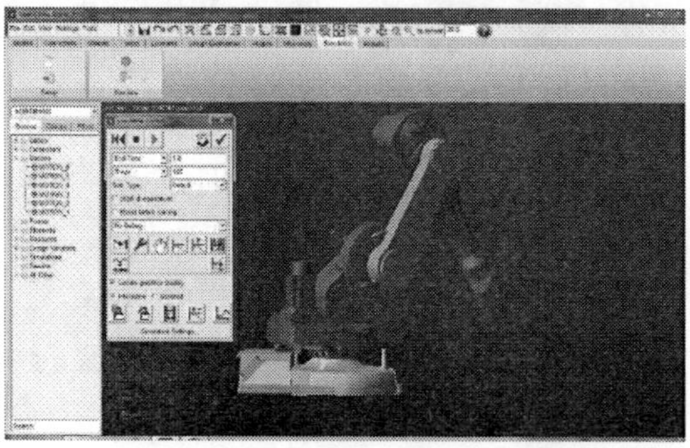

Fig. 9. The changed position of the robot

Using the animation module, the trajectory of the Marker 32, belonging to the end effector, is observed after a simulation of 1s, in Fig. 10.

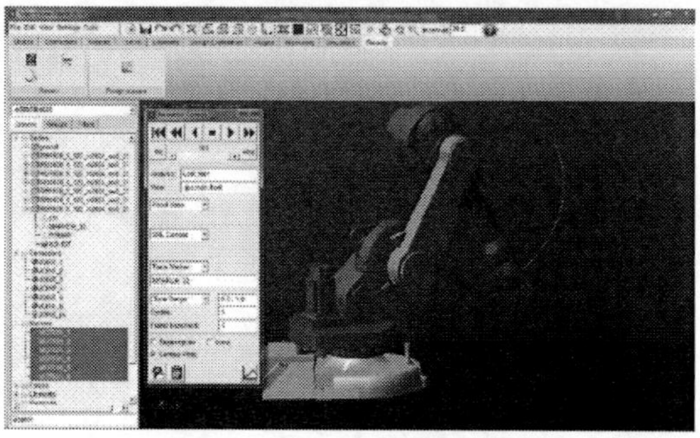

Fig. 10. Trajectory of the end effector

Now, the model of the robot is prepared for kinematic analysis using ADAMS MBS software.

4 Kinematic analysis of the robot

The goal of the kinematics is to describe the motion of the robot without to take into consideration the forces and torques causing the motion.

There are two types of kinematic analysis, forward kinematic analysis, and inverse kinematic analysis. In this work, we will do a simple forward kinematic analysis, in which we will plot the position and orientation of the end effector, its velocity and acceleration, under the imposed motion generators of those six revolute joints. So, through the forward kinematic analysis, we will determine the cumulative effect on the end-effector of the entire set of joint variables. In contrast with the forward kinematic analysis, the objective of the inverse kinematic analysis is to determine the values for the joint variables, given the position and orientation of the end effector [10].

In the next five figure, we present the charts of variation of some kinematic parameters of the end effector after the above simulation run, in fact position (Fig. 11), velocity (Fig. 12), acceleration (Fig. 13), angular velocity (Fig. 14) and angular acceleration (Fig. 15) of the mass center of the end effector.

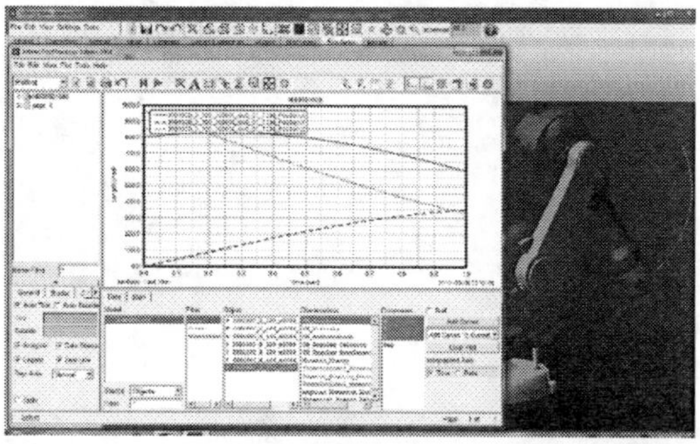

Fig. 11. Position of the end effector

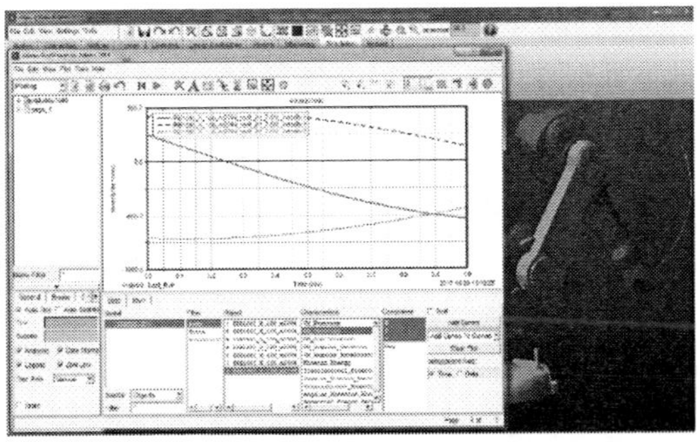

Fig. 12. Velocity of the end effector

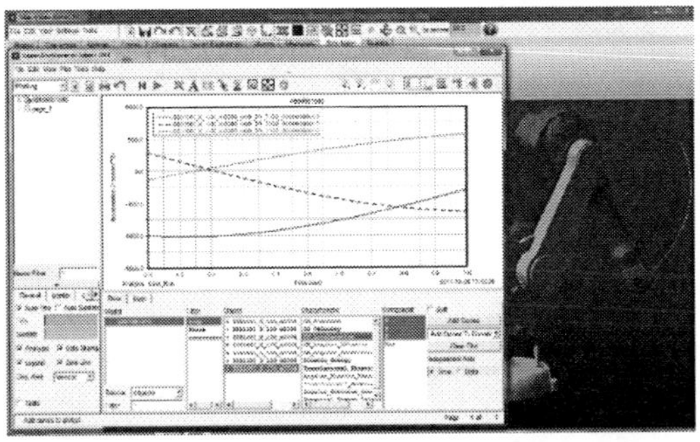

Fig. 13. Acceleration of the end effector

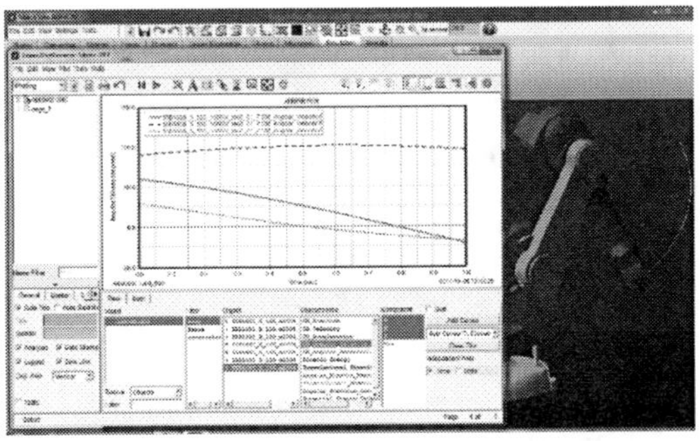

Fig. 14. Angular velocity of the end effector

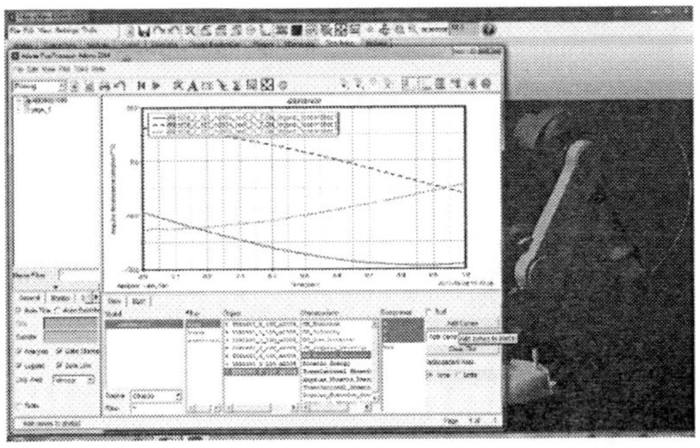

Fig. 15. Angular acceleration of the end effector

5. Conclusion

We conducted this work, which represents a first and small step in modeling in ADAMS MBS software of a 6R articulated robot, in order to help us in our future research, which will consist of kinematic and dynamic analysis and optimization of the robot motion. We hope that this example will be useful to students or other interested people who want to start learning to model such mechatronic systems in ADAMS.

References

1. M. Ratiu, Short literature review on the kinematics and dynamics of the industrial robots, Annals of the University of Oradea, Fascicle of Management and Technological Engineering, ISSUE #2, pp. 49-52 (2016)
2. M. Ratiu, A. Rus, Modeling of the trajectory-generating equipments, IEEE publication, 14th International Conference on Engineering of Modern Electric Systems (EMES), DOI: 10.1109/EMES.2017.7980418 (2017)
3. www.abb.com/robotics
4. Getting started using ADAMS/View, MSC Software (2014)
5. Adams Tutorial Kit for Mechanical Engineering Courses (Third Edition), MSC Software
6. C. Alexandru, Modeling in mechatronic concept and virtual testing of orientation mechanisms used in solar panels, Mechanism and Manipulators Magazine, vol. 2, nr. 2, pp. 49-54 (2006)
7. M. Enescu, C. Alexandru, Modeling and simulation of a 6 DOF robot, Proceedings of 2011 International Conference on Optimization of the Robots and Manipulators (OPTIROB), Sinaia, Romania (2011)
8. M. L. Enescu, C. Alexandru, Parametric optimization in virtual prototyping environment of the control device for a robotic system used in thin layers deposition, 7th International Conference on Advanced Concepts in Mechanical Engineering, IOP Publishing Ltd, Series: Materials Science and Engineering, 147, doi:10.1088/1757-899X/147/1/012087 (2016)
9. V. Singh, S.S. Dhami, Robot Position Optimization for a Pick and Place Operation, International Journal of Emerging Technologies in Engineering Research (IJETER), Volume 5, Issue 12, pp. 63-70 (2017)
10. M. W. Spong, S. Hutchinson, M. Vidyasagar, Robot modeling and control, Wiley (2006)

Android application for ocular training in amblyopia

Daniela Mariana Barbu[1,*]

[1]Transilvania University of Brasov, 29 Eroilor Bd., 500036 Brasov, Romania

Abstract. This paper propose a mobile application that runs on the Android system, an app that aims to help persons with amblyopia by running recovery exercises in a very simple and convenient way, right on their own phone. It is very important to know that training the amblyopic eye even five minutes a day can result in vision acuity recovery, so it is recommended to use this application every day. Moreover, it is very easy to use, with an intriguing and intuitive menu with very simple commands. The subject with amblyopia can do the exercises anywhere, anytime. Another advantage is the accessibility of the application, which is used by the phone, can be used at any time of the day; the user has to give only five minutes of his daily activities.

1 Introduction

The visual analyser is effectively built to provide optimal visual function. The light penetrates the eyes, crosses the transparent and refractive media, formed by cornea, aqueous, crystalline and vitreous, eventually reaching the retina where the image is formed [1]. The amount of light that enters the eye is regulated by the pupillary opening [2]. In the emetropic eye, transparent and refractive media contribute to the convergence of luminous rays and their focusing on the retina. It is a selective receptor for electromagnetic radiation between 375 and 760 nm [3]. Its photoreceptors are the ones that convert electricity into light energy, thanks to photo pigments that contain a protein role, rhodopsin [4]. On the other hand, the eyeball has the property of orienting the visual axis in the direction of the object fixed by the extrinsic muscles that rotate it around it. The muscles are commanded by oculo-motor nerves so that the two eyes act simultaneously as one eye to achieve binocular and stereoscopic vision [5].

Amblyopia is the decrease in monocular visual acuity (measured with the best optical correction) undetermined by apparent organic causes of the posterior eye or the visual pathways. After Cernea [6] amblyopia is the diminution of visual acuity in the absence of visible organic lesions or organic lesions that do not adequately explain the decrease in vision. Vancea et al. [7] define amblyopia as a decrease in visual acuity having as a substrate the primary or secondary damage to neuronal circuits. Amblyopia is an alteration of the central area of the visual field, while its peripheral area is usually unaffected [3]. The more amblyopia appears earlier in life (interfering with the process of developing normal binocular vision), the worse it is. To the same extent, the earlier it is treated, the better the prognosis treatment. Yet, recovery exercises for amblyopic eye are recommended at any age at least to stop the degree of amblyopia already installed [5].

2 The mobile application Amblioptix

During this age of technology becomes increasingly shorter and are more and more work to be done, we divide between professional and personal life. Fortunately, technology makes stunning progress and gives modern man the means to make things much faster in the online environment that would otherwise have consumed him for quite some time. One of the most interesting aspects of technology development is the ability to intuitively relate to the wishes and needs of users, enabling them to optimize daily activities [8].

Now, according [9] the challenge for software developers is to create the smartest programs that mimic the human mind in terms of functioning, perform more and more complex tasks, and even self-optimize, similar to the process by which people constantly learn and memorizes new information every day.

Android is a software platform and operating system for mobile devices and phones, based on the Linux kernel originally developed by Google, and later by the Open Handset Alliance commercial consortium [10]. Android allows developers to write managed code in the Java programming language, controlling the device through the Java libraries developed by Google [11].

Amblioptix is a mobile application, developed with student Alexandru Spătaru, running only on the Android system, an application that aims to help persons with amblyopia by running recovery exercises in a very simple and convenient way, even on their own phone. It is very important to know that training of the amblyopic eye even for only five minutes a day can result in its disambiguation, so it is recommended to use this application every day. Moreover, it is very easy to use, with an attractive and intuitive menu with very simple controls. The subject with amblyopia can do the exercises anywhere, anytime. Another plus is the accessibility of the

* Corresponding author: dbarbu@unitbv.ro

© The Authors, published by EDP Sciences. This is an open access article distributed under the terms of the Creative Commons Attribution License 4.0 (http://creativecommons.org/licenses/by/4.0/).

application, which is used by the phone, it can be used at any time of the day, and the user has to give only five minutes of his daily activities.

Table 1. Minimum phone requirements

Requirement	Description
Android version	4.4.2 or later
Phone processor	Quad Core Processor, 1.0 GHz
RAM Memory	500 MB minimum
Memory space	4.25 MB
Screen resolution	480x800 minimum

2.1 Application design

The application was developed in the Android Studio 2.3.3 development environment by using the Java and XML programming language, which is an extensible mark-up system, that is, it is simpler, a mark-up system similar to HTML, but it is much better and more dynamic, the key difference being that tags are not defined, the programmer being free to experiment. [12]

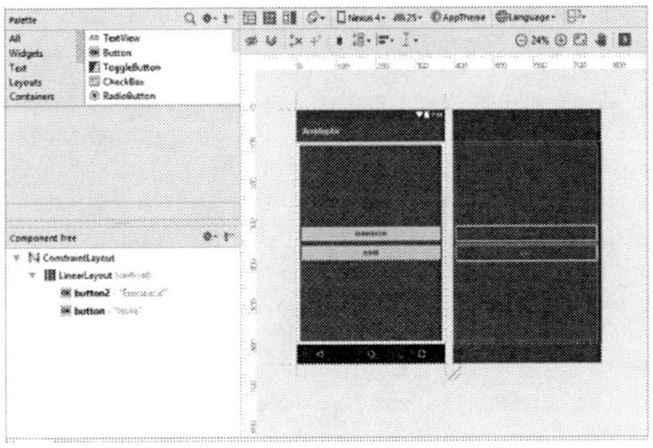

"Exit" button

```
public void onIesireBtnClick(View view) { finish(); }
}
```

"Practice" button

```
public void onExerseazaBtnClick(View view) {
    Intent myIntent = new Intent(this, MeniuExercitii.class);
    startActivity(myIntent);
}
```

Fig. 1. Android Studio presentation

To begin with, the main menu of the application was created, where the two buttons, "Exersează" (Exercise) and "Ieşire" (Exit), intuitive names for their use, was introduced. Behind each button there is a method represented by a piece of code that gives meaning and representation in the application. For example, the "Ieşire" button will close the application, represented simply by finishing it, and the "Exersează" button will also open a new menu consisting of the actual recovery exercises.

By navigating through the exercise menu, you can find out how each button for each exercise has been implemented (Figure 2).

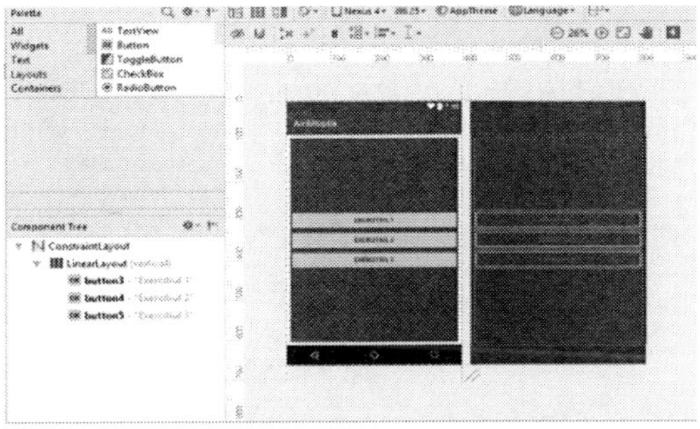

```
public void onExercitiul1BtnClick(View view) {
    Intent myIntent = new Intent(this, Game.class);
    myIntent.putExtra("type", Exercitii.EXE1);
    startActivity(myIntent);
}

public void onExercitiul2BtnClick(View view) {
    Intent myIntent = new Intent(this, Game.class);
    myIntent.putExtra("type", Exercitii.EXE2);
    startActivity(myIntent);
}

public void onExercitiul3BtnClick(View view) {
    Intent myIntent = new Intent(this, Game.class);
    myIntent.putExtra("type", Exercitii.EXE3);
    startActivity(myIntent);
```

Fig. 2. Recovery exercises in Android Studio

Also, behind each exercise button were written recovery methods, which are designed to start the recovery protocol as efficiently as possible.

2.2 Ambliopitx implementation

The application consists of two menus: principal menu (Figure 3) and exercises menu (Figure 4).

Fig. 3. Main menu

Fig. 4. Exercises menu

The "Exersează" button will send us to another menu where we will retrieve the recovery exercises, while the "Ieşire" button will result in closing the application in a simple and direct way.

All three exercises have the role of strengthening the external muscles of the eyeball along with the effective transmission of visual impulses from the retina to the visual cortex.

The use of this application is especially for people with amblyopia, and in order to have a remarkable result, it is necessary to strictly observe a daily program of use.

Exercise 1

By accessing the first exercise of this application, one can see the black-and-white background (like chess table) and a red dot, a point that must always be watched by the amblyopic eye, the congener eye being in occlusion all this time.

In this exercise, the subject follows the red point, which carries a circular motion whose radius decreases to a certain limit, and then returns to its original value. Practically, the point moves initially to the inside, and then vice versa, returns to the shape of the original circle. This exercise can work for a maximum of five minutes, a value set in the program.

Fig. 5. Exercise No. 1

Exercise 2

Using the second exercise of this application, it can be notice the background represented by black and white background, and a red dot, a point that must always be watched by the amblyopic eye. In the second exercise, the subject strengthens the point, which, this time, intersects the two horizontal sides of the screen, following the motion of the point in the form of zigzag shapes (Figure 6). This exercise works for a period of five minutes, a value imposed by the program.

Fig. 6. Exercise No. 2

Exercise 3

By accessing the third exercise of this application, one can see the background in black and white, and the same red dot, a point that must always be watched by the amblyopic eye. This exercise is intended for sharper, more difficult to follow by the amblyopic eye, an exercise where the level of difficulty is high. The red dot rises to the vertical line of the screen until it touches the corner of the screen, and then suddenly changes its position by moving diagonally to the opposite corner of the screen (Figure 7). This movement is repeated over a period of five minutes, also a value imposed by the program.

Fig. 7. Exercise No. 3

Instructions for using the application:

- The app installs on a phone whose operating system must be Android. Any version of it is supported for this application to run. After the application is installed, the Amblioptix icon will launch it, where it can be seen its main menu, along with the two buttons for Exercise or Exit.
- Before starting exercises, it is extremely important to occlude the director eye, which is the eye with a much better visual acuity. A professional occlusive devices can be used or the eye it can be occluded by a handy method (for example, using a tissue). After that, the exercises for the amblyopic eye can be started;
- Pressing the "Exersează" button will access a new menu showing three exercises for eye recovery. The amblyopic subject can work with three exercises. They are numbered from 1 to 3, on different levels of difficulty, from the most accessible to the most difficult.
- To begin with, exercise 1 is recommended, and the training time should be five minutes. To avoid "cheating", it can also call for additional people for support. In one training session, it is recommended to use the entire set of exercises.
- For a notable and positive result, daily exercise is continued and it is very important for the subject to follow the recovery protocol.
- It is also recommended to periodically submit to a specialist consultation during the use of this software.

2.2 Application testing

The application developed and presented above was tested on a group of students, two of them having amblyopia confirmed by specialized tests.

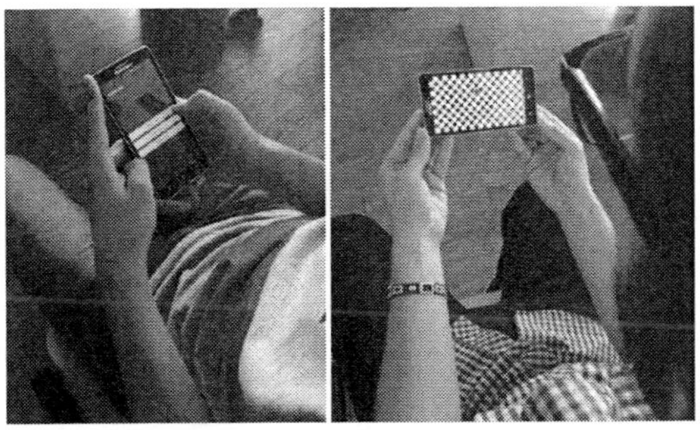

Fig. 8. Amblioptix testing

Application testing was conducted over three weeks, during which the subject used the app installed on your phone. The application was used for 5 minutes each day. Subjects with amblyopia began to work on difficulty levels from exercise 1 to exercise 3. Each exercise was used over a period of one week, a period of time necessary for the eye with amblyopia to get used to the work environment. The subjects felt better and easier to work with the lunch-time application, as opposed to the morning when they had difficulties in exercising.

After a period of six weeks, re-testing of amblyopia was performed by a specialist medical consultation and an obvious improvement in visual acuity was achieved by about 10%.

3 Conclusions

The application proposed in this paper can become a very accessible tool for anyone who needs training for visual recovery. Its main advantage is that it can be installed on any device that runs using the Android operating system, which gives it great mobility. It takes a few minutes of use per day and can be used in school, office or any service that can provide a short break. For better results, the application should be accessed at different times of the day for visual application under different conditions of eye fatigue.

The development possibilities would consist in the fact that this app could reach the Play Store and so could be accessed by all children and young people, and not only, who have visual problems. As the application is specifically intended for those with amblyopia, it could be used every day by people with ocular problems to be able to perform recovery exercises in a simple and efficient way. Particularly, the application could also be used in optician and optometry offices, ophthalmic and paediatric clinics, as well as in educational institutions.

A real possibility of future development would be to design a more advanced application that could use access to the front camera phone to avoid cheating. Thus the front camera could detect eye movement and the subject would be warned whenever his eyes do not follow the target. Also a great advantage would be the possibility of timing, registration of the recovery program and the results obtained.

Still, performing these exercises is extremely important in visual recovery, and specialist medical advice is strictly recommended to be performed periodically.

Thanks to the students from the Optometry study program of Transilvania University of Brasov, and especially Alexandru Spataru, for the contribution made in this study.

References

1. M.I. Baritz, The 7th International Conference on Computational Mechanics and Virtual Engineering, COMEC 2017, 182-187 (2017)
2. B. Braun, The 13th International Scientific Conference eLearning and Software for Education, ELSE 2017, **1**, 500-505 (2017)
3. M.I. Baritz, The International Scientific Conference eLearning and Software for Education ELSE 2016 **1**, 498-502 (2016)
4. M.I. Baritz, Conference on Sustainable Energy CSE 2017, 720-729 (2017)
5. B. Braun, Bulletin of the Transylvania University of Brasov, **9 (58)**, 57-62 (2016)
6. P. Cernea, Treaty Ophthalmology, Medical Press, Bucharest (1997)
7. P. Cernea, F. Constantin, M. Aconiu, Strabismus, Medical Press (1981)
8. A. Pringle, S. Somerville, Mental Health Practice **17**, 34–37 (2013)
9. D. Smite, C. Wohlin, T. Gorschek, R. Feldt, Journal of Empirical Software Engineering **15 (1)**, 91-118 (2010).
10. P. Kaur, S. Sharma, 2014 Recent Advances in Engineering and Computational Sciences (RAECS), Chandigarh, 1-5 (2014)
11. M.A. Anwar, D. Gangodkar, Communication Control and Intelligent Systems (CCIS) 2015, 120-123 (2015)
12. A. Singh, S. Sharma, S. Singh, International Journal of Computer Applications (0975–8887), 5-8 (2016)
13. Android Official Website http://www.android.com/ (2018)

MATEC Web of Conferences **184**, 02008 (2018)
Annual Session of Scientific Papers IMT ORADEA 2018
https://doi.org/10.1051/matecconf/201818402008

Device for persons with low vision

Vlad Horațiu Sîrbu[1,*], *Daniela Mariana* Barbu[1]

[1]Transilvania University of Brasov, 29 Eroilor Bd., 500036 Brasov, Romania

Abstract. As defined by http://www.lowvision.org, "low vision is a bilateral impairment to vision that significantly impairs the functioning of the patient and cannot be adequately corrected with medical, surgical, therapy, conventional eyewear or contact lenses". The device is a keyboard that can be connected to any computer, phone or tablet via cable or Bluetooth. The keyboard will have eight buttons, two of them are predefined, and the other 6 form the Braille language cell.

1 Introduction

According to National Eye Institute (NEI) [1], low vision is a visual impairment that usually cannot be corrected by standard eyeglasses, contact lenses, medication, or surgery. It is often a loss of sharpness or acuity but may present as a loss of field of vision, light sensitivity, distorted vision or loss of contrast. Usual activities like reading, writing, watching TV or walking may be hard to do. Further, the mainly consequences of vision loss can cause anxiety, loneliness, helpless or even depressions of several kinds.

American Optometric Association consider low vision as a disorder caused by an ocular disease where the visual acuity is 20/200 or less for the better eye and / or a field of vision of 20 degrees or less and can`t be corrected or improved with regular glasses, surgery or medications [2].

The World Health Organization uses the following classifications of visual impairment [2]: viewing in the best eye with the best optical correction is:

- 20/20 to 20/60, it is considered a slight loss of vision or an almost normal view;
- 20/70 to 20/160, is considered to be moderate visual impairment or low moderate vision;
- 20/200 to 20/400, is considered a severe visual impairment or a severe low vision;
- 20/500 to 20/1000, is considered to be almost total visual impairment or almost total blindness;
- no perception of light, it is considered a total visual deficiency or total blindness or cecity.

The most common causes of low vision are [3] (figure 1):

- Macular degeneration - occurs due to circulatory disorders (which occur more often in the elderly) or factors not known primarily heredities (which we meet in children and young people).
- Diabetic retinopathy - occurs during diabetes and manifests as ophthalmic lesions (haemorrhages and white spots).
- Retinitis - it is a group of retinal lesions, due in part to circulatory disorders, plus a toxic factor.
- Amblyopia is a syndrome characterized by a decrease in visual acuity, whether or not associated with the modification of the foveal fixation reflex.
- Retinal detachment - in this disease, the retina separates from the underlying layer. This can be described as a curtain on one side of the field of vision.
- Cataracts - is the disorder that is manifested by the disorder of the lens that is subject to metabolic, physical, chemical, traumatic or age-related aggression.
- Glaucoma - under the name of glaucoma, there are a number of pathological conditions, the common symptom of which is the increase in eye pressure and the impairment of visual function.
- Acquired (traumatic injuries) - following a study, eye trauma can be of several kinds, namely: work accidents, accidents in everyday life, accident at birth, self-injury. Or, by their nature, trauma to the visual organ can be: contusions, actinic lesions, (physical or chemical), cuts cut or stung, with or without a foreign body.

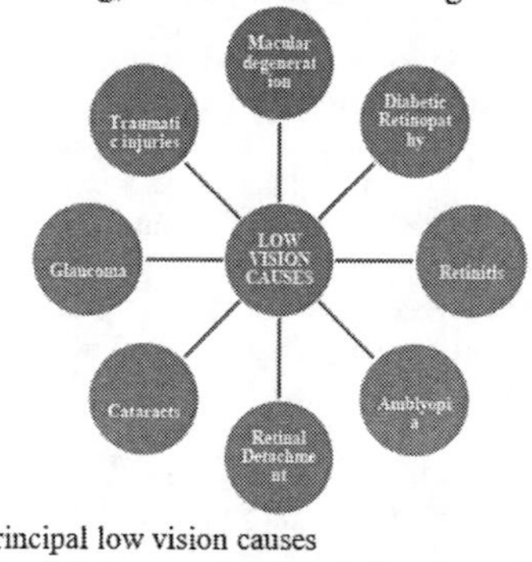

Fig. 1. Principal low vision causes

Some of these may be detected early and treatments can be made to stop disease progression (as macular degeneration or diabetes) or may have partial or total vision treatments such as retinal detachment or cataracts [4-5]. People with low vision may experience the following symptoms: loss of central vision, night blindness, loss of peripheral vision, blurred vision and hazy vision (figure 2) [6].

Email: sirbuvladhoratiu@gmail.com

© The Authors, published by EDP Sciences. This is an open access article distributed under the terms of the Creative Commons Attribution License 4.0 (http://creativecommons.org/licenses/by/4.0/).

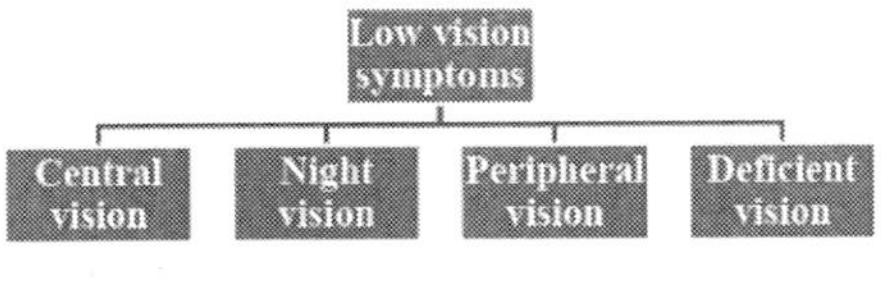

Fig. 2. Low vision symptoms

2 Assisting systems for low vision and blindness

2.1 The cane

This is the most distinctive sign of a man with a low vision or blindness. It is used for travel and is a very important tool in crossing the street because a person affected can cross a road without traffic signs only by lifting this cane [7]. There are several models available [8]:

- Long cane: is primarily designed as a mobility tool used to detect objects in a user's path. The length of the cane depends on the height of a user and extends traditionally from the floor to the user's sternum.
- Guide cane: This is a shorter cane that generally extends from the floor to the hip of the user - with a more limited mobility function. Used to scan borders and steps.
- Support cane: It is white and designed primarily to provide physical stability to a visually impaired user. Through its color, the cane also functions as a means of identification for someone with low vision or blindness.
- Green cane: Used in some countries to indicate a reduced vision, while the white cane indicates blindness.

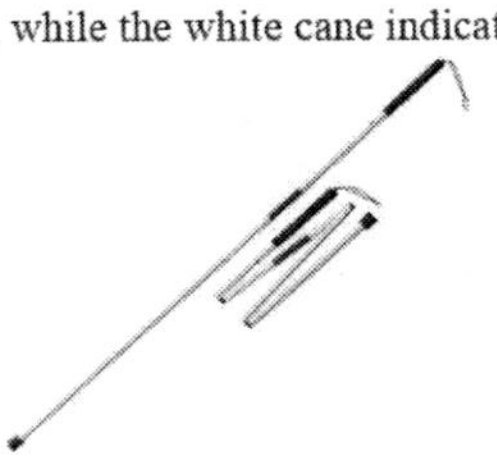

Fig. 3. The cane for blindness [8]

2.2 Using Braille alphabet

Braille is a write and print system for visual impairment people in which letters and numbers are identified by various arrangements of dots that are tactile [9]. It is a technique by which people with low or blindness can read and write. Each "cell" has 6 points, which are placed in a rectangle on 2 columns and 3 rows.

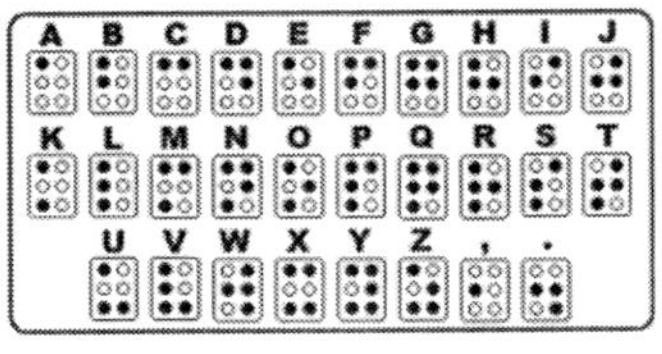

Fig. 4. Braille alphabet [9]

Braille can be "handwritten" - from right to left - with a paper puncher. When the sheet is inverted, the points are raised and the text can be read from left to right. It can also be achieved with the help of classical instruments, a slate and stylus with typewriters or access technology (Braille printers analogous to standard).

Therefore, the print / display equipment of Braille alphabet text, images, is based on a special paper that embosses the forms of the 3x2 matrix points that form the letters of the alphabet.

2.3 Ocutech VES-AutoFocus-Glasses

This device is an autofocus telescope, prescribed for the eye with a better sight [10]. It is mounted on special glasses frames, available in many colours and sizes. Also, the normal distance and bifocal prescription can also be included.

Fig. 5. Ocutech VES-AutoFocus [10]

VES-AutoFocus incorporates a 4x Keplerian telescope (the two lenses of the telescope are positive, the resulting image is real and reversed, prisms are used to reverse the image of the good field of view) coupled to an automated infrared computerized autofocus system [10]. It measures the focusing distance more than 30 times per sec to give a clear picture almost immediately and at any distance as close as 12 inches.

Operated with a separate, rechargeable battery, VES-AutoFocus can be used for over 8 hours continuously. This method to reduce low vision was adopted by practitioners and people with visual impairments around the world.

2.4 Smartwatch for the Blind

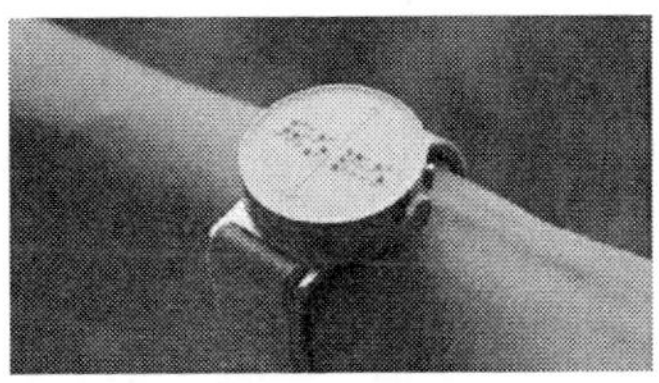

Fig. 6. Dot smartwatch [11]

Unlike most assisted smartwatches for blinds that rely on audio communication, Dot is the world's first Braille intelligent device that allows blind people to read messages on the screen. It has all the facilities a blind would need from a 21st century device, such as social or email notifications, Google Maps instructions, or weather updates. All information is displayed on smartwatch using 4 Dynamic Braille cells and its users can select the speed at which characters are updated [11].

Just like any other smartwatch the Dot connects to a smartphone via Bluetooth, when connected it can receive any text information from any app [11].

https://doi.org/10.1051/matecconf/201818402008

3 Device for persons with low vision

This project is about creating a device for people with low vision or blindness.

Specifically, this device will be a simplified acquisition system that will act as a Braille keyboard. This will allow people with low vison and / or blind people to write various SMS, E-MAIL or even WORD documents without having to carry a large keyboard.

The device will simplify the lives of visually impaired people for whom the 21st Century flat screens are not "communicative". The importance of the project is given by the constant evolution of devices for people with normal vision (from button phones to screen phones), a development that limits communication for people with low vision. Through this device, a new world opens for those with low vision to written communication and to what is "normal".

The first step in creating this project is designing the electrical system of the device, followed by a connection with Arduino Mega 2560 equipment and two keypads of four buttons each.

Fig. 6. Arduino Mega 2560 [12]

The MEGA 2560 is designed for more complex projects. With 54 digital I/O pins, 16 analogue inputs and a larger space for your sketch it is the recommended board for 3D printers and robotics projects. This gives your projects plenty of room and opportunities [12].

This keypad has 4 buttons, and since every key has its own wire line, no matrix code is required - just treat these like every day switches. The membrane is soft and has a removable paper backing to expose a strong adhesive so you can stick this on an enclosure and feed the cable through a slot.

The software will be loaded onto the board and tested for functionality. The results will be displayed on a LCD screen and transmitted through Bluetooth, Wi-Fi or cable.

The two keyboards each with 4 buttons will consist of a Braille cell (this consists of two columns of three buttons), and the other two buttons will have predefined space and delete functions [13].

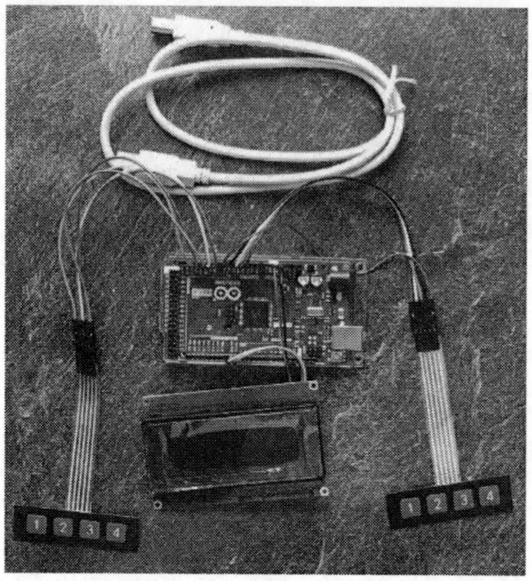

Fig. 7. Electronic components

The second step is designing the case with „Computer-aided design" (CAD) software and then to be produced from plastic (PLA) with a Creality 3D printer.

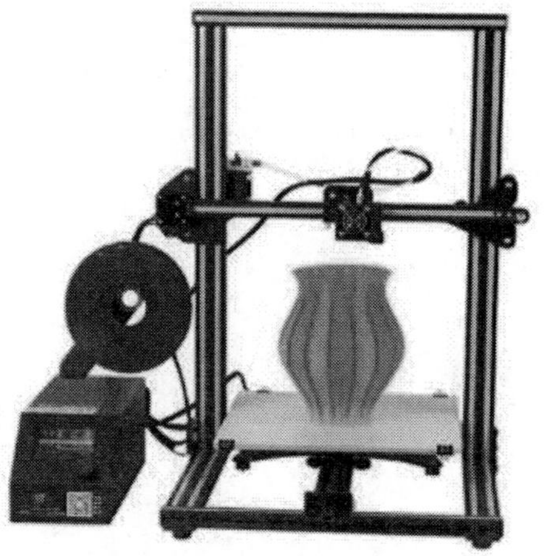

Fig. 8. Creality 3D CR-10S-3 [14]

This is an open-frame printer and heated print pad that can use a wide range of thermoplastic materials including PLA, ABS, TPU, Copper PET and composite materials [14].

Important specifications:

- High accuracy
- Print speed max. 200 mm / s
- Max. Print pad temperature 60 °C
- Layer thickness 0.05-0.35 mm
- Extrusion temperature max. 250 ° C

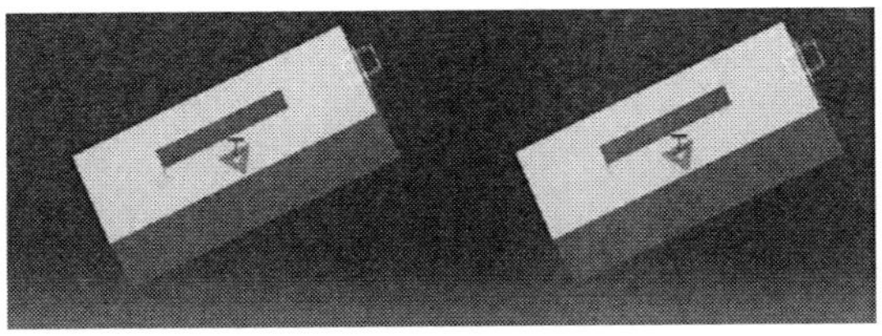

Fig. 9. CAD prototype of the case

Before the actual printing, the CAD model will be introduced into software called Cura 3.2.1., where the necessary print settings will be made.

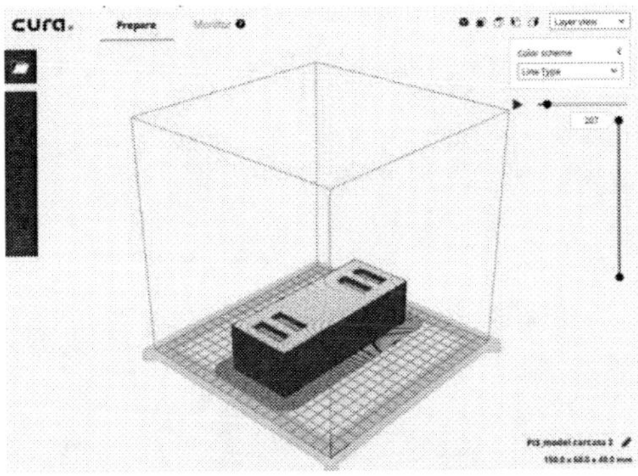

Fig. 10. Preparation of the casing printing and related settings

The final device will be assembled from the electrical part and the casing specially designed to implement all the necessary components.

The construction of the system has certain risks, such as applying over voltages on certain electrical components and implicitly damaging them, or creating multiple carcases in order to get the most ergonomic option, but these risks are part of testing a prototype, which means that broken components will be replaced and are assumed.

Once the system has been fully assembled, tests will be performed to verify that the device returns the desired characters. A series of instructions for commissioning and using the device will be prepared

4 Conclusions

In conclusion, this keypad prototype designed for people with low vision is designed to improve their communication capabilities by writing with technology. With its small size, lightweight materials, the device will be easy to carry and handle which opens their world to SMS, E-mails and is a step towards a common communication in the century of devices.

One of the main goals of the project is the low cost of the final product, because many devices designed for the visually impaired people have high prices which is why not many people use them.

This study was funded through a winning project in a university competition for final study projects.

References

1. Official Site of National Eye Institute (NEI) - https://nei.nih.gov/
2. Official Site of American Optometric Association, https://www.aoa.org/documents/optometrists/CPG-14.pdf
3. STL Chung, IL Bailey, G Dagnelie, et al., Optometry and vision science: official publication of the American Academy of Optometry, **89(9)**, 1244-1245 (2012)
4. HR Taylor, Investigative Ophthalmology & Visual Science, **44**,1413–1418 (2003)
5. MI Baritz, Cyberpsychology & Behavior **8(4)**, 304-305 (2005)
6. EM Vingolo, V De Rosa, D Domanico, F Anselmucci, Clinical Optometry, **7**, 53-58 (2015)
7. QK Dang, Y Chee, DD Pham, YS Suh, Sensors (Basel, Switzerland); **16(1)**, 95 (2016)
8. SM Bradley, CR Hernandez, American Family Physician. **84 (4)** (2011)
9. J Jiménez, et al., Survey of Ophthalmology, **54(1)**, 142-149 (2009)
10. Official Site of Ocutech, https://www.ocutech.com/product/ocutech-ves-autofocus/
11. Official Site of Steemit, https://steemit.com/technology/@penguinpablo/the-dot-world-s-first-braille-smartwatch-for-the-blind
12. Official Site of Arduino, https://store.arduino.cc/arduino-mega-2560-rev3
13. Official Site of Adafruit, https://www.adafruit.com/product/1332
14. Official Site of Formwerk, http://formwerk.ro/ro_imprimante.html

MATEC Web of Conferences **184**, 02009 (2018)
Annual Session of Scientific Papers IMT ORADEA 2018
https://doi.org/10.1051/matecconf/201818402009

Internet of things and machine parts laboratory

Dan Ioan Țarcă[1*], *Radu Cătălin* Țarcă[2], *and Ioan Constantin* Țarcă[3]

[1]Țarcă Dan Ioan, PFA, Oradea, Romania
[2]University of Oradea, Mechatronics Department, Oradea, Romania
[3]University of Oradea, Mechanical Engineering and Automotive Department, Oradea, Romania

Abstract. Internet of Things (IoT) is a relatively new approach. Due to its wide spreading and simplicity it has a various number of application, practically in every domain. This paper aims to present the way in which a classic stand can be transformed into a microprocessor-controlled system. The chosen stand is related to experimental determination of the friction coefficient and efficiency for a threaded mechanism.

1 Theoretical considerations

Same as threaded joints, in threaded mechanisms the friction torque between threads in nut and bolt is proportional to axial load, thread diameter and helix angle, number of threads and friction coefficient between threads. The relationship between axial load and friction torque is presented in eq.1.

$$M_{f1} = \frac{1}{2} F \cdot d_2 \cdot \tan(\beta_2 + \varphi') \quad (1)$$

Where
M_{f1} is the friction torque
F is the axial load
d_2 is the mean diameter of the thread
β_2 is the helix angle of the thread measured on the mean diameter (eq.2)

$$\beta_2 = arctg \frac{p}{\pi \cdot d_2} \quad (2)$$

p is the thread pitch
φ' is the reduced friction coefficient (eq.3)

$$tan\varphi' = \mu' = \frac{\mu}{cos\frac{\alpha}{2}} \quad (3)$$

μ is the friction coefficient for the bolt and nut materials
α is the angle of the thread profile.

In Mechanical Engineering student laboratory activities this torque is measured by the mean of an elastic system based on a spring and a device used to measure the elongation of the spring, for example a dial indicator. In this paper the authors present a way to develop this stand using affordable components with small modification in the mechanic system of the stand.

2 Stand analyses

2.1 Existing test stand [1]

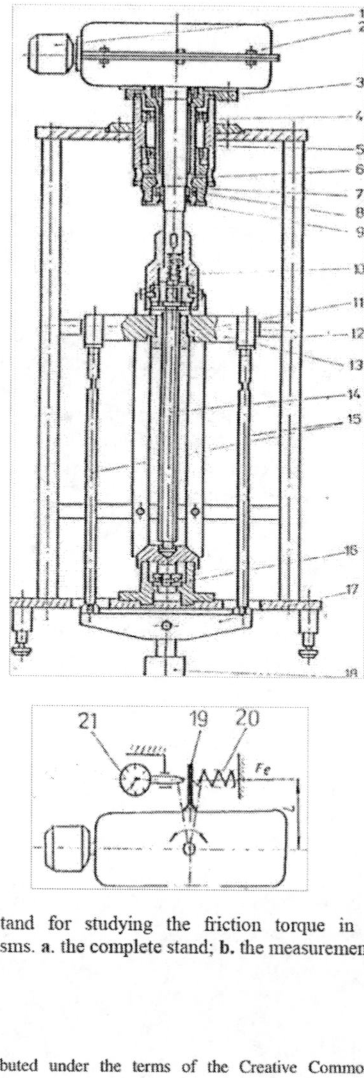

a

b

Fig.1 Stand for studying the friction torque in threaded mechanisms. **a.** the complete stand; **b.** the measurement system

[*] Corresponding author: bh90tdi@gmail.com

© The Authors, published by EDP Sciences. This is an open access article distributed under the terms of the Creative Commons Attribution License 4.0 (http://creativecommons.org/licenses/by/4.0/).

The original test stand is presented in fig.1 and is composed by DC motor (1) which rotates the threaded bolt (14) by the mean of a worm gear mounted inside a housing (2), a shaft (6) and a clutch (10).

The radial bearing (5) ensures that the shaft is correctly positioned against the fixed crosspiece of the device during operation.

The threaded bolt rotation generates a translational movement of the nut (11) and also of the crosspiece (13) which is guided by the columns (12). Axial load is created through the weights (18) attached to the crosspiece (17) which load the nut through tyrants (15).

During the lifting of the weights (18) axial load creates a frictional torque between the bolt threads and nut threads which acts as a resistance in the rotating movement of the bolt. Currently the fictional torque is measured by the mean of an elastic system which uses a spring (20) and a dial indicator (21) which measures the displacement of the indicator (19). The spring and the dial indicator are placed at the certain distance (L) from the rotational axis. When the friction torque equals the torque generated by elastic forces the system is in equilibrium. In this case one can calculate the value of the torque generated by the elastic force:

$$M_t = F_e \cdot L = kxL \quad (4)$$

Where:

M_t is the torque generated by the elastic forces (spring)

F_e is the elastic force developed in spring; $F_e = kx$

k is the characteristic of the spring

x is the spring deformation, measured by the dal indicator

L is the distance between the dial indicator and bolt axis

Tests are made for different weights.

2.2 Modernized test stand

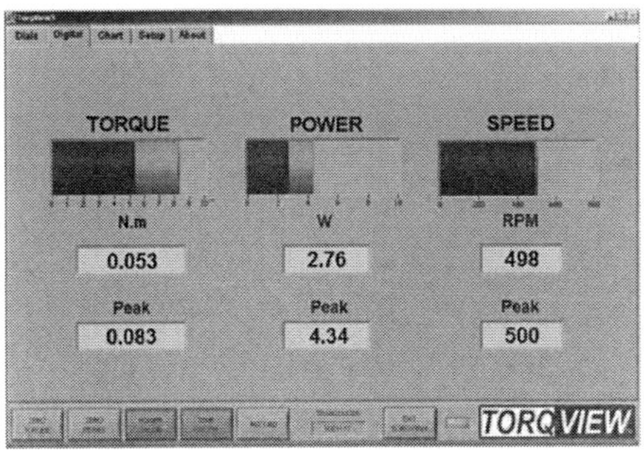

Fig. 2 Interface for data acquisition of torque

Another way of measuring the resistant torque is by the main of a torque sensor which will transmit to a PC both the values of the measured torque and the motor speed. Also, power absorbed by the motor can be measured. The interface for data acquisition is presented in fig.2 together with the results measured by the sensor. For this case an expensive E200 ORT optical torque sensor from Sensor Technology is used, on a scale of 1 Nm, placed between the shaft and the threaded bolt. From the original stand measuring elements 19 to 21 are no longer needed.

2.3 Internet of Things modified stand

A simpler and cheaper way for measuring the resistant torque is presented in fig.3. The measurement principle is similar to the first stand i.e. torque is calculated by the mean of applied forces. A load cell sensor (20) mounted in a fixture (21) is used as a force sensor, without the need of a spring. The sensor is pushed with the force F_e by an arm (19) provided with a tip placed at a distance L from the centre of rotation of the threaded bolt which generates the force to be measured. Usually a load cell sensor is used in equipment to gauge weight and force. In this case the friction forces generated by the rotation of the bolt are measured by the mean of the cell sensor. The sensor unit voltage signal is converted into a digital signal using a 24-bit analogue-digital converter (ADC) and transferred to a microcontroller unit (MCU). The force logger schematic is presented in fig.4. The supply voltage values are indicated by (E+) and (E-) while load cell output voltage is indicated by (S+) and (S-). In fact, the load cell sensor is a classic Wheatstone bridge following the classic rules presented in equation 5.

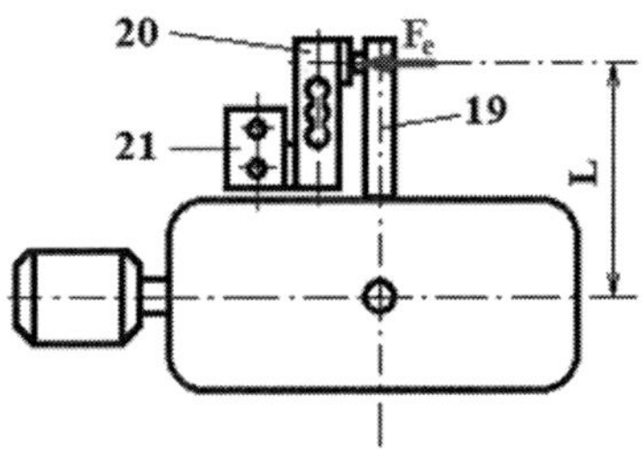

Fig. 3 Stand modified for load cell sensor

$$V_{R2} = I_U \cdot R_2 = \frac{V_{ref}}{R2+R1} R2; \; V_{R4} = I_B \cdot R_4 = \frac{V_{ref}}{R3+R4} R4$$

$$V_{OUT} = V_{R2} - V_{R4} = \frac{R2R3 - R1R4}{(R2+R1)(R3+R4)} V_{ref} \quad (5)$$

Where:

V_{R2} and VR4 are the voltage across R2 and R4 respectively (fig.4)

R_1 and R_2 respectively R_3 and R4 are the left and right branches of the Wheatstone bridge

I_V and I_B are currents passing on the left and respectively right branches.

V_{OUT} is the output voltage of the Wheatstone bridge

V_{ref} is the excitation (or supply) voltage

For this stand a 5 V power supply is set as the reference or excitation voltage (V_{ref}). Descriptions of the load cell sensor is presented in table 1.

Maximum output force measured on the first presented stand, achieved for a loading weight of 6.95kg (79.99N) was 1.359N, so that a maximum output of 3N which can be measured by the load cell sensor is more

https://doi.org/10.1051/matecconf/201818402009

than sufficient to achieve usual laboratory values, which are shown in table 2.

Table 1 Load sensor cell properties

Rated output	1.0±10%mV/V
Excitation	3-15V
Zero balance	±5% RD
Nonlinearity	±0.02% RD
Hysteresis	±0.015% RD
Repeatability	±0.015% RD
Creep(30min)	±0.03% RD
Safe overload	150% F.S.
Ultimate overload	200% F.S.
Compensated temperature	-10...+40°C
Operating temperature	-20...+60°C
Temperature shift zero	±0.05% RD/10°C
Temperature shift span	±0.03% RD/10°C
Input resistance	405±10 ohms
Output resistance	350±3 ohms
Insulation resistance	>2000M ohms
Ingress protection	IP65
Material of element	Aluminium alloy
Cable	Ø4*250mm
Wiring code	Red--E+ Black--E- Green--S+ White--S-

Note:
%RD = accuracy as percentage of reading
%F.S. = accuracy as percentage of full scale

Table 2 Data achieved during tests for classic stand

Specification	U/M	Value 1	Value 2	Value 3
Bolt /nut materials	-	Steel/Bronze		
Thread type	-	Tr24		
Load	N	56.408	65.727	77.990
Load	kg	4.75	5.70	6.95
Spring elongation	mm	1.75	2.05	2.47
Friction torque	N·mm	96.250	112.750	135.850
Force in spring	N	0.963	1.128	1.359
Friction coefficient	-	0.080	0.081	0.084
Efficiency	-	46.64%	46.39%	45.68%
Distance from measurement system to axes	mm	100		

The force logger scheme is presented in fig.4, while in fig.5 the wiring scheme can be seen. The voltage for the load cell is supplied by the 24bit HX711 ADC and amplifier, and input values of the voltage difference are also acquired and amplified by the HX711 module (fig.4), then transmitted to the Arduino Uno microcontroller system which converts data into useful information on a display. With a rated output of 1mV/V, for an excitation voltage of 5V the load cell will produce at maximum load (full scale, which corresponds to 0.3kg x 9.81m/s^2=2.943N), 5mV.

The resolution of the scale is given by the AD converter, that is:

$$Resolution = \frac{5V}{2^{24}} = 0.298\mu V \quad (6)$$

This specification is not usable, since the noise generated by wire and the environment have greater values.

The sensitivity of the load cell is calculated based on relation [2]:

$$Sensitivity \left[\frac{mV}{N}\right] = \frac{exc\ voltage\ (V) \times rated\ output\left[\frac{mV}{V}\right]}{max\ force\ [N]} \quad (7)$$

With actual values sensitivity is 1.698mV/N, which is quite satisfying.

Conversion of the output value of the load cell to force values is based on the formula [3]:

$$Force\ [N] = K \times \left(Rated\ output\left[\frac{mV}{V}\right] - Offset\right) \quad (8)$$

Calibration of the load cell is made in two steps. Firstly, the offset of the load cell is measured, by putting the load cell on a plane surface, unloaded, while measuring the output value.

K factor (load cell factor) is computed by comparison to known forces applied to the cell while adjusting the output to correct value. After knowing the *K* value, equation 8 can be used to compute the force measured by the load cell.

If needed, the value of the friction torque can be directly displayed by multiplying relation 6 with the distance from the threaded bolt to the point where the force is applied. The 16x2 I^{2}C (Inter-integrated Circuit) liquid crystal display (LCD) outputs the values of the force (torque) on the first row and eventually, the value of threaded bolt rotational speed, if needed.

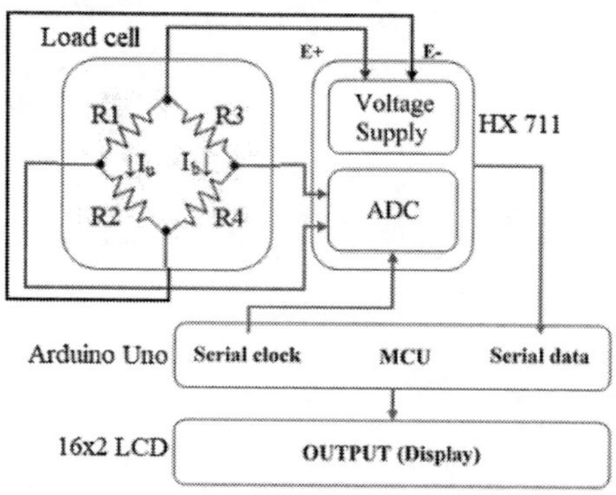

Fig. 4 Force logger principle

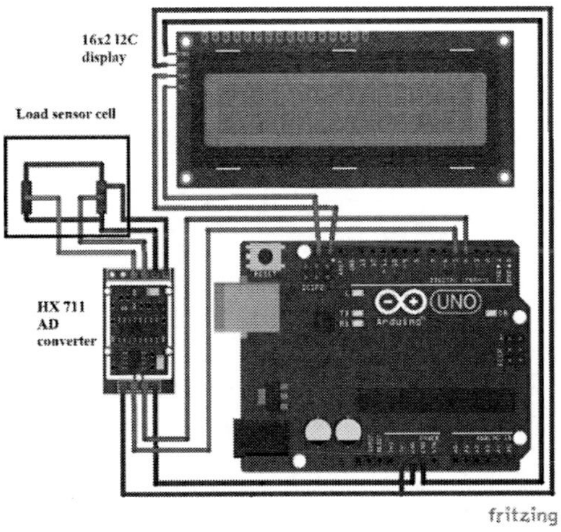

Fig. 5 Wired scheme of the force logger

As Microcontroller Unit (MCU), the Arduino UNO capabilities are exploited, due to its affordability, ease of use and simplicity in writing code [4]. Also, Arduino Nano is suitable for this modification, if miniaturization is one of the desired aspects of the designer.

Other modules can be attached to the Arduino, such as analogue inputs for continuous data reading, digital ports either for input (used with digital sensors) or output (for example modification of the motor speed using power width modulation (PWM) facilities of the board) and so on.

3 Conclusions

Laboratory activities for mechanical engineering students might be at a crucial point. Due to rapid development of Mechatronics and their spread in all field of activity, Internet of Things emerges, giving a new breath to classic assumption of mechanical engineering.

Many of the stands based of force measurements can be adapted as presented in the paper, with almost no effort, both financial and material.

Three types of stands were presented in the paper:

The classic mechanical stand, which uses the deformation of an elastic system deformation (spring) to measure forces and thus, torques, and all other physical dimensions. Even though is an obsolete method, it has the advantage of making students to find all kind of means (direct or indirect) by which one can measure needed dimensions.

The second method allows, with small changes in the configuration of the stand, to get more accurate results, near to those acquired in specialized laboratories, which, for research purposes is mandatory. For didactic activities though, these professional or semi-professional systems are too expensive for their purpose, which mainly is to offer students ways to find correlations between certain physical/mechanical dimensions.

Finally, the last method uses common, low priced materials. Not only that the new concept of Internet of Things offers a large variety of solutions to daily problems, but, as a matter of fact, this concept allows the students to easily access the world of experiment and innovation.

In this paper, the modification made to the laboratory consisted in changing the obsolete mechanical way of measuring with a combined one, in which microcontrollers together with microelectromechanical systems (MEMS) play an important role. The elastic system (spring) has been substituted with another elastic system (load cell), but sensors (resistive gauges) took the place of the dial indicator. Other problems related to the design of the IoT system were the selection process of the load cell, the problems related to transducers selection, creating the software for measurement and also for the control of the system, and so on.

Also, differences between the second method presented in the paper and the IoT one resides in the fact that beside the fact that IoT is much cheaper, it sustains the students' creativity, motivating them to find functional solutions with a great level of confidence in respect to the results.

References

1. C. Cornea, *Machine Parts. Laboratory Handbook (Organe de Masini. Îndrumar pentru lucrări de laborator)*, (1996)
2. W. K. Lee, et al. *Physiological Signal Monitoring Bed for Infants Based on Load-Cell Sensors*. Sensor Journal, 390-409, (2016).
3. https://www.phidgets.com/docs/Load_Cell_Primer
4. A.G. Smith, Introduction to Arduino: A piece of cake!, www.introtoarduino.com (2011)
5. M.R. Thakur, Measurement made simple with Arduino, https://docs.google.com/viewer?a=v&pid=sites&srcid=ZGVmYXVsdGRvbWFpbnxwYXRyaWNpYXN1c2FuYW5pbmRpdGF8Z3g6NTE5NjU2MTRlZjM0ODA1Zg (2016)

MATEC Web of Conferences **184**, 02010 (2018)
Annual Session of Scientific Papers IMT ORADEA 2018
https://doi.org/10.1051/matecconf/201818402010

Mechatronical system used for the recovery of motor functions

Roxana Andreiaş[1,*], and *Daniela Mariana* Barbu[1]

[1]Design Product, Mechatronics and Environment Department, Transilvania University of Brasov, Brasov, Romania

Abstract. Robotics of medical recovery represent a subject of great interest among various research centers that are trying to develop different complex systems dedicated to the recovery of motor functions through fulfilling the required tasks imposed by applied therapy programs. In order to implement the concept of 'robotical therapy' it is required to use a robotic system specially designed to adhere to medical protocols and to meet the requirements of therapeutic exercises by programming the command and control unit. This paper is in the field of mechatronic systems used in medicine because the purpose of this kind of equipment is to help patients who underwent surgery in the lower limbs regain the mobility of the affected joints as soon as possible. The proposed objective is to improve an exising concept so that a patient may be able to use it in the comfort of their own house, but still being 24/7 monitorized using various sensors and even being able to contact medical care assistance in case of emergency.

1 Introduction

Rehabilitation robotics is a field of research dedicated to understanding and augmenting rehabilitation through the application of robotic devices. Rehabilitation robotics includes development of mechatronic systems specially created for assisting different sensorimotor functions(e.g. arm[1], hand[2], leg[3], ankle[4]), development of different schemes of assisting therapeutic training and assessment of sensorimotor performance (ability to move) of patient.[5]

The main objective is to use robots as therapy aids instead of assistive devices. Rehabilitation using robotics is generally well tolerated by patients and has been found to be an effective adjunct to therapy in individuals suffering from motor impairments, especially due to stroke or after surgery recovery.

Although popular culture presents the image of a robot as a mechanical, humanoid device, often evil in intent, in attempt to follow the development of rehabilitation robotics, it should all start from an official definition. The Robot Institute of America defined a robot as

„A reprogrammable, multifunctional manipulator designed to move material, parts, tools or specialized devices through variable programmed motions for the performance of a variety of tasks."[6]

It is obvious that this definition was intended for industrial robots but it identifies the key features of programmability, flexibility and movement.

In 1987 the Department of Trade and Industry in the UK had an initiative in encouraging the usage of robots in areas other than the industrial field. Using the definition of Advanced Robotics

„The integration of enabling technologies and attributes embracing manipulators, mobility, sensors, computing (IKBS, AI) and hierarchical control to result ultimately in a robot capable of autonomously complementing man's endeavours in unstructured and hostile environments." [6], it can be stated that the aim of the integration of advanced technologies is to develop a device that can both operate autonomously and in an enviroment that may be unstructured and/or hostile.

The use of robots in rehabilitation is often associated with assistive devices, some used in special dedicated areas, some as aids to daily living. There are some other areas where robotic technology has been applied such as in mobility, where traditional wheelchairs, prosthetics and orthotics are being developed to increase the usefulness. A major area of concern is represented by the robot mediated therapy, where robots are used to increase the effectiveness of traditional therapies and even to develop new therapy protocols.

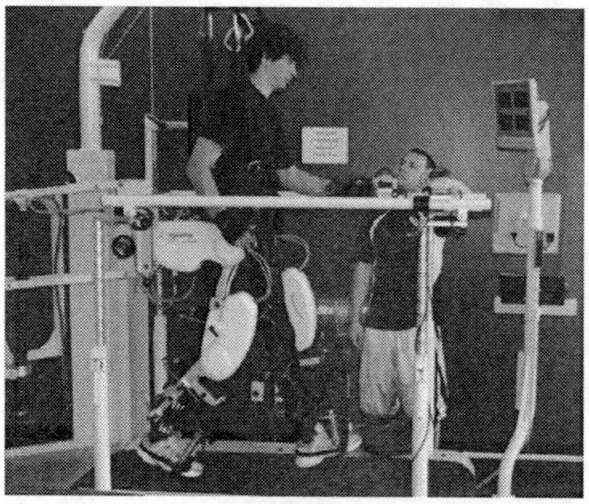

Fig. 1 – Lokomat therapy robot [7]

Researches in neuroscience had shown that the brain and spinal cord retain a remarkable ability to adapt, even after injury, through the use of practiced movements. The first

* Corresponding author: andreias.roxana@yahoo.com

© The Authors, published by EDP Sciences. This is an open access article distributed under the terms of the Creative Commons Attribution License 4.0 (http://creativecommons.org/licenses/by/4.0/).

robot used in that way, MIT-Manus [8], helped stroke patients to reach across a tabletop if they were unable to perform the task by themselves. Patients who received extra therapy from the robot improved the rate of their arm movement recovery. Another therapy robot, the Lokomat (presented in figure 1), supports the weight of a person and moves the legs in a walking pattern over a moving treadmill, with the goal of retraining the person to walk after spinal cord injury or stroke.[9]

The major limiting factor in the development of rehabilitation robots is that researchers do not know what exactly needs to happen in order for the nervous system to adapt to overcome a physical impairment. Researchers are developing rehabilitation robots that assist in movement, resist movement when it is uncoordinated, or even make movements more uncoordinated in an attempt to trick the nervous system into adapting. Advances have been made in the development of robotic exoskeletons, which are lightweight wearable devices that assist in limb movement. Other types of rehabilitation robots could play a role in assisting the nervous system to regenerate appropriate neural connections following stem cell and other medical treatments.[10]

Robotic systems have the ability to take precise measurements of kinematic and dynamic values, which are far more reliable than those achieved with human error, and have the potential to be very useful for assessment purposes. This being said, there still remains a need to develop standardized procedures and protocols in order for this data to be useful. Currently some examples of the data used from the robotics systems during assessments are ROM, walking distance, gait velocity and other various dynamic measures, but we do not yet have a standardized measure for assessment, like those seen in other gait related assessments (Barthel Index, Dynamic Gait Index, etc.).

Furthermore, the effectiveness of robotics has not been shown to be highly superior to that of typical manually-assisted therapy provided by therapists and this is a driving reason as to why it has not yet been implemented into regular practice.[11]

One of the less obvious examples of incorporating robotics into rehabilitation is that of its assessment capabilities. While machines and receptors have been used for years in the rehab field (eg. EMG, force plates), their primary purpose has been to just collect raw data. With the right robotics, measures such as joint angles, velocities, amplitudes can be observed and recorded easily. This information can then be used as an outcome measure to see the progress a patient can make during a treatment plan. [12]

The purpose of this work is to develop a mechatronic system that will serve as an aid for patients who underwent surgery in the lower limbs in order to regain the mobility of the affected joints as soon as possible.

2 AID device

Gait training represents one of the main purposes for lower limb robotic rehabilitation. The current methods of improving gait require the assistance of multiple therapists to help the patient move each joint and leg appropriately for best results. This represents an expensive and very difficult approach for rehab because it is extremely labour intensive for the therapists to provide such care. [12]

Using robotics, the patient will be strapped onto the robotic device, allowing the therapists attention to be concerned only for providing supervision and for setting up of the device. This type of therapy should deliver the same mechanical movements as the traditional approach in a shorter period of time.

2.1 CPM Machine

Studies suggest that continuous passive motion (CPM) can improve range of motion (ROM) in those patients undergoing surgical release of arthrofibrosis of the knee or manipulation of the knee under anesthesia. In these settings, CPM provides for early post-operative motion and is considered a substitute for active physical therapy (PT).[13]

The method of rehabilitation with the use of the CPM device affects the acceleration of the process of post-operative knee rehabilitation after arthroscopy and arthroplasty.It allows, among others, the improvement of metabolism within the joint, accelerates healing of damaged parts of the joint and reduces increased soft tissue tension. Furthermore, one can observe a faster absorption of hematoma, a better blood supply to the limb and increased strength of the ligament tension.[14]

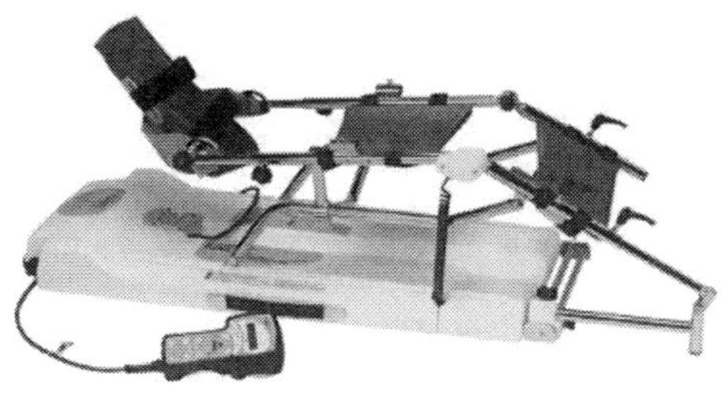

Fig. 2 – Commercial example of CPM machine (Kinetec Performa knee CPM) [15]

Figure21 presents an existing CPM machine for the purpose of studying its characteristics in order to adapt and develop an improved version.

Criteria usually considered in the design and construction of these machines are as follows:

- Adjustability to a wide range of patient sizes;
- Comfort, particularly for long uninterrupted application;
- Use in a variety of settings: in bed while sleeping, while ambulating during the day, at home or in hospital;
- Durability;
- Light weight;
- Low cost;
- Full anatomic range of motion capability;

https://doi.org/10.1051/matecconf/201818402010

- Multi-axis motion capability;
- Minimum loading of joints;
- Precise and variable range of motion (ROM) control;
- Speed control;
- Safe and reliable operation;
- Simple and straightforward operation;
- Protection for unstable joints.

There is also a significant creative and innovative element to the process that can not be ignored. The art of creative product design lies precisely in being able to find simple and creative solutions to constantly broaden and redefine the range of what is considered possible.[16]

2.2 Overview of Device

The proposed device should be able to provide different additional functionality beyond ensuring the proper movement of the rehabilitation limb.

Mechanical components used to elaborate this device should be rigid enough to resist working under maximum load (min. 90 kg or greater for adult men). Also, the materials from which this device is to be made are lightweight, so the final resulted equipment could be portable and easy to manipulate. The most important aspect is that this device will meet all the safety requirements (it is to be used by patients by themselves at home).

It is designed as a combined mechanism, with an assembly of three links made from aluminum (Figure 3 : (a) Femural link ; (b) Tibial link ; (c) Plantar link) which will sustain the patient's leg that will be strapped onto the device. The femural and tibial links will be conjoined using a mobile joint (figure 3-(d)) in order to execute motions of flexion and extension of the affected knee joint. There are also four links attached to stabilize the whole process (one on each side of the assembly).
The elaborated design will also incorporate two 12V servo motors (figure 3-(e)) that will ensure the movements of the linkage mechanism along the base board.

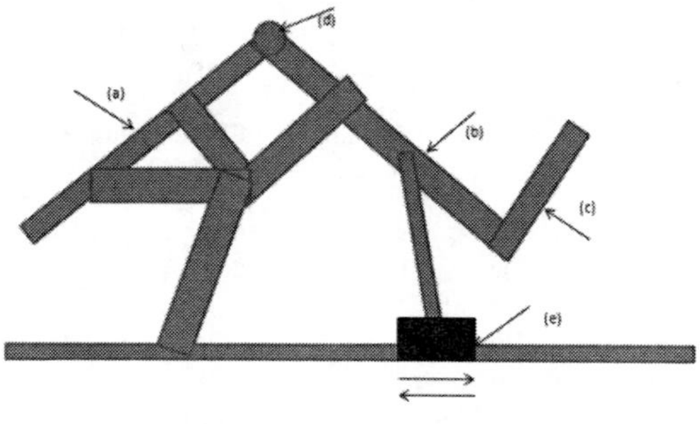

Fig. 3 – Proposed structure of own CPM machine

The device has open-loop control with a constant power supply. The outer current of the power supply can adjust the speed of the device, so that the user can control and adapt the needed motion.
This device will be adjustable for several leg lengths because the plantar link will be provided with the possibility of extension for a range of 10 to 20 centimeters.

Also, it will incorporate several soft covers for all the metallic components that come in contact with the patient's leg in order to make any procedure as comfortable as possible. Nevertheless, a fixed foot ankle orthesis can be added for more confort and better support.

2.3 Software and Programming

The micro controller used for the custom CPM device is an ATmega2560 Arduino board. Using the Arduino dedicated application, the source-code containing all the parameters involved and controlled during therapy programs can be uploaded and runed in order to set the motors in motion.
The application allows to upload several programs and to access any of it before or during the procedure. It will allow to adjust the amplitude of the movement, the angles needed for the knee flexion motion, the total duration of a procedure etc.
As an additional feature, several sensors can be added to the Arduino board in order to monitor various functions during the therapy procedures. Adding two-three muscular sensors will give the superviser/user of this device the possibility to monitor the muscular activity and even to avoid any overstressing of the affected joint.

3 Results and Conclusions

The purpose of this project is to try and create a CPM custom machine that can be used to complete a rehabilitation routine on patients that are recovering after knee surgery using mechatronic knowledge and minimal medical information.

Continuous passive motion is a common rehabilitation therapy for post-operative knee surgery patients. However, speed, duration, maximum flexion angle, and maximum extension angle have no standard settings for physical therapy methodology. Adjusting these input parameters to maximize the benefits of the post-operative rehabilitation treatment may reduce the treatment needed time.

The study or observation of the biomechanical properties of a knee undergoing passive motion can be beneficial for several surgical procedures in addition to post-operative applications. For example, this testing device will be used in our university's institute by other fellow researchers to simulate and investivate the effects of different parameter values and motion programs.

Also, we can observe passive motion for various types of prosthetics and study the effects on the biomechanical properties in the knee.

In addition to the study of the passive motion effects on the knee, this project may lead to future studies involving active motion. Passive motion in the knee does not require muscle activation.

Simulation of muscle contraction during knee flexion and extension by pulling on the tendons may give further insight into the biomechanical properties of a knee undergoing active motion, which is important in determining biomechanical properties during everyday activities such as walking, running, or stair climbing.

The experimental test setup will motivate further research studies that may improve the surgical techniques, reduce the treatment time for physical therapy, or improve the outcome of rehabilitation programs for post-operative knee surgery patients.

References

1. R.B. Brewer, S.K. McDowell., L.C. Worthen-Chaudhari, *Poststroke Upper Extremity Rehabilitation: A Review of Robotic Systems and Clinical Results* (Topics in Stroke Rehabilitation. 14 (6): 22–44, 2007)
2. S. Balasubramanian, J. Klein, E. Burdet, *Robot-assisted rehabilitation of hand function* (Current Opinion in Neurology. 23, 2010)
3. Y. Kang, D. Jeon, *Rehabilitation robot control using the VSD method* (System Integration(SII) IEEE/SICE International Symposium: 192., 2012)
4. K.P. Michmizos, S. Rossi, E. Castelli, P. Cappa, H.I. Krebs, *Robot-Aided Neurorehabilitation: A Pediatric Robot for Ankle Rehabilitation* (". IEEE Transactions on Neural Systems and Rehabilitation Engineering. 23 (6): 1056–67, 2015)
5. L. Marchal-Crespo, D.J. Reinkensmeyer, *Review of control strategies for robotic movement training after neurologic injury* (Journal of NeuroEngineering and Rehabilitation. 6: 20, 2009)
6. S. Balasubramanian, R. Colombo, I. Sterpi, V. Sanguineti, E. Burdet, *Robotic Assessment of Upper Limb Motor Function After Stroke* (American Journal of Physical Medicine & Rehabilitation. 91 (11 Suppl 3): S255–69., 2012)
7. Center For Neuro Recovery. Lokomat Pro Gait Training (20 April 2018)http://www.centerforneurorecovery.com/wp-content/uploads/2015/06/Lokomat-Pro-Robotic-Gait-Training
8. H. Krebs, M. Ferraro, P.S. Buerger, M.J. Newbery, A. Makiyama, M. Sandmann, D. Lynch, B.T. Volpe, N. Hogan, *Rehabilitation robotics: pilot trial of a spatial extension for MIT-Manus* (Journal of NeuroEngineering and Rehabilitation. 1 (1): 5., 2004)
9. R. Riener, M. Frey, M. Bernhardt, T. Nef, G. Colombo, *Human-Centered Rehabilitation Robotics* (International Conference On Rehabilitation Robotics (ICORR), 2005)
10. Rehabilitation robot (20 April 2018) https://www.britannica.com/technology/rehabilitation-robot
11. I. Diaz, J. Gil, E. Sanches, *Limb Robotic Rehabilitation: Literature Review and Challenges* (J Robotics.;2011:1-11., 2011)
12. S. Maggioni, A. Melendez-Calderon, E. van Asseldonk, V. Klamroth-Marganska, L. Lünenburger, R. Riener et al., *Robot-aided assessment of lower extremity functions: a review* (. J Neuroeng Rehab;13(1)., 2016)
13. M. Hillman, *Rehabilitation Robotics from Past to Present – A Historical Perspective* (Advances in Rehabilitation Robotics, LNCIS 306, pp. 25-44, 2004)
14. R. Trochimczuk, T. Kuźmierowski, *Kinematic analysis of CPM machine supporting to rehabilitation process after surgical knee arthroscopy and arthroplasty* (Int. J. of Applied Mechanics and Engineering, vol.19, No.4, pp.841-848, 2014)
15. Kinetec Performa knee CPM machine (20 April 2018) http://www.thelivingcentre.co.uk/kinetec-performa.html
16. J. Saringer, *Bridging theory with practice: practical considerations in the design of CPM machines* (Continuous Passive Motion: A Biological Concept for the Healing and Regeneration of Articular Cartilage, Ligaments, and Tendons : From Origination to Research to Clinical, 1992)

Automatic fuel tank monitoring, tracking & theft detection system

Komal D/o Shoukat Ali Khuwaja[1,*], *Brohi* Arif Ali[2], *Vlad O.* Mihalca[1] and *Radu Cătălin* Ţarcă[1]

[1]Mechatronics Department, University of Oradea, Romania
[2]Electronics Department, MUET Jamshoro, Pakistan

Abstract. This paper presents the design, implementation and characterization of a hardware platform for Fuel Management System (FMS) for fuel carrying vehicles. The primary design goal is to devise a system capable of monitoring the fuel level in real time. This system is proposed for fuel carrying road tankers which carry fuel from oil depots to end users such as petrol-stations. This system is based on hardware as well as software. The hardware part consists of fuel level circuits, on-board Arduino, GSM and GPS modules. While the software part consists of LabView for Database purposes. This system offers an Automatic Electronic System (AES) which replaces manual monitoring of fuel. This system measures fuel volume and sends measured volume to the owner's mobile as well as owner's computer through GSM network. It also provides a technique for detecting theft or fraud incidents in case of fuel theft or fuel leakage. Additionally the system can track the location of each fuel carrying tanker in order to locate the place of any incident that could have occurred. This system allows for an automated analysis and monitoring of fuel level, having a reduced cost due to affordable and easy-to-acquire electronic components.

1 Introduction

Fuel Management Systems (FMS) are used to maintain, control and monitor the fuel consumption in any type of organization that uses transport, including: rail, road, air and water as means of business. Currently the organizations are facing a serious problem of managing the fuel transportation due to manual monitoring. This manual monitoring provides an inefficient way of calculating and analysing the fuel ingestion and can lead to financial losses for the company.

In the typical scenario used by most organizations there is no logging or auditing mechanism to check that the number of litres written on the receipt correspond to the actual amount of fuel in the tank. The driver takes the vehicle to the fuel depot, fills the tank and gets the receipt from the pump manager, mentioning the number of litres of fuel inserted into the vehicle tank, date and the amount. The driver gives this receipt to the person in charge of the organization, which keeps up the record and sends the amount to the pump-owner at the end of the month.

This scenario is affected by many possible flaws: the driver does not know whether the amount of fuel inserted in the fuel is the same as the amount declared on the receipt. Another flaw of the system is caused by misleading information created by the pump station manager. A third example would be the involvement of the staff from company's management personnel which commit fraud by falsifying information regarding the amount of fuel inserted.

Many companies must keep all the above mentioned factors into account, therefore the need for an automated electronics system (AES) that not only checks the fuel ingested but also digs out the theft of fuel in order to save company from loss [10]. Furthermore, this system provides an efficient way of tracking vehicles in order to localize the fuel Tankers as well as a software database for fuel records, residing on end user's workstations or mobile phones [11]. The database contains the number of fuel litres at the time of insertion and number of fuel thefts when the volume of tanker changes a lot before reaching the destination.

Fuel transportation or fuel delivery to the sites takes place through various ways such as approved highway tanks, heavy ships, trains and through pipelines. This system only focuses on highway fuel tankers for the transportation of fuel on the subject of management. The System offers the most comprehensive solution on the market for knowing exactly how much fuel is inserted and how much is lost. From complete fuel accountability to total reconciliation and protection against fuel theft, this keeps our assets covered.

This system provide an efficient, cost effective and automated fuel management system that is especially customized for the companies of Pakistan that view stealing fuel as a serious problem but due to costly existing solutions, it is difficult for them to afford the systems already available on the market.

1.1 Fuel Management System (FMS)

GPS Fleet Tracking is an all-in-one management solution that can be used to monitor, manage, and recover assets. By tracking valuables, we can better manage and monitor

* Corresponding author: komalali_89@yahoo.com

© The Authors, published by EDP Sciences. This is an open access article distributed under the terms of the Creative Commons Attribution License 4.0 (http://creativecommons.org/licenses/by/4.0/).

their whereabouts, cutting down on wasted time and fuel consumption. With GPS Fleet Tracking information, we can evaluate performance and cut down on extra costs. This enhances the company's value by improving the level of service provided, leading to higher profits and customer satisfaction.

This automatic Fleet Tracking reports are easy-to-read graphs that let us identify operational trouble spots and improve fleet management [1].

1.2 Automatic Fuel Management System (AFMS)

Automatic Fuel Management System (AFMS) consists of basic electronic components which are then mounted on highway tankers for the purpose of measurement and monitoring of its volume. It measures the number of fuel litres when inserted into the tank as well as when fuel is stolen or leaked and then it sends that measured data for analysis and additional statistics.

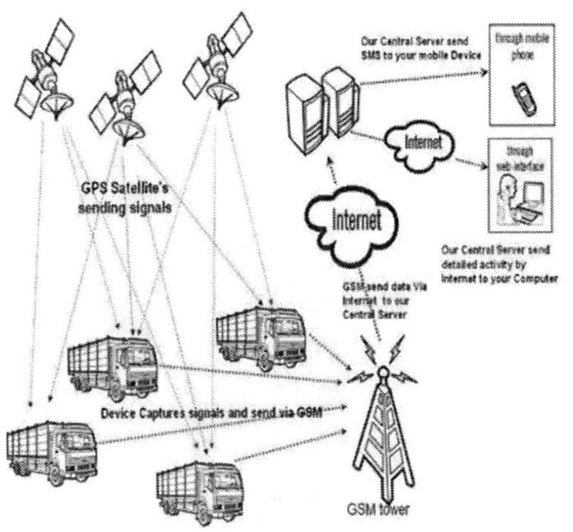

Fig. 1. Automatic Fleet Management System (AFMS)

According to Wikipedia a "GPS Fleet Tracking Is An All-In-One Management Solution That Lets You Monitor, Manage, And Recover Assets". By tracking your valuables, you can better manage and monitor their whereabouts, cutting down on wasted time or unnecessary fuel [1].

M. Saraswati in "Design and construction of water level measurement system" states that water measurement data is very important in some water related fields. An automatic water measurement system is needed which prevent the difficulties when one does measurement manually [2].

M. Saravanan, in his paper "Tracking Anomalies in Vehicle Movements using Mobile GIS" stated that detecting fraud activities and anomalies in the movement of vehicles are the two major concerns for the vehicle fleet industry. Some of these which incur heavy losses to the organization include fuel theft, loading and unloading related frauds, route variations, inefficient fuel maintenance and frequent breakdowns [3][4].

2 Methodology

The System may consist of hardware part as well as software part. Hardware part is used for sensing number of litres and used for communication of data processing whereas software part is used for the representation of data. The suggested system mainly consists of 3 modules namely embedded system module, GSM communication module and GPS tracking module. This structure is similar to the one presented in [5] and [7]. The first module contains fuel level sensor with a microcontroller which is placed in each fuel tankers to detect the number of litres and to interact with the GSM communication network.

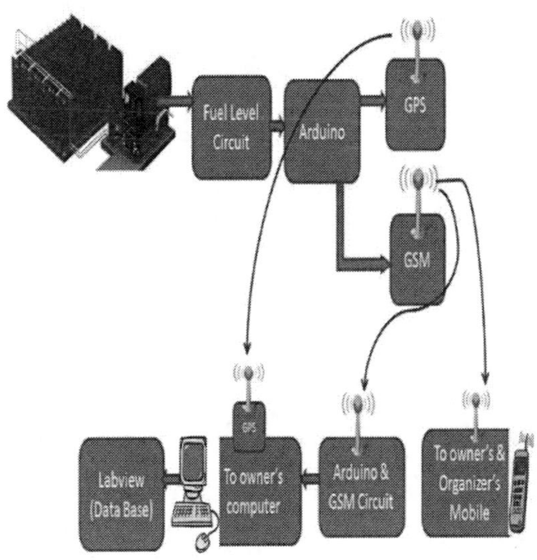

Fig. 2. General Block Diagram

The embedded system module acquires information from the fuel vehicles, maintains record and sends messages to the monitoring agents through communication module. The GPS tracking module is used for the real time tracking of fuel tankers [6] .

2.1 Embedded System Module

Embedded system module is composed of fuel level sensor and a processing unit known as Arduino Mega 2560. In order to measure tanker volume the distance sensor HC-SR04 is used which works by sending out a pulse of ultrasonic sound and measuring the amount of time it takes for the sound to come back after hitting the fuel surface [2]. The sensor is implemented with Arduino controller for the purpose of measuring and processing the distance above fuel level.

2.2 Communication Module

This system mainly depends on the communication module for sending and monitoring of data wirelessly. The measured data is sent remotely for further analysis [8]. A similar UDP-based communication scheme can be seen in [9]. Our system makes use of AT commands sent through the Arduino GSM module.

2.3 The prototype Container

The container is made up of plastic which has dimensions 18 inch length, 12 inch width and 12 inch height. It has holes at upper portion of container for placing ultra-sonic sensors for measuring fuel level within the container. The container is shown in figure below.

Fig. 3. Container measurement

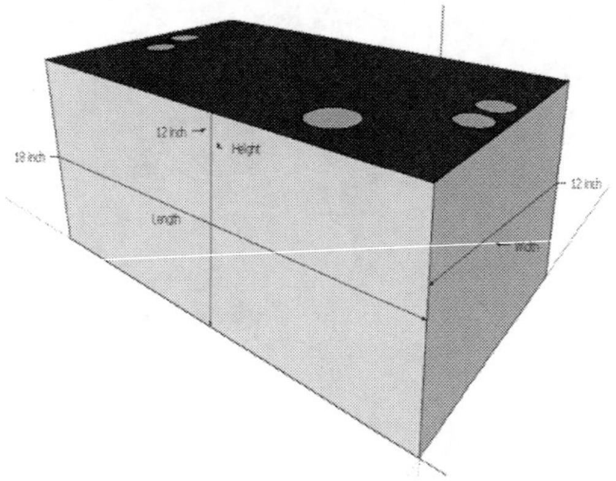

Total volume= ((1296.08) × height)/1000) litters.......(1)

Note: The above formula calculates the total volume of container whereas the height is measured by the Ultra-sonic sensor and multiplied with the constant which is the overall dimensions of the container

2.4 System software

Software part is comprised of LABVIEW. It is a system design platform and development environment for a visual programming language from National Instruments. It comprises of database which is holding vehicle's number plate, total volume of the tanker, number of fuel litres inside the vehicle's tanker, number of lost fuel litres, total number of fuel litres carried in a month, total price of the fuel in a month and total loss/profit in a month.

LABVIEW integrates seamlessly with thousand of different hardware devices, and helps save development time with convenient features and a consistent programming framework across all hardware [2]. The flowchart depicts the procedure for receiving message through GSM modem into LABVIEW environment and hence extracting the data from it.

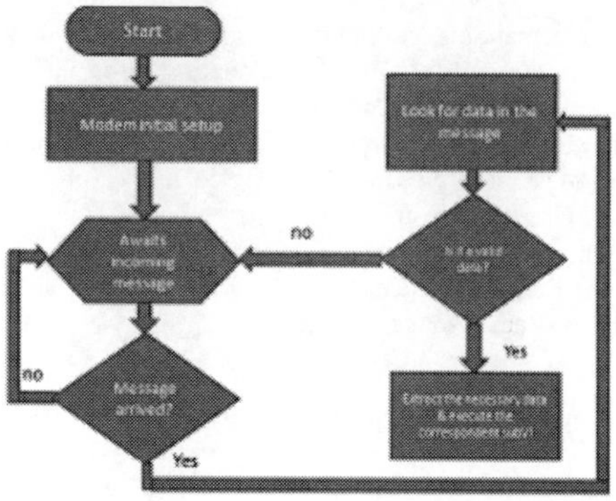

Fig. 4. Flow chart of message receiving

The database contains the data of all specified fuel tankers, each having individual system for the monitoring of fuel. The tanker truck and fuel within it is automatically displayed in LABVIEW just after receiving a message. The level indicator shows level of fuel inserted into the fuel tanker and map shows the location of the truck. This is shown below.

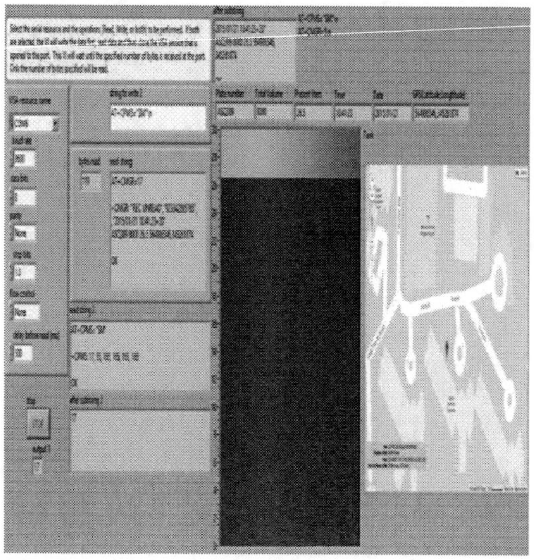

Fig. 5. Final view

3 Survey

We created a survey from different Petrol pumps and some other organizations and these are some questions and the results we found. In Fig. 6 first row is the answer to the first question and so on. According to the survey results, most of the people want this system in their practical life in order to obtain accurate information.

- Are you sure whether the number of fuel litres mentioned on the receipt is filled into the tanker or not?
- Have you ever caught the pump-employer doing fraud by filling the tanker with less fuel than required?

- Do you want an automatic system for your organization which could keep record of fuel being transported in an automatic manner?

- Do you want an automated system which could calculate and send on your cell phone as well as personal computer the measured litres within the tank along with the amount specified on the receipt and the vehicle plate?

- Do you want an automated system which could track all your fuel tankers and inform you about incidents involving any of them on your computer?

- Do you want an automated system which could prevent the fuel theft/fraud or leakage and could calculate and analyse the number of wasted litres of fuel in a month incidents?

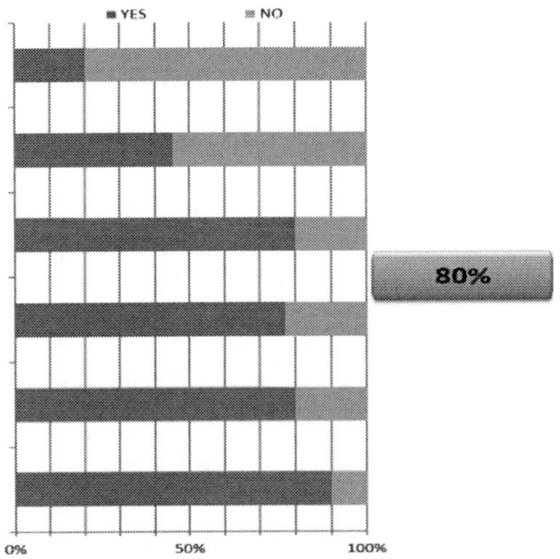

Fig. 6. Survey chart

4 Conclusion

This investigation presents and discusses the problems related to fuel management systems and overcoming these problems through an automated electronic system. The article presents considerations about both the hardware and software part of the system which involves measurement of fuel litres through fuel level circuit and sending these measurements through GSM to the Organizer's mobile as well as to the computer wirelessly.

The theoretical results show that 90% people are facing problems in managing and analyzing the fuel consumption through manual monitoring. Hence, there is a need of such a system which could minimize the problems occurring in fuel management and which could also analyze and calculate fuel consumption in an automated way. Although some systems are available regarding such applications, they are expensive hence organizations do not consider investing into them after comparing the fuel consumption and the price of such systems. The proposed system is designed by using easily available and cheap electronic components hence beneficial for any organization or industry having vehicles as key resources.

Transportation being an important resource for an Organization can have great impact on the Organization's economy. Management of the fuel involved by the logistics resource is of great importance. As a current practice it is done by Manual Monitoring which consists of Organizations having an agreement with nearby Petrol/Diesel suppliers on a monthly basis. Sometimes Manual Monitoring leads the Organization to financial losses by printing the wrong amount of fuel litres on the receipt. Such a scenario endorses an automated system which could allow for automated Monitoring and Analyzing of fuel.

References

1. Fleet Management Systems from http://gpsintegrated.com/services/fleet-management-system
2. M. Saraswati, E. Kuantama, P. Mardjoko, Eur. Symp. Comp. Model. Sim., *Design and Construction of Water Level Measurement System Accessible through SMS,* **Vol. 6.**, 49-53, (2012)
3. M. Saravanan, S. Aishwarya, L. N. Aravindan, Sci & Inf. Conf., *Tracking anomalies in vehicle movements using mobile GIS*, (2013)
4. T. Le-Tien, V. Phung, The IEEE International Symp. on Es. Design, Test & Apps., *Routing and Tracking System for Mobile Vehicles in Large Area*, **Vol.5** 297-300,(2010)
5. M. Popa, B. Suta, TELFOR, *A solution for tracking a fleet of vehicles*, **Vol.19**,1558 –1561, (2011)
6. G. Yuan, Z. Zhang, W. ShangGuan, International Conference on Internet Computing in Science and Engg., *Research and Design of GIS in Vehicle Monitoring System*, 223-228, (2008)
7. H. Hu , L. Fang , Inter. Sympo. on Inte. Infor. Techno. Apps., *Design and Implementation of Vehicle Monitoring System Based on GPS/GSM/GIS*, **Vol.3**, 278 – 281, (2009)
8. P .Chen, S. Liu , ICIE, *Intelligent Vehicle Monitoring System Based on GPS, GSM and GIS*, **Vol.4**, 38-40, (2010)
9. D. Liu, Z. Liang, Y. Li, H. Yang, FITME, *Research and design of a high performance GPS vehicle monitoring system*,**Vol.3**,59 – 62,(2010)
10. T. Enomoto, M. Sekine, S. Chikara, K. Kubota, Intelligent Transportation Systems, *A freight status management system with restriction to record loading/unloading information by location*, **Vol.3**, 548 – 553, (2003)
11. A. Tahat, A. Said, F. Jaouni, W. Qadamani, ISCE, *Android-based universal vehicle diagnostic and tracking system*, **Vol.16**, 137-143, (2012)

MATEC Web of Conferences **184**, 02012 (2018) https://doi.org/10.1051/matecconf/201818402012
Annual Session of Scientific Papers IMT ORADEA 2018

Analysis of gait recognition system developed in NI LabVIEW

Dániel Salánki[1] , *Kornél* Sarvajcz

[1]University of Debrecen, Faculty of Engineering and Mechatronics, 4028 Debrecen, Hungary

Abstract. Nowadays, the world of biometric identifiers is improving one of the most significantly in security technology. Within the biometric identifiers, the research group is working with the gait recognition speciality. The research group realized a complex gait recognition system in NI LabVIEW, that can detect more reference points simultaneously with a universal camera and can suit a function on the detected points. Moreover, the program can compare the functions of the reference curve with the functions of the actual curve and evaluate if the two gait images are the same or not. The program can save the actual gait data and reload them. The leg/foot ratio analyser program before the gait recognition is designed to improve accuracy. The research group tested the program in different ways: walking only in socks in shorts or in trousers, walking in slippers, walking in the opposite way. These facts have affected the gait recognition's accuracy, the lowest False Rejection Rate (FRR) was in the case of walking in socks in shorts.

1 Introduction

In today's world, I find it important to protect our values with the mechanical and electronic tools at our disposal. As a mechatronics engineer student, I sought to find a theme that plays a decisive role in everyday life and is closely related to mechatronics. Numerous biometric identification methods exist today (e.g. face, voice, palm, iris, retina, fingerprint authentication and DNA – Deoxyribonucleic Acid – analysis), but gait recognition is less common.

This is how we came up with the idea of wanting to implement a camera-based gait recognition program using the Vision module of National Instruments LabVIEW software, which is capable of real-time image processing to recognize that the gait image of a person passing by in front of the system is the same as that of the previously stored person's gait image. If this condition is met, the system gives him / her permission. The gait recognition is preceded by a foot examination that examines the leg/foot ratio of the person. If the test is passed, the actual gait recognition begins.

2 Gait recognition – References

People can often feel that a familiar person is already recognized far from their walk. This common experience is the idea behind the fact that district recognition systems can be used in security technology. You can easily make an image of a person's walk, even in public places, without even getting observed by that person. Walking is also influenced by several factors: footwear, soil, fatigue, current state of mind or any injury [1].

Walking can be defined as a series of cyclic and coordinated movements that result in human displacement [1]. Gait is defined as "a manner of moving on foot; way of walking or running" in Webster's New Collegiate Dictionary. However, human gait is more than that: it is an idiosyncratic feature of a person that is determined by, among other things, an individual's weight, limb length, footwear, and posture combined with characteristic motion [2]. Compared to physiological biometric identification (fingerprint authentication, iris and face identification, earlobe geometry identification), the gait recognition has the advantage of identifying persons with low resolution or hidden camera shots [3]. Physiological biometric methods cannot reliably detect non-cooperating individuals, especially not at distances, under the changing environmental conditions of the real world [4]. However, the accuracy of walking recognition can be greatly influenced by several factors, such as the viewing angle that changes the visual attributes [5].

3 Leg/foot ratio analysis

The essence of this analysis is that we determine the ratio of the leg's and foot's lengths with a program developed in LabVIEW. We first measured the ratio of one of our legs, which ranged between 2.42 and 2.43, so we selected these two numbers as limit values. So, if the ratio of the leg and foot measured in pixels is between these two numbers, the program will give a green signal.

Fig. 1 shows a measurement in trousers and in slippers. The proportional value (2.26314) did not correspond to the set limits, so the green light did not flash. Fig. 2 shows another measurement. In this case, the person was sitting in socks, without any footwear. In this

[1] Corresponding author: danisalanki@yahoo.com

© The Authors, published by EDP Sciences. This is an open access article distributed under the terms of the Creative Commons Attribution License 4.0 (http://creativecommons.org/licenses/by/4.0/).

case, the proportional value was adequate, and the green light flashed. In shorts or without socks the green light also flashed. It can be stated that the clothes have minimal influence on the results, but the footwear has an important role in this measurement.

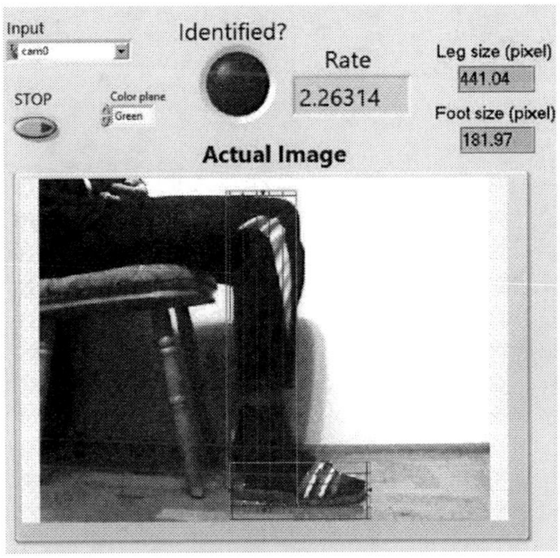

Fig. 1. Measuring in slippers.

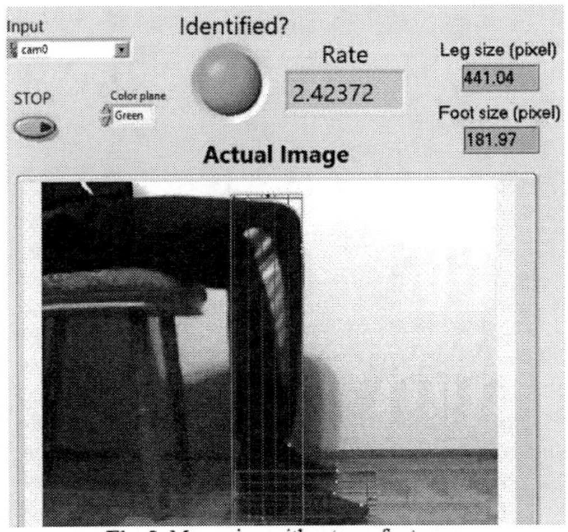

Fig. 2. Measuring without any footwear.

4 The presentation of gait recognition software

We attached a point to the knee and another to the leg of the person, about 35 centimetres from the ground. The background of the points is white, the point itself is black. It is important to have a light or white background to avoid false point detection, and to have your foot closer to the camera where the reference points are located. The measurements were made with artificial light, with normal room lighting, beside the darkened drapery avoiding sunlight, creating approximately the same lighting conditions for each measurement.

Fig. 3 shows the front panel while the program is working. On the right side of the figure, the coordinates of the detected points will be stored in blocks after running. X1 and Y1 are valid for the first point (suited on the knee), while X2 and Y2 are valid for the second point. In the middle it is the actual image, the two red rectangles frame the reference points. On the second and third tabs, the gait image is processed. On the second tab the actual gait image is processed, while on the third tab the saved and reloaded gait image is processed. If the actual gait image is like the saved one, the green light below the actual image will flash. The program suits a function on the curves drawn by the two reference points. You can choose between four function models (Linear, Polynomial, Exponential, Power) and you can choose the order of the function too.

On Fig. 4. a) and b) the blue points represent the detected points and the red curves are the functions suited on them. According to the tests, the most accurate fit is the fourth-degree (biquadratic) polynomial function (Fig 4. a)). However, for biquadratic polynomials, the third and fourth coefficients take very small values (up to 10^{-10}), and the second-order (quadratic) function also follows well the curve drawn by the points (Fig 4. b)), so the use of the fourth-degree polynomials becomes unnecessary. Working with quadratic function is simpler than working with biquadratic ones.

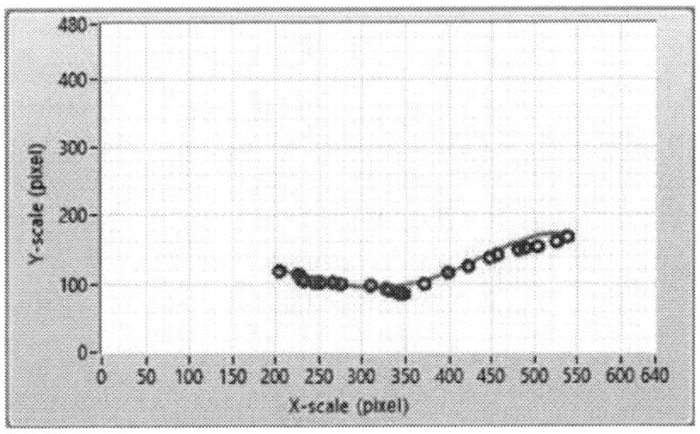

Fig. 4. a) Suiting a fourth-degree polynomial function on the gait curve.

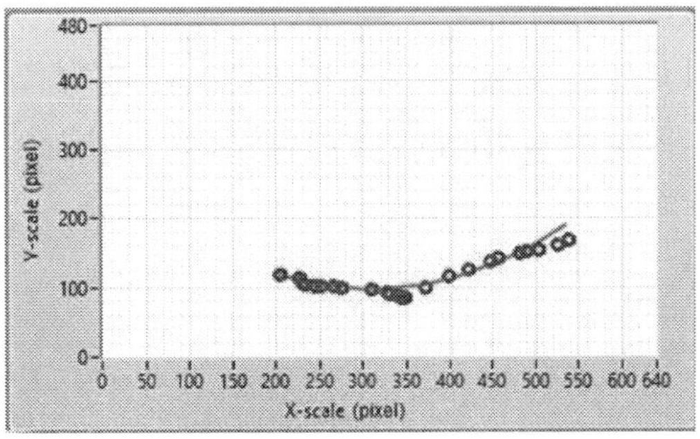

Fig. 4. b) Suiting a second-degree polynomial function on the gait curve.

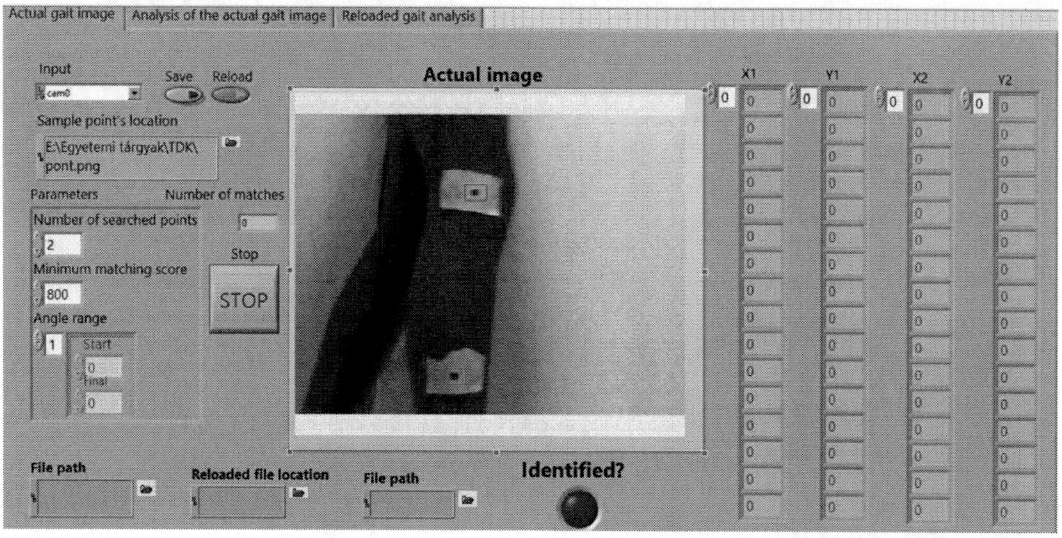

Fig. 3. The front panel of the program while working.

5 Gait recognition results in different conditions

Fig. 5. shows the coefficients of the functions suited on the curve drawn by the first (blue columns) and the second point (green columns). During this series of measurements, the tested person wore shorts and socks without any shoes or slippers, and the first step was made with the right leg (the reference points were also fixed on the right leg). According to the results, in 8 cases out of 10 the system recognized the tested person. The unsuccessful attempts are written with red characters.

In shorts, 1. point - no footwear			In shorts, 2. point - no fw		
X^0	X^1	X^2	X^0	X^1	X^2
399.7463	-0.0469	-3.46E-06	185.8495	0.5765	0.0011
396.7941	-0.0554	3.05E-05	213.5363	-0.8573	0.0017
404.3768	-0.134	2.00E-04	156.7454	-0.3408	0.0007
400.0114	-0.0578	3.05E-05	227.1697	-0.8753	0.0016
393.8913	0.0003	-9.64E-05	226.7635	-0.9018	0.0017
404.5176	-0.1228	1.00E-04	234.8857	-0.9617	0.0018
406.8849	-0.1174	1.00E-04	212.0161	-0.7867	0.0015
390.3333	0.0225	-1.00E-04	209.1956	-0.7424	0.0013
395.9785	-0.0232	-4.19E-05	220.0219	-0.7801	0.0013
397.4828	-0.0574	3.27E-05	179.0767	-0.546	0.0011

Fig. 5. Results in shorts, no footwear.

On Fig. 6. you can see the coefficients of the functions received from measurements made in shorts wearing slippers. The results are very similar to the first measurement, the False Rejection Rate (FRR) increased a bit: the tested person was recognized 7 times out of 10 measurements.

In slippers, 1. point			In slippers, 2. point		
X^0	X^1	X^2	X^0	X^1	X^2
413.5568	-0.0701	6.55E-05	217.3839	-0.751	0.0015
405.3897	-0.0107	-5.83E-05	225.0628	-0.7645	0.0014
417.8884	-0.093	7.29E-05	262.8625	-1.1075	0.0021
413.9129	-0.0565	3.49E-05	181.0427	-0.5	0.0011
414.6489	-0.0667	3.15E-05	229.4339	-0.8247	0.0016
423.2002	-0.1287	1.00E-04	234.7215	-0.928	0.0018
412.7888	-0.0507	1.94E-05	192.342	-0.5711	0.0012
413.5032	-0.065	3.16E-05	211.6114	-0.6993	0.0013
424.9448	-0.1551	2.00E-04	238.6543	-0.9568	0.0019
415.4354	-0.0592	3.53E-05	210.8684	-0.6811	0.0014

Fig. 6. Results in shorts, in slippers. The unsuccessful attempts are written with red characters.

Fig. 7. shows the data received from measurements made in trousers, without shoes. The FRR increased to 50%, so only half of the measurements were successful. These results can be explained with the fact that the reference points fixed on the trousers were moving with the cloth (the trousers are moving on the leg while walking), while fixed directly on the tested person's knee and leg, they followed exactly the movement of the knee and the leg.

In trousers, 1. point - no fw			In trousers, 2. point - no fw		
X^0	X^1	X^2	X^0	X^1	X^2
401.4465	-0.1794	0.0002	270.2624	-1.285	0.0023
389.2778	-0.134	0.0002	152.8095	-0.4528	0.0009
384.0136	-0.0881	0.0001	169.2744	-0.7754	0.0018
377.0618	-0.0694	0.0001	161.3756	-0.6282	0.0015
382.4782	-0.0582	0.0001	198.1475	-0.8582	0.0016
401.3688	-0.2588	0.0004	256.0288	-1.4425	0.0032
381.9576	-0.1289	0.0002	167.4292	-0.6569	0.0014
385.261	-0.2386	0.0002	203.0616	-0.9265	0.0019
373.701	-0.0971	0.0002	174.136	-0.7995	0.002
375.3779	-0.1076	0.0002	171.3232	-0.8775	0.0024

Fig. 7. Results in trousers, no footwear. The unsuccessful attempts are written with red characters.

 https://doi.org/10.1051/matecconf/201818402012

Fig. 8. shows the results from a series of measurements made in shorts, wearing no shoes, the reference points fixed on the left leg, and going from the right of the camera to the left. The previous measurements were made from the left of the camera to the right. With these attempts, the FRR became 90%, only one time out of 10 was the test person recognized. This result was caused by the fact that the software was calibrated to start from the left of the camera. The curves drawn by the reference points fixed on the left or on the right leg are different.

Opposite direction - 1. point			Opposite direction - 2. point		
X^0	X^1	X^2	X^0	X^1	X^2
343.794	0.2355	-0.0002	215.5685	-0.8344	0.0013
326.8067	0.3227	-0.0003	234.3298	0.9133	0.0014
365.491	0.1453	-0.0002	225.2881	-0.8438	0.0013
328.3563	0.3319	-0.0004	283.1634	-1.2319	0.0019
338.6368	0.2772	-0.0003	211.0794	-0.7592	0.0012
365.3066	0.1254	-0.0001	265.4828	-1.1291	0.0019
342.1156	0.2323	-0.0002	193.6888	-0.6821	0.0011
368.3189	0.1154	-0.0001	214.2421	-0.7694	0.0011
388.4625	**0.0199**	**2.85E-05**	**225.5323**	**-0.949**	**0.0016**
353.3414	0.2108	-0.0002	253.5832	-1.0858	0.0017

Fig. 8. Results in shorts, walking in the opposite direction. The unsuccessful attempts are written with red characters.

The measurements presented above were made with starting the walking with the right leg. We have made some measurements with the same conditions as in the first measurement series (reference points on right leg, wearing short and no footwear), but the first step was made with the left leg. The results are shown on Fig. 9. Because the coefficients were very different from the initial values, we made only three attempts, certainly without any success. The curves drawn by the two reference points, that are shown on Fig. 10. a) and b) were far from the initial curves because of the cyclical shift and because the camera can view only a few steps. So, it is normal that the program didn't make any recognition in this measurement.

Starting with left - 1. point			Starting with left - 2. point		
X^0	X^1	X^2	X^0	X^1	X^2
358.0921	0.1561	-0.0001	150.8065	0.2241	-0.0005
355.9683	0.0869	-1.36E-05	150.1471	0.1522	-0.0004
323.1626	0.259	-0.0002	164.19	0.1417	-0.0004

Fig. 9. Unsuccessful results starting with the left leg.

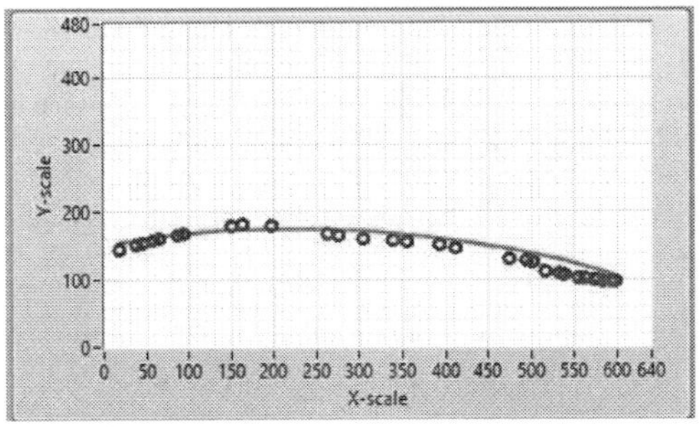

Fig. 10. a). Curve drawn by the second reference point starting the walking with left leg

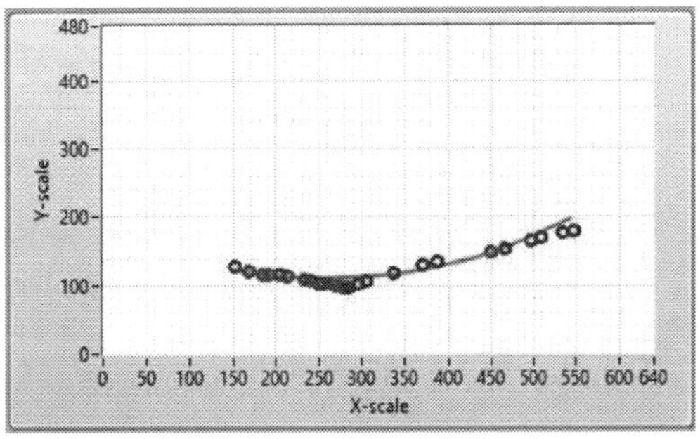

Fig. 10. b). Curve drawn by the second reference point starting the walking with right leg

6 Conclusion

With these tests, we have showed how our gait recognition system is working in different situations. According to the result, the footwear has the least influence on the system, but wearing trousers, or walking in the opposite direction can have serious impact on the successfulness of the test. It is also important to start the measurement with the right leg, because otherwise the curves will be far different from the reference curves, and the program will not recognize even the person with permission.

Acknowledgement

The work/publication is supported by the EFOP-3.6.1-16-2016-00022 project. The project is co-financed by the European Union and the European Social Fund.

References

1. Jeffrey E. Boyd, James J. Little: Biometric Gait Recognition, **19-42** (Springer - Verlag Berlin Heidelberg, 2005)
2. Lee, Lily, W. Eric L. Grimson. Gait analysis for recognition and classification. Automatic Face and Gesture Recognition, 2002. Proceedings. Fifth IEEE International Conference on (**pp. 155-162**). IEEE, 2002.
3. Mark S. Nixon, T. N. Tan, R. Chellappa: Human Identification Based on Gait, (Springer Science + Business Media, Inc., 2006)
4. Ju Han, Bir Bhanu: Individual Recognition Using Gate Energy Image, **28.2: 316-322,** (IEEE transactions on pattern analysis and machine intelligence, 2006)
5. Wei Zeng, Cong Wang: Neurocomputing - View-invariant gait recognition via deterministic learning, **175: 324-335** (2016)

Designing a delta tripod based robot fused deposition modelling 3 dimensional printer using an open-source Arduino development platform

Tamás Templom[1], *Timotei István* Erdei[1], *Zsolt* Molnár[1,*], *Edwin* Shaw[2] and *Géza* Husi[1]

[1]University of Debrecen, Mechatronics Department, 4028 Ótemető street 2., Debrecen, Hungary
[2]London South Bank University, School of Law and Social Sciences, London, United Kingdom

Abstract. The pinnacle of 3D printing is its effect on the field of rapid prototyping. The major advantage comes from the fact that designers can quickly materialize a part or object, which then could be tested in practice, and can be effortlessly modified if needed. This obviously cuts the development expenses and time by a significant percent. Moreover, it's possible to create complex and precise shapes with the technology, which would take more time and would be resource intensive if done by older methods, for example manual or automatic machining.

1 Introduction

The University of Debrecen's Building Automation Research Faculty houses numerous projects of the Faculty of Engineering [1]. These projects always inspire us for further research, like our IoT (Internet of Things) enabled FDM (Fused Deposition Modelling) 3D printer [2]. During studies, smaller projects always required a modest amount of mechanical parts, which were often needed to be of a special size, or hard to manufacture, and would cost a great deal amount of money. The explosion of CAD (Computer-Aided Design) systems in 1980 drastically changed the field of engineering, and enabled us to quickly create and manufacture parts at a negligible cost and time. This whole principle made us consider the project. In comparison to the "classical" Cartesian 3D printers, the decision was made to use a less common, Delta mechanism, as an essential challenge. The problem comes from the much more complex kinematical and calibration issues of said parallel delta systems. The project started in the 2nd semester of year 2016 to 2017, within the framework of the University of Debrecen's Program for Gifted Students (DETEP).

2 Design points

One of the most crucial design point was cost efficiency, that's why Open-Source Arduino platform was chosen. [3]. For the Control Software, Arduino firmware Marlin was selected, which is specially designed for Delta 3D printers [4] The other control and slicing software communicates via USB (Universal Serial Bus). Delta robots can have two types of movement, linear and radial, linear was favourable. Furthermore, the aim was to create a workspace capable of manufacturing parts of standard sizes and general uses, not only example of concept smaller ones. Delta tripods consist of three vertical axes, each rotated from 60 [°] from one another. By comparing different present designs, toothed belt was approved as a method of moving the system, as it's fast, and precise enough for our demands [5].

The drive consists of a toothed belt of 2 [mm] pitched, GT2 belt and a gear with 20 teeth driven by hybrid stepper motors. The choice was a Nema 17 type bipolar stepper motor, with the dimensions of 42,3 x 42,3 x 34 [mm] [6]. The sled structure consists of 2-2 pieces of Ø8 [mm] diameter circular steel pieces connected with linear bearings.

To reach the entire working area with the hotend, 5 parameters needed adjustments: height of axis(H), radius of delta (DR), consisting of the circle included by the 3 axis, length of pushrod (L), that connects the sled and the effector, offset of the effector (EO), and the offset of the sled (SZO).

The sled structure is downloaded from an Open-Source STL (stereolithography) file sharing site, printed from ABS (Acrylonitrile Butadiene Styrene), and designed for LM8UU type bearings and GT2 toothed belt [7]. The pushrod is made out of carbon fibre, and has ball joints on each end. The heating solution of the hotend is a E3D V6 type element. To connect the pushrods to the hotend, an adapter is needed, a so-called effector, this connects the 6 pushrods to the hotend. As the MK2Y type was chosen, d = 220 [mm] diameter heated bed, printing work area was designed as well. On the Z axis, 110 [mm] maximum height was aimed for.

Fig 1. shows our 3 variables (H, DR, L) are heavily dependent on each other. The larger the delta, the longer

* Corresponding author: zsolt.molnar94@gmail.com

© The Authors, published by EDP Sciences. This is an open access article distributed under the terms of the Creative Commons Attribution License 4.0 (http://creativecommons.org/licenses/by/4.0/).

pushrods we need to reach the correct area on the X-Y plane. This brings the need to lengthen the axis as well, to keep the movement range on the Z axis as well. The pushrods' angle with the bed at the robot's HOME position is: α. However, during movements pushrod's angle with the bed cannot go under 20°, so the angle in the endpoint is: γ.

- $SZO_z = 20$ [mm]
- $HEO_z = 60$ [mm]

$$[L * \cos(\gamma) + SZO + EO] - [L * \cos(\alpha) + SZO + EO] = 110 \quad (1)$$

$$[L * \cos(\gamma)] - [L * \cos(\alpha)] = 110 \quad (2)$$

$$Z\_max = H - SZO\_Z - HEO\_Z - (L * \sin(\alpha)) \quad (3)$$

Table 1. Specified physical parameters

Component	Specified size
Pushrod length [mm]	336
Radius of Delta [mm]	260,75
Height of Axis [mm]	460

After laying down design points, free Autodesk Inventor was used to create the 3D model of our system [8], which was necessary for further planning. The 3D model enables the verification of the fitting and usability of any part that will be added to the system later. It's also beneficial that collisions in a simulation can be checked, preventing real accidents.

Components on Fig 1.:

1. Bottom Tray
2. Top Tray
3. Bottom Axis Support
4. Top Axis Support
5. Heated Bed
6. Round Bar
7. Pushrod
8. Hotend
9. Effector
10. Sled
11. Stepper Motor
12. Extruder motor
13. Pulley
14. Control Circuit
15. Hollow Section Legs

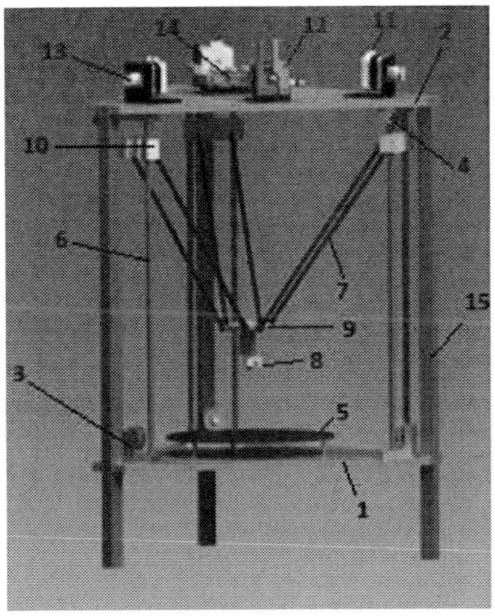

Fig. 1. The model 3D rendered in Autodesk Inventor

A Bowden solution was used for the feeder unit, which means that the extruder motor is not directly above the hotend, but connected to the top tray and connected via a Teflon tube, which guides the thermoplastic material [9]. On the plus side, the total moved weight is less. However, it requires a motor with higher torque, some leftover filament will remain in the pipe, this is not suitable for feeding flexible materials.

During the assembly, tensioners to fasten the toothed belts were used, and it is advantageous to use some kind of elastic connecting element in between the pushrod pairs.

3 The completed circuit

Next task was to install the electronic components. The servo drivers and the other necessary components were connected to an Arduino Mega with a Ramps 1.4 as a shield, which is a specialized instrument for 3D printers. The shield runs on 12 [V] DC (Direct Current), provided by a 280 [W] ATX PSU (Power Supply). It possesses two 5 [A] outputs, one for the hotend, which is in charge of heating the thermoplastic material, and one for the fan, to cool down the already extruded layers. Additionally, it includes an 11 [A] output, which provides the required power for the heated bed. Up to 3 thermistors can be connected to it, in this case 100 [kΩ] resistance ones were used. Moreover, 6 limit switches were added, of which 3 were used as the top endpoints, as basic mechanical NO (Normally Open) switches. Ramps 1.4 can support up to five stepper drivers, currently 3 are used for movement, and 1 for the extruder, so in the future an extra extruder can be added for a second hotend.

The chosen stepper motors technical data: 1,3 [A], 0,22 [Nm] torque, 1,8° per step. We used DRV8825 (Pololu) drivers to control the motors [10]. This IC (Integrated Circuit) also houses the necessary resistors and potentiometers to operate the motors.

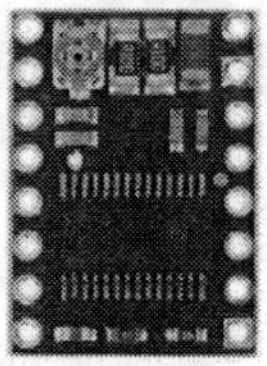

Fig. 2. Pololu DRV8825 module [10]

One of the greatest benefit of the IC is the so-called microstep function, which enables the user to split the steps of the motor to even more steps. In microstep, the user can distinguish 1/2, 1/4, 1/8, 1/16, 1/32 step angles, depending on the setting of the circuit. The correct microstep on the jumper on the driver can be set. Because DRV8825 is capable of 1/32 microstepping, this option was chosen. The maximum amperage and voltage is 2,2 [A], and 8-45 [V] over the motors in the case of Pololu.

The following formulas were used to calculate the movement resolution: [11]:

$$\frac{Steps}{mm} = \left(\frac{\left(\frac{360°}{step\ angle°}\right) * \left(\frac{1}{microstep}\right)}{belt\ pitch * dial's\ number\ of\ teeth} \right) \quad (4)$$

$$\frac{Steps}{mm} = \left(\frac{(360°/1{,}8° * (1/\frac{1}{32}))}{2 * 20} \right) = 160 \quad (5)$$

4 Calibration

During calibration, differentiation must be made between Software and Hardware calibration. During the Software calibration, the Arduino program is set up, and the slicing software to the correct configuration. In the firmware Marlin, physical parameters of the printer have to be submitted, the data for the driver, limit switches, and temperature data and also the pinouts. The free slicing software Ultimaker Cura even possesses an official manual, to set the correct work area, correct communication, and the diameter of the nozzle [12]. The parameters for the current print are set here as well, for example: fill factor, print speed, and the filament's specified heat values.

Before the physical calibration of the printer, individual part calibration is needed as well. Such parts are like the limit switches, that need an individual check-up for each component. Moreover, it's important to set the current limit for the stepper motors, thus to achieve the highest torque, a driver with higher current flow than the maximum current needs to be used on the stepper. Overpowering of the motors is not desirable, the maximum set current should be the highest possible power draw of the motor. The correct calibrations result in cooler motors, and no loss of steps. After the parameters are set correctly, the units can be connected to the system. For the correct settings, reference voltage of the instrument must be measured. The DRV8825 method follows the equation:

$$I_{limit} = 2 * U_{ref} \quad (6)$$

Because the chosen motors maximum current per phase is 1,3 [A], formula no. 6. enables calculation of the reference voltage: $U_{ref} = 650$ [mV].

The first step of calibrating a Delta-Tripod type printer is to set the P(0,0,0) mechanical null point in advance, manually. The robot set in the HOME position will be moved along the Z axis in the negative direction (down) until the hotend reaches the glass plane covered with a sheet of ordinary paper, but it shouldn't press it onto the glass. Because a sheet of paper usually has about ≈0,1 [mm] thickness, the correct parameters can be entered into the Arduino program from the total distance corrected with the girth of the sheet. This is the so-called paper test [13].

These corrections are followed by refining the limit switches. This is done by printing a one layer, but thick test prop, to measure the difference between the thickness of the layer approaching to the axes from the centre. If the measured values do not correspond to the pre-set thickness of the layer, the difference can be used for correction purposes again.

The latest task is to set the dimensions. This can be done with two methods. In the first one an equilateral triangle based prism was used. By measuring the height of the object on different edges and sides, some errors can be concluded, such as loose belts.

On the other hand, a cube can be used as our test print as well, but all in all, the main goal is to measure the dimensions, correct the erroneous calibrations, then print once more, and measure once more as well, repeatedly, until the same values on the physical manifestation are corresponding with those given in the program.

5 Reference prints

With the calibration prints, the following conclusions can be made:

- X and Z axes had marginal differences, but on the Y axis at 20 [mm] had -0,1 [mm] difference.
- At 80 [mm/s] print speed, tear-free fill factors can still be used.

After accomplishing the correct print conditions a nozzle was implemented to cool down the extruded material, to further increase the print quality. Tensioners were also fabricated for the belts, and a filament drum holder with bearings, to ease the feeding process.

The opportunity was given to print out a three finger gripping tool designed by Viktor Varga, an undergraduate colleague, for the KUKA KR5, located at the Mechatronical Faculty of the Engineering Department, University of Debrecen. 1,75 [mm] diameter PLA (Polylactic Acid) filament was used, with its correct heat values visible in the table bellow

The print consisted of the three fingers, and the pushrods and rotary elements as well. Additionally, a holder for the tool was printed for secure storing when not in use. All parts have been made with 0,2 [mm] layer thickness, 30 [%] fill factor and 50 [mm/s] print speed.

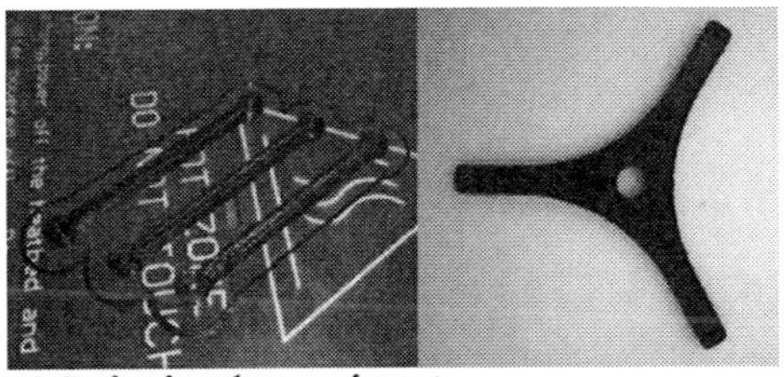

Fig. 3. Pushrods and rotary element

The Pushrods' print time was: 3 minutes per piece, and the rotary element took 10 minutes total.

The total print time per finger was: 85 minutes, and required supporting material, the additional holder took another 163 minutes. The total print time did not go over 8 hours and their total weight is only 52,2 [g].

Fig. 4. The finished griping tool on the holder

6 Conclusion

After carefully designing the mechanical frame, installation of the electronic components proceeded as well. After setting up the chosen programs, the greatest challenge of configuring, calibrating, and correcting the robot in every aspect was met. In addition to theoretical knowledge of the field, practical experience was acquired in the field of FDM 3D technology. The final task of printing the aforementioned 3-finger gripping tool, gave presentation of the ability to quickly and efficiently use the printer for rapid prototyping.

For future development, removal of human factor as a constant surveillance for the printing process, implementation and Open-Source Linux-based remote surveillance system accessible via Internet can be added, enabling the user to print or get real-time footage of the printing process ongoing from anywhere.

Fig. 5. The finished 3D Printer

Acknowledgment

The publication was supported by the project EFOP-3.6.1-16-2016-00022. This project was co-funded by the European Union, from the European Social Fund.

References

1. Kalmár F, Kalmár T, Csáki I, Husi G „Interrelation between ACH and air temperature distribution in a room," 2009: Proceedings of the 11th International ROOMVENT Conference. Busan, Dél-Korea, 2009.05.24-2009.05.29. Busan: [s. n.], 2009. pp. 911-917.
2. T. I. Erdei, ZS. Molnár, G. Husi, „Selecting Equipment and Supplies for Self–Replicating 3D Printer," Acta Technica Corviniensis - Bulletin of Engineering, Hunedoara9.1, Jan-Mar 2016, pp. 59-62.
3. S. Glaves, „Rocholl (Rostock) Style Delta Robot Kinematics".
4. „Marlin," [Online]. Available: http://marlinfw.org/. [Access Date: 27 11 2017].
5. R. Santos, J. James, T. Chris, S. Marshall, P. Maalouf, „Delta 3D Printer," 2015
6. „Reprap," [Online]. Available: http://reprap.org/wiki/Main_Page. [Access Date: 27 11 2017].
7. „Thingiverse," [Online]. Available: https://www.thingiverse.com/thing:1495442. [Access Date: 27 11 2017].
8. „Autodesk Inventor," [Online]. Available: https://www.autodesk.com/products/inventor/overview. [Access Date: 27 11 2017].
9. T. I. Erdei, Zs. Molnár, N. C. Obinna, G. Husi, „Cyber physical systems in mechatronic research centre," MATEC Web Conf. Volume 126, 2017.
10. „Pololu," [Online]. Available: https://www.pololu.com/product/2133. [Access Date: 27 11 2017]
11. J. Prusa és N. Zadoks, „RepRap Calculator V3," [Online]. Available: https://nathan7.eu/stuff/RepRapCalculator/RepRapCalculator.html. [Access Date: 27 11 2017].
12. „Ultimaker Cura," [Online]. Available: https://ultimaker.com/en/products/ultimaker-cura-software. [Access Date: 27 11 2017].
13. C. T. Brittain, „Ladvien," 2014. [Online]. Available: https://ladvien.com/robots/kossel-mini-calibration/. [Access Date: 27 11 2017].

MATEC Web of Conferences 184, 02014 (2018) https://doi.org/10.1051/matecconf/201818402014
Annual Session of Scientific Papers IMT ORADEA 2018

Redesigning of the gripping system on KUKA KR5 by utilizing 3 dimension printing technologies and Arduino

Viktor Varga[1], *Timotei István* Erdei[1,*], and *Géza* Husi[1]

[1]University of Debrecen, Mechatronics Department, 4028 Ótemető street 2. Debrecen, Hungary

Abstract. During this short summary of the project, we'd like to go through all the steps of the project, from the very first planning stages, up to the 3D modelling, and finishing up with the actual printing itself. We've also had the chance to install our finished tool onto the KUKA KR5 itself, and evaluate the quality of our tool using sensors, measurement, and by analysing our Arduino code.

1 Introduction

During Robotics course, the chance was given to study the KUKA KR5 robot, with many other robots [1] and automated systems [2], both theoretically, and practically, from the basic toolset to the robot's inner code and structure. At the moment, the robot uses a two finger pneumatic gripper as its tool. This gripper can hold objects ranging for 30 to 45 [mm] in diameter, and currently is used to grip on a tool used for educational purposes. This method is sufficient, but far from optimal, considering its small dynamic range, two-state (open or closed) nature only enables small objects to work with, and moreover they cannot be fragile as a result of either maximum or zero gripping strength. Because the Laboratory does not possess any more gripping tools, it was an obvious initiative to create more varied gripper both to increase the variety of tools, and to challenge us with this endeavour. We choose a 3 finger parallel gripper. As a result of the spread of 3D printing technologies, we could physically manifest the tool in real life as well. The supervision and control has been done with a Bluetooth module using an Arduino. Because the fingers are under constant surveillance by the sensor, we can get accurate measurements on the object held onto on the fly. We used a FSR (Force Sensitive Resistor) dynamometer to get the current clamping force on the fingers. After integrating the tool into the test environment, we proceeded to carry out some measurements and testing, to make sure our project was successful.

2 Presenting the mechanism

As our first step, before we started the 3D modelling, we had to specify the method of moving the fingers. The fingers are 120° form each other, and are being moved by a rotary component (*Fig. 1., Fig. 2.*).

Fig. 1. Presenting the mechanism

This enabled us to convert rotational movement to linear. Same method was used in steam engines, or being used in modern reciprocating (piston) engines, compressors or pumps.

Next step was to define the greatest angle which grants us safe and accurate, collision free movement. After the measurements, we got the precise angle of 106° to be used as the maximum range, which enables us to have a scope of 50 [mm] for gripping, which proved to be adequate. By knowing these values, we can use a sketch drawn in Fusion 360 [3] to accurately measure the upcoming components' sizes, and also to keep track of the angles, and mechanical model of our system. On the left side of the figure, we can see the motor at 0°, thus it's fully open position, while on the right we can see maximal 106°, approximately 50 [mm] long contraction.

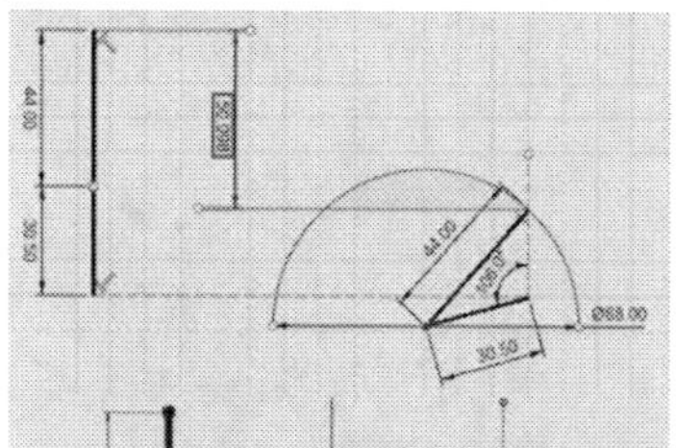

Fig. 2. Presenting the mechanism

* Corresponding author: timoteierdei@gmail.com

© The Authors, published by EDP Sciences. This is an open access article distributed under the terms of the Creative Commons Attribution License 4.0 (http://creativecommons.org/licenses/by/4.0/).

3 Presenting the 3D model

After we successfully drawn out our parts, and measured them correctly, we can get to 3D modelling the parts of our gripper. We used Autodesk Fusion 360 to create our 3D models. In the upcoming figures, we can see the parts of our gripper separately. On *Fig. 3.* we can see the frame of the gripper, which houses our motor, and the circular sliders placed 120° apart, to enable the fingers to move.

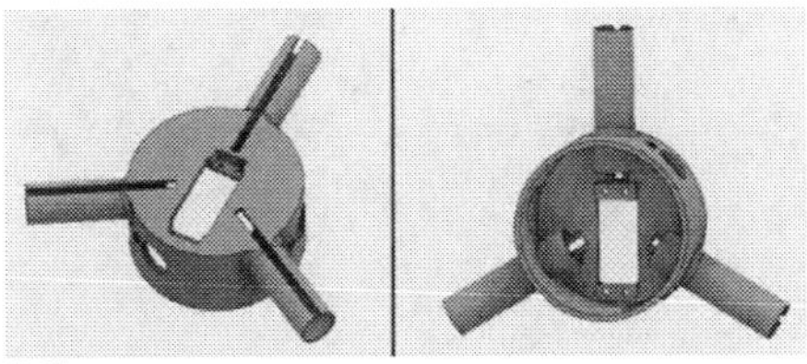

Fig. 3. Frame

The images down below (*Fig. 4.*) are the already presented 44 [mm] long pushrods (left), and the rotary component mounted to the motor later (right).

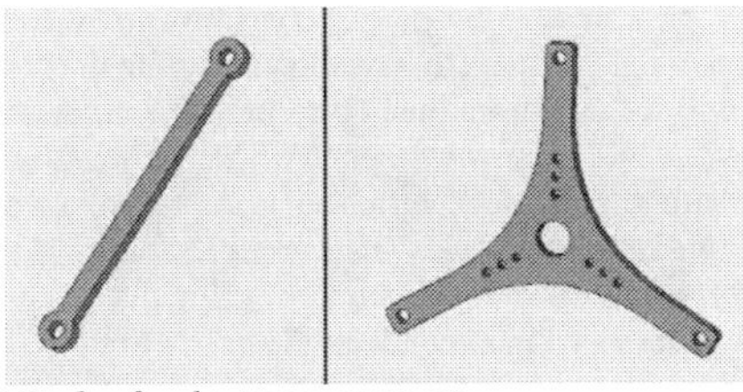

Fig. 4. Pushrod and rotary component

Fig. 5. shows our 3D modelled finger, which is made so that it can slide in the circular slider of the frame effortlessly.

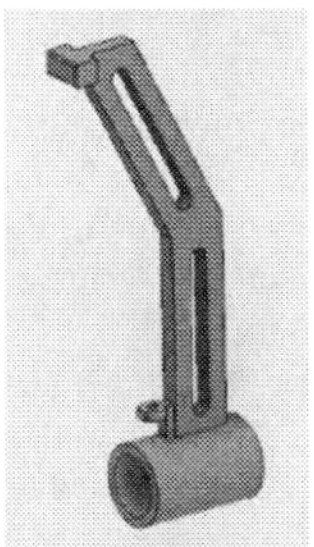

Fig. 5. Pushrod and rotary component

Another task was to create correct a fitting mechanism in between the robot and the tool itself. The 6th axis of the KUKA KR5 possesses 4 mounting points where we can attach tools. By knowing the exact position and size of these bores, we made an adapter (*Fig. 6.*) so we can fit the tool onto the robot easily.

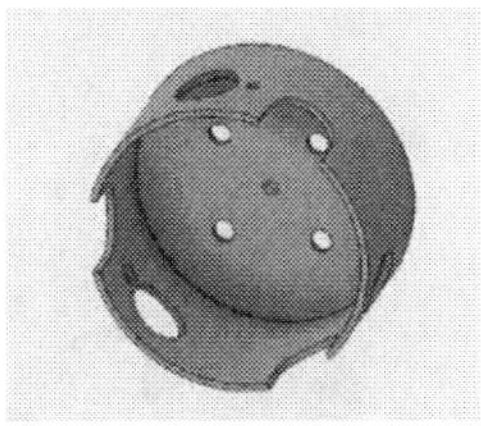

Fig. 6. The adapter

4 Assembling the parts in Fusion 360

As the first step of the assembly, we connected the TowerPro MG995 servo to our frame, then we mounted the rotary component onto it. The chosen servo has great range and great torque compared to its size. The table below shows some technical data about it:

Table 1. Specifications of the servo [4]

Servo type	Torque [Nm]	Speed [s/60°]	Weight [g]
TowerPro MG995	1	0,16	55

Coming up, we added the fingers, sled in from the side into the frame's sliders, connected with the pushrods of which thereafter connected to the rotary component. This concludes our 3D model of the gripper (*Fig. 7.*).

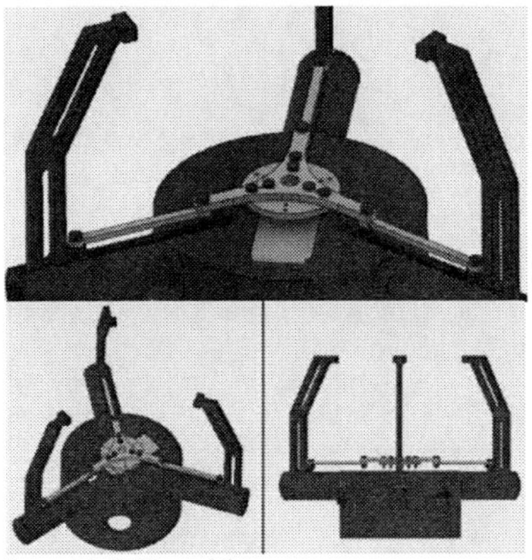

Fig. 7. Assembled 3D model

The grip range is the difference in between the fully open and closed positions, which lets us grab objects ranging from 9 [mm] to 109 [mm]. The adapter and the gripper itself can be sled together, and then secured together with screws (*Fig. 8.*).

https://doi.org/10.1051/matecconf/201818402014

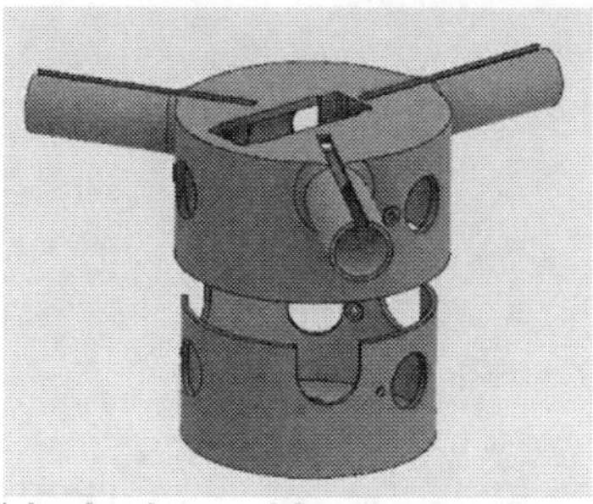

Fig. 8. Joining the adapter and the gripper together

5 Printing out the components in 3D

The printing has been done by a third party, after we agreed in the terms and materials of the print (*Fig. 9.*). Generally, ABS is more ductile, heat resistant and less prone to shattering, but since we won't need these advantages, since we won't move anything heavy or hot, PLA was a cheaper alternative. It's also preferred by hobbyists and designers also endorse it for its ease of use and larger colour variation [5].

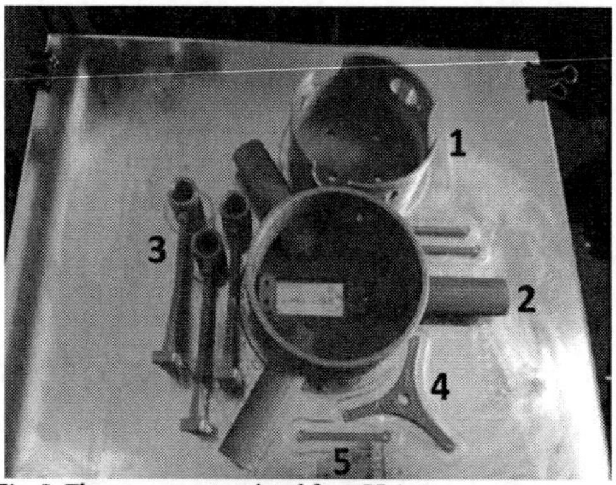

Fig. 9. The components printed from PLA

Ultimaker Cura takes the burden of setting all the parameters for each print from us, since it automatically sets it for us, saving a lot of time. The details can be found in the next table [6].

Table 2. Printing details per components

Component	Fill factor [%]	Total print time [hour:min]	Weight [g]	Cost [Ft]
1	30	2:48	31,1	165
2	30	6:54	82,1	435,1
3 (3 pieces)	30	2:54	23,7	126,6
4	30	00:10	1,6	8,5
5 (3 pieces)	30	00:06	1,2	6,3
Σ	N/A	12:52	139,7	741,5

The total weight of the gripper without the screws is only 194,7 [g], which takes up the 3,89% of the maximum of 5 [kg] load on the robot's last axis. After we manufactured the parts, we assembled in the same way as we planned it out in the virtual world (*Fig. 10.*). To further increase the grip strength, we used rubber ends on the fingertips.

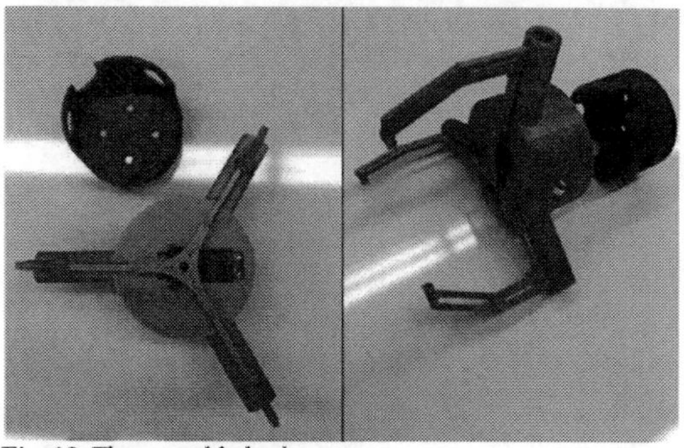

Fig. 10. The assembled gripper

6 Controlling the gripper with Arduino

As the basic methodology of servos, even if they hit an obstacle while turning, they won't stop, quite the contrary, they will try to continuously turn until the potentiometer inside reaches the given angle [7]. This can cause serious overheating, and later the failure of the servo. To project this issue onto our design, its not preferable to drive the servo for 0°to 106° as a method of gripping, since it will damage the servo in a short time. Thus, we tried to achieve the optimal state of gripping on the object, then locking the servo at that angle, preventing further movements, but also gripping on it steadily.

The issue was solved by analysing the potentiometer inside the servo. After we soldered a pin from the Arduino to the signal line of the potentiometer, we could easily, and accurately measure the state of the servo at any time in a scale of 0 to 1024 units. Then we used command "Map" to map out the potentiometer values for each angle, 122 for 20° and 302 for 126° (this +20 shift will be beneficial later).We saved this to the val variable, which will now always show +20 from the real value, which should be accounted for.

The program seen on *Fig. 11.* basic premise is to drive to motor from 0° to 106° inside a loop, incrementing the angle in each loop. As stated above, the val integer will keep increasing with the loop, as the motor is turning about, but if it gets stuck, it will cease to increase. As the grip happened, the loop will continue to run, and as It equals with the val integer, we can safely say that the grip has happened, so we set the servo to the loop's local variable x-17. The subtraction is important, since as val equals to x we are already overturning by 20°, so we step that down to only 3°which produces a firm grip, yet its light on the servo.

```
void loop() {
  if (Serial.available()) {        // is serial communication available?
    bluetooth = Serial.read();     // reading bluetooth signal
    Serial.println(bluetooth);     // writing bluetooth signal

    if (bluetooth == '+') {                  // if the received signal from bluetooth is '+'
      for (int x = 0; x < 107; x++) {
        myservo.write(x);                    // move the servo from 0 degrees to 106 degrees
        int val = analogRead(A5);            // read the value of the servo potentiometer
        val = map(val, 122, 382, 20, 126); // mapping the 'val' variable between 12 and 118 degrees
        Serial.print(x);                       // writing the value of 'x'
        Serial.print("    ");
        Serial.println(val);                   // writing the value of 'val'

        if (x == val)   // if the servo gets stuck, the 'x' value reaches the 'val' value
        {
          myservo.write(x - 17);    // stop the servo at the stuck point
          Serial.println(x - 17);
          break;
        }
        delay(35);
      }
    }
    else {                  // if the bluetooth signal is '-'
      myservo.write(0); // servo goes to 0 degrees, the gripper opens
    }
  }
}
```

Fig. 11. A snippet from the Arduino code

The gripper can be controlled via Bluetooth, with the Module HC-06, along with a free Android app downloadable online [8]. The App can send two commands to the Module, on the first press, it starts the contraction, by sending a + signal, and then we can press it again to reverse the contraction, loosening it, with the - signal. We've drawn the wiring diagram in Fritzing, shown on *Fig. 12.* [9][10].

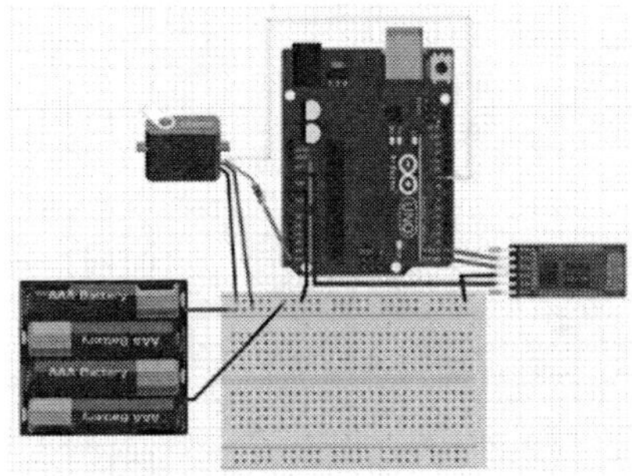

Fig. 12. The Arduino wiring diagram

The power for the motor is supplied by external source, since the Arduino cannot give it enough current to run smoothly. We also had to wire in a 1 [kΩ] resistor, so the signals would not fluctuate so much, and we grounded the whole contraption to a common ground so we achieve the desired results.

7 Calculating the diameter of the held object

By stopping the servo on the grip position, we can precisely calculate the diameter of the object held onto, using the position of the servo. The code presented it is obvious, that we can use the motor's present angle stored in x to calculate the diameter of said object. We will go through the calculation in a parametric fashion. On the figure below (*Fig. 13.*) we can see the symbols, and for added methodical approach, we added the indications for both left and right figures.

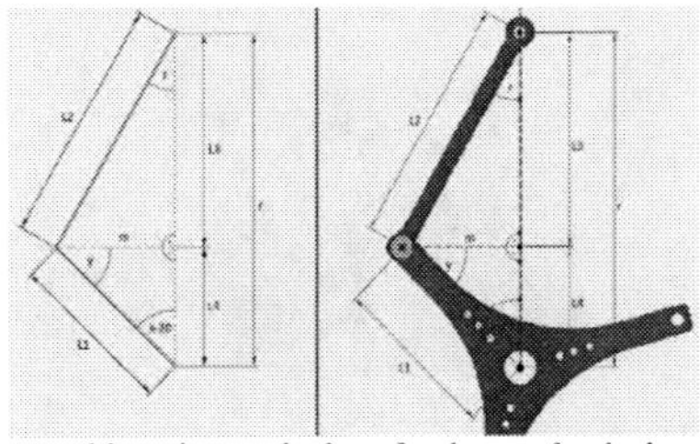

Fig. 13. Notable points and edges for the mechanical model

$$sin(x-20) = \frac{m}{L1} \tag{1}$$

As we've already stated x is always shows 20° more than the real value, so we subtract it.

$$m = sin(x-20) * L1 \tag{2}$$

$$sin(z) = \frac{m}{L2} \tag{3}$$

$$z = arcsin\left(\frac{m}{L2}\right) \tag{4}$$

$$L3 = cos(z) * L2 \tag{5}$$

$$L4 = cos(x-20) * L1 \tag{6}$$

$$r = L3 + L4 \tag{7}$$

$$d = (L3 + L4 - 20) * 2 \tag{8}$$

L1 and L2 are constants.

The 8th formula represents the 20 [mm] shift from the calculated to the real position. The multiplication of the radius with 2 equates to the diameter of the object. After we declared the necessary variables, we simply implemented the formula into the Arduino code (*Fig. 14.*), sometime consolidated together for saving space (and memory).

```
m = sin(radians (x-20))*L1;
w = m/L2;
z = asin(w);
d = (((cos(radians(x-20)) * L1) +( cos(z) * L2)) - 20) * 2;
Serial.print("The diameter of the gripped object = ");
Serial.println(d);
```

Fig. 14. Calculating the diameter of the held object

8 Determining the clamping force using the FSR 400 dynamometer

We've chose the FSR 400 (Force Sensitive Resistor) dynamometer for this task. The sensor is basically a strain gauging resistor increasing its resistance with pressure in between 0,2 [N] and 20 [N] [11]. The measurement can be carried out easily, visible on this simple wiring diagram:

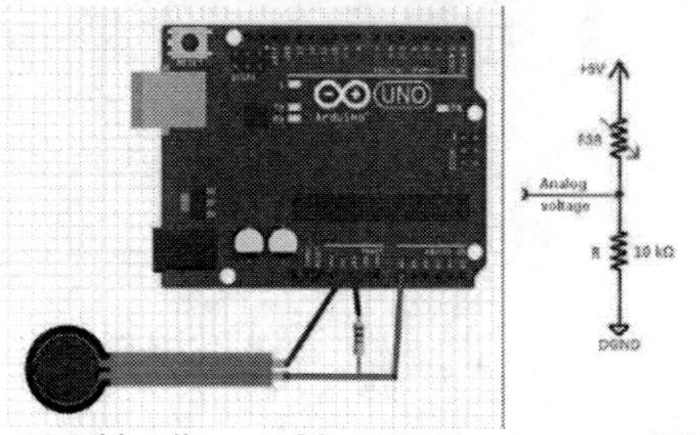

Fig. 15. Wiring diagram of the FSR 400 measurement [12]

One pin of the sensor goes to the 5 [V] power supply, the other goes to the common ground after a scraper resistance. Additionally, we connected the analog pin in between the scraper resistance and the FSR 400 (*Fig. 15.*). After overcoming the minimal threshold load (pressure), the FSR acts like a resistor of 100 [kΩ]. Proportionally with the pressure, the resistance drops from 100 [kΩ] to a 10 [kΩ] minimum. This means an increasing voltage on the line, equating to data for us on the analog pin. After the wiring had been done, we created a program to acquire the results from raw voltage (*Fig. 16.*).

```
int fsrAnalog; int fsrVolt; unsigned long fsrResistance;

void setup() {
  Serial.begin(9600);
}

void loop() {
  fsrAnalog = analogRead(A0);
  Serial.print("Analog value = ");
  Serial.println(fsrAnalog);

  fsrVolt = map(fsrAnalog, 0, 1023, 0, 5000);
  Serial.print("fsrVolt value in [mV] = ");
  Serial.println(fsrVolt);

  if(fsrVolt == 0) {
    Serial.println("No pressure");
  }
  else {
    // FSR = ((Vcc - V) * R) / V
    fsrResistance = 5000 - fsrVolt;
    fsrResistance *= 10000;
    fsrResistance /= fsrVolt;
    Serial.print("FSR resistance in ohm = ");
    Serial.println(fsrResistance);
  }
  Serial.println("------------------------");
  delay(1000);
}
```

Fig. 16. The code for the FSR 400 dynamometer

Firstly, as usual, we declared our variables. The main loop first prints the raw value of the analog pin onto the serial plotter. This equates from 0 [V] to 5 [V] proportional of the pressure, to increase the resolution, we used the command map to convert this from 0 [mV] to 5000 [mV]. We printed out the voltage to the serial monitor as well, in millivolts. Without any load, the program simply writes out: "No Pressure". We can calculate the voltage on the scraper resistance with the following formula:

$$Ur = U\left(\frac{R}{R + FSR}\right) \tag{9}$$

By sorting the equation, we can get the formula:

$$FSR = \frac{(U - Ur) * R}{U} \tag{10}$$

U = 5 [V], R = 10 [kΩ]

We implemented this formula into our Arduino code, resulting the FSR sensor's resistance in [Ω] (*Fig. 17.*). To measure the clamping for of the gripper, we use a 30 [mm], a 60 [mm] and a 90 [mm] diameter test prop, printed out in 3D [13]. We secured each prop with the gripper, while having the FSR 400 on one of the fingers. We can see the results of the 60 [mm] object on the image below.

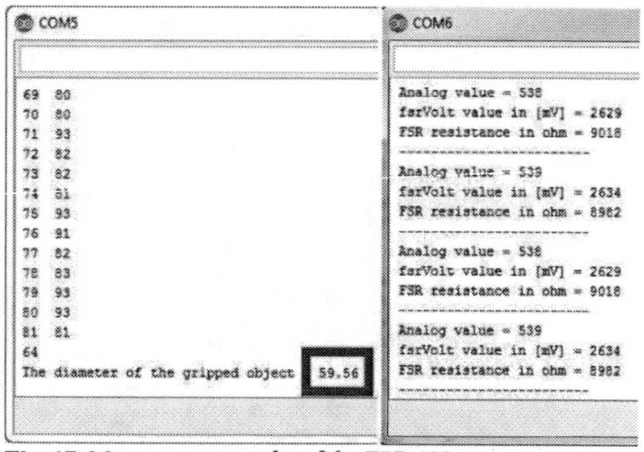

Fig. 17. Measurement results of the FSR 400

On the left side, the already presented diameter calculation is visible, and on the right, we can see the measurement results the FSR resistance's results are necessary to measure the force acting upon the sensor. Visible on the diagram below, where we display the results on the y axis, then intersecting that at 30,60 or 90 (diameter), then intersecting those points from the arc to the x axis, then divide these values by 100, we can get the correct force in [N].

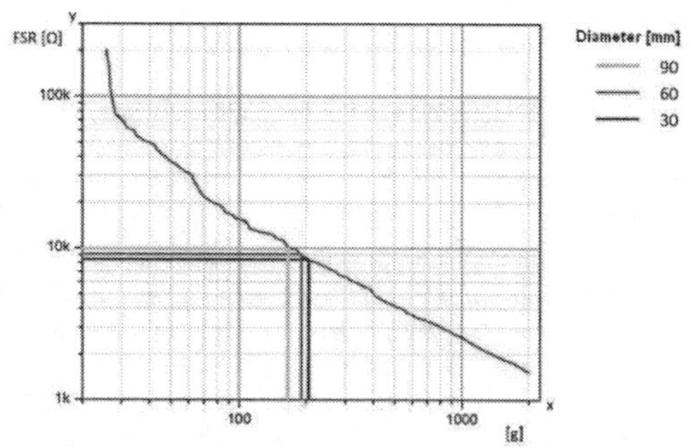

Fig. 18. Diagram of the FSR dynamometer

Interpreting the values, we can see that the finger acts with 1,7 [N], 1,9 [N] and 2,05 [N] on different diameters (*Fig. 18.*).

9 Reference work

After completing the setup, we installed the gripper with the adapter mentioned above to the robot (*Fig. 19.*), also, we rigged the Arduino onto the robot as well, for simplicity's sake.

Fig. 19. Rigging the gripper onto the robot

After installing, we manipulated different objects, lifting them, or gripping them alike (*Fig. 20.*).

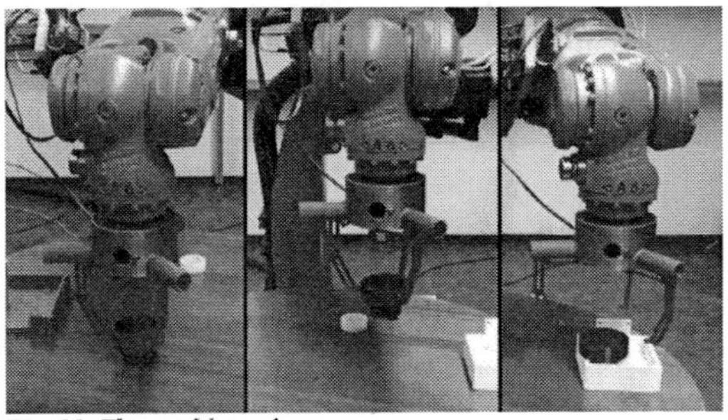

Fig. 20. The working robot

9 Conclusion

We can confidently say that we created a working gripping tool for the KUKA KR5, and acquired useful and practical knowledge on the way. The code and sensor working together can measure diameter and clamping force, adding another layer of complexity of our system. In this paper we went over the detailed creation of the whole project, and we are considering further improvements for the tool.

Acknowledgment

The publication was supported by the project EFOP-3.6.1-16-2016-00022. This project was co-funded by the European Union, from the European Social Fund.

References

1. Zs. Molnár, T. I. Erdei, N. C. Obinna, G. Husi, „*A novel design of an air-cushion vehicle and its implementation,*" MATEC Web Conf. Volume 126
2. T. I. Erdei, Zs. Molnár, N. C. Obinna, G. Husi, „*Surveillance and Security System in the Building Mechatronics Research Center,*" Electrical Engineering and Mechatronics Conference EEMC'16 - 7th-18th-19th March, 2016.
3. „*Autodesk Fusion 360,*" [Online]. Available: https://www.autodesk.com/products/fusion-360/overview. [Access Date: 27 11 2017].
4. „*TowerPro MG995 Servo,*" [Online]. Available: https://servodatabase.com/servo/towerpro/mg995. [Access Date: 27 11 2017].
5. „*ABS or PLA: Which 3D printing filament should you use?,*" [Online]. Available: https://www.digitaltrends.com/cool-tech/abs-vs-pla-3d-printing-materials-comparison/. [Access Date: 27 11 2017].
6. „*Ultimaker Cura software,*" [Online]. Available: https://ultimaker.com/en/products/ultimaker-cura-software. [Access Date: 27 11 2017].
7. F. Reed, „*How Do Servo Motors Work,*" [Online]. Available: https://www.jameco.com/jameco/workshop/howitworks/how-servo-motors-work.html. [Access Date: 27 11 2017].
8. „*Arduino Bluetooth controller,*" [Online]. Available: https://play.google.com/store/apps/details?id=com.giumig.apps.bluetoothserialmonitor. [Access Date: 27 11 2017].
9. „*Arduino,*" [Online]. Available: https://www.arduino.cc/. [Access Date: 27 11 2017].
10. „*Fritzing,*" [Online]. Available: http://fritzing.org/home/. [Access Date: 27 11 2017].
11. „*Force Sensitive Resistor (FSR), Overview,*" [Online]. Available: https://learn.adafruit.com/force-sensitive-resistor-fsr/overview. [Access Date: 27 11 2017].
12. „*Force Sensitive Resistor (FSR), Using an FSR,*" [Online]. Available: https://learn.adafruit.com/force-sensitive-resistor-fsr/using-an-fsr. [Access Date: 27 11 2017].
13. T. I. Erdei, Zs. Molnár, N. C. Obinna, G. Husi, „*A Novel Design of an Augmented Reality Based Navigation System & its Industrial Applications,*" 15th IMEKO TC10 – Technical Diagnostics in Cyber-Physical Era Budapest, Hungary, 6 – 7 June, 2017 - Organised by: MTA SZTAKI – Hungarian Academy of Sciences - Institute for Computer Science and Control.

Modelling and simulation of a series wound direct current motor using ANSYS

István Kovács[1,*], *Kornél* Sarvajcz[1]

[1]Mechatronics Department, Faculty of Engineering, University of Debrecen, Hungary

Abstract. In this paper we will discuss the process of making the model of a series wound DC motor and performing simulation and analysis using the ANSYS products. The geometry model was done in RMXprt, and the electromagnetics simulation in MAXWELL. Using the data acquired from those analyses a thermodynamical simulation was also done on the geometric model.

1 Introduction

The goal of the project was making a geometric model and performing electromagnetics simulation on an electric motor using ANSYS. The electric motor in question was used as a traction motor in an electric competition vehicle built by the students of the Engineering Faculty of University of Debrecen. The first steps of the job were learning how the software works, doing research on the subject of direct current (DC) motors. Once that has been done, we can model the motor's geometry in RMXprt, and do analysis in the MAXWELL electromagnetics module. The ANSYS products provide great solutions for mechanical, electrical and fluid dynamics problems using finite element methods.

2 Prologue

2.1 A short introduction to electric machines

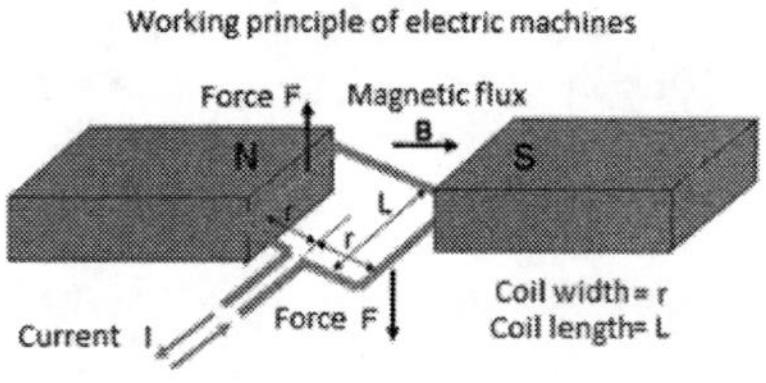

Fig. 1. Basic working principle of electric machines.

The basic working principle of electric motors is the electromagnetic induction. Around every conductor, in which current flows through, there will be a magnetic field created. The change in magnetic field will produce a change in the electric field. The force that acts upon the conductor in the magnetic field is called Lorentz-Force [2]. This force creates the moving torque in the electric motor, that we can use for different purposes, like lifting loads or propelling a vehicle [1] (Fig. 1).

3 Properties of the modelled electric motor

The first step towards the goal of the project was the disassembly of the motor. We had to take measurements on the different parts, because a detailed datasheet was not available for the product. The motor is a series wound direct current motor, and it was used on a voltage of 12 [V] powered by batteries, propelling the electric vehicle by a chain drive.

Fig. 2. The rotor (left) and stator (right) of the electric motor.

The data plate on the stator housing provided some nominal or peak values of the electric motor, the model number is ZC4-48C, the nominal current is 103 [A], nominal voltage is 48 [V], nominal speed 1200 [rpm] (with a maximum of 4000 [rpm]) and nominal torque of 14,7 [Nm]. The number of poles in this model is 4, the carbon brushes are placed according to the placement of the poles, there are 4 brushes divided by 90° from each other, under the poles. The conductor used for the stator coils has 8 strands of wire, with a diameter of 1,75 [mm] each, and the coils on each pole consist of 1,5 turns. The rotor has 36 slots, in which the rotor coils are placed with wave winding pattern [3]. The conductor used for this winding consist of 2 strands of rectangular wire with 0,8 [mm] height and 4 [mm] width. These are the most important values that describe the winding on the stator and the rotor

* Corresponding author: k.ist96@mailbox.unideb.hu

© The Authors, published by EDP Sciences. This is an open access article distributed under the terms of the Creative Commons Attribution License 4.0 (http://creativecommons.org/licenses/by/4.0/).

4 Creating the model

One of ANSYS's electrical modules is the RMXprt, it can be used for setting up geometry of electric machines. Also it gives a quick performance and operating overview of the electric machine with a simulation.

The software uses prebuilt templates for different electric machine designs and these can be adjusted with the measured (or calculated) values of the dimensions and winding properties.

In the software window (Fig. 3.) on the left hand side there is a project tree containing every part of the geometry. Here are the settings for the type of the electric machine (alternating or direct current, and the different designs of each type). Also the operation of the machine can be set, either motor or generator simulation can be done. Each part of the geometry can be opened and under the „Properties" window the dimensions or other properties of the part can be set to the desired values. Based on these values and parameters the software creates a 2D axial cross section of the motor. Here the winding and the coils are not shown, however they are included in the calculations. The most critical properties for an accurate operating model are the windings on the stator and rotor, and the poles of the stator.

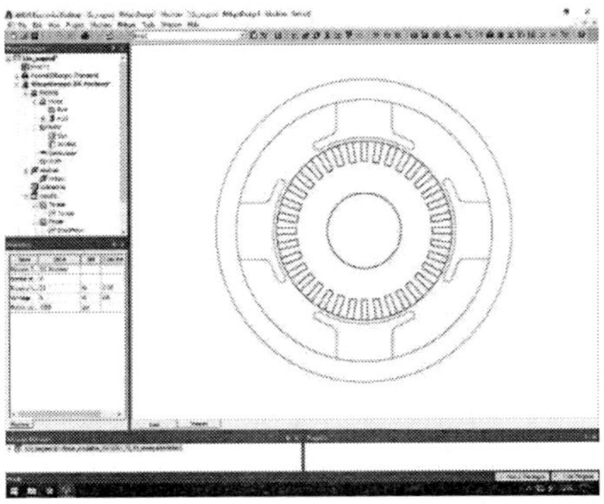

Fig. 3. 2D model of the electric motor in RMXprt.

At the analysis settings we can set the load of the motor to the desired type, these can be constant power/speed/torque, linear torque or fan load (where we increase the torque proportional to the speed). During the RMXprt analyses we used the fan load setting. We can set the nominal values of the electric machine in the properties window [4].

An important note on these simulations is, that the RMXprt module performs only static analysis. The different parameters are not displayed in function of the time (as in a transient analysis), instead they are displayed in function of a different parameter.

5 Evaluation of the results

According to the results of the RMXprt simulation, the operation of the motor is close to the nominal values. The current flowing through the coils is 108 [A] at 48 [V] voltage. However the produced torque does not meet the nominal 14,7 [Nm] in fan load operation, these results can be improved with further simulations in the MAXWELL module.

6 Modelling in MAXWELL3D

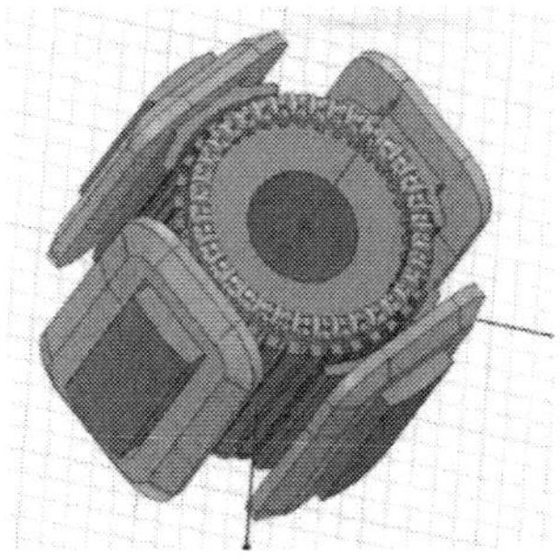

Fig. 4. 3D model of the electric motor in MAXWELL (without the outer frame).

The next part of the analysis was done in the MAXWELL module, here we can observe the operation of the motor considering different losses, and also the change of electromagnetic fields during the operation. Using the 2D model created by the RMXprt module, the MAXWELL module can extend the cross section to a 3D model (Fig. 4.). The MAXWELL module can operate with only a segment of the geometry to decrease the resources needed for the calculations. The MAXWELL module also has 3D drawing tools, so we can set up the whole geometry in this module (using primitive shapes), although it would require more time and effort.

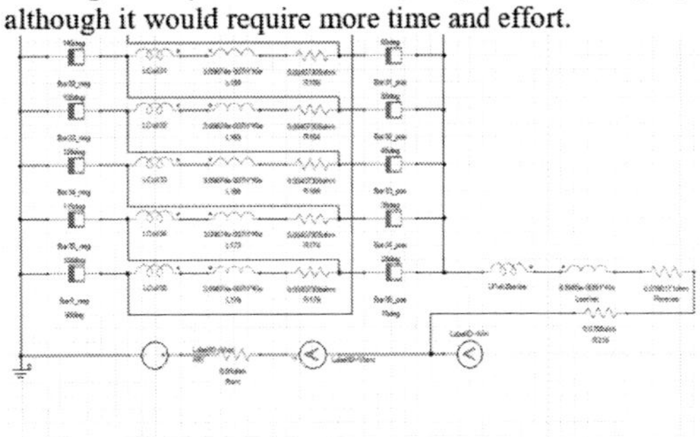

Fig. 5. A section of the windings' circuit diagram in MAXWELL

After we export the geometry model, the software creates the corresponding electrical circuit from a template, this comes with an advantage that different elements of the geometry become automatically connected to each other. We can also edit the circuit diagram and set the values on the different components.

If we want to do thermodynamical analysis, we have to adjust the thermal coefficients of the conductors. By default the simulation will not calculate with the thermal coefficient. The resistance of the conductors changes with the temperature and this can affect the performance of the motor. Also this result can be the base of a thermodynamical simulation

$$\sigma' = \sigma \frac{1}{1+0.0039*(T-22)} \quad (1)$$

The 1. equation defines the relation between the temperature and the conductivity. Unless we allow the addition of the „Thermal Modifiers", the software won't compute thermal losses. Aside from thermal losses, we can compute other type of losses - eddy currents and core losses.

For the simulations we set the duration to 2 [s] and the time step at which the calculations occur to 60 [μs].. Fig. 6. shows the current-time graph. The red coloured line means the current on one coil of the rotor winding, and the purple line on the field winding. The peak current on the field winding is 222.21 [A], and 55.63 [A] on the rotor coil, this occurs at the start of the motor. These values drop top 103.36 [A] and 22.26 [A] respectively after 2 [s] of running time.

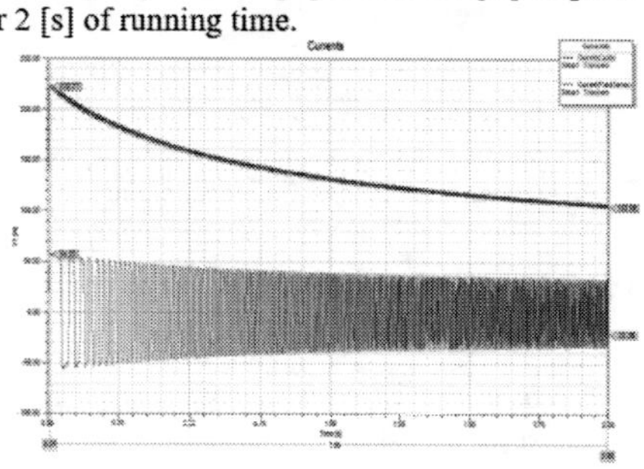

Fig. 6. Current-time graph of the stator (purple) and rotor (red) windings

The following Fig. 7. shows the torque-time graph. At the start of rotation the motor produces 14.84 [Nm] torque, which diminishes to 3.54 [Nm] after 2 [s].

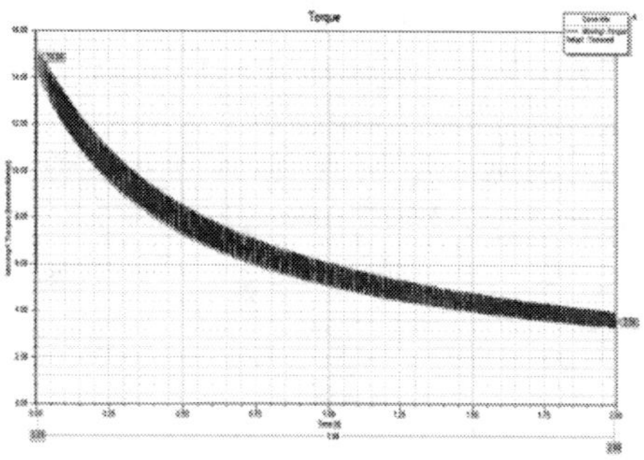

Fig. 7. Torque-time graph of the electric motor

Fig. 8. shows the speed-time graph of the motor, which links the torque and current graphs. After 2 [s] the motor reaches 4000 [rpm] which is it's highest rated speed. The results show the relation between current and torque, which is proportional, and the current and torque values are inversely proportional to the speed. This relation is fitting for a series wound DC motor.

Creating graphs is not the only way we can display the results of the analysis. We can create animations that show how the electromagnetic field occurs and changes during the operation of the motor. The save of the data corresponding to these fields does not happen by default.

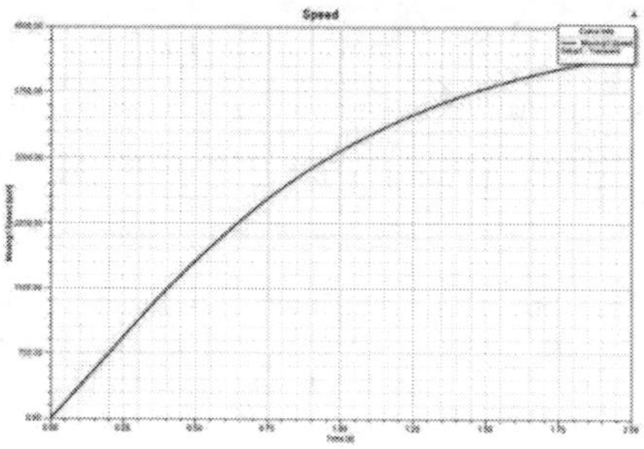

Fig. 8. Speed-time graph of the electric motor

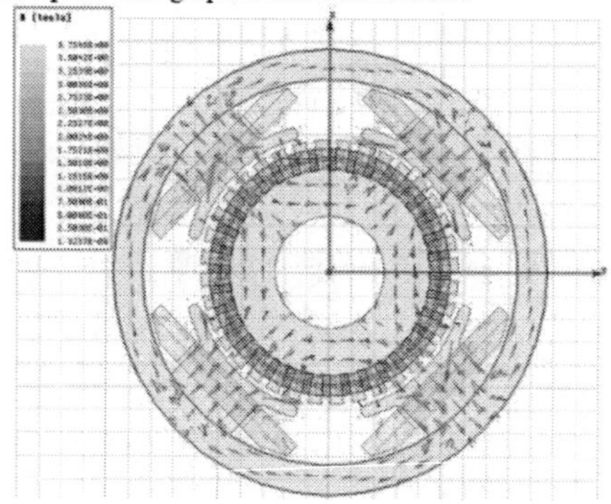

Fig. 9. Magnetic flux density, B vector shown in the motor

Fig. 9. shows the flux density measured in [tesla], also called as B vector. This gives the amount of magnetic induction lines crossing a given surface. In most cases we describe the strength of magnetic field with the flux density, because we can measure the magnetic flux easier than the intensity of the magnetic field. Also the direction of the vectors show the direction of the flux lines and the poles of the field. We can see how the magnetic lines go from magnetic pole to pole, according to the poles of the stator where the field windings are placed.

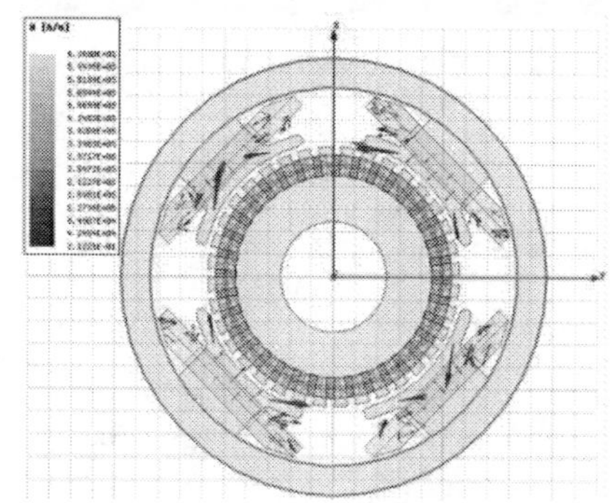

Fig. 10. Magnetic field strength, H vector shown in the motor

Fig. 10. shows the intensity of the magnetic field, measured in [A/m], also called H vector. This quantity can be related to the direction of the current. It shows the direction and amount of force the magnetic field exerts on a magnetic material. The highest value of field intensity can be found on the poles of the stator.

Fig. 11. shows the current density in the conductors, also known as J vector, measured in [A/m^2]. This shows the current flowing through a given cross section of conductors. The figure shows the direction of the current during the commutation of rotor coils.

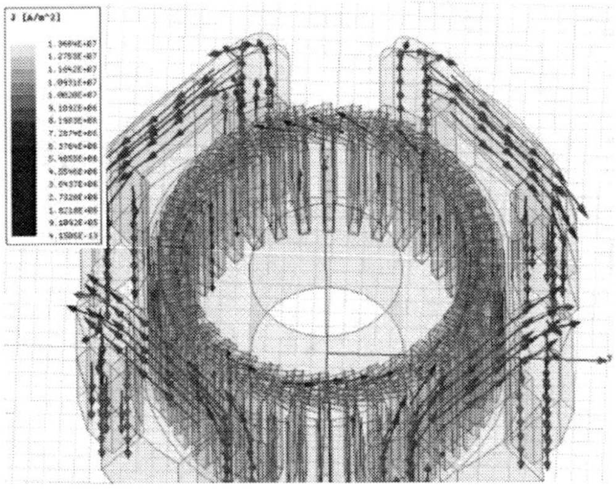

Fig. 11. Current density of conductors, J vector shown in the motor

7 Thermodinamics analysis

To perform a thermodynamics simulation, we have to use the Workbench. We can connect the MAXWELL module with the thermodynamical modules of ANSYS Mechanical. We can conduct both transient and steady-state analysis, we used both simulations.

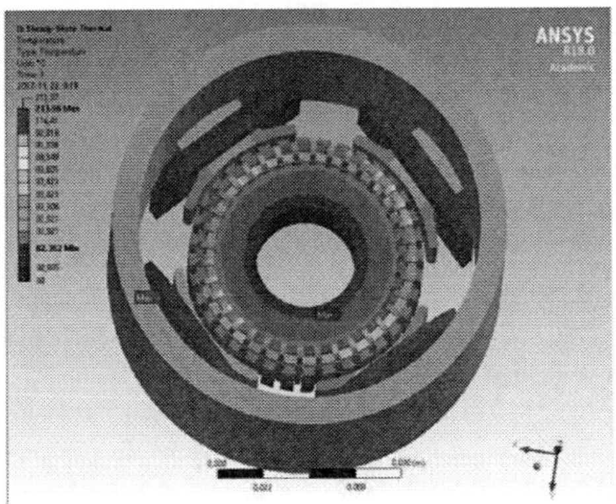

Fig. 12. Results of a steady-state thermal analysis

Fig. 12. shows the steady-state result of the analysis. In this case we take the greatest thermal load and assume that it affects the geometry during the whole simulation. This is the reason why the elements reach a much higher temperature than in the transient analysis (and the real operation of the motor). The peak temperature is 213.96 [°C], which is too high for the motor to operate normally, it would cause damage to the winding.

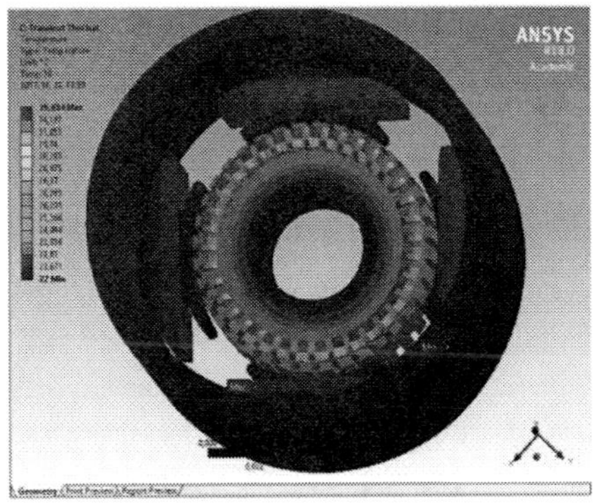

Fig. 13. Results of a transient thermal analysis

Fig. 13. shows the transient result of the simulation, which represents the real operation. The peak temperature is 39.434 [°C] after 30 [s]. According to the insulation class F on the nameplate of the motor, the maximum internal temperature without causing any damage can go up to 155 [°C]. This means that this operation would not cause damage to the motor.

8 Summary

9.1 Evaluation of the project

We created a 2D and 3D model. Performed transient simulations using MAXWELL. The results were evaluated in graph and animation forms. Based upon the results a thermodynamical simulation was also done. These results can be made more accurate with the involvement of further modules and the use of smaller time steps in the simulations. Using the results of the thermodynamics analysis, we can create a fluid dynamics analysis, which can be linked back to the thermal module to produce the airflow created by the heat to get better results. These steps can be repeated to get more accurate results each time. Also we can use the MAXWELL module to optimize the parts of the electric motor.

Acknowledgement

The work/publication is supported by the EFOP-3.6.1-16-2016-00022 project. The project is co-financed by the European Union and the European Social Fund.

References

1. Chapman, Stephen J. *Electrical Machinery Fundamentals*. 4th Ed. (2005)
2. Simonyi Károly, *The history of physics, A fizika kultúrtörténete*, 2nd Ed. (1981).
3. Pálfi Zoltán, *Electric drives, Villamos hajtások*, 1978.
4. Páczelt István, Szabó Tamás, Baksa Attila – *Fundamentals of the finite element method, A végeselem-módszer alapja* (lecture)

MATEC Web of Conferences 184, 02016 (2018) https://doi.org/10.1051/matecconf/201818402016
Annual Session of Scientific Papers IMT ORADEA 2018

Measuring drone flight-stability using computer vision

Dávid Dezső [1,*], and *Kornél* Sarvajcz [1]

[1] University of Debrecen, Faculty of Engineering, 4028 Debrecen, Hungary

Abstract. In this project we made an image processing software that we used to examine drone flight stability. One can read about the closed loop control of a quadcopter, thé computer vision technics and the results of the test under real circumstances.

1 Introduction

In the last few years the use of quadcopters expanded, more people use it as a hobby and its presence can be seen in professional enviroment. Their use as a video recorder device is the most common.

We work with these kind of quadcopters in the Drone Club of the Faculty of Engineering. The size of these are much larger than a racer drone and its main attribution is stability. This means they can hold their position with the less movement possible. To reach this we need multiple sensors and their precise cooperation.

The closed-loop control needs a lot of tuning to work as precisely as it should. After an adjustment we only can be sure if the change went into the right direction when we test the position hold flight mode under normal circumstances. To see this we can only use our eyes and the impression of the pilot. To fly we need good weather conditions and this can mean that between two flights too much time goes by and we cannot be sure that the tuning is better or worse than it was last time.

We want to give a solution to this problem in this paper. To reach this goal, we have to make a testing method that gives exact data on the stability of the position holding so that multiple versions of the same quadcopter or different drones can be compared.

Our plan is to write a program that processes a recording of a drone flight using image processing algorithms. It detects a marker and shows the route of a drone in the test time. After that it makes diagrams on these we can see the results.

The full test stands of the recording, the process of the recording with the analyser software and last the evaluation of the results.

2 A QUADCOPTERS SENSORS AND CONTROL

2.1. Sensors

So that a multicopter can fly stable without special pilot abilities it needs closed-loop control. In a perfect world where all the parts the drone stands of are perfect and where there are no disturbing external forces this control system would not be needed. But in the real life there is no such thing as two perfectly similar motor or propeller. This is why we need sensors to „keep an eye" on the conditions of the quadcopter and electronics to control the actuators according the datas they get.

The most necessary sensors take place in the IMU (Inertial Measurement Unit). One of these is the gyroscope. The task of the gyroscope is to give information about the angle of the UAV (Unmanned Aerial Vehicle) compared to horizontal. For this it uses the angular momentum. In the everyday life they use it in many fields such as on airplanes in the so called artificial horizon. Two gyroscope measures the horizontal and also the vertical angles, the three axes they give information of are:

- Yaw = roll on the vertical axis
- Roll = roll on the vertical perpendicular to forward movement axis
- Pitch = roll on the vertical parallel to forward movement axis

Despite the angular momentum a drift appears in the gyroscope. This is the moving of the rotation plane. Here gets part the next part of the measurement unit, the accelerometer. It can compensate the forces that cause this drift. The principle of operation of the accelerometer is the same as the behaviour of a mass fastened to springs, the forces moves the mass so the lengths of the springs change. These two sensors are cooperating with each other and we need both to get the right datas.

Knowing the position is just as important as the angle. Primary is the height. For this a barometer is integrated to the IMU. The barometer measure the atmospheric

* Corresponding author: dezso.david15@gmail.com

© The Authors, published by EDP Sciences. This is an open access article distributed under the terms of the Creative Commons Attribution License 4.0 (http://creativecommons.org/licenses/by/4.0/).

pressure. The vertical position of a multicopter specifies the height of the air pillar above it and the weight of that. Naturally the pressure of the air around the drone is highly affected by the temperature. That is why we also need a temperature sensor at the side of the barometer. With the information of these two can be calculated the exact altitude of our quadcopter.
Staying at the calculation of the flight altitude a few words about a sensor that is used in the middle-class and higher priced UAV as a standard equipment in the last few years. This is the ultrasonic distance sensor. This is placed at the bottom of the multicopters vertically downwards. A sensor like that send an ultrasonic wave from the „trigger" transceiver, this wave is reflected from an object, in this case the ground, and the receiver called „echo" detects it. The time difference between the released wave and the received one shows how much time th ultrasonic wave needed to go back and forth. We can calculate the altitude if we know the speed of the wave that is 343.2 m/s in 20 degree celsius. Of course the ultrasonic wave cannot go large distances. That is why it is only useful while take off and landing or during low altitude flights as a completion of the barometer.
The fully autonomous position holding cannot be achieved without a GPS receiver. This sensor with triangulation calculates the position of the UAV using its own clock signal, the time code and orbit data received from the satellites, to do this the GPS sensor needs minimum 4 satellite signals [1].

2.2 Control system

After the introduction of the sensors used in control let's talk about the closed loop control itself. The UAV needs closed loop control to keep its state stable while flying. Stability here means that despite the environmental effects and despite the imperfection of the structural elements the drone stays in a balanced state. A system like that consists of a controller and a controlled section.

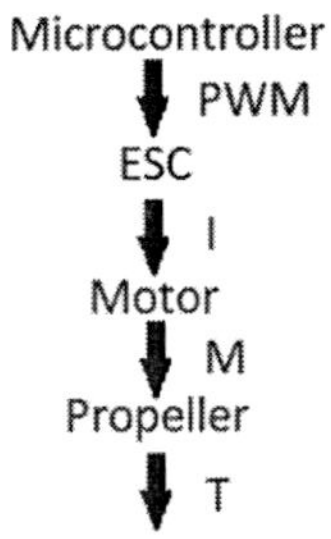

Fig. 1. : Controlled section of an UAV

On Fig. 1. the controlled section of a quadcopter can be seen. The microcontroller acts according the controller section's commands by sending Pulse Width Modulated (PWM) signals. The electronics of the motors translates the PWM signal to the current the brushless DC motor needs. This electronics called ESC (Electronic Speed Controller). The motors rotates the propellers with the momentum needed to reach the thrust force we want to archive [2].

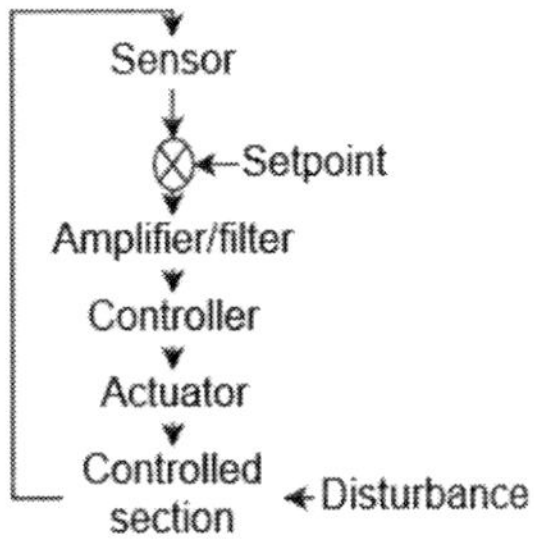

Fig. 2. : Controller section of an UAV

In quadcopters, PID (Proportional–Integral–Derivative) control units are used to create the command needed for the intervention unit, that is, the microcontroller. After the sensor signal has been merged, a differential generator compares the set signal that is sent by the autopilot or the remote control operator. After amplifying and filtering, it reaches the PID unit. Such a unit consists of three parts: The P, proportional member output signal is proportional to its input signal. The integrating I, which output signal is proportional to the integrity of the input signal. Finally, the D, differentiator part, its output signal is proportional to the change in the input signal.
Most quadcopters' central computers are pre-programmed to self-tune the closed-loop control but we may achieve better results with self-adjusting. The custom method, as well as the built-in, compensating parameter setting is done using the Ziegler-Nichols method. The loop gain is increased until a permanent swing is obtained. With drones this is a continuous tremor in the floating. Then the period of time for the critical open loop gain can be determined. These data can be used to determine the value of each of the three parts by using the given relations [1].

3 OBJECT FOLLOWING PROGRAM

3.1 Importing libraries

The list and short description of the libraries needed to operate the program:

- CV2: The OpenCV (Open Source Computer Vision) library gives thousands of algorithms for image processing and computer vision [3].
- Numpy: This is a indispensable algorithm package if you use OpenCV. Python's basic library for scientific calculations. It contains multi-dimensional array elements creating functions. Tools for integrating C / C ++ scripts. It can be used to calculate linear algebra, Fourier transformation and random number generation [4].
- Matplotlib: The Python's two-dimensional drawing library. It allows you to create diagrams for presentations and evaluations. Interactive environment for communication between platforms [5].

- Imutils: A free to use library made by Adrian Rosebrock directly for OpenCV. With it you can transform and resize images [6].

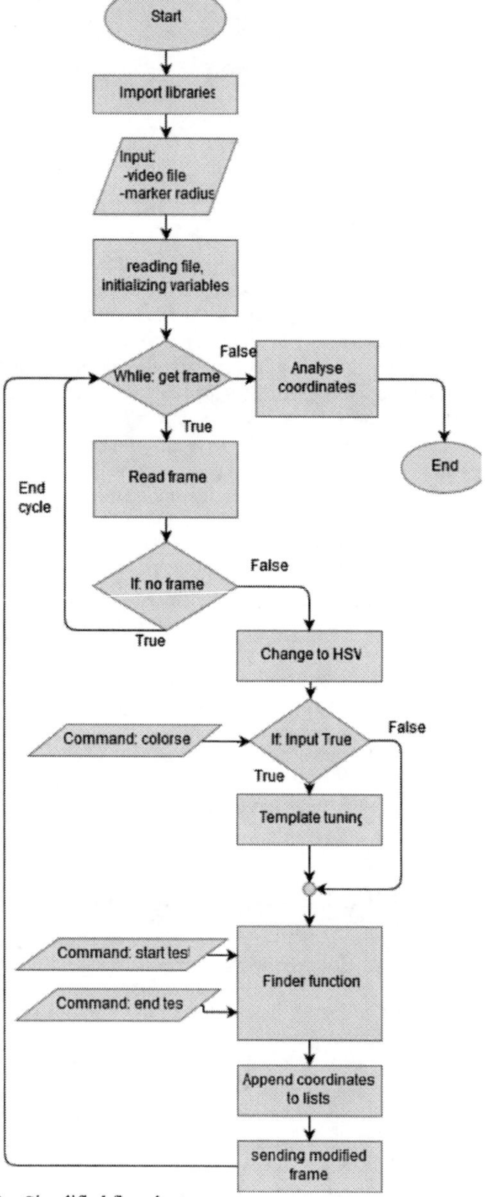

Fig. 3. : Simplified flowchart

3.2 Data input

When a Python program is started, an operation window pops up immediately. After the startup, the following message appears:
„Insert name of the video file:"
That is, the program needs the name and extension of the recording you want to test. In case the file is not in the same folder as the program, the name of the file must be entered along with its path. Since the OpenCV's primary punctuation is real-time image processing, of course this is possible. If you want to process continuous stream, the video file name must be replaced by the serial number of the connected recorder, ie. "0". After the filename is read, another initial variable is required, and this is the circle radius of the object marker in millimeters:
„Radius of the circle (in mm):"
For this input I have set a default value of 25 mm. By sending this you will also start processing and the following message will appear in the keypad window: „Press 'C' to set the colour (esc if done) S' to start test and 'Q' to quit."
The flowchart shown in Fig. 3. has three inputs in its while loop, which are used to control the software. The message above provides information on this.

3.3 Detecting cycle

After the video file has been called in and the necessary variables and lists initialized, the program enters its main cycle. This is where the object is detected and the necessary data for motion analysis is recorded. As a first step, the next frame of the recording is called with the "getframe" function, which returns a boolean value with the frame that indicates there is another frame or not. If you have an empty image, the variable is "False". That follows the test, so if the program runs out of frames, it interrupts the cycle and jumps to the coordinate analysis part. Then the frame size is reduced by using the "imutils" library, thus accelerating the processing. Then the "cvtColor" function changes the color space. RGB encoding switches into HSV (H = hue, S = saturation, V = value) color space. With this encoding, the boundaries can be defined, which means that if only the marker remains, detection will be successful in the color search function. Values are defined in the next section. At this point, the video file is displayed from frame to frame and the program is waiting to receive one of the commands from the keypad. Pressing the "C" key sends the "colorset" command. At this point, the video stops in a given frame. New window pops up, you have three windows for the current frame, the mask, and one where six sliders represent the color boundary values. Once they appear, the program enters a while cycle. The positions on the sliders are scanned continuously. Using these, you create a new mask that is updated in the previous window. It is a good idea to start the border search by adjusting the hue, as it can only slightly change and thanks to the homogeneous marker it can be within narrow limits. After that, it is worth refining the picture by shrinking the boundaries of saturation. The final set is the brightness, in this case it is not fortunate to set it for only that frame, since the movement of the drone can change the brightness, and if it is outside the boundaries, wrong coordinates will be given. When we are done with setting the boundaries, pressing the "ESC" key accepts the current HSV values and the main cycle of the script continues. After that comes the "finder" function written by me, which detects the marker on the frame, changes the displayed image and returns the X and Y coordinates and the radius of the marker in pixels.

3.4 Color finder function

The "finder" function mentioned in the previous chapter gives the root of the software. When calling, three arguments are required:

- HSV: The frame in new colorspace.
- MIN: The minimum boundaries set on the „colorset" sliders.
- MAX: The maximum boundaries set on the „colorset" sliders.

These are all needed in the first step. As in the mask setting, it also uses the "inRange" function to highlight the marker to the template image. This examines all the pixels if the color values are within the defined limits. It gives the corresponding pixels "High" value on the template and all other receives logical "0" values, so the template itself can be interpreted as a binary array. Our next task would be to find the contours, but in this case the operation of the program would not be impeccable, it would make mistakes several times, making it impossible to perform the test. This is because the edge of the mask in the template is uneven and overall too noisy from the scattered "High" pixels. The problem is solved by morphological transformations. The collective name of multiple interconnected image processing tasks in computer vision. In our case erosion and dilation are applied by the software. During erosion, it slides the template below itself and only the pixel remains "1" which a high value is found on both layers. Thus the tiny noises from the template disappears, but the contour that is important for us will also be smaller. Although the program would reduce the size of the reduced mask, the radius values collected for the later process of analysis would be altered, causing the accuracy of the test result to fall. That's why dilation is also needed. What makes the inverse process, that is, every pixel gets a high value with a value of "1" on either layers. Here you get the original size of the contour you want, but the lost noise will not come back. It can be seen that the two processes are the same, the difference lies in the fact that eroding is the logical "and" the dilatation acts as the logic "or". Subsequently, the template is now suitable for finding contours. This means that you can determine the visible shapes by searching for "0" - "1" borders. This is done by CV's "findContour" function. Creating an array of all the found contour's datas, the task is to choose only the largest of these, after all, the boundary setting was made so that only the marker would appear on it. The software runs on all frames to this point. I made it like this because these transformations are the most complex operations that require the most time. Therefore, if the test starts, there will not be a noticeable deceleration in the processing, and until the beginning of the test, the frames won't be shown too fast.

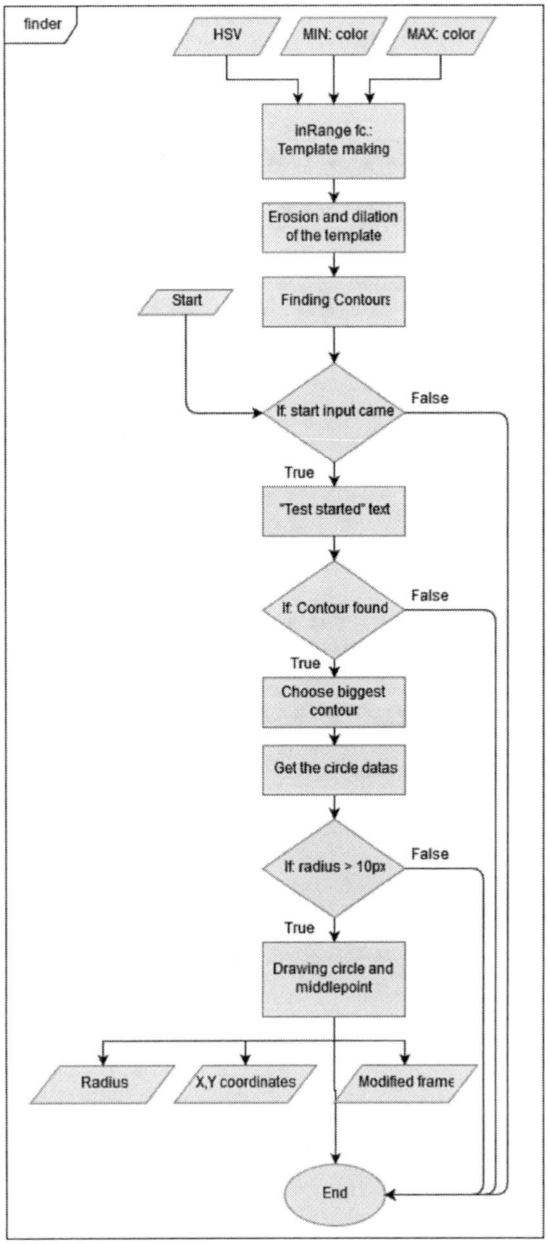

Fig. 4. : Color-finder function

The start of the test begins with the sending of the start signal as it is seen in the flowchart (Fig. 4), as it was explained in the explanatory message mentioned above, by pressing the "S" key. As a primary feedback, a yellow "TEST STARTED" text appears in the upper left corner of the window. It will then be checked if it has found a contour. If yes, it uses the OpenCV's "minEnclosingCircle" algorithm. Entering the largest contour as an argument returns three values, all of which can get from the smallest writable circle around the contour. They are in the order: the X coordinate of the circle's center, the Y coordinate, and finally the radius. If the radius exceeds a minimum value, the circle and the center are displayed as feedback on the image. This is the end of the function. Returns the coordinate and radius values as a closing.

3.5 Analyzing coordinates

The main cycle produces three lists of Y, X coordinates and radiuses. They contain an element for each frame, before the test is started "0" values are given, so it is necessary to filter them. For this task, I defined a simple function to shorten the script. The original list contains the image coordinates, those are the positions of the object on each frame. As a first conversion, a relative movement list is developed. The first element of the filtered coordinate lists is named for the origo, then it replaces all the others with the difference between them and the origo value. This is still only suitable for comparison of percent deviation, the general use requires real movement determination. This is achieved by the list of radiuses. Because the software recognizes the true radius of the marker, since the user specified it at the beginning of processing. Calculates the actual movement of the visible axes in the frame with the following equation:

$$\frac{Radius\ in\ pixel}{real\ radius} = X \quad (1)$$

$$X = Millimeter\ pixel\ coefficient \quad (2)$$

$$\frac{Movement\ in\ pixels}{X} = Real\ movement \quad (3)$$

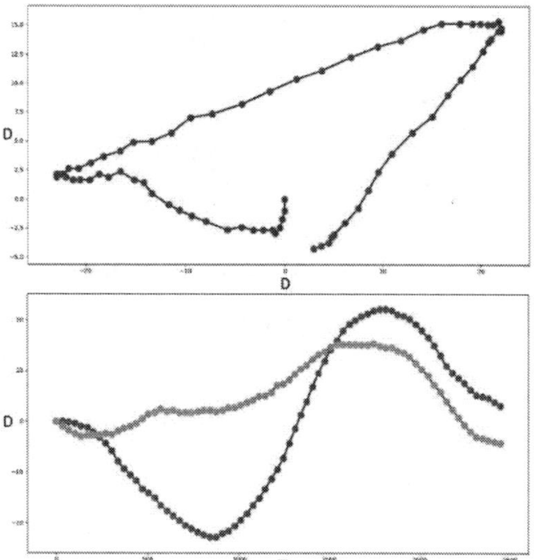

Fig. 5. : End result diagram

The Fig. 5. above shows the first result of the image processing. The X and Y axes return the full motion route. Both axes are in millimetres. This diagram shows rather the success of tracking than it is needed for evaluation. Two functions are displayed on the next diagram. With those we are capable to evaluate the motion. The vertical axis shows "d" movement in millimetres, and the horizontal "t" axis is the time. When the program generates a chart that can be magnified, it can be read up to 10 millimetres and milliseconds. .

4 Position hold test

4.1 Phantom 1 test

The test was originally designed for laboratory conditions, that is, adequate GPS signals, good lighting, and windless environment. They were not available, but it is possible to observe the effects of lack of these. At the time of this test, there was a constant 7 km / h wind and heavy occasional gusts of 15 km / h during the flight while the air temperature was 12 ° C. The drone was unable to hold its position, but apparently worked against the wind, without the control even the droneframe's horizontal close angle would not be possible. A persistent, slightly accelerating sliding can be observed. Within this 3.5 seconds, there have been some outbreaks in this measurement, but they can still be evaluated. After that, the functions are not as smooth as before, there is an explanation for this, the quadcopter has been moving away more than five meters from the camera. I limited the motion evaluation to the first three seconds.

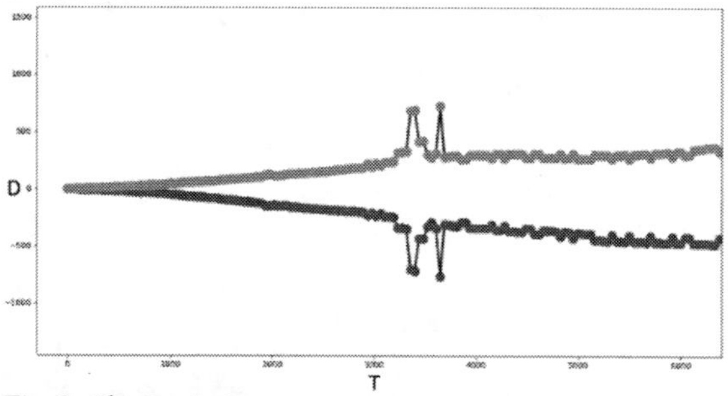

Fig. 6. : Phantom test

Table 1. Phantom test

t [s]	0,5	1	1,5	2	2,5	3
d_X [mm]	-17,4	-49,9	-86,8	-136,7	-178	-208,3
v_X [m/s]	-0,0348	-0,065	-0,0738	-0,0998	-0,0826	-0,0606
a_X [m/s^2]	0,0696	0,0604	0,0176	0,05	-0,0344	-0,044
d_Y [mm]	21,7	49,9	78,1	117,2	154,1	200
v_Y [m/s]	0,0434	0,0564	0,0564	0,0782	0,0738	0,0918
a_Y [m/s^2]	0,0868	0,026	0	0,0436	-0,0088	0,036

The horizontal motion on the "X" axis is negative, meaning that it is floating towards the left edge of the frame, which is to the right of the drone's point of view, the same as the wind direction. The speed accelerated to varying degrees for two seconds, then began to slow slightly, but the entire diagram showed that the direction of the speed did not change.

Movement along the vertical axis shows a steady increase which, with the exception of two small overthrow, is also steadily accelerating. On both examined axes, the movement of the drone was approximately 20 centimetres from the initial.

4.2 Inspire 1 test

I tried to reverse the use of program with an inverted approach during the test of the DJI Inspire 1 drone. In this case, the marker is not attached to the droneframe, but to a stable point, specifically on the ground. Thus, the program captures the relative motion of the video maker to a reference point.

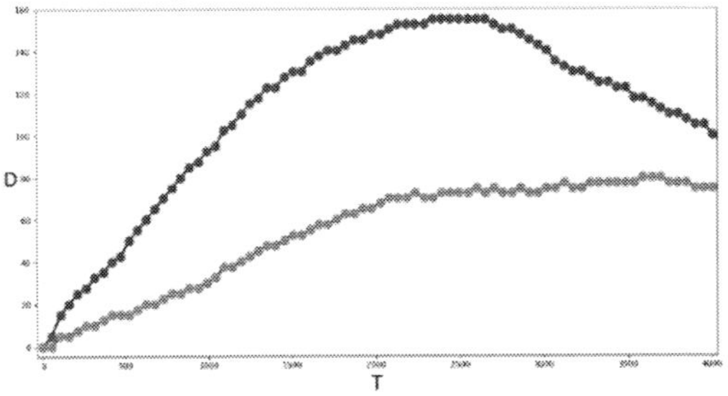

Fig. 7. : Inspire test

Table 2. Inspire test

t[ms]	400	800	1200	1600	2000	2400	2800	3200	3600	4000
d_X [mm]	39	77,2	110,5	133,8	147,5	155	150	130	117	100
v_X [m/s]	0,195	0,191	0,167	0,117	0,069	0,038	-0,03	-0,1	-0,07	-0,09
a_X [m/s²]	0,98	-0,02	-0,12	-0,25	-0,24	-0,16	-0,31	-0,38	0,18	-0,1
d_Y [mm]	14,5	25,2	40,3	54,6	66,3	72,6	72,6	75	80	75
v_Y [m/s]	0,073	0,054	0,076	0,072	0,059	0,032	0	0,012	0,025	-0,03
a_Y [m/s²]	0,363	-0,095	0,11	-0,02	-0,07	-0,14	-0,16	0,06	0,065	-0,25

During the test, it was windless and about 12°C. One possible mistake was that it was surrounded by a densely built environment that could disturb GPS signals. There is a slow slide here, but it is also captured by these sensors. Axis "X" shows the movement perpendicular to the direction of motion. Beginning at the starting point to the right at a speed of 0.195 m / s. The control reacted to that and slowed down smoothly, and by the time it reached the maximum 15.5 cm shift the drone started moving backwards. There is a very similar tendency along the "Y" axis. It moves forward at a slower starting speed than on the other axis. The response speed declaration is also smaller and even though it stops for 2.8 seconds, it reaches its maximum movement after 3.6 seconds and then corrects itself back.

5 Summary

During this project we were able to create a software that uses color detection method to follow a marked object on the video file of its movements. We then completed it with properties that make it suitable for use in a measurement to determine the stability of a quadcopter during the position hold mode. These properties include customizing the boundaries of the searched color, starting and ending the analysis of the recording at a command, determining the true motion and displaying the result of the test in an evaluable form. Once all these have been completed, the test of real conditions has taken place. Measurements were made using two commercially available UAVs. In addition to the appropriate conditions, the software can be used during the development of quadcopters to analyse the effects of changes in hardware or hardware.

6 Acknowledgement

The work/publication is supported by the EFOP-3.6.1-16-2016-00022 project. The project is co-financed by the European Union and the European Social Fund.

References

1. „hobbielektronika.hu,"[Online]. Available: https://www.hobbielektronika.hu/cikkek/multicopterek_nullarol_az_uav-kig_i.html?pg=1. (2015)
2. Phd. J. Tóth , Author, Measurement and Control Technology I. . [Performance]. University of Debrecen, Faculty of Engineering Debrecen, Hungary (2015)
3. „Opencv introduction," OpenCV, [Online]. Available: https://opencv.org/about.html (2017)
4. „Numpy," Numpy, [Online]. Available: https://www.numpy.org/ (2017)
5. „Matplotlib," Matplotlib,. [Online]. Available: https://matplotlib.org/ (2015)
6. A. Rosebrock, „Pyimagesearch," [Online]. Available: https://www.pyimagesearch.com/ (2017)
7. Phd. J. Tóth , Author, Measurement and Control Technology II. . [Performance]. University of Debrecen, Faculty of Engineering Debrecen, Hungary (2015)

Homes of the Future

Bence Kovács[1*], *János* Tóth[1]

[1]University of Debrecen, Mechatronics Department, 4028 Debrecen Ótemető street 2-4., Hungary

Abstract. The industry is being revolutionised again. It is the fourth time, that is why this revolution is often called Fourth Industrial Revolution. One of the definitions of this process says that the aim is to digitalize the machines and resources and those should communicate with each other in the cyber-physical space via sensor networks and mobile communication. This automation process starts to appear in the households. People have the need to own fully automated houses and intelligent homes. In this article you can read about how a house can be intelligent.

1 What does intelligent mean?

If we use this phrase to humans, it is easy to tell what an "intelligent person" means. However, it is not that easy to define when we talk about houses and homes. Since it is the concept of the future, we cannot find exact definition of it. As I have mentioned in the Abstract, there are several definitions of the fourth industrial revolution and I had to choose my ideal one [1]. The case is the same with "intelligent houses". During my research I have found thoughts about being "intelligent" based on energy efficiency or the usage of renewable energy resources. Others say that a house can be called intelligent if it has an internal informatic network. Some say that an automated house is "Smart", but only when it is controlled by a central control unit, not separated, individual controls [2-4].

I do not agree with all of the statements, that is why I have decided to make my own criteria of "intelligent" and will build a model that meets the requirements and simulates real-life like conditions. I do not think that being "Smart" depends on energetic issues or renewable energies, but internal informatic network, automatic processes controlled by a central control unit and IoT.
The automation should:

- make the owners life easier and more comfortable,
- provide security,
- have positive effects on (energy) consumption,
- and provide advanced fire alarm system.

2 Selection of the central control unit

I had to choose a control unit that could fulfil all four requirements of automation on its own. A Programmable Logic Controller (PLC) would have been a good choice, because it is usually manufactured to create automations. It usually has 24 VDC inputs and outputs, so it can directly control actuators and process sensor signals. However, it requires specific knowledge to create complex controls and it is hard to fulfil IoT requirements. With the usage of a processor it is much easier to connect the whole house to the internet, but those chips are not designed for controlling power-hungry devices. And those have serial control.

My choice was a product of National Instruments. The control unit is a myRIO 1900. It has an integrated processor and a Field-Programmable Gate Array (FPGA) module. Thanks to the processor it is easy to connect the device to the internet and have real-time supervision, and to the FPGA I can achieve massive parallel control. It is a stand-alone application. The chosen controller can be seen on Fig. 1.

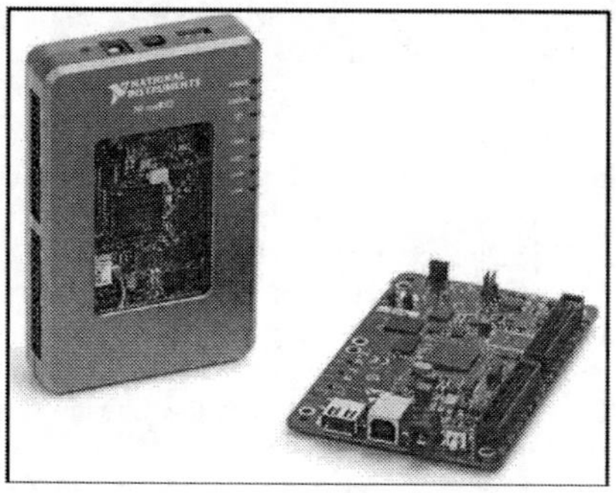

Fig. 1 - myRIO 1900 control unit

With this choice I am able to use the LabVIEW development platform provided by National Instruments. I can create my own programming code, the block diagrams and the user interface at the same time. Also, programming in g-code makes it more effective.

3 Automatic controls

My aim was to create:

- key-free access to the house,

* Bence Kovács: kovacs.bence0730@gmail.com

© The Authors, published by EDP Sciences. This is an open access article distributed under the terms of the Creative Commons Attribution License 4.0 (http://creativecommons.org/licenses/by/4.0/).

- controlled lighting whit the option of mood-lighting,
- motorized controls (Direct Current (DC) and Stepper),
- heating and cooling control,
- controlled sprinklers,
- security system with door/window opening and motion sensors,
- and an advance fire alarm system.

The most important aspect of an intelligent house for me is that the house should be aware of the presents of the owner and should create user-specific processes based on their needs.

I have achieved it by using Radio Frequency Identification (RFID) access control system. Each owner has an own card. The house senses their entry to the house and by using Universal Asynchronous Receiver-Transmitter (UART) communication protocol, it can recognise the individuals and perform pre-programmed user specific controls such as setting a specific room temperature or switching on mood-lightings. I had to use a hierarchical system among the users. If more persons are present and they have different needs, the control unit has to decide what to do. This decision making is based on my pre-programmed hierarchy. I have used a While loop with 10 ms timing for this part of the program. Each RFID card contains a 10 digit identification character. The program waits till all 10 of the characters arrive and then examines the code. I have used American Standard Code for Information Interchange (ASCII) character coding with 9600 baud rate, without parity bit. One stop bit registers the end of the communication. A portion of the RFID program can be seen on Fig. 2.

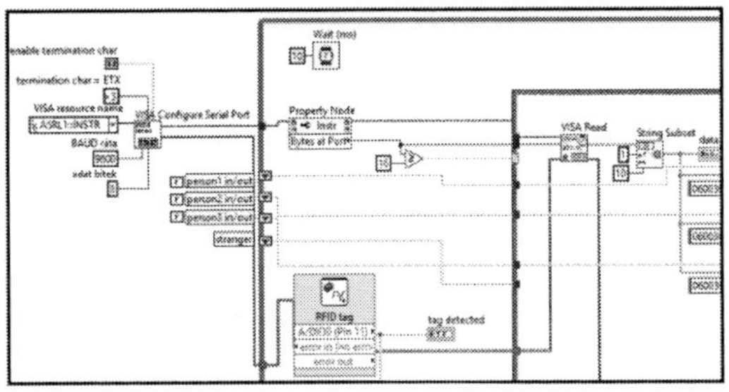

Fig. 2 - RFID program portion

Controlling lightings, motors, actuators and sensors was not that easy. Unfortunately, I had to make extra control circuits with amplifiers, because the myRIO can only provide 3.3 VDC and 4 mA. I used transistors, logic-level metal-oxide-semiconductor field effect transistors (MOSFET) and relays for amplifying. I soldered everything together on universal printed circuit boards. An example of lighting control circuit with motion control can be seen on Fig. 3.

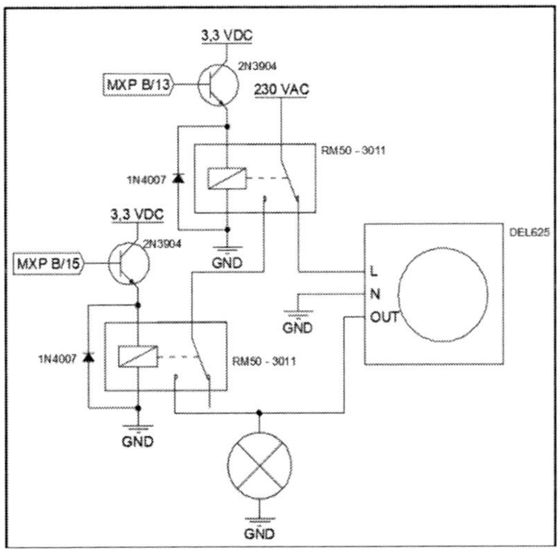

Fig. 3 - Additional control circuti for lighting

I have installed a wide variety of sensors to automate lighting. I have used ambient light sensors to detect the need of the lights to switch on. I have also used motion sensors and timers so that the house owners do not need to use switches. However they are able to overwrite any decision made by the controller. You can find several types of lights on the model, e.g. outdoor lights, indoor lights, standard bulbs, 12 VDC spot lights, light emitting diodes (LED), LED strips, etc. When I used MOSFET-s to control the brightness and the colour of my red-green-blue LED-s, it was important to use heatsinks as well. During the usage of lots of LED-s, amperes can go through the MOSFET-s and those produce significant heat. When applying heatsinks, I have to deal with electrical isolation between the Gate of the MOSFET (which is connected to the back area of the MOSFET) and the heatsink. Otherwise it would be dangerous to use.

The lights can be controlled manually (by switches) and via the Front Panel. A portion of the control program can be seen on Fig. 4.

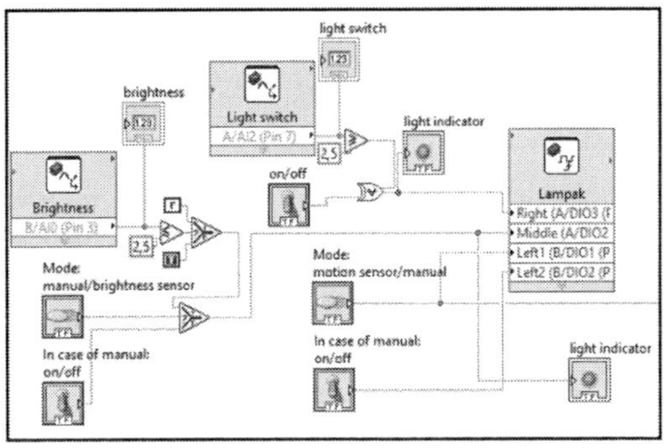

Fig. 4 - Portion of the lighting control program

There can be found an automatic sprinkler system in my intelligent home to make the user's life easier. It is time-controlled but the control unit detects the moisture of the soil and if it is wet, sprinkling will not happen.

I have used two kind of motors to achieve controlled motions. For places where positioning is important, I have used stepper motors, otherwise I have used simple DC motors. The two kind requires different control methods, which I was able to program. The DC motor represents an automatic garage door. Two button is used on the user interface to "open" and "close" it. There is a light sensor for protection. If it signs during the closure of the door, the process is terminated.

The control of the stepper motor is more complicated. The owner of the house have the opportunity to vary the direction and the speed of the motor.

4 Security functions

For security purposes I have installed door and window opening sensors. If there is a robbery, a loud and unique alarm sound is activated. I have created the sound by programming Pulse Width Modulations (PWM). The frequency of the PWM is responsible for the tone, while the duty cycle is responsible for the uniqueness of the sound.

I have created other sound profiles for the RFID system. The access control system gives different sounds when it recognises and accepts a card and different ones when refuses them. This system also logs every movement. It logs when someone enters or leaves the home and saves it to a text file. This text file can be processed by any spreadsheet program. You can apply filtrations to get valuable information from the data with e.g. Excel. Fig. 5 shows and example.

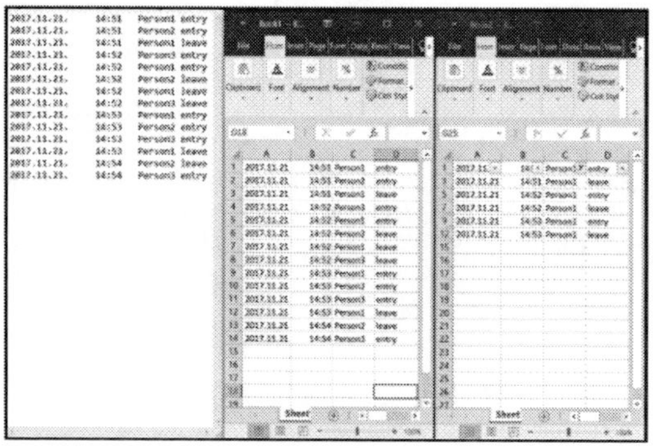

Fig. 5 – Movement logging and processing of the data

You can find an "intelligent" fire alarm system in my Smart home model. I have programmed the following functions to make it "intelligent":

- It has drift compensation. That means that by using counters, shift registers and logic values in my program, it can compensate the dust pollution of the sensor. However, after a while cleaning and maintenance of the sensors are compulsory.
- It will not provide false alarms. The signal of the sensors must be continuous for 30 seconds otherwise it is not accepted as fire and the system will not alarm.
- On the other hand, the system uses group decisions. If two or more neighbour sensors signs at the same time, there is no delay in the system, it is immediately activated. It is not the case when two, but not neighbour sensors signs.
- The users have the opportunity to check the fire as soon as possible, because they are notified via e-mail, and they receive a pop-up window on the front panel of the program, when there is a fire alarm.

A portion of the fire alarm program can be seen on Fig. 6.

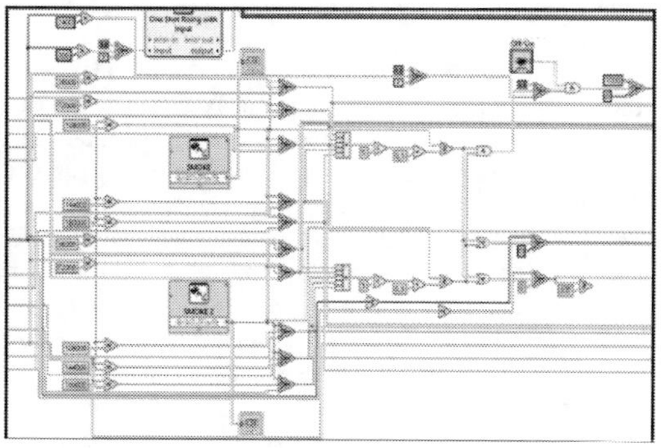

Fig. 6 - Protion of the intelligent fire alarm program

5 Remote control

I have connected my system to the internet. The myRIO has a built in Wireless Fidelity chip, so it can connect to a router wirelessly.

I have created a front panel in LabVIEW which can be seen on Fig. 7. The colour coding (e.g. yellow for lightings, blue for sprinkling, etc.) and the labels help the users to know what they are controlling. Also, indicators shows what is happening currently in the house.

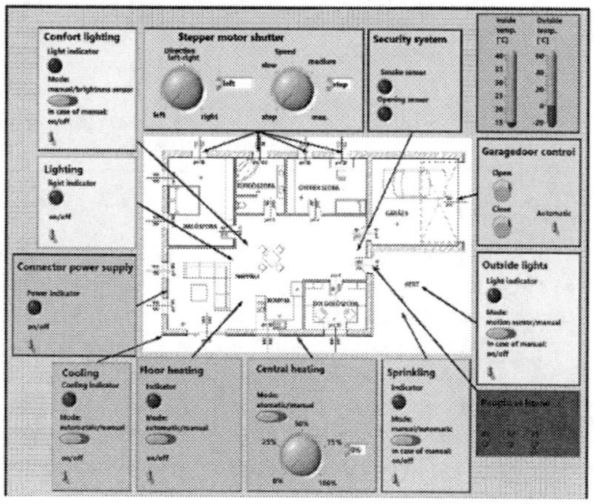

Fig. 7 - User interface / control panel

I have converted this front panel into a hypertext markup language (HTML) page and created a web server on the myRIO.

With this technique, I am able to connect to the controller through the Internet. The only thing needed is a web browser.

6 Built home model

Last but not least I have built my home demonstration. I have used a big Oriented Standard Board as a base of my model. I wanted to have real-life like conditions, so I have used 230 VAC and real sensors or actuators where I was able to. Every part of my automated processes can be found on the front face of the board, while the electronic components and the wiring is located on the back. On Fig. 8. you can find on these specific locations:
① lightings,
② controlled consumer electronics,
③ security system,
④ motorized controls,
⑤controlled sprinklers,
⑥ key-free access control system,
⑦ heating and cooling control.

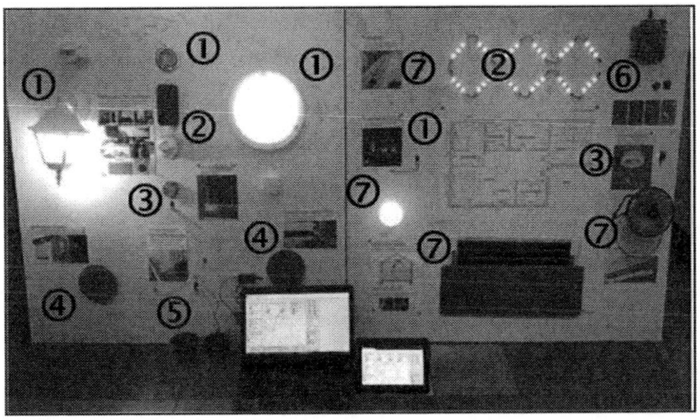

Fig. 8 - My home model that simulates real-life like conditions

7 Conclusions

I have created a model of a smart home that simulates real-life like conditions. This model meets the criteria that I have made about an "intelligent" house. There are automatic lighting, heating, cooling and sprinkling systems. There are also motorized controls and an advanced protection against robbery or fire.
The users can control the home with a user interface. This user interface is accessible through the web.

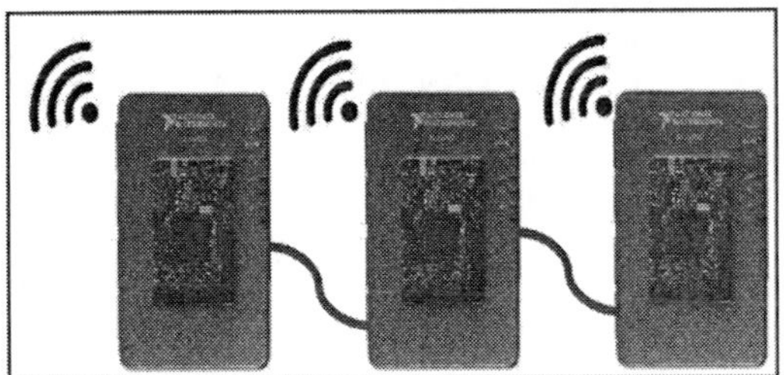

Fig. 9 – Usage of more myRIO and wireless communication

In the future I would like to work on improvements and more functions. With my technique any standard need can be satisfied. However there can be extreme needs. myRIO has 40 digital input-output, 10 analogue input and 6 analogue output ports. If a user has such specific needs that require more ports than that, I would like to find the way using more myRIO-s in serial or parallel configuration.

Other solution would be using wireless sensors and actuators. In this way I need to learn in details about wireless protocols and how to use them effectively.

References

1. G. András, Industry 4.0 and Hungary, Ipar 4.0 és Magyarország, lecture record, Eger, (2017)
2. P. Kádár, The Smart home, A Smart lakás, Óbuda University, Budapest, (2012) ISBN: 978-615-5018-49-7
3. Prof. Dr. L. Monostori, I. Lepényesi, I. Ács, A, Oláh, M. Petrik, Prof C. Hassan, Industry 4.0 National Technological Platform, Ipar 4.0 Nemzeti Technológiai Platform, Abailable: https://www.i40platform.hu/
4. R. Haddad, How can our home be Smart Home ?, Hogyan lehet otthonunk Smart Home?, Óbuda University, Budapest, (2012) ISBN: 978-615-5018-49-7

MATEC Web of Conferences **184**, 02018 (2018) https://doi.org/10.1051/matecconf/201818402018
Annual Session of Scientific Papers IMT ORADEA 2018

Comparison between proportional, integral, derivative controller and fuzzy logic approaches on controlling quarter car suspension system

Ahmet Mehmet Karadeniz[1,*], *Ammar* Alsabbagh[1], and Dr. *Geza* Husi[1]

[1]University of Debrecen, Department of Mechatronics, 4028 Debrecen Otemeto utca 2.-4., Hungary

Abstract. Developing and constantly changing technologies, efforts to achieve maximum efficiency with minimum fuel consumption, as well as the development of comfort and safety systems, have become very essential topic in car manufacturing and design. Whereas comfort and security were not given a high importance in the first produced cars, they are indispensable elements of today's automobiles. Since public transportation uses road in large scale, the need for safety and repose is also increasing. Nowadays, vehicles have better security and comfort systems, which react very quickly to all kinds of loads and different cases of driving (braking, acceleration, high speed, cornering), where the tires can keep the road at its best, utilizing an advanced suspension system. In this study, a quarter-car model was fulfilled using LabVIEW (Laboratory Virtual Instrumentation Engineering Workbench) software. The control of this model has been realized by applying two different controllers. PID (proportional, integral, derivative) controller which is a common and conventional control method and the Fuzzy Logic controller which is considered as an expert system that is becoming more and more widely used. In both control approaches, controlling the suspension system was achieved successfully. However; It has been determined that controlling the system using Fuzzy Logic controller gave better dynamic response than applying the PID controller for the quarter car suspension model that has been used in the direction of this study.

1 Introduction

In recent years, improving of vehicle suspension systems has been an intensive research and development topic. These studies are carried out in two different forms, scientific and commercial. Commercial work is carried out by vehicle and parts manufacturers aiming to enhance the quality of their products. Scientific studies are carried out by researchers, it concentrates more on controlling the system in order to provide automatic control of vehicle suspensions relying on the new developments of actuators, sensors and low cost electronics. The features expected from a good suspension system make suspension control more interesting. The desired features are:

- Adjustment of body movement: Ideally, a suspension should protect the body from road irregularities and from irregularities caused by inertia at the start, stop and during cornering.
- Adjustment of the suspension movement: Extreme vertical movement of the wheel will cause the tire to be in an improper position with respect to the road, which will also result a poor roadholding. In other words, the surface between the tires and the road should be maximized in order to increase the engine efficiency.
- Force distribution: Proper tire and road contact must be provided on all four wheels to ensure a good roadholding and handling.

These features overlap each other according to the road and vehicle situation, therefor, traditional suspension systems cannot provide all these features. A design that reduces the sideways tilt of the torso, for instance, will have too much stiffness, which cause the vibrations from the road to be transmitted to the vehicle body.

This can be shown on the quarter car model in which the single wheel is considered. The amplitude of the transfer function, which relates the vertical changes on the curve to the changes in the vertical velocity of the vehicle body, is seen. The straight line is the answer of the traditional vehicle suspension system. There are two resonance states. Lower one; the vibrations of the body (spring mass), and the other one; that is related to the vibrations of the wheel hub (unspring mass). Improving the low frequency characteristics of the suspension may be possible by increasing the damping constant of the suspension. The dotted line indicates the case where the damping constant is increased ten times. The response at low frequencies can be improved by shifting the resonance region. Note that the resonance amplitudes of the body and the wheel increase while the gain at high frequencies decreases.

* Corresponding author: ahmetm.karadeniz@gmail.com

© The Authors, published by EDP Sciences. This is an open access article distributed under the terms of the Creative Commons Attribution License 4.0 (http://creativecommons.org/licenses/by/4.0/).

The simple quarter car model mentioned above clearly demonstrates the dilemma between low and high-frequency driving comfort. Furthermore, the control of the side tilt of the nose and the rear part will be a function of the interaction between each suspension [1].

2 Quarter Car Model

The quarter car model, which is one of the most preferred, most useful and simplest car models, is shown in Fig. 1. The parameters of the model used in this study are obtained by extracting the mathematical model of the system and using the reference [2].

The vibrations of a vehicle subjected to a vertical axis or as result from dynamic behaviours can be modelled as M_1 and M_2 double masses. M_1 represents quarter of the total mass of the vehicle, and M_2 represents the mass of the wheel of the vehicle.

The equations for the quarter car suspension model are shown in the both equations 1 and 2:

$$M_1\ddot{x}_1 = -b(\dot{x}_1 - \dot{x}_2) - K_1(x_1 - x_2) + u \qquad (1)$$

$$M_2\ddot{x}_2 = -b(\dot{x}_1 - \dot{x}_2) - K_1(x_1 - x_2) + b_2(\dot{w} - \dot{x}_2) + k_2(w - x_2) - u \qquad (2)$$

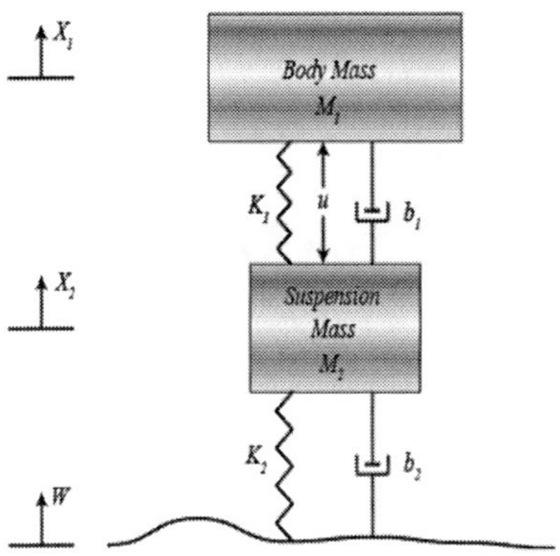

Fig. 1. Quarter Car Suspension System.

2.1. Modelling Quarter Car Suspension System

LabVIEW (Laboratory Virtual Instrumentation Engineering Workbench) allows modelling of systems that can be expressed as mathematical equations using block diagrams. The modelling in the LabVIEW environment using the equations for the quarter vehicle suspension model is shown in Fig. 2. The parameters and values used during this modelling are shown in Table 1.

Table 1. System Parameters

Parameter	Value	Unit
M_1	2500	Kg
M_2	320	Kg
K_1	80000	N/m
K_2	500000	N/m
b_1	350	N.s/m
b_2	15020	N.s/m

As minimising the vibration is one of this research most significant aims, controlling the vertical movement and forcing it to goes to zero is what is important. For non-feedback systems, the desired parameter cannot be controlled or only the open-loop response, which is the uncontrolled response of the system, can be obtained. Fig. 3 shows the open loop response obtained for the quarter car suspension system. As shown in Fig. 3, it has an oscillated dumbed behaviour because it has no feedback nor controller (open loop response).

2.2. Controlling System Utilizing PID

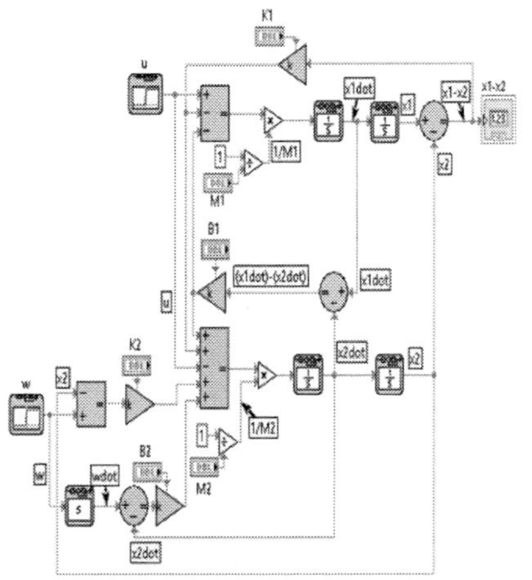

Fig. 2. LabVIEW Model of the System.

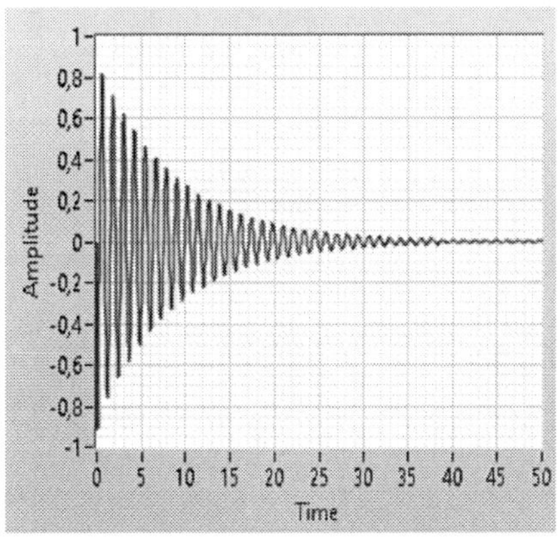

Fig. 3. Open Loop Response of the System.

PID control is a proportional integral plus derivative controller whose transfer function is:

$$G_{PID}(s) = K_p + \frac{K_i}{s} + K_d s \qquad (3)$$

PID controller is considered as one of the simplest controllers, easy to be applied and it does not require high performance microprocessors [5]. The PID gains can be obtained using many different methods like zero-pool cancelation, Ziegler-Nichols, and trial and error. In this paper, Trial and error method has been used in order to tune the PID controller. K_p, K_i, K_d gains are called proportional PID gain, integral PID gain, and derivative PID gain, respectively. Error signal is the input for proportional, derivative and integral gains of the PID controller to obtain desired output. The block diagram of the LabVIEW PID controller with the subsystem which represents quarter car suspension system for this study model is shown in Figure 4.

Where the PID (proportional, integral, derivative) controller is applied to the quarter car suspension system as series with a feedback. Table 2 shows the values used for PID parameters which are K_p, K_i, K_d gains.

Table 2. PID Parameters

PID Parameteres	Values
K_p	146975.92
K_i	890807. 1099
K_d	956.8692

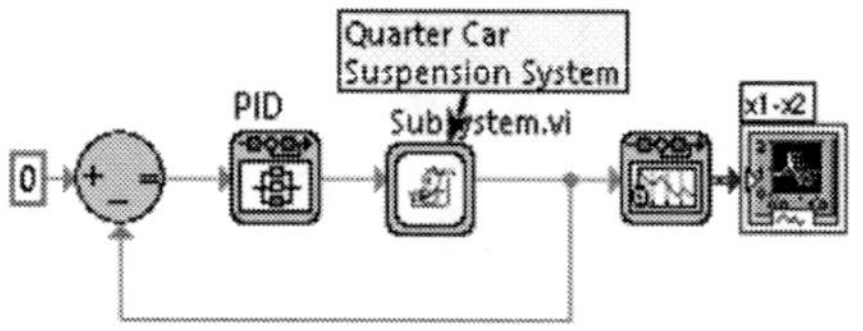

Fig. 4. Controlling the suspension system using PID

Fig. 5 shows the obtained result when the PID controller is applied with the parameters given in Table 2. Looking at this result, we can see that the system respond is zero after 1 second. The settling down of the system to zero means that the vibrations along the vertical axis in the vehicle will not affect the passengers.

2.3. Controlling System Utilizing Fuzzy Logic

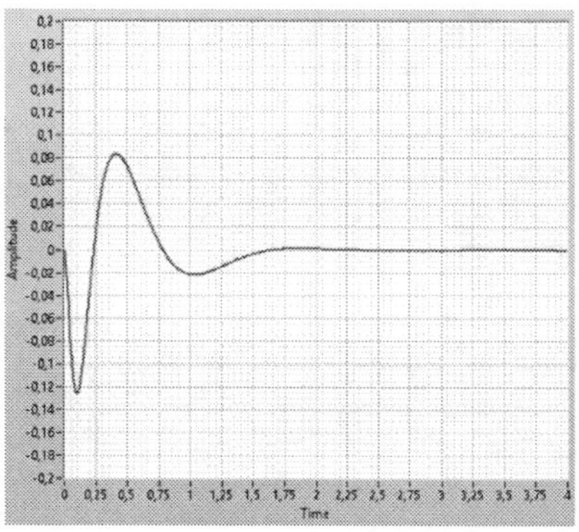

Fig. 5. The system response to the PID controller

Fuzzy logic, unlike classical logic, does not have sharp distinctions. That is, one level does not only have only 0 or 1 values, but it can also have values in between. NB (negative big), NS (negative small), Z (zero), PS (positive small), PB (positive big) levels, which are commonly used in fuzzy logic controllers. Fuzzy System Designer is used to transfer these levels to the LabVIEW environment. With Fuzzy System Designer, the membership functions of the system were defined using the table 3, which was created by experts and the values of membership functions can be determined. After the decisions, the control model is designed with the fuzzy logic controller in LabVIEW. The fuzzy logic controller model designed for this system is shown in Fig. 6. Since identification of membership functions the fuzzy logic controller is important, the type, number and normalization factors of the membership functions should be carefully determined at this step. In this study, the general trapezoidal membership function was chosen however, only the Z membership function was chosen as the triangular membership function because it is desired to be highly sensitive to make the error goes to zero as precise as possible. In the normalization process, the input function is kept in the range of -1 to +1.

The rule tables that show the relationship between the input and output in the control with the fuzzy logic controller are important factors. The rule table used for this study is shown in Table 3 and membership functions are shown in Fig. 7 on Fuzzy System Designer [3,4].

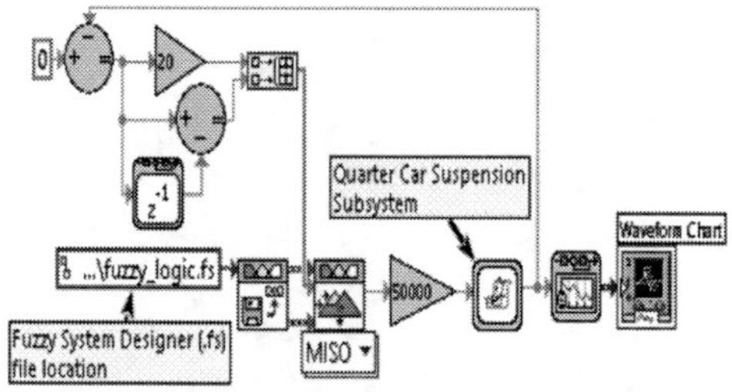

Fig. 6. Controlling the system using Fuzzy Logic

Table 3. Rule Table

du / u	*NB*	*NS*	*Z*	*PS*	*PB*
NB	NB	NB	NB	NS	Z
NS	NB	NB	NS	Z	PS
Z	NB	NB	Z	PS	PB
PS	NS	NS	Z	PB	PB
PB	Z	Z	PS	PB	PB

In order to evaluate the performance of the fuzzy logic controller designed in accordance with the rules and membership functions specified in Table 3 and Fig. 7, it is necessary to compare the output values with the desired values. The closed-loop response of the system controlled using the fuzzy logic controller for this study is demonstrated in Fig. 8.

The choosing of the controller to be applied in order to control any system is performed as a result of comparing the responses of the specified controllers to this system. Fig. 9 shows the response of both controllers for this system.

The behaviour of the vehicle's suspension system without any controller is shown in the open loop response in Fig. 3. To conclude from this, the uncontrolled vehicle

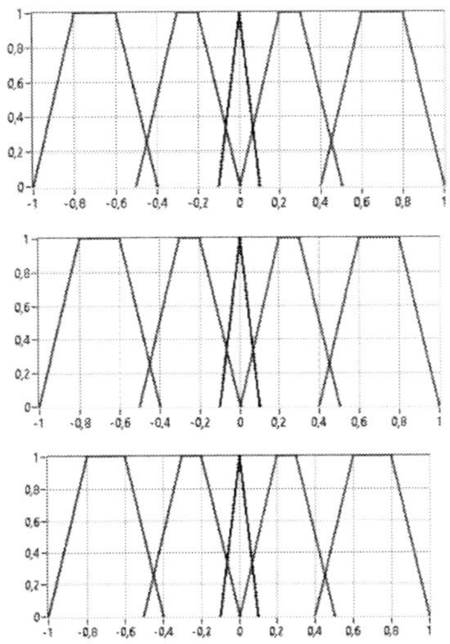

Fig. 7. Membership Functions 'u', 'Δu', 'output'.

suspension system is not desirable considering material life and passenger health and safety. A control method

2.4. Comparison of PID and Fuzzy Logic

should be selected to remove such undesirable situations and developed in such a way that the control parameters are applied to the system to provide the optimum level. The control method that should be applied to the quarter car suspension system will be clarified by comparing the results obtained in this study.

As can be seen form fig. 9, the response using fuzzy logic controller is tree time less downshoot than the utilizing PID (about 4 centimetres downshoot utilizing fuzzy logic controller against 13 centimetres downshoot using PID controller). Moreover, control the quarter suspension system takes about half of consumed time by PID to reach the steady state (the fuzzy logic controller leaded the system to reach the stable state in approximately 0.75 second, while the PID controller took around 1.5 seconds to make the quarter suspension system reach the steady state).

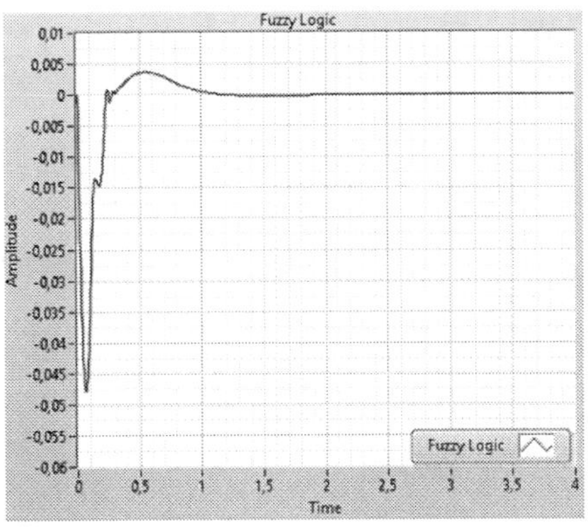

Fig. 8. The system response to Fuzzy Logic controller

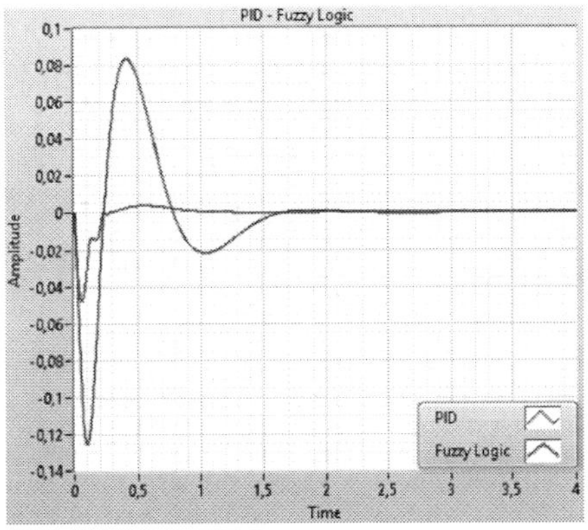

Fig. 9. Comparison of PID and Fuzzy Logic Controllers.

3 Conclusions

The performance of the PID and Fuzzy Logic controllers for the quarter car suspension system was tested. The model used in this study is taken from [1]. Fuzzy Logic and PID controllers were applied to the system separately and the obtained results were compared with each other. In this comparison, it has been observed that the Fuzzy Logic controller gives better results than the PID controller for the control of the designed model in respect of both time that was taking to reach the steady state, and in respect of overshoot and downshoot, which means, that the fuzzy logic controller was able to reduce the oscillation amplitude of the system and enhance the safety of mechanical parts of the vehicle as well as the passenger safety.

Acknowledgment

The publication is supported by the EFOP-3.6.1-16-2016-00022 project. The project is co-financed by the European Union and the European Social Fund. Moreover, thanks to Adewale Oriyomi Oseni for being encouraging, supporting, and making this work more efficient and easier.

References

1. R.K. Pekgöz, M.A. Gürel, M. Bilgehan, M. Kısa, *Active Suspension of Cars Using Controller Optimized by Genetic Algorithm*, Int. J. of Eng. and Appl. Sci. (IJEAS) **2**, Issue 4 27-37, (2010)
2. http://ctms.engin.umich.edu/CTMS/index.php?example=Suspension&section=SystemModeling, (retrieved on 25.11.2017)
3. J. Jantzen, *Foundations of fuzzy control*, Chichester, England; Hoboken, NJ: John Wiley & Sons, 2007.
4. N. E. I. S. U. Manual, *NI Educational Laboratory Virtual Instrumentation Suite II Series (NI ELVISTM II Series) User Manual*, (2009)
5. Ammar Alsabbagh, Péter T. SZEMES, Abdulkader Baki, *Brushless DC Motor Modeling Using Bond Graph Method and Control using LabVIEW.* (RIim) Recent Innovation in Mechatronics. Vol. 5, No 1, 2018.

Three phase induction motor modelling and control using vector control in LabVIEW

Wadee Khoury[1,*], *Aram* Nasser[1], and *Péter* Tamás Szemes[2]

[1]Masters Student, Debrecen University, Mechatronics Department, 4028 Debrecen, Hungary
[2]Associate Professor, Debrecen University, Mechatronics Department, 4028 Debrecen, Hungary

Abstract. In this paper, a speed control system for a three-phase induction motor was modelled and designed within LabVIEW software environment. After structuring the dynamical model of the motor, a current controller was developed to stabilize the system and avoid a wind-up situation. Then, a speed controller was designed, using PID and Field Weakening techniques, to generate the reference current values. The field-weakening algorithm is used to achieve stability at speeds that are higher than the nominal one. An estimator is used to calculate the flux angle and the electro-mechanical speed of the motor. As a result, the system achieved the desired speed with good transient and steady state responses. In addition, the system proved to be robust when the torque load is applied in all cases.

1 Introduction

Three phase induction motors (IM), particularly squirrel cage, are rugged, efficient, and cheap compared to other motor types. Because of this, they are widely used in industry for many applications. In early days, due to the high cost and low efficiency of the IM speed control techniques, the use of the three-phase induction motor was limited to the applications that have constant speeds, such as driving fans and pumps. Following to the advancements in power electronics, IMs had become known as the industry horses and been more widely used in applications that require variable speed control [1, 2].

Simulating the control system behaviour before implementing it into a real world environment is convenient, and gives information about the control system efficiency and performance, as well as the system sensitivity against disturbances and noise signals.

In order to simulate the system and test it, the motor was modelled in LabVIEW using the T-model and the mathematical equations in the stationary reference frame, αβ. The vector control method, with PI and PID controllers, is used after converting the stationary variables αβ into a rotating dq reference frame using the Park transform. The Park transformation needs the rotor flux angle as input. This angle is difficult to be measured, therefore, a current estimator was used to calculate it. Thanks to the field-weakening algorithm, described in [3], the system is able to operate on speeds higher than its nominal one and with good response. The system's simulation was tested on two different speeds (1400) [rpm], which is the nominal speed, and (2000) [rpm], with and without the torque load, where the system stability was evident.

2 Induction Motor Modelling

Several circuit representations are used to describe the motor's behaviour for transient and steady state analysis [4]. In this paper, the simple T-form of linear equivalent circuit in the stationary reference frame, shown in figure 1, is used. This representation helps getting the state space that describes the IM's behaviour. By doing this, the complexity in the voltage equations of the IM can be simplified, resulted from the time varying mutual inductances between the stator and the rotor circuits [5]. To get the T-model, The IM's variables, currents, voltages, and fluxes, needed to be transferred into the stationary reference frame using Clarke Transform [6], given by:

$$\begin{bmatrix} i_\alpha \\ i_\beta \end{bmatrix} = \frac{2}{3}\begin{bmatrix} 1 & -\frac{1}{2} & -\frac{1}{2} \\ 0 & \frac{\sqrt{3}}{2} & -\frac{\sqrt{3}}{2} \end{bmatrix}\begin{bmatrix} i_a \\ i_b \\ i_c \end{bmatrix} \quad (1)$$

Now, the dynamic model of the IM can be described in the stationary reference frame αβ, using the T-model shown in figure 1.

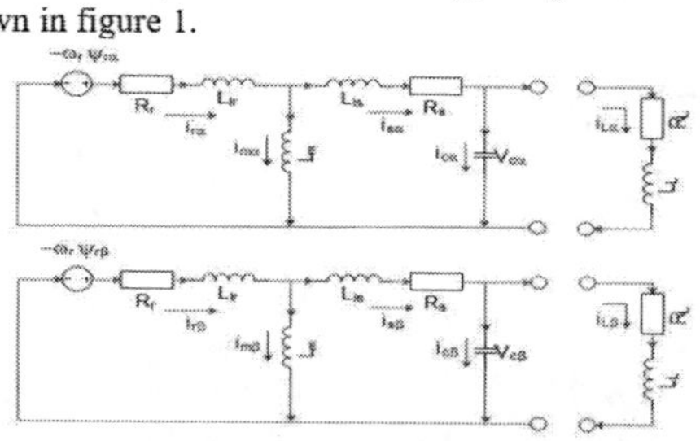

Fig. 1. The T-model equivelant circuite of the IM [1].

* Corresponding author: w.khoury@outlook.com

© The Authors, published by EDP Sciences. This is an open access article distributed under the terms of the Creative Commons Attribution License 4.0 (http://creativecommons.org/licenses/by/4.0/).

From figure 1, the dynamical equations can be derived to get the IM's model [1].

$$v_{s\alpha} = R_s i_{s\alpha} + L_s \frac{d}{dt} i_{s\alpha} + L_m \frac{d}{dt} i_{r\alpha} \quad (2)$$

$$v_{s\beta} = R_s i_{s\beta} + L_s \frac{d}{dt} i_{s\beta} + L_m \frac{d}{dt} i_{r\beta} \quad (3)$$

$$0 = R_r i_{r\alpha} + L_r \frac{d}{dt} i_{r\alpha} + L_m \frac{d}{dt} i_{s\alpha} + w_r \varphi_{r\beta} \quad (4)$$

$$0 = R_r i_{r\beta} + L_r \frac{d}{dt} i_{r\beta} + L_m \frac{d}{dt} i_{s\beta} - w_r \varphi_{r\alpha} \quad (5)$$

For a magnetically linear machine, the stator and rotor flux linkages are given by [4]:

$$\begin{bmatrix} \boldsymbol{\varphi_{s,\alpha\beta}} \\ \boldsymbol{\varphi_{r,\alpha\beta}} \end{bmatrix} = \begin{bmatrix} L_s & L_m \\ L_m & L_r \end{bmatrix} \begin{bmatrix} \boldsymbol{i_{s,\alpha\beta}} \\ \boldsymbol{i_{r,\alpha\beta}} \end{bmatrix} \quad (6)$$

Where:

- $v_{s\alpha}, v_{s\beta}$ –Stator voltages in the stationary reference frame [V].
- $i_{s\alpha}, i_{s\beta}, i_{r\alpha}, i_{r\beta}$ –Stator and rotor currents in the stationary reference frame respectively [A].
- $\varphi_{r\alpha}, \varphi_{r\beta}$ –Rotor magnetic flux in the stationary reference frame [Wb].
- R_s, R_r –Resistances of the IM's stator and rotor respectively [Ω].
- L_s, L_r –Self-inductances of the IM's stator and rotor respectively [H].
- L_m –Mutual inductance between the stator and the rotor [H].
- w_r –Rotor speed [rad/s].

As the self-inductance of the stator and rotor are composed of the leakage and mutual inductances as [7]:

$$L_s = L_r = L_{ls} + L_m \quad (7)$$

The mechanical behaviour can be expressed as:

$$\frac{J}{n_p}\frac{dw_r}{dt} = T_e - T_L \quad (8)$$

$$\frac{d\theta}{dt} = w_r \quad (9)$$

And the electrical torque developed by the IM:

$$T_e = \frac{3n_p}{2} L_m \left(i_{r\alpha} i_{s\beta} - i_{r\beta} i_{s\alpha}\right) \quad (10)$$

When considering the coefficient of friction, $B \neq 0$, the load is expressed as:

$$T_L = B\frac{w_r}{n_p} + T_{L,extra} \quad (11)$$

Where:

J: Moment of inertia [Kg. m^2]

θ: Angular position [rad]

n_p: Number of pole pairs

By combining equations (2-10), the comprehensive equation of the IM [7] is obtained and can be represented in the following form:

$$u = Rx + L\dot{x} \quad (12)$$

$$\begin{bmatrix} V_{s\alpha} \\ V_{s\beta} \\ 0 \\ 0 \end{bmatrix} = [\mathrm{R}] \begin{bmatrix} i_{s\alpha} \\ i_{s\beta} \\ i_{r\alpha} \\ i_{r\beta} \end{bmatrix} + [\mathrm{L}] \begin{bmatrix} \frac{di_{s\alpha}}{dt} \\ \frac{di_{s\beta}}{dt} \\ \frac{di_{r\alpha}}{dt} \\ \frac{di_{r\beta}}{dt} \end{bmatrix} \quad (13)$$

Where:

$$[R] = \begin{bmatrix} R_s & 0 & 0 & 0 \\ 0 & R_s & 0 & 0 \\ 0 & w_r L_m & R_r & w_r L_r \\ -w_r L_m & 0 & -w_r L_r & R_r \end{bmatrix}$$

$$[L] = \begin{bmatrix} L_s & 0 & L_m & 0 \\ 0 & L_s & 0 & L_m \\ L_m & 0 & L_r & 0 \\ 0 & L_m & 0 & L_r \end{bmatrix}$$

In order to get the IM state-space model, equation (13) is rearranged to be in the form [7]:

$$\dot{x} = -L^{-1}Rx + L^{-1}u = Ax + Bu \quad (14)$$

3 Control Method

3.1 VECTOR CONTROL

Two major control methods are used to control the IM's speed: the scalar control, referred to as V/Hz, and the vector control, also known as field-oriented control, referred to as FOC. The first method is cheap, easy to implement, and gives good steady state results. However, this method gives slow transient response; therefore, it is not efficient in controlling dynamical systems [5]. On the other hand, although it is more complex, FOC has fast and excellent transient response [8]. Because of this, vector control method has become more popular in industry for the speed and torque control of the IM.

The idea behind the vector control method is to identify the stator currents of an IM as two orthogonal components that can be visualized with a vector, which gives the ability to deal with the IM as a separately excited two-phase DC machine. One component (i_q) refers to a virtual current that controls the torque and the other component (i_d) refers to a virtual current that controls the flux [9]. As they are decoupled or orthogonal in space, the torque and the flux can be controlled separately without affecting one another.

To apply the FOC, the rotor flux angle is needed, and since this angle is hard to be measured, a flux estimator is used to calculate it.

3.2 Flux Estimator

There are two basic ways to determine the angle of the rotor flux. The direct scheme, which needs field sensors to measure the rotor flux position directly. This scheme is difficult, or maybe impossible to implement [5]. Instead, the indirect scheme can be used, which uses equations to estimate the rotor flux angle mathematically from other measurable components. For the indirect estimation, either the current based flux or voltage based flux can be used [7]. In this paper, the measured current is taken to estimate the rotor flux angle, as well as the synchronous speed of the IM.

By introducing a new constant, Leakage coefficient (σ), as in equation (15), and rearranging equations (2–5), the stator and rotor flux equations can be written as [10]:

$$\sigma = 1 - L_s i_{sa} \quad (15)$$

$$\varphi_{sa} = \int (v_{sa} - R_s i_{sa})\, dt \quad (16)$$

$$\varphi_{s\beta} = \int (v_{s\beta} - R_s i_{s\beta})\, dt \quad (17)$$

$$\varphi_{ra} = \frac{L_r}{L_m}(\varphi_{sa} - \sigma L_s i_{sa}) \quad (18)$$

$$\varphi_{r\beta} = \frac{L_r}{L_m}(\varphi_{s\beta} - \sigma L_s i_{s\beta}) \quad (19)$$

The rotor and stator flux angles can now be calculated as:

$$\theta_{\varphi s} = tan^{-1}\frac{\varphi_{s\beta}}{\varphi_{sa}} \quad (20)$$

$$\theta_{\varphi r} = \tan^{-1}\frac{\varphi_{r\beta}}{\varphi_{ra}} \quad (21)$$

From equation (20), the synchronous speed is:

$$w_1 = \dot{\theta_{\varphi s}} = \frac{\varphi_{sa}\dot{\varphi_{s\beta}} - \varphi_{s\beta}\dot{\varphi_{sa}}}{\varphi_{sa}^2 + \varphi_{s\beta}^2} \quad (22)$$

3.3 Park Transformation

The estimated rotor flux angle is used to transform the stationary reference frame into the rotating reference frame using the Park transform [6]:

$$\begin{bmatrix} i_d \\ i_q \end{bmatrix} = \begin{bmatrix} \cos\theta & \sin\theta \\ -\sin\theta & \cos\theta \end{bmatrix} \begin{bmatrix} i_\alpha \\ i_\beta \end{bmatrix} \quad (23)$$

Or, in case of designing using complex numbers, the following formula can be considered:

$$v_{dq} = v_{\alpha\beta} e^{-j\theta}$$

Where $j = \sqrt{-1}$ is the complex number.

3.4 Speed control model

3.4.1 Generation of the q current component

The i_q component is controlled using a parallel PID controller after getting the error signal, which is the difference between the measured mechanical speed and the desired one. To satisfy the design requirements, the controller parameters were found as follow:

TABLE I. PID PARAMETERS

Controller Parameters	Value
k_p	30
k_i	0.35
k_d	0.04

3.4.2 Generation of the d current component

To generate the desired i_d component, the field weakening technique is used. This technique allows to operate the IM on speeds higher than its nominal one by reducing the flux in the air-gap between the stator and rotor, which in turn reduces the back emf constant, and hence the torque. From the field weakening equations (24) and (25), the reference rotor flux is generated using the inverse-Γ-model, described in [3], as follow:

$$\varphi_{r,ref} = \int_{\varphi_{r,min}}^{\varphi_{r,rated}} k(v_{base}{}^2 - v_{d,ref}{}^2 - v_{q,ref}{}^2)\, dt \quad (24)$$

$$k = \frac{\propto_f L_m}{2\, w_f\, L_\sigma\, v_{s,rated}} \quad (25)$$

$$L_\sigma = L_s - \frac{L_m}{L_r} \quad (26)$$

Where:

- v_{base} –Maximum peak voltage that the power electronics can put out (326.6 [v]).
- $v_{d,ref}, v_{q,ref}$ –Reference voltage values from the voltage controller.
- $\propto_f$–A constant bandwidth of the field-weakening algorithm [0.5].
- L_σ –Leakage inductance variable from the inverse-Γ-model.
- w_f is given by:

$$w_f = \begin{cases} w_{1,rated} & if \quad |w_1| \le w_{1,rated} \\ |w_1| & if \quad |w_1| > w_{1,rated} \end{cases} \quad (27)$$

The output of the field-weakening algorithm, which is the reference i_d component, is calculated using the following equation:

$$i_{d,ref} = \frac{\varphi_{r,ref}}{L_m} \quad (28)$$

The Field-Weakening method is illustrated in figure 2.

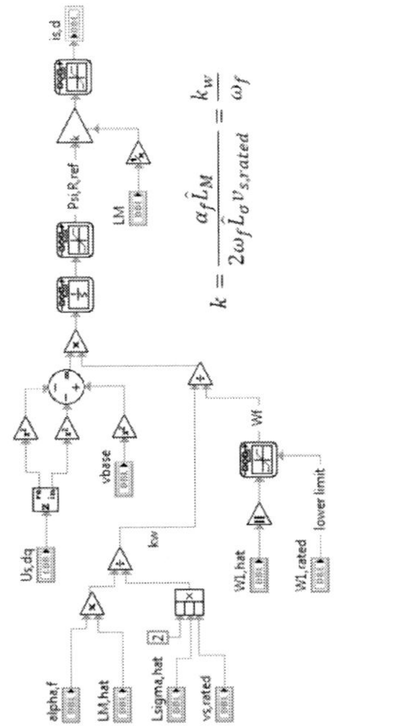

Fig. 2. Field wakening block diagram.

3.4.3 Current Controller model

The resulting currents $i_{s.dq}$ from the Park transformation will be compared with the reference $i_{s.dq,ref}$; the output of the speed controller, and the error signal will be fed to the current controller. Using a normal PI controller might cause the system to wind-up when the voltage is saturated by its maximum value. If the reference voltage gets higher than the voltage limit, there will be an error between the limited voltage and the actual voltage. This causes the current to increase slower, which makes the integrator integrates too much. When the current reaches the reference value, the accumulated error will be too large, causing an overshoot. To negate this, a back calculation algorithm is used by introducing a new error, ē, which will be fed to the integrator, using the following equations [7]:

$$\bar{e} = e + \frac{\bar{v}_{s,lim} - v_s}{k_{pc}} \tag{29}$$

Where:

$$\bar{v}_{s,lim} = \begin{cases} v_s & if \quad |v_s| \le v_{s,max} \\ |v_{s,max}| < v_s & if \quad |v_s| > v_{s,max} \end{cases} \tag{30}$$

$$v_{s,dq,out} = ek_{pc} + \int \bar{e}\, k_{ic} \tag{31}$$

k_{pc}, k_{ic} –PI controller gains, and their values are 27 and 6400 respectively.

Finally, the output voltage is transferred to the three-phase equivalent reference frame using the inverse Clarke-Park transformations to feed the motor with the proper signals.

4 LabVIEW Simulation

LabVIEW simulation loop was used for modelling and controlling the three phase induction motor, where the dynamical model, represented in state-space form, is implemented as shown in figure 3.

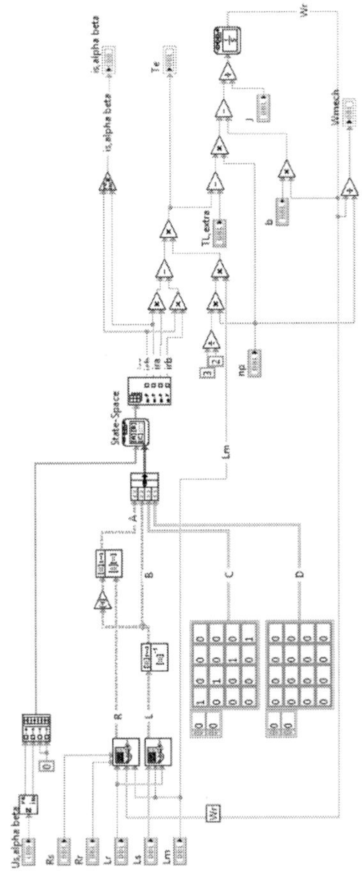

Fig. 3. AC Induction Motor block diagram.

The IM model parameters used in the simulation are described in the following table.

TABLE II. MOTOR PARAMETERS

Parameter [unit]	Value
Stator inductance L_s $[H]$	0.1276
Rotor inductance L_r $[H]$	0.1276
Mutual inductance L_m $[H]$	0.1186
Stator resistance R_s $[\Omega]$	2.4
Rotor resistance R_r $[\Omega]$	2.3
Number of pole pairs n_p	2
Coefficient of friction B	0.003
Moment of inertia J $[Kg.m^2]$	0.00529

The complete LabVIEW implementation of the control system is presented in figure 4.

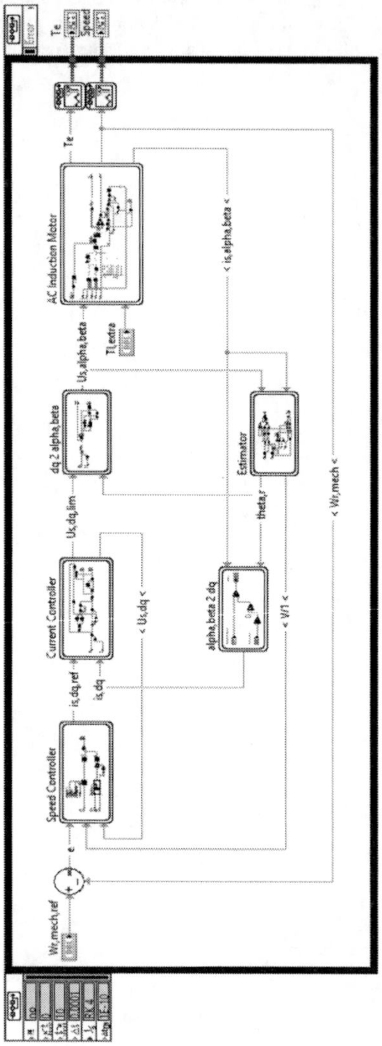

Fig. 4. Full control system block diagram.

First, the system is simulated with a desired speed of 1400 [rpm], which is the rated speed of the motor. The speed response is shown in figure 5.

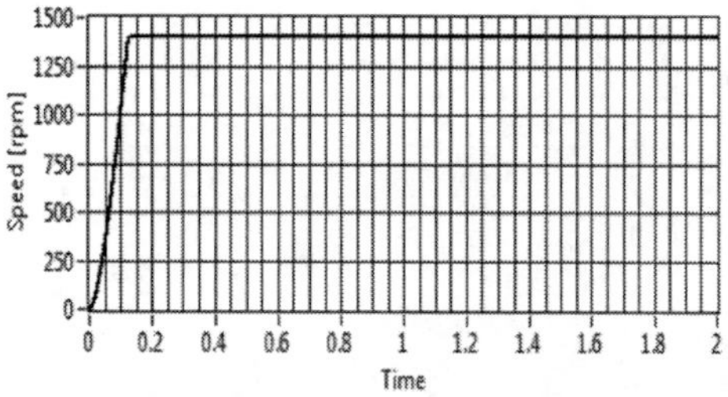

Fig. 5. Mechanical speed at $w_{mech,ref} = 1400\ [rpm]$.

As a result of this simulation, the speed reaches its final value without any overshoot and only within 0.1 second, which means that the control system is fast and has a good transient response. As the load torque is applied at the first second, the system behaviour remains the same without any change in the speed value, allowing the motor to work at a constant speed with or without the load, proving the robustness of the system.

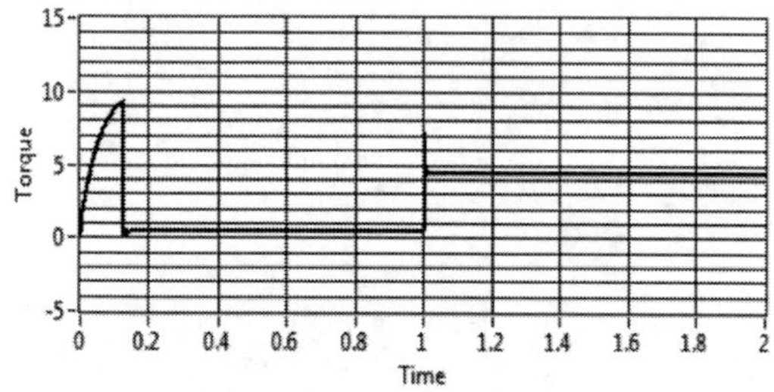

Fig. 6. Electromechanical torque at $w_{mech,ref} = 1400\ [rpm]$.

To keep the motor's speed constant, current components change while working on the transient state as shown in figure 7, to become stable at steady state. Then, they change again, especially i_q, when applying the torque at the first second. These components affect the motor's torque and flux, as shown in figure 6, and therefore, maintaining a constant motor speed.

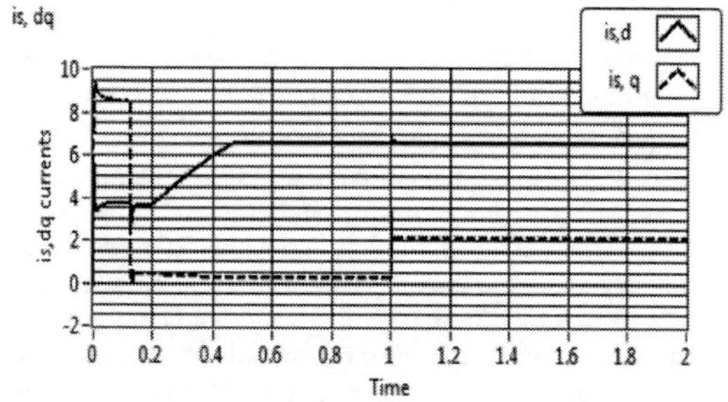

Fig. 7. Current components at $w_{mech,ref} = 1400\ [rpm]$.

In order to test the effectiveness of the field weakening technique, the system is simulated with a desired mechanical speed of 2000 [rpm], higher than the rated speed, 1400 [rpm]. The field weakening will reduce the d current component, which will decrease the flux in the air-gap, causing higher speeds. The speed response is presented in figure 8.

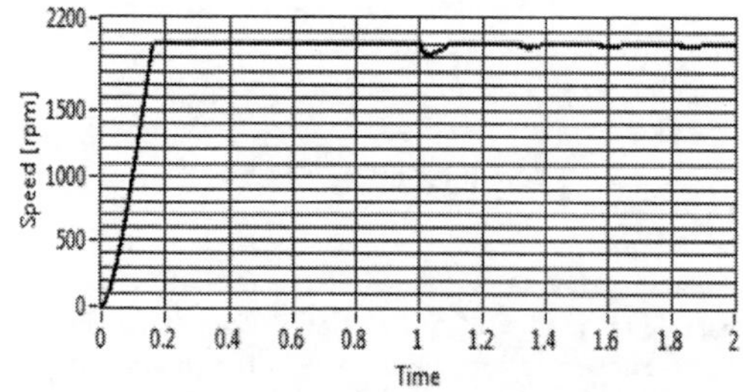

Fig. 8. Mechanical speed at $w_{mech,ref} = 2000\ [rpm]$.

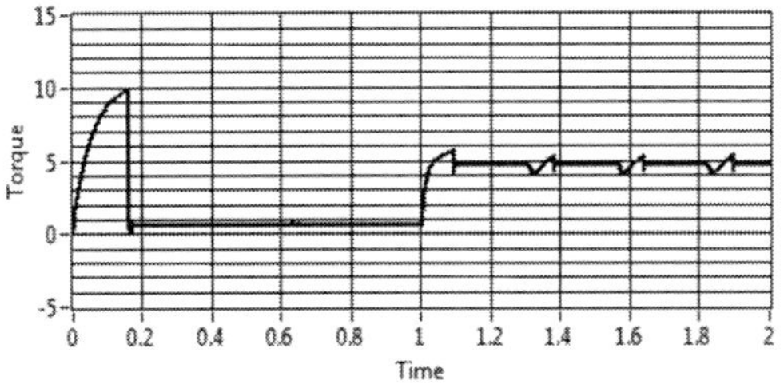

Fig. 9. Electromechanical torque at $w_{mech,ref} = 2000$ $[rpm]$.

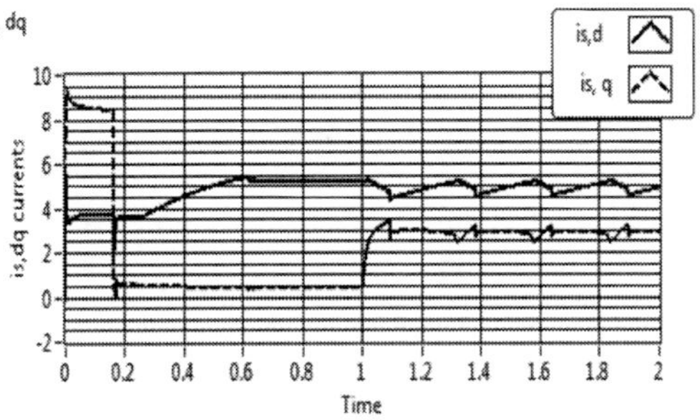

Fig. 10. Current components at $w_{mech,ref} = 2000$ $[rpm]$.

As a result, in figure 8, the speed reaches its desired value within 0.4 sec that is larger than the one in the previous simulation. This response delay is due to the fact that the field weakening block is reducing i_d, as shown in figure 10, comparing with the nominal value in figure 7. This in turn will affect in reducing the flux, back EMF constant, and torque, resulting in achieving the high speed.

Additionally, the existence of small fluctuations in figure 8 is due to applying a high load, 7 [N.m], when the motor was working in the field weakening region. Therefore, the field weakening technique should be used in applications where the applied load is expected to be small. Nevertheless, although a high torque was applied in the field weakening region, the control system tried to keep the desired speed constant, which reflects the robustness of the system. With real motors, however, this might not be totally true, where the motor's heating problems should be taken into consideration while designing and implementing the control system.

5 Conclusion

As discussed earlier, LabVIEW software is used to model and simulate the speed control of a three-phase induction motor with and without the extra load. At normal speeds, the system achieved good response and proved to be robust when the load was applied. At speeds higher than the nominal one, the field weakening technique was able to achieve good response with being carful not to operate the system in applications that require high toque. The simulation results show the efficiency of using the vector control and field-weakening methods.

Acknowledgment

The work/publication is supported by the EFOP-3.6.1-16-2016-00022 project. The project is co-financed by the European Union and the European Social Fund.

References

1. K. Sundararaju, R.S. Kumar, and I.G.C. Raj, Modeling and simulation of neural based speed controller for direct torque control of three phase induction motor. In *Region 10 Conference, TENCON*, 1439-1444 (2017). IEEE.
2. S. Hegde, S. Angadi, A.B. Raju, Speed control of 3-phase induction motor using volt/hertz control for automotive application, *I4C*, 17214371, (2017). IEEE.
3. L. Harnefors, K. Pietilainen, L. Gertmar, Torque-maximizing field-weakening control: design, analysis, and parameter selection, *IEEE, Transactions on Industrial Electronics*, **48(1)**, 161-168, (2001).
4. G.R. Slemon, Modelling of induction machines for electric drives, *IEEE Transactions on Industry Applications*, **25(6)**, 1126-1131, (1989).
5. M.R.B. Fathima, P.M.J. Princy, S. RamPrasath, Mathematical modeling of SVPWM inverter fed 3 phase induction motor vector control in MATLAB/Simulink environment, In *Circuit, Power and Computing Technologies (ICCPCT), 2017 International Conference on*, 1-8, (2017). IEEE.
6. J.M. Peña, E.V. Díaz, Implementation of V/f scalar control for speed regulation of a three-phase induction motor, In *ANDESCON*, 1-4, (2016). IEEE.
7. J. Bj¨orkqvist, D. Stockman, Implementation and simulation of sensorless control and field weakening for an induction machine Using the the Statically Compensated Voltage Model, Department of Energy and Environment Division of Electric Power Engineering, Chalmers University of Technology, *Master's thesis*, (2014).
8. G. Kohlrusz, D. Fodor, Comparison of scalar and vector control strategies of induction motors, *Hungarian Journal of Industry and Chemistry*, **39(2)**, 265-270 (2011)
9. M.S. Aspalli, S.J. Patil, Study of AI and PI controller using SVPWM technique for induction motor speed control, In *Electrical, Electronics, Communication, Computer, and Optimization Techniques (ICEECCOT), 2017 International Conference on*, 1-6, (2017). IEEE.
10. P. Parida, A Sliding Mode Controller For Induction Motor Drives. National Institute of Technology in Rourkela, *Doctoral dissertation* (2009).

Robust control system design for a brushed direct current motor using LabVIEW simulation loop

Wadee Khoury[1,*], and *Péter* Tamás Szemes[2]

[1]Masters Student, Debrecen University, Mechatronics Department, 4028 Debrecen, Hungary
[2]Associate Professor, Debrecen University, Mechatronics Department, 4028 Debrecen, Hungary

Abstract. This paper presents the design of a robust control system for a PMDC motor using LabVIEW software. At first, the mathematical equations of the motor were modelled and simulated. Then, a proportional integral controller (PI) was used as a speed controller, and to guarantee the robustness of the system against the disturbances, the PI parameters were tuned by applying a sensitivity analysis. Finally, a proportional gain (P) was used as a position controller to guide the system towards the desired angle. The simulation was performed using LabVIEW Simulation Loop, and the sensitivity analysis was carried using LabVIEW MathScript tool.

1 Introduction

The brushed DC motors are widely used in various robotic applications and industrial areas in fields ranging from toys to spacecraft due to their flexibility, high reliability, and relatively low cost. One of the most common techniques used to control the position and speed of a DC motor is the PID controller, because of its simple structure and comprehensible control algorithm.

Many papers address the design of a DC motor control system using different methods. For example, Yolchan et. al. [1] compared PID and State Feedback methods to control real-time position, trajectory, and speed of a brushed DC motor. Sahin et. al. [2] conducted a research on position control of a DC motor using Artificial Neural Network as the main controller of their system. Namazov et. al. [3] used fuzzy logic control. Mao et. al. [4] also researched precision positioning of a DC motor, but in the presence of aerostatic friction, and they used a PID controller as well. Although many advanced control algorithms were used, PID control method has proved its efficiency, robustness, and considered to be fully effective for various DC Motor applications.

2 DC Motor Model

2.1 Mathematical Model

In armature control of Permanent Magnet DC motors, the voltage applied to the armature of the motor is adjusted without changing the voltage applied to the field. Figure 1 shows the Permanent Magnet DC Motor equivalent model.

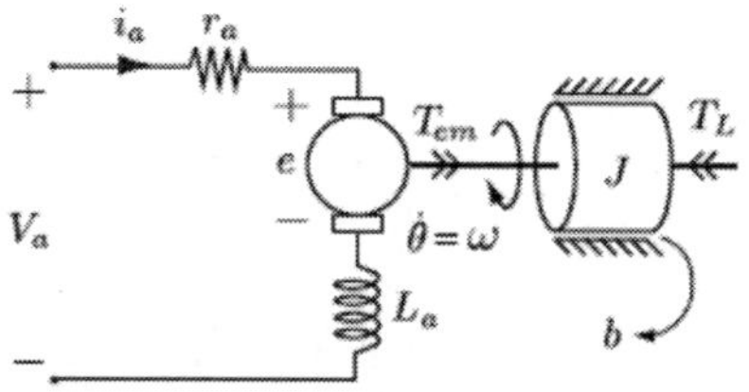

Fig. 1. Permanent magnet DC motor model [5].

Under the assumption of a homogeneous magnetic field, the direct current (DC) motor is modeled as a linear transducer from motor current to electrical torque. The classical model of the DC motor is composed of a coupled electrical and a mechanical subsystem.

The angular velocity is controlled by the input voltage V_a with a constant voltage drop attributed to the brush and rotor resistance, and a back-electromotive force (back emf) caused by the rotary armature. The motor inductance contributes proportionally to the change in the motor current. The motor current couples the electrical component with the mechanical one as it generates the driving torque. This torque is antagonized by the motor inertia, structure damping, friction, and the external load.

2.1.1 Electrical Characteristics

A differential equation of the equivalent circuit of a DC motor, illustrated in figure 1, can be derived by using Kirchhoff's voltage law around the electrical loop. Kirchhoff's voltage law states that the sum of all voltages around a loop must equal to zero, or:

$$V_a - V_{Ra} - V_{La} - V_c = 0 \tag{1}$$

* Corresponding author: w.khoury@outlook.com

© The Authors, published by EDP Sciences. This is an open access article distributed under the terms of the Creative Commons Attribution License 4.0 (http://creativecommons.org/licenses/by/4.0/).

According to Ohm's law, the voltage across the armature's resistor V_{Ra} can be represented as:

$$V_{Ra} = i_a R_a \tag{2}$$

Where $R_a[ohm]$ & $i_a\ [A]$ are the armature resistance and current respectively. The voltage across the inductor is proportional to the change of current through the coil with respect to time as:

$$V_{La} = L_a \frac{d}{dt} i_a \tag{3}$$

Where $L_a\ [H]$ is the inductance of the armature coil. Finally, the back emf can be written as:

$$V_c = k_v \omega_a \tag{4}$$

Where $k_v\ [V/rad/sec]$ is the velocity constant determined by the flux density of the permanent magnets, the reluctance of the iron core of the armature, and the number of turns of the armature winding. $\omega_a\ [rad/sec]$ is the rotational velocity of the armature.

Substituting equations (2-4) into equation (1) gives the following differential equation:

$$V_a - i_a R_a - L_a \frac{d}{dt} i_a - k_v \omega_a = 0 \tag{5}$$

2.1.2 Mechanical Characteristics

By balancing the energy of the system, the sum of the torques must be equal to zero. Therefore,

$$T_e - T_{\omega'} - T_{\omega} - T_L = 0 \tag{6}$$

Where $T_e\ [Nm]$ is the electromagnetic torque, $T_{\omega'}\ [Nm]$ is the torque due to rotational acceleration of the rotor, $T_w\ [Nm]$ is the torque produced by the velocity of the rotor, and $T_L\ [Nm]$ is the torque of the mechanical load. The electromagnetic torque T_e is proportional to i_a, and can be written as:

$$T_e = k_t i_a \tag{7}$$

Where $k_t\ [Nm/A]$ is the torque constant and, similar to the velocity constant, is dependent on the flux density of the fixed magnets, the reluctance of the iron core, and the number of turns in the armature windings. $T_{\omega'}$ can be written as:

$$T_{\omega'} = J \frac{d}{dt} \omega_a \tag{8}$$

Where $J\ [kgm^2]$ is the inertia of the rotor and the equivalent mechanical load. The torque associated with the velocity is written as:

$$T_{\omega} = B \omega_a \tag{9}$$

Where $B\ [N/rad/sec]$ is the damping coefficient associated with the machine mechanical rotation system.

Substituting equations (7-9) into equation (6) gives the following differential equation:

$$k_t i_a - J \frac{d}{dt} \omega_a - B \omega_a - T_L = 0 \tag{10}$$

2.2 Transfer Function and Block Diagram

A block diagram of the system is developed based on the differential equations (5) and (10). Taking the Laplace transform of each equation gives:

$$sI_a(s) - i_a(0) = -\frac{R_a}{L_a} I_a(s) - \frac{k_v}{L_a} \omega_a(s) + \frac{1}{L_a} V_a(s) \tag{11}$$

$$s\omega_a(s) - \omega_a(0) = \frac{k_t}{J} I_a(s) - \frac{B}{J} \omega_a(s) - \frac{1}{J} T_L(s) \tag{12}$$

Where s is the Lablace variable (complex number).

If permutations around some steady-state value are considered, the initial conditions go to zero and all variables will have some change around a reference state. Then, equations (11) and (12) can be expressed as:

$$I_a(s) = \frac{-k_v \omega_a(s) + V_a(s)}{L_a s + R_a} \tag{13}$$

$$\omega_a(s) = \frac{k_t I_a(s) - T_L(s)}{Js + B} \tag{14}$$

The above equations can then be easily put into a block diagram that represent the mathematical model of the PMDC Motor, as shown in figure 2.

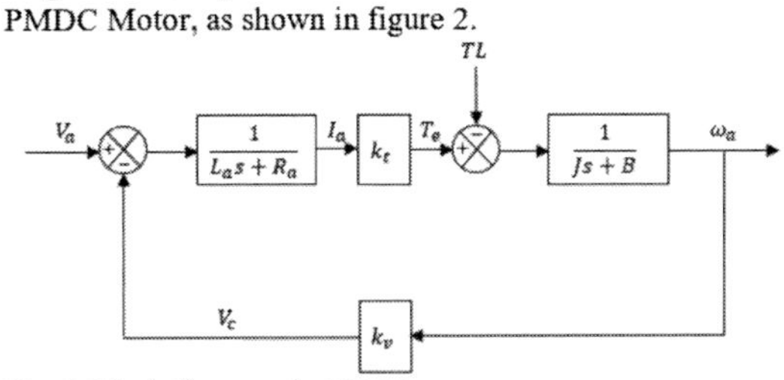

Fig. 2. Block diagram of a PMDC motor system.

From the block diagram, and assuming that $T_L = 0$ [Nm], the equivalent transfer function is found to be:

$$\frac{\omega_a(s)}{V_a(s)} = \frac{k_t}{L_a J s^2 + (R_a J + L_a B)s + R_a B + k_t k_v} \tag{15}$$

The AXEM DC motor F9M2 [6], which has the characteristics shown in table 1, is considered for the simulation in this article.

Table 1. Motor Characteristics.

Parameter [unit]	Value
Total motor resistance $\boldsymbol{R_a}$ [Ω]	0.98
Torque constant $\boldsymbol{k_t}$ $[\boldsymbol{Nm/A}]$	0.0274
Back emf constant $\boldsymbol{k_v}$ $[\boldsymbol{V/rad/sec}]$	0.0297
Damping constant $\boldsymbol{B}$ $[\boldsymbol{N/rad/sec}]$	7.2×10^{-5}
Total system inertia $\boldsymbol{J}$ $[\boldsymbol{kgm^2}]$	3.2×10^{-5}
Inductance $\boldsymbol{L_a}$ $[\boldsymbol{H}]$	25×10^{-6}
Rated torque $[\boldsymbol{Nm}]$	0.282
Rated speed $[\boldsymbol{RPM}]$	3000
Rated volt $[\boldsymbol{V}]$	14

For the selected motor, the transfer function of the speed $[\omega_a(s)/V_a(s)]$, assuming $T_L = 0$, is:

$$G(s) = \frac{0.0274}{8 \times 10^{-10} s^2 + 3.136 \times 10^{-5} s + 0.00088434} \tag{16}$$

3 Robust sensitivity analysis

In reality, T_L is not zero, and its value is considered as a disturbance to the system. Therefore, to get the best results and make the system robust against the disturbances and noise signals that can affect the response, as shown in figure 3, the sensitivity of the open loop system of the angular speed is analysed in the presence of the PI controller.

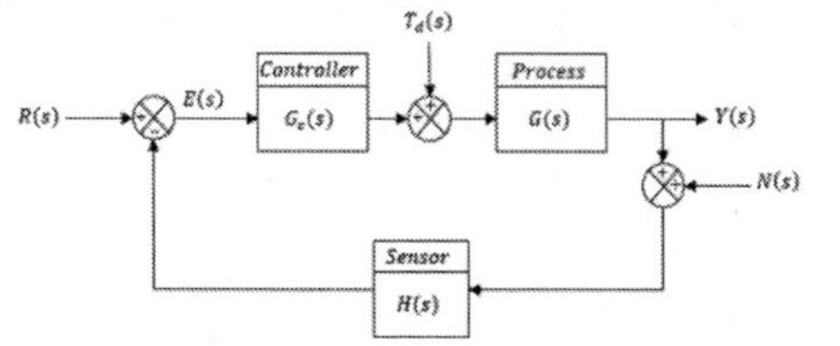

Fig. 3. Closed loop system in the presence of disturbances and noise signals [7].

Generally, most of the disturbances occur at low frequencies, which can affect the response, and most of the noise signals occur at high frequencies, which can affect the measurement. Mathematically, the sensitivity function, $S(s)$, is proportional to the disturbances, and the complementary sensitivity function, $T(s) = 1 - S(s)$, is proportional to the noise signals. Therefore, the sensitivity function must be of a low magnitude at low frequencies and high magnitude at high frequencies, so the effects of the disturbances and noise signals will be minimized. As a result, a robust system is obtained, that is capable of withstanding un-modelled dynamics, disturbances, noises ...etc.

Since the open-loop function, $L(s) = G(s)G_C(s)$, is inversely proportional to $S(s)$, it must be of a high magnitude at low frequencies and low magnitude at high frequencies. In other words, the open-loop function should behave as an integrator, and the controller should be designed to achieve this behaviour.

For this purpose, a parallel PI controller is chosen:

$$G_C(s) = k_p + \frac{k_i}{s} = 10 + \frac{20}{s} \tag{17}$$

The open-loop transfer function is given as:

$$L(s) = G(s)\, G_C(s) \tag{18}$$

$$L(s) = \frac{0.274s + 0.548}{8 \times 10^{-10}s^3 + 3.136 \times 10^{-5}s^2 + 0.00088434s} \tag{19}$$

The sensitivity function of the open loop system is:

$$S(s) = \frac{1}{1 + L(s)} \tag{20}$$

$$S(s) = \frac{s^3 + 39202.2s^2 + 1.10543 \times 10^6 s}{s^3 + 39202.2s^2 + 3.436 \times 10^8 s + 6.85 \times 10^8} \tag{21}$$

Where Bode diagrams of both the open-loop and sensitivity transfer functions were drawn using LabVIEW MathScrip tool, which resulted in the following plots:

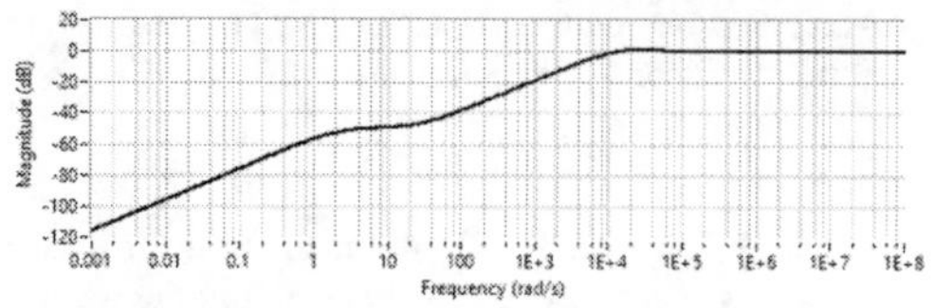

Fig. 4. Magnitude diagram of the sensitivity function.

As shown in figure 4, there are no peaks in the magnitude of the sensitivity, which means that the effects of the disturbances and the noise signals on the system are minimized at all frequencies.

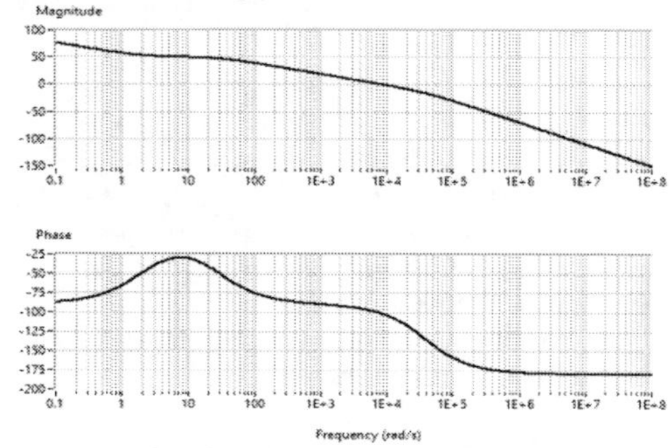

Fig. 5. Bode plot of the open-loop transfer function L(s).

It can be observed form the magnitude plot in figure 5 that the loop function behaves like an integrator, which, in this case, will increase the robustness of the system depending on the earlier discussion. Additionally, the phase margin is about 80 degrees, and the gain margin is infinity. This means that the system is stable at every value of the gain, and the phase margin is large enough to ensure that the system is not close to -1 in the Nyquist plot. However, the sensitivity analysis is good enough to ensure that some amount of uncertainties might not have effects on the proposed system, which, in turn, ensures that the system is robust under some boundaries. Nevertheless, since the speed loop system does not have any right half zeroes or time delays, it does not suffer the "lack of robustness", but it remains for the unknown uncertainties to determine how much the system is vulnerable, or robust.

4 LabVIEW implementation

To design the system within LabVIEW Simulation loop, several simulation Sub-VIs were developed, and two controllers were used, one for the speed (PI), and another for the position (P). The implementation process is detailed and shown in figures 6-8.

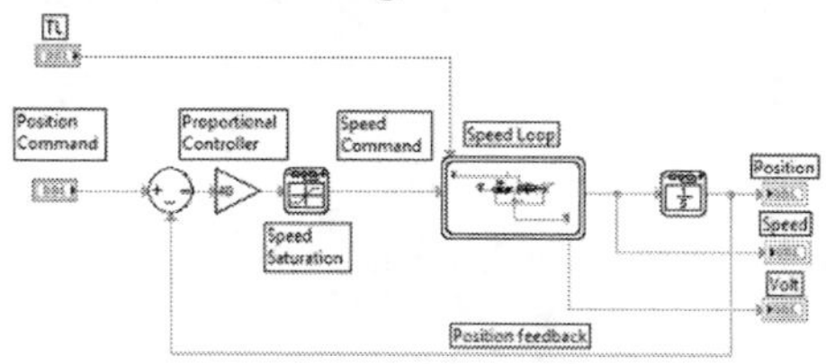

Fig. 6. Position loop subsystem architecture.

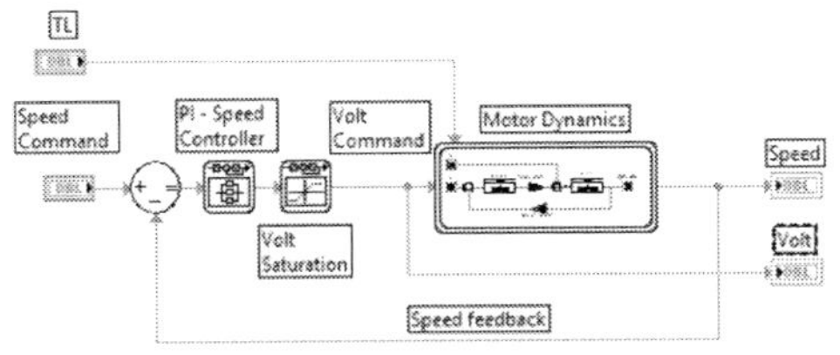

Fig. 7. Speed loop subsystem architecture.

Finally, the motor dynamics subsystem, having the same architecture as in figure 2, is shown in figure 8.

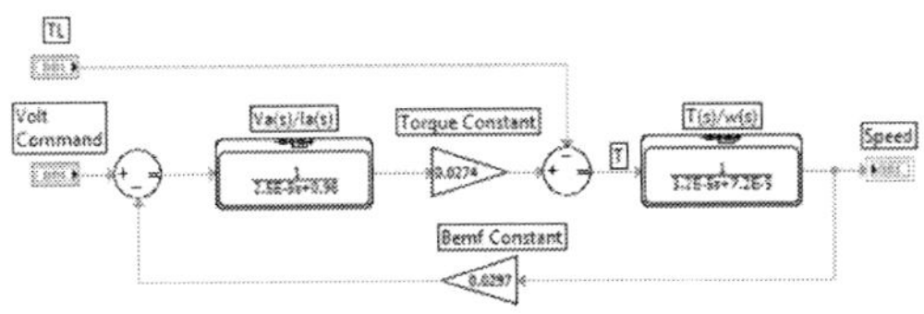

Fig. 8. Motor dynamics subsystem architecture.

5 Results

According to the sensitivity analysis and the system requirements, the final values of the controllers were found as $k_p = 10$ & $k_i = 20$ for the PI speed controller, and a proportional gain of 40 for the position controller.

To test the effectiveness of the system, the system is simulated by applying a step signal of 90 degrees, and when the system is stabilized, a load of 0.282 $[N.m]$ is added at $0.4\ sec$ in order to test the robustness of the system. The results are shown in figure 9.

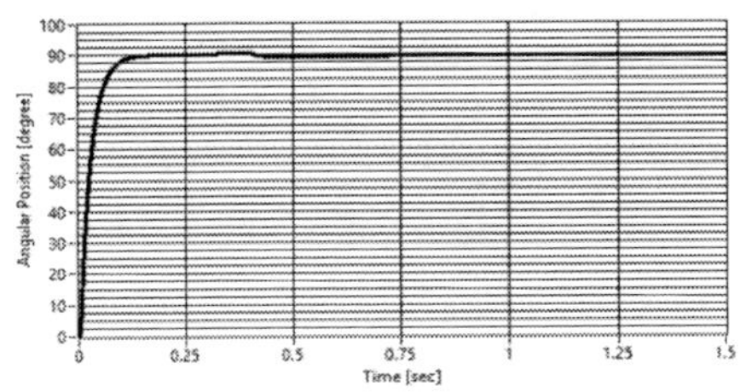

Fig. 9. Angular position response, with applying T_L at 0.4 *sec*.

The system reaches the desired position in $0.15\ sec$ with no overshoot or oscilations, and the system is well stable even when the torque load is applied. Moreover, the PI speed controller's zero succeeded to cancel out one of the poles of second order characteristics equation of the motor's dynamics. The resulting system acted like a first order system, with no oscillation in the final response as shown in figure 9.

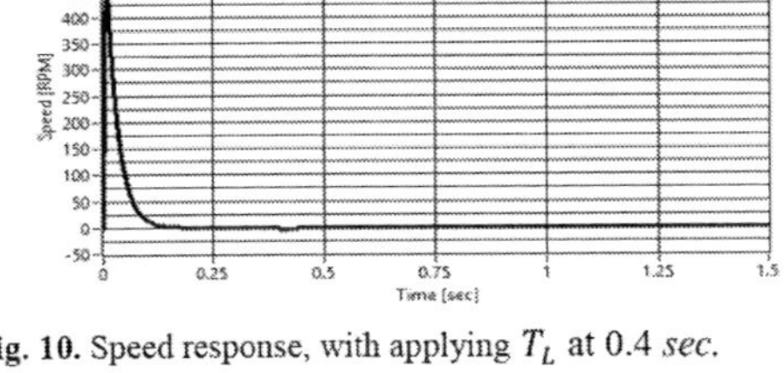

Fig. 10. Speed response, with applying T_L at 0.4 *sec*.

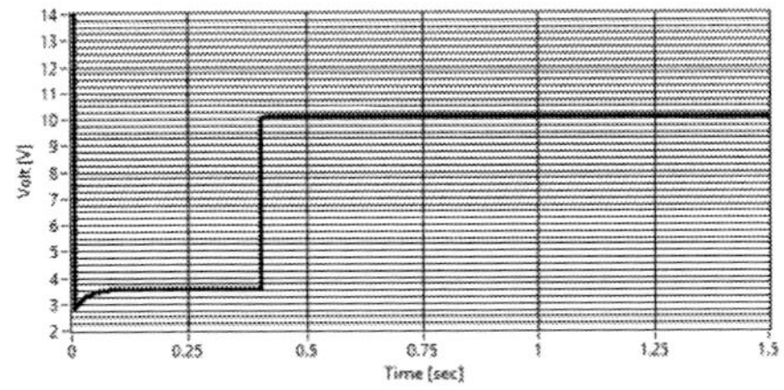

Fig. 11. Voltage response, with applying T_L at 0.4 *sec*.

It is seen that all the motor's parameters are within the maximum limits, this is achieved by adding the nonlinear function, saturation, to the control system. Additionally, the robustness of the system was fully able to overcome the uncertainties (caused by the torque load) and nonlinearities (caused by the saturation). However, the position response could be achieved in a shorter time by choosing different parameters of the PID, but it would cause oscillations in the magnitude of the sensitivity function, which in turn decreases the robustness.

6 Conclusion

LabVIEW Simulation Loop was used to build the PMDC motor dynamical model. The sensitivity analysis allowed to find the best parameters of a robust controller with no oscillations or overshoot in the final response, even when the maximum load is applied.

Acknowledgment

The work/publication is supported by the EFOP-3.6.1-16-2016-00022 project. The project is co-financed by the European Union and the European Social Fund.

References

1. E. Yolacan, S. Aydin and H.M. Ertunc, Real time DSP based PID and state feedback control of a brushed DC motor. In *Information, Communication and Automation Technologies (ICAT), 2011 XXIII International Symposium on*, 1-6, (2011). IEEE
2. E. Yolacan, S. Aydin, and H.M. Ertunc, Position control of a DC motor used in solar panels with artificial neural network. In *Machine Learning and Applications (ICMLA), 2012 11th International Conference on*, **2**, 487-492, (2012). IEEE
3. M. Namazov, and O. Basturk, DC motor position control using fuzzy proportional-derivative controllers with different defuzzification methods, *An Official Journal of Turkish Fuzzy Systems Association*, **1**, 36-54, (2010).
4. J. Mao, H. Tachikawa, and A. Shimokohbe, Precition positioning of a DC-motor-driven aerostatic slide system, *Precition Engineering*, **27(1)**, 32-41, (2003).
5. W. Borutzky, Bond graph modelling of engineering systems. *New York: Springer*, (2011).
6. M. Ruderman, J. Krettek, F. Hoffmann, T. Bertram, Optimal state space control of DC motor. *IFAC Proceedings Volumes*, **41(2)**, 5796-5801, (2008).
7. R.C Dorf, and R.H. Bishop, Modern control systems, *Pearson*, 12th Edition, 235-241 (2011).

The maintenance, commissioning of FANUC M-1iA 0.5A type selecting robot and implementation of iRVision picture-recognizing procedure

Gyula Korsoveczki[1,*], *Géza* Husi[2], and *Timotei István* Erdei[3]

[1]Mechatronics Department, Engineering Faculty, University of Debrecen, Hungary
[2]Mechatronics Department, Engineering Faculty, University of Debrecen, Hungary
[3]Mechatronics Department, Engineering Faculty, University of Debrecen, Hungary

Abstract. The topic of the given project is based on the maintenance, commissioning and implementation of iRVision picture-recognizing procedure, and the operation of the associated program of the FANUC M-1iA 0.5A type delta tripod designed selecting robot that can be found at the Mechatronics Department, Engineering Faculty of the University of Debrecen. Keywords — FANUC; M-1iA 0.5A; delta tripod; robot; iRVision; picture-recognizing

1 Introduction

The permeation of robots is more considerable in every area of industry. With the development of the technic and growing of demand there is need for automated production lines. The velocity and precision of these lines gives industrial robots. With the opportunity of this project I could work with that kind of industrial robot I have always wanted to. That is why I chose the mentioned FANUC M-1iA 0.5A type delta tripod designed selection robot that can be found at the Mechatronics Department, Engineering Faculty of the University of Debrecen. After I had completed the maintenance operations, I measured the tool used by the robot. Next, I optimizationed the sphere of the workplace and I measured the user frame that I used during the project. After this operation, I designed several tools and elements for the robot. These metal elements make the work desk stronger and ensure a higher level of stability for the construction that supports the work desk. The subjoined CD contains the complete plans. Furthermore, I wanted to know the working of the iRVision system of FANUC and the implementation of picture-recognizing procedure that I wanted to show in my task. As the result of these operations, the sphere of work of the robot was broadened and the functioning became more efficient and competent in educational purposes.

2 Description of the system

2.1. FANUC M-1iA 0.5A type robot

The robot I used during my task is a delta tripod designed robot, that possesses 6 degrees of freedom. This model is versatile and nimble, and it can achieve excellent work rates on complex component assembly operations. The robot is capable of working at any angle, so it is good choice for those applications that demand high level of repeatability and flexibility.

The scope of the application is electronics, food products industry, manufacture of medical instruments optic and medical industry and precision mechanics [1].

Fig. 1. FANUC M-1iA 0.5A [1]

The technical specifications of the robot:

- Number of controlled axes: 6
- Maximal reach: 280 mm
- Repeatability: +/- 0,02 mm
- Mechanical weight: 23 kg
- Maximal load capacity at wrist: 0,5 kg
- Maximal velocity: 4000 mm/s [1]

2.2. FANUC R-30iB Mate controller

The robot is controlled by FANUC R-30iB Mate Open-Air type of controllers. It is compact and easy to integrate into single robot production cells. Unlike Windows-based controllers, the R-30iB types use FANUC's own propriety

* Corresponding author: korsoveczki.gyula@gmail.com

© The Authors, published by EDP Sciences. This is an open access article distributed under the terms of the Creative Commons Attribution License 4.0 (http://creativecommons.org/licenses/by/4.0/).

software. This allows that the system is more stable and it is protected against the most cyber virus and attacks of hackers. Furthermore, its controller advantage is that there is no need frame system to run [2].

The controller also suits the requirements of industrial environment, because it has suitable protection against wetness and dust. To use the iRVision system easily, an Ethernet port can be found on the controller. The main parts are a switch with two position (ON-OFF), an emergency toggle with red colour and an operation mode changer green toggle that can be actuated by a small key. This green toggle is the responsible for the running of the cycle program [2].

Fig. 2. FANUC R-30iB [3]

2.3. FANUC iPendant Touch

The robot can be programmed in two ways. One of these ways is the Karel program language. To use this one, a PC is needed. The other opportunity is the iPendant Touch hand controller that is powered by FANUC. As this hand controller uses block style program language, there are more options to program the robot and it is faster than in case of Karel language.

The iPendant Touch has ergonomic shape, touch screen and its programming environment is user-friendly. To further enlargements it contains function toggles and there can be opened two windows on the big colourful screen at the same time. The hand controller has a 4D designer option and USB input to connect external store or mouse [4]. The main parts are the big colourful screen, the keyboard, the safety Deadman switch with three positions, an emergency toggle and a switch with two positions that can enable the iPendant Touch.

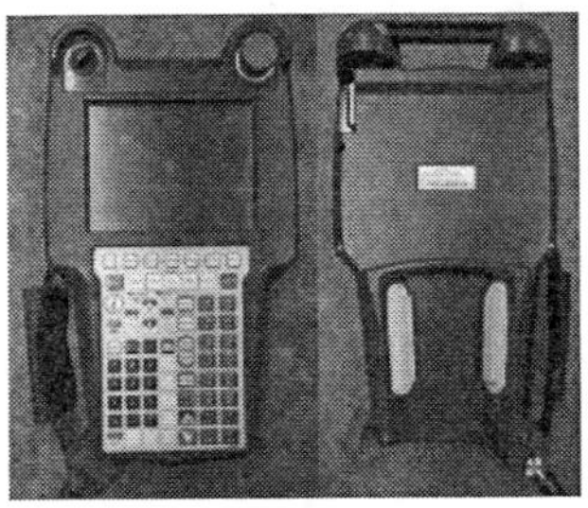

Fig. 3. FANUC iPendant Touch [Own figure]

2.4. SONY XC-56 type camera

The system includes an industrial monochromatic camera that is powered by SONY. This camera is compatible with the iRVision picture-recognizing system. It has the duty to accomplish the visual part of the implemented procedure.

The main technical specifications of the camera:

- High resolution: VGA resolution (648 x 494)
- High rate scanning: 120 frames / second
- Minimum illumination: 0,5 lux
- High shock and vibration resistance [5]

Fig. 4. SONY XC-56 camera [5]

3 The maintenance operations

3.1. Reset of the serial pulsecoder

During the task, my first problem was related to the batteries of the serial pulsecoder. The robot has 6 moveable axes. These axes are moved by 6 AC servo motors. To give their actual position, there are incremental pulsecoders in the robot. These kinds of pulsecoders need stationary electric supply. Without this, the result of the counting perishes [6]. That is why the robot includes 4 Industrial ALCALYN batteries. However, these batteries have gone flat since the last use of the robot. In this case a special alarm turns up, namely SRVO-062 BZAL alarm (Group: i Axis: j). It means Battery Zero Alarm. To solve this problem, changing the batteries is not enough, because there is need to delete and unlock this one in the system. With using of the iPendant Touch, we need to enter into the menu of the robot with pushing of the MENU button. After that, there is need to select NEXT (0) and the SYSTEM (6) options. Finally, we can reach the internal library of the robot with pushing of the VARIABLES (2) button. With the ITEM function we can search according to serial numbers. In this case, the number of the searched item is 308. This is $MRC folder. In this folder we need to find the $SPC_RESET variable with the serial number of 14. It means SERIAL PULSECODER RESET. When it is found, we have to change the value FALSE to value TRUE. Completing this operation, there is need to push the Deadman switch into middle position and the RESET button at the same time. As the next step, there is need to shut down the robot and disconnect to the electric circuit. When we turn on the robot, the axes will be moveable and we can use the JOINT coordinate-system [7].

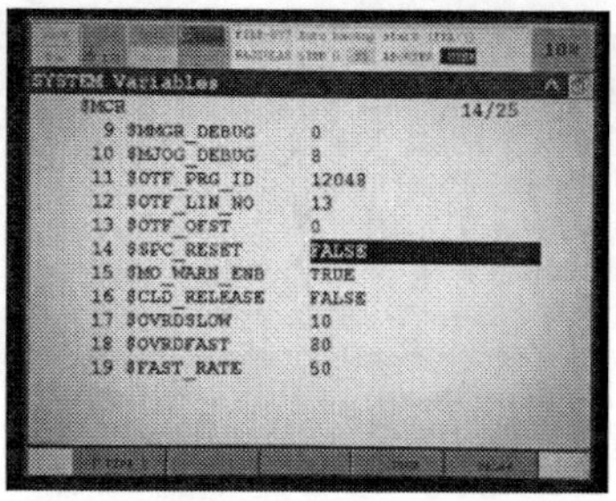

Fig. 5. The $SPC_RESET [Own figure]

3.2. The MASTERING and the CALIBRATION

Following the reset of the serial pulsecoder, it was necessary to perform the MASTERING and the calibration procedures. The actual position of the robot depends on the values of the pulsecoders. The corresponding count value is obtained from the rotational speed of the pulsecoder connected to the relevant motor and the rotation angle within one rotation. By right of this principle we can give values to the pulsecoders with the MASTERING operation.

This procedure is accomplished at the factories, users do not need to perform at every use. However, it is necessary after motor, pulsecoder and cable replacing, moreover, in case of battery problems [8].

During my task I accomplished the QUICK MASTERING operation on the robot. This procedure is implemented in a position that is defined by the user. QUICK MASTERING uses the fact that the absolute value of a rotation angle within one rotation will not be lost. This kind of MASTERING can be used, if the pulsecoder count value is lost, for example a low voltage detects on the backup battery for the pulse counter. However QUICK MASTERING cannot be used, after the serial pulsecoder is replaced or the mastering data is lost from the robot controller [8].

To accomplish QUICK MASTERING, there is need to select the SYSTEM option and then we have to choose the MASTER/CAL opportunity with pushing the button 3. Here we can reach all of the MASTERING options. First of all, it is necessary to move the robot axes into a random position. To complete this operation, we need to select the JOINT coordinate-system with the COORD button. The next step is to push the RESET button and the Deadman switch into the middle position at the same time. With this operation, we can unlock the FAULT alarm.

To move the robot axes, it is needed to push the Deadman switch, the SHIFT button and one of the direction buttons at the same time. These direction buttons can be found on the right side of the iPendant Touch. In this case, the TOOL CENTER POINT (TCP) moves into the required direction. The direction buttons are the following:

- +X; -X; +Y; -Y; +Z; -Z: Moving along the axes of the Descartes coordinate-system.
- +X ↷; -X ↷; +Y ↷; -Y ↷; +Z ↷; -Z ↷: Rotating around the axes of the Descartes coordinate-system.

To select the QUICK MASTERING, we need to use the navigation buttons and to complete it, we have to press ENTER.

After performing this operation, pulsecoders get count values. However, it is not enough to give values to the pulsecoders, because we have to save these values. There is the CALIBRATING option for this purpose. To perform this operation, it is important to choose the CALIBRATE opportunity [8].

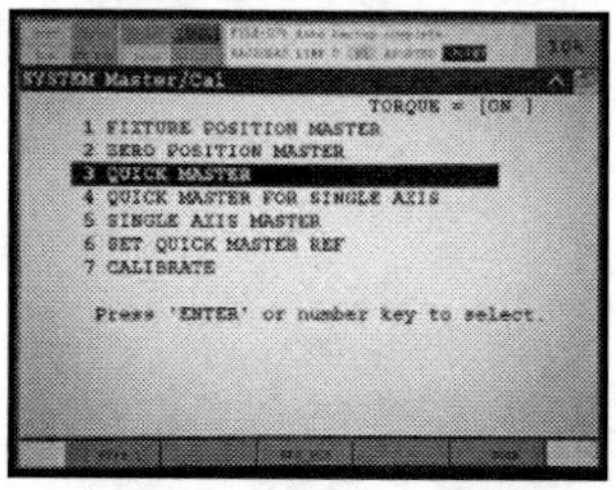

Fig. 6. The MASTER/CAL menu [Own figure]

During my task I completed all of the described operations. As a result, I achieved the aim that the robot can move in WORLD coordinate-system. To define and present the directions, I created an illustration that contains these directions. The illustration can be found on the construction that supports the robot. Knowing of the directions enables faster and easier robot programming for the users.

4 The commissioning of the robot

4.1. Measuring of the tool frame

Finishing with the maintenance, I had the opportunity to commission the robot. The first step of commissioning was the measuring of the tool frame.

To work the robot properly, there is necessary that we perform this operation. With measuring of the tool frame, we teach the robot the TOOL CENTER POINT. Knowledge of this information is essential for the robot because the robot gets information about the geometric parameters of the tool, which is found on it.

To accomplished the measuring, it is necessary to enter into the MENU of the iPendant Touch. After that, we need to select the SETUP (6) option and within that the FRAMES possibility with the button 3. Here there is need to open the [OTHER] function to reach the TOOL FRAME platform. Following this procedure, the system gives us a list about the previously measured tools. We have the opportunity to modify these tools but we can also define new tools. To measure a new tool, we have to move the cursor on an empty place and then with the CREATE order we are start the operation. Opening of the [METHOD] function helps us to choose the method of the tool measuring. The system offers to us 4 possibilities. These are the Three Point, Six Point, Two Point + Z and the Direct Entry.

While I was dealing with my task, I used a 3D printed tool and I measured it with the Three Point method. The essence is of this method, that we touch with the target tool the chosen reference point from three different directions and with three different orientations. To complete the procedure, there is need to define 3 points which are APPROACH POINT 1, 2 and 3. These points are the mentioned touching points. We have to move the TCP into the demanded points and then it is necessary to save them. We can do this with pushing the SHIFT button and selecting the RECORD order at the same time. We have the possibility to comment the measured tool into the COMMENT space.

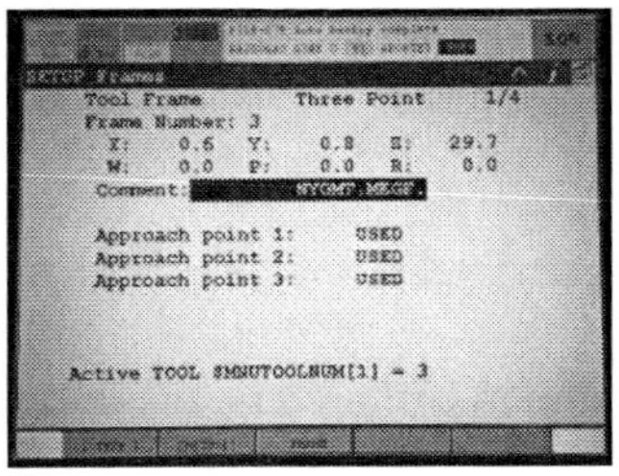

Fig. 7. Measuring of the tool frame [Own figure]

4.2. Optimization of the work sphere

The next purpose of my task was the optimization of the work sphere. Thanks to this operation, the scope of the work of the robot broadened and with the setting of the work sphere the precision has increased.

Before the procedure, the robot had possessed a work sphere but from the point of the view of my project it had been unsuited.

The first problem I discovered that the work table was not horizontal, because it possessed a thereabouts 20 degrees. This problem made the robot unsuited for the task solution because I wanted to implement a 2D viewed picture-recognizing procedure. To solve this problem, I created a construction made of performed aluminium. It was compatible with the supporting construction of the robot. To fix the fittings, I used small metal components that have screw-threads. With constraining or slacking the components the work table proved moveable along the splints.

As the work table I used a metal practise table powered by KUKA GmbH. It is stable, it includes height differences and there are practise figures on the surface. Another problem I discovered that the maximal reach of the robot is 280 mm, so it was not able to reach the surface of the work table. To solve this problem, I have lifted up the new work table. Being stable and adjustable are very important criterions against the work table. I accomplished these viewpoints with 4 shafts that possess screw threads. Their diameters are M6 and they are 120 mm long. The screws are compatible with the small metal components and with the whole construction. They warrant X and Y direction mobility by the WORLD coordinate-system. To fix the practise table, I used screw-nuts and vapid spacers. Thanks to that, the position of the practise table is variable vertically. It means that the Z position is adjustable.

As a result of the performed changeovers, I could set the work sphere of the robot optimal. The directions of the mobility are presented in figure 8.

Fig. 8. The new work sphere [Own figure]

4.3. Measuring of the user frame

After the suitable constructing of the work sphere, I had the opportunity to measure the user frame that was used by the robot. In addition, I laid down the used coordinate-system and I defined the directions of this.

The measuring of the user frame can be reached from the menu of the iPendant Touch with the SETUP (6) option. To get this, there is need to open the FRAME (3) opportunity and then it is needed to select the USER FRAME possibility with the opening of the [OTHER] function. Here we can modify before measured frames or we can measure a new one. The [METHOD] function offers the Three Point, Four Point, and the Direct Entry operations.

During the soluting of the task I performed the Three Point method. In this case we need to define the origo of the user frame and two other points which are in X and Y directions. These directions are the same with the directions of the WORLD coordinate-system.

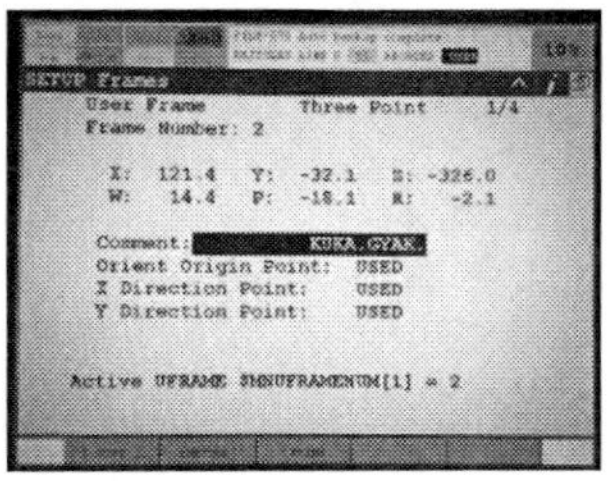

Fig. 9. Measuring of the user frame [Own figure]

5 Creating of the iRVision procedure

The iRVision is the FANUC's own PLUG & PLAY visual sensor system. The robot possesses a camera and a picture-analysing process that is connected to the camera. These items provide vision for the robot. It is completely integrated into the R-30iB Mate controller. As a result of this, the iRVision can be used easily and it offers a high level of flexibility to the users. Thanks to the iRVision system, the robot is suitable for selecting, complex component assembly operations and it is able to recognize things with optional scale, colour and position in 2D and 3D as well [9].

Meanwhile I was dealing with my project, I wanted to implement a picture-analysing procedure that recognizes a given 2D coordinate-system with the use of the camera. After that, it touches the center point of the coordinate-system and it draws around the wholesale triangle which is placed in this one. However, this triangle is invisible for the robot. To accomplish these operations, we need a robot program that connects the iRVision system to the robot.

The speciality of the iRVision is that it has a direct connection to the controller. In case of computer connection, there is need an Ethernet cable and a PC that can run Microsoft Internet Explorer.

The first step of the implementation is connecting the PC to the controller of the robot with Ethernet cable. For the communication, we need to overwrite the IPv4 address of the PC to 192.168.0.11. Following this operation, it is necessary to open the Internet Explorer and we can reach the main software of the controller with the address of 192.168.0.1. There are special components for the Internet Explorer which are essential to work the system. These components can be downloaded with clicking to the iRVision Setup menu.

From the appearing menu we should choose the CAMERA SETUP TOOLS opportunity with the help of [TYPE] option. After that, with using of the CREATE command we can create a new one. Here we can define the settings of the used camera. In addition, we have to select fix camera option and set the exposition time.

The next step is to complete the calibration of the camera with the CAMERA CALIBRATION TOOLS function. To do this, we need the FANUC's own calibration grids. Without this, we cannot accomplish the operation. The main point of the calibration is that the robot optimizes the settings of the camera itself to get the maximal efficiency. Since I didn't possess with calibration grids, I used the basic settings of the camera.

After all these operations, there is need to create a new picture-recognizing procedure, namely VISION PROCESS TOOLS and within this a 2-D SINGLE – VIEW VISION PROCESS. The name of the process I created is "KRESHAROMSZOG".

In this opportunity, we have to define the camera calibration, the exposition time and the criterions of the finding of the searched components. Furthermore, there is need to select the previously measured user frame and we have to define a reference position.

The menu contains a GPM LOCATOR TOOLS option. With using of it, we can teach various objects to the robot. To teach them, we have to take a picture with the SNAP command. Following this operation there is need to trace the object that we want to teach to the robot. In addition, we have the possibility to mask the parts which are unimportant from the point of view if searching, as I did in case of the triangle. This operation can be completed with the TRAINING MASK command.

Important tools are the CONTRAST TRESHOLD, ORIENTATION AND SCALE OPTIONS. With the first one, we can search by colours. In case of the second one, we have the opportunity to find objects in various orientations. The third one enables us to search objects by their scales.

With the end and saving of the edition comes into existence the picture-recognizing procedure that can be testing straight. Meanwhile the testing, the software shows visually the progress and the result of the searching.

With completing of the presented steps, my procedure could be accomplished successfully. I taught the coordinate-system to the robot as a searchable object.

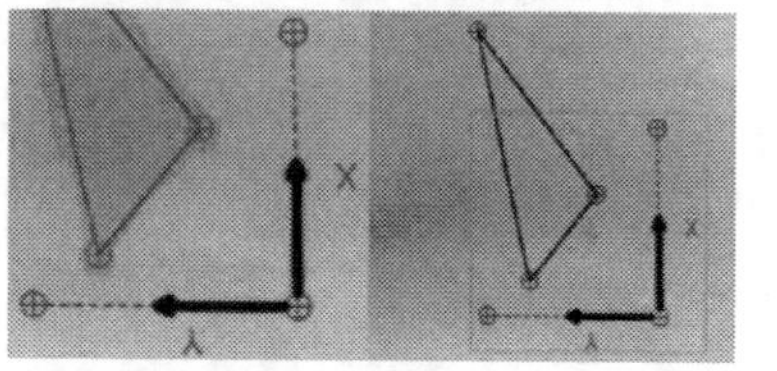

Fig. 10. The searched and found object [Own figure]

6 The associated robot program

6.1. Description of the program code

As the last step of my task, I wanted to create an associated robot program to the picture-recognizing procedure. Thanks to that, the robot can use the mentioned procedure in practice. With running the program, the robot is able to recognize the taught coordinate-system which is can be found on the work table of the robot in any position but in given orientation. Furthermore, the robot can touch the origo of the coordinate-system and after that it can draw around the triangle in the coordinate-system. In this case the triangle is invisible for the robot, so the position of the triangle depends on the found coordinate-system.

The programming happened with the using of iPendant Touch. It is much more suitable and faster than the usage of Karel program language.

In pursuance of this as the first step, I defined the previously measured tool frame and user frame in line 1 and 2.

As the third step, I implemented a view point for the procedure. In this point the TCP do not disturb the vision of the camera. The type of the motion is linear (L), the velocity is 100 m/s and at the end of the motion, the robot stops (FINE).

In line 4, I connected the picture-recognizing procedure to the robot program with the VISION RUN_FIND command. The purpose of the searching, I gave the name of the visual process which is (KRESHAROMSZOG). The camera of the robot checks the work table according to the picture-recognizing procedure to find a taught object.

In case of successful detecting, the robot saves the coordinates of the object into an OFFSET file in line 5. However, if it does not find anything, it creates a jump (JMP) onto a marked label (LBL [1]) to skip over the part of the program code which belongs to the finding.

With the commands which can be found in lines 6-8, the robot touches the origo of the coordinate-system. After that, with the commands in lines 9-14, the robot draws around the triangle according to P [4], P [5], P [6], P [7], P [5] and P [4] points.

Finally the LBL [1] can be found in line 15, and with line 16 and the [END] command the robot returns into the P [1 : VIEW] point and finishes the program. The program code is the next:

```
 1 :   UFRAME_NUM=2
 2 :   UTOOL_NUM=3
 3 : L   P [1 : VIEW]  100mm/sec  FINE
 4 :   VISION RUN_FIND 'KRESHAROMSZOG'
 5 :   VISION GET_OFFSET
   :   'KRESHAROMSZOG'  VR [1]  JMP  LBL [1]
 6 : L   P [2 : APRCH]  80mm/sec  FINE
   :   VOFFSET , VR [1]
 7 : L   P [3 : PICK]  80mm/sec  FINE
   :   VOFFSET , VR [1]
 8 : L   P [2 : APRCH]  80mm/sec  FINE
   :   VOFFSET , VR [1]
 9 : L   P [4]  100mm/sec  FINE
   :   VOFFSET , VR [1]
10 : L   P [5]  100mm/sec  FINE
   :   VOFFSET , VR [1]
11 : L   P [6]  100mm/sec  FINE
   :   VOFFSET , VR [1]
12 : L   P [7]  100mm/sec  FINE
   :   VOFFSET , VR [1]
13 : L   P [5]  100mm/sec  FINE
   :   VOFFSET , VR [1]
14 : L   P [4]  100mm/sec  FINE
   :   VOFFSET , VR [1]
15 :   LBL [1]
16 : L   P [1 : VIEW]  100mm/sec  FINE
[END]
```

6.2. The flowchart diagram

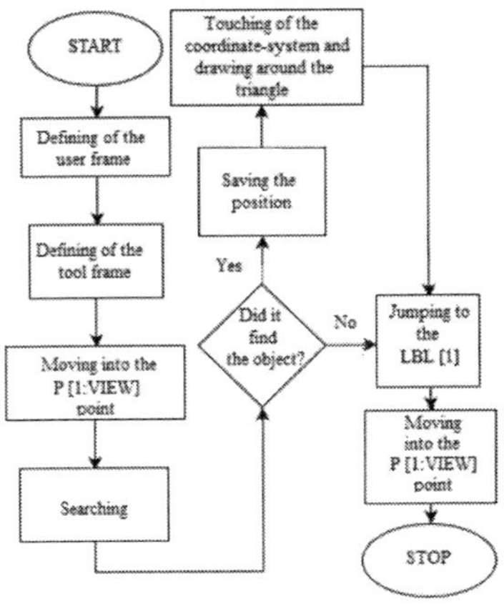

Fig. 11. The flowchart diagram [Own figure]

7 Conclusion

The purpose of my task was to accomplish the maintenance and the practice oriented commissioning of the robot. Furthermore, I wanted to test and submit the iRVision system of the robot with the help of the associated program. As a result, I got a system that works well and I accomplished the purposes. There sre other possibilities, for example creating various target specific tools or implementation of external light source and conveyor line. Moreover, the recalibration of the camera is more important to make the precision and efficiency better and to improve the scope of the work. However, the software part can be broadened without any problem with new part of program codes and functions.

Acknowledgement

The publication was supported by the project EFOP-3.6.1-16-2016-00022. This project was co-funded by the European Union, from the European Social Fund.

References

1 FANUC Hungary Kft., „M-1iA/0.5A," [Online]. Available: http://www.fanuc.eu/hu/hu/robotok/robotsz%c5%b1r%c5%91-lap/m1-sorozat/m-1ia-05a. [Accessed: 27.11.2017.].
2. FANUC Hungary Kft., „R-30iB controller." [Online]. Available: http://www.fanuc.eu/hu/hu/robotok/tartoz%c3%a9kok/r-30ib-vez%c3%a9rl%c5%91. [Accessed: 27.11.2017.].
3. RobotWorx, „FANUC R-30iB Open Air Cabinet," [Online]. Available: https://www.robots.com/fanuc/controller/r-30ib-open-air-cabinet. [Accessed: 27.11.2017.].
4. FANUC Hungary Kft., „iPendant Touch," [Online]. Available: http://www.fanuc.eu/hu/hu/robotok/tartoz%c3%a9kok/r-30ib-vez%c3%a9rl%c5%91/ipendant-touch. [Accessed: 27.11.2017.].
5. SONY, „XC56," [Online]. Available: https://pro.sony.com/bbsc/ssr/product-XC56/. [Accessed: 27.11.2017.].
6. Dr. P. Boza, Dr. J. Pintér, *Manufacturing automatization (Gyártásautomatizálás)* Győr, 2011.
7. FANUC Robotics America, Inc., „M-1iA Operator Manual," [Online]. Available: https://www.scribd.com/document/350205121/M-1iA-Operator-Manual-B-83084EN-08. [Accessed: 27.11.2017.].
8. FANUC Robotics America, Inc., „R-30iA Mate CONTROLLER," [Online]. Available: http://www.msamc.org/aimss/documentation/pdf/manuals/lr_mate_manuals/R30iA%20Mate%20Controller.pdf. [Accessed: 27.11.2017.].
9. FANUC Hungary Kft., „Visual sensor system," [Online]. Available: http://www.fanuc.eu/hu/hu/robotok/tartoz%c3%a9kok/l%c3%a1t%c3%a1s. [Accessed: 27.11.2017.].

The influence of profile angle on forces distribution on silent chain transmissions

Radu Velicu[1*], *Lenard* Jurj[1] and *Radu* Săulescu[1]

[1]Transilvania University of Brasov, Product Design Mechatronics and Environment Department, Eroilor 29, 500069 Brasov, Romania

Abstract. The aim of this paper is to define and to analyze the theoretical contact forces that appear in a sprocket - silent chain joint. The contact force between the links and the teeth flanks of the sprocket can be considered as a reaction to the axial force (tensioning force) transmitted throughout the pins. The centrifugal force is considered to be constant and the friction between the elements is not considered. This paper is presenting the steps to obtain the relationships for the forces from the sprocket-link contact depending on the tensioning force component between the links and the centrifugal force. These forces are computed in order to establish their distribution on links.

1 Introduction

Chain drive system's dynamics became of great interest to the automobile industry with the increase in importance of noise, vibration, and harshness (NVH - noise, vibration and harshness). Chain drives are used because of the non-slip, reliability and compact power transmission compared to belt transmissions, but usually with the disadvantages of increased noise and vibration. A brief history of chain drive systems and the important milestones in their practical use and development through the 20th century can be found in [1].

Timing chain transmissions are used more and more in combustion engines. Their robustness ensures the transmission of higher power, respectively higher torque for a longer maintenance-free period aggregate, [2].

Silent chains as timing chains have a great interest regarding performance improvements of automobile engine systems, which can generate benefits like: reduced consumption of fuel and oil, higher engine power and torque, reduced harmful gas emissions, longer life and improved reliability of engines [3]-[5].

Silent chains are usually composed, as seen in Figure 1. of: 1 – inner plates, 2 – middle plates, 3 – outer plates and 4 – pin. The inner plates and the outer plates are fixed on the pin, while the middle plates are mounted with clearance on the pins. Only the inner and the middle plates are engaging with the teeth of sprockets. The outer plates are used for guiding. This type of chain is not standardized, there are different shapes and sizes of the plates; different number of inner and middle plates.

References of theoretical studies on silent chain transmissions come from early 20th century, [6]. A geometrical study of chain transmissions with two sprockets and a guide is presented in [7]. An FEM study of forces distribution on links of a silent chain, at the contact with a circular guide, is presented in [8]; pressure distribution between a silent chain and a guide, resulted from a tribological study, is presented in [9]. A comprehensive approach on general chain transmissions, including aspect of silent chains is in [10]; experimental results on load distribution between bush chains and sprockets is presented in [11]. This paper intends to study the theoretical contact which appears between the silent chain's elements and the sprockets.

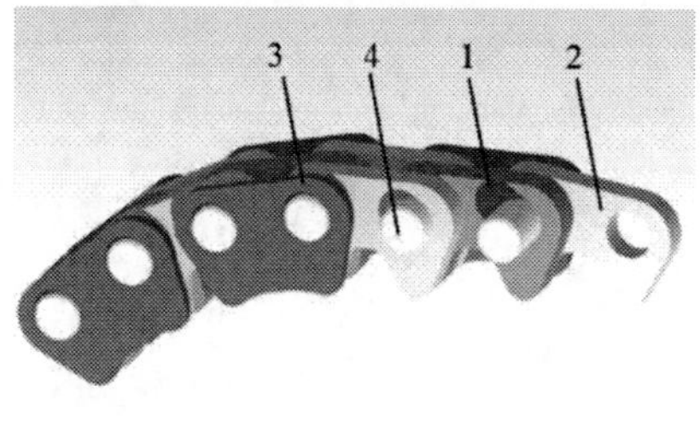

Fig. 1. Silent chain.

2 Theoretical approach

Figure 2. presents the theoretical model of the silent chain transmission. The middle or the inner plates of each link are coming in contact directly with the linear flank of the tooth, without rolling or sliding.

F is the tensioning force applied to the first pin on the first link (1) of the chain. The theoretical force distribution takes into account the normal forces that appear between the sprocket tooth and chain plate of link 1 - F_{z1}, the tensioning force component between the silent chain links 1 and 2, that is transmitted by the second plate-pin joint – F_1, and the centrifugal force acting on link 1 that appear during the rotational speed of the chain transmission - F_c.

* Corresponding author: rvelicu@unitbv.ro

© The Authors, published by EDP Sciences. This is an open access article distributed under the terms of the Creative Commons Attribution License 4.0 (http://creativecommons.org/licenses/by/4.0/).

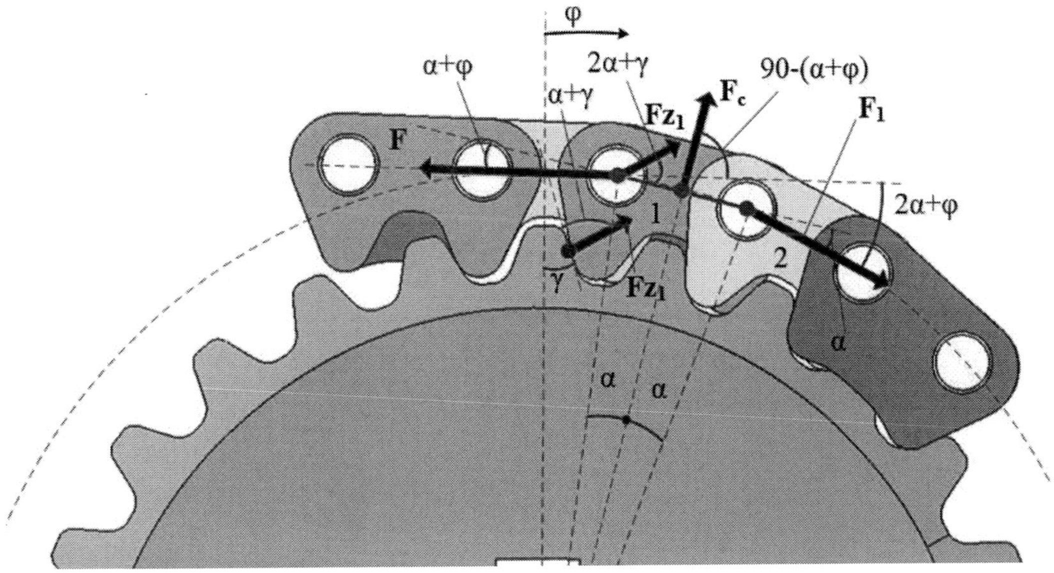

Fig. 2. Force distribution on silent chain transmission.

In this case, the force distribution is varying within an angle φ from -α to α, where α represents the half of the pitch angle on the sprocket. This angle depends on the number of teeth of the sprocket. The profile angle γ appears between the axis of a tooth and its flank line (plate's exterior flank). There will be taken into account three values for γ to find out what the effects of this angle are over the overall force transmission between the sprocket and the chain.
The theoretical model will be established using a constant rotational speed of the system, thus constant centrifugal force on the plates.
The calculus starts with the equilibrium of the forces written on two directions:

- on the direction of the tensioning force F:

$$F = F_{z1}\cos(\alpha+\gamma-\varphi) + F_c\sin(\varphi+\alpha) + F_1\cos(\varphi+2\alpha); \quad (1)$$

- on perpendicular direction over the tensioning force:

$$F_1\sin(\varphi+2\alpha) = F_{z1}\sin(\alpha+\gamma-\varphi) + F_c\cos(\varphi+\alpha). \quad (2)$$

Extracting from (2) the force from the plate-sprocket contact F_{z1} we have:

$$F_{z1} = \frac{F_1\sin(\varphi+2\alpha) - F_c\cos(\varphi+\alpha)}{\sin(\alpha+\gamma-\varphi)}, \quad (3)$$

Using (3) in (1) we get the equation for the component of the tensioning force between the chain links:

$$F_1 = \frac{F\sin(\alpha+\gamma-\varphi) + F_c\cos(2\alpha+\gamma)}{\sin(3\alpha+\gamma)}, \quad (4)$$

Based on Figure 2, the following equations can be drawn:

$$\alpha = \frac{360}{2z} \quad (5)$$

$$v_t = \frac{\pi\cdot D_p\cdot n}{1000\cdot 60} \quad (6)$$

$$F_c = \frac{m_{chain}\cdot v_t^2}{\frac{D_p}{2}\cdot 1000} \quad (7)$$

where z is the number of teeth on the sprocket, Dp is the pitch diameter of the sprocket, n is the rotational speed of the chain transmission, v_t is the tangential speed at the pitch diameter and m_{chain} is the mass of the silent chain.

For the equilibrium of link 2, the input force is the tensioning coming from link 1 F_1, which will replace the input F force used for equilibrium of link 1. The relative angle between the links 1 and 2 being constant, angle φ in (3) and (4) will have the value of the pitch angle α.

The force F1 for the second link will become F_2, while F_{z1} will become F_{z2}. All these equivalent replacements of the forces will be computed with the following observation: F_1 and F_{z1} are varying in function of the angle φ of link 1.

The tensioning force component for the second link can be described as:

$$F_2 = \frac{F_1\sin(\gamma) + F_c\cos(2\alpha+\gamma)}{\sin(3\alpha+\gamma)}. \quad (8)$$

The forces acting between the sprocket and the second link can be written as:

* Corresponding author: rvelicu@unitbv.ro

$$F_{z2} = \frac{F_2 \sin(3\alpha) - F_c \cos(2\alpha)}{\sin(\gamma)} . \qquad (9)$$

For the next links, the forces can be computed with the following generalized formulas:

$$F_m = \frac{F_{m-1} \sin(\gamma) + F_c \cos(2\alpha + \gamma)}{\sin(3\alpha + \gamma)} \qquad (10)$$

$$F_{zm} = \frac{F_{m-1} \sin(3\alpha) - F_c \cos(2\alpha)}{\sin(2\gamma)} \qquad (11)$$

where m represents the number of the corresponding link of the silent chain.

3 Numerical analysis – Results

The model for the force equilibrium presented above has been applied for z =27 and three values of profile angle γ (20°, 25°, 30°). Table 1 presents the input parameters used to compute the theoretical forces that appear between the silent chain and the sprocket, calculated according to [12].

Table 1. Gear parameters, z = 27, γ (20°, 25°, 30°).

Symbol	Unit	Value
z	-	27
n	rpm	5000
F	N	500
α	deg	6.66666
P	mm	8
D_{pitch}	mm	68.91032
M_{chain}	kg	0.0034
L_{chain}	mm	518
v_t	mm/s	9.51745
$F_{c/plate}$	N/plate	5.082994

The diagrams of tensioning force F_m depending on rotation angle φ, for three values of the contact angle γ, are presented for links 1, 2 and 3 in Figure 3., Figure 4. and Figure 5. They are showing the continuous decrease with rotation angle and sequent number of links, with smaller values for smaller contact angle. The diagrams of contact forces F_{zm} depending on rotation angle φ, for three values of the contact angle γ, are presented for links 1, 2 and 3 in Figure 6., Figure 7. and Figure 8. The contact force on the first contact increases with rotation angle and with decrease of contact angle, but for the following contacts there is a continuous decrease with rotation angle.

For the sequent number of links, the trend on the contact force is a faster decrease for smaller values of the contact angle. The decrease of contact angle determines a decrease of contact force starting with link 3. The second link is just a transition situation between the trend of the first link and the trend of the next links.

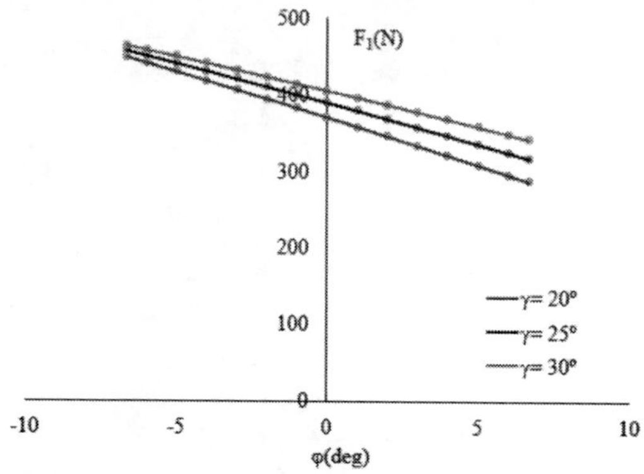

Fig. 3. Tensioning force distribution on the first link for three values of the profile angle γ (20°, 25°, 30°).

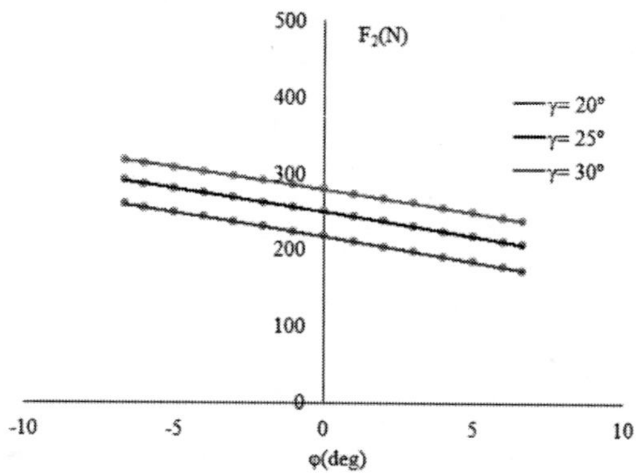

Fig. 4. Tensioning force distribution on the second link for three values of the profile angle γ (20°, 25°, 30°).

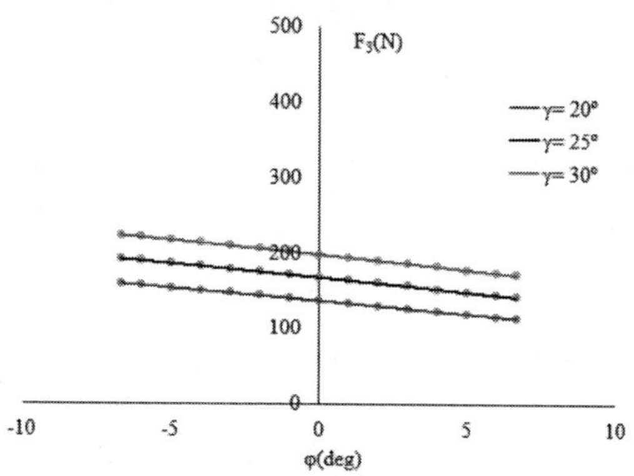

Fig. 5. Tensioning force distribution on the third link for three values of the profile angle γ (20°, 25°, 30°).

The analysis of the theoretical force distribution shows that the tensioning force in the silent chain transmission is distributed on more than the first couple of chain links.

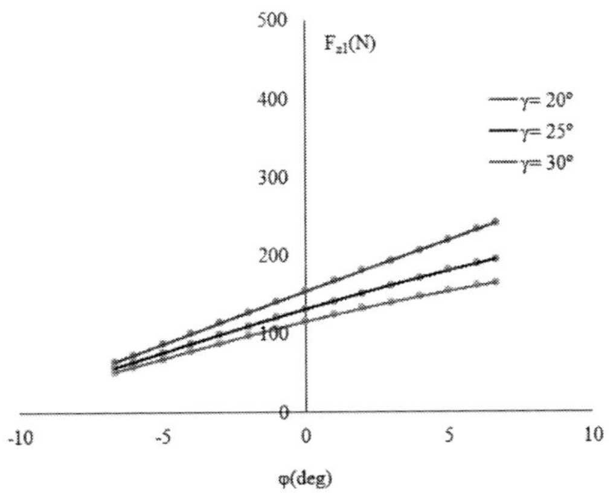

Fig. 6. The contact force on the first link for three values of the profile angle γ (20°, 25°, 30°).

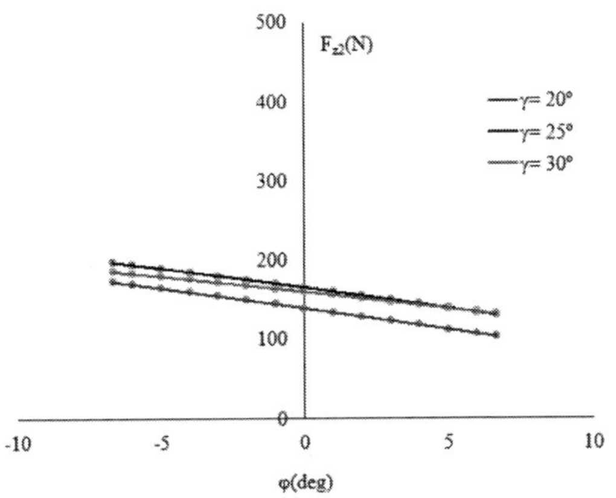

Fig. 7. The contact force distribution on the second link for three values of the profile angle γ (20°, 25°, 30°).

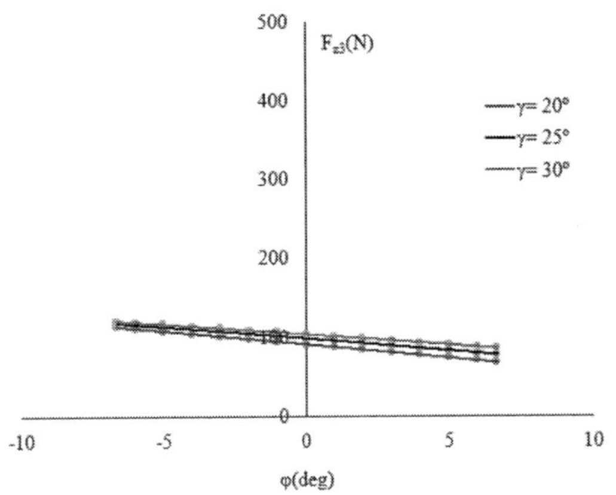

Fig. 8. The contact force distribution on the third link for three values of the profile angle γ (20°, 25°, 30°).

The results of the theoretic analysis for the silent chain-sprocket shows that:

1) When the forces get to negative values the chain gets off the sprocket – until that moment the force are transmitted between the chain links and the sprocket.

2) The profile angle of the silent chain plate (middle plate, inner plate) has a great effect on the number contacts between the chain and the sprocket. As the profile angle increases the number of links in contact increases.

3) The F_z component from the chain transmission, as a reaction from the sprocket to the chain link, tends to move off the links from the sprocket.

3 Conclusion

The main conclusion from theoretical analyse of the silent chain contact forces is that the number of chain links in contact is influenced by the profile angle of the tooth flank.

However, this conclusion must be proved when the friction forces are also considered.

References

1. J. C. Conwell, *An Examination of Transient Forces in Roller Chain Drives, PhD Thesis,* Vanderbilt University, Nashville (1989).
2. S. Belmer, T. Fink, , I. Lorenz, H. Neukirchner, MTZ **66**, 6 (2005).
3. K. Holmberg, P. Andersson, A. Erdemir, Trib Int **47**, (2012).
4. T. Hyakutake, M. Inagaki, M. Matsuda, N. Hakamada, Y. Teramachi, JSAE Review 22:5 (2001).
5. M. Schwaderlapp, F. Koch, J. Dohmen, FISITA (2000).
6. J. R. Cautley, SAE Technical Paper 140021 (1914).
7. R. Papuc, R. Velicu, C. C. Gavrila, Annals of the Oradea University, Fascicle of Management and Technological Engineering XXIII (XIII) (2014).
8. M. T. Lates, R. Papuc, C. C. Gavrila, Annals of the Oradea University, Fascicle of Management and Technological Engineering XV (XXV)/1 (2016).
9. R. Papuc, C. C. Gavrilă, M. Lates, Bulletin of the Transilvania University of Brasov 9 (58) 2 Prasic (2016).
10. J. H. Pfalzgraf, *Experimental research on the chain transmission from timing system of combustion engines (in German), PhD thesis* Technical University Munchen (2011).
11. M. R. Naji, K. Marshek, Journal of Mech., Trans. and Autom. in Design **105** (1983).
12. Schaeffler Group Automotive, *SCDS Catalogue* (2009).

3D end milling finite element thermal analysis

Andjelija. Mitrović [1], *Pavel.* Kovač [2,*], *Nenad.* Kulundžić [2], *Borislav.* Savković,[2] *Ildiko.* Mankova[3]

[1] Visoka škola tehničkih strukovnih studija Čačak, Svetog Save 65, 32000 Čačak, Serbia
[2] UNS Faculty of Fechnical Sciences, Trg Dositeja Obradovića 6, 21000 Novi Sad, Serbia
[3] Technical university in Kosice Kosice, Slovakia

Abstract. The paper presents a modern approach to the phenomenon of thermal analysis in end milling by the finite element method. 3D model of the end mill and work-piece was created in the software package SolidWorks. In order to predict the occurrence of thermal phenomena in milling process software package Third Wave AdvatEdge was used. Influence of cutting speed on the temperature in cutting zone was modelled and analyzed.

1 Introduction

Computer application has brought revolutionary changes in the field of different engineering and scientific disciplines, and one of the first was the mechanics of the solid bodies. In the number of computer calculation methods, created in the past four decades, the most widely accepted and developed is the finite element method, which, in its present form, dates from the late 1950's.

The latest period of development is marked by the appearance of the FEM system working with discrete elements, thus enabling the simulation of gas flow with high speeds. For the reason of relative complexity of these kinds of problems, those systems are still the subject of interest for the small number of specialists.

Past years, the finite elements analysis becomes the main method in the simulation of metal cutting. Recently, programming systems are developed based on the finite elements method, which are only intended for the simulation of cutting process [1, 2]. Among scientists, that have examined this problematic couple of them have made quite interesting results [3, 4].

Wang and Liu [5] have estimated maximum temperatures in both tool and workpiece by a thermal method based on Green's function and a microstructure-based method in orthogonal turning. Korkut et al. [6] reported an average temperature of the chip leaving the cutting zone. Jaspers et al. [7], have also investigated the average temperature of the chip leaving the cutting zone but in this case with infrared method. Ueda et al. [8] and Tanaka et al. [9] have performed experimental investigation of the cutting tool edge temperature.

2 Methodology and material

Third Wave AdvantEdge is a commercial program for designing, improving and optimizing the machining process. The program structure consists of three modules: pre-process module, simulation module and post-process module. The pre-process module is the starting point that allows users to set the whole simulation, define the geometry of the tool, the properties of the materials, and the processing parameters. Simulation control data is also entered therein. The simulation module is a module where the simulation is actually performed. When the modeling data is entered and the solver simulation is started (part of the software that solves the mathematical problem) for calculations that are based on the finite element method. Since the data are completed, the post-processing module processes the results and displays them in various forms, such as graphs and images. Among the other results that can be obtained are the following: shape of the chip, temperature on the tool-chip interface, cutting forces, wear, damage and so on [10].

The AdvantEdge software package has a library with a large number of standard materials that can be used to define the material of the tool, tools and coatings. In the Tools menu, a selection of tool material and coatings is selected, and in the Workpieces menu, a selection of the material of the workpiece. The chemical composition and properties of standard materials are listed in the menu of the tool and workpiece material window. When using the coated tool, the thickness and number of coating layers (up to three layers) are defined and there is also the possibility of adding new materials to the library [11].

In the AdvantEdge software package, the simulation can be run in a demonstration or standard mode. Demonstration mode reduces time for simulation processing but provides less reliable results. Standard simulations are more reliable, more accurate and more stable than demonstration simulations.

After selecting the processing type by cutting in AdvantEdge, it is necessary to select the appropriate tool

* Corresponding author: pkovac@uns.ac.rs

© The Authors, published by EDP Sciences. This is an open access article distributed under the terms of the Creative Commons Attribution License 4.0 (http://creativecommons.org/licenses/by/4.0/).

for the appropriate machining technology. In order to machining the stepped surfaces with end milling cuter (in the case of the opposite movement of the tool and the workpiece), the input parameters such as the number of rotations, step per tooth, width and depth of cutting are entered in the Process Parameters window.

For 3D simulations in AdvantEdge, under the Create / Edit Standard Workpiece / Advanced Options / Meshing Parameters option, there is a default processing that contains the corresponding network parameters. The initial network parameters define the preliminary network before the simulation starts, capture the original geometry, and define the starting points for the parameters of the custom re-networking. The Tecplot subroutine allows users to define the output parameters of the simulation to be displayed. To start the program, in order to process the simulation, select the Simulation / Submit / Submit Current Job option.

In this study, simulations of end milling were performed in AdvantEdge with high-speed steel milling tool, and the steel for the improvement of 30CrNiMo8 was selected for the material of the workpiece.

3 Results of the simulation of end milling process and analyze

3D simulations were created for the three corresponding machining regimes (Table 1) with the end milling toolmade of the high-speed steel, based on which the influence of cutting speed on the cutting temperature in the cutting zone was analyzed.

Table 1. Machining regimes during simulations.

Cutting regime	Simulation 1	Simulation 2	Simulation 3
cutting speed v, m/s	0.25	0.29	0.34
Feed per tooth s_z, mm/t	0.045	0.045	0.045
Depth of cut a, mm	1.00	1.00	1.00

The representations of comparative simulations was recorded at the same time intervals (by selecting the same zone being displayed) for 3 different cutting speeds are given in Figures 1, 2 and 3. A minimum temperature value of 50 °C and a maximum value of 250 °C reach(with 11 steps levels selected).

In Figure 1, the zone shortly before the outflow of the tooth from the contact with the workpiece was selected, and due to the chip that has not yet been separated, no good temperature zones near the cutting edge are seen.

In Figure 2, the display of the tool is turned off in software, in order to see better the FEM network of the workpiece in contact with the tool tooth.

Figure 3 shows the temperature of the tool from comparative simulations for three cutting speeds where it can be seen that immediately before the tooth exits from the contact, the higher tooth temperatures in the regime with a lower cutting speed are observed.

Fig. 1. Comparison simulation with different cutting speeds.

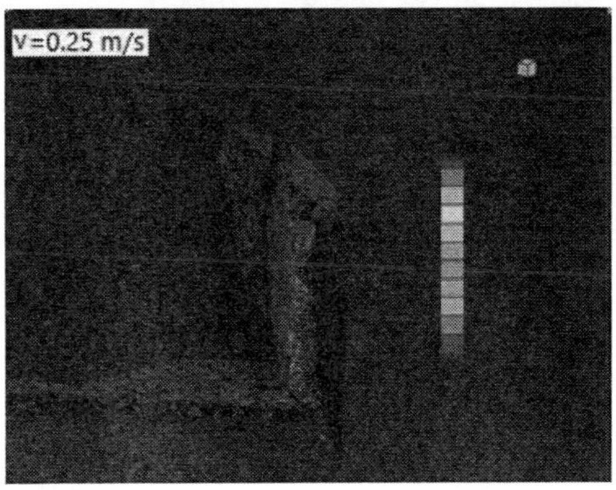

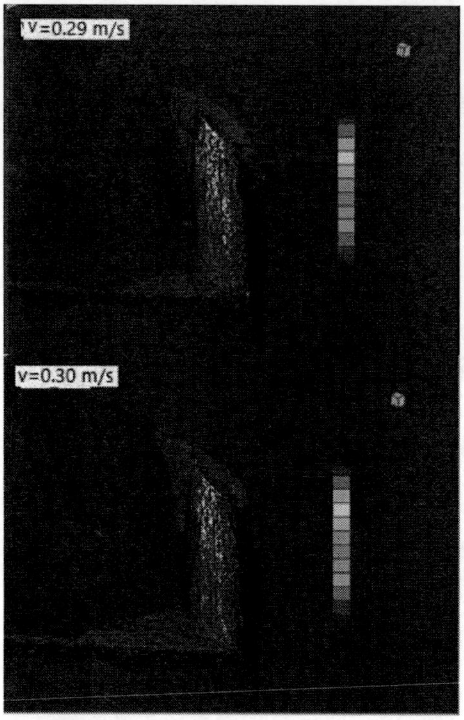

Fig. 2. Comparative simulation of temperature without tool display.

Figure 4 shows the comparative temperature diagrams in the cutting zone (the highest value) versus the machining time. From the diagram it can be noticed that at the beginning of machining the highest temperature is achieved with the highest cutting speed and that at a lower cutting speed, more time should be allowed to achieve the maximum temperature value.

Figure. 5 shows a temperature dependence diagram of an angular processing length. As with the previous diagrams for an angular processing length of 90 °C, it can be seen that the temperature is rapidly decreasing at half the machining length.

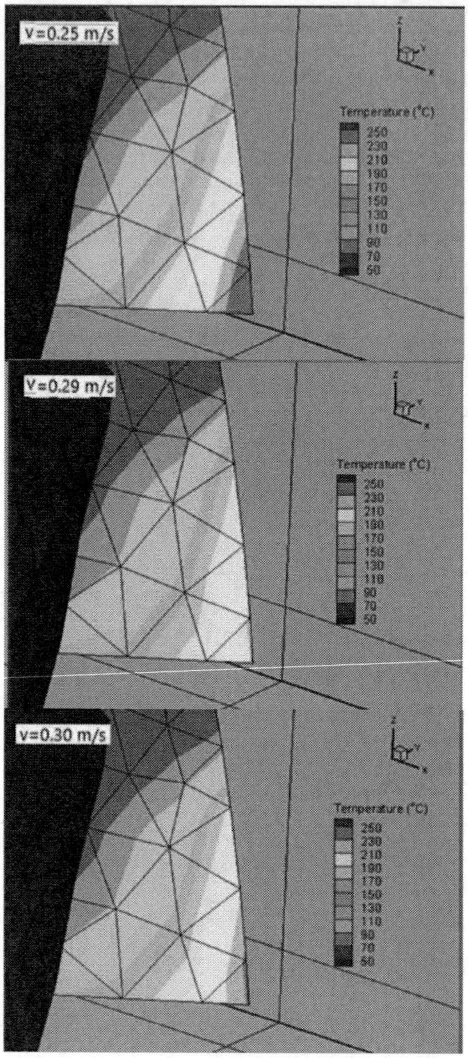

Fig. 3. Temperature on tool display on comparative simulations

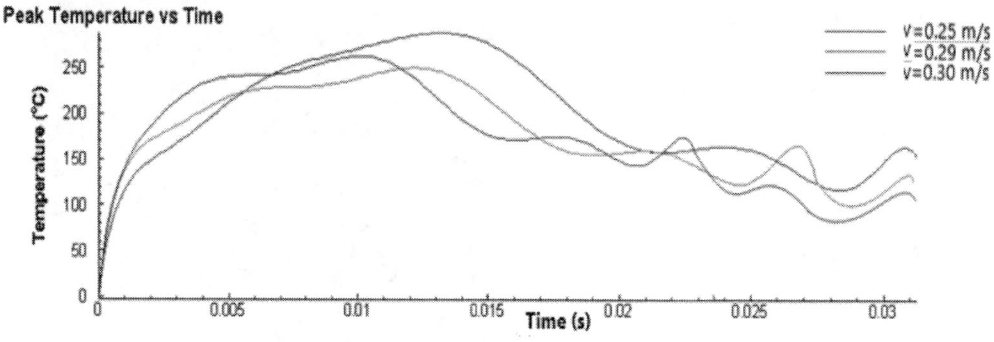

Fig. 4. Diagram of temperature-time simulation for different cutting speeds.

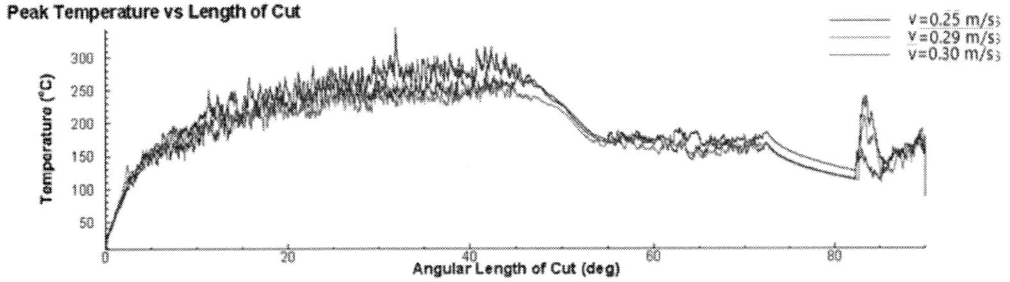

Fig. 5. Temperature dependence diagram versus angular processing length.

3 Conclusion

On the basis of the obtained results, it can be concluded that in the AdvantEdge software package, temperatures in the cutting zone for machining simulations can be analyzed with great reliability. For the appropriate treatment regimes, the influence of cutting speed on the temperature in the cutting zone can be conveniently analyzed. Precise geometry of the tool allows obtaining more reliable results in predicting temperature in the cutting zone, although this significantly influences the increase in the time required to process the simulation.

The results obtained by testing needs to be taken with the material irregularities that are formed during the production of the material, as well as the vibration on all machines.Equations should be centred and should be numbered with the number on the right-hand side.

References

1. P. Kovac, M. Gostimirovic, M. Sekulic, B. Savkovic A review to advanced modeling and simulation of machining process, Journal of Production Engineering, **12**, 1 (2009), 17- 22.
2. P. Niesłony, W. Grzesik, W. Habrat, Experimental and simulation investigations of face milling process of Ti-6Al-4V titanium alloy, Advances in manufacturing science and technology, **39**, 1 (2015)
3. C. Constantin, S. Croitoru, G. Constantin, E. Străjescu, FEM tools for cutting process modelling and simulation, UPB Scientific Bulletin, Mechanical engineering, (2012)
4. K. Kadirgama, M. Noor, M. Rahman, W. Harun, C. Haron, Finite Element Analysis and Statistical Method to Determine Temperature Distribution on Cutting Tool in End-Milling, European journal of scientific research, (2009)
5. J. Y. Wang, C. R. Liu, A new concept for decoupling the cutting forces due to tool flank wear and chip formation in hard turning, Mach SciTechnol **2**, 1 (1998), 77-90.
6. I. Korkut,M. Boy, I. Karacan, U. Seker, Investigation of chip back temperature duringmachining depending on cutting parameters. Mater Des 28, (2007), 2329-2335.
7. S. Jaspers, J. Dautzenberg, D. Taminiau, Temperature measurement in orthogonal metal cutting. Int J Adv Manuf Technol **14**, (1998), 7-12.
8. T. Ueda, M. Huda, K. Yamada, K. Nakayama, Temperature measurement of CBN tool in turning of high hardness steel. CIRP Ann **48**, 1 (1999), 63-66.
9. R. Tanaka, H. Motishita, Y. Lin, A. Hosokawa, T. Ueda, T. Furumoto, Cutting tool edge temperature in finish hard turning of case hardened steel. Key Eng Mater **407/408**, (2009), 268-272.
10. A. Mitrovic, P. Kovač, N. Kulundžić, B. Savković, 3D finite element simulation of milling, Journal of Production Engineering, **19**, 1 (2016), 31-34.
11. Third Wave Sistems AdvantEdge 7.1, User's manual, (2015)

MATEC Web of Conferences **184**, 03002 (2018) https://doi.org/10.1051/matecconf/201818403002
Annual Session of Scientific Papers IMT ORADEA 2018

Study of functional parameters of a conveyor by modeling and simulation

Florin Blaga[1,*], and *Iulian* Stanasel[1]

[1]University of Oradea, Industrial Engineering Department, Universităţii 1, Oradea, Romania

Abstract. The transfer systems conveyor type are an important component of the modern manufacturing systems. Designing and constructing a conveyor involves evaluating its performance for the adoption of the most effective constructive solutions. The paper shows how modelling and simulation can be used in assessing the performance of a conveyor. The studied parameters refer to the kinematic characteristics (the speed of the conveyor) and to the functional indicators of the system components (input rate, service rate).

1 Introduction

The conveyors are transfer systems which are used to carry different products in successive steps up to the point of destination. Also, with the aid of conveyors is ensured buffer volumes of products to balance the flow when it is necessary [1]. The study of the performance of these systems through modelling and simulation is the subject of several researches.

In [2] is presented a method for designing the driveshaft of a monorail conveyor, using CAD-CAE applications.

In paper [3] a model comparison approach based on material flow systems is investigated That is divided into a microscopic and a macroscopic model scale.

The paper [4] studying existing conveyor system and optimize the critical parts like: Roller, shafts, c-channels for chassis and support, to minimize the overall weight of assembly and material saving.

An applicable scheme for production design in manufacturing, and provides a valuable tool to conclusively obtain the optimal profit of a given production quantity for operations research engineers in today's manufacturing with profound insight is presented in [5].

This paper highlights how modeling and simulation have been used to improve the solutions adopted in the implementation of a transfer system (of) conveyor-type.

2 Modeling the conveyor using the Taylor ED program

Figure 1 shows the model of the transport system conveyor-type developed in the Taylor ED program. The conveyor is composed of four sections: 4 and 6 are linear segments, 5 and 7 are circular segments . Also in Figure 1 are shown the other components of the system in which the conveyor is integrated. Thus, the following "atoms" (name specific to the TaylorED program) are highlighted: 1- Product; 2- Bringing installation (source); 3, 8 - Operators; 9 - The Team Block is specific to Taylor ED (it interconnects the operators); 10- Exhaust system.

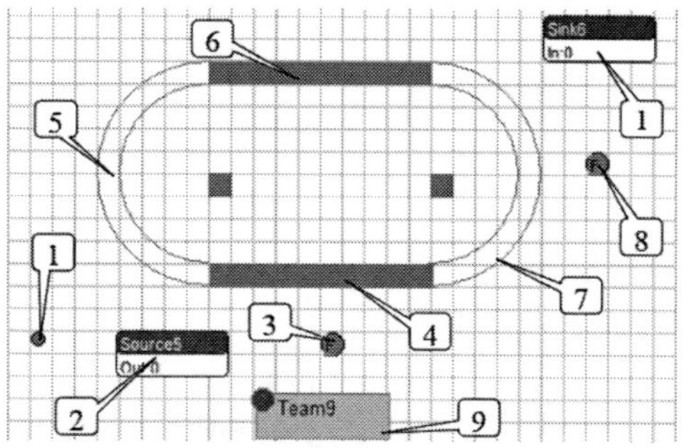

Fig. 1. The model of the conveyor system

In Figure 2 are highlighted the connections between the components of the model. These connections are established between the output ports and the input ports of the elements (atoms) of the model.

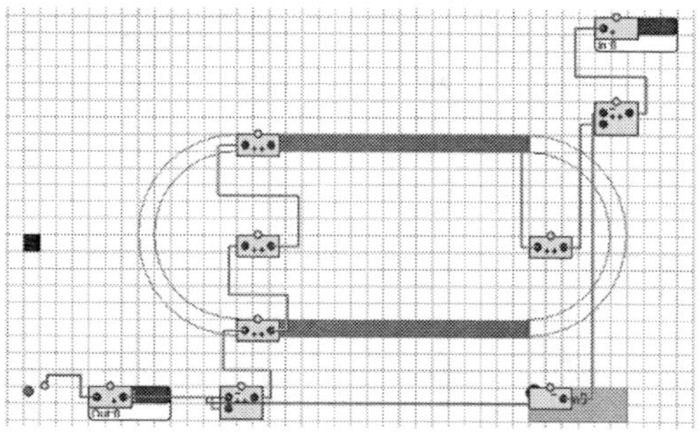

Fig. 2. The model of the system with the connections between atoms.

For each component of the model are made the settings which correspond with to the characteristics of the real system. Thus, in the case of the bringing

* Corresponding author: fblaga@uoradea.ro

© The Authors, published by EDP Sciences. This is an open access article distributed under the terms of the Creative Commons Attribution License 4.0 (http://creativecommons.org/licenses/by/4.0/).

installation (source), is set the time interval between two successive inputs. In the first variant, the interval between two successive entries is considered to be a deterministic parameter with a value of 60 seconds (1 minute) (Fig. 3).

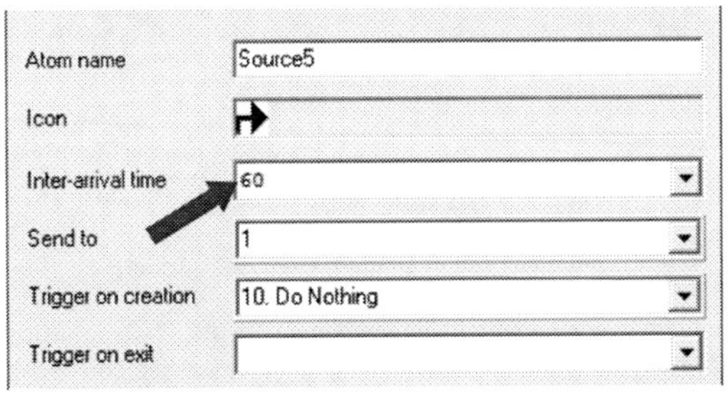

Fig. 4. Setting Inter-arrival time. Variant 1.

The parameters (capacity, length, speed) of the linear sections of the conveyor are shown in Figure 6. In Figure 7 is shown the window setting of specific parameters of the circular sections of the conveyor (speed, radius).

The seeted speed is proportional to the real speed

$$Speed_{model} = 5 * Speed_{real\ system} \quad (1)$$

Speed multiplication was required because in the model also the distances are five times higher compared to the real system.

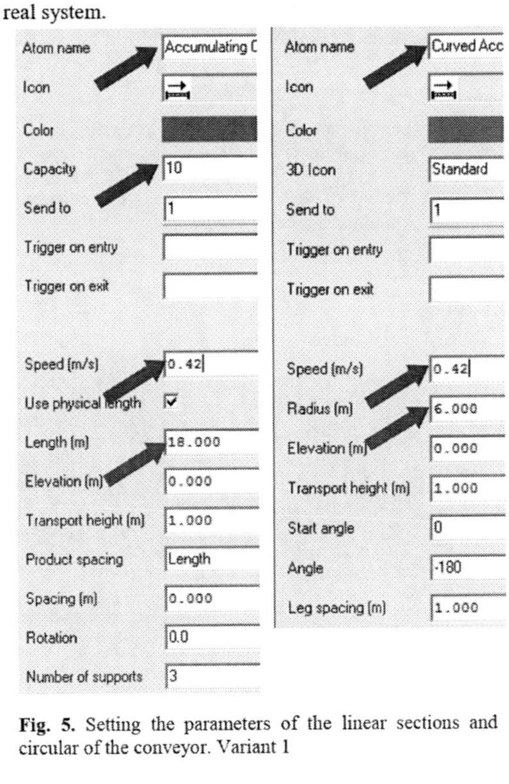

Fig. 5. Setting the parameters of the linear sections and circular of the conveyor. Variant 1

3 Simulation of conveyor system operation

To simulate the operation of the conveyor system, a few adjustments must be done in the main menu - *Simulate*.

The reference interval of time for the simulation *Set Stop Time*- 8 hours (Fig. 6). Will be recorded information relating to all the components of the model History- All on) on the reference interval of time for which the simulation is done (Fig. 6). To start the simulation is enable the option *Reset+ Run until Stoptime* (Fig. 6).

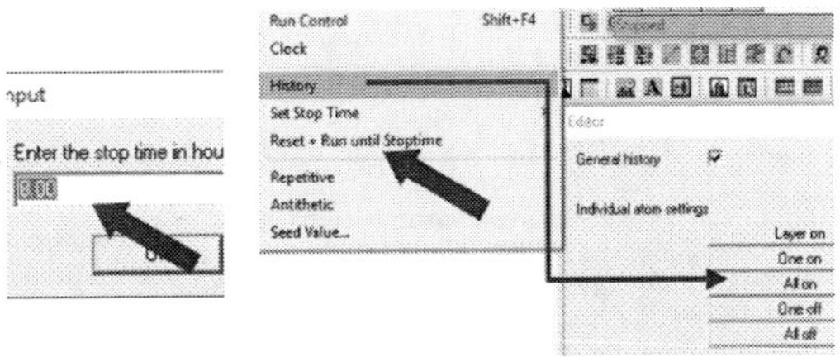

Fig. 6. Setting the reference time and the option to record data

After the simulation, it is generated a summary report which shows the performance of the system components. This report is shown in Figure 7. It may be observed that the system has entered 480 parts and was evacuated 472 parts. The report also shows how many parts there are in each component of the system when the simulation is finished.

For example, on the first linear segment there is a part. Another information refers to the average number of parts, (considering the reference time 8 hours) which can be found on a specific component of the system. Thus, the average number of parts on the first curved section of the conveyor is 3.442.

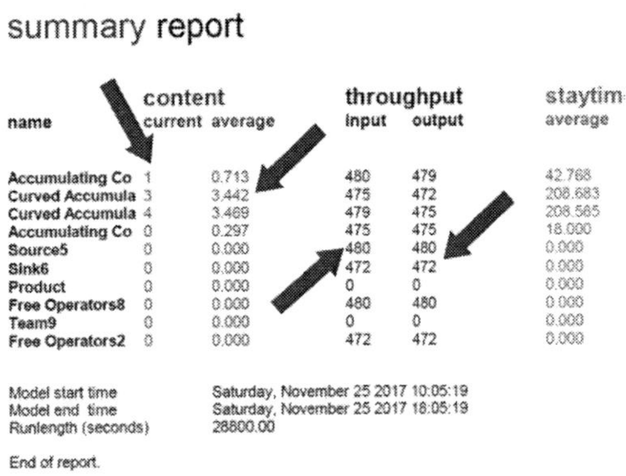

summary report

	content		throughput		staytim
name	current	average	input	output	average
Accumulating Co	1	0.713	480	479	42.768
Curved Accumula	3	3.442	475	472	208.683
Curved Accumula	4	3.469	479	475	208.565
Accumulating Co	0	0.297	475	475	18.000
Source5	0	0.000	480	480	0.000
Sink6	0	0.000	472	472	0.000
Product	0	0.000	0	0	0.000
Free Operators8	0	0.000	480	480	0.000
Team9	0	0.000	0	0	0.000
Free Operators2	0	0.000	472	472	0.000

Model start time Saturday, November 25 2017 10:05:19
Model end time Saturday, November 25 2017 18:05:19
Runlength (seconds) 28800.00

End of report.

Fig. 7. Summary report of the simulation. Varinat 1

In addition to the synthetic report, the Taylor ED program provides detailed information about each component of the system. Thus, in Figure 8 is shown the filling degree of the source (Status overview). It was busy in the proportion of 100%.

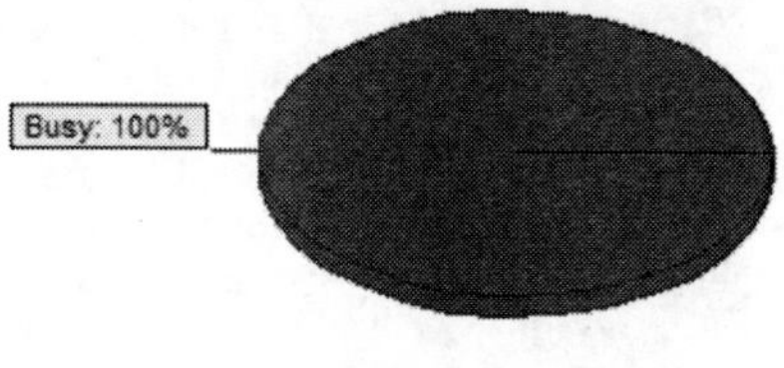

Fig. 8. The degree of loading of the source. Varinat 1

Figure 9 shows the time progress of the number of parts in the waiting string of the first curved segment. So in this string are found, alternatively, 3 or 4 pieces.

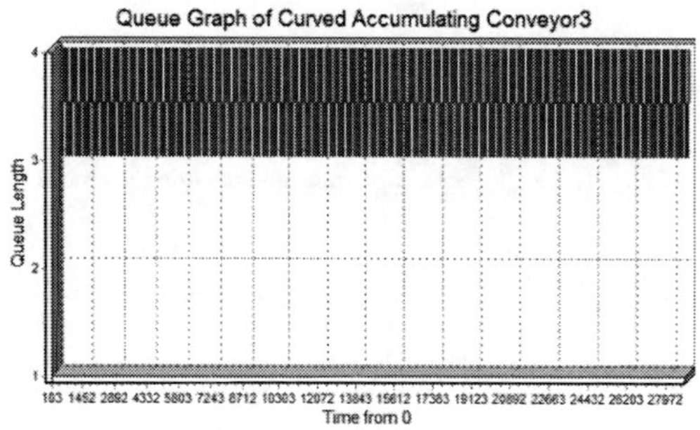

Fig. 9. The evolution in time of the number of parts in the waiting string of the first curb segment

The built model allows simulation taking into account other functional parameters. Thus, if it is considered the time interval between two entries in the system, for a random variable it can be assigned a probability function of the negative exponential type. (Fig.10). The average value for the time interval between two successive inputs is 30 seconds.

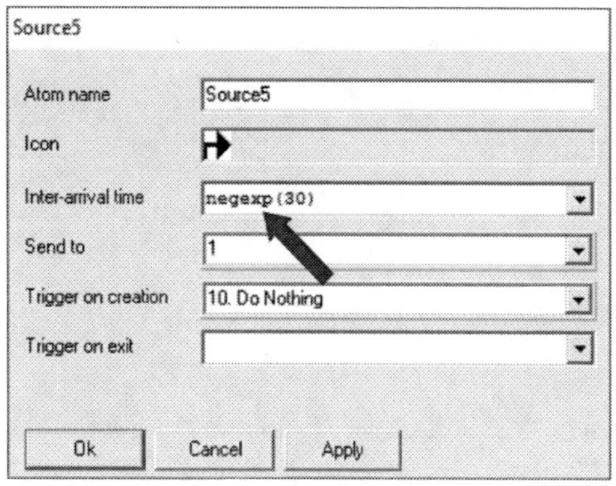

Fig. 10. The Setting of random entries

In figure 11 are presented the results of the simulation shown in the summary report. Thus, have entered in the system 984 parts and 982 come out.

The average time that a part spends on the different sections of the conveyor varies between 31,032 seconds (linear segment) and 32,499 seconds (curve section) (Fig.11)

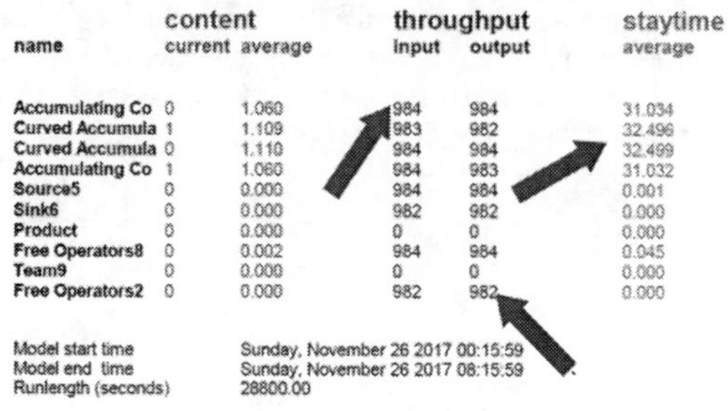

	content		throughput		staytime
name	current	average	input	output	average
Accumulating Co	0	1.060	984	984	31.034
Curved Accumula	1	1.109	983	982	32.496
Curved Accumula	0	1.110	984	984	32.499
Accumulating Co	1	1.060	984	983	31.032
Source5	0	0.000	984	984	0.001
Sink6	0	0.000	982	982	0.000
Product	0	0.000	0	0	0.000
Free Operators8	0	0.002	984	984	0.045
Team9	0	0.000	0	0	0.000
Free Operators2	0	0.000	982	982	0.000

Model start time Sunday, November 26 2017 00:15:59
Model end time Sunday, November 26 2017 08:15:59
Runlength (seconds) 28800.00

Fig. 11. Summary report of the simulation. Variant 2

Figure 12 shows the histogram of the waiting string corresponding to the first linear segment of the conveyor. It is highlighted that the waiting string contains 0, 1, 2, 3, 4, 5 pieces at different intervals of time.

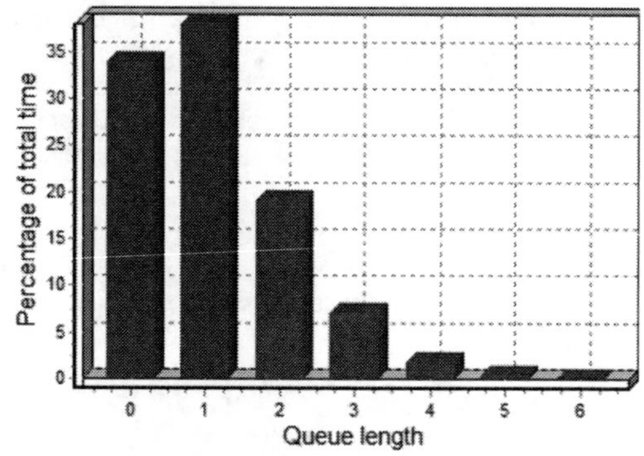

Fig. 12. The histogram of the waiting string corresponding to the first linear segment of the conveyor.

The variation of the number of parts on the second circular segment is shown in Figure 13.

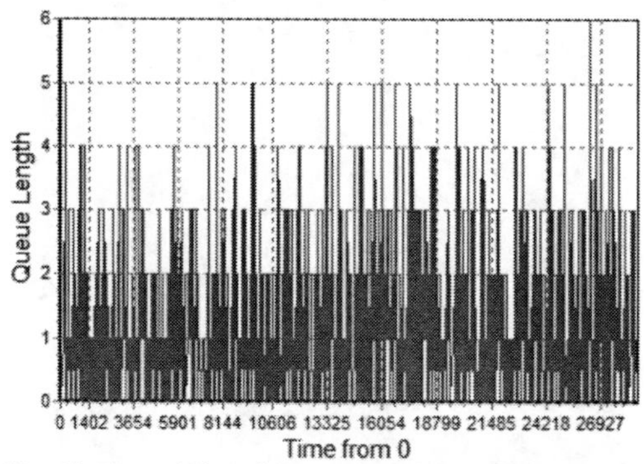

Fig. 13. The variation of the number of parts on the second circular section.

4 The integration of the transfer system conveyor type into a complex manufacturing system

The transfer system of conveyor type can be integrated into a complex manufacturing system (Figure 14) with two machines tools (servers) (positions 11 and 14) and two linear conveyors (12 and 13). In Figure 20 are presented the components of manufacturing system interconnected between them.

In the case of this model time of processing have random sizes. The function of the probability is negative exponential type [6] expressed by:

$$f(t)=\beta e^{-\beta t} \tag{2}$$

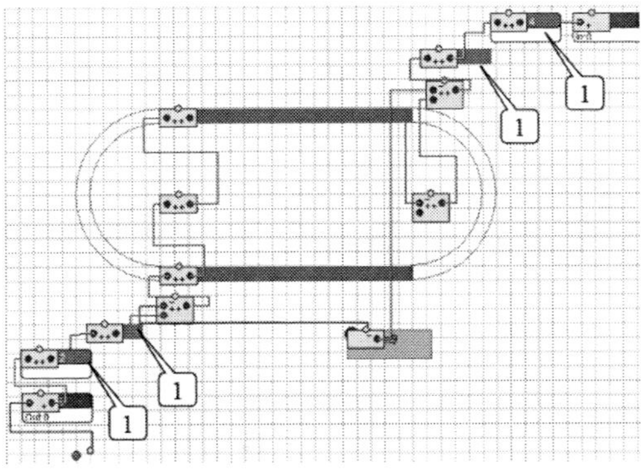

Fig. 14. The transfer system conveyor type integrated into a complex manufacturing system.

In figure 15 is presented the synthetic report. It highlights the fact that 685 parts have been entered into the system and 677 were evacuated.

	content		throughput		staytime
name	current	average	input	output	average
Accumulating Co	1	0.737	684	683	31.033
Curved Accumula	2	0.958	681	679	40.528
Curved Accumula	1	0.770	683	682	32.453
Accumulating Co	1	0.734	682	681	31.012
Source5	0	0.307	685	685	12.893
Sink6	0	0.000	677	677	0.000
Product	0	0.000	0	0	0.000
Free Operators8	0	0.000	684	684	0.018
Team9	0	0.000	0	0	0.000
Free Operators2	0	0.152	679	679	6.466
Accumulating Co	0	0.095	684	684	4.000
Server12	1	0.617	685	684	25.924
Accumulating Co	1	2.068	679	678	87.896
Server14	1	0.809	678	677	34.383

Model start time Sunday, November 26 2017 18:37:19
Model end time Monday, November 27 2017 02:37:19
Runlength (seconds) 28800.00

End of report.

Fig. 15. Summary report of the simulation. Variant 3.

The Taylor ED program provides information on the performance parameters of the system components. Thus, Figure 16 shows the occupancy degree of the machine 1. It is found that the machine processes 62% of the reference time interval (8 hours).

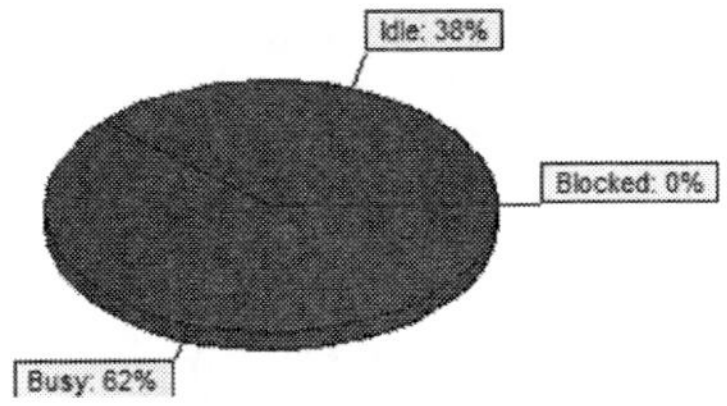

Fig. 16. The degree of loading of the machine 1 (position 11).

Figure 17 highlights the histogram of the first linear section of the conveyor. Thus, 37% of the time on this section there is only one part into the string, 15% of the time they are two machines and there are three parts in 2.5% the reference time.

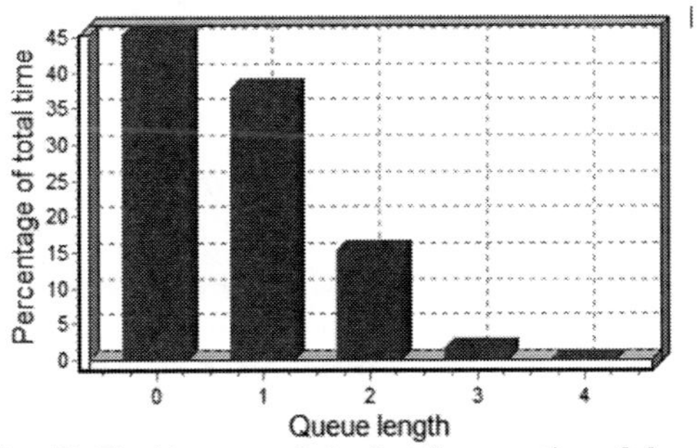

Fig. 17. The histogram of the first linear section of the conveyor.

5 Conclusions

Modeling and simulation are appropriate tools for evaluating the performance indicators of manufacturing systems and their components. In the case of the conveyor system, the Taylor ED program allows obtaining detailed information about the components of the manufacturing system in which the conveyor is integrated. Through simulation, the model allows the evaluation of the operation of the conveyor with different sets of data, so that the decision making factor has information to optimize the system form the constructive design and functional point of view.

References

1. M. Hapenciuc, *Transport equipment in the food industry* (in romanian) (Ed. Fundaţiei Universitare "DUNĂREA DE JOS" Galaţi, 2004)
2. I. Stănăşel, F. Blaga, D. Crăciun, C. Bungău, T. Buidoş, *AN. UO. FMTE*, **3**, 27 (2016)
3. S. Göttlich, S. P. Schindler, V. Schleper, A. Verl, *Apl. Math. Mod.* **38**, 3295 (2014)
4. S. M. Shinde, R.B. Patil, *Int. J. Emer. Techno.*, **3** (1): 168, (2012)
5. Chun-Hsiung Lan, *Int. J. of Adv. Man. Techno.* **22** (7), 510, (2003)
6. I. Abrudan, *Flexible manufacturing systems. Concepts of design and management* (in romanian) (Ed. Dacia, Cluj- Napoca, 1996)

The work is a part of project MANUNET ERA-NET, Contract nr.19/2015 "Monorail conveyor system for automated production lines".

A brief overview of the evolution and trends in maintenance

Eugen Pămînţaş[1*], and *Felicia veronica* Banciu[1]

[1]Politehnica University of Timişoara, Materials and manufacturing d Department, 300222, Timişoara, Timiş, Romania

Abstract. From the mistakes of industrial past, from the needs of today's manufacturing industry, from the knowledge accumulations of theoreticians, researchers and practitioners in the field presented in scientifical papers and reviews, the future directions of R & D in the field of maintenance were outlined in this paper.

1 Introduction

For the production of goods the industry has widely used different processes that have known various classifications. For example, the simplest classification of manufacturing processes is made by Swift and Booker [1]: casting, cutting, forming and manufacturing, while Kalpakjian and Schmid [2] provide a more comprehensive classification in six subcategories of casting, machining and finishing, joining, sheet metal, polymer processing and deformation processes. It is noted that both traditional classifications have certain limitations on the identification of newly developed technologies (e.g., rapid prototyping, die casting and injection molding) [3]. Regardless of the category they belong to, machine tools, equipments or machinery have been stepping up with technological advances but also with their needs and especially the scientific knowledge they have embedded. Regardless of their degree of complexity, throughout their technological evolution, all machines and installations, regardless of the type of drive and the working environment, needed repair work due to natural damage or the effect of work wear.

Technological developments and the current economic context have created major changes in maintenance systems. These developments need to be analyzed to determine their contributions and their negative effects. For this it is necessary to know the new achievements in the field, evaluation techniques and methods of controlling their effects. An ample bibliographic synthesis limited to a specific topic within a particular manufacturing area provided by [4, 5] has shown that:

- As new concepts are introduced into the manufacturing industry, the research community is trying to identify future frontiers of research;
- As new manufacturing paradigms are revealed, the component subsystems of the manufacturing systems should be carefully re-examined to explore how new technologies influence the current infrastructure;
- To date, it is noted that there are relatively few advanced manufacturing concepts syntheses published;
- In the literature reviewed, manufacturing systems are classified differently according to their evolution and therefore, it is a real challenge to try to establish a type or a subsequent category of a manufacturing system. It is therefore necessary to develop an appropriate taxonomy to investigate the various aspects of its production and maintenance systems. Most researchers in the field are concerned with studying new advanced manufacturing trends that are already developing and less with systematizing the knowledge gained in the field.
- few studies have addressed the negative effects that ICT may induce in a manufacturing company through changes in labor patterns and due to their poor integration and utilization.

This paper presents a view of maintenance, analyzing a wide range of publications covering many sub-areas and topics. Some of the arguments that lead the industry to take new initiatives in enterprises, processes, production systems and equipment levels are also underlined.

The purpose of this paper is to report the state of the art of maintenance related to structural changes of machine tools by presenting a clinical review of the literature currently available on the subject.

The paper aims to bring attention to the concepts of self-learning, the summative analysis of becoming so far, the emphasis on previous features and the pre-configuration of the directions of the future as well as examples of new approaches to the self-maintenance concept along with future research challenges..

2 From simple repair to maintenance

It is widely known that industrial maintenance is a knowledge intensive field based on different disciplines covering a wide range of technical sciences involved in the various technologies included in modern industrial equipment. Therefore, until recently, the complexity of the equipment and the need to have a knowledge bag corresponding to a very large area of science and technology was the main limitation of substantial improvements to the discipline of maintenance. In fact, innovative advances in support technologies are needed to enable maintenance engineering to fully develop its potential. In this respect, new ICT-based and microelectronic technologies - such as intelligent data

* Corresponding author: eugen.pamintas@upt.ro

© The Authors, published by EDP Sciences. This is an open access article distributed under the terms of the Creative Commons Attribution License 4.0 (http://creativecommons.org/licenses/by/4.0/).

capture, advanced viewing, wireless, IoT (internet of things) and intelligent sensors - provide maintenance with the environmental intelligence needed to renew the way forward maintenance [6].

The evolution of the field to which the task of preserving machinery and installations has come to be seen, in a general picture, as having three well-defined areas (see Fig. 1): management and management of activities (underground), workers maintenance and repair work (First Level) and Maintenance Intervention Objects (Workshop). The results of the company from the operation of machinery and installations are linked to the hopes of those who have their eyes and ears at every moment on how they "fight": the productive but also the related and support sector, the administrative management compartments but more chosen patronage.

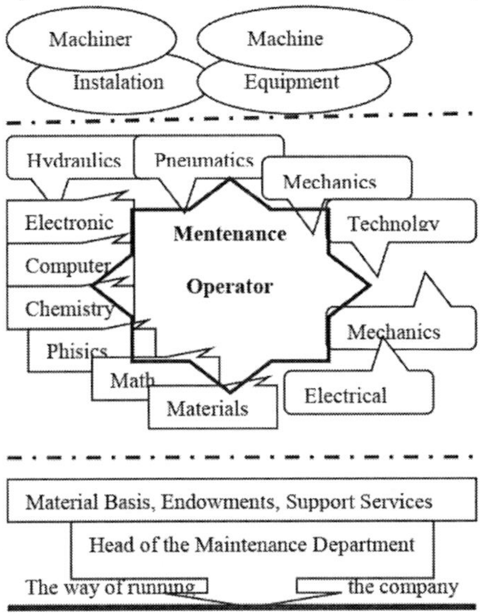

Fig. 1. General picture of maintenance field.

Regarding the first two aspects of the present study field evolution, the literature highlights several approaches that chronologically systematize the stages or ample and detailed the content. Under these circumstances, just to avoid hiatus in the logical development of the subject, we will continue to make a brief and narrative review of them.

The maintenance operator, the one who had to repair and re-install the fallen equipment, has accumulated constant knowledge, has acquired skills and has perfected his technique continuously. This happened until he has not dealt with the constructive and functional complexity of the subject of its activity. The need to carry out the tasks led to the transition from a worker to a team of workers. First, the team was made up of at least two operators of the same profession (say mechanic or electro-mechanical locks as examples) and then the team members got different qualifications. The path to the distinct specialization of operators in the work team has passed through the transient phase of cumulative qualifications, e.g. mechanic and welding or electromechanical and welding or electrician and electronics etc. Then he continued with qualifications centred on the types of actuation and control (mechanical, electrical, hydro-pneumatic, electronics) and more rarely on machine type and their subassemblies (gearboxes, transport systems and tracks, frames and elements of the resistance structure, spindles etc.).

At the same time, there has also been a change in the role of the teams: intervention, maintenance and supervision, repairs, etc. and the coordination of the work of the teams has gone from the task of maintenance and repair to the maintenance department which, with the change of name, has also received new tasks: planning, management, monitoring, control, prediction, design and development strategies modernization, etc. Synthetically and schematically, the evolution of the maintenance domain in the manufacturing industry can be replayed by the schemes of Fig. 2

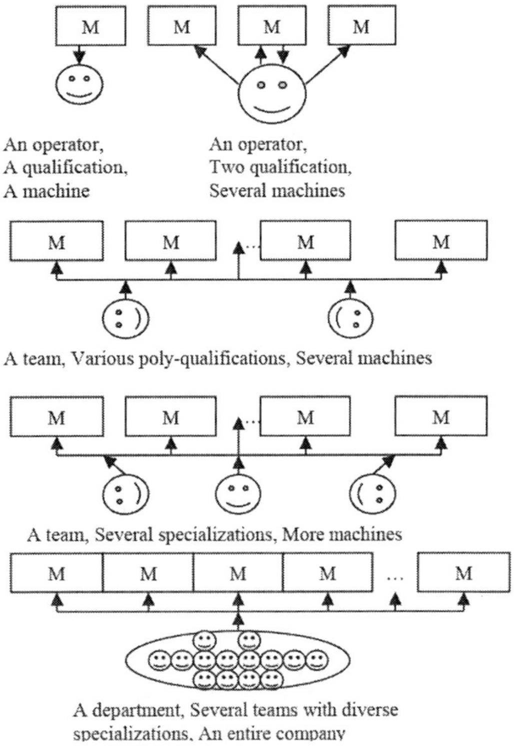

Fig. 2. Maintenance tasks in the manufacturing industry

The third area, the object of maintenance activity, will be given a wider attention in several respects, in order to argue the choice of the subject of the present paper and possible solutions for this.

3 On the evolution of manufacturing machinery and equipment

As machines and installations have increased their performance, the need to increase production flexibility, reduce energy consumption, increase operational safety, reduce to complete elimination of downtime and unplanned stops has increased the attention paid for their

constructive-structural conception. Here are some remarkable results:
- to counteract the wear effect of the moving parts: antifriction alloys, oils and greases with superior and long-lasting properties, screws, slides and guides with rolling elements, etc.;
- for shortening the kinematics drive chains: replacement of gearboxes with electric speed changers;
- for monitoring the technical condition of the machine: built-in sensors for measuring of temperature, vibration amplitude and frequency, working fluid pressure, etc.

If we take into account only how the machine designers have responded over time to the requirements imposed by the need for maintenance, the main solutions consisted of the constructive choices presented in Fig. 3.

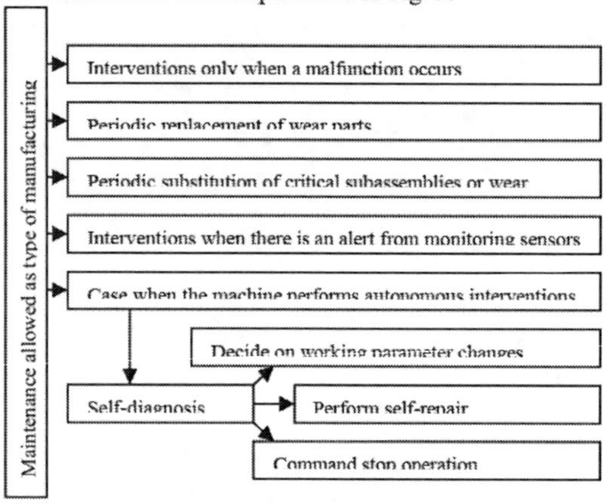

Fig. 3. Type of maintenance interventions

From the mistakes of the industrial past, from the needs of today's manufacturing industry, from the accumulations of theoreticians, researchers and practitioners in the field, the future directions of R & D in the field of maintenance were outlined:
- The key issue that needs to be overcome in order to achieve the full potential of unmanned machining centers (but, we say, also for self-diagnosis and self-maintenance) is the development of reliable and effective sensors for monitoring machine operation, ensuring efficient metal removal rates and taking corrective action in the event of accidents or breakages ;
- Machines should include dimensional and proximity sensors to dimension and geometric corrections, cutting force sensor, spindle force and power load sensors, torque sensor and speed sensor to monitoring tool wear and tool breakage, acoustic emission sensors to monitoring entire manufacturing process [7];
- Use of intelligent maintenance systems to monitor, diagnose and predict the health of machines and their integrated components. Interfaces that mix the virtual reality can provide valuable support to maintenance operators to isolate and repair components of the equipment that might fail. The decision will be based on the analysis of the degradation behaviours of these components. The advantages of such a way of working consist in reducing maintenance time and costs caused by unpredictable failures of critical industrial equipment, where failures can seriously compromise the production process. For example, an interface that combines real and virtual using mixed/augmented reality techniques [8] has already been tested to facilitate replacement of spare parts [9, 10], as well as for technical inspections or repair [11, 12]. Using reality mixed with virtual in maintenance is not new [13÷15], although previous research works do not deal with how to manage maintenance information or what data it needs to be presented;
- Diagnostic interface design have major implications on: complete redesign of the troubleshooting approach so as to allow for a smooth transition from failure mode to the failure cause, rethinking the reasoning behind which fault/damage messages are displayed, the maintenance operator will be fully immersed in the troubleshooting process, being in fact an active part of the local or remote computer system managing the diagnostic process of the manufacturing systems;
- Embedding in the construction of device fabrication systems such as sensors, embedded systems and computing devices have given a level of intelligence to system maintenance. This has led to a change from "fail and fix" to "anticipate and prevent" approach. This has irreversibly transformed future maintenance strategies [13] with regard to the complexity and uncertainty of the equipment, the trends being linked to new maintenance management methods such as resilient systems and self-maintenance systems that are able to cope more good debris – base for developing these solutions as a CBM (Control Bazed Maintenance) paradigm [14]. Such systems, based on the CBM, have an information system [15] that acquires, stores and processes the information provided by a multitude of different sensors to monitor the state of operation of the production machines [16,17]
- - several standards (e.g. ISO-13374 for Condition Monitory and Diagnostics of Machines [18], which provides a specification for information flow and data processing for monitoring and diagnosis of equipment health) were developed to assist engineers in the development of platforms and architectures to implement CBM on the factory floor (see a complete review of the state of the art, concepts, standards and platforms for maintenance in [19].

4 Conclusions

An overview of the evolution of tandem means of production - efficient maintenance techniques is presented in this study. Conventional and relevant modern methods used in the largest industrial sectors of the aeronautical industry, automotive industry, etc., have begun to provide information on the effectiveness of implementing artificial intelligence and electronic communication techniques in predictive maintenance.

Today's maintenance function is focused on e-maintenance solutions, considering artificial intelligence techniques for decision-making, appropriate work patterns, and accentuating collaborative phenomena that are becoming more and more prevalent in manufacturing firms through systems distributed. The emphasis in the paper, whether positive or poor, is based on practical experience and provides recommendations for the further development of research in areas such as machine tools, sensors, monitoring and diagnostics, decision-making techniques with embedded AI, e-maintenance strategies with self-diagnosis and self-maintenance. Regardless of what is missing at the moment or is still inappropriately done, the potential of these techniques could be implemented through integration between data from multiple processing subsystems is huge.

From the point of view of practical implications:

- Self-diagnosis and self-maintenance are not a panacea. However, it is a concept of significant potential value, providing ways for businesses to exploit production activities with minimum wastage of materials, energy, human strength and time. Unfortunately, so far there have been few practical achievements that can be used to help students and practitioners alike.

From the point of view of originality

- This paper provides a useful insight into machine tools viewed from the point of view of its maintenance and an approach on which to base more in-depth research on the broader subject of the competitive strategy based on the compatibility of manufacturing systems with IT and AI techniques offered by results in related research fields.

Based on the deployment of the faulty condition and related equipment information from the maintenance system, the visualization interface will be able to suggest/plan the repair/recovery solution. In this way, a computer-aided diagnostic system is supported by the interface which should be more intuitive interface and non-invasive for maintenance operator which must have mobility and free hands to perform the needed maintenance operations.

References

1. K.G. Swift, J.D. Booker, *Manufacturing process selection handbook: from design to manufacture*. Waltham, MA, USA: Butterworth-Heinemann, p. 456, 2013.
2. S. Kalpakjian, S.R., Schmid, *Manufacturing engineering and technology*. NewYork, Toronto, Prentice Hall, 2010.
3. Z. Zhu, V.G., Dhokia, A. Nassehi, S.T. Newman, *A review of hybrid manufacturing processes-state of the art and future perspectives*. Int J Comput Integr Manuf, **26** (Feb (7)), pp. 596÷615, 2013
4. D. Espındola, L. Fumagalli, M. Garetti, C.E. Pereira, S.S.C. Botelho, R.V. Henriques, *A model-based approach for data integration to improve maintenance management by mixed reality*, Computers in Industry **64**, pp. 376÷391 (2013)
5. B. Esmaeiliana, S. Behdadb, B. Wang, *The evolution and future of manufacturing: A review*, JMS **39**, pp. 79÷100 (2016)
6. L. Fumagalli, D. Elefante, M. Macchi, B. Iung, Evaluating the role of maintenancematurity in adoption of new ICT in the process industry, in: In IMS'08: 9th IFAC Workshop on IMS, Szczecin, Poland, October 9–10, 2008, (2008)
7. J.Tlusty, G.C.Andrews, *A Critical Review of Sensors for Unmanned Machining*, CIRP Annals, Volume **32**, Issue 2, pp. 563÷572, (1983)
8. P. Milgram, F. Kishino, *A taxonomy of mixed reality visual displays*, in: IEICE Transactions on I.S., vol. **E77-D**, No. 12, December, (1994)
9. H. Regenbrecht, G. Baratoff, W. Wilke, *Augmented Reality Projects in the Automotive and Aerospace Industries*, Published by the IEEE Computer Society, IEEE, November, (0272-1716 / 05, (2005).
10. S. Henderson, S. Feiner, *Evaluating the benefits of augmented reality for task localization in maintenance of an armored personnel carrier turret*, in: IEEE International Symposium on Mixed and Augmented Reality 2009, ST Proceedings, (2009)
11. S. Henderson, S. Feiner, *Opportunistic tangible user interfaces for augmented reality*, IEEE Transactions on Visualization and Computer Graph. **16** (1) (2010)
12. J. Song, Q. Jia, H. Sun, X. Gao, Study on the perception mechanism and method of virtual and real objects in augmented reality assembly environment, in: The 5th IEEE Conference on Industrial Electronics and Applications, (2009)
13. J. Lee, *Design of self-maintenance and engineering immune systems for smarter machines and manufacturing systems,* Invited Keynote Paper for IMS 2010, Lisbon, Portugal, July (2010)
14. M. Lebold, M. Thurston, *Open standards for condition-based maintenance and prognostic system*, in: Proceedings of MARCON 2001 – 5th Annual MRC, Gatlinburg, USA, (2001)
15. L. Fumagalli, M. Macchi, M. Rapaccini, *Computerized maintenance management systems in SMEs: a survey in Italy and some remarks for the implementation of Condition Based Maintenance,* in: IFAC Proceedings Volumes (IFAC-Papers Online), vol. **13**, Issue Part 1, 13th IFAC Symposium INCOM'09, Moscow June 3–5, pp. 1615–1619, (2009)
16. J. Lee, J. Ni, *Infotronics-based intelligent maintenance system and its impacts to closed-loop product life cycle systems*, in: Invited Keynote Paper for IMS'2004, Arles, France, (2004)
17. J. Lee, L. Liao, E. Lapira, J. Ni, L. Li, *Informatics platform for designing and deploying e-manufacturing systems*, in: Collaborative Design and Planning for Digital Manufacturing,, pp.1–35, (2009)
18. ISO, 13374 – International Organization for Standardization ISO 13374-1*: condition monitoring and diagnostics of machines: data processing, communication and presentation, Part 1: general guidelines*, Geneva, (2003)
19. A. Muller, A.M. Crespo, B. Iung, *On the concept of e-maintenance: review and current research, in: Engineering and System Safety*, Elsevier, http://dx.doi.org/10.1016/j.ress.2007.08.006, pp. 1165÷1187 (2007)

MATEC Web of Conferences **184**, 03004 (2018)
Annual Session of Scientific Papers IMT ORADEA 2018
https://doi.org/10.1051/matecconf/201818403004

Reverse engineering and CAD-CAM approach for manufacturing of spare parts. Case study

Iulian Stănăşel[1,*], *Florin* Blaga[1], *Traian* Buidoş[1] and *Dan* Crăciun[2]

[1]University of Oradea, Faculty of Management and Technological Engineering, Industrial Engineering Department, Romania
[2]University of Oradea, Faculty of Management and Technological Engineering, Mechatronics Department, Romania

Abstract. Reverse engineering represents a method to retrieve information from an existing product, information which are then used for designing a new product or to reproduce it in order to make spare parts. In this paper is presented a case study on making a spare part of complex shape from a worn part. There are presented the stages of acquisition of data by scanning, the use of reverse engineering for the realization of the 3D model, the virtual manufacturing and then the machining of the workpiece CNC machine tools.

1 Introduction

In the present time due to factors such as increased competition, the market pressure, the development cycle of new products tends to become shorter. The process of achieving a product consists of several stages Fig. 1.

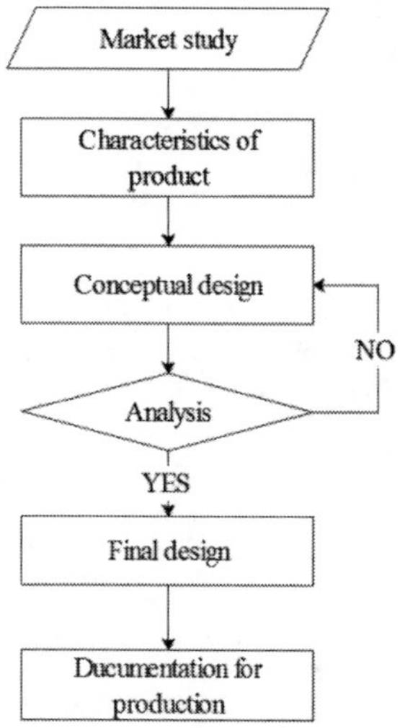

Fig. 1. The main steps of reverse engineering process.

The amount of information in the field of engineering knows an unprecedented development which is why software applications have become increasingly more complex and faster, allowing the integration of activities of design, analysis, and manufacturing of the virtual prototype.

New software technologies make possible to design and study the behavior of a new product in the virtual environment with high accuracy, which has a high impact on reducing the development cycle.

In accordance by [1], Reverse engineering is a process by which digital data are obtained in the form of a cloud of points belonging to the surfaces of a physical object, which are processed and transformed into a 3D model of the scanned object.

Digital data acquisition can be done by measuring or 3D scanning. Although numerically controlled machines are more accurate than the 3D scanners, they offer the advantage of having higher productivity under right precision conditions [2].

The large amounts of data needed to be processed make it necessary to use high-performance computers equipped with dedicated applications with the aim to increase the working speed and to achieve a high quality of the 3D model.

Reverse engineering is used for various purposes [3]:

- for the reproduction of a piece (as a spare part) in the situation where there is no longer any documentation of execution, when the producer no longer exists, or the workpiece is no longer produced, but there are requests from the customers.
- analyzing some features and exploring possibilities to enhance the performance of an existing product.
- to create a new product with improved performances, by redesigning an existing product
- to achieve the 3D model of a product based on a physical model made by artists in clay, wood or plaster.
- to create3D models of objects from nature or to reproduce on a specific scale in order to produce them.
- for protheses design and manufacturing in the medical field
- in archeology to rebuild missing parts of certain objects by using data from existing parts

* Corresponding author: stanasel@uoradea.ro

© The Authors, published by EDP Sciences. This is an open access article distributed under the terms of the Creative Commons Attribution License 4.0 (http://creativecommons.org/licenses/by/4.0/).

2 Obtaining the 3D model through reverse engineering

Obtaining the 3D model of the object which must be carried out, involves the following steps (Fig. 2.)

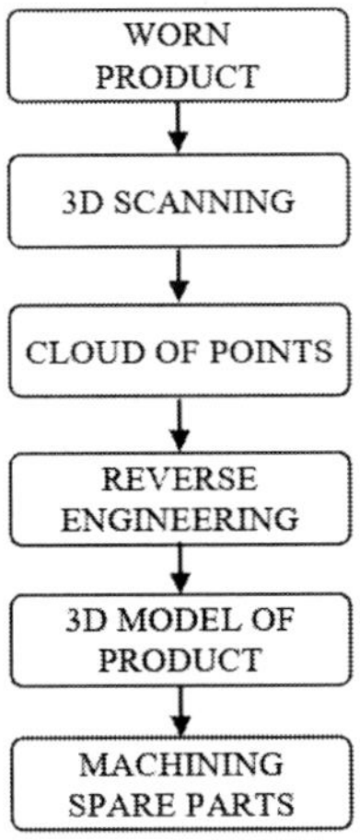

Fig. 2. The main steps of reverse engineering process.

For scanning the physical object it was used a numerically controlled rotary table that can be programmed for different angles, and a 3D scanner Optim II supplied with blue light and two video cameras.

Fig. 3. The 3D scanner

The physical object that had to be reproduced is shown in Fig. 4. From its analysis, it can be seen that it is delimited by several curved surfaces. For this reason the determination of dimensions by measuring with traditional instruments is practically impossible.

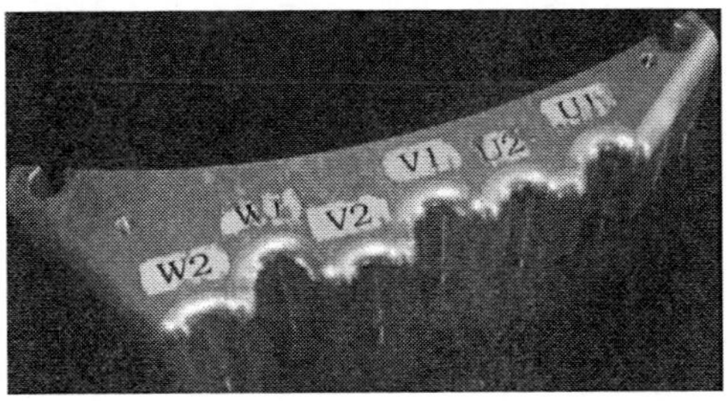

Fig. 4. The worn part.

On the other reason, the accuracy required to the achieved workpiece is within ± 0.1, which is ensured by the used scanner.

They have performed ten scans at equidistant angles until a complete rotation of the workpiece. The data resulting from scanning are shown in Fig. 5. These were saved, and for their processing, it was used reverse engineering application Shiningform XOR.

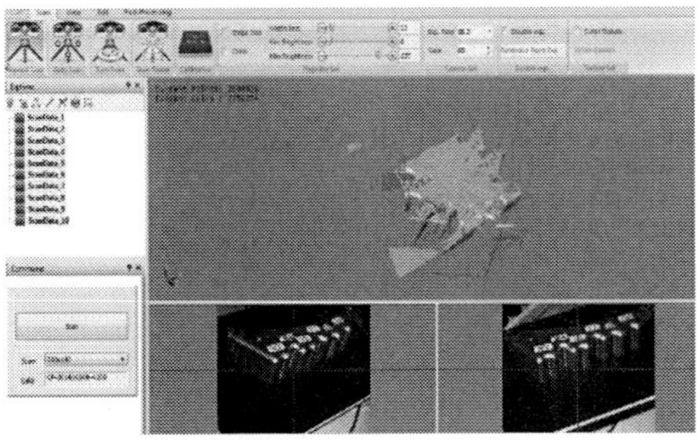

Fig. 5. The unaligned scanned data

For the construction of the workpiece surfaces, the points resulting from the scan are subjected to several operations performed in several stages. In the first phase, each set of data corresponding to the successive scans was cleaned by the parasitic points which come from the objects located in the visual field of the two cameras of the 3D scanner.

Then was done the alignment of the data for each scan, the operation of which depends on the accuracy with which the scanned object was reproduced. The application allows you to determine the accuracy of the cloud of points alignment by displaying a coloured map, in which green colour corresponds to surface that is in interval ± 0.1 mm which is prescribed by the beneficiary of the parts (Fig. 6).

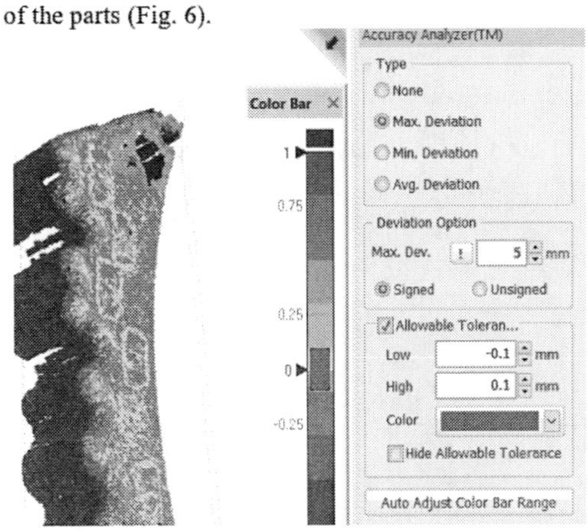

Fig. 6. Accuracy analyser.

The cloud of points of the scanned object is shown in Fig. 7. Because the beneficiary did not allow the application of a layer of matte paint on the surface of the piece, the mesh shows some discontinuities due to the metallic gloss. However, the surface shows enough points which allow determining the correct shape of the object.

The cloud of points is exported to the Siemens PLM NX application, which allows using the scanned data to create the sketch needed to extrude the 3D model of the object.

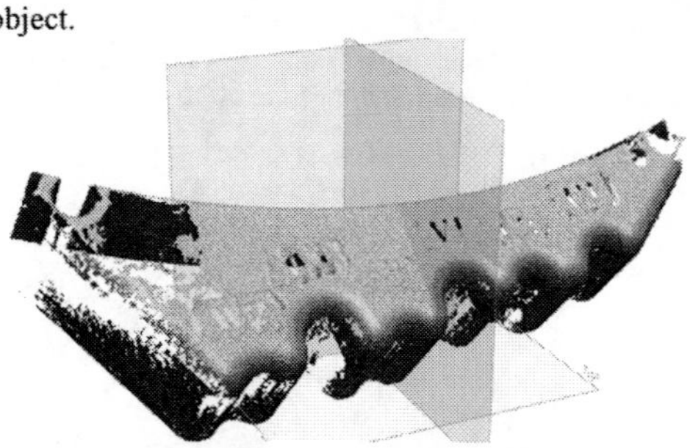

Fig. 7. The cloud of points of the scanned object.

In order to obtain the sketch required to extrude the solid model, the mesh is intersected with several parallel planes to get as many points on the contour in order to represent correctly the surfaces with discontinuities. (Fig. 8).

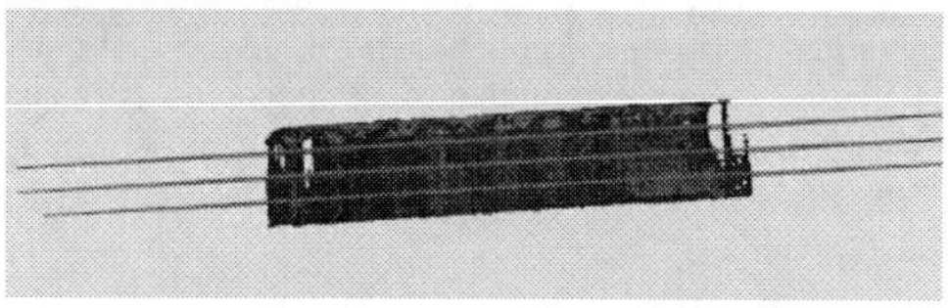

Fig. 8. Intersections with parallel planes

The points obtained were projected onto the plane on which the sketch is built and were joined by using specific commands to define the contour of the workpiece (Fig. 9).

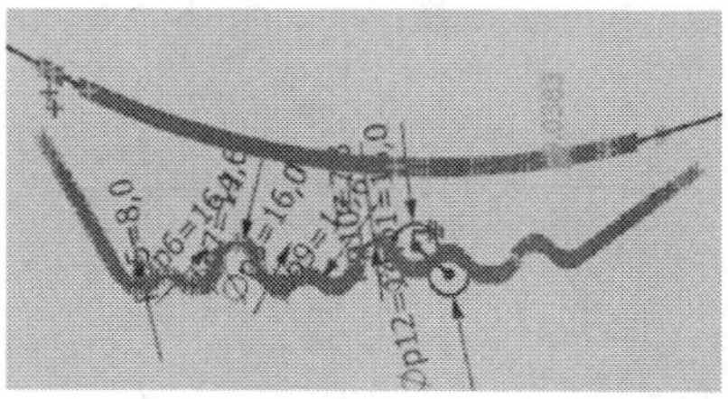

Fig. 9. The sketch of the part

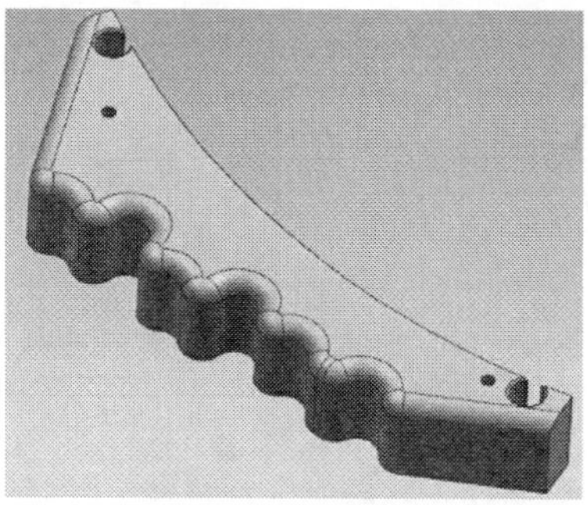

Fig. 10. The 3D model of the scanned object

By extruding the outline and applying chamfering and filleting operations, they obtained the 3D model [4] of the scanned object (Fig. 10).

3 Manufacturing of the spare part

Taking into account the complexity of the workpart, it is proposed to be done on numerically controlled machine tools. There were determined the machining operations, the needed cutting tools, and were calculated the cutting parameters.

The cutting parameters were determined by using an application which contains a database of the different materials and cutting tools. In Fig. 11 there are shown the parameters of the drilling operation.

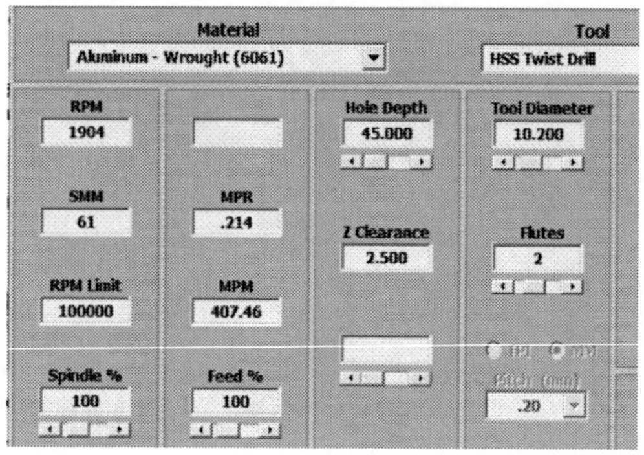

Fig. 11. The geometrical parameters of the drilling tool.

Table 1 shows the machining operations, the cutting parameters and the cutting times calculated for each operation.

Table 1. The necessary operations for manufacturing the part.

Operation	Tool Number	Speed	Feed	Time
		rot/min	mm/min	min
Spot_drilling	1	1618	196	00:00:17
Deep_hole_drilling_d4.8	2	4000	450	00:00:20
Deep_hole_drilling_d10.2	3	1900	400	00:00:16
Countersinking	4	950	300	00:00:03
Reaming_d5	5	2900	300	00:00:20
Tapping_m12	6	800	300	00:00:20
Milling_contour	7	4000	1200	00:11:34
Fillet_mill	8	4000	1000	00:07:25
				00:22:13

To determine the paths for each tool used to make the workpiece, it was used a Computer Aided Manufacturing (CAM) application [5].

Using the application for digital production involves going through several stages. It was established the origin of the coordinate system for manufacturing, there were indicated the geometry of the blank and the geometry of the workpiece. For each operation, were defined the cutting tool, the surfaces to be machined, cutting parameters, cutting strategy.

As an example, in Fig. 12 are shown the geometrical parameters for the drill Ø10.2 used to make the threaded hole.

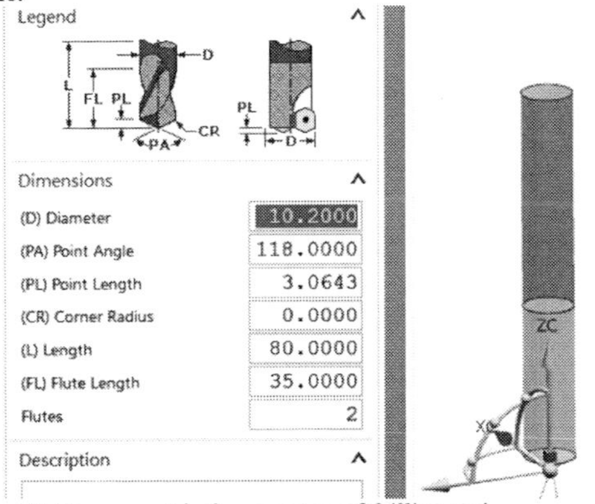

Fig. 12. The geometrical parameters of drilling tool.

All the tool paths for manufacturing the workpiece are shown in Fig. 13.

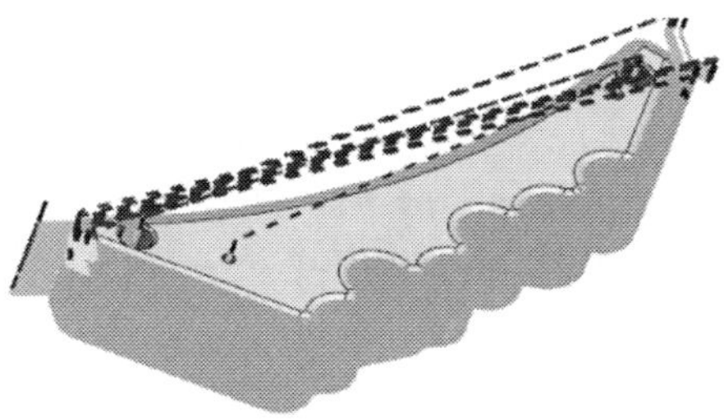

Fig. 13. The tools paths for manufacturing

After generating the tools paths, it is possible to check them by 3D simulation. By measuring of the resultant piece after virtual manufacturing, it is possible to highlight collisions or to identify the areas which are inappropriately processed (Fig. 14).

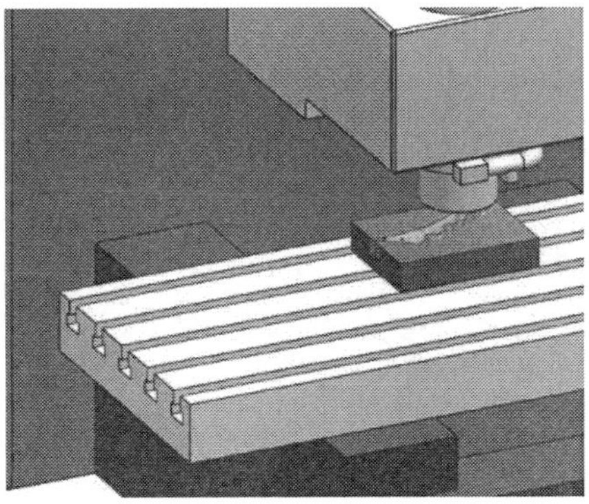

Fig. 14. The 3D simulation of the tools paths.

The postprocessing was done for a HAAS VF3 computer numerically controlled (CNC) equipment on which the workpiece was machined. The running of the program on the CNC machine tool is shown in Fig. 15, and in Fig. 16 it is shown the workpiece obtained after machining.

Fig. 15. Aspect during the processing of the piece

Fig. 16. The machined part

4 Conclusions

The case study shows the use of reverse engineering and CAM techniques to obtain a spare part of complex shape for which there is no documentation nor the manufacturer is known, starting from a worn part.

The digital data obtained through 3D scanning was used to get the 3D model of the part.

It was done the virtual machining of the3D model using a CAM application.

The CNC program obtained after post-processing was tested and back plotted before running on the HAAS VF2 equipment.

The work is a part of project MANUNET ERA-NET, Contract nr.19/2015 "Monorail conveyor system for automated production lines".

References

1. R. Abella, J. Daschbach, R. McNichols *Reverse engineering applications*. Comput Ind Eng, (1994).

2. D. Nedelcu, R. Avasiloaie, D. Florea, A. Rajic, Applications of the Rapid Prototyping Technology to Manufacture the Pelton Runners, Anale Univ. "Eftimie Murgu" Reşiţa, anul XX, **1**, (2013).

3. W. Wego, *Reverse Engineering: Technology of Reinvention*, CRC Press, Taylor & Francis, (2010).

4 G. Manole, E. Oprea, M. Iosip, *Concept and design of products (Conceptia si proiectarea produselor)*, Qual Media, (2009).

5. M. Iosip, E. Oprea, D. Boricean, *Performing digital manufacturing of products unsing virtual prototyping (Realizarea fabricaţiei digitale a produselor folosind prototipul virtua)l*, Qual Media, (2010)

MATEC Web of Conferences 184, 03005 (2018) https://doi.org/10.1051/matecconf/201818403005
Annual Session of Scientific Papers IMT ORADEA 2018

Considerations on determining the explosion risk and the areas with explosion hazard in a natural gas compressor station

Alina Maria Gligor[1,*], and *Valentin* Petrescu[1]

[1]Lucian Blaga University of Sibiu, Romania, Department of Industrial Engineering and Management, Emil Cioran str., 4, Sibiu, Romania

Abstract. Compressor stations are important components within natural gas transportation networks. The current paper analyses some of the safety aspects involved in the operation of such facilities – on the one hand related to the explosion risk for employees working in compressor stations and on the other hand related to determining to the areas with increased explosion hazard in such a station, in order to help reducing work-related accidents and the occurrence of professional illnesses. The explosion risk is determined based on a method devised by the National Institute for Research and Development for Environmental Protection Bucharest, while the explosion hazard areas are determined and displayed with the help of the software package AutoCAD Plant 3D.

1 Introduction

Natural gas is a very important energy source, but also an energy source that requires special attention with regard to the safety features of the production, storage, treatment, transportation and distribution facilities.

The assessment of explosion and fire hazards is in this case of essential importance for the selection of equipments that are to be used in such facilities and for their configuration. [1] For example, the usage of electrical equipments in areas with explosion hazard mandates an approach that differs from the case of normal areas, in several aspects.

Firstly, even before implementing any technical solution, for ensuring the explosion safety of the personnel and of the equipment operating in such demanding environments it is necessary to analyse in detail the possibility of flammable substance accumulations that may create explosive atmospheres. [2]

Secondly, it is necessary to determine the probability of an explosive atmosphere forming within the facility, taking into account and identifying all types of employed flammable substances so as to facilitate the placing of the various areas within one of the standardised types of explosion hazard areas.

While various authors have discussed the possibilities of assessing the explosion risk in various other parts of a natural gas extraction, storage and transportation system [1, 3, 4], in the current paper, the authors have on the one hand used a consecrated method to determine the explosion risk in a natural gas compressor station and on the other hand they also employed a specialised 3D computer-aided design software, AutoCAD Plant 3D, to help indicate the main problem areas.

2 Presentation of the analysed natural gas compressor station

Natural gas compressor stations are facilities for the processing of natural gas coming from the extraction wells, used to increase the gas pressure in order to facilitate its transportation over large distances through main pipeline systems and for the subterraneous storage of the gases from the transportation and distribution network. [5]

In Romania, the first compressor station, with an installed capacity of 8000 hp and used in the transportation of natural gases, started to function in the year 1965.

The compressor station chosen by the authors for the analysis comprises following elements: the installation for separating and filtering gases; the technological piping system consisting of aspiration and evacuation collector pipes, bypasses, interconnections, gas stacks, evacuators; the installation for measuring basic technological parameters (such as flows, pressures, temperatures); compressor units; cooling systems for the compressed gases, oil circuits, compressor cylinders and engines; installations for separating the oil used for lubricating the compressor cylinders; the electric energy system; compressor station rooms; systems for detecting the presence of hazardous atmospheres; fire prevention systems, systems for isolating and depressurising in the event of an accident, video surveillance systems. [5]

The analysed compressor station has been modelled using the computer-aided design software AutoCAD Plant 3D (figure 1).

* Corresponding author: alina.gligor@ulbsibiu.ro

© The Authors, published by EDP Sciences. This is an open access article distributed under the terms of the Creative Commons Attribution License 4.0 (http://creativecommons.org/licenses/by/4.0/).

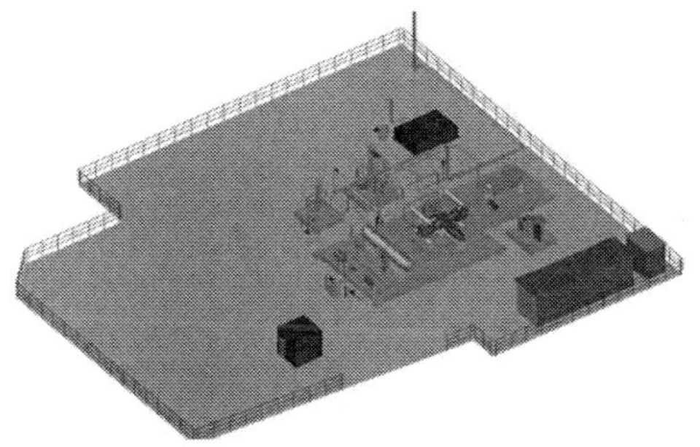

Fig.1. Schematic presentation of the analysed natural gas compressor station

Fig. 2. Detail of a modelled equipment from the natural gas compressor station

3 Risk Assessment Methodology

The method employed by the author to determine the levels of explosion risk within the analysed natural gas compressor station is based on the one conceived by the National Institute for Research and Development for Environmental Protection (I.N.C.D.P.M.) Bucharest. This method involves a quantitative risk/security level determination for a workplace, area or facility, based on a systemic analysis, as well as the risk assessment for accidents and professional illnesses.

This method requires the identification of all explosion risk factors within the analysed workplace or facility, using predetermined checklists and then quantifies the risk level according to the gravity and frequency of the maximal predictable consequence. A workplace's security level is then inversely proportional to the determined risk level.

When applying the method, following steps should be followed:

- defining the analysed system (workplace);
- identifying the risk factors in the system;
- assessing the explosion risks;
- ranking the risks and determining the prevention priorities;
- suggesting prevention measures.

The above-mentioned steps require the employment of following work instruments:

- List for the identification of risk factors;
- List of possible consequences of the action of risk factors on the human body;
- Scale for ranking the gravity and probability of consequences;
- Risk assessment grid;
- Risk level scale and safety level scale, respectively;
- Workplace assessment form – centralising document;
- List of proposed measures.

A workplace's global risk level (Nr) is then determined as weighed average of the risk levels determined for the identified risk factors. The rank of each risk factor is used as weighing element (being equal to the risk level), so that the result will reflect reality as closely as possible.

The formula for calculating the global risk level is given in (1):

$$N_r = \frac{\sum_{i=1}^{n} r_i * R_i}{\sum_{i=1}^{n} r_i} \quad (1)$$

where:
- N_r is the global risk level for a workplace;
- r_i is the rank of the risk factor „i";
- R_i is the risk level for the risk factor „i";
- n is the number of risk factors identified for the considered workplace.

4 Assessing the explosion risk in the analysed natural gas compressor station

In the case of the natural gas compressor station, the explosion risk factors were determined taking into account the tasks indicated in the job description files for the personnel operating within this facility, as well as the corresponding technological processes. The risk factors were then identified and grouped by the generating element within the facility (operator, work task, production means and work environment).

For the next step, the authors employed the previously mentioned work instruments in order to determine the partial risk level for each risk factor.

The global risk level for the workplace "Natural gas compressor station" has been calculated using formula (1) and the partial risk levels indicated in figure 3, leading to value of $N_r = 2.58$.

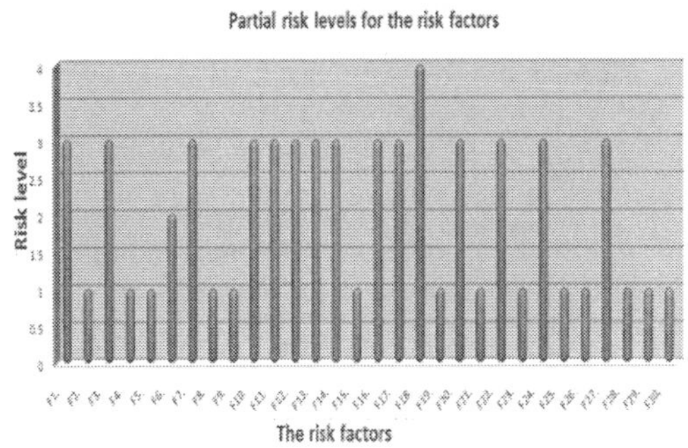

Fig. 3. Partial risk levels for the risk factors at the natural gas compressor station

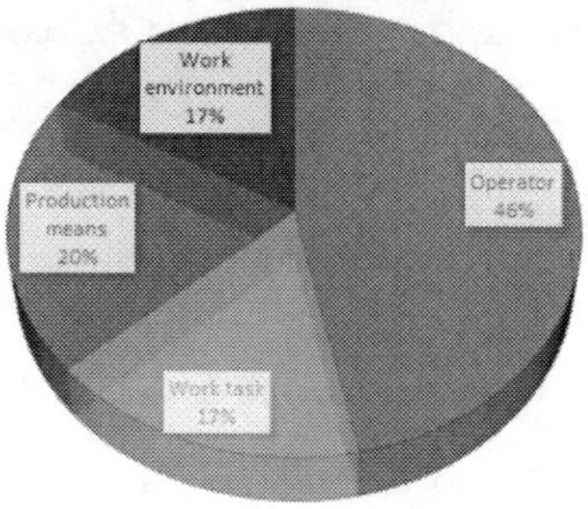

Fig.4. Distribution of the risk factors by generating sources

Table 1 presents some examples of risk factors, corresponding to the 4 types of explosion risk generating sources specific for the natural gas compressor station.

Table 1. Examples of risk factors taken into account

Risk factor generating source	Risk factors
Operator	F1. Carrying out tasks not included in the job requirements
	F11. Entering the compressor area without adequate protection equipment, or with equipment that can lead to electrostatic discharges
Work task	F17. Wrong succession of operations or absence of operations required by technological documents
	F18. Employment of personnel who do not have the necessary competence or qualification for the assigned work task
Work equipment	F24. Contact with moving elements of the compressor
	F25. Lack of equipotential and grounding connections on the instruments or on other electrical equipments
Work environment	F26. Loud noise level (above the maximally admitted level) near the compressor and during the opening of the pressure relief valves
	F30. Inadequate lighting level during works carried out during the evening or night

Thus, the global risk level calculated for the workplace "Natural gas compressor station engineer" is 2.58, a value that places it among the workplaces with small to medium risk levels.

From figure 2 it can be noticed that of the total of 30 identified risk factors, only 1 (F18) surpasses a partial risk level of 3, thus being considered medium to high risk factors.

By determining the explosion risk level, it is possible to determine the priorities in prevention measures for the workplace, but also to optimise the usage of resources allocated for this purpose.

The explosion risk in installations where explosive atmospheres may form can be reduced in two ways:

- continuous checking of the concentration of flammable in the area of equipments and initiating their decoupling from the electric power network if approaching the explosion limit;
- construction and usage of electric equipment with specific constructive and functional characteristics, that cannot become ignition sources for the explosive atmosphere that surrounds them.

An advantage of the I.N.C.D.P.M. method is the fact that it does not require an assessed system to be physically present. Also, it can be employed for all stages of a system's (or of an element of this system's) lifecycle: conception and design, physical creation, production, actual functioning, dismantling.

5 Indication of the explosion risk areas

An explosion risk area represents an industrial space in which, under normal functioning conditions, there can accumulate, permanently or accidentally, gases, vapors of flammable liquids, dusts or powders in sufficient amounts to give birth to an explosive atmosphere. Explosion risk areas are consequently very important from the point of view of the work safety legislation [7, 8] and require a special training of the personnel working within or near them. [9]

Explosion hazard areas can be classified into several types, function of the length of the periods during which the explosive atmosphere is present: [5, 10]

- area 0 - is the area in which the explosive atmosphere is present during the normal functioning of the technological installations, permanently or for a period of more than 1000 hours per year.
- area 1 - is the area in which the explosive atmosphere can occur during the normal functioning of the technological installations for intermittent periods with durations comprised between 10 and 1000 hours per year.
- area 2 - comprises the industrial areas within which an explosive atmosphere cannot occur during the normal functioning, but only in cases of predictable defects or for periods between 0.1 and 10 hours per year.

The zoning of industrial spaces represents the process of classification of spaces within an industrial installation function of the probability of occurrence of an explosive atmosphere and is important for the subsequent correct selection of the electrical and mechanical equipment that will be used in that area.

An area's extent depends mainly on the physical and chemical parameters of the flammable material and of the production process, such as:

- Discharge flow of gases and vapours
- The lower explosion limit (LEL)
- Amount of ventilation
- Relative density of the gas or of the vapours at the moment of discharge
- Climatic conditions
- Local topography

The result of the zoning process is a set of documents that comprises maps with precise indications regarding the different areas within facilities, to which there are associated specific parameters for the explosion hazard (explosion group and subgroup, temperature class, area index etc).

The electrical apparatuses for normal environments cannot be used, in general, in areas with explosion hazard due to the risk of the explosive atmosphere igniting due to electric sparks or arcs, hot surfaces of equipments, electrostatic discharges etc. [5]

Therefore, at the execution of the electrical apparatuses and equipments for potentially explosive areas, there have to be applied certain supplemental constructive measures that would offer them safety during functioning and in the presence of the explosive atmosphere.

For the zoning of the natural gas regulating and metering station, the author has used the software package AutoCAD Plant 3D, that allowed to clearly emphasise the areas with explosion hazard, as shown in figure 5.

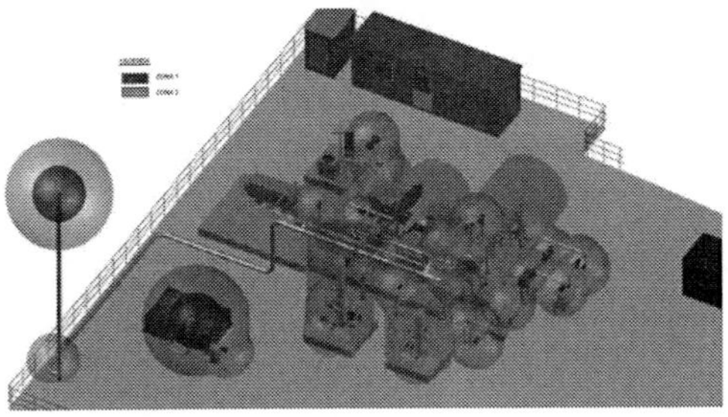

Fig. 5. 3D zoning plan showing the physical location of areas with explosion hazard

The computer-aided zoning allowed to notice that for the analysed natural gas regulating and metering station, area 1 is found for the following equipments: odorisation room, exhauster, safety valve, inner technological installation, while area 2 is encountered for the outer technological installation.

Figure 6 shows the extent of the explosion hazard areas in vertical cross-section.

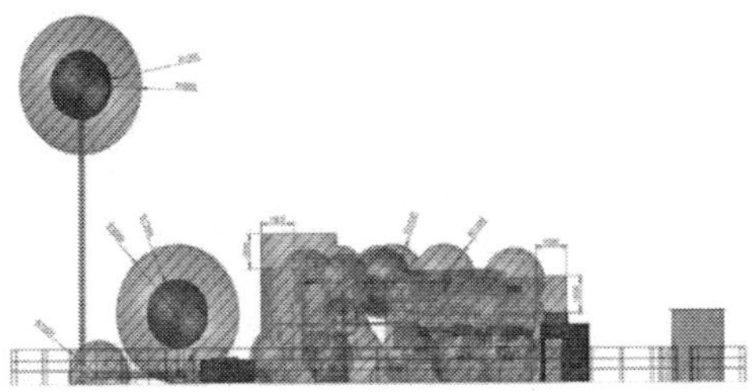

Fig. 6. Vertical zoning plan showing the physical location of the areas with explosion hazard

6 Conclusions

One of the main advantages of the method presented in this paper is that its application is not limited by the condition of a physical existence of the system to be assessed. It can be used in all stages related to the lifecycle of a work system or of an element of such a system: conceiving and design, physical creation, putting into production, unfolding of the work processes.

The application of the gravity/probability method for the analysed workplace has allowed the identification of explosion risk factors and to determine the dimension of risks for the occurrence of an explosion at this workplace.

For this, for each identified risk factor there will be indicated technical and organisational measures that need to be taken in order to keep the risk level in the „area of acceptable risks".

After the division of the space into specific hazard areas, the electrical installations can be conceived either through the usage of electrical equipments certified for such areas, or, if no such equipment is available, through conceiving new equipments and the certifying it.

In order for an explosion to happen, it is necessary that an explosive atmosphere and an ignition source coexist. Therefore, in order to reduce the explosion hazard, there need to be taken measures for the prevention of explosions, by:

- avoiding explosive atmospheres. This goal can be achieved mainly by modifying either the concentration of the flammable substance to a value outside of the explosion range or the oxygen concentration to a value below the limit oxygen concentration (LOC);
- avoiding all possible ignition sources;

As a general rule, the equipments with potential ignition sources have to be, if at all possible, placed in non-hazardous areas or, if that is not possible in the least hazardous area.

References

1. D.P. Nolan, *Handbook of fire and explosion protection engineering principles for oil, gas, chemical and related facilities* (Elsevier, 2014)
2. D.M. Duse, C.S. Duse, C. Deac, Wiert Naft Gaz, **25**, 2 (2008)
3. Y.-D. Jo, D.A. Crowl, J Loss Prevent Proc, **21**, 8 (2008)
4. Moraru, R.I., Proc. 18th Conf. Envir. Min Proc., 2014
5. A.-S. Huidan, *Electrical equipment for automation and control in environments with explosion hazard* (Translation from Romanian) (Technical Publishing House, 2008)
6. I. Foidas, *Testing and production of natural gas deposits* (Translation from Romanian) (Publishing House of the „Lucian Blaga" University of Sibiu, 2014)
7. L. Tarnu, *Elements of Law and Legislation* (Translation from Romanian) (Publishing House of the „Lucian Blaga" University of Sibiu, 2014)
8. L. Tarnu, *International protection of human rights* (Translation from Romanian) (Tehnomedia Publishing House, 2014)
9. D.M. Duse, C.S. Duse, C. Deac, M.I. Rusu, Trans Adv Eng Edu, **3**, 9 (2006)
10. ASRO, *Explosive atmospheres. Part 10-1: Classification of hazardous areas* (ASRO, 2008)

Contribution for increasing the manufacturing companies productivity using a tracking and planning production program

Iaroslav Francisc Pişec[1,*], and *Alin* Pop[1]

[1] University of Oradea, Department of Industrial Engineering, Universităţii 1, Oradea, România

Abstract. In this paper is presented roughly a tracking and planning production program. The major challenge for manufacturing companies is to have the highest efficiency regarding the productivity. For this reason, the best way to achieve this is to track all elements from production process and to plan all phases as well as possible. The method used for planning production, is the Critical Path Method (CPM) because this kind of planning method determinates in advance the time of fabrications of complex parts. Using this tracking and planning production program, the companies can check the machine tools workload, can analyse the capacity, can assign the proper equipment for part's fabrication, can estimate the delivering products time and also the cost of it. One of the advantages is that this information is provided on real time.

1 Introduction

In this days, most of the manufacturing companies are involved in a dynamic challenge [1] because the need of a short product and technology life cycle [2] and a high numbers of product variants as the demand for individualized products rises [2; 3].

An important obstacle which manufacturing companies have to cross, when is needing to concept a new manufacturing system to produce several products, is the right selection of the optimal quantity and portfolio of product-dedicated and flexible capacities.

Production planning is a time bound activity, in terms of years, months weeks and days. It is an integral part of the companies planning process and depicts the organization planning process [4].

The support activities of the tracking and planning production program can be broken roughly into three-time horizons: long term, medium term, and short term.

In the long term, the system is responsible for providing information to make decisions on the appropriate amount of capacity (including equipment, buildings, suppliers, and so forth) to meet the market demands of the future.

In the intermediate term, the fundamental issue addressed by the tracking and planning production program is matching supply and demand in terms of both volume and product mix. Although this is also true in the longer term, in the intermediate term, the focus is more on providing the exact material and production capacity needed to meet customer needs. This means planning for the right quantities of material to arrive at the right time and place to support product production and distribution.

In the short term, detailed scheduling of resources is required to meet production requirements. This involves time, people, material, equipment, and facilities. Key to this activity is people working on the right things. As the day-to-day activities continue, the tracking and planning production program must track the use of resources and execution results to report on material consumption, labour utilization, equipment utilization, completion of customer orders, and other important measures of manufacturing performance.

Moreover, as customers change their minds, things go wrong, and other changes occur, the tracking and planning production program must provide the information to managers, customers, and suppliers on what happened, provide problem-solving support, and report on the resolution of the problems. Throughout this process, communication with customers on production status and changes in expectations must be maintained.

The specific requirements for the tracking and planning production program design depend on the nature of the production process, the degree of supply chain integration, customers' expectations, and the needs of management. As the program is required to integrate with other company systems in the supply chain and/or with the tracking and planning production programs of the firm, additional design parameters are introduced. Moreover, these tracking and planning production program requirements are not static. As competitive conditions, customer expectations, supplier capabilities, and internal needs change, the tracking and planning production program needs to change.

2 The need of creating a tracking and planning production program

In order to manage the production process easily and efficiently, it is necessary to implement a program that closely follows all the elements that make up the process. It is built on rigorous methodology and intends to bring

* Corresponding author: pisec.iaroslav@gmail.com

© The Authors, published by EDP Sciences. This is an open access article distributed under the terms of the Creative Commons Attribution License 4.0 (http://creativecommons.org/licenses/by/4.0/).

 https://doi.org/10.1051/matecconf/201818403006

together all available resources within the company to make the product as efficient as possible.

Getting real-time information (even on-line) for all areas of the company determines maximum control, which reduces the risk of the occurrence of some disturbing factors.

First, the program is used to assign a machine tool which will perform the mechanical processing. Each production part passes through at least two machining operations, cutting the blank and machining, drilling or turning. In this way, it will be assigned a cutting machine and a milling machine. Some parts, such as BLOCK LOCATOR, also have electro-erosion machining, and PIN parts have a grinding process. These mechanical processes will be assigned to specific machines.

Assignment to a machine tool is required for generating the CNC program. This differ depending on the machine's working parameters, the tools available in the machine's storage tool, fixing devices, etc. It is not possible to assign a machine tool that is in reparation or maintenance because the processing time adds the waiting time until the machine can be used again.

The next step, after the machine is assigned, is to generate the CNC program. Once generated, it is transferred online, and the processing process starts.

The tracking and planning production program manages the machine tools activity, the parts processing part stages. An important aspect of this program is the ability to stop a machining process caused by the modification of the 3D model of the part. It is possible for the part to suffer shape changes in the design phase even at the time of its machining. The machining process is stopped, is checked the stage where the blank is and if it is possible to insert the changes into the actual piece, the CNC program is changed.

Improving the efficiency of the machining department, cost reduction and process assignment, continuous control over the entire activity, real-time analyses and reports are some of the benefits of implementing the tracking and planning production program.

3 The method used by production planning program

The method used for planning production, is the Critical Path Method (CPM) because this kind of planning method determinates in advance the time of fabrications of complex parts.

The principle of critical path analysis is to divide a project (complex actions) into component parts at a level that allows for their logical and technological correlation, that is, to make it possible to establish interactions between the component parts. These components are the activities of complex actions.

When defining the list of activities, the specialist involved in this operation uses his experience to answer, for each activity, the following questions:
"What other activities do or necessarily precede this activity?";
"What is the duration of the activity?".

The CPM method is a critical path analysis process in which the only parameter analysed is the time and the representation of the network graph takes into account the following conventions:

- each activity is associated with an orientated segment
- called the arc, defined by its ends, each activity being identified by a spring;
- each arc is associated with a value equal to the duration of the activity it represents;
- the conditioning of two activities is represented by the succession of two arcs.

Bellow is an example of how the CPM method works.

If two or more activities have the same precedent activity, for example A precedes B and A precedes C, the representation in the network graph will take the form of Figure 1 (a). Arcs B and C symbolize two activities that can only begin after activity A is completed. Activities B and C can be executed simultaneously. Also, the execution of an activity may depend on the completion of several previous direct activities, for example A precedes C and B precedes C as in Figure 1 (b). In this situation, activity C cannot begin, logically, until A and B have finished.

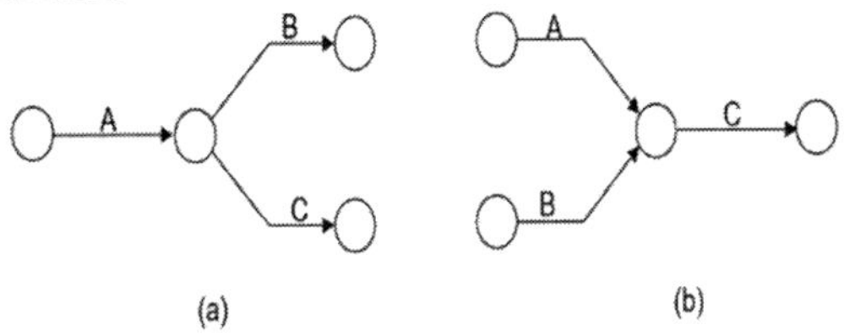

Fig. 1. Examples of network graph

The most important values to be computed after the network has been traced are:

- the earliest start time of an event - is the closest (time) moment at which a node can be reached;
- the latest moment of realization of an event - is the

most remote (time) moment at which a node must be reached for the project to finish on the set date.

Among the advantages of the CPM method (and in general the critical path analysis) we highlight:

- anticipating the execution time of the complex projects;
- during the running of the project allows for permanent control of its execution;
- explaining the logical and technological links between activities;
- highlighting critical activities;
- highlighting non-critical activities, which have time reserves;
- allows periodic updates to be made without redoing the graph;
- offers the possibility to carry out calculations to optimize the duration of a project, based on the cost criterion;
- represents a streamlined and rational method that allows timely programming of activities considering resources.

4 Input data / Output data

The operation of the program for tracking and planning the production requires the introduction in good conditions of the input data as can be seen in the figure below.

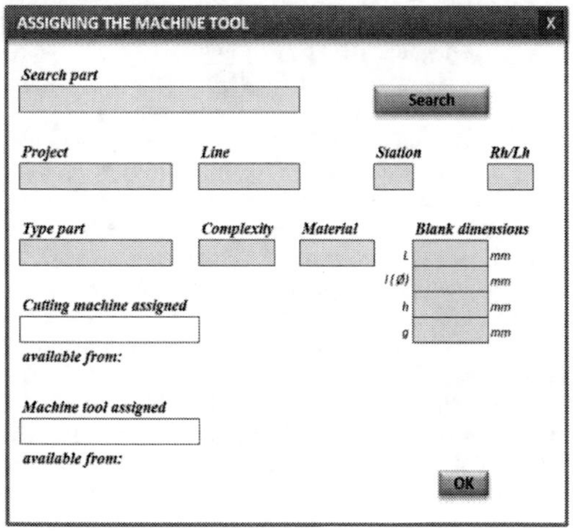

Fig. 2. Entering part details for machine tool assignment

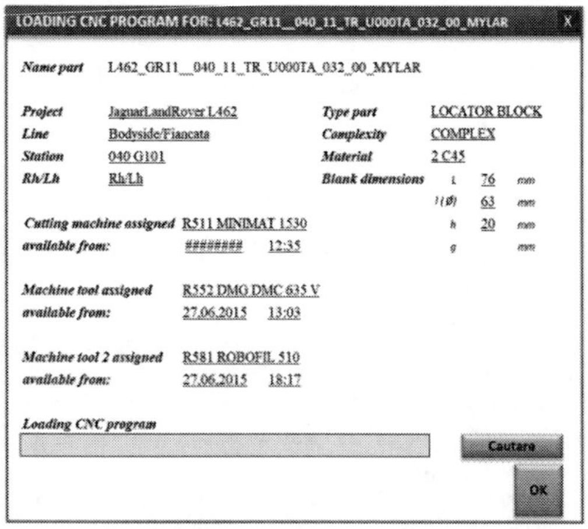

Fig. 3. Loading the CNC program for the given part

In Figure 2, it can be seen with yellow colour, which part information should be entered for assignment of the machine tool. The first step involves loading the 3D model of the part into the program. In this way is named also the part, the name which will be used in the program. Some inputs are just to know the system from which the part is part of (project, line, station). The "Rh / Lh" box determines the amount of pieces needed to be manufactured. For example, if "Rh" or "Lh" is selected, only one piece will be processed. If "Rh / Lh" is selected then two pieces will be processed, one of them will be symmetrical, and if "Rh=Lh" is selected then two identical pieces will be processed.

The type of part and its complexity is the main factor in the assignment of the machines tool. Below is a table in which is showed the predefined assignment of existing machine tools to processing different parts.

Table 1. Predefined assignment of machine tools to different parts

		R510 MINIMAT 1530	R511 MINIMAT 1530	R512 BRRANCO 500	R520 DMG CTX beta 2000	R521 DMG CTX beta 2000	R530 CORREA A30	R531 CORREA A30	R540 HARTFORD HV35	R541 HARTFORD HV35	R542 HARTFORD HV35	R550 DMG DMC 635 V	R551 DMG DMC 635 V	R552 DMG DMC 635 V	R560 DMG DMU 70	R561 DMG DMU 70	R562 DMG DMU 70	R563 DMG DMU 70	R570 DMG DMF 260/11	R580 ROBOFIL 190	R581 ROBOFIL 510
Locator block	simple	X	X				X	X	X	X	X									X	X
	medium	X	X						X	X	X	X	X	X						X	X
	complex	X	X								X	X	X	X	X	X				X	X
L Block	simple	X	X				X	X	X	X	X										
	complex	X	X								X	X	X	X	X	X					
Plate	simple	X	X				X	X	X	X	X										
	medium	X	X				X	X	X	X	X	X	X	X							
	complex	X	X						X	X	X	X	X	X							
Blade	simple	X	X									X	X	X	X	X	X	X			
	medium	X	X												X	X	X	X			
	complex	X	X												X	X	X	X			
Pin	simple			X	X	X															
	complex			X	X	X															
Weldment	simple	X	X												X	X	X	X			
	simple-medium	X	X														X	X	X		
	medium	X	X														X	X	X		
	medium-complex	X	X																X		
	complex	X	X																X		

The reports about machine tools, about the parts processed or which will be processed are the output data of the program for tracking and planning production. They can be saved as files in Microsoft Office Excel format or can be viewed online within the program.

After loading the CNC program, it is automatically transferred to the machine tools assigned via the digital data management system, Product Life Management (PLM). The PLM system is primarily a database of all the information about the projects that a company have realized or is realizing it.

Other outputs are:

- the actual start date of each operation;
- the actual end date of each operation;
- the duration of the operation in the programming process (i.e., operational or non-operational time);
- the gap between each operation and its starting date at the latest;
- delaying the execution of the production task with its starting date at the latest.

The generation of reports can be determined by their own information and control requirements or can be set as follows:

- daily reports;
- monthly reports;
- quarterly reports;
- annual reports.

These reports can be used to calculate the cost of mechanical processing, to calculate machine tool efficiency, to calculate service life, etc.

Based on the reports, it can be found whether a machine tool does not work at full capacity due to the small number of machined parts. Then it must be equipped with tools and devices suitable for the processing of other types of parts. It is important that the machine tool does not have a rest time because it is a disadvantage in the production efficiency.

5 Presentation of the functioning and usage of the program

The tracking and planning production works well and generates accurate information with the condition that it is used correctly. That is why the operator of this program should be trained and then supervised for a determined period of time.

Below is a broad overview of how to use the program.

Figure 4 illustrates the start-up page of the program after opening it from the computer start-up bar.

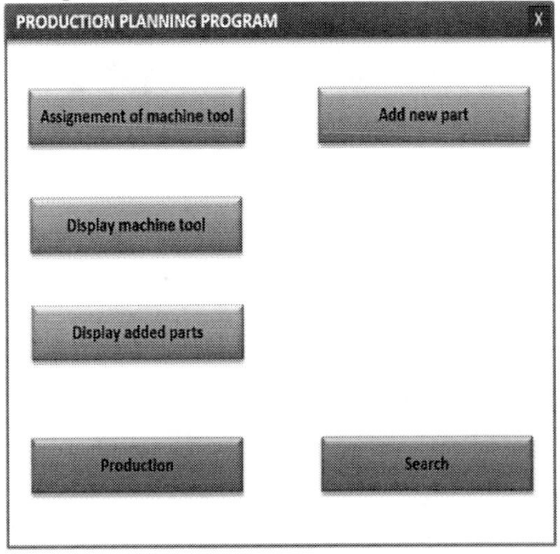

Fig. 4. Program start page

The Assignement of machine tool button opens the window where you enter the 3D model of the part and where the details about it are selected.

Loading the 3D model is done by activating the Search button in the newly opened window. In this way is named also the part, the name which will be used in the program. Some inputs are just to know the system from which the part is part of (project, line, station). The "Rh / Lh" box determines the amount of pieces needed to be manufactured. For example, if "Rh" or "Lh" is selected, only one piece will be processed. If "Rh / Lh" is selected then two pieces will be processed, one of them will be symmetrical, and if "Rh=Lh" is selected then two identical pieces will be processed.

The type of part and its complexity must be known by the operator because these are the main factors in assigning the suitable machine tool. Machine tools are automatically selected based on their availability. For example, a part that can be machined on five machines, for it will be assigned the machine that has the shortest amount of time to complete the current processing. At the same time, the tools and devices available in a machine tool can determine its assignment.

After entering all part information, it can be seen a preliminary assignment of the machine tool, Figure 5.

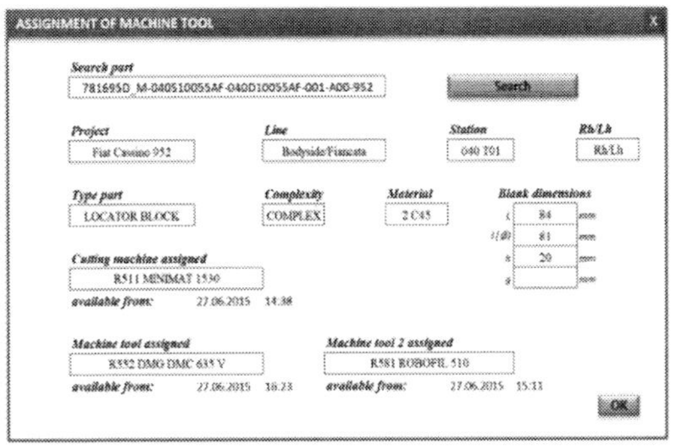

Fig. 5. Entering part information and preliminary view of assigned machine tool machines

If all the information is entered correctly, press the OK button and generate the file with the assigned machines, a file that is saved in the same location as the 3D model of the part. The program operator views the contents of this file, as illustrated in Figure 6.

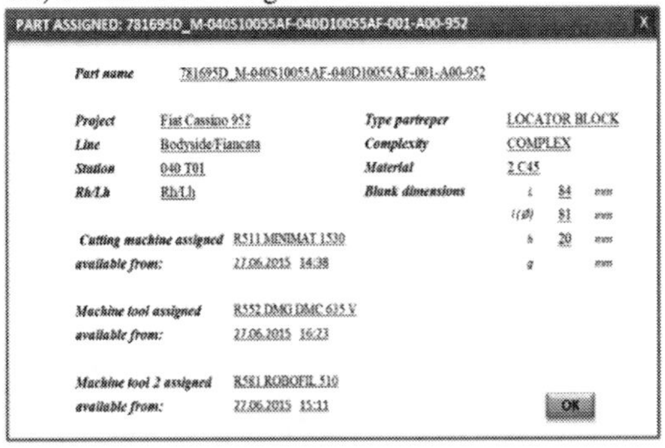

Fig. 6. Final view of machine tools assigned for the given part

The Add new part button opens the window where the CNC program of the given part is loaded. The file containing the CNC program is called "*CNC-x_y*" where "*x*" is the CNC program number created and "*y*" is the part name (Figure 7).

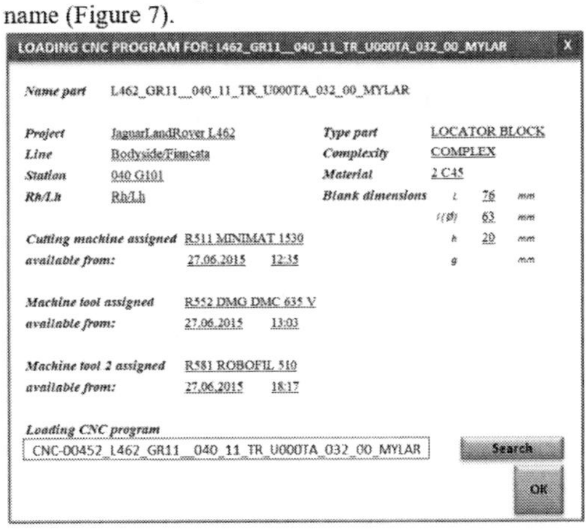

Fig. 7. Loading the CNC program for the given part

The Display machine tool button opens the machine tool tracking window. Here are all the machine tools available in the mechanical processing site, which parts are processed, and which parts will be processed on these machines, their availability, their capacity, as shown in Figure 8.

A machine tool cannot be assigned for the machining of a part if its processing takes longer than the remaining time until the machine enters into maintenance. That's why each machine has an Activate / Deactivate button that stops it or starts it for various operations, either maintenance, repair or improvement by adding tools, devices.

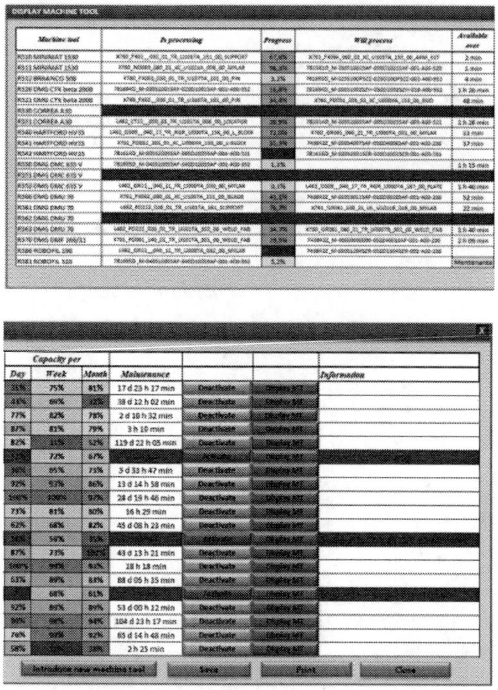

Fig. 8. Displaying the machine tools and the information about them

By pressing the Display MT button, it can be seen all details about the tool machine: the workpiece which is process on it, the workpiece which will be process, the available tools and devices, the machine capacity, the machine efficiency, etc.

The Display added parts button opens the window with the parts loaded in the program (Figure 9).

The parts can be processed on maximum 3 type of machine tools. In this window it can be seen which machine tools was assigned for the given part. The "*Processing time*" column contain the total time of processing part (in all phases: cutting, machining and electro-erosion) if the progress is 100% and the time spent until now if the progress is less than 100%.

Pressing the Display part details button, a new window is opened where is displayed all information about the given part. Information like who have added this part in program or how much it cost, can be printed as a report file.

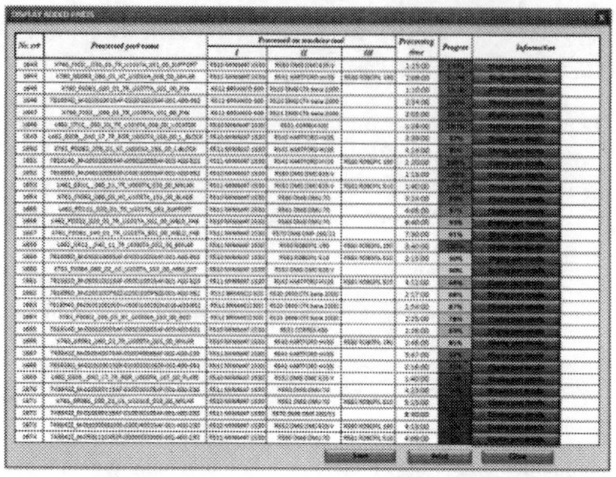

Fig. 9. Displaying the added parts in program and information about them

6 Diagrams and graphs generated by the program

The following figures show some examples of diagrams generated by the tracking and planning production program.

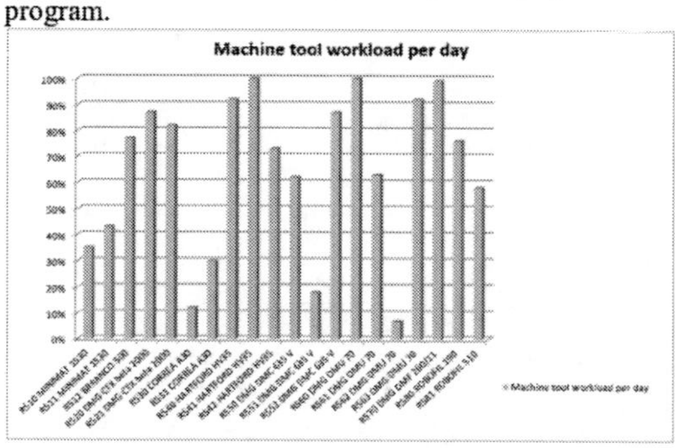

Fig. 10. Diagram of machine tool workload per day

As it can be seen from the daily workload diagram, Figure 10, some machine tools have a workload per day of 100%. These machine tools are productively efficient because they work 24 hours a day.

The vast majority of machine tools have a workload between 50% and 80%. These machine tools have an average workload that can be increased, but the fact that there are several machine tools of the same type can be considered that they have a good workload. The machine tools which have a low workload, such as the R530 CORREA A30 and the R562 DMGDMU 70, don't produce enough parts to be efficient.

Analysing this diagram, it is the decision to discover the elements that contribute to low efficiency. Some of the reasons may be: customer ordering of parts that can be produced on this machine is insufficient, the machine is in the maintenance or repair phase, the stages preceding the machining of the parts on these machines last longer than was planned.

For a better analyse of machine tools efficiency, is needed to see also the weekly and monthly workload diagrams, Figure 11 and 12.

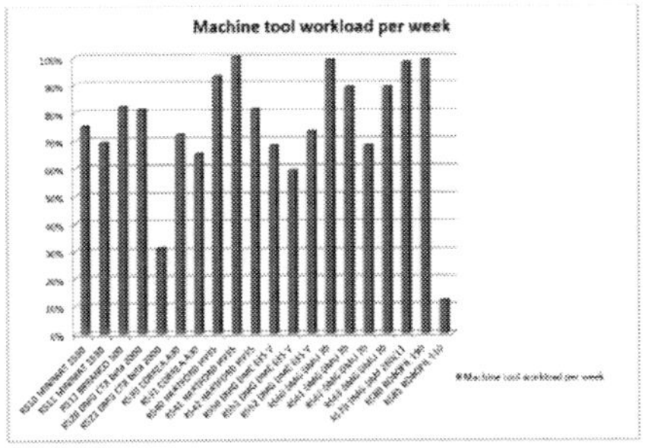

Fig. 11. Diagram of machine tool workload per week

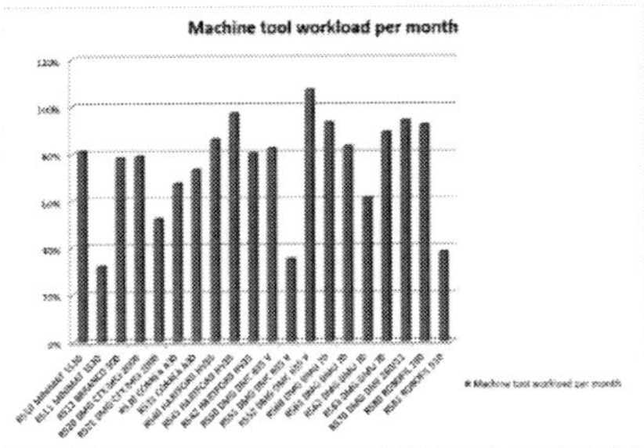

Fig. 12. Diagram of machine tool workload per month

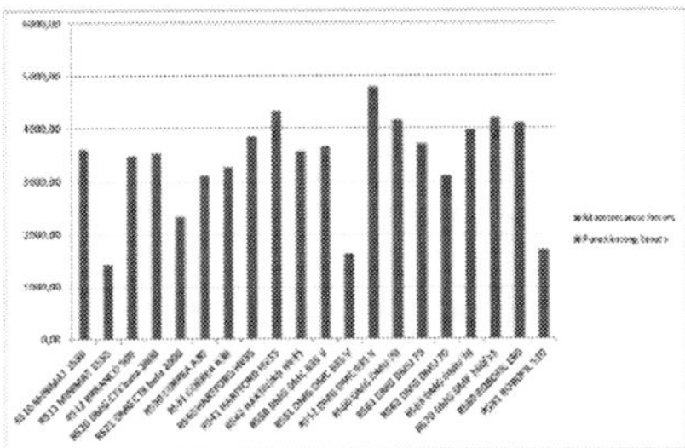

Fig. 13. Number of hours for functioning of machine tool and number of maintenance hours

If some machine tool has a lower workload per month is a disadvantage for production. This mean that this machine which was a cost for the company, now I not produce enough to amortize investment. If after the check, the machine is working in normal conditions, and is equipped properly for manufacturing specific part, is necessary to increase the order from clients. If not, it will be an extra cost for the company.

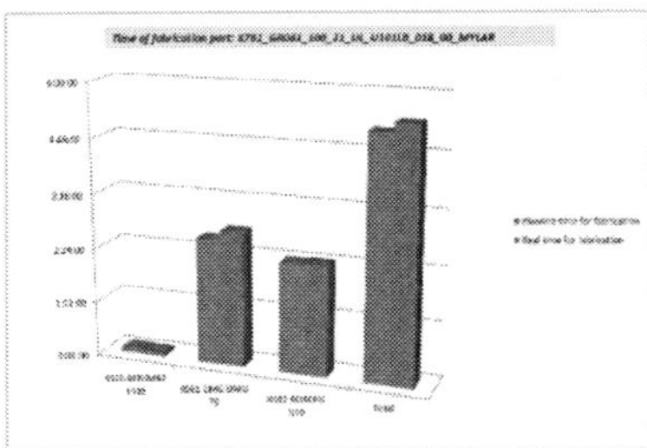

Fig. 14. Difference between the planned and real fabrication time for the given part on all machine tools where it is being processed

Another diagram provided by the tracking and planning production program, is the duration of mechanical fabrication for different parts diagram. This show how long time take the all operation to produce the specific part. The different operations are marked with separately colour to be more visible.

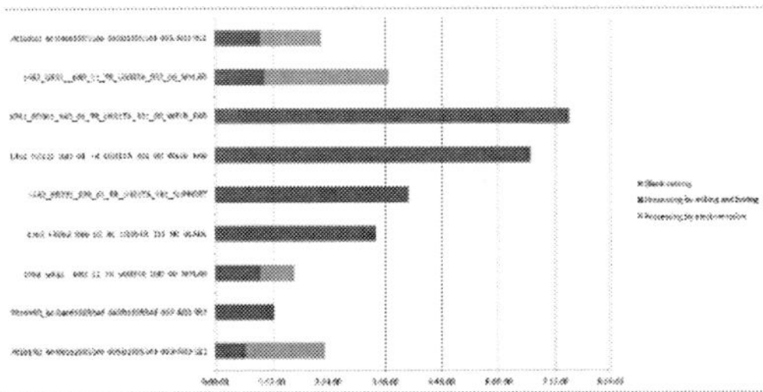

Fig. 15. Duration of mechanical fabrication for different parts, within 8 working hours

7 Conclusions

The program for tracking and planning production can be improved by changing the user interface in a more interactive way so that it is as easy to use. The program's databases must be as detailed as possible, to contain the correct information, which can be done by adding a procedure that checks all the information entered into the system.

With the increase in production capacity, this program needs to be adapted so that more users can enter, process, and save information in the system at the same time without affecting each other's work.

In conclusion, the program presented is appropriate to a manufacturing company's production by its simplicity of use, by generating reports on the actual situation in the department and by its efficiency in assigning machine tools suitable for processing the given parts.

References

1. H. P. Wiendahl, H. A. ElMaraghy, M.F. Zaeh, M. Kolakowski, Annals of the CIRP **56**, 783-809 (2007)
2. E. Abele, J. Elzenheimer, T. Liebeck, T. Meyer, Reconfigurable manufacturing systems, Berlin : Springer, 3-13 (2006)
3. F. Jovane, E. Westkämper, D. Williams,Berlin : Springer, (2009)
4. S. V. Kumar, V. G. S. Mani, N. Devraj, Proc. Mat. Sci. 5, 1710-1715 (2014)
5. M. T. Tabucanon, B. Sasiwong, Applied Mathematical Modelling, **15**, 3 (1991)

MATEC Web of Conferences **184**, 03007 (2018) https://doi.org/10.1051/matecconf/201818403007
Annual Session of Scientific Papers IMT ORADEA 2018

Considerations on the automation of a natural gas compressor station

Alina Maria Gligor*

"Lucian Blaga" University of Sibiu, Romania, Department of Industrial Engineering and Management, Emil Cioran str., 4, Sibiu, Romania

Abstract. Natural gas compressor stations are very important components of a natural gas life cycle. Unfortunately, in Romania most of the existing natural gas compressor stations are equipped only with old-type automation systems, with a high degree of physical and moral wear. The current paper describes the main elements and functions of an automation system that was implemented in a natural gas compressor station, offering an improved pressure control function, but also the possibility to control the compressors from a distance, for example from a central dispatching station.

1 Introduction

In many parts of the world and especially in the central part of Romania, where there are important reserves, natural gas represents an essential source of energy. However, being a fossil fuel with finite supply, it is of great importance to efficientise all aspects of its extraction, processing, transportation, storage, distribution and usage.

Natural gas compressor stations are processing units of the natural gas provided by the extraction wells, having the task to raise the gas pressure in order to allow either its transportation over large distances through main pipelines or the underground storage of the transported gases.

The main components of a natural gas compressor station are: the installation for separating and filtering gases; the technological piping system consisting of aspiration and evacuation collector pipes, bypasses, interconnections, gas stacks, evacuators; the installation for measuring basic technological parameters (such as flows, pressures, temperatures); compressor units; cooling systems for the compressed gases, oil circuits, compressor cylinders and engines; installations for separating the oil used for lubricating the compressor cylinders; the electric energy system; compressor station rooms; systems for detecting the presence of hazardous atmospheres; fire prevention systems, systems for isolating and depressurising in the event of an accident, video surveillance systems [1].

In natural gas compressor stations, the gas pressure can be increased by means of specific compressor units – either turbocompressors, motocompressors or electro-compressors [2].

The compressor units in operation in Romania are mostly still fitted with older types of automation systems that display a high degree of physical and moral wear and most importantly lack the possibilities to transmit data over longer distances and to process numerically the data related to the compression process.

The current labour legislation and regulations demand better working conditions for operators in potentially hazardous environments, but also enhanced possibilities for the direct or remote control of equipments by operators and a minimal reaction time [3, 4].

Therefore, in the current paper, the author presents the implementation of a solution for a modern automation system for a natural gas compressor station that can overcome the shortcomings mentioned above and does not require a direct monitoring by an operator or can be remotely monitored by a central dispatcher.

2 Industrial automation systems

In many cases, the unfolding of a production process requires physical parameters such as pressure, flow or temperature to either be maintained at a constant level or to vary over time according to specific laws.

In practice, this can be achieved by using specialised automation equipments called automated regulators that maintain the parameter at the desired level in an automated manner, without human intervention [5].

Such regulators can be used for controlling parameters as diverse as the feed rate and precision for a contouring equipment [6, 7] or the pressure in a natural gas transportation or compression system.

A regulator can work either in manual mode or in automated mode.

A first step in starting a regulation process is the tuning of the regulator, by successively modifying the work parameters so that the system reacts according to the requirements of the process [5].

Function of the type of dependence between the control parameter and the error, automated regulation systems can be proportional (P), proportional-integrative (PI) or proportional-integrative-derivate (PID).

© The Authors, published by EDP Sciences. This is an open access article distributed under the terms of the Creative Commons Attribution License 4.0 (http://creativecommons.org/licenses/by/4.0/).

Figure 1 shows the standard configuration of an automated regulation and control system [5].

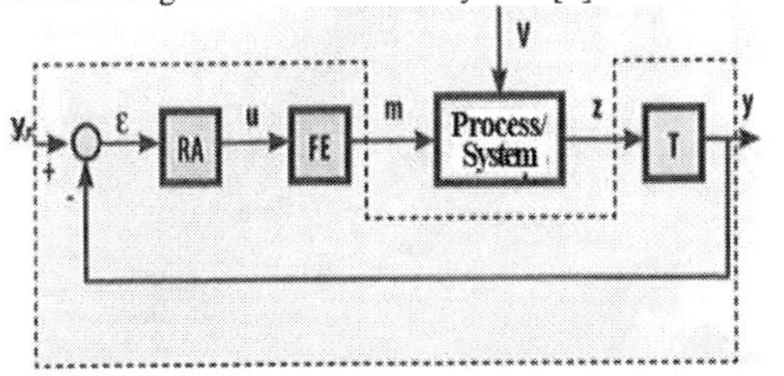

Fig. 1. The basic structure of an automated regulation and control system [5]
RA = automated regulator; EE = execution element; T = transducer
y_r = reference value; y = output value; ε = error; u = control parameter; m = execution parameter; v = disturbance; z = quality

The most common type in use today is the PID regulator. The control parameter can be determined with a function of the type [5]:

$$u(t) = K_R*\varepsilon(t) + K_i*\int \varepsilon(t)dt + K_D*\varepsilon(t)/dt \quad (1)$$

PID regulators are employed especially when there are larger disturbances in the system. The P component reduces the overregulation and leads to a short transition time, but also introduces a large stationary error. The I component cancels the stationary error but introduces a larger overregulation and a longer response time. In turn, the D component improves the dynamic behaviour, introduces small values of the overregulation and of the transition time but introduces a large stationary error. Their combination therefore leads to better performance both in terms of the overregulation, of the transition time and of the size of the stationary error [1, 5, 6].

Figure 2 [5] shows the response of a PID automated regulator to a step-type input.

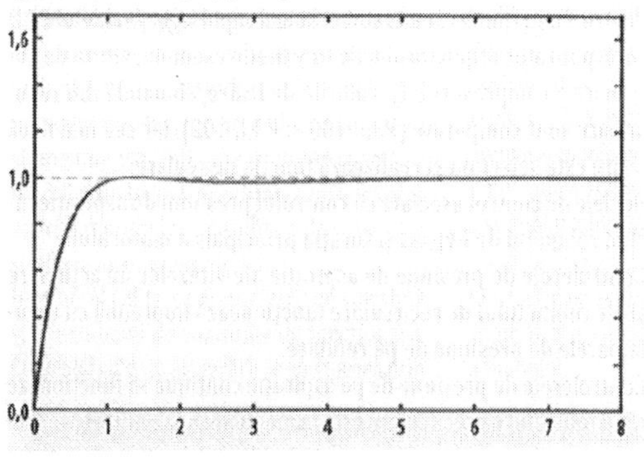

Fig. 2. Response of a PID regulator to a step-type input.

3 The new automation system implemented in the natural gas compressor station

Given the fact that the existing equipment within the natural gas compressor station does not allow for the automated regulation of the gas pressure, it was considered as necessary to implement a new automation solution to solve this problem.

One of the main problems that needed to be solved was which component to use as execution element for the automation (control) system, or more precisely on what type of valve to fit it. The compressor systems comprise two main types of valves – slide valves and spherical shutter valves. However, the automation system cannot be fitted directly on a spherical shutter valve, because this type of valve is not designed to work in an intermediate position, as this would subject the valve's sphere to intensive corrosion, which in turn would lead to a deteriorated surface and to leaks. Therefore, the automation system is fitted on a slide valve that is better suited for this type of operation.

The main elements of the implemented automation system are as follows:

- a BIFFI ICON 2000 electric actuator fitted on the slide valve (figure 3) and a programmable automation and monitoring panel that will be installed in the control room of the natural gas compressor station;

Fig. 3. BIFFI ICON 2000 electric actuator fitted onto a disassembled valve.

- a process computer equipped with a power source, input-output and communication modules and with clips for electrical connections;
- HMI (human machine interface) operation panel with an adequate display.

- automatic circuit breaker for overload and short-circuit protection for the general power circuit and for the power circuit of the controlled valve.

Also, in order to provide for the automated regulation of the gas pressure, it is necessary to install some pressure transducers, each of which will transmit to the automated regulation system a signal of 4-20 mA, proportional to the value of the pressure in the measuring point:

- a pressure transducer for the aspiration part of the compressor, installed ahead of the regulation valve, with a measurement range of 0-60 bar – PT1;
- a pressure transducer for the aspiration part of the compressor, installed below the regulation valve, with a measurement range of 0-60 bar – PT2;
- a pressure transducer for the exhaust part of the compressor, with a measurement range of 0-100 bar – PT3.

All three transducers will provide the regulation system (the automation panel) with the realtime values of the three pressures, values that represent the input for the calculation of the control parameter of the regulation system, parameter that will be also sent through an unified 4-20 mA current signal to the electrical actuator BIFFI ICON2000 (figure 4 [8]), fitted on the slide valve.

Fig. 4. Electric actuator BIFFI ICON 2000 [8]

The electric actuator BIFFI ICON 2000 has following characteristics [6]:

- Power supply: either 3 phase from 208 V to 690 V at 50/60 Hz or 1 phase from 110 V to 240 V at 50/60 Hz
- Torque output: From 30 to 57,000 Nm
- Speed range: From 12 to 173 RPM at 50/60 Hz
- Work temperature range: -20°C to +85°C
- Non-intrusive configuration;
- User-friendly push-button panel for operation, setting and diagnostics;
- Bluetooth™ wireless connectivity;
- Advanced maintenance data and alarm reports;
- Valve condition monitoring;
- Configurable 'data logger' function for maintenance and diagnostic programs in recorder or event modes;
- Customized numeric and graphic displays with 8 language options;
- Single enhanced terminal block;
- Digital contactless torque and position sensing;
- Watertight and explosionproof PDAs available;
- Suitable for use in SIL 2 applications.

Also, the automation panel will be connected to the spherical shutter valve, that can be controlled manually from the HMI part of the automation panel, should this become necessary during the compression process.

Each of the two valves can be controlled manually only after introducing a user-specific password and any of these manoeuvres will be automatically stored and included in a report that can be visualised later on the HMI display.

Also, all essential information from the regulation process can be remotely visualised by means of mobile devices connected through GPRS to the process controller.

In order to ensure the system's cybernetic security there will be used a GSM data card with fixed IP and the access to the information provided by the controlled will require the usage of user names and of the corresponding passwords. It will not be possible to issue a command for the modification of the valves' position by GPRS communication, since this connection will be dedicated exclusively to the reading of data from the regulated process. The only possibility for controlling the valves or for cancelling the controlled pressure will be by means of the HMI installed on the frontal panel of the automation system or by using the buttons of the local panels for the operation of the electric actuators BIFFI ICON 2000 fitted on the slide valve and on the spherical shutter valve.

In turn, the controller will communicate with a remote process computer by GPRS for the transfer of acquired data and for acquiring the targeted values of the regulated process variables. In order to visualise the parameters, the prescription of the regulated process variables and the control of the execution elements connected to the controller, independently from the process computer, there will be used a 15" touchscreen installed on the front panel of the automation block. This panel will allow the operation of the two valves, the monitoring of the regulation process from the compressors' aspiration part, as well as the visualisation of data on mobile devices by means of GPRS connections.

As can be seen in figure 5, the HMI display allows the visualisation of the three pressure values in both analogical and digital form, as well as the position of the valve (100% open). The visualisation of the functioning parameters and of the valves' position allows a parametrisation of the control function. By means of the HMI, the operator can supervise all process parameters and can compare the values of the various displayed parameters, so that he can then make quick decisions function of the indicated pressure value.

The situation presented in figure 5 in terms of the displayed values requires some further discussion. The value of the set pressure is higher than that of the pressure registered ahead of the control valve (PT1). In this situation, it can be assumed that the controller will be unable to bring the pressure to the targeted value, as even in the case in which the valve is 100% open, the pressure ahead of the valve can be at most equal to that behind the valve. This is a particular case for which the control system has to be programmed in such a manner that it alerts the operator if the value is too high.

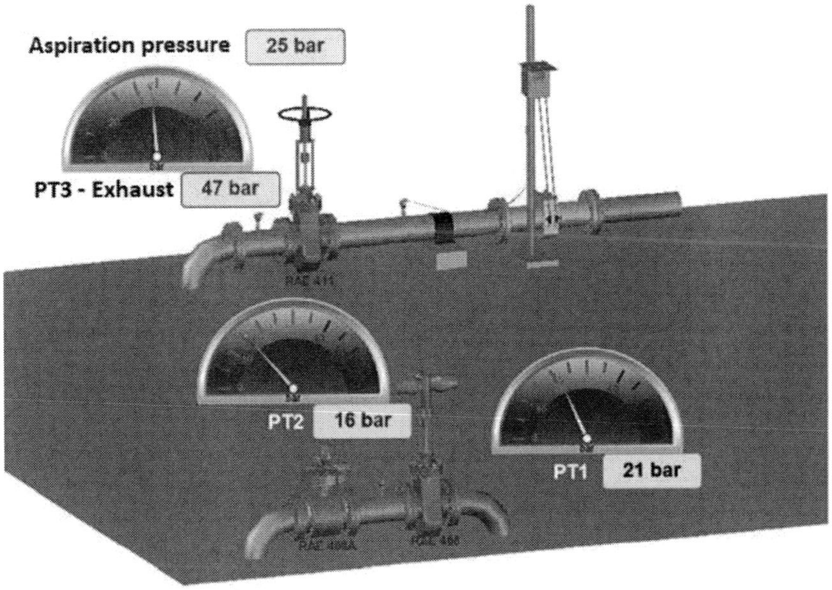

Fig. 5. The graphical interface developed on the HMI of the implemented automation system, showing the valves, the transducers and the controlled parameters.

4 Conclusions

The paper has described the implementation of an automation system for pressure control in a natural gas compressor station.

By implementing such a control system, the compressor station can work more efficiently and the gas injection process will be much easier to monitor, since the operator will have quick access to all information and can make informed decisions. Furthermore, this information can get directly to the decision factors in the company by means of the process computer that communicates with the controller.

All data are stored in the process computer and there can be made real time analyses of the functioning of the control system and of the process of injecting natural gas into a gas storage facility.

The new control system therefore represents a step towards achieving a higher efficiency and lower exploitation costs for natural gas compressor stations.

In future, the author will expand her researches in order to improve this automation solution and to apply improved automation solutions also to other elements within the natural gas life cycle.

References

1. I. Foidas, *Testing and production of natural gas deposits* (Translation from Romanian) (Publishing House of the „Lucian Blaga" University of Sibiu, 2014)
2. A.-S. Huidan, *Electrical equipment for automation and control in environments with explosion hazard* (Translation from Romanian) (Technical Publishing House, 2008)
3. L. Tarnu, *Elements of Law and Legislation* (Translation from Romanian) (Publishing House of the „Lucian Blaga" University of Sibiu, 2014)
4. L. Tarnu, *International protection of human rights* (Translation from Romanian) (Tehnomedia Publishing House, 2014)
5. O. Tanase, Rev. Automatiz. Instrum., **27**, 1 (2018)
6. R. Breaz, O. Bologa, M. Tera, C. Deac, Proc. 9th IEEE Intl. Conf. Ind. Inf., 1, (2011)
7. C. Biriş, C. Deac, M. Tera, Annals DAAAM, 1 (2008)
8. BIFFI, ICON 2000 Electric actuators, https://biffi.it/en-us/biffi-products/biffi-electric-actuators/biffi-icon-2000, 2018

The lubricants' parameters monitoring and data collecting

Gavril Grebenişan[1,*], *Nazzal* Salem[2], and *Sanda* Bogdan[3]

[1]University of Oradea, Faculty of Manag. and Technological Eng., Industrial Engineering Department, Universitatii 1, Romania
[2]Zarqa University, Faculty of Engineering Technology, P.O. Box 132222 Zarqa 13132, Jordan
[3]University of Oradea, Faculty of Manag. and Techn. Eng., Mechanical Eng. and Automotive Department, Universitatii 1, Romania

Abstract. This approach is focused on Machine Intelligence for Diagnosis Automation, a research program, which promotes « preventative maintenance in manufacturing plants through the development of a fully automated prototype for oil analysis and fault prediction. The prototype is based on Artificial Intelligence (A.I.) software and online hardware ». Monitoring the condition of lubricants requires the examination of the physical, chemical and additive states, which maintain the quality of the lubricants, which is necessary for the proper functioning of the equipment. A lubricant monitoring program, especially from a qualitative point of view, will need to focus on both machine tool wear and degradation of lubricants, as well as on detecting and describing abnormal working conditions for both lubricants and machine parts. This goal can be satisfied by examining all the oils used in a company by completing laboratory tests to generate steps and acceptance classes, as well as unplanned contingency analyzes. These laboratory tests can be concentrated and classified into technology-based data sheets based on test-based information and test results, ultimately constituting consistent databases needed to generate monitoring and prevention reports. Data on the condition of the oil parameters used in the hydraulic system for lubricating machine tools have been collected during six months. The data as matrix organized, with 258648 rows (observations) and 21 columns (parameters).

1 Introduction

One of the major causes of premature wear of machines parts, moving or working, under load on the contact surface is the practice of improper lubrication. This includes contamination with foreign particles, the use of improper lubricants, the use of lubricants in excessive quantities or in insufficient quantities. In addition to the quality of lubricants to ensure technological processes under friction, wear and temperature control, lubricants also become diagnostic tools. A lubricant status monitoring program can provide information on additive inefficiency, change in viscosity, water content, coolant, and emulsion losses, and particulate matter contamination [1], [2]. Depending on the objectives and size of the monitoring system with the lubricant analysis, one can observe: control and analysis of the contaminating fluids; fluid properties analysis, metal wear analysis. Typical tests are:

- Number of foreign particles;
- Viscosity;
- Water content;
- Spectroscopy of chemical elements;
- Infrared Spectroscopy (FTIR);
- Measurement of Total Acid Number;
- Wear density (excluding measured particles);
- Analytical Ferrography (exception of the density of wear);

One of the key elements in the lubricant analysis is the ISO Code (ISO 4406). This code for solid contaminants expressly provides for the purity levels of liquids and is internationally recognized. The code provides a reference: the number of particles per volume unit, between 5 and 15 microns.

Table 1. The ISO codes.

Range Code	Particles per milliliter	
	More than	Up to / Including
24	80000	160000
23	40000	80000
22	20000	40000
21	10000	20000
20	5000	10000
19	2500	5000
18	1300	2500
17	640	1300
16	320	640
15	160	320
14	80	160
13	40	80
12	20	40
11	10	20
10	5	10
9	2.5	5
8	1.3	2.5
7	0.64	1.3
6	0.32	0.64

These two dimensions have some significance, because lower concentration gives an accurate appreciation of particle deposits in oil, and the upper level of 15 microns reflects the type of particle concentration

* Corresponding author: ggrebenisan@gmail.com

© The Authors, published by EDP Sciences. This is an open access article distributed under the terms of the Creative Commons Attribution License 4.0 (http://creativecommons.org/licenses/by/4.0/).

from wear. The ISO code is represented by a combination of $R_{5\ microns}$ / $R_{15\ microns}$, according to the reference data in the Table1.Thus, if the analysis resulted in the following data:-400 particles> 5 microns / ml; -65 particles> 15 microns / ml; then the ISO code is: ISO 16/13.

Contamination of lubricants with water is a subject of their analysis. The degree of contamination is expressed in PPM (parts per million) of water content. A percentage (1%) of water is 10000 PPM. The low percentage of water in lubricants means good quality.

Viscosity is the critical property to follow in the case of lubricants. An increase in viscosity is more tolerable than a decrease in viscosity. Viscosity does not have an exact value; it varies within certain limits. This is guaranteed by the manufacturer.

The Neutralization Number, or the Total Acid Number (TAN), corresponds to its acidity. Acidity is primarily due to oil degradation. Acidity varies according to oil level.

Most experts in tribology recommend that oil analyzes be made mandatory for all new oils. These must include:

- Quantification of the number of particles in the workshop using porous sites (10 microns, characteristic mesh size);
- Determination of viscosity in the unit of work;
- Water content, using fluid separators;

Another set of laboratory analyzes should be performed on "critical" fluids, such as lubricants used in equipment tested at intervals of 60 to 90 days. Tests performed outside the workshop are:

- Karl Fischer Specialized Laboratory Test;
- Special Spectroscopy Test of Elements;
- Infrared Spectroscopy (FTIR);
- Total Acid Number (TAN).

New oils must have an ISO code as well as a water content equal to or better than prescribed in the technical documentation of the machine tool.

Efficiency and cost-effectiveness calculations have highlighted the savings of up to 10% of operating costs if contaminated oils are changed in time, i.e., before the affected equipment is introduced into the service, repair or repair process. Regarding the benefits of investment and reliability, one can say that the key elements of a program for the monitoring, evaluation, and analysis of the state of the oils in a company with specific machining are:

- Acceptance criteria for new lubricants
- Acceptance criteria for servicing oils
- Defining storage conditions
- Clear storage procedures
- Handling and transport procedures
- Oil change procedures
- Ensure the filtering process
- Ensure the filtering process on the tool machine
- Specific airing conditions

Worldwide, the yearly damage to machinery due to wear is equivalent to the destruction of about 20% of their total. Tribology, science dealing with the study of the phenomena and processes of friction, lubrication and wear occurring in the superficial surfaces of the machine organs in contact with relative motion, seeks to prolong the duration of operation of machines and installations by combating or eliminating wear, both by knowing the causes (friction phenomenon) and by preventing or reducing wear by using proper lubrication.

Some preliminary considerations on how to behave in a relative movement of the machine parts with direct reference to the immediate surroundings of the couples formed in the manufacturing process and machine tools:

After surface lubrication, the friction may be:

- dry - in case of direct contact between the two coupling elements without lubrication (friction coefficient $\mu > 0,3$);
- at the limit (unctuous) - in this case, the interposition of molecular layers of lubricant occurs, the oil film reduces the contact between the elements of the coupling, but does not remove it, the lubrication exists but is at the limit ($0,1\ \mu < 0.3$);
- mixed (semifluid) - when the lubricating film has a thickness corresponding to the leakage of the fluids, but due to the roughness of the surfaces it breaks and recovers, at high frequencies, part of the normal load is taken over by the fluid film. Percentage, this assumption of normal forces, is estimated at several percentages, which are significant in the case of components with large, non-dimensional masses. In this case, the lubrication is called partial ($0.05 < \mu < 0.1$). This type of friction is encountered when starting and stopping sliding bearings, hydrostatic assembly of tightening assemblies, etc.
- fluid - when the joint surfaces of the coupler are separated, through a continuous lubricant film. The lubrication is fluid ($0.01 < \mu < 0.05$).

It has been found in industrial manufacturing practice that most machine parts are not destroyed by both mechanical and thermal stresses, especially wear, due to sources of uncontrolled friction and inappropriate lubrication. Not always, inappropriate lubrication is due to design mistakes or non-conforming operations. These can also occur due to correct operation, but the composition of the machine tool materials in contact, as well as oils or lubricants, is inappropriate for various reasons. Some of these reasons may also be the deterioration of the properties and technical characteristics of the lubricants, but also of the materials of the machine parts in contact. By definition, the wear finds its physical expression in the loss of material on the contact surfaces of the relative moving bodies, resulting in the modification of the dimensions, the geometric shape and under certain temperature conditions, even leading to structural changes in the superficial layers [3] . The wear in lubricated friction contacts is influenced by:

1.1 - Rheological lubrication regime

The characteristic process values of the rheological lubrication regime are the gs viscosity parameter and the elasticity parameter g_e:

$$g_v = \frac{\alpha \cdot \omega}{R} \cdot \left(\frac{\omega}{\mu_0 \cdot V}\right)^{\frac{1}{2}} \qquad (1)$$

$$g_e = \left(\frac{\omega}{E \cdot R}\right)^{\frac{1}{2}} \cdot \left(\frac{\omega}{\mu_0 \cdot V}\right)^{\frac{1}{2}} \qquad (2)$$

in which: α is the viscosity-pressure coefficient; ω is the normal force per unit of contact width; R is the radius of curvature of the contact surface; E represents the effective elastic modulus; μ_0 is the viscosity under normal conditions, and V is the relative speed between the contact surfaces [1-4]. The values of these parameters must be within the ranges: g_v <1.5, and g_e <0.6. If these conditions are not complied with due to the operating process, this is called the elastohydrodynamic lubrication process, characterized by the g_s speed parameter and the g_l load parameter, which must be within the following limits:

$$1.8 < g_s = \alpha \cdot \left(\frac{E^3 \cdot \mu_0 \cdot V}{R}\right)^{\frac{1}{2}} < 100,$$

$$1.0 < g_l = \alpha \cdot \left(\frac{\omega \cdot E}{2 \cdot \pi \cdot R}\right)^{\frac{1}{2}} < 100$$

respectively.

1.2 - Functional lubrication regime

This is characterized by the minimum thickness of the oil film in the case of the pressed surfaces and is calculated by the formula:

$$h_0 = 4.9 \cdot \frac{\mu_0 \cdot V \cdot R}{\omega} \quad (3)$$

where the coefficient "4.9" refers to a solid rigid with a lubricant isoviscosity. In the case of elastohydrodynamic conditions, the minimum thickness of the lubricant film shall be calculated with the formula:

$$h_0 = 2.65 \cdot \frac{\alpha^{0.54} \cdot (\mu_0 \cdot V)^{0.7} \cdot R^{0.43}}{\omega^{0.13} \cdot E^{0.03}} \quad (4)$$

In the case of point-like contact of smooth surfaces, the minimum thickness of the lubricant film is:

$$h_0 = 0.84 \cdot (\mu_0 \cdot \alpha \cdot V)^{0.74} \cdot R^{0.41} \left(E/_{\omega}\right)^{0.074} \quad (5)$$

If the process takes place by sliding the contacts on a thin film of lubricant, it is necessary that this regime is not considered a rheological one and the ratio lambda: $\lambda = {h_0}/_{S}$, where $S = \frac{R_1+R_2}{2} = 1.11 \cdot R_a$ and R_a, is the height of the surface roughness. If λ> 3 then it can be considered that there is a fairly consistent likelihood that this contact of asperities is insignificant, so adhesive bonding is impossible to achieve. Otherwise, if λ <1, then the operating mode takes place at the extreme limit of lubrication, which implies the presence of wear and tear wear, respectively. Therefore, in this situation, lubrication conditions must be reconsidered. If this is practically impossible, the contact mode at the asperity limit must be determined by examining the plasticity index:

$$\psi = \frac{E}{P_y} \cdot \left(\frac{\sigma}{\beta}\right)^{\frac{1}{2}} \quad (6)$$

where E is the modulus of elasticity; P_y is the normal pressure on the contact surface; β- the asperity dispersion factor, σ -the mean standard deviation of the asperity height.

Thus, in the case of the mixed lubrication regime in which 1 <λ <3, in which most of the machine tools exhibit sliding or sliding / rolling contacts, the total load is divided between the asperity load and the lubricant film, respectively in this case only the load on the roughness contributes to the generation of wear. If ψ <0.6, contact between the asperities can be considered as elastic under any type of loading, and if this plasticity index ψ> 1 then the contact can be considered partially elastic, even at low loading forces. If the plasticity index is in the range of 0.6 <ψ <1, the contact mode is a mixed one, and an increase in the loading force can generate a change of the elastic-to-plastic contact. If ψ <0.6, there is a predominant metal-metal contact due to the fluctuation of the absorbed lubricant, and the idea of approaching the defects of the fractionated lubricant film should be considered.

1.3 -The fractional lubricant film defect

A property with measurable influence, which has a critical effect on wear in lubricated contacts, is the heat absorbed by the lubricant. This feature is valid in the case of adhesion, resulting, in practice, from direct metal-metal contact between roughness. If the lubricant molecules remain attached to the contact surfaces, then the probability of forming a pronoced wear is extremely low. In physical terms, the film fractionation and the occurrence of fractionation defects of the oil film, respectively, b, can be defined as the ratio between the number of unrealized contacts, between the roughness, and the actual total number of contacts between the possible asperities:

$$\beta = \frac{A_m}{A_r} \quad (7)$$

in which A_m represents the contact area, metal to metal, and A_r represents the real contact area.

1.4- Dispersion of loads in the contact

For the discussion of this situation, it is considered a lubricated contact, in which only adhesion wear is present. The solution to the problem is based on the theory of partial elastohydrodynamic lubrication. According to this theory, both the contact between the asperities and the lubricant film contributes to the absorption of the dynamic loads. Thereby:

$$W_c = W_s + W \quad (8)$$

in which W_c represents the total load on the contact surface, W_s represents the charge taken up by the lubricant film, and W represents the load assumed by the asperities. According to experimental data, only part of the total load is considered responsible for adhesion. Analytically, the size of the contact pressure, empirically given by the relationship:

$$p = {}^{2}/_{3} \cdot E \cdot (Nr\sigma) \cdot ({}^{\sigma}/_{r})^{\frac{1}{2}} \cdot F_{\frac{3}{2}}\left({}^{d_e}/_{\sigma}\right) \quad (9)$$

The total pressure resulting from the total load W_c is given by:

$$p_c = 0.39 \cdot E^{\frac{2}{3}} \cdot W_c^{\frac{1}{3}} \cdot R_e^{-\frac{2}{3}} \quad (10)$$

What determines the following ratio calculation:

$$p/_{p_c} = 1.7 \cdot R_e^{\frac{2}{3}} \cdot W_c^{-\frac{1}{3}} \cdot E^{\frac{1}{3}}(Nr\sigma) \cdot$$

$$\cdot ({}^{\sigma}/_{r})^{\frac{1}{2}} \cdot F_{\frac{3}{2}}\left({}^{d_e}/_{\sigma}\right) \quad (11)$$

in which $R_e^{\frac{2}{3}}$ is the relative radius of curvature of the contact surfaces, E- is the effective elastic modulus, N- the relative density of the asperities, r- is the mean radius of curvature of the asperity peak, σ- is the standard deviation of the peaks, $F_{\frac{3}{2}}\left({}^{d_e}/_{\sigma}\right)$ is a statistical function, and d_e is the mean distance of separation between the tip of the asperities and the smooth surface.
The ratio between the lubricant pressure taken and the total pressure on the contact surfaces is given empirically by:

$$p_s/_{p_c} = \frac{1}{\lambda}\left(\frac{h_0}{h}\right)^{6.3} \quad (12)$$

where λ is the specific thickness of the lubricating film, h is the average height of the oil film in the case of rough surfaces, and h_0 is the height of the oil film in the case of smooth surfaces.

1.5- A mathematical model of adhesion wear

Theoretically, the amount of adhesion wearing should be expressed by a function of the metal to the metal contact surface and the slip distance:

$$V = k_m \cdot A_m \cdot L \quad (13)$$

Where k_m is a non-dimensional constant specific to friction materials and independent of contaminants or lubricants. Expressing the actual contact surface, as a function of loading W and P, the flow pressure, taking into account the oil film fractionation defects, β, obtains:

$$V/_{L} = k_m \cdot \beta \cdot ({}^{W}/_{P}) \quad (14)$$

This equation has a special meaning for the terms it contains. Thus k_m characterizes the tendency of shrinkage of the surfaces by wear, and the parameter β indicates the ability of the lubricant to reduce the metal to the metal contact surface, taking values in the interval [0,1].

2 Operational Objective: Integration of the whole system, prototype development, testing

The first of the directives of this operational objective is certainly to build a hardware system that can support the new requirements of the intended use range. In industrial practice, there is obviously a major requirement, namely that technology is structured so that it is "safe", without any risk to the work environment, and at the same time offers the opportunity to store and transmit in a way efficient information retrieved from sensors located on industrial equipment.

The designed system, which has begun this research, has some clear drawbacks, especially as regards the speed of adder acquisition, as well as its transmission and processing. This system was totally lacking in the possibility of data transmission, i.e., communication, of the local area networks(LAN) type, which revealed a fundamental deficiency in the intention of interfacing the system in the current industrial context. It was obvious that the system had to be rethought in terms of a new neural network approach, with the application of genetic algorithms, with the avoidance of traditional, high-risk switches or connectors. It was obvious the need to reorient the power supply to a mains or power source, max. 24 VDC, through LEMO connectors (fiber optic). It would have been desirable to involve an external touch screen, a smartphone model, through which the basic operations of the activation functions, display parameters, etc. could be performed. Considering these requirements, using specialized literature and specific search engines, a rigorous selection of appropriate component manufacturers was needed, so that the model of this system, which is prefigured, would allow a wide range of options and facilities to suit the intended purpose, and objectives listed.

A testing phase of a working model took place at the Mecoil (Italy) headquarters, in the company's lab, on a demo system with controlled parameters. The elements that were tested (physicochemical parameters) were evaluated in a first stage, on a single level. As the data was collected, based on predefined scenarios, "event / error modules" were interleaved on one level to test the system's response speed. The data was then processed and interpolated, at which point the contribution of "IA" became essential to evaluate the infinitesimal transformations (those that can escape the eye of a technician) of the multiparametric elements. The system was thus tested to detect between the "n" elements to be taken into account simultaneously, and to evaluate in real time both the evolution and the trend of the monitored parameters, being "learned" to detect essential anomalies or possible failures . The overcoming of this early phase allowed the decision to install the system designed and physically installed on industrial installations (pilot systems) at EMSIL Techtrans SA-Romania, as well as in four other locations in Italy, to Mecoil partners, [6].

2.1 Activities carried out

In order to remove the obvious limitations shown by the first version of Monitoil's (Fig. 1.) monitoring system, it was decided to restructure the equipment by approaching a reasonable option for industrialization in the idea of efficient and logical improvement of the electronic components that would validate the project from the point of view of its operation under an environment of real, industrial thing.

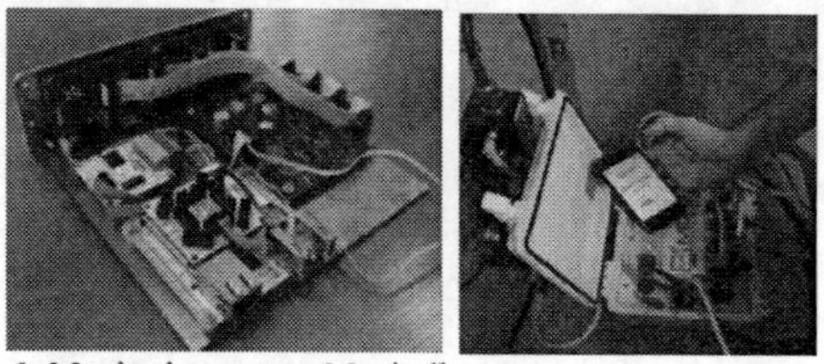

Fig. 1. Monitoring system Monitoil

Using advice with specialists in preventive maintenance, networking and robotics design (DM Electronics and Rodolfo Facchini), the planned activities were channeled towards the preliminary design of the acquisition board and the sensor work pattern, paying attention all possible causes of "distortion" of the signal and environmental and electrical interference. In this respect, consideration was given to the evaluation of a Raspberry type interface, with emphasis on data acquisition and sensing response checking. In addition, TecService Italy's TecService Italy consultancy has been used to build a first-ever version of the Machine Intelligence for Diagnosis Automation (MIDA), with both industrial and IP67 protection restrictions for these work environments, [6], (Fig. 2).

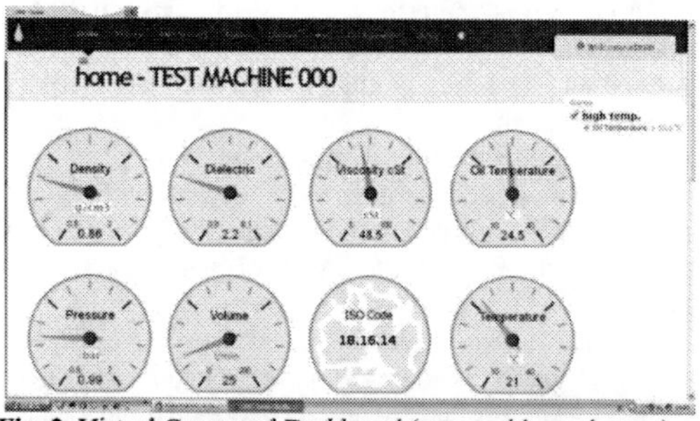

Fig. 2. Virtual Command Dashboard (www.mida-project.eu)

Fig. 3. The installation of MIDA on a machine tool at EMSIL Techtrans SA Oradea

Testing and tuning of the final phase algorithms were made possible by collecting and making available to the Project partners a set of observations data from the installation of MIDA on a machine at EMSIL Techtrans SA Oradea (Fig. 3.). Starting from the scheme of the first prototype, the Mida's miniseries were made. In addition to the test prototype test at EMSIL Techtrans SA Oradea, 4 other units were sent to Mecoil customers and partners to validate the device's operation under industrial conditions and obtain new test databases for artificial intelligence. The exploitation and testing of this equipment for many users simultaneously played an essential role in identifying problems, but also in evaluating the datalogger design phase(Fig. 4.).

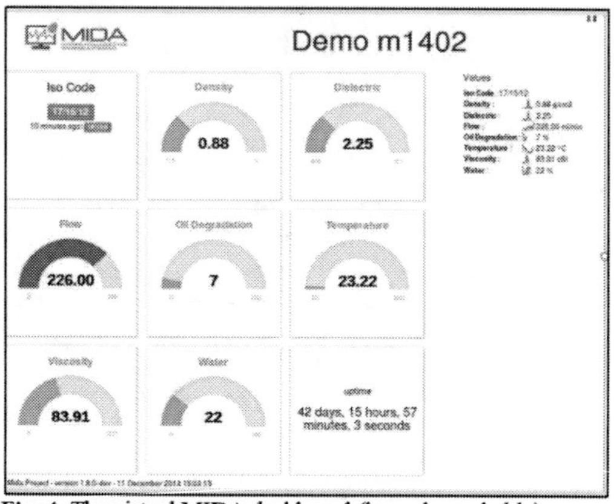

Fig. 4. The virtual MIDA dashboard (by web reachable)

3. Results

The data are matrix organized, with 258648 rows (observations) and 21 columns (parameters). The data refer to the analyzes made on the manufacturing line on the following parameters: density, viscosity, dielectric, water, oil degradation, temperature, ISO4, ISO6, ISO14, with obvious importance for the oil condition, but also for the operation of the industrial equipment being tested .

Particle sensors (Fig. 5) may be of several types: particle count or particle detector. The two are listed below:

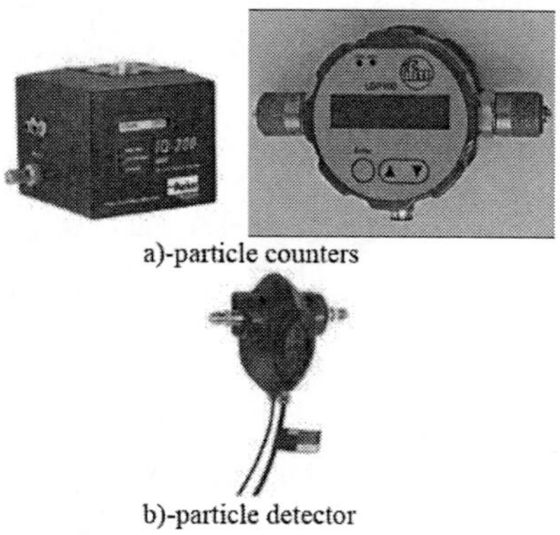

a)-particle counters

b)-particle detector

Fig. 5. Sensors used

Water sensors or moisture sensors can be of several types. Some are mounted on the pressure pit at others on the return circuit of the installations. Water or humidity sensors continuously monitor water / humidity in hydraulic and grease oils. They were designed especially for preventive maintenance of installations and machines. Reporting is done in% relative humidity of the water content, giving the user how close is the saturation point oil. Data on the condition of the oil parameters used in the hydraulic system for lubricating machine tools have been collected for six months. The data are matrixed, with 258648 rows (observations) and 21 columns (parameters). The parameters are presented in Table 2:

Table 2 The parameters collected

id	measure identifier
postgresql_timestamp	writing time
time	writing time
raw_data	Maybe neglected, it is not important (data before process starts)
icm_isocode	ISO code 4406 (icm_iso4/ icm_iso6/ icm_iso14)
icm_rh	water saturation%
icm_flow	oil flow through particle counter
icm_temp	particle counter electronic temperature
icm_iso4	ISO 4406 code class particles >4 μm
icm_iso6	ISO 4406 code class particles >6 μm
icm_iso14	ISO 4406 code class particles >14 μm
icm_pc4	number of particles >4 μm
icm_pc6	number of particles >6 μm
icm_pc14	number of particles >14 μm
fps_vcst	viscosity [cSt]
fps_v	viscosity [cP]
fps_density	density
fps_dielectric	dielectric constant
fps_temp	oil temperature
oh_temp	Oilhealth electronic temperature
oh_parama	Oilhealth paramA (not important)
oh_paramb	Oilhealth paramB (not important)
oh_paramc	Oilhealth (not important)
oh_od	Index of degradation OD%
Date	day of analysis
time	hour/time of analysis

Table 3. Data collected- the first 30/ 258648 rows and all columns of the database, are showed

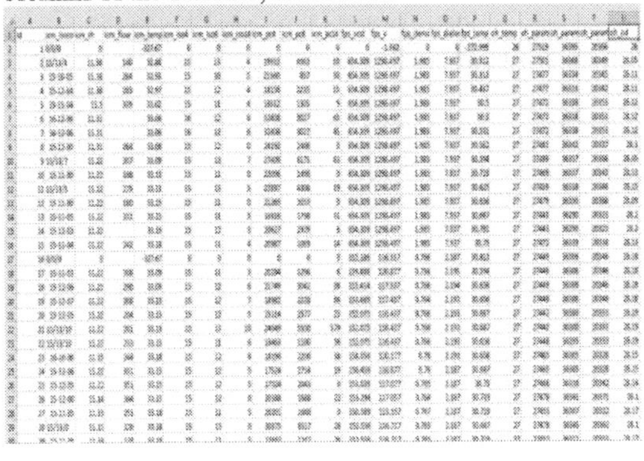

4. Conclusion

The activities carried out at this stage by the project partners were focused on the development and testing of prototype variants; analyzing continuous signals to identify the relevant signal characteristics and determining the useful data collection (Table 3) space to be able to provide an accurate estimate, or as close as possible to accuracy; Support Vector Machines training algorithms were developed to read the signal characteristics based on "learning". Numerical experiments have been conducted in different ways of using data through on-line analysis[5], [7]. Alternatively, the signal was divided into a larger number of frames that overlap partially (at the limit, NK-1 frames, provided that the first contains the data between 1 ... K, the second zone 2 ... K + 1, etc.). In this way, the data is partially superimposed on a given order, but the advantage is that the data set becomes a time series of parameters in a smoother regression and therefore easier to extrapolate.

Acknowledgment

This work is supported by Machine Intelligence for Diagnosis Automation (MIDA), Research Program, National Program PN II, ERA MANUNET: NR 13081221/13.08.2013

References

1. Dennis H. Shreve-Integrated Condition Monitoring Technologies, IRD Balancing LLC, 2003.
2. Stolarski, T. A. –Tribology in Machine Design, Butterworth Heinemann, 1990.
3. Constantin, V., Palade, V.-Machine Tools Parts and Mechanisms (Organe de maşini, şi mecanisme), Ed. Fundaţiei Universitare Dunărea de Jos, Galaţi, 2004.
4. Prashad, H. -Solving tribology problems in rotating machines, Woodhead Publishing Ltd., Cambridge, 2006, England.
5. Grebenişan Gavril, *Experimental research data processing by wavelet families transforms*, Nonconventional Technologies Review (indexată ProQuest) – no. 1/2009, ISSN 1454-3087, Oradea, 2009.
6. G. Adriani, M. Campatelli, M. Paoli, -Monitoil® - Online sensors for efficient and cost-saving oil conditions' monitoring, Mecoil Diagnosi Meccaniche s.r.l., Florence, Italy, http://www.mecoil.net/wp-content/uploads/2013/04/20120914-Articolo-colonne_Monitoil_eng.pdf
7. Grebenişan Gavril, Sanda Bogdan-*Parameterized Finite Element Analysis of a Superplastic Forming Process, Using ANSYS*, MATEC Web of Conferences, Volume 126, 2017, https://doi.org/10.1051/matecconf/201712603003

An approach of classification and parameters estimation, using neural network, for lubricant degradation diagnosis

Gavril Grebenişan[1,*], *Nazzal* Salem[2], and *Sanda* Bogdan[3]

[1]University of Oradea, Faculty of Manag. and Technological Eng., Industrial Engineering Department, Universitatii 1, Romania
[2]Zarqa University, Faculty of Engineering Technology, P.O. Box 132222 Zarqa 13132, Jordan
[3]University of Oradea, Faculty of Manag. and Techn. Eng., Mechanical Eng. and Automotive Department, Universitatii 1, Romania

Abstract. This paper addresses a delicate problem, namely the diagnosis of the state of the oils in the industrial systems, namely the machine tools. Based on measurements (the data set contains over five million records), within a Machine Intelligence for Diagnosis Automation (MIDA) project funded by the National Program PN II, ERA MANUNET: NR 13081221 / 13.08.2013, several applications of MATLAB toolbars are being developed in the field of artificial intelligence, specifically using the Support Vector Machine algorithms and neural networks. The tests were carried out on several distinct situations, followed by validation and verification tests on the devices designed and developed within the project (MIDA, Monitoil).

1 Introduction

A lot of methods in the fault (error) diagnosis literature are based on linear or accurate methodologies, or with a high dose accuracy. It is known that industrial processes are often difficult to model. These are complex and not precisely known; noise and uncertain, error-prone sensors corrupt measurements. Therefore, some researchers have perceived artificial neural networks as an alternative way to represent knowledge about malfunctions, but also to find ways to solve technical problems with infinitesimal errors, but sometimes also abruptly influenced by errors approach, or design. Neural networks can filter noise and disturbance; they can provide a stable diagnostic, failures without traditional types of models, extremely sensitive, and economic efficiency due to insignificant computing and design effort. Another desirable feature of neural networks is that exact patterns are not required to reach the decision stage [1], [2], [4], [5], [7], [8], [11]. In a typical operation, the process model can only be approximate and critical measurements may be capable of internally crunching functional relationships that represent processes, filtering noise, and managing correlations. Although there are many promising examples of neural network simulations in diagnosing errors in specific literature, real applications are still quite rare. There is a great need to carry out more detailed scientific investigations on the application of neural networks in real industrial installations to achieve full use of their attractive features.

Using a neural network with appropriate parameters and valid statistical architecture and weights, any continuous non-linear relationship can be approximated with arbitrary precision, a significant property that a relatively complicated but controllable algorithm as a mathematical process can be called by functions or specific applications in MATLAB, or dedicated software.

Neural networks are composed of simple elements that work in parallel. The biological nervous systems inspire these elements. As in nature, the network function is primarily determined by the connections between the elements. We can train a neural network to perform a specific function by adjusting the values of the connections (weights) between the elements, [3], [6]. Typically, conventional neural networks are adjusted, or trained, so that a particular entry leads to a specific target output. Such a situation is presented in Fig. 1.

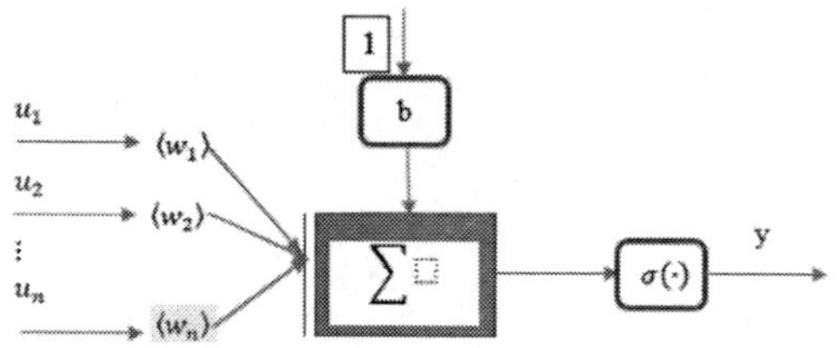

Fig. 1. Neuron scheme with n inputs(u_i), w_i −weighted, biased(b), $\sigma(\cdot)$ is the non-linear activation function and one output (y)

Here, the neural network is adjusted, based on a comparison between output and target, until the network output matches the target. Typically, many such input/ target pairs are used to train a network, this type of neural network is called a supervised learning network. The classical mathematical model of such a neural network is described by the following equation, [1]:

$$y = \sigma(\sum_{i=1}^{n} \omega_i u_i + b) \qquad (1)$$

* Corresponding author: ggrebenisan@gmail.com

© The Authors, published by EDP Sciences. This is an open access article distributed under the terms of the Creative Commons Attribution License 4.0 (http://creativecommons.org/licenses/by/4.0/).

where u_i are the system inputs, w_i – the weights coefficients, b-the bias(treshold), $\sigma(\cdot)$ is the non-linear activation function and the output y. In Fig. 2. a single-layer neural network, with S neurons, is presented. Note that each of the network inputs is connected to each of the network's neurons, for this reason the weight matrix now has S rows. The layer includes the $W_i, i = 1 \ldots n$- weight's matrix, the summers, the bias vector, $b_j, j = 1 \ldots s$, the transfer functions,$\sigma(\cdot)$, and the output vector, $y_j, j = 1 \ldots s$. Each element of the input vector u_i is connected to each neuron through the weight matrix, $W_i, i = 1 \ldots n$. Each neuron has a bias vector, $b_j, j = 1 \ldots s$, a summer, a transfer function, $\sigma(\cdot)$, and an output vector, $y_j = \sigma(W \cdot u_i + b_j), j = 1 \ldots s$.

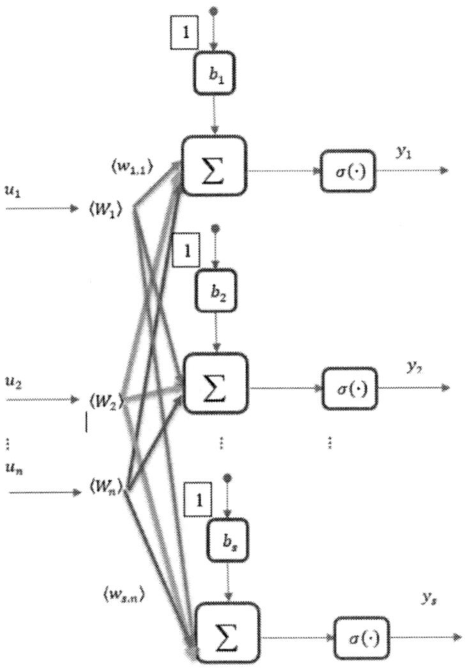

Fig. 2. A single layer network, with n inputs and S neurons

Not always the inputs vector size is similar to the number of neurons, it is quite no common to be equality between these characteristics. Also, not all neurons in a layer may have the same transfer function. In this case, it is common to define a single (composite) layer of neurons that have different transfer functions by combining two or more networks in parallel, depending on the number of different transfer functions. Thus the neural networks could have the same inputs and each network would create their own outputs. The weight matrix, W, trough which the inputs vector enter to the network is:

$$W = \begin{bmatrix} w_{1,1} & w_{1,2} & \cdots & w_{1,n} \\ w_{2,1} & w_{2,2} & \cdots & w_{2,n} \\ & \vdots & & \\ w_{S,1} & w_{S,2} & \cdots & w_{S,n} \end{bmatrix} \quad (2)$$

Now we consider a multilayer neural network(Fig. 3.), [2]. Each layer has its own weighting matrix, its own bias vector, a vector with n inputs, and a vector of outputs, y(the layers are identified, symbolically, by the superscript, which indicates its number, eg the weight matrix, in layer 3 is W^3). Thus, the output of the neural network is given by the relationship:

$$y^3 = \sigma^3\left(W^3 \cdot \sigma^2\left(W^2 \cdot \sigma^1\left(W^1 \cdot u_i + b^1\right) + b^2\right) + b^3\right) \quad (3)$$

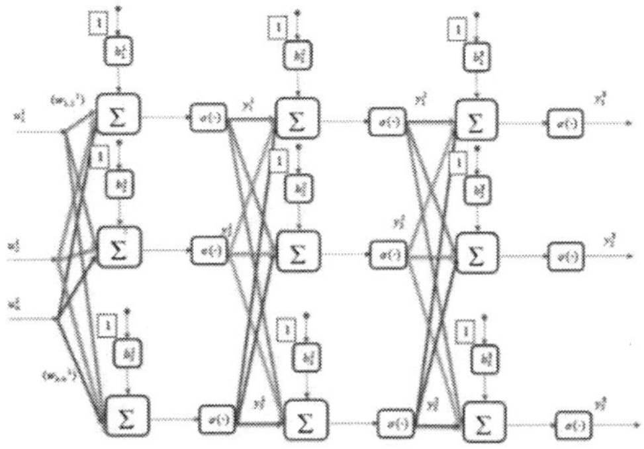

Fig. 3. The 3-layer neural network, with n inputs and S neurons, [9]

Some definitions shall be presented in order to use a unified terminology, [8]:

- The fault is an unaccepted deviation of at least one characteristic property or system variable from that characteristic of the acceptable / standard / standard behavior of the system.
- The failure is a permanent interruption of the system's ability to perform a function under certain operating conditions.
- The fault detection is an identification of the defects present in the system and the time required to detect them.
- Isolation of the fault is an identification of the type, location and time of detection of an error / malfunction. Follow fault detection.
- Identifying defects is a determination of behavioral deviation, size, and variance over time of a malfunction. The isolation of the fault follows.
- The fault diagnosis is a determination of the type, size, location, and detection time of a malfunction. Follow fault detection. Includes both the isolation of defects as well their identification. The main objective of fault diagnosis is to detect defects in each subsystem and their causes early enough so as to avoid global system failure and provide information about their size and sources.

Even if the elements used in the current industrial systems are of high reliability, there are always failures of the components of the technological components and of the automatic devices as well as faults due to the mistakes of the operator. According to some specialists, the degradation states, which occur frequently during the maintenance of the technological components, are quite natural. They cause significant and long-term disturbances during the manufacturing process, which leads to a decrease in productivity and can sometimes result in complete, definitive damage to these systems. In these cases, economic losses are very high. Some malfunctions may lead to reactions that may result in environmental hazards or damage to production facilities. They can also be dangerous to human life. Identifying the failure state is usually a delicate problem, especially in the case of complex automated systems. The human operator may often be overwhelmed in short periods of time (such as accidents or successive chemical reactions). These systems are irremediably affected by the lack of a prompt decision taken by the human operator, or by the effects of delayed decisions, and in the absence of correct decisions, systems can evolve into defects or physically irreparable states. Problems of diagnosis and process protection are constantly developing. The importance of this problem increases with economic growth, with the degree of automation and justifies the development of maintenance and technological monitoring compartments. Therefore, computational systems that allow either operators to perform a correct diagnosis in the event of an alarm or fault detected or to establish a rational diagnosis in an automatic mode, which is essential.

Compared to diagnostic operations performed by an operator, the automation of diagnostic operations makes it possible to shorten the response times significantly, and to narrow the time elapsed between the time of identification and complete isolation of the malfunctions. This has the effect of increasing the reliability, the economic efficiency of protected systems and the development of automation in the field, [8].

2 Databases used for classification using neural network

Data on the condition of the oil parameters used in the hydraulic system for lubricating machine tools have been collected for six months. The data are matrixed, with 258648 rows (observations) and 21 columns (parameters). The parameters are presented in Table 1, [12].

Developing and testing prototype variants of experimental data monitoring and collection systems, analyzing continuous signals to identify the relevant features of the investigated parameters, and determining the "useful" data collection space to be able to provide an estimate accurate, or as close as possible, were designed to create the conditions and the mathematical environment that are conducive to the development of Support Vector Machines and Classification and Parameters Estimation (using neural network based modeling) algorithms suitable for reading the signal characteristics on the basis of "learning".

Table 1 The parameters collected, [12]

id	measure identifier
postgresql_timestamp	writing time
time	writing time
raw_data	Maybe neglected, it is not important (data before process starts)
icm_isocode	ISO code 4406 (icm_iso4/ icm_iso6/ icm_iso14)
icm_rh	water saturation%
icm_flow	oil flow through particle counter
icm_temp	particle counter electronic temperature
icm_iso4	ISO 4406 code class particles >4 μm
icm_iso6	ISO 4406 code class particles >6 μm
icm_iso14	ISO 4406 code class particles >14 μm
icm_pc4	number of particles >4 μm
icm_pc6	number of particles >6 μm
icm_pc14	number of particles >14 μm
fps_vcst	viscosity [cSt]
fps_v	viscosity [cP]
fps_density	density
fps_dielectric	dielectric constant
fps_temp	oil temperature
oh_temp	Oilhealth electronic temperature
oh_parama	Oilhealth paramA (not important)
oh_paramb	Oilhealth paramB (not important)
oh_paramc	Oilhealth (not important)
oh_od	Index of degradation OD%
Date	day of analysis
time	hour/time of analysis

Table 2. Data collected- the first 30 from 258648 rows and all columns of the database, are showed, [12]

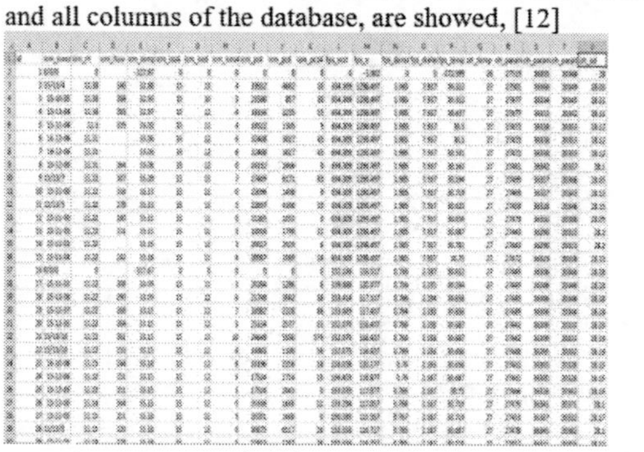

Experiments and in-situ trials have been carried out considering the basis of data usage using on-line analysis, which means efficiency and reproducibility, [10].

3 Results and validation discussion

Testing, training, and validation of the collected data set (N observations) were possible through mathematical and experimental statistical operations, by generating/ separating K different time frames (each consisting of N /

K data) and analyzing the signal characteristics on each of these frames (which have the advantage of being composed of data that is not repeated and therefore not linearly dependent). Alternatively, the signal was divided into a larger number of frames that overlap partially (at the limit, NK-1 frames, provided that the first contains the data between 1 ... K, the second zone 2 ... K + 1, etc.). In this way, the data is partially superimposed on a given order, but the advantage is that the data set becomes a time series of parameters in a smoother regression and therefore easier to extrapolate.The data set consists of 5431608 observations collected every minute from a work machine. The data refer to the analyzes made on the manufacturing line on the following parameters: density, viscosity, dielectric, water, oil degradation, temperature, ISO4, ISO6, ISO14, with obvious importance for the oil condition, but also for the operation of the industrial equipment being tested (partialy showed on Table 2, and graphical represented in plot figure, Fig. 4.) .

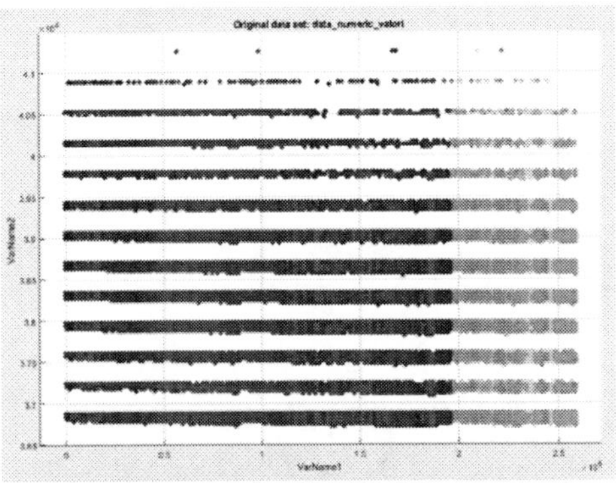

Fig. 4. Original data set (the all 5431608 parameters)

Once the dataset was received, the project coordination team, conducted data analysis, customized software development, and data set configuration to "clean up" the missing data set. Later, the "learning" of the machine was performed using the Support Vector Machine for a class (Fig. 5.).

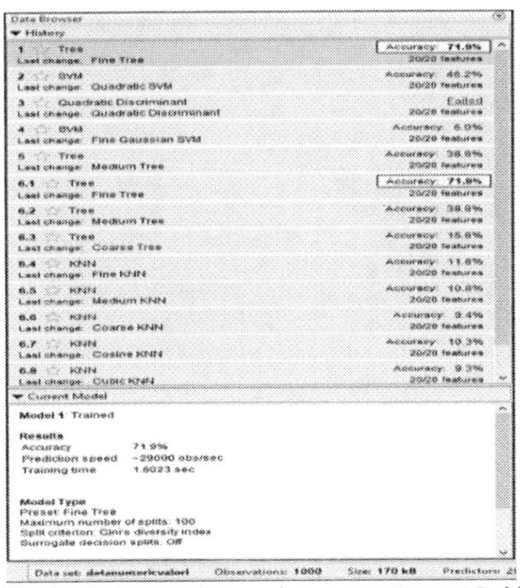

Fig. 5. The training phase, using Support Vector Machine

At the end of experimentation on this first set of data, it was decided to perform the training phase using 50 minute superimposed frames (the first frame consists of the data collected between minute 1 and minute 50, the second frame consists of data collected between minute 2 and minute 51, and so on). With the data in each frame the best fit is based on a model based on the ARIMA model whose parameters are the characteristics used for the training. The training phase, including a cross-validation part, generated SVM machine, resulting in machine training in a class with an accuracy of more than 70%(Fig. 5.). Support Vector Machine is used for training on the computing units available, capable of analyzing a 50-minute observation sample and performing the test using SVM in less than one second by the computing processor (Artificial intelligence).

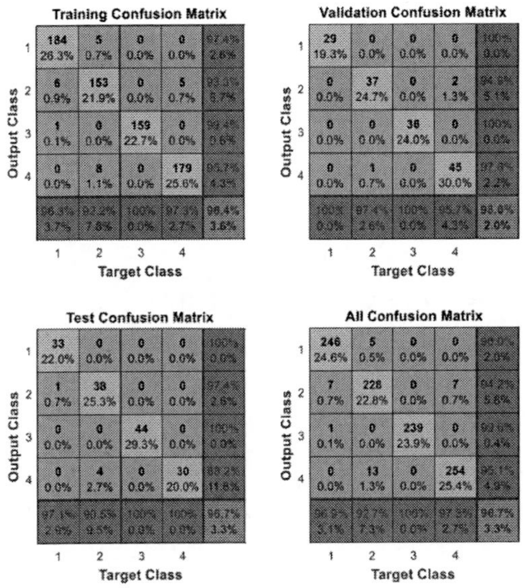

Fig. 6. Training Confusion Matrix, Testing Confusion Matrix, Validation Confusion Matrix, the summary of Confusion Matrix

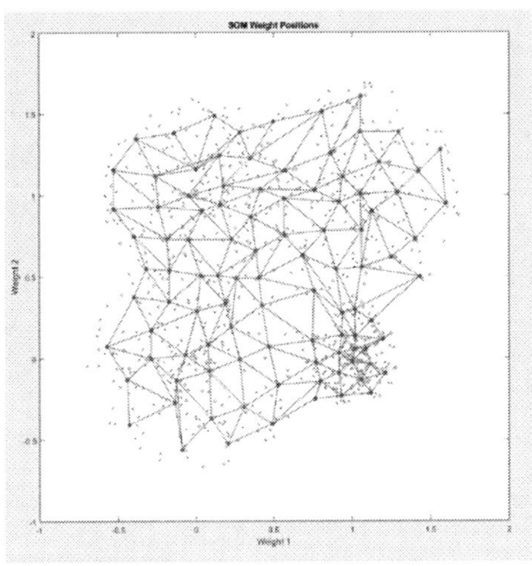

Fig. 7. The regression learner for selected data

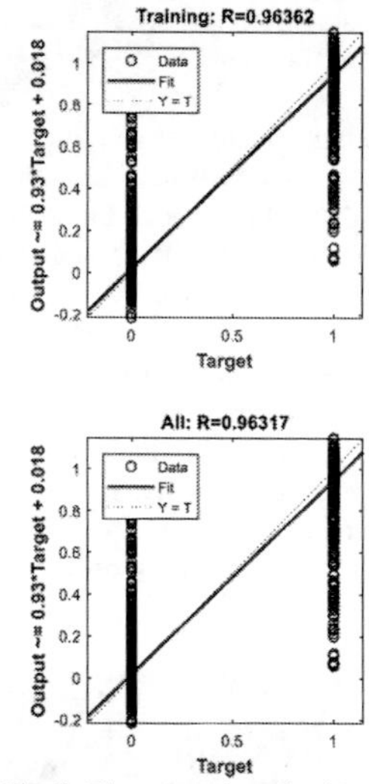

Fig. 8. The regression plot (SVM)

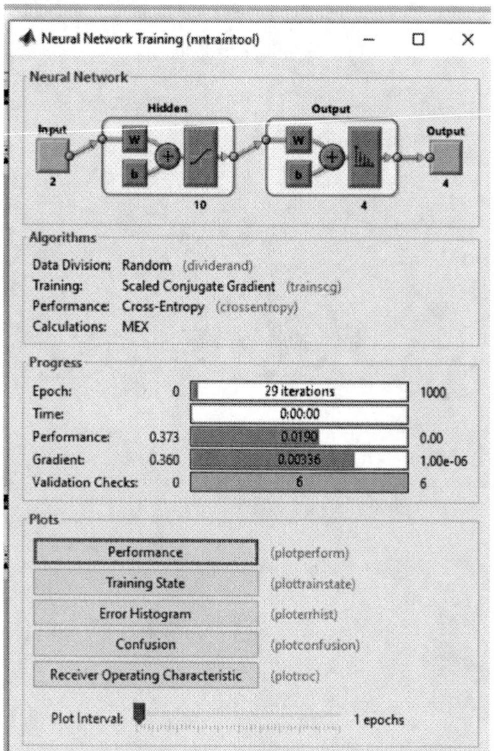

Fig. 9. The neural network training results and coordinates

The results of the training using the neural network algorithm are presented in Fig. 9. Here you can notice the architecture of the network as well as the numerical parameters of the network. In Fig. 10, respectively in Fig. 11. The results of the tree classification are represented graphically in two variants: complex tree, and simplified tree, respectively, [9], [10].

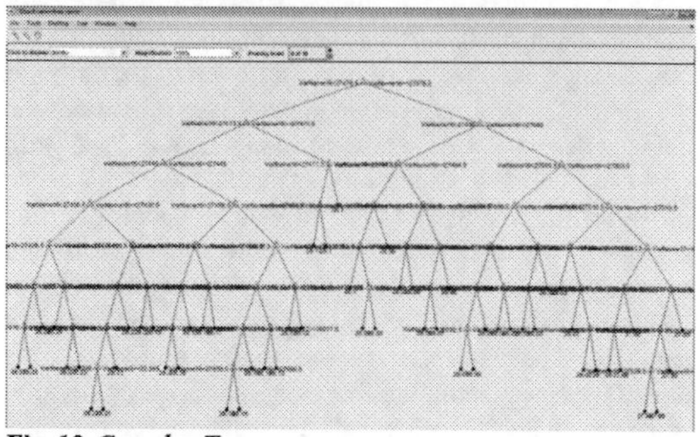

Fig. 10. Complex Tree

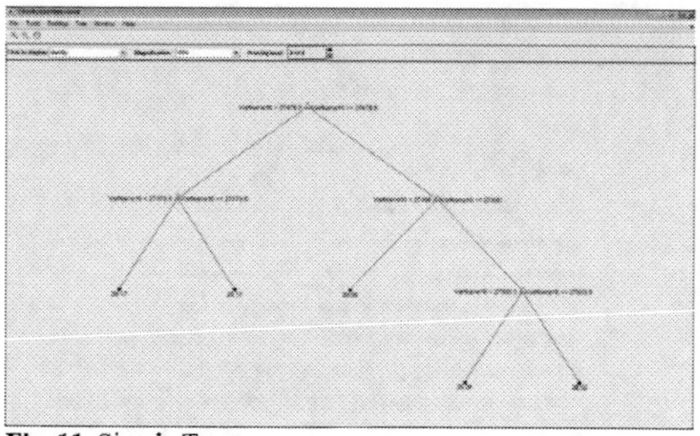

Fig. 11. Simple Tree

In Fig. 12. the distribution in two dimensions of the oil status parameters analyzed in the research project is represented, ie the influence of the content of hydrological contaminants and the impurities on the viscosity of the lubricant. Classification accuracy was 63% for a SVM-Complex Tree algorithm.

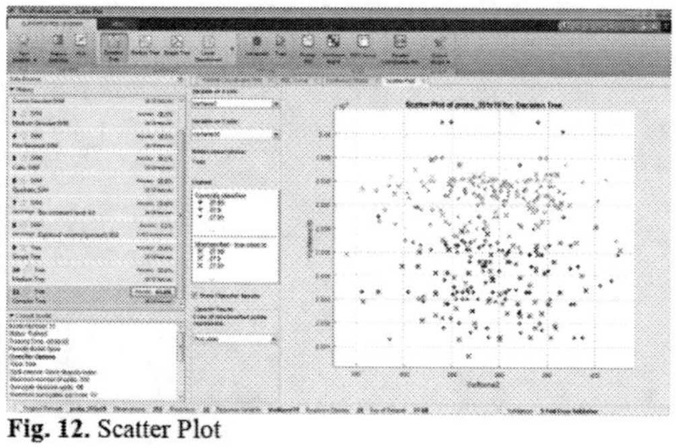

Fig. 12. Scatter Plot

Perhaps a more eloquent and deeper exposure to the behavior of the classifier is revealed by the so-called matrix of confusion. The i, jth element of this matrix represents the number of ω_i samples in the validation set to which the ω_j class is assigned. The confusion matrix is graphically presented in Fig. 6 and Fig. 13.

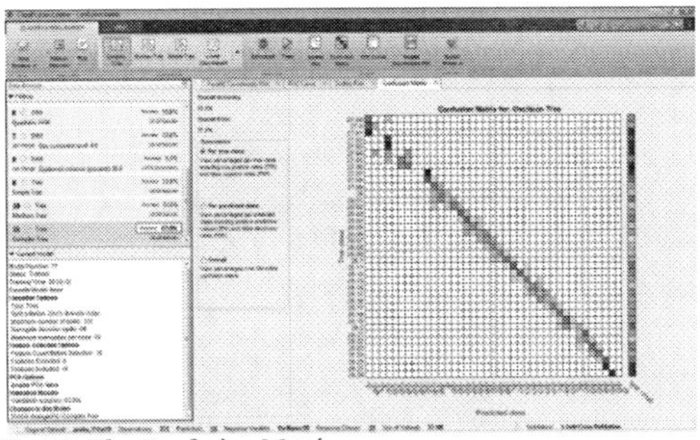

Fig. 13. The Confusion Matrix

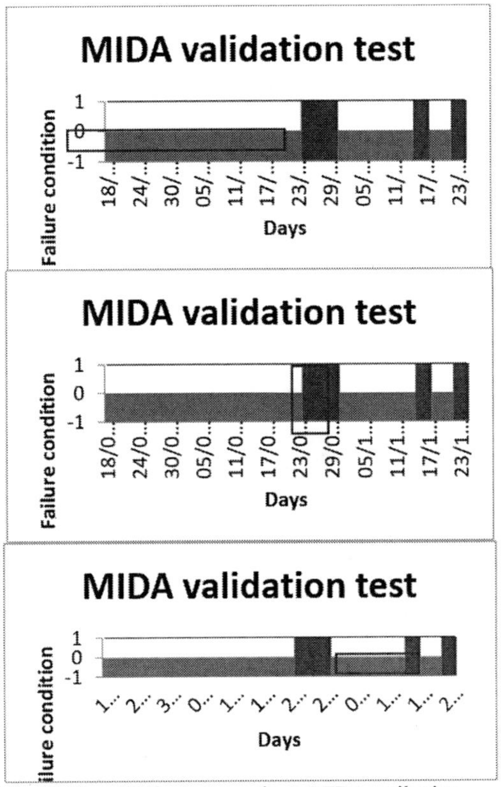

Fig. 15. The validation tests, results on MIDA application

4. Conclusion

The algorithm on which the system operates is stable, not influenced by variations in operating parameters. The system has an extremely simple operating principle and does not require additional staff training. The system works on-line: the speed of response to errors in measurement is very high. Alarms are generated to exceed the preset (learned) values of the parameters. The system can be customized for any industrial equipment. The system is adaptable as it can be "taught" to determine the errors that occurred during the monitoring of any process. The system is safe, with no possibility of interference during data collection or transmission of collected data. The system is replicable. System improvements are possible: as regards the computing system (more efficient processor); as regards the speed of data acquisition; regarding the speed of data transmission. Application of the system to other types of equipment in other fields is always possible (in aviation, auto, etc.). The system can be certified.

Acknowledgment

This work is supported by Machine Intelligence for Diagnosis Automation (MIDA), Research Program, National Program PN II, ERA MANUNET: NR 13081221/13.08.2013

References

1. K. Patan- *Artificial Neural Networks for the Modelling and Fault Diagnosis of Technical Processes*, LNCIS 377, Springer-Verlag Berlin Heidelberg 2008.
2. H. Demuth, M. Beale-*Neural Network Toolbox For Use with MATLAB®*, The MathWorks, Inc., 2014.
3. M. T. Hagan, H. B. Demuth, M. H. Beale, O. De Jesús- *Neural Network Design*, Second Edition, 2014, ebook.
4. P. M. Frank, B. Koppen-Seliger *New Developments Using AI in Fault Diagnosis*, Engng. Applic. Artif Intel. Vol. 10, No. 1, pp. 3-14, 1997 © 1997 Elsevier Science Ltd.
5. H.N. Kolvo, *Artificial Neural Networks In Fault Diagnosis And Control*, Control Eng. Practice, Vol. 2, No. 1, pp 89-101, 1994, Printed in Great Britain.
6. J. R. Rabunal and J. Dorrado *Artificial Neural Networks In Real-Life Applications*, Idea Group Publishing, 2006, London.
7. L. H. Chiang, E. L. Russell and R. D. Braatz, *Fault detection and diagnosis in industrial systems*, Springer-Verlag London Limited, 2002.
8. J. Korbicz : J.M. Koscielny - Z. Kowalczuk : W. Cholewa, *Fault Diagnosis -Models, Artificial Intelligence, Applications*, Springer-Verlag Berlin Heidelberg GmbH, 2004.
9. B. Lei, D. de Ridder, D. M. J. Tax, F. van der Heijden, G. Xu, M. Feng, Y. Zou - *Classification, Parameter Estimation and State Estimation An Engineering Approach Using MATLAB*, Second Edition, 2017 JohnWiley & Sons, Ltd.
10. A. Gelman and J. Hill, *Data Analysis Using Regression and Multilevel/Hierarchical Models*, Cambridge University Press, 2006.
11. J. Korbicz, M. Mrugalski: *Confidence estimation of GMDH neural networks and its application in fault detection systems*, International Journal of Systems Science (2007) DOI: 10.1080/00207720701847745
12. G. Grebenişan, S. Bogdan, N. Salem: *The lubricants parameters monitoring and data collecting*, Annual Session Of Scientific Papers, "IMT ORADEA 2018", Volume XXVII (XVII), ISBN 978-606-10-1537-5, Oradea, 2018

Development of the method of direct temperature measurement for gear cutting of cylindrical gears by hob milling

Sandra Sovilj-Nikić[1,*], *Bogdan* Sovilj[2], *Dušan* Ješić [3], *Vladimir* Blanuša [4], *Vladimir* M. Tonkonogyy[5], *Liubov* Bovnegra[6]

[1]Iritel a.d. Beograd, Batajnički put 23, 11080 Belgrade, Serbia
[2]University of Novi Sad, Faculty of Technical Sciences, Trg Dositeja Obradovica 6, 21000 Novi Sad, Serbia
[3]International Technology Management Academy, Trg Dositeja Obradovica 7, 21000 Novi Sad, Serbia
[4]High Technical School of Professional Studies, Školska 1, 21000 Novi Sad, Serbia
[5]Odessa National Polytechnic University (ONPU), Shevchenko Avenue 1, 65044 Odessa, Ukraine
[6]Odessa National Polytechnic University (ONPU), Shevchenko Avenue 1, 65044 Odessa, Ukraine

Abstract. The availability, adaptability, reliability and productivity of modern industrial complexes, maintenance costs, energy costs and industrial production costs depend also on the intensity of wear of the elements of tribo-mechanical systems. Gear cutting operations are usually a bottleneck of production, especially in higher types of production. Therefore, the necessity of researching the process of gear cutting by hob milling is obvious. In the paper a part of the results related to the possibility of direct measurement of the temperature in a very complex process of gear cutting of cylindrical gears by hob milling is given.

1 Introduction

The rapid development of industrial production in the world has caused the need for extensive research in all areas, even in the field of machining.

Hob milling, as one of the most complex machining processes, has the widest application in the gear cutting of cylindrical gears due to the high productivity of the process. In Fig. 1 [1], the principle of hob milling and kinematics are shown.

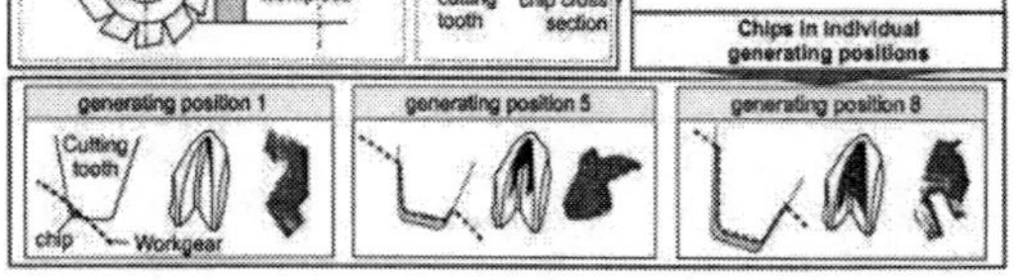

Fig. 1. Hob milling principle and kinematics

The complicated kinematics and geometric relations between the hob milling tool and the work-piece create a series of difficulties and problems that prevent the optimal use of tool and machine (Fig 2 [2]).

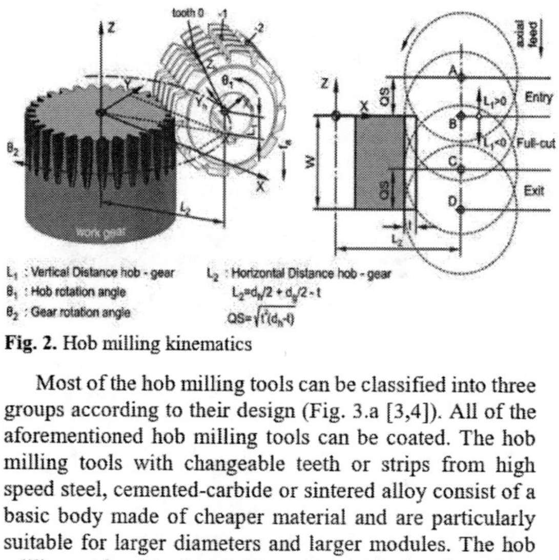

Fig. 2. Hob milling kinematics

Most of the hob milling tools can be classified into three groups according to their design (Fig. 3.a [3,4]). All of the aforementioned hob milling tools can be coated. The hob milling tools with changeable teeth or strips from high speed steel, cemented-carbide or sintered alloy consist of a basic body made of cheaper material and are particularly suitable for larger diameters and larger modules. The hob milling tools with changeable teeth beside that they allow larger constructive back angles, they also have a relatively large, usable cutting length.

Various variants of hob milling tools that combine different processes such as roughing and finishing and machining of chamfers for removing irregularities (Fig. 3 b [3,4]) have also been developed. In order to increase the productivity of the hob milling, gear cutting process is very close to the limit of the technological capabilities of the machine tools and the hob milling machines. Therefore, the optimal choice of the hob milling tool material is important in achieving this goal.

* Corresponding author: sandrasn@eunet.rs

© The Authors, published by EDP Sciences. This is an open access article distributed under the terms of the Creative Commons Attribution License 4.0 (http://creativecommons.org/licenses/by/4.0/).

Fig. 3. Modern hob milling tools

In Fig. 4 [5], the influential parameters on the process of wear of hob milling tool are given. Hob milling is a multi-parametric and complex method for the gear cutting of cylindrical gears.

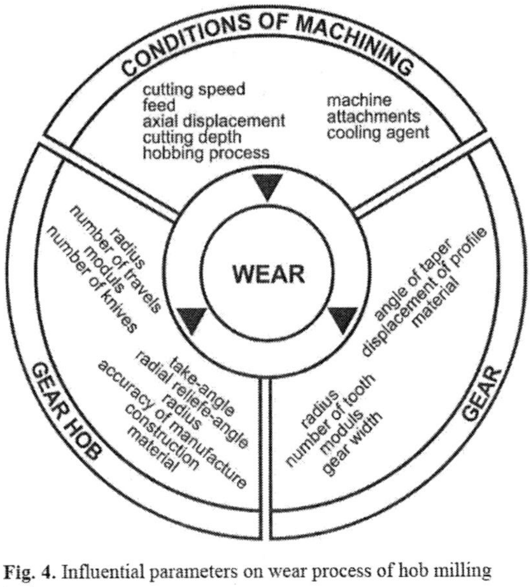

Fig. 4. Influential parameters on wear process of hob milling tool

A comprehensive knowledge of the gear cutting process is crucial for solving the demands of the global market. The world market today is being tackled with rapid and continuous changes. Fig. 5 [6] shows a production environment in the 21st century. A number of factors affecting the world market can be seen in the Figure.

Production operations and methods of gear cutting in mass production depend on various parameters such as cost of gear cutting, production philosophy, practical experience, etc. A common strategy is to achieve the highest possible accuracy during rough gear cutting and to perform thermal treatment with the necessary tolerance to completely avoid finishing the serration [4].

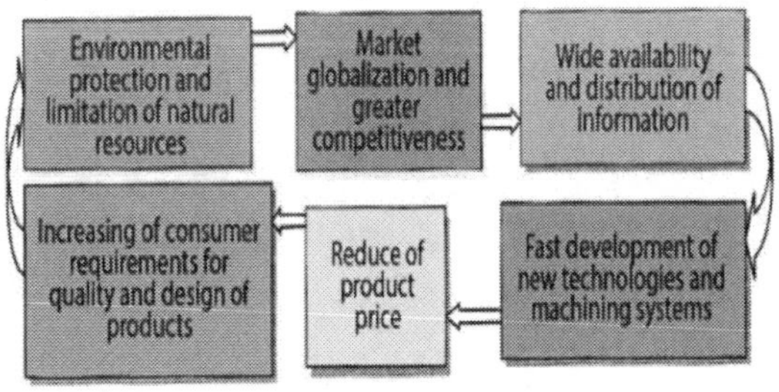

Fig. 5. Manufacturing environment in 21st century

Many physical and chemical reactions that occur in the process of gear cutting by hob milling and are directly related to the wear of hob milling tool, are largely dependent on the temperature at which they occur, whether it is the speed of their advancement or achieving the balance. It is obvious that the reasons for measuring the temperature are numerous [7].

2 Thermal phenomena in hob milling

The research of the thermal phenomena due to mentioned in previous section is one of the key problems in the gear cutting by hob milling. Knowledge of the basic principles on which the layout of the heat sources and the character of the temperature field in the hob milling tool, work-piece and chip are based, enables a better formulation of the basic theory of the chip formation and other phenomena characteristic for the process of gear cutting by hob milling.

The scope of the method for the research of the chip formation in the gear cutting by hob milling is given in Fig. 6 [4,8].

Fig. 6 also shows areas in which the knowledge about the chip formation in the hob milling can be applied. The fields of application of the knowledge of the chip formation are widespread according to the direction of the evaluation of the cutting properties of the tools for gear cutting and materials of the hob milling tools, the machining economy, the conditions in the gear cutting on modern hob milling machines, the use of cutting fluids in the gear cutting, etc.

The theory of gear cutting, especially the part dedicated to the chip formation, is always conceived of knowledge and experimental procedures. Experimental methods are the source of the a priori data and information used to verify the set modeling procedures, which are shown in Fig. 7, 8 and 9 [4,8].

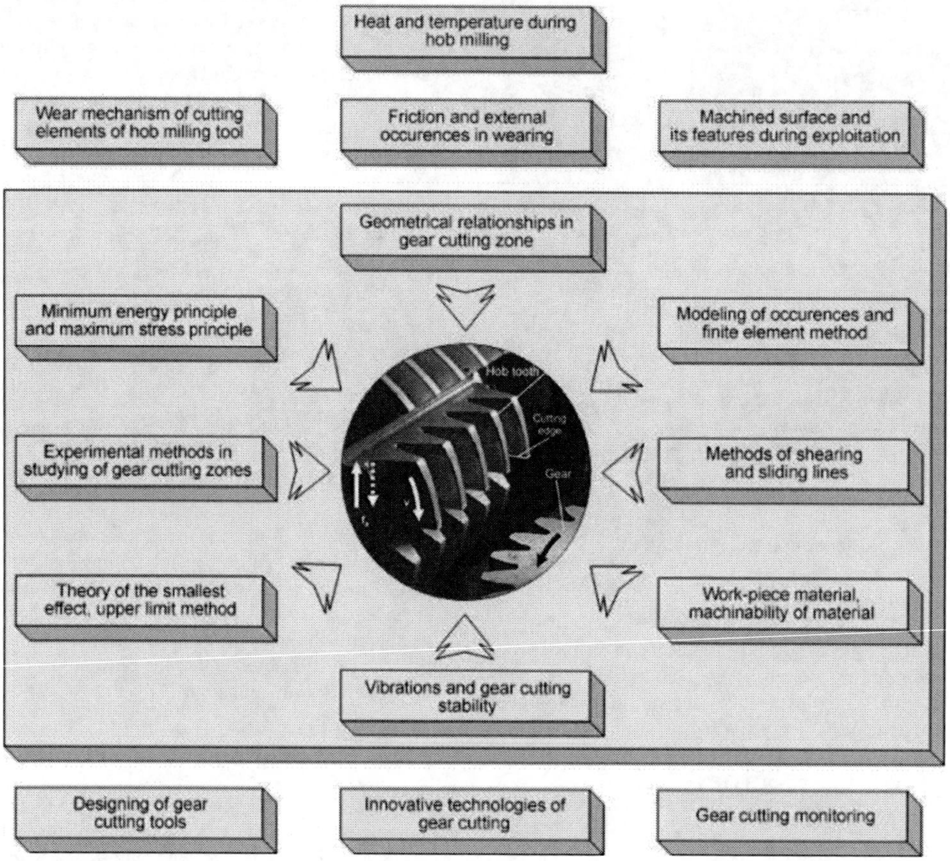

Fig. 6. Methods and procedures in the scientific approach to the research of the chip formation in gear cutting by hob milling

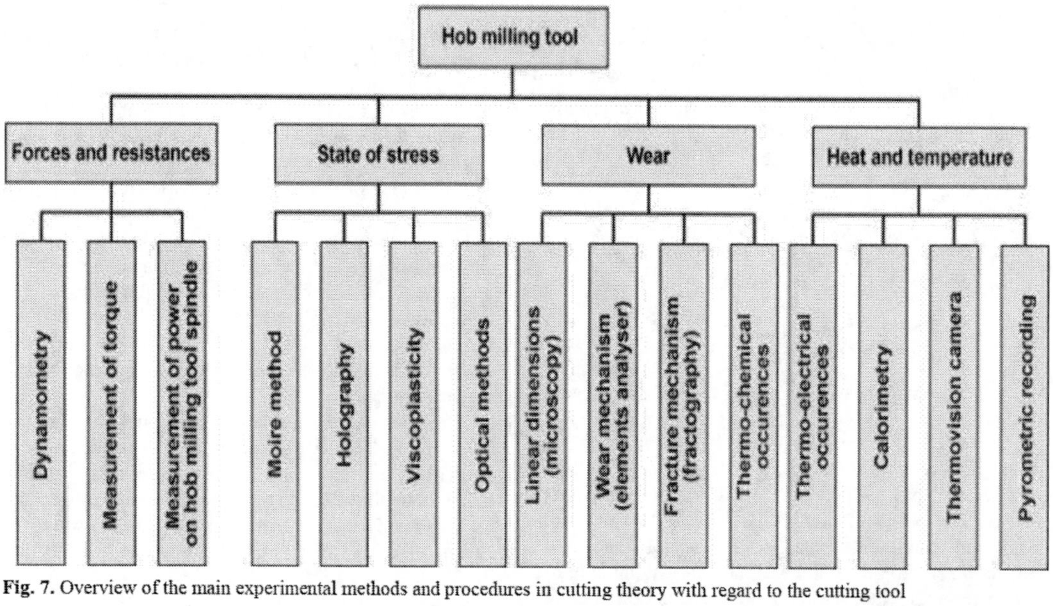

Fig. 7. Overview of the main experimental methods and procedures in cutting theory with regard to the cutting tool

 https://doi.org/10.1051/matecconf/201818403010

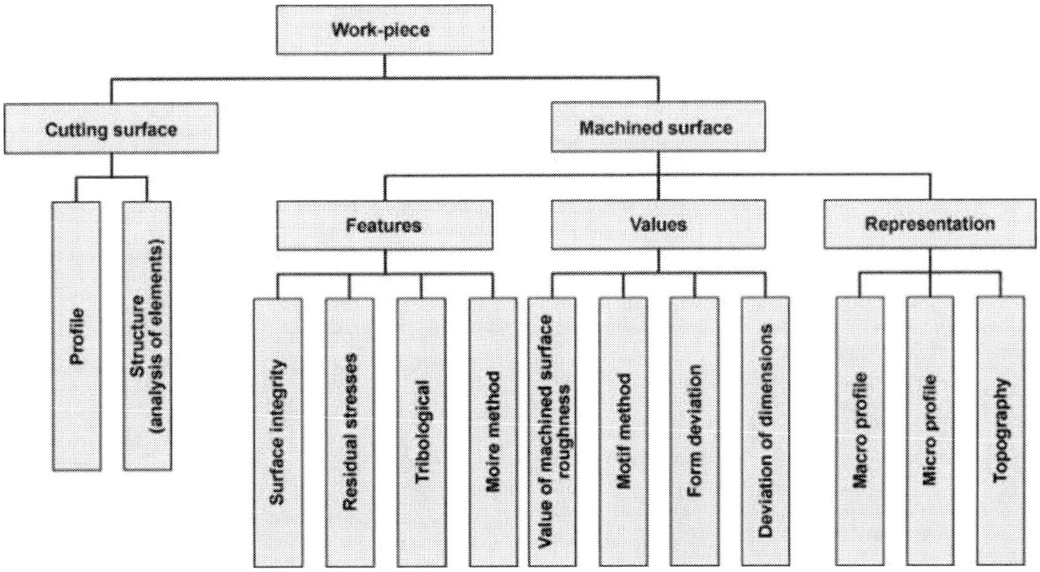

Fig. 8. Overview of the main experimental methods and procedures in cutting theory with regard to the work-piece

The scale of the experimental methods used is wide. In their application, the goal and availability decide. More procedures require special devices (for example, cutting failure) that are most often protected by patents.

Standardized methods for measuring surface roughness, stiffness of hob milling machines nodes, ISO standards and testing of cutting tools abilities are known, etc., but there are requirements for acquiring new knowledge that lead to new and unique measurement procedures. The overview shown in Fig. 7, 8 and 9 can be considered as orientation and can be further addressed in iterative procedures. They always refer to pairs of observation objects (for example a chip-hob milling tool, for example: for examining a surface contact, i.e., a real surface of the contacts, a shear surface of chip, etc.). A similar situation occurs in the case of a pair of hob milling tool- work-piece, where it is possible to test the dependence of the tool life, the machinability of the material and the cutting ability for gear cutting.

Contemporary gear cutting technology includes also finishing processes, which give the final shape of gear serration, determine the successive position of the surfaces, the accuracy and quality of their manufacture. For these reasons, it is very important to know all the processes and phenomena that have a significant impact on the resulting accuracy and quality of products to ensure effective production.

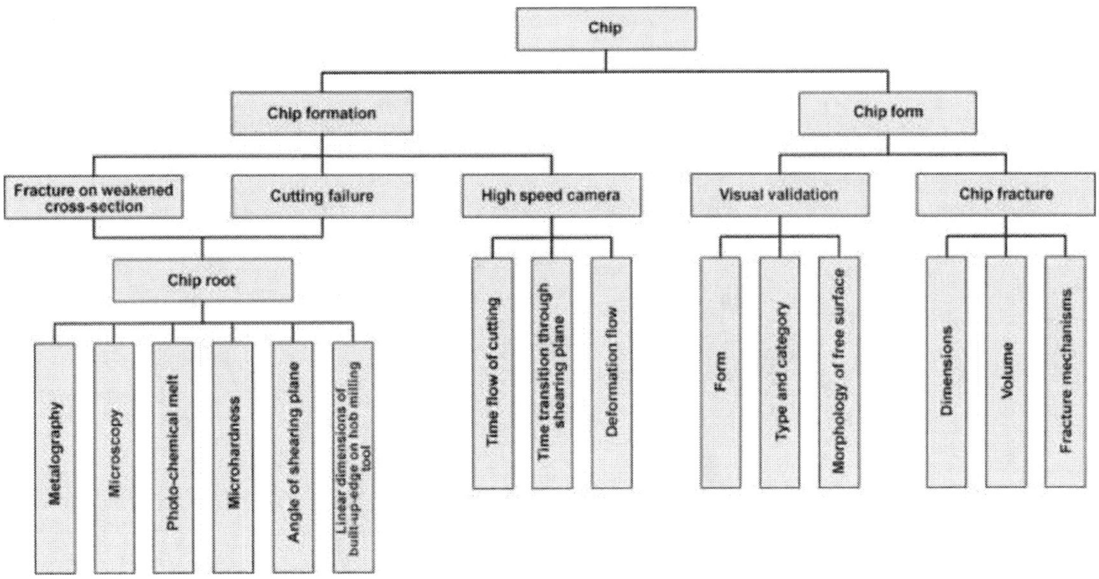

Fig. 9. Overview of the main experimental methods and procedures in cutting theory with regard to the chip

In addition to the above, the knowledge of the basic principles on which the layout of the heat sources and the character of the temperature field in the hob milling tool, work-piece and chip are based enables: the correct formulation of the basic theory of the tool life of the hob milling tool, as well as predicting the wear character and the degree of its intensity under different machining conditions; the correct formulation of the requirements for new materials of the hob milling tool that would allow the application of optimal parameters of the cutting regime; the evaluation of the cutting fluids in the gear cutting process, and the proper formulation of requirements for new more efficient cutting fluids, including their proper application; the application of the
results of the study of thermal phenomena in order to reduce the harmful effect of temperature on the work-piece, especially on its surface layer; determining the heat stresses in the cutting elements of the hob milling tool and their changes in time, in order to could evaluate their impact on the sudden fracture of the cutting elements of the hob milling tool; the improvement of the existing construction of the hob milling tools from the point of better drainage of the thermal energy, ie, the reduction of wear; the development of sensors for adaptive control of the hob milling machines in order to optimize the machining process using the influence of temperature on the tool life of the hob milling tool and the quality of the machined surface.

In this paper, the basis for the method of direct measurement of temperature in the gear cutting of cylindrical gears by hob milling is given.

3 The method for direct temperature measurement during hob milling of serration

Due to the great importance that knowledge of the temperature has in the process of gear cutting by hob milling, the need for the development of a new simpler, more reliable and accurate measuring method was imposed, which would operate on a direct principle. This can be achieved using miniature thermocouple whose external diameter of the protective coating is 0.5 mm and less.

In Fig. 10 a schematic diagram showing the principle of measuring the cutting temperature when machining by hob milling according to the design solution is given. A sample of 20CrMo5 was used as a work-piece with the same characteristics as well as samples that were used to research optimal cutting regimes and optimal cutting geometry of the hob milling tool, but additional preparation was performed. The preparation consisted of drilling four holes $\Phi 15$ at a distance of 15.5 mm from the outer edge of the work-piece. On the outside, and at a distance of 2 mm from the forehead, the hole the diameter $\Phi = 0.6$ mm was drilled normal to the axis of already drilled hole $\Phi 15\,\mathrm{mm}$. The next hole was drilled at a distance of 30 mm from the first, i.e. 32 mm from the forehead, which is the place where the model hob milling tool completed the entire machining of gullet at a distance of 2 mm from the forehead. The first thermocouple is indrawn into the first hole, and the second thermocouple in the second hole. Therefore, the model hob milling tool proceeds completely out of

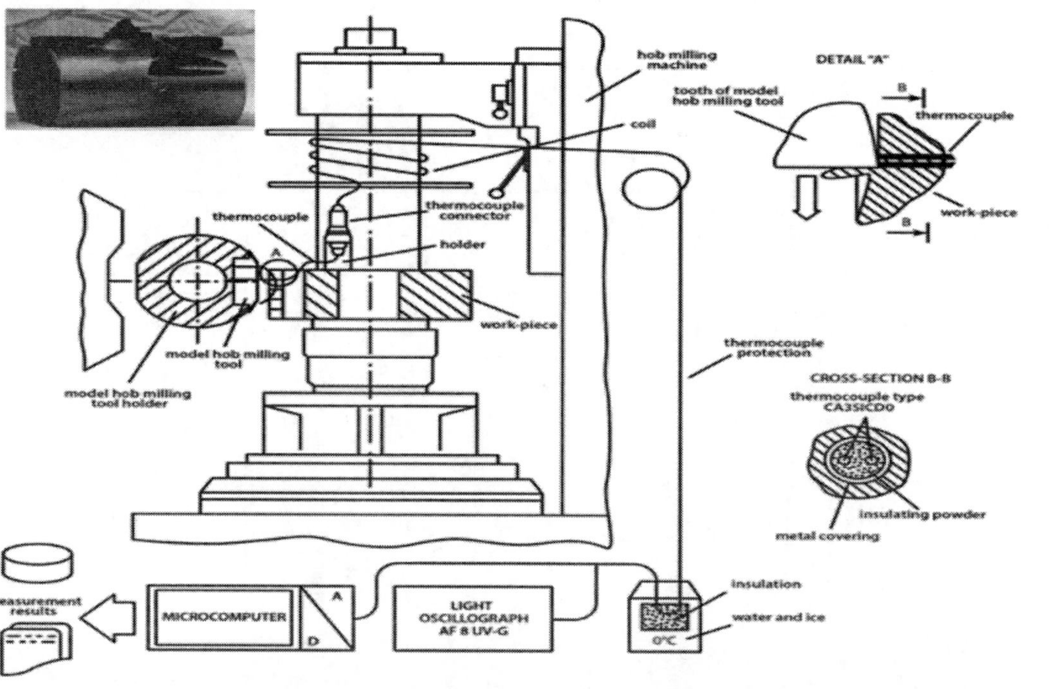

Fig. 10. Scheme of instrumentation used in measurement of temperature during hob milling

contact with the first thermocouple and then, after its rotation, it contacts with another thermocouple.

From the "Thermocoax" series, a miniature thermocouple type CA 3 S 1000 has been selected with the following characteristics: thermocouple wire: NiCr-Ni (Chromel-Alumel); polarity: (C 5000 S) NiCr (+) Ni (-); the voltage characteristic and tolerance; thermo stress cca $41\,\mu V/^{\circ}C$; insulation: Magnesium Oxide MgO; external diameter: 0.5 mm; diameter of the thermocouple wire: 0.08 mm; protective sheath: stainless steel; thickness of protective sheath: 0.06 mm; temperature range: 200 to 1200; minimum bending diameter: 0.5 mm.

The length of the thermocouple is 1000 mm, the length of the corresponding cable duct C5000S for the connection with the instrument is 5000 mm, and the connection between the thermocouple and the cable duct is carried out with a special plug LBE1/LSA1. In Fig. 10, detail A, a thermocouple with a protective sheath of Philips Company is shown.

Two holders are placed on the sample, in which plugs are installed (thermocouple connection). The cable duct is connected by copper conductors connecting the thermocouple with the measuring instrumentation or the microcomputer system. The connection of cable duct and copper conductors is placed in a bowl with ice in order to maintain zero.

The AF8 UV-G light oscillograph was used as an instrument (Fig. 10). It is necessary for the new research to develop software support that will be compatible with the process of hob milling, since the existing one, which is used for the measurement of the mean temperature in the turning process, is not satisfied due to specific conditions.

The light oscillograph has a very sensitive galvanometer with electromagnetic silencer type GA 1,1/8H (Fig. 10).

Calibration was performed in order to translate the size of the deflections registered on the paper into the values for the thermo stress or temperature. It is achieved by means of the compensating instrument "Thermocontrol" type YHK, by which it can be very precisely achieved the DC voltage from 0 to 100 mV. On the characteristic record, the measurement temperature was $670^{\circ}C$ (Fig. 11).

Conditions of measurement:

- Work-piece material: 20CrMo5,
- Model hob milling tool material: HS6-5-2-5,
- Cutting speed: v_c=98,25 m/min,
- Axial feed: f_a=2 mm/rev,
- Tangential feed: f_t=1,25 mm.

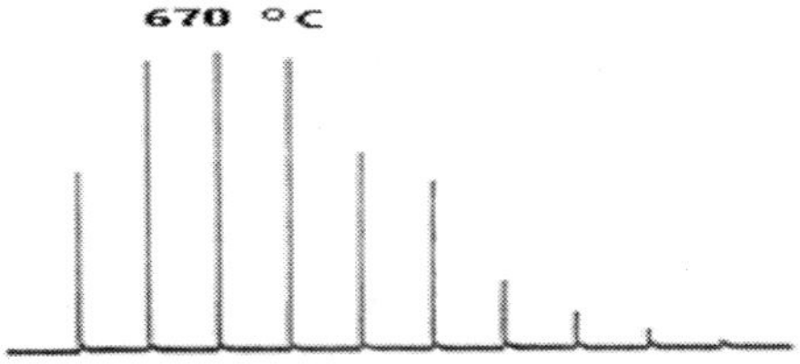

Fig. 11. Characteristic record of measured cutting temperature

In the future, the described method will be further improved and it will be possible reach to the basic principles on which the layout of the heat sources and the character of the temperature fields in the tooth of the model hob milling tool, the work-piece and the chip is based.

4 Conclusion

Improving the hob milling process is important and useful both for gear manufacturers and for manufactures of hob milling tools. Increasing the efficiency of the gear cutting of cylindrical gears is one of the basic directions for the development of modern gears. Hob milling is an efficient process for the gear cutting of high quality cylindrical gears.

The developed method of direct measurement of temperature in the gear cutting of cylindrical gears with straight teeth by model hob milling tool enabled the signal recording.

Acknowledgments

The research was funded by the Ministry of Education, Science and Technological Development of the Republic of Serbia, within the projects TR 35015 and III 43008, and it is also the result of the cooperation within CEEPUS project CIII-RO-0058-07.

References

1. A. Antoniadis, N. Vidakis, N. Bilalis, Fatigue fracture investigation of cemented carbide tools in gear hobbing, *J. Manuf. Sci. Eng.* **124** (4), 784-791 (2002)
2. V. Dimitriou, N. Vidakis, A. Antoniadis, Advanced computer aided design simulation of gear hobbing by means of three-dimensional kinematics modeling, *J. Manuf. Sci. Eng.* **129** (5), 911-918 (2007)
3. K. D. Bouzakis, E. Lili, N. Michalidis, O. Friderikos, Manufacturing of cylindrical gears by generating cutting processes: A critical synthesis of analysis methods, *CIRP Ann.* **57**, 676-696 (2008)
4. I. Sovilj-Nikić, *Modeling and optimization of hob milling process (Modelovanje i optimizacija procesa odvalnog glodanja)*, Fakultet tehničkih nauka, Novi Sad, Serbia, Doctoral Thesis (unpublished)
5. I. Sovilj-Nikić, B. Sovilj, M. Kandeva, V. Gajić, S. Sovilj-Nikić, S. Legutko, P. Kovač, Tribological characteristics of hob milling tools from economic aspect, *J. Balk. Tribol. Assoc.* 18 (4), 577-585 (2012)
6. M. Čohodar Husić, Đ. Begić-Hajdarević, A. Čekić, *Modern machining systems and industrial robots (Savremeni obradni sistemi i industrijski roboti)*, Sarajevo: Mašinski fakultet (2014)
7. D. Ješić, *Measuring techique (Merna tehnika)*, Univerzitet Banja Luka, Mašinski fakultet Banja Luka, Republic of Srpska (2004)
8. J. Beno, *Theory of metal cutting (Teoria rezania kovov)*, Srojnicka fakulta TU Košice, Slovakia (1999)

MATEC Web of Conferences **184**, 03011 (2018) https://doi.org/10.1051/matecconf/201818403011
Annual Session of Scientific Papers IMT ORADEA 2018

Analysis of the tool life of coated hob milling tools for gear cutting of cylindrical gear

Sandra Sovilj-Nikić[1,*], *Bogdan* Sovilj[2], *Gyula* Varga[3], *Nicolae* Ungureanu[4] and *Vladimir* Blanuša[5]

[1] Iritel a.d. Beograd, 11 000 Belgrade, Serbia
[2] University of Novi Sad, Faculty of Technical Sciences, 21 000 Novi Sad, Serbia
[3] University of Miskolc, Faculty of Mechanical Engineering and Informatics, 3515Miskolc, Hungary
[4] North University of Baia Mare, Engineering Faculty, 430311 Baia Mare, Romania
[5] High Technical School of Professional Studies, 21000 Novi Sad, Serbia

Abstract. The improvement of the gear cutting process by hob milling is also significant for the producers of gears as well as for the producers of hob milling tools. Due to the complexity of the process, the high values of the gears, and in particular the hob milling tool, the research of the bases for optimizing the process of gear cutting by hob milling requires significant financial resources and significant efforts of the researchers. In the modern production of gears there is a continuous demand for increasing the economy. In the gear cutting process improvements of the production equipment and materials for the cutting elements of the hob milling tools are necessary. In this paper an analysis of the tool life of coated hob milling tools for gear cutting of cylindrical gears is given. The experimental results have shown that the hob milling tools with inserted combs coated with TiAlN outperform the hob milling tools with inserted combs coated with TiN. They have higher resistance of 60.6 %.

1 Introduction

The occurrence of the power transmitter and movement dates back to ancient times. In China, Mesopotamia and Egypt, parts and assemblies of irrigation devices with toothed portable elements have been found, which according to the kinematic principle are still used today. In addition to agriculture, such devices have been developed for war and civil engineering purposes [1].

Modern gear cutting of cylindrical gears is most often performed by relative rolling methods, i.e. hob milling and hob planning. Thanks to the high productivity of the process, the hob milling has the widest application in the gear cutting of cylindrical gears. The process of hob milling of cylindrical gear serration is one of the most important operations in the technological process of machining, since the productivity, the final geometric accuracy and the quality of serration surface depend on the machining methods and the applied tools [2, 3].

Tribo-mechanical power transfer systems, known in the engineering as mechanical transmissions, are used to transmit energy of mechanical motion at a distance, as well as to modify its parameters. At the same time, they transfer revolving motion. The general purpose of the transmission is firmly associated with the performance of elementary functions, among which the most characteristic are: energy distribution, reduction or increase of speed, change of motion type, speed control, start, stop, direction change, etc. Functionality of mechanical transmission is based on a specific triboelement form and the corresponding law of meshing the contact surfaces. This group of transmissions includes: transmission gears, chain transmissions, timing belt drives, cardan shafts, variator transmissions and all of them found their application in modern technology.

2 Gear cutting and materials for manufacturing of hob milling tools

The problem of the production of gears is analyzed in science and practice in various ways, identifying it once as an element of the machine and another time as a part of the production, that is, a final product. The basic types of serration, the basic types of tools and the basic methods of gear cutting are shown in the Figure 1 [3,4,5,6].

Designed and properly manufactured gears are key components in most complex systems. Gears with involute teeth are among the most important components of efficient modern transmission techniques, in accordance with relevant legislation on pollution and energy saving. Figure 2 [6] shows how to get involute.
In the hob milling process, the teeth and hollows between teeth are formed by removing the material from the work-piece. Problems that arise in the process of material removal are due to the presence of tribological processes on the contact surfaces of the elements of tribo-mechanical system.

* Corresponding author: sandrasn@eunet.rs

© The Authors, published by EDP Sciences. This is an open access article distributed under the terms of the Creative Commons Attribution License 4.0 (http://creativecommons.org/licenses/by/4.0/).

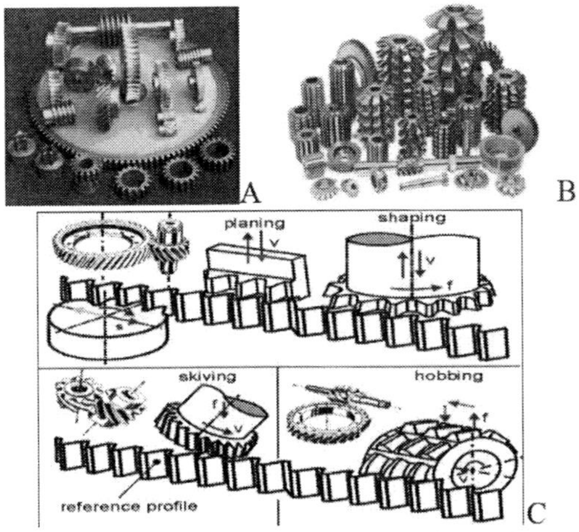

Fig. 1. Basic types of gears, hob milling tools and gear cutting methods

In the Figure 1c [3,6] examples of tribo-mechanical systems with basic tribo-mechanical elements and basic methods of gear cutting are given.

A fundamental knowledge of the process of producing transmission gears is crucial to oppose the demands for continuous improvement in a globalized market. Related development is continually involved in the development and production of machine tools and tools for gear cutting.

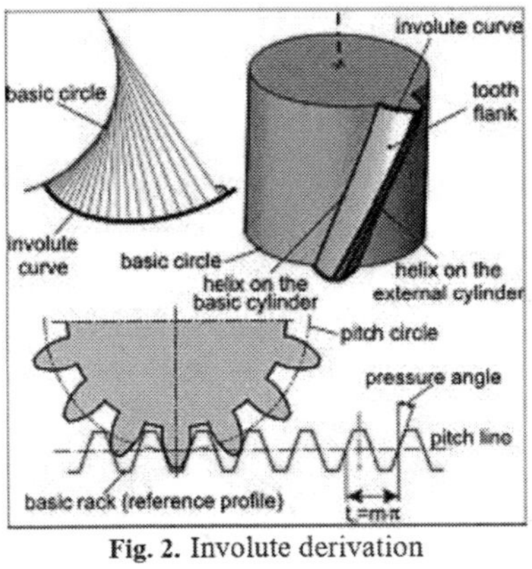

Fig. 2. Involute derivation

Manufacturers of involute cylindrical gears during their production process, regardless of their size, from a few millimeters to several meters (Figure 3 [6]), face common technological problems. The predictability of machining parameters, such as geometry and chip flow, tool wear, etc., with regard to work-piece, tool and machining data is of great interest for research and industry. The changeable chip formation during the manufacturing process of serration of cylindrical gears causes different wear laws.

Production operations and methods in mass production depend on various parameters such as the cost of machining, production philosophy, practical experience, etc. A common strategy is to achieve the highest possible accuracy during roughing and to perform thermal treatment with the necessary tolerance to completely avoid finishing [3].

Fig. 3. Possible gear sizes

Hob milling is a multi parametric and complex method for gear cutting of cylindrical gears. Hob milling is an efficient machining process for the production of serration of high quality cylindrical gears. This process is related to complicated kinematics, the formation and the flow of chip, as well as to the hard-to-describe tool wear mechanisms.

The hob milling process is one of the most important links in the chain of gear machining, having in mind that the productivity, the final geometric accuracy and the quality of the serration depend to a great extent on the hob milling process.

Gear cutting operations most often represent a bottleneck in the production of cylindrical gears, especially in higher types of production. Therefore, the necessity of research and development of optimal tool constructions, as well as research of the optimal conditions of these processes, especially hob milling, is obvious.

The optimization of the production conditions for the gear cutting of the cylindrical gears depends to a large extent on the correctness of the selection and maintenance of the hob milling tool, as well as on the selection of the appropriate cutting regimes.

The modern technical development of the production of cylindrical gears is directed in two basic directions: increase the productivity of the appropriate process and gear cutting while improving the quality of the machined serration. Achieving these, mostly opposite, demands is a very complex task.

The productivity and cost-effectiveness of the hob milling is determined by: the characteristics of the hob milling machine, the construction characteristics of the hob milling tool and the material of the cutting edge [3]. The enhancement of the hob milling of serration of cylindrical gears can be achieved by increasing the productivity and increasing the quality of the serration (Figure 4 [2,3]).

In industrial countries today, losses in billions of Euros are generated through wear and corrosion when manufacturing and using mechanical gears. For this reason, the design of hob milling tools, that will provide minimal losses of this type, is of great importance expressed through the construction and cutting geometry of the hob milling tool, and also through the selection of the optimal tool material.

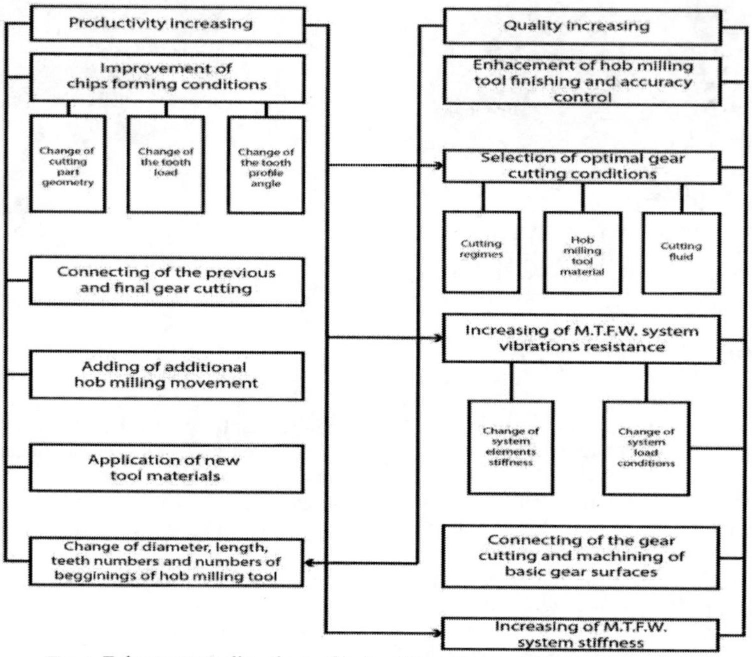

Fig. 4. Enhancement directions of hob milling of cylindrical gear serration

The influence of the structure and microstructure of the materials and technological processes of manufacture of blank and work-piece and the properties and behavior of the material, as well as their interaction are given in [7,8]. Improving and finding more perfect materials has played the most important role in the development of gear cutting tools of cylindrical gears by hob milling [3]. Hob milling tools are made of high-speed steel, cemented-carbide, sintered alloys, etc.

The development of new and improvement of the existing materials for the production of hob milling tools in order to minimize the total cost of gear cutting and retain the same quality of gear cutting is of great importance in view of the current economic situation. The steadily growing demands placed on the hob milling tools require the continuous development and finding of new methods for making combined materials. Significant progress has been made in that field in the last fifteen years. A large number of hard materials coating methods have been developed in order to protect against the wear of the hob milling tools, but due to insufficient knowledge and experience, they slowly find application in the industry [3,6].

The process of the gear cutting by hob milling is characterized by different thicknesses and lengths of the chips, and the materials from which the cutting edge of hob milling tools are made should have significant tenacity and hardness. Based on the previous significant attention is paid to the strengthening of the hob milling tool by applying a permanent layer to the working surfaces of the hob milling tool. The application of coated hob milling tools is conditioned by their high efficiency. Until now there is no correlation between the strength of the hob milling tool and the properties of the coating layer, and the issues of optimizing the properties of the complex layer and the tool base remain open. There are also difficulties in the industrial application of progressive surface reinforcement technology and the rational application of a reinforced hob milling tool [3,6,9,10].

The tendency of the material development for the hob milling tools and the technology of their surface reinforcement are conditioned by the need directed towards improving the physical and mechanical properties of the material for the hob milling tools, among which strength, thermostability, resistance and brittleness have long been considered to be important, while other properties are considered to be less important. Ignoring the importance of certain properties of the hob milling tool materials is one of the main reasons for the lack of understanding of the main functions of the layer's stability on the hob milling tool.

There are more than one hundred different methods of surface coating of metals and alloys, by which it is possible to increase the efficiency of the tool to a certain extent. An overview of basic techniques and surface engineering (surface modification and application of a layer) is shown in Figure 5 [3,11,12]. In Figure 6 [3,12], coating techniques based on the physical state of the applied material are given. The surface coating technique is perfected by a rapid rhythm, which greatly contributes to the development of theoretical concepts in the field of surface resistance of materials.

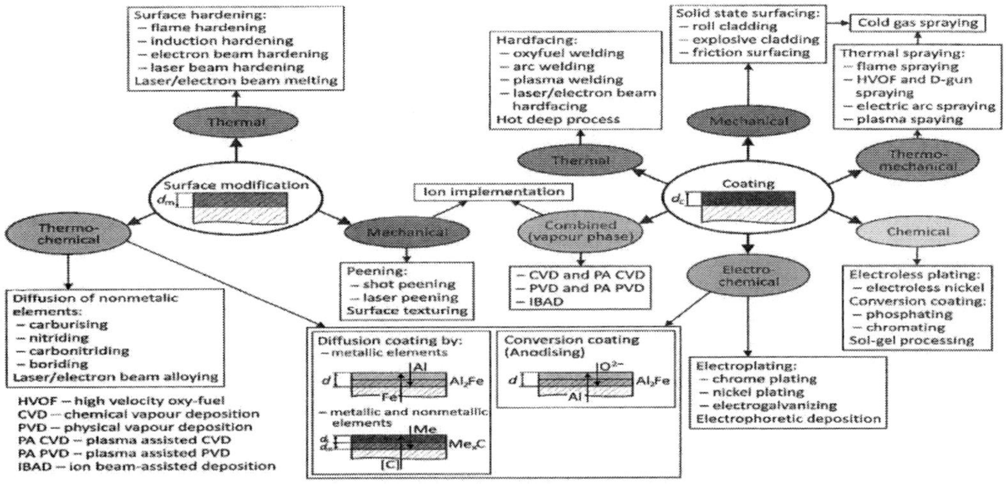

Fig. 5. Overview of main surface engineering(surface modification and coating deposition)techniques

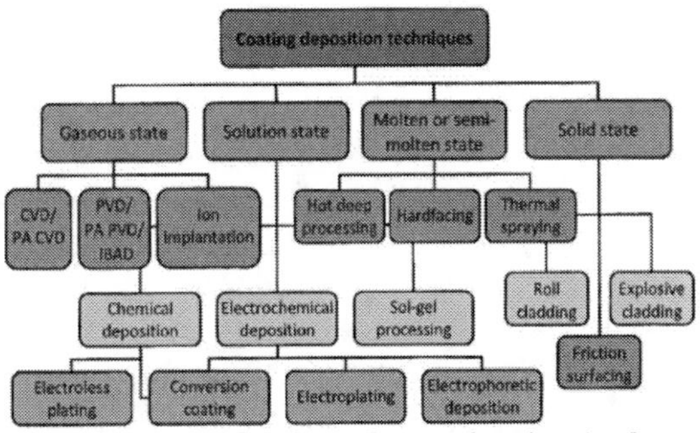

Fig. 6. Overview of main surface engineering (surface modification and coating deposition) techniques

Today, high-speed steel hob milling tools continue to be used primarily for gear cutting in the automotive industry, especially in the production of truck changers. The further evolution of metallurgy of powdered high-speed steel in combination with the application of PVD layers has contributed to this application and the development of high-speed steel. Coated hob milling tools made of high-speed steels are also used for dry gear cutting of cylindrical gears.

Cemented-carbide hob milling tools are also used in gear cutting of cylindrical gears. These tools have a high wear resistance. With these tools, there is the possibility of increasing the cutting speed in order to increase productivity and reduce production costs. Mechanisms affecting the reliability of the process of gear cutting by hob milling tools of cemented-carbide are caused by a stochastic occurrence of a built-up-edge on the cutting edge and a deterioration of the cutting ability after the sharpening of the hob milling tools [3,6].

Hob milling tools can also be made from sintered alloy. In the comparison with conventional cemented-carbide based on tungsten carbide, they have shown greater hardness and wear resistance, including high blade stability. This ensures the invariability of the form and dimensions of smaller tool cutting edge and better machined surface of the work-piece. Good cutting edge stability allows the use of this material for hob milling tools with smaller module. The rounded cutting edge of hob milling tools from the sintered alloy to prevent damage does not produce better results. The reason for this is the fact that the sintered alloy cutting edges have already increased stability. The cracks that occur on the tools of this material and represent one of the wear criteria are not the result of an unstable cutting edge and do not arise accidentally. They are largely the result of damage of the structure. The rounded cutting edge results in an increase in the cutting force, which leads to occasional damage of the face surface and decrease tool life.

The productivity of the hob milling of cylindrical gears serration has not been able to increase until now using sintered alloys instead of cemented-carbide. Cemented-carbide hob milling tools have greater tool life than sintered alloy tools. Based on the aforementioned, it can be concluded that today, sintered alloys cannot be economically applied for hob milling tools. As previously mentioned, the design and the material of the hob milling tool are also important. Hence, an attention will be paid to this in the experimental part of this paper.

3 Material and method

In the framework of the research on project entitled The Modeling and Optimization of the Hob Milling Process [3], one part is dedicated to hob milling tools. In the paper [13], the advantages of the hob milling tools with inserted cutting elements-combs in relation to the integral hob milling tools are highlighted. The experimental results have shown that the hob milling tools with inserted coated combs have a greater resistance of 27.28 % in comparison to the hob milling tools with inserted uncoated combs [8]. This paper presents only a part of experimental research. The experiments were carried out in several positions with hob milling tools with inserted coated combs. The combs are coated with TiN and TiAlN. The work-piece material is 20MnCr5. The serration of straight cylindrical gear was machined. Its characteristics are: the module m = 4.75 mm, the number of gear teeth $z_2 = 39$, the angle of the contact line $\alpha = 20^\circ$, the angularity of the gear tooth $\beta = 0^\circ$, the width of the serration b=66 mm, the diameter of the work-piece $\Phi 194.75$. Tools with the diameter $d_{ht} = 145\,\text{mm}$, length $l_{ht} = 180\,\text{mm}$, module m = 4.75 mm with one beginning, 12 teeth at one comb and 14 teeth per circumference were used on the hob milling tool. In the experiments, the cutting speed $v_c = 101.94\,\text{m/min}$, $n = 224\,\text{r/mm}$ and the feed $f_a = 2\,\text{mm/r}$ were used, which were determined in the previous researches as optimal. The hob milling tool is moving for $f_t = 1.25\,\text{mm}$ after gear cutting on one work-piece in order to equalize the wear of as many teeth as possible on the hob milling tool. The body and rings of hob milling tool are made of 42CrMo4, and the comb of HS6-5-2-5. Coating of the combs was done with TiN and TiAlN. Measuring and checking of the hob milling tools and gears were carried out on a special measuring machine KLINGEBERG. Monitoring and measurement of wear were carried out using a special magnifier and tool microscope manufactured by KARL ZEISS. The gear cutting was carried out by (up) conventional milling.

4 Results and discussion

In tribological processes, the very small mass of the tool material and the large mass of the material of the work-piece are involved in the machining of serration of the gear. The consequences of the development of tribological processes in the zones of contact are the friction and wear of the cutting elements of hob milling tools [14].

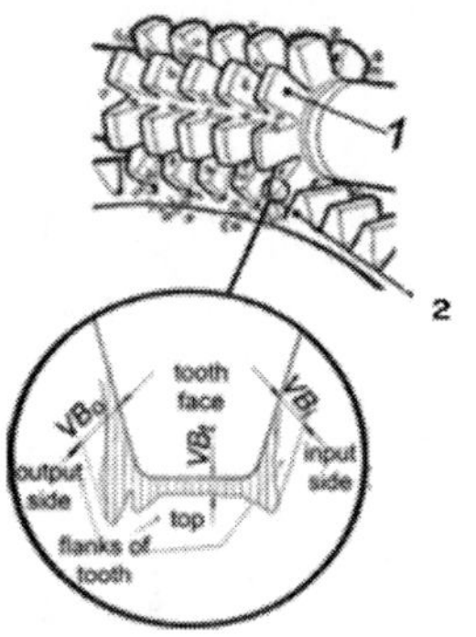

Fig. 7. Tribo-mechanical system for gear cutting

The wearing process takes place on the teeth in contact with the work-piece. The hob milling tool and work-piece make a tribo-mechanical system (Figure 7 [3]). The Figure shows the detail of wear on one tooth of the hob milling tool. The primary processes of wear are the most important. In the experiments shown in this paper, the flank wear on the back surfaces of the hob milling tool teeth are the primary wear zone. The back surface of the teeth of the hob milling tools wears on the input lateral back surface cutting edge, the tip back surface cutting edge and the output lateral back surface cutting edge. The development of the wear process on one tooth of the hob milling tool is given in the Figure 8. The Figure shows the wear on all three back surfaces. The wearing of the output lateral back surface of cutting edge $VB_o = 0.7\,\text{mm}$ in roughing has been adopted as a wear criterion. At the cutting speed of $v_c = 101.94\,\text{m/min}$, $n = 224\,\text{r/min}$ and axial feed $f_a = 2\,\text{mm/r}$, the hob milling tool with inserted TiN coated combs made of high speed steel manufactured by powder metallurgy had a tool life L = 2178 mm on machined serration of straight cylindrical gear (Figure 8a).

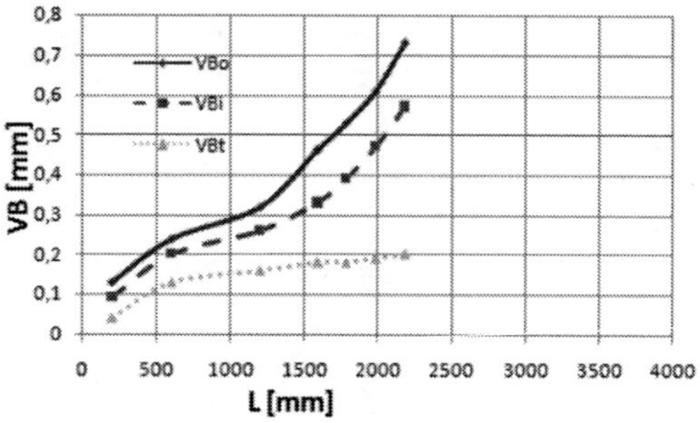

A

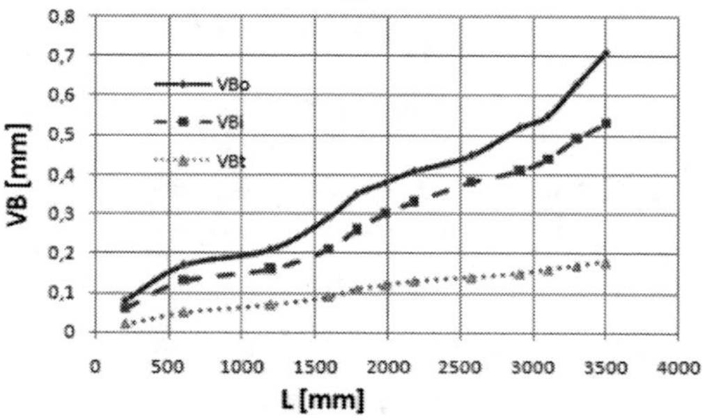

B

Fig. 8. The development of the wear process on hob milling tools coated with TiN and TiAlN

The development of the wear process on one tooth of the hob milling tool with inserted combs of high speed steel manufactured by powder metallurgy and combs coated with TiAlN is shown in the Figure 8b. The tool life of this tool with the same cutting regimes and the same wear criterion was L = 3498 mm of machined serration of straight cylindrical gears.

Comparative studies of hob milling tools with inserted TiN coated combs made of high speed steel, manufactured by powder metallurgy and hob milling tools with inserted combs of the same material and coated with TiALN were performed in real production conditions. These experimental researches once again confirmed that there is a concentrated wear on the cutting elements of the hob milling tools when gear cutting of straight cylindrical gears. The morphology of the flank wear is characterized by the appearance of most commonly two grooves on the output lateral back surface and one groove at the input lateral side back surface. The grooves in the cutting elements of the hob milling tool are located in the transition areas of the tip back surface into the output and input lateral back surfaces.

The experimental researches have shown that the hob milling tools with inserted TiAlN coated combs have a higher resistance of 60.6 % compared to the hob milling tools with inserted TiN coated combs. Previous and these studies have shown that the design, that is, the construction of the hob milling tools and the coating with TiAlN, have great significance for increasing the tool life of the hob milling tools and the quality of the machined serration. It is important to note that the operating modes were the same for all the hob milling tools used in experimental research. Based on the knowledge in the available literature and own research with tools coated with TiAlN, it is possible to apply larger cutting regimes and thus achieve even more techno-economic effects. In further research, comparative testing and application of hob milling tools coated with two or more layers of different manufacturers should be carried out as well as their more rational application. It is necessary to research the technological and economical characteristics of hob milling tools with several beginnings and a new construction of hob milling tools.

5 Conclusion

On the basis of theoretical and experimental researches, the necessity for further researching the hob milling process of the serration of straight cylindrical gear is obvious.

The experimental results have shown that the hob milling tools with inserted combs coated with TiAlN have a greater resistance of 60.6 % in comparison to the hob milling tools with inserted combs coated with TiN.

The effects of the application of hob milling tools coated with TiAlN would be considerably better if cutting speeds greater than 180 m/min were applied.

References

[1] M. Opalić, *Mechanical transmissions and motion (Prijenosnici snage i gibanja)*, Sveučilište u Zagrebu, FSB, Zagreb, Croatia (1998)

[2] I. Sovilj-Nikić, *Application of genetic algorithm in optimization of geometric parameters of hob milling tool (Primena genetskog algoritma u optimizaciji geometrijskih parametara odvalnog glodala)*, Fakultet tehničkih nauka, Novi Sad, Serbia, Master Thesis (2007)

[3] I. Sovilj-Nikić, *Modeling and optimization of hob milling process (Modelovanje i optimizacija procesa odvalnog glodanja)*, Fakultet tehničkih nauka, Novi Sad, Serbia, Doctoral Thesis (unpublished)

[4] http://w.w.w.indiamart.com

[5] http://w.w.w.ags-industrial.com

[6] K. D. Bouzakis, E. Lili, N. Michalidis, O. Friderikos, Manufacturing of cylindrical gears by generating cutting processes: A critical synthesis of analysis methods, *CIRP Ann.* **57**, 676-696 (2008)

[7] T. Filetin, M. Franz, Đ. Španiček, V. Ivušić, *Properties and characteristics of material (Svojstva i karakteristike materijala)*, Sveučilište u Zagrebu, Fakultet strojarstva i brodogradnje, Croatia (2012)

[8] B. Sovilj , S. Sovilj-Nikić, G. Varga, V. Blanuša, Analysis of the possibility of improving the characteristics of hob milling tools for gear cutting of cylindrical gears, *Proceedings of International Scientific Conference KOD 2018,* 6-8 June 2018, Novi Sad, Serbia (2018)

[9] K. D. Bouzakis, G. Skoridatis, M. Michalidis, Innovative methods for characterizing coatings mechanical properties, *Proceedings of International Conference 'The Coatings 2008'*, October 2008, Kallithea, Greece (2008)

[10] I. Sovilj-Nikić, S. Sovilj-Nikić, B. Sovil, R. Đokić, Analysis of roughness parameters of the tooth-face of model hob milling tool, *Mach. Dsgn* **9**(3), 99-106 (2017)

[11] M. Stupnišek, B. Matijević, Overview of procedures for modifying and coating of metal (Pregled postupaka modificiranja i prevlačenja metala), Proceedings of International Scientific Conference "Thermal treatment of metal and surface engineering" (Znanstveno stručni skup s međunarodnim učešćem 'Toplinska obradba metala i inženjerstvo površina'), Zagreb, Croatia, Jun 8, 53-62 (2000)

[12] M. Kandeva-Ivanova, A. Vencl, D. Karastoyanov, *Advanced Tribological Coatings For Heavy-Duty Applications: Case Studies*, Prof. Marin Drinov Publishing House of Bulgarian Academy of Sciences, Bulgaria (2016)

[13] B. Sovilj, A. Popov, D. Ješić, I. Kolev, I. Sovilj-Nikić, Treatment in Tribological Processes of Spur Gear in Modern Agricultural Engineering, *Con. Agricul.Eng.* **37**(3), 305-313 (2011)

[14] I. Sovilj-Nikić, S. Sovilj-Nikić, B. Sovilj, V. Blanuša, Topography of Characteristics Surfaces of Model Hob Milling Tools for Machining of Gear Serration, *Proceedings of International Scientific Conference* IRMES 2017, 7-9 September 2017, Trebinje, Bosnia and Herzegovina, 241-246 (2017)

Acknowledgments

The research was funded by the Ministry of Education, Science and Technological Development of the Republic of Serbia, within the projects TR 35015 and III 43008, and it is also the result of the cooperation within CEEPUS project CIII- RO-0058-07-1415.

 https://doi.org/10.1051/matecconf/201818404001

Aspects of Human Behaviour in Negotiation Process

Dan Săvescu[1]*

[1]Transilvania University of Braşov, Romania

Abstract. Paper presents some aspects regarding negotiation human behaviour process. There are presented some ideas about entrepreneurship, negotiation being an important part of this "job", negotiating as art, negotiation for your own advantage, barriers of negotiation, objections' management, conclusions and references.

1 Introduction

One of the factors that makes the difference between a successful entrepreneur and an average entrepreneur is thinking.

A successful entrepreneur understands that the world is full of great possibilities and ideas. A successful entrepreneur understands that failure is a step towards even greater success, and not just an obstacle to overcome. A good entrepreneur needs to know how to develop creative ideas. Creativity is present in every lucrative success. A good entrepreneur needs to know quickly to make the right decisions. It has been discovered that the speed of decision-making is a feature of many millionaires. Business thinker James Allen [1] said, "Your previous thoughts brought you to where you are today. Tomorrow you will be where your thoughts will take you today. "And indeed, at work, the quality of the way of thinking determines the quality of the gain.

Entrepreneurship is the basis for a successful business in the long run. In the process of material exchanges between market players, there are exchanges, with already existing patterns, standards, requiring or offering products / technologies (these are TTO - Technology to offer and TTR - Technology to request, known in EU policy, 2020). All these product exchanges are based on negotiation. In other words, negotiation is an integral part of the competitive entrepreneurial process.

2 Negotiating as art

As in every moment of our life we sell, we also negotiate something in any day. A leader will be put in a number of times in the face of a situation where he has to convince a manager of the need to allocate resources or the need for a certain decision. Also, a leader can be part of the team negotiating the customer contract or can lead negotiations with a raw material supplier or simply negotiate with a team member the assignment and the characteristics of tasks within the project. Negotiation is an organized social process, a set of techniques used by at least two partners to reach a consensus; it requires a dialogue, a permanent communication between the two sides in which it is sought (at least theoretically and declaratively) the avoidance of confrontations.

3 What is negotiation?

After François de Collieres (1716): the science of agreeing on the benefits and interests of the parties concerned.

Arthur Lall: a dispute resolved by peaceful means other than legal.

Smith Simpson: a part of the vast and complex managerial activity through which management promotes its own objectives.

F. Ikle: links the need for negotiation with the presence of conflicting interests that can be achieved through explicit proposals presented ably for the obvious purpose of reaching an agreement.

John Illich, in 1980, said that negotiation, in its purest form, is the intelligence opposed to another intelligence [2].

According to Gerhard I. Nierenberg, in his work "The Art of Negotiation" [1-3] "whenever people change ideas with the intention of changing relationships, whenever they prefer a deal, then they negotiate ... Negotiation is not led either to open or to limit a relationship, but form a new and different configuration. "

4 The results of negotiation process

The negotiation process is under the risk. It is hard to assume that "the home account matches the market one."

Any negotiation ends in one of four possible variants:
- Both sides have to lose;
- One part is worth winning, the other has to lose;
- Both sides have a win;
- No results - no positive or negative consequences.

The results in which both parties have to lose appear when none of them get what it needs or wants. One example: a customer wants to buy a card processing

* Corresponding author: dsavescu@unitbv.ro

© The Authors, published by EDP Sciences. This is an open access article distributed under the terms of the Creative Commons Attribution License 4.0 (http://creativecommons.org/licenses/by/4.0/).

machine with the features. A vendor is presenting the machine at a price that was real with the one on the market. After a while, the buyer changes his mind and cancels the contract, moving on a more tempting offer. The seller lost a customer and the buyer purchased a non-performing machine, being in litigation, in court, with the seller. Therefore, the buyer lost the transaction, lost time, lost customers. Another example is related to disagreements between employer and union on wages. The end of the negotiation is loss-loss, the strikers did not get the salary increases, the company lost time, customers and money. Usually the failure is the poor preparation of the negotiation session, somewhat unprepared participants!

The second variant is interesting, it depends only on what part you are: winner or defeated. The taste of defeat is bitter, hard to bear, and it is hard to assume there will be a similar situation between the two, the buyer and the seller. One example: a client calls a bank for a short-term loan (60 days). Receives the credit but after 45 days the bank changes the terms of credit repayment (reasons, inflation, legislation, etc.). Being only a small loss, for the remaining days - 15, the customer accepts the new conditions. The bank won, the client lost (as usual!). It's obvious that in a similar situation the client will not call that bank anymore.

The ideal outcome of the negotiation is mutual gain, basically both sides are pleased with this move. It is a pleasure to see satisfaction on the faces of the participants after finalizing the negotiations, the needs and goals of both parties are fulfilled. Negotiation is a cooperative action; it is not a game, in a good bargain, when everyone has a win.

In the fourth situation, the one of the null result, the parties, can be said to have lost time, were not sufficiently motivated or simply did not prepare for negotiation. These situations are often encountered at "rigid" positions in the negotiation process or do not accept that there is a "competitive market" with many actors involved in the same field.

5 Negotiation itself

Negotiation itself is a creative process that involves imagination, flexibility and rapid adaptability to the partner, how to think and act. All these involve a strategic culture that takes on several aspects: setting individual priorities and objectives, negotiating orientation or overall vision, finding change or replication solutions in a potentially dangerous situation.

Arsenal of questions used in negotiation:

- Investigation questions (clarification, details);
- Confirmation questions (verification, summary, rewording);
- Return questions (the question is answered with a question);
- Alternative false questions (control of discussion by limiting options);
- Questions to relaunch (diverting the discussion).

An efficient negotiation begins with answers to some questions like (Table 1):

Table 1. Questions of an efficient negotiation

Efficient negotiation bazed on quetions	-What do I want to get from the negotiation? -What does the partner / opponent want to achieve? -How can we both achieve our interests? -What does the man in front of me want to know? -What questions will I ask? -What questions to ask? -How will my partner react to my questions? -How can we reach consensus?

Major strategic decisions at the beginning or during the negotiation are the choice of an offensive or defensive strategy, a conciliatory attitude or, on the contrary, intransigence, an open or closed approach to the process, etc.

Information is also in the negotiation process the main instrument available to the negotiator. Going into the door of negotiations with "Lessons Made" is a sometimes decisive advantage in achieving the goals.

Another strong point is the ability to transpose instead of the one in front of you; Understanding your negotiating partner as to the motivations and constraints to which it is subject is another key factor in success. The "read" techniques of non-verbal signals are here at home. Even if their interpretation in an isolated context can lead to false ideas, most often a corroboration of these gestures and posts with "snapped" information from others can reveal aspects that make the difference between success and failure.

The way is to admit that:

- The one in front of us has a different reality from ours;
- We can understand the other's point of view;
- The idea that "different" does not mean "worse" or "bad".

In the negotiation, the actual submission of an initial offer must be made firmly, without reservations, hesitations or justifications of any kind. It's more convenient to submit your offer after you hear one of your partners. Every aspect of the offer must be realistically addressed and there is the possibility of its real argument. Opening bids must be the best for their point of view (setting high goals leads to better results) with rigorous and solid arguments. Successful negotiators make much less concessions than their opponents because of an indispensable quality: they are unpredictable in terms of concession size. Also a competent negotiator responds to the offer of others through clarifications, not through justifications.

The features of negotiating offers in order to achieve maximum satisfaction are:

- each item of the offer received is analysed;
- no opinions or reasons are speculated;
- does not contradict directly, openly;
- the answers received will be noted before commenting on them;
- its own position does not occur very quickly;

- does not immediately come into direct problem analysis;
- it does not advance on a wide front.

6 Negotiating for your own advantage

If you can not convince them, get them confused.
(Murphy)

Succession of problems in negotiation process, in case of looking for your own advantage is presented in Table 2.

Table 2. Succession of problems

1	It starts with a problem that is not very important to us, from which concessions can be made and the willingness to make concessions.
2	Another problem is equally unimportant, to test „their negotiation" mode and to analyse the concessions they are willing to make.
3	The critical issue for us (but which will not be presented) is being addressed for which we will seek special concessions from others.
4	There are other major, then minor problems. For the end, a minor problem for which significant concessions are granted is retained.

7 Barriers to negotiation

The main obstacles to negotiation are represented by:
- Frustration and anger;
- Perceiving negotiation as a win-loss game ("What I earn will have to lose you and vice versa");
- Clear distinctions of status and power between the two negotiators.

The worst negotiation is done when you are frustrated or angry ... Ancestral wisdom indicates that nothing in this world is catastrophic for at least two reasons:
- Most events are not as serious as we think;
- if something really serious happened by going out of the way we make it even worse.

Frustration is the situation where you do not get what you want or when you get forced what you did not want. Frustration does not necessarily lead to disorder and aggression (if we know how to control it), but it often generates anger. Angry people actually conceal the fear of failure, and anger raises a reaction to anger (for example, a dictatorial attitude generates frustration and anger). Anger can also be born out of self-pity.

Anger is a self-healing and self-sustaining process. The first phase is because I want something. Not getting what I wanted to become frustrated. Then I come to the idea that it is terrible that I did not get that "something" and end up in a depression phase. At this stage you have to look for guilty ... every time someone else is to blame; that one is bad because his attitude led me to depression and must be punished. How can I punish him? Simply: by shouting at him, wounding him, wiping him, watching his balance and dignity, etc. Accusation (personal or other) has an essential role in the appearance of anger.

Generally, the three reasons for inappropriate behaviour are: stupidity, ignorance, and disorder. If we consider any unwanted behaviour as a natural result of one of these three reasons, we will certainly be able to avoid (at least partially) anger. People's selfishness is a fact that can be treated as a simple and natural meteorological phenomenon: they cannot blame the phenomenon but can do something to avoid its consequences.

How can we avoid the emergence of nervousness in the negotiations? Albert Ellis [2] suggests an answer to the above question: "treats with gratitude those who act ugly with you; so they become ridiculous. " By controlling your anger you force the other not to break into a conflict; the Bible parable tells us "after a slap that gives you back and the other cheek". If you are attacked by insulting remarks, becoming more courteous, you can create consternation through your behaviour and even a radical change in the attitude of the one in front of you who is "suddenly" quiet.

The harmful consequences of anger are clear: it increases frustration, prevents you from solving problems, it is a poor example in the eyes of the collaborators and can lead to physical illness.

Combating rage starts from accepting the idea that we are upset ourselves because of the irrational ideas that we impose on ourselves. We can counter anger by learning how to identify these irrational thoughts and what to replace them.

Affective - rational anger therapy has some clear features, as seen in Table 3:

Table 3. Anger therapy

Do not be discouraged	When you try to reject the neurotic ideas and still feel angry, continue to discuss them with yourself until you stop believing in them.
Discipline yourself	You have to be able to shut up when it is very hard not to say something to someone
You do not have to aggravate the situation	The greatest suffering is not caused by others, but by what we allow our own anguish; should be taken from the "look what they did" approach to the "look what I did from what they did to me"

To perceive negotiation as a place where there is a defeated and a winner does not mean negotiation or manipulation, it is to make the one in front of you have a totally inappropriate attitude with its principles and values. In the short term, such an approach seems beneficial, because you can get the best price or you can intimidate and annoy the "opponent"; but in the long run you have won a battle, but you will never win the war. The "defeated" often has profound resentment and is, in most cases, really disgusted with the "winner" negotiation. Especially when handling techniques are applied in a constant way to collaborators this attitude will have a demotivating effect; in the medium and long term (from just over 2-3 months) your collaborators will leave you, the team spirit begins to become a declared but unapplied thing and the project begins to squeak ...

When negotiating a team leader from Section 2 with the group's general manager, who is surrounded by 5 counsellors, 2 lawyers, and 3 executives, the outcome of the negotiation seems to be out of the box; in order not to impose solutions and to find the best option to convince everyone involved, it is necessary for the negotiations to take place between persons at a relatively close hierarchical level.

8 Attitude to objections

Objections must be allowed to appear on the surface; they need to be searched for as a source of information about the person, his needs and his intellectual capacity. In addition, the partner appreciates the other's interest in his problems. A good negotiator is not afraid of objections because, through a good argument, they lose their strength; the negotiator will gain the prestige in the partner's eyes.

One of the basic rules of diplomacy is to always allow the partner the possibility of an honourable exit ("Allow the deceased to be worthy" - Fisher said); this behaviour leaves the door open for future communication. It should not be interrupted but listened with attention and respect and the answers given should not immediately follow the question but after a few seconds, even if you do not need to meditate. Trust, do not want to have the last word at all costs, do not panic or dramatize. A crisp attitude, a serious tone and a long response reflect in the partner's mind another dimension of the problem, which is often not true.

A nodal point in the objection technique is the detection of false objections; generally a true, "honest" objection hides a need for additional information. False objections are those that rise in a shameless way.

When you get a false objection you can answer in several ways:

- Silence - just let a tense silence wait over your conversation while staying firm and insistent on your partner; most of the time it will yield and withdraw or reformulate its objection (here is the principle of "who speaks first, loses")
- You can ask for additional clarifications, especially if you are not sure that this is a mischievous objection; by clarifying the objection and by considering the logic of argumentation, you can see whether it is a sincere or not objection.
- You can use the hanging technique - the words: "If you respond to this objection, we will finish the discussion (contract)?". In the negotiation process, when seeking a solution, each party is required to make concessions.

The courtesy rules assume very strictly that a concession made by one of the parties is followed by a concession of the same level made by the other party. It is advisable that the pace of concessions made be similar - it begins by giving up a little more then more and more to end concessions when the other side does not respond in the same rhythm. Another rule is not to make concessions if they are not needed.

9 Conclusions

A solid basis in the negotiation process is trust. The more honesty, integrity and credibility, the more trust your business partner has, the more successful the negotiation is. If for some reason your partner considers you inferior, unreliable, it is difficult to obtain even minor concessions. When it comes to gaining someone's trust, actions say more than words. Here is a succession of things on which to build trust in the negotiating partner.

- Demonstrate your competence;
- Make sure that the non-verbal signals you send fit with your statements;
- maintain a professional image;
- Communicate your good intentions
- Do what you say you want to do
- Go beyond the conventional relationship;
- Listen;
- Communicate as much as possible;
- Discuss what is not to be discussed;
- Provides accurate information without hidden thoughts;
- Be honest, even if it costs you something;
- Be patient;
- Attention to fairness;
- Negotiate for abundance and not for poverty;
- Assume calculated risks.

"You never get what you deserve, you get what you negotiate" *(Bill Scott)*

References

1. P. B. Stark, J. Flaherty. Practical guide in negotiation. ED Amaltea (2008)
2. A. Pekar Lempereur, A. Colson. Méthode de négociation. DUNOD, Paris (2004)
3. D. Săvescu. Project management. ED LUXLIBRIS, Braşov (2017)

Innovative architectural design development: the Arcbazar creative crowdsourcing contests perspective

Adriana Bujor[1,*], and *Silvia* Avasilcai[1]

[1]"Gheorghe Asachi" Technical University of Iasi, Department of Engineering and Management, Romania

Abstract. Growth and value creation have become the dominant preoccupations for today's managers/companies. That is why they have increasingly embraced crowdsourcing and co-creation as business practices, especially online. If most of them aim to increase their performance with the help of these practices, there have been cases where they have been approached in critical moments, when various businesses have found themselves on the verge of collapse, as is the case with Legos. If crowdsourcing is about getting the services, ideas or content needed, by asking for contributions from a large group of people and especially from an online community rather than from traditional employees or suppliers, co-creation is considered to be the strongest business strategy, at the moment, to involve consumers in co-creating products, services, and experiences with a unique value for both parties. In this paper, architectural design crowdsourcing will be presented as an alternative method to the acquisition of traditional design, which can overcome the potential deficiencies, faced by this sector of the creative industries, and may bring smaller projects in the field of architectural practices.

1 Introduction

The term crowdsourcing, representing the combination of crowd and outsourcing, was first mentioned by author Jeff Howe in his article *The Rising of Crowdsourcing* in 2006 [1] and represents a fashionable business practice. Therefore, firstly, it is necessary to understand what outsourcing means.

According to the author James Bucki, outsourcing, also called contracting out, is a business practice used by businesses to reduce their costs or improve their efficiency by transferring tasks, operations, jobs, or processes to an outsourced (external) third party for a significant, established, period of time. The outsourced functions may be performed by third parties, whether on-site or off-site the business [2].

Since the beginning of the 21st century, society, in general, and the structure of the market, in particular, has undergone a major transformation process, according to Prahalad, et al., with a shift in power from those businesses that have determined which are their customers' needs, in order to manipulate the demand to satisfy their offers to networks of customers and manufacturers that create value together [3].

Thus, co-creation represent another business strategy, as crowdsourcing, that managers and entrepreneurs have embraced and adopted for their business growth, using mostly creative online platforms.

Once implemented, both crowdsourcing and co-creation, as open innovation strategies for a business performance growth, offers many benefits to manufacturers.

2 Crowdsourcing – definition and meanings

Crowdsourcing, in short, is defined as a method that uses people around the world to help accomplish a particular task. The masses are invited, in various, projects to propose and even help develop a new product or service, refine a project, calculate or obtain different algorithms, or assist in providing, organizing, or evaluating significant amounts of information in viable data [4].

Nowadays, crowdsourcing is used by almost all industries in order to face the changes in the market, to align with the new trends, and to register a better performance, in the end.

As highlighted by Pripic at.al., most recently research on crowdsourcing considers that there are three types of crowdsourcing applications that can be used by organizations: virtual labor markets, tournament-crowdsourcing, and open collaboration [5]. Each of these types can be compared along the following universal characteristics: *the cost of using a crowdsourcing technique, the anonymity of crowdsourcing participants, the scale of crowd size, the IT structure of crowdsourcing application, the time required to implement crowdsourcing, the magnitude of crowdsourcing tasks, and the reliability of the crowdsourcing technique.*

One of the most popular examples of crowdsourcing, Wikipedia, is a platform developed through the voluntary collaboration of a large number of

* Corresponding author: adbujor@gmail.com

© The Authors, published by EDP Sciences. This is an open access article distributed under the terms of the Creative Commons Attribution License 4.0 (http://creativecommons.org/licenses/by/4.0/).

people, which are the creators and editors of its extensive web content.

3 Co-creation – definition and meanings

Seen as an art, an initiative of management, or an economic strategy, co-creation successfully brings different parties together: a company and a community of customers, in order to jointly create a mutually valued good/service, as C.K. Prahalad, and V. Ramaswamy underlines [6].

C.K. Prahalad and Venkat Ramaswamy describe co-creation as "viewing markets as forums for organizations and active customers to share, combine and renew each others resources and capabilities in order to create value through new forms of interaction and learning" [3].

According to Piller, through the open line created for their customers by those manufacturers that have embraced co-creation, businesses have access to ideas for new products or for improving others, even to complete models/designs. Thus, helping customers build networks, organize themselves in communities, and express their commitment to a particular project, enables companies to transform market research spending into sales [7].

Table 1. The concept of co-creation [7].

What Co-creation IS NOT	What Co-creation IS
Customer focus Customer is king or customer is always right	Co-creation is about joint creation of value by the company and the customer. It is not the firm trying to please the customer.
Delivering good customer service or pampering the customer with lavish customer service	Allowing the customer to co-construct the service experience to suit her context
Mass customization of offerings that suit the industry's supply chain	Joint problem definition and problem solving
Transfer of activities from the firm to the customer as in self-service	Creating an experience environment in which consumers can have active dialogue and co-construct personalized experiences; product may be the same (e.g. Lego Mindstorms) but customers can construct different experiences
Product variety	Experience variety
Segment of one	Experience of one
Meticulous Market research	Experiencing the business as consumers do in real time Continuous dialogue
Staging experiences	Co-constructing personalized experiences
Demand-side innovation for new products and services	Innovating experience environments for new co-creation experiences.

However, it does not mean that co-creation of customers is always beneficial. The development of conventional products and customer co-creation should be seen as complementary rather than substitute, and very important, according to C.K. Prahalad, and V. Ramaswamy, is to understand what is and what is not co-creation [6].

4 Crowdsourcing and co-creation in creative industries

Creative industries are those industries which combine the creation, production, and commercialization of creative elements [8], comprising all that is produced by artistic and scientific creativity and has industrial potential, enters in the economic circuit, produces value, generates public income through taxes, creates jobs and profit, and was proved that helps to national and regional development [9].

During the years, different researches have proven that creative businesses can save the economy at the national or regional level, mainly where the traditional economies based on agriculture, industry or commerce has failed. Representing the happy combination of both qualities: that of the artist and that of the businessman [], creative businesses are increasingly becoming important components of modern post-industrial knowledge-based economies [9].

Even though they usually are based on the individual talent, according to the DCMS (UK) definition of creative industries, and on creative content, they are also adopting new business strategies, such as open innovation. By using these new innovative strategies these industries demonstrate that they have also understood the importance of the crowd, the power of its voice and role in delivering outputs to meet the demands of increasingly competitive markets.

Thus, more and more, creative businesses have also started to gather ideas for innovation from customers or stakeholders through their involvement into the early stages of the adopted open innovation process, either through crowdsourcing, or co-creation. In fact, it has been demonstrated that their ideas outline their needs, desires, and have been described as *need for information* [10]. Customers, and stakeholders generally also offer ideas that have been called *solution information*, which is not just need information but also customer-oriented proposals that describe how ideas can be transformed into marketable products [10].

In the field of architectural design, sector that belongs to the creative industries, the traditional model acquisition is mainly for all smaller-scale projects ineffective, unaffordable, and unattainable. There are different reasons why clients are not attracted to the idea to pursue architectural design services, namely:

• the way of finding an architectural designer: via word of mouth, or through friends and family,

• the fees of service providers which can be quite high (can range between 10 and 15% of the construction cost),

• the predominant fear that in the end, after all the hard work and cost, the client might not be happy with the singular design they get from only one designer [4].

Thus all of the above make room for disruptive business models in order to make it all easier for both clients and designers. Arcbazar is the first architectural design crowdsourcing platform, that right from the beginning it has registered success, but also critics from professionals in the field, which consider that by offering free ideas the company placed no weight on value [11].

5 Arcbazar case study – an example of a design crowdsoucing platform

Access to architectural design services has been historically excessively high. As a result, 89% of the annual housing improvement projects are completed without a project, according to Modern Ventures [12]. This often leads to inappropriate expectations and costly mistakes by all parties to trading.

5.1. About Arcbazar

Considering these prohibitive costs and taking into account the needs and wishes of those interested in design and architecture services, Arcbazar was founded in 2010, in Cambridge, Massachusetts, with the aim to help designers and architectures, especially the young ones, to find exciting and challenging design opportunities, according to the platform [13].

Being the first-of-its-kind online competition website, as the company claims [13], Arcbazar offers a market for real estate professionals and consumers to present their home and property redevelopment project, where thousands of architects compete for to provide the best performance in a fraction of what traditional architectural services are taxing, as mentioned by Modern Ventures [12].

The idea for arcbazar.com was developed after the founders, both executives and web developers, noticed that:

• too often customers could not find a simple and affordable way to find competitive architectural design services, and, at the same time,

• designers have a hard time getting in touch with customers and using their design talent [13].

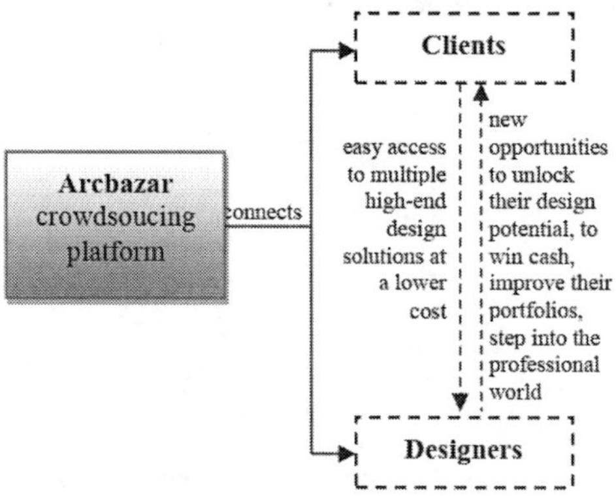

Fig. 1. Arcbazar's main benefits to its stakeholders [14].

The name of the platform, Arcbazar, is inspired by the idea of an architecture bazaar, wanting to be an energetic, dynamic space for changing and negotiating design and architecture services, for which the founder, Imdat As, and his team work constantly to match designers with exciting projects of site customers, as declared on the platform. [13].

5.3. Arcbazar business model

Anyone can start a design competition using the Arcbazar crowdsourcing platform. Competitions can be of various types, from small to medium scale, including: remodeling jobs (of an attic, basement, bathroom, and kitchen, or of an entire apartment), office spaces, additions (e.g. a room addition to the house), retail, landscape (for backyard, front yard, or other open areas/spaces) or any other architectural project (new construction) [14][15].

The arcbazar.com is user-friendly built, with an easy and intuitive interface, and diverse competition tools in order to help clients better describe their dream design challenges. All competition projects are verified by the company, safety, and privacy being its top priorities [16].

There are three important steps to be followed by both clients and designers in order to participate to a design competition, starting from the client's dream (design competition's input) to the final chosen project (design competition's output).

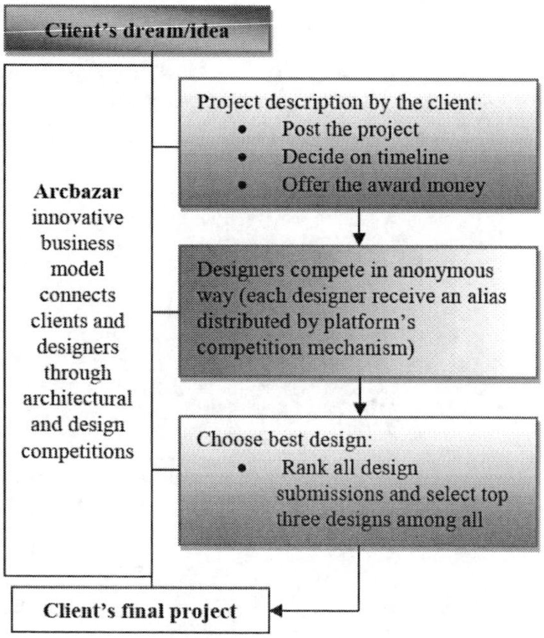

Fig. 2. Arcbazar's design competition steps [16].

Arcbazar registration is free to all stakeholders. As they begin to upload details about their projects, clients are helped by the company to establish the award money, by suggesting how much the prize should be. Some clients offer less, according to Ana Batista Company's spokesperson, up to the $ 250 which is the minimum required by the site. Others decide to offer triple of what they are being suggested, hoping to get a variety of high-quality designs [17]. The company's suggestions on the award money are based on different indicators, such as the complexity, scale, type, deliverables, and duration of a competition [15].

The platform is provided with a client-centered selection process that has the goal of controlling entries. In the process of choosing the final offer, the client must evaluate the 10 average proposals based on cost, schedule and proposal quality. The client performs a ranking of submissions, the first three being the one that collects the prize for the project, as it follows: 60% for the first place, 30% for the second and 10% for the third [18].

The process ends with ArcBazar automatically completing a set of documents and deliverables needed to complete the project.

6 Conclusions

The advantage of ArcBazar consists in the existence of the possibility to choose from architectural or quality design firms, lower or competitive costs, and faster delivery of architectural design.

Despite criticism, ArcBazar promotes enhanced access to high-quality/professional architecture or design services as a benefit to platform users.

All the data offered by the platform, together with the testimonials, proves that the decision to use crowdsourcing as an alternative for the traditional business model in creative industries, architecture, and design, in this case, was a right one to take by the founder. Thus, at the moment, the platform has managed to attract:

- 1,000,000 total views
- 1,000+ competitions
- 50,000+ projects,
- 190+ participant countries.

The potential for growth still exists and is becoming more and more exciting once the platform becomes known through different advertising channels.

Acknowledgment

This work was supported by a grant of the Romanian National Authority for Scientific Research and Innovation, CNCS/CCCDI – UEFISCDI, project number PN-III-P2-2.1-PED-2016-0689, within PNCDI III.

References

1. M. Rouse, Retrived from TechTarget web site, www.searchcio.techtarget.com/definition/crowdsourcing (2017)
2. J. Bucki, Retrived from The Balance web site, https://www.thebalance.com/what-is-outsourcing-2533662 (2018)
3. C.K. Prahalad, and V. Ramaswamy, *The Future of Competition: Co-Creating Unique Value with Customers*, Boston, MA: Harvard Business School Press (2004).
4. A. Imdat, and A. Maria, *Crowdsourcing Architecture: A Disruptive Model in Architectural Practice*, ACADIA 12: Synthetic Digital Ecologies, San Francisco 18-21 October, 2012, pp. 439-443.
5. J. Prpic, and P. Shukla, *Crowd Science: Measurements, Models, and Methods*, https://arxiv.org/ftp/arxiv/papers/1702/1702.04221.pdf
6. C.K. Prahalad, and V. Ramaswamy, *Co-Creation Experiences: the Next Practice Value in Creation*, Journal of Interactive Marketing, vol. 18, no. 3 (2004)
7. F.T. Piller, *Open Innovation with Customers: Crowdsourcing and Co-creation at Threadless*, SSRN Electronic Journal (2010)
8. A. Bujor, and S. Avasilcai, *Creative industries as a growth driver: an overview*, Proceedings of MakeLearn: Management, Knowledge and Learning International Conference, pp. 1725-1731, 25 – 28 Iunie, Portoroz, Slovenia (2015).
9. A. Bujor, and S. Avasilcai, *Creative Entrepreneurship in Europe: A Framework of Analysis*, Annals of The Oradea University, Fascicle of Management and Technological Engineering, B+, Vol. XXIII, nr. XIII, issue 1, pp. 151 – 156 (2014).
10. A. Bujor, S. Avasilcai, *Creativity and Innovation by Empowering the Customer: The Case of Mulino Bianco. Part I: Organisational Innovation Context*, IOP Conference Series: Materials Science and Engineering, Vol. 95, pag. 012146, DOI: http://dx.doi.org/10.1088/1757-899X/95/1/012146, indexed ISI Web of Science (2015).
11. The Architects's Journal, https://www.architectsjournal.co.uk/news/daily-news/architecture-crowd-sourcing-website-criticised/8620481.article (2011)
12. Modern Ventures, Arcbazar Pilot Program https://www.moderneventures.com/arcbazarpilot/ (2018)
13. Arcbazar, About us, https://www.arcbazar.com/about (2018)
14. Crunchbase, Organization: arcbazar.com, https://www.crunchbase.com/organization/arcbazar-com#section-overview (2018)
15. Arcbazar, FAQ for Clients, https://www.arcbazar.com/faq-clients (2018)
16. Arcbazar, How it works, https://www.arcbazar.com/howitworks (2018)
17. The Orange County Register, https://www.ocregister.com/2014/02/22/architects-miffed-at-website-that-offers-low-cost-bids/ (2014)
18. Digital Transformation and Innovation, https://digit.hbs.org/submission/arcbazar-crowdsourcing-design/ (2017)

MATEC Web of Conferences **184**, 04003 (2018) https://doi.org/10.1051/matecconf/201818404003
Annual Session of Scientific Papers IMT ORADEA 2018

Sensitivity of indicators used in cost-volume-profit analysis

Carmen-Elena Stoenoiu[1,*]

Technical University of Cluj-Napoca, Department Electrical Machines and Drives, Cluj-Napoca, România

Abstract. The need for cost-volume-profit analysis (CVP) starts from the need to optimize and manage costs due to unforeseen events that accompany economic activity in all areas. In this paper, the dependency relationship between the three cost-volume-profit indicators was analyzed to highlight the need to permanently track and optimize these variables so that managers' decisions can be supported by adaptive analyzes to concrete needs. The study also includes a sensitivity analysis that has shown that the magnitude and meaning of changes occurring when changes occur to one or more variables may be different due to direct and inverse relationships that are established between these variables.

1. INTRODUCTION

In the economic activity of any one enterprise, there is the operational risk due to dependent and independent variables that influence the result of the exploitation activity and ultimately the profitability. Adaptation of the enterprise to the present is thus a condition for survival but also for development, and the ability of the enterprise to cope with the economic risk is the condition to reach profitability when the lowest costs are obtained [1]. When firms want to maintain their production capacity, variable costs are those on which managers will focus because fixed costs are mostly committed to becoming uncontrollable [2], [3].

For this, there are several studies and researches demonstrating the importance of anticipating costs as well as the advantages and disadvantages of costing models [4] - [6]. This is the case for researchers who come up with general arguments about how to understand costs [7], which supports the role of complexity in cost-increasing operations [8], or who argue that increasing varieties and heterogeneity in the product mix will most often lead to a negative impact on costs and on operational performance [5], [9].

Organizational agility and dynamic capabilities of the enterprise (adaptation to change, innovation and creating a favourable market for customers and unfavourable to competition) are two topics debated in recent studies and research, this being seen as helping to prioritize the consistency between strategy, structure and business environment [10], [11].

Various studies show that managers are inclined to use transfer prices that are not at the total cost level [12] while others advocate their alignment to total costs [13]. Although the former affects tax levels and managerial decisions, accounting and cost management have shown that the use of transfer pricing at total costs affects efficiency not stimulating the increase in production [12]. Different levels of production capacity utilization and arbitrary cost allocations often impede the correct cost allocation, which may affect the sales level, ie the expected profit [14] - [17]. Cost behaviour is different depending on the product, and when costs increase more than decreases in volume production, managers feel they are entitled to limit production [18] – [20]. Cost behaviours can be studied through cost – volume - profit analysis (CVP) that allows highlighting changes that may occur in the mix of product mix, sales volume, or profits through the margin of safety [21].

The purpose of this research was to identify the variables that affect a company's profit and then to observe the dependency relationship between these variables. CVP analysis allowed, through 21 cases studied, to demonstrate the existence of causal relationships between variables, and sensitivity analysis determined the magnitude and meaning of variation of a variable when changing another variable.

2. MATERIALS AND METHODS

The CVP analysis has identified relationships of dependence that exist between cost, volume or level of activity and profit. The values taken in the study are presented in Table 1.

This analysis is required because, in a company's activity, volume fluctuations of activity occur that cause changes within fixed and variable costs or sales price changes (only we can sell at the same price or the same amount), which ultimately can affect the profit of companies.

The equation of profit used in this article is given by the following calculation formula, (1):

$$P = SP \times Q - VC \times Q - TFC \qquad (1)$$

* carmen.stoenoiu@emd.utcluj.ro

© The Authors, published by EDP Sciences. This is an open access article distributed under the terms of the Creative Commons Attribution License 4.0 (http://creativecommons.org/licenses/by/4.0/).

Where: P - profit; SP - sales price per unit of product; Q - the number of units produced and sold; VC - variable cost per unit of product; TFC - total fixed cost.

This CVP analysis was needed to calculate the breakpoint (zero or critical point), that is, the quantity (number of units) to be produced and sold for the company to fully cover its costs without profit or loss.

Table 1. Quantitative and valuable state of CVP indicators.

No. cases	Quantity [Pcs.]	Selling price per unit [RON/Pcs.]	Variable cost per unit [RON/Pcs.]	Total fixed cost [RON]	Profit [RON]
1	1,000	40.00	24.40	10,000.00	5,600.00
2	1,200	40.00	23.42	10,000.00	9,900.00
3	1,400	40.00	23.07	10,000.00	13,700.00
4	1,600	40.00	24.88	10,000.00	14,200.00
5	1,800	40.00	25.72	10,000.00	15,700.00
6	2,000	38.00	25.60	10,000.00	14,800.00
7	2,200	38.00	27.82	10,000.00	12,400.00
8	2,400	38.00	29.08	10,000.00	11,400.00
9	2,600	38.00	28.92	10,000.00	13,600.00
10	2,800	38.00	28.36	10,000.00	17,000.00
11	3,000	36.00	27.80	10,000.00	14,600.00
12	3,200	36.00	32.13	10,000.00	2,400.00
13	3,400	36.00	33.03	10,000.00	100.00
14	3,600	36.00	33.19	10,000.00	100.00
15	3,800	36.00	32.95	10,000.00	1,600.00
16	4,000	34.00	32.50	10,000.00	-4,000.00
17	4,200	34.00	32.81	10,000.00	-5,000.00
18	4,400	34.00	32.36	10,000.00	-2,800.00
19	4,600	34.00	32.22	10,000.00	-1,800.00
20	4,800	34.00	32.38	10,000.00	-2,200.00
21	5,000	32.00	32.00	10,000.00	-10,000.00

Since the managers are concerned to achieve a higher sales level than the critical amount, the margin of safety has been calculated (how close it is to the equilibrium level). The safety margin was obtained by the difference between the expected sales level and the critical point sales, (2).

$$MS = ES - BES \tag{2}$$

Where: MS - The margin of safety; ES - Expected sales; BES - Break-even sales.

The equation of profit (1) can be rewritten according to (3):

$$P = (SP - VC) \times Q - TFC \tag{3}$$

From rewriting the profit equation using Eq. (3) we get the marginal contribution per unit of product as the difference between the selling price and the unitary variable cost, (4).

$$P = Cm \times Q - TFC \tag{4}$$

Where: Cm - Contribution margin per unit.

The marginal contribution per unit measures the value of the incremental profit generated by the sale of an additional unit. This may not be obvious at first glance, but with this indicator, we can see what happens when sales and production grow with one unit.

The company may benefit from incomes equal to the sales price, but it often involves increased costs equal to the variable cost per unit. Fixed costs are not affected by changes in production volume, so they do not affect the incremental profit associated with the sale of an additional unit. If we multiply the contribution margin per unit by the number of units sold, we get the total contribution margin.

If we solve the profit equation for the number of sales in units (Q), we obtain a formula that allows us to calculate the unit sales (quantity) needed to attain specified profit (USNASP) or profit target, according to (5):

$$USNASP = (P + TFC)/Cm \tag{5}$$

Finally, it was necessary to carry out a sensitivity analysis to determine the effect of variation of an indicator on another indicator, both in size and level of influence (direct or inverse). The coefficient of variation used in the sensitivity analysis is within ±1%, ±5%, ±10%. The sensitivity analysis was performed by studying the variance of all existing variables in case 10 (see Table 1, when the profit is the maximum). For this, the relative diminution and the relative deviation of the variables were calculated. Relative diminution was obtained using (6):

$$X' = X\,(100\% + a) \tag{6}$$

Where: X '- the variable that suffered a relative decrease from baseline; X - the variable in the initial state; a - coefficient of variation.

Later, the relative change of a variable from itself was determined, according to (7):

$$X'' = (X' - X)\,/\,X' \tag{7}$$

Where: X"- the relative change of a variable from the value of that variable obtained from the relative decrease.

By sensitivity analysis, it was desired to know the meaning and magnitude of the variation that occurs when changing a variable over the others. Thus the case 16 was studied where the quantity of 2,800 pieces was recorded, respectively the highest profit. Finally, a comparison was made between two cases A and B, the first one in which the effect of changing a variable was calculated, while the other was the effect of the simultaneous change of two variables. This allowed knowledge of the magnitude and intensity of the effect due to these changes.

3. RESULTS AND DISCUSSION

Using the data in Table 1 was represented in Fig. 1 the evolution of profit, total cost and sales revenues according to the quantities taken into study.

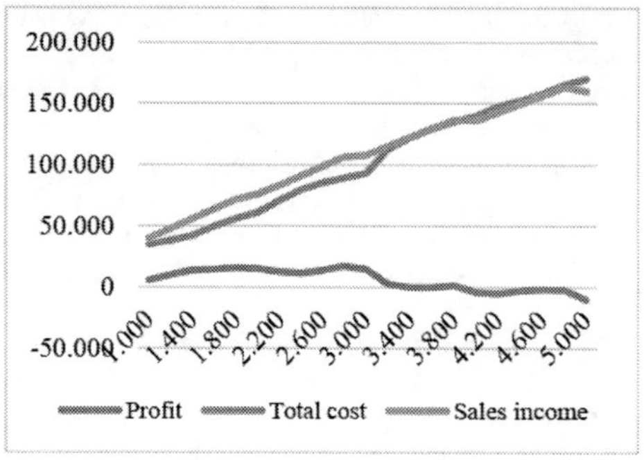

Fig. 1. Evolution of indicators according to the quantity.

From Fig. 1 analysis it is noted that up to 3,800 pieces of positive profit (higher incomes than costs) and from 4,000 pieces it becomes negative (loss).

Subsequently, taking into account the data in Table 1, it was observed that in case 16 there was a negative profit, and this value of 4000 RON was taken as a break-even (i.e. the critical point). Figure 2 shows graphically expected sales and break-even sales.

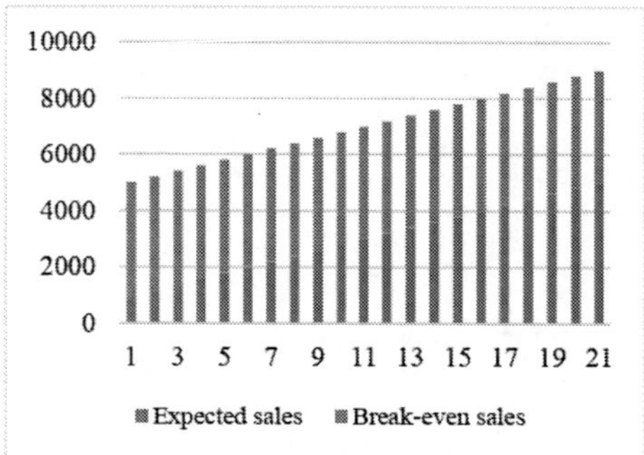

Fig. 2. Breaking point delimitation.

Using Eq. (2), the margin of safety was calculated and plotted in Figure 3.

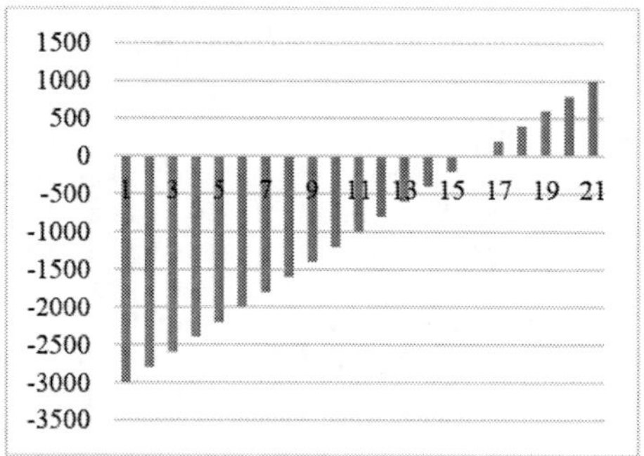

Fig. 3. Evolution of the margin of safety.

From the Fig. 3 analysis, it is noted that when the safety margin is negative (the expected sales are lower than the sales at the point of breaking) we say that the company is making profit.

Fig. 4 shows the evolution of the unit sale price, unit variable cost, and unit marginal contribution.

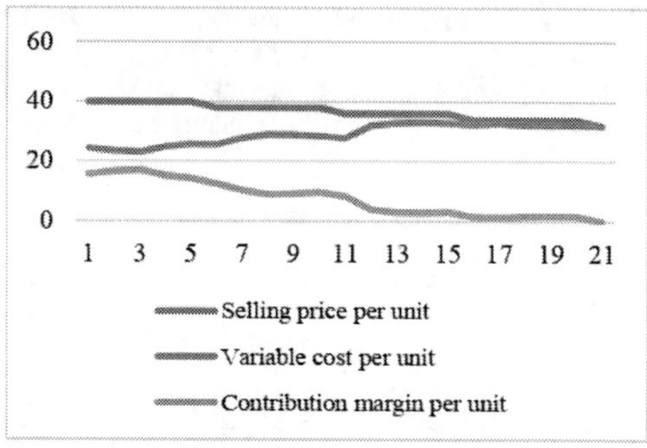

Fig. 4. Evolution of variables.

From the Fig. 4 analysis, it is observed that when the quantity increases (according to Table 1) the unit sale price and the unitary contribution margin decreases (reverse link) and the unitary variable cost increases (direct link).

In Table 2, the amount needed to be sold was calculated in order to obtain the expected profit.

Table 2. The distribution of the quantity according to the expected profit.

Profit [RON]	Total fixed cost [RON]	Cm [RON/Pcs.]	USNASP [Pcs.]
5,600	10,000	15.60	1,000
9,900	10,000	16.58	1,200
13,700	10,000	16.93	1,400
14,200	10,000	15.13	1,600
15,700	10,000	14.28	1,800
14,800	10,000	12.40	2,000
12,400	10,000	10.18	2,200
11,400	10,000	8.92	2,400
13,600	10,000	9.08	2,600
17,000	**10,000**	**9.64**	**2,800**
14,600	10,000	8.20	3,000
2,400	10,000	3.88	3,200
100	10,000	2.97	3,400

From Table 2 we can see that if we know the unitary contribution margin (the difference between the unit sale price and the unitary variable cost) and the fixed costs we can get the quantity to be sold to obtain the expected profit. It is noticed that the biggest profit is recorded in the quantity of 2,800 pieces (case 10).

To observe the influence of a variable on the other variables, the relative diminution of a variable, Case A (Table 3), and the relative diminution of two variables in the Case B (Table 4), were calculated.

Because for managers in economic and financial analyzes the profit is the indicator that determines the performance of a company, in Table 5 we obtained the relative deviation of the unitary variable cost on the profit.

Table 3. The relative change of indicators according to the coefficient of variation (Case A).

Coefficient of variation [%]	Variable cost [RON]	Total cost [RON]	Sales income [RON]	Profit [RON]
-10%	72,886.15	82,886.15	106,400.00	23,513.85
-5%	76,935.38	86,935.38	106,400.00	19,464.62
-1%	80,174.77	90,174.77	106,400.00	16,225.23
0%	80,984.62	90,984.62	106,400.00	15,415.38
1%	81,794.46	91,794.46	106,400.00	14,605.54
5%	85,033.85	95,033.85	106,400.00	11,366.15
10%	89,083.08	99,083.08	106,400.00	7,316.92
Coefficient of variation	**Profit**	**Variable cost**	**Total cost**	**Sales income**
-10%	15,300.00	80,984.62	90,984.62	106,284.62
-5%	16,150.00	80,984.62	90,984.62	107,134.62
-1%	16,830.00	80,984.62	90,984.62	107,814.62
0%	17,000.00	80,984.62	90,984.62	107,984.62
1%	17,170.00	80,984.62	90,984.62	108,154.62
5%	17,850.00	80,984.62	90,984.62	108,834.62
10%	18,700.00	80,984.62	90,984.62	109,684.62
Coefficient of variation	**Sales income**	**Profit**	**Total cost**	**Variable cost**
-10%	95,760.00	17,000.00	78,760.00	68,760.00
-5%	101,080.00	17,000.00	84,080.00	74,080.00
-1%	105,336.00	17,000.00	88,336.00	78,336.00
0%	106,400.00	17,000.00	89,400.00	79,400.00
1%	107,464.00	17,000.00	90,464.00	80,464.00
5%	111,720.00	17,000.00	94,720.00	84,720.00
10%	117,040.00	17,000.00	100,040.00	90,040.00

Table 4. The relative change of indicators according to the coefficient of variation (Case B).

Coefficient of variation [%]	Total cost [RON]	Sales income [RON]	Profit [RON]
-10%	82,886.15	95,760.00	12,873.85
-5%	86,935.38	101,080.00	14,144.62
-1%	90,174.77	105,336.00	15,161.23
0%	90,984.62	106,400.00	15,415.38
1%	91,794.46	107,464.00	15,669.54
5%	95,033.85	111,720.00	16,686.15
10%	99,083.08	117,040.00	17,956.92
Coefficient of variation	**Profit**	**Total cost**	**Sales income**
-10%	15,300.00	72,886.15	82,886.15
-5%	16,150.00	76,935.38	86,935.38
-1%	16,830.00	80,174.77	90,174.77
0%	17,000.00	80,984.62	90,984.62
1%	17,170.00	81,794.46	91,794.46
5%	17,850.00	85,033.85	95,033.85
10%	18,700.00	89,083.08	99,083.08
Coefficient of variation	**Sales income**	**Profit**	**Total cost**
-10%	95,760.00	15,300.00	80,460.00
-5%	101,080.00	16,150.00	84,930.00
-1%	105,336.00	16,830.00	88,506.00
0%	106,400.00	17,000.00	89,400.00
1%	107,464.00	17,170.00	90,294.00
5%	111,720.00	17,850.00	93,870.00
10%	117,040.00	18,700.00	98,340.00

Table 5. Relative deviation of the variables on profit.

ΔVariable cost	ΔTotal cost	Case A		Case B	
		ΔSales income	ΔProfit	ΔSales income	ΔProfit
-10%	-8.90%	0	52.53%	-10%	-16.49%
-5%	-4.45%	0	26.27%	-5%	-8.24%
-1%	-0.89%	0	5.25%	-1%	-1.65%
0%	0.00%	0	0.00%	0%	0.00%
1%	0.89%	0	-5.25%	1%	1.65%
5%	4.45%	0	-26.27%	5%	8.24%
10%	8.90%	0	-52.53%	10%	16.49%

From the analysis of Table 5 for Case A it is observed that when the variable cost changes, it will change in the same direction but with a smaller weight and the total cost (the difference is given by the fixed cost) and the profit will change in the sense the opposite (lower costs increase profit and vice versa) but in a much bigger share. Table 5 for the Case B shows that the change in variable costs and sales revenue causes a change in the same sense of profit, which is close to size.

Table 6 shows the percentage changes in sales revenue for Case A and Case B when the total cost does not change or change with the same percentage as the profit.

Table 6. The relative deviation of the variables on sales income.

ΔProfit	Case A		Case B	
	ΔTotal cost	ΔSales income	ΔTotal cost	ΔSales income
-10%	0	-1.57%	-10%	-9.07%
-5%	0	-0.79%	-5%	-4.54%
-1%	0	-0.16%	-1%	-0.91%
0%	0	0.00%	0%	0.00%
1%	0	0.16%	1%	0.91%
5%	0	0.79%	5%	4.54%
10%	0	1.57%	10%	9.07%

The analysis of Table 6 can be seen in the Case A that the change of profit when the total cost is maintained at the same level leads to the same change of the sales revenues. In the Case B when the same changes in profit and total cost occur, they will cause a change in sales revenue of approximately the same level.

Table 7 shows the percentage changes in the total cost in in the Case A and in the Case B when we change

Table 7. Relative change of variables on total cost.

ΔSales income	Case A		Case B	
	ΔProfit	ΔTotal cost	ΔProfit	ΔTotal cost
-10%	0	-11.90%	-10%	-11.26%
-5%	0	-5.95%	-5%	-5.63%
-1%	0	-1.19%	-1%	-1.13%
0%	0	0.00%	0%	0.00%
1%	0	1.19%	1%	1.13%
5%	0	5.95%	5%	5.63%
10%	0	11.90%	10%	11.26%

 https://doi.org/10.1051/matecconf/201818404003

to only one variable (sales income) or when there is a change to two variables (sales and profit).

The analysis of Table 7 shows that in the Case A, when sales revenue changes and profit remains the same it will cause a change in the same sense of the total cost but at a higher level. In the Case B we have two variables that change in the same direction and at the same level that cause a variation of profit at a lower level.

Fig. 5 (Case A) shows the evolution of profit when the variable cost changes and the amount (2,800 pieces) and sales revenue (106,400 RON) remain unchanged.

Fig. 5 (Case B) shows the evolution of profit when both variables (unit variable cost and sales revenue) change with the stated percentages.

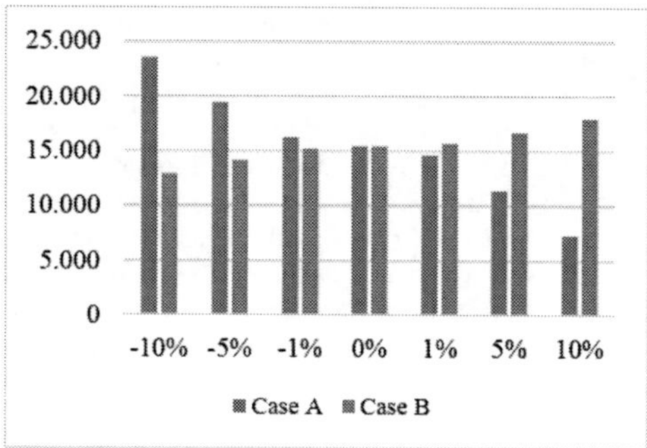

Fig. 5. Evolution of profit

From the analysis of Case A, it can be seen that when the unitary variable cost increases the profit decreases. From the analysis of Case B, it can be seen that when the quantity, unitary cost and sales revenue increase with the same percentages and the profit will increase the same. From the comparison between Case A and Case B, we can see that the effect produced by an indicator (unitary variable cost) has higher consequences for the total cost and implicitly for the profit than for the total variable cost product and the quantity on the profit.

Fig. 6 shows the evolution of sales revenue when the quantity and profit vary or when the quantity, profit, and total cost varies with the specified percentages. From Case A, it is noted that when profit changes and total quantity and cost remain unchanged then sales revenue is the difference. The analysis of Case B shows that when all variables change with the same percentages it causes a change in the same direction and with the same percentage of sales revenue. By comparison, it is noticed that the effect of changing a variable is smaller than the order of magnitude than the one produced by all the variables.

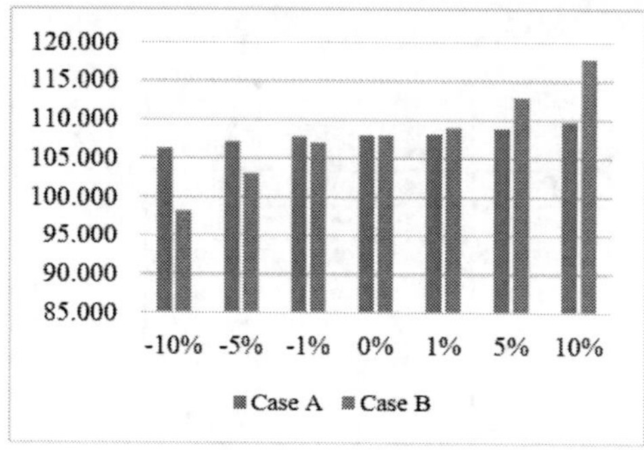

Fig. 6. Evolution of sales income

Fig. 7 shows the evolution of total cost when sales quantity and sales (Case A), quantity, sales revenue and profit varies (Case B).

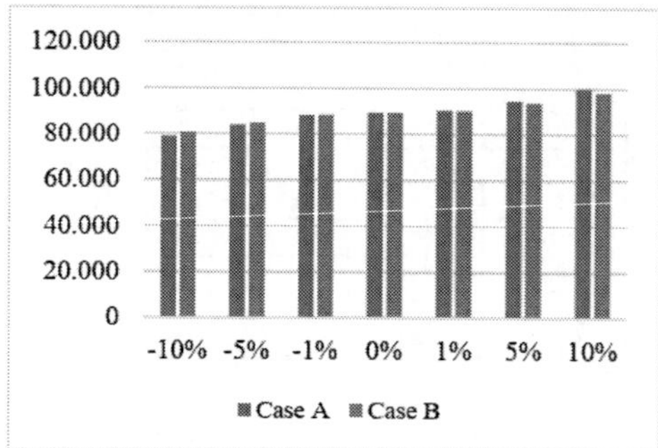

Fig. 7. Evolution of total cost

From the analysis of Case A, it can be seen that the change in the sales revenues determines the total cost changes in the same direction when it is desired that the profit (17,000 RON) and the quantity (2,800 pieces) remain at the same level. Case B shows that the increase in total cost is determined by the increase in quantity, unitary cost, and profit. The effect determined by the variation obtained on the total cost is higher in Case A than in Case B.

4. CONCLUSIONS

The Cost-Volume-Profit Analysis has allowed you to calculate the breaking point, safety margin and sales expected. Subsequently, the evolution of some indicators (unitary variable cost, unit selling price, unitary marginal contribution) to the volume change of the production, according to some cases taken into the study, was calculated and observed. Activity-related fluctuations have demonstrated chain changes in unit costs and implicitly in sales prices.

In fact, the changes are not only related to the volume of activity and these are also given by the other variables: production costs, the selling price accepted by the market, the profit obtained differently from the expected, etc.

The sensitivity analysis allowed the change of variance to be determined when a change occurs at the level of a single variable or when it affects two variables. Although production costs decrease as a result of economies of scale and can be achieved as the quantity increases (due to fixed costs and less of the variable costs), in practice it is found that they do not propagate in the same proportions. Similarly, if there is a change in the sales mix, the earnings will show changes in the amount sold.

The disadvantage of these analyzes can be given by the uncertainty that exists in certain activities of correctly dividing fixed and variable costs. In reality, some fixed costs may be semi-fixed or may be transformed from a certain level of activity into variable costs. Splitting costs into fixed and variable costs is sometimes difficult. Profit can be predicted on the basis of variable cost, or vice versa if a certain level of profit is desired, then either the sales revenue in the sense of maximizing or the minimization costs.

I consider that the use of CVP analysis and sensitivity analysis in forecasting, tracking, and control activity will result in an efficient optimization of cost-benefit, profit and profitability indicators as a result of both anticipative and participatory management. Knowing activity through these analyzes will be a useful tool for managers, providing the enterprise with the ability to adapt and respond to changes in the environment.

References

1. S. Liebesman, *How to manage risk in a global economy*, QP, **41 (3)**, p. 58-60 (2008).
2. R. Balakrishnan, E. Labro, N.S. Soderstrom, *Cost structure and sticky costs*, J. of Manag. Acc. Res., **26 (2)**, 91-116 (2014).
3. T. Groot, F. Selto, *Advanced management accounting*, Pearson Education, Amsterdam (2013).
4. S.W. Anderson, *Measuring the impact of product mix heterogeneity on manufacturing cost*, The Acc. Rev., **79**, 363–387 (1995).
5. R.S. Kaplan, R. Cooper, *Cost and effect: sing integrated cost systems to drive profitability and performance*, Harvard Business School Press, Boston (1998).
6. J. Glomazić, *Influence of operating conditions to the method of calculating cost prices*, J. of Acc. and Manag., 2016, **6 (2)**, 43 - 54.
7. J.G. Miller, T.E. Vollman, *The hidden factory*, Harv. Bus. Rev., **63**, 142–150 (1985).
8. H.T. Johnson, R.S. Kaplan, *Relevance lost: The rise and fall of management accounting*, Harvard Business School Press, Washington, DC (1987).
9. R.D. Banker, G. Potter, R.G. Schroeder, *An empirical analysis of manufacturing overhead cost drivers*, J. of Acc. and Ec., **19**, 115–137 (1995).
10. D. Teece, M. Peteraf, S. Leih, *Dynamic capabilities and organizational agility: Risk, uncertainty, and strategy in the innovation economy*, California Manag. Rev., **58 (4)**, 13-35 (2016).
11. C. Drury, M. Tayles, *Cost system design and profitability analysis in UK companies*, Chartered Institute of Management Accountants, London (2000).
12. C.T. Horngren, S.M. Datar, and M. Rajan, *Cost Accounting: A Managerial Emphasis*, 15th edition Harlow, U.K.: Pearson Education Limited (2015).
13. Ernst & Young, *Global transfer pricing survey - Addressing the challenges of globalization*, New York, NY: Ernst & Young Ltd. (2010).
14. R.S. Kaplan, A. Atkinson, *Advanced management accounting*, 3rd edition Englewood Cliffs, NJ: Prentice Hall (1998).
15. S. Sahay, *Transfer pricing based on actual cost*, J. of Man. Acc. Res., **15**, 177–192 (2003).
16. K.A. Merchant,d W.A. Van der Stede, *Management control systems: Performance measurement, evaluation and incentive*, 3rd edition Harlow, U.K.: Financial Times Prentice Hall (2012).
17. J.L. Zimmerman, *Accounting for decision making and control*, 8th edition New York, NY: McGraw-Hill Companies Inc. (2013).
18. M. Anderson, R. Banker, S. Janakiraman, *Are selling, general, and administrative costs „sticky"?*, J. of Acc. Res., **41**, p. 47–63 (2003).
19. K. Calleja, M. Steliaros, D.C. Thomas, *A note on cost stickiness: Some international comparisons*, Manag. Acc. Res., **17**, 127–140.
20. R.D. Banker, L.T. Chen, *Predicting earnings using a model based on cost variability and cost stickiness*, The Acc. Rev., **81 (2)**, 285–307 (2006).
21. M. Potkany, L. Krajcirova, *Quantification of the volume of products to achieve the Break-Even Point and desired profit in non-homogeneous production*, Proc. Ec. and Fin., **26**, 194 – 201 (2015).

Comparative analysis for estimating production costs

Carmen-Elena Stoenoiu[1,*], *Ciprian* Cristea[2]

[1, 2] Technical University of Cluj-Napoca, Department Electrical Machines and Drives, Cluj-Napoca, România

Abstract. The analysis of the cost of production is a complex problem, being very useful in the planning and control process as well as in production and sales. In this study, this was done using three methods: accounting analysis, high-low and regression analysis that allowed cost elements to be identified through a cost equation that can be used in planning and control analysis or in sales activity. At the same time, the evolution of the cost elements was achieved using two estimates used by the accounting method. The High-Low method allowed the determination of the minimum and maximum activity level at one year's level, which was then analyzed through the company's available activity range. The multivariate regression analysis determined the analysis of the obtained model, being linear, thus establishing the production line.

1. INTRODUCTION

The increasing complexity of economic activity as well as its variety are increasingly requiring an effective management of resources. In this context, cost management tools play an important role because they allow costs to be aligned to the strategy of the company [1], [2]. Because in the economic activity the behaviour of production costs is different, managers are often put in the position to make estimates based on past or present information provided by financial or management accounting, through financial reports or through dashboard [3], [4].

The decision or the managerial act depends increasingly on the information provided by the financial-accounting activity, the quality and the speed of the information, the way it is processed and transformed into clear, pertinent and complete information [5]. The future of a company is often dependent on the speed with which it manages to make decisions, respectively to adapt to the concrete conditions existing on the market. So, most of the informational system plays an essential role because it allows to know the costs, how they change to the volume of production, allowing component identification, fluctuation observation and the successful completion of the managerial act [6], [7].

Cost management is a widely discussed topic in the literature as it is considered to provide support in the decision-making process [8], [9], allows the development of performance planning and management systems [10], [11], advising managers on the formulation and implementation of corporate strategy [12], [13], and when flexible they represent a resource to create new scenarios within the organization [14], [15]. Fixed costs, typically linear, are those expected costs because they are already engaged (in progress) and can not be changed in a short period of time [16]. Variable costs, in turn, are fluctuating (direct material costs, direct wages, etc.), including various aspects related to the characteristics of manufactured products [17], [18]. Variable cost tracking is useful because it allows the manager to know the costs which can be manipulated and controlled [19]. In the literature, it is considered that when fixed costs have a significant weight in total costs, managers need to find solutions for sales growth [20]. The way of tracking the information provided by accounting and subsequently encompassing the financial reports is also debated by various specialists [21].

Since managers make many decisions through production costs, precision is important because a small distortion can have a major influence on the firm's profit [22]-[24].

In this paper, an analysis of production costs was made, starting from the information provided by the accounting. The goal was to exemplify through three different ways (accounting analysis, High-Low and regression analysis) the calculation, estimation and cost analysis based on the information from financial accounting.

2. MATERIALS AND METHODS

Three methods were used to analyze production costs: accounting analysis (AA), High-Low method (HLM) and regression analysis (RA), starting from the idea that small and medium enterprises do not always have sufficient financial resources to acquire performance software capable of providing detailed cost analyses.

The information on the production costs is in the accounting records of any company, irrespective of its size, it only requires processing, and i.e. their take over from the balance of monthly checking of the accounts,

* carmen.stoenoiu@emd.utcluj.ro

© The Authors, published by EDP Sciences. This is an open access article distributed under the terms of the Creative Commons Attribution License 4.0 (http://creativecommons.org/licenses/by/4.0/).

the distribution by cost types and then the summation of each type of cost. The starting point was the identification and grouping of all production costs in a company's accounting records, according to Table 1, by type of costs (fixed and variable).

Table 1. Correspondence costs from financial accounting into managerial accounting.

Production costs	Accounts groups (Account)
1. Fixed costs, from which:	
- Utilities (electrical energy, water, telephone, gas energy, etc.)	60. Stock expenditure (acc. 605); 62. Expenses with other service provided by third parties (acc. 626).
- Services performed by third parties (rent, maintenance and repairs, insurance, studies and research)	61. Expenditure related to third-party services (acc. 611÷614)
- Fixed asset depreciation (equipment, machinery, buildings, means of transport, etc.)	68. Expenses, provisions and adjustments for depreciation or loss of value (acc. 6811)
2. Variable costs, from which:	
- Consumables	60. Stock expenses (acc. 601 ÷ 604, 608, 609)
- Assembly / processing salaries	64. Staff costs (acc. 641-645)

The equations used was defined by (1-6):

$$PC= FC + VC \quad (1)$$

$$FC = U + TPS + FA \quad (2)$$

$$VC = M + S \quad (3)$$

$$EPC = CV/ \quad (4)$$

$$EC = (MAC - MC)/(MaxQ - MinQ) \quad (5)$$

$$TC = FC + (UVC \times Q) \quad (6)$$

Where: PC - production cost; FC - fixed cost; VC - variable cost; U – utilities; TPC - third party services; FA - fixed assets; M – materials; S – salary; EPC - estimated poor cost; CV - cost variation; AV – activity variation; EC - estimated cost; MAC - maximum activity costs; MC - minimum costs; MaxQ - maximum quantity; MinQ - minimum quantity; TC - total cost; UVC - unit variable cost; Q – quantity.

The AA is based on the use of the professional judgment of the analyst and involves the grouping of costs in fixed and variable, and then the establishment of the unitary cost of the activity. This analysis seeks to know: what are the possible variations, in the unit cost of production, when changing the quantity produced. The way in which the calculations were made are presented in Table 2, using the month of May as the reference month.

Because of the complex nature of these costs, sometimes the assignment is difficult, therefore two estimates for accounting analysis were used in this study:

The first estimate involves the delimitation of the production costs in fixed and variable ones, considering that the specificity of the activity allows their clear delimitation (Table 3).

Table 2. Production costs related to May [RON].

Cost elements	Value
Quantity	1,950
Production costs, from which:	
- Consumables (C)	115,900
- Assembly / processing salaries (AS)	83,100
- Utilities (U)	7,100
- Services performed by third parties (SPTP)	18,000
- Fixed asset depreciation (FAD)	50,900
Total	275,000

The second estimate of fixed and variable production costs takes into account that some of the utility costs change with the volume of production and therefore cannot be considered totally fixed, the appreciation being 50% fixed and 50% as variables (Table 4).

After identifying costs by cost type for each month, the data was centralized to analyze: fluctuations that exist from one month to the next at fixed cost, variable cost and total cost, physical output level (quantitative) and fixed unit costs and variables. It was considered that the company's business is not constant, with some changes from one month to the next. Subsequently data was centralized at the level of a calendar year.

It was considered that the analyzed activity evolved linearly over twelve months and what is outside this behavior is an exception that needs to be identified and analyzed separately.

The HLM was used to see the variation of production cost components (fixed and variable) and their level at various work volume changes. The High-Low method allows us to observe the maximum and minimum points, that is, the maximum and minimum level of activity performed. Starting from (4) we obtained (5), which allows us to determine the high and low cost variables through the High-Low method.

The RA was based on the following assumptions: the independent variable is quantitative or dichotomic (in this variable study is the quantity of products obtained in one year); the variable is measured without error; the independent variable has a nonzero variant; there is no multicollinearity; the mean value of the variable e (error) is zero for any set of values of the independent variables; each independent variable is uncorrelated with the variable ε; the variant of ε is constant-homoscedasticity; for any two observations, the errors are uncorrelated; for any values of the independent variables, ε is normally distributed.

Taking (x_i, y_i), as a set of data, it results a function f: R → R. The variable ε was defined by (7):

$$e_i = y_i - f(x_i) = y_i - \dot{y}_i \quad (7)$$

Where: i = 1 ÷ m, m, is called the residual variable, and the values of this variable are called residual values.

To solve it, we need the absolute value of the residual values, but to work with expressions we use their squares. The sum of squares of residual values is called

https://doi.org/10.1051/matecconf/201818404004

the residual quadratic sum. By replacing "y" with its value, the relationship becomes (8):

$$\Sigma\ (y_i - a - bx_i)\ 2 = \text{minimum} \qquad (8)$$

Deriving in relation to "a" and "b" by canceling the partial derivatives, the system of normal equations is obtained (9-10):

$$na + b\ \Sigma x_i = \Sigma y_i \qquad (9)$$

$$a\ \Sigma x_i + b\ \Sigma x_i = \Sigma x_i y_i \qquad (10)$$

Where: n - the number of units observed, ie the number of pairs (x, y).

It follows that y = a + bx, which means that the regression right passes through the mean point (x, y). By solving the system of normal equations, parameters "a" and "b" are obtained. The coefficient a, which can take both positive and negative values, is ordered at origin, that is, the value of y when x is equal to zero. The coefficient b is referred to as the regression coefficient which shows the extent to which the dependent characteristic varies when the independent feature changes with a unit.

3. RESULTS AND DISCUSSION

Through the AA, fixed costs (FC) and variable costs (VC) were calculated using the two estimates (cases A and B), which are presented in Table 3.

Table 3. Estimated production costs for May through accounting analysis [RON].

Cost elements	Case A	Case B	Cost elements	Case A	Case B
1. FC	76,000	72,450	2. VC	199,000	202,550
- U	7,100	3,550	- U	-	3,550
- SPTP	18,000	18,000	- C	115,900	115,900
- FAD	50,900	50,900	- AS	83,100	83,100
			- Q	1,950	1,950
			Unit variable cost	102.05	103.87

Table 4. Distribution of monthly production costs [RON].

Month	Quantity [Pcs.]	Case A		Case B	
		FC	VC	FC	VC
January	800	76,000	104,000	72,450	107,550
February	1,100	76,000	114,000	72,450	117,550
March	1,250	76,000	134,000	72,450	137,550
April	1,650	76,000	169,000	72,450	172,550
May	1,950	76,000	199,000	72,450	202,550
June	2,250	76,000	209,000	72,450	212,550
July	2,900	76,000	339,000	72,450	342,550
August	2,750	76,000	299,000	72,450	302,550
September	2,450	76,000	224,000	72,450	227,550
October	1,200	76,000	132,000	72,450	135,550
November	1,000	76,000	104,000	72,450	107,550
December	500	76,000	74,000	72,450	77,550
Total	20,000	912,000	2,101,000	869,400	2,143,600

From the analysis of Tables 3 it can be noticed that the different estimation of a cost of production component (utilities) led to the obtaining of different values of the unitary and total variable cost and of the total fixed cost.

The process used for May continued for all the months of the year, thus obtaining a monthly cost of production, as shown in Table 4.

The evolution of fixed and variable costs using the two estimates at the level of the 12 months is shown in Fig. 1 and Fig. 2.

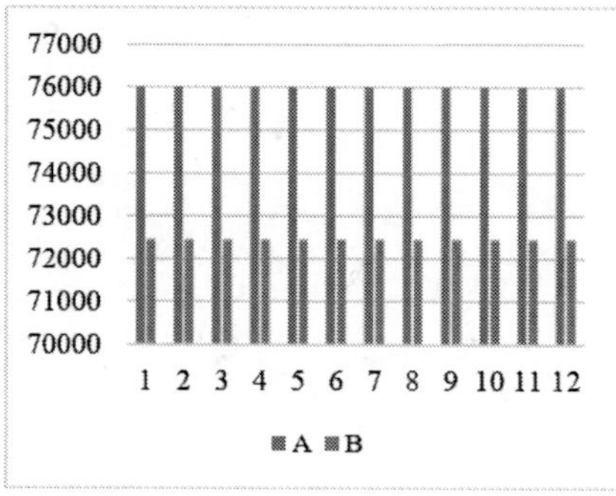

Fig. 1. Evolution of the Fixed costs

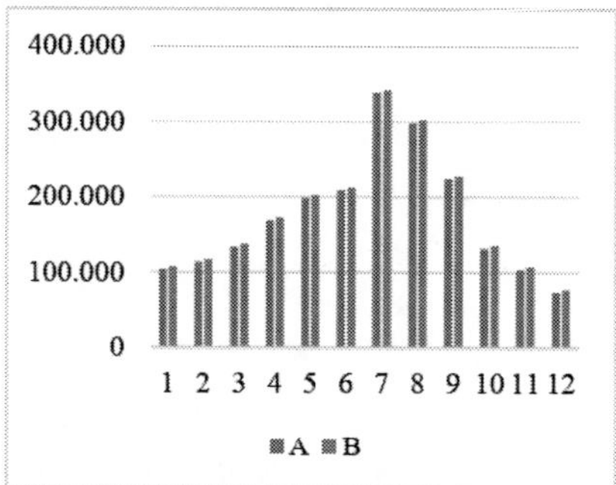

Fig. 2. Evolution of the Variable costs

From the analysis of Fig. 1 and Fig. 2 it can be noticed that the evolution of the fixed and variable costs at the level of the two production cost estimates (Cases A and B) is slightly modified. The evolution of fixed and variable costs per unit using the two estimates at the level of the 12 months is presented in Fig. 3:

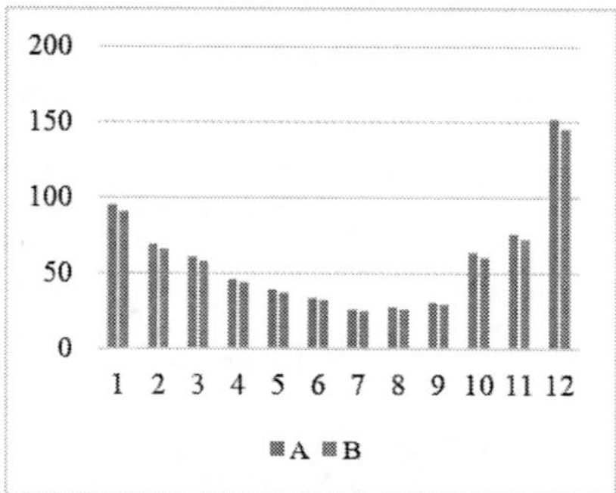

Fig. 3. Evolution of the Fixed costs per unit

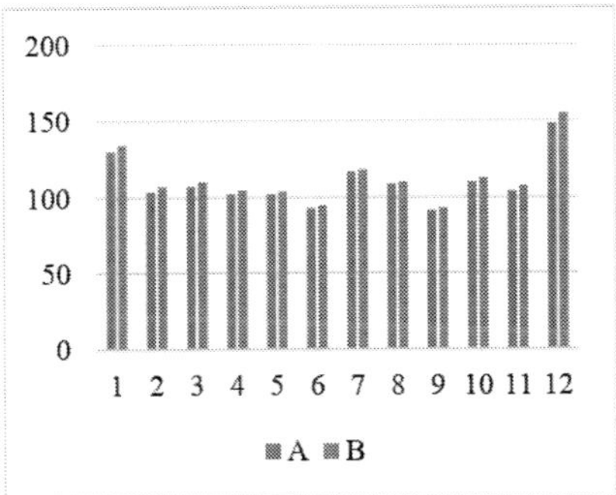

Fig. 4. Evolution of the Variable costs per unit

The AA allows us to observe the minimum level of unitary cost versus the physical volume of production (Fig. 4).

From the AA using the two estimates presented in Fig. 5 we can see that the evolution of the unitary cost variables, according to the quantity shows a small difference between the two estimates (Cases A and B).

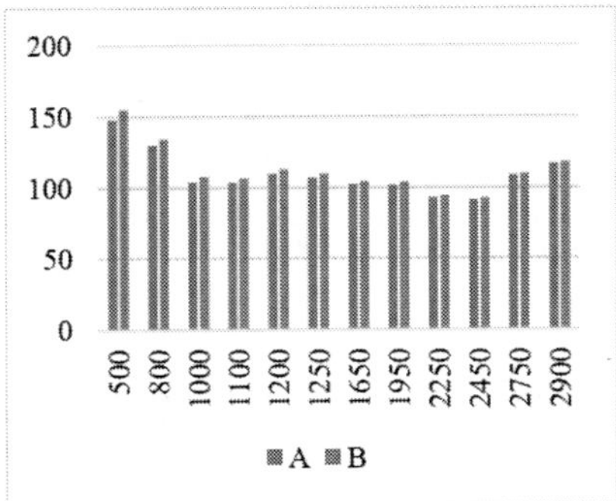

Fig. 5. Evolution of the unit variable costs according to the quantity

Since the distribution of total costs in fixed and variable is sometimes subjective on the one hand due to their variation over time and, on the other hand, due to the way these costs are accepted or understood, the use of this method does not always provide us with surety about costing.

Using the HLM, it was possible to estimate the level of the variables in the production cost component to the lowest level activity (LLA) or the highest level of activity (HLM), as shown in Table 5.

Table 5. Estimated High-Low method [RON]

Indicators	LLA	HLA
Total cost	131,208.33	396,208.33
Variable cost estimate	55,208.33	320,208.33
Fixed Cost Estimate	76,000	76,000

The HLM is useful after the managers have set the possible production range that can be achieved taking into account the production capacity. According to the example taken in the study, if at 50% we are profit and not loss and this means producing 500 pieces using a production range between 500-3,200 pieces, it means we have 100% at maximum. Through this analysis we can delimit both the degree of loading and the monthly production capacity recorded at the entity level, ie the position against the minimum or the maximum taking into account the technical level, as well as the minimum acceptable quantities for the company to make profit.

For the RA, the following variables listed in Table 6 were taken into account.

Table 6. Annual state

Month	Quantity [Pcs.]	Total cost production [RON]
January	800	180,000
February	1,100	190,000
March	1,250	210,000
April	1,650	245,000
May	1,950	275,000
June	2,250	285,000
July	2,900	415,000
August	2,750	375,000
September	2,450	300,000
October	1,200	208,000
November	1,000	180,000
December	500	150,000

Table 7 presents statistical indicators computed through regression analysis.

Table 7. Statistical indicators

Statistic	Value
Multiple R	0.97
Multiple R^2	0.95
Adjusted R^2	0.94
F(1,10)	179.93
p	1.02×10^{-7}
Std.Err. of Estimate	19,735.39

The RA led to the identification of unit variable cost and total fixed cost when the total cost of production and the quantity for each month of the year are known, as shown in Table 8.

Table 8. Regression analysis

N=12	B
Intercept (a)	85,930.01
Quantity (b)	100.09

From the analysis of Table 8 we can see the value of the regression analysis for the fixed cost and for the unitary cost.

Following the calculations of the total fixed cost and the unitary cost of each of the 3 methods taken in the study, the equation of production cost was obtained, as shown in Table 9:

Table 9. Equation of production cost

Methods	General equation	Equation of production cost
AA (Case A)	$Cp = CF + Cv_u \times Q$	$Cp = 76,000 + 106.11 \times Q$
AA (Case B)		$Cp = 72,450 + 108.26 \times Q$
HLM		$Cp = 76,000 + 110.41 \times Q$
RA		$Cp = 85,930.01 + 100.09 \times Q$

For each total value of your monthly (Q) production, the total cost of production also recorded a different value. The unitary variable cost, using the AA, was obtained by means of the simple arithmetic mean, both for case A and case B. The unit-variable cost of HLM was obtained by reporting the total variable cost to the quantity (min. or max.) and the RA mathematically calculated this cost based on the monthly quantity and total cost of production. Analysing the four equations there are some differences from the initial situation, which are presented in Table 10.

Table 10. Comparison between the initial cost of production and the values obtained by using the methods [RON]

Month	CTPI	CTPACA	Dif. CTPACA	CTPACB	Dif. CTPACB
0	1	2	3 = 2-1	4	5=4-1
01	180,000	160,888.88	-19,111.12	159,060.08	-20,939.92
02	190,000	192,722.21	2,722.21	191,538.86	1,538.86
03	210,000	208,638.88	-1,361.13	207,778.25	-2,221.75
04	245,000	251,083.32	6,083.32	251,083.29	6,083.29
05	275,000	282,916.65	7,916.65	283,562.07	8,562.07
06	285,000	314,749.98	29,749.98	316,040.85	31,040.85
07	415,000	383,722.19	-31,277.81	386,411.54	-28,588.46
08	375,000	367,805.53	-7,194.48	370,172.15	-4,827.85
09	300,000	335,972.20	35,972.20	337,693.37	37,693.37
10	208,000	203,333.32	-4,666.68	202,365.12	-5,634.88
11	180,000	182,111.10	2,111.10	180,712.60	712.60
12	150,000	129,055.55	-20,944.45	126,581.30	-23,418.70
Total	3,013,000	3,012,999.78	-0.22	3,012,999.48	-0.52

Month	CTPMHL	Dif. CTPMHL	CTPAR	Dif. CTPAR
0	6	7=6-1	8	9=8-1
01	164,328.00	-15,672.00	166,002.01	-13,997.99
02	197,451.00	7,451.00	196,029.01	6,029.01
03	214,012.50	4,012.50	211,042.51	1,042.51
04	258,176.50	13,176.50	251,078.51	6,078.51
05	291,299.50	16,299.50	281,105.51	6,105.51
06	324,422.50	39,422.50	311,132.51	6,132.51
07	396,189.00	-18,811.00	376,191.01	-38,808.99
08	379,627.50	4,627.50	361,177.51	-13,822.49
09	346,504.50	46,504.50	331,150.51	31,150.51
10	208,492.00	492.00	206,038.01	-1,961.99
11	186,410.00	641,000	186,020.01	6,020.01
12	131,205.00	-18,795.00	135,975.01	-14,024.99
Total	3,098,118.00	85,118.00	3,012,942.12	-57.88

CTPI - total initial production cost; CTPACA - total cost of production obtained through the accounting analysis variant (a); CTPACB - total cost of production obtained through the accounting analysis variant (b); CTPMHL - total cost of production obtained by the High-Low method; CTPAR - total production cost obtained by regression analysis.

Table 10 shows that the total differences recorded for the accounting analysis (Cases A and B) per total year are very small (to be neglected), which are followed (as order of magnitude) by the RA. HLA does not help us estimate the cost of production because the difference is very high. The strong point of the RA is that this and if it is not known, the values of the total fixed and unitary cost can establish the equation of the total cost of production, thus establishing the relationship between the fixed cost, the variable cost and the quantity, very useful for managers in the decision-making process as it allows the sale price to be determined and participation in negotiations with potential customers before the production is in stock, meaning to know in detail all the cost elements.

4. CONCLUSIONS

The AA made it possible to know the production costs and their evolution. This analysis is useful for managers in small and medium businesses because it allows to determine the quantity of products that can be produced at the lowest cost and to adapt it to the economic conditions existing on the market. Using billboards that take account of accounting information, managers can get to know these production costs and track them both on a total and per product basis. Thus, they can establish or adjust their production volume and, implicitly, the sales price in terms of profitability, taking into account the production costs, the changes that occur at the cost level of both the material component and the wage component.

HLM is useful because it allows managers to know where the costs are high and small, and by delimiting a possible production range, depending on the capacity of the firm, to know the variance of these costs within that range. It is an analysis of the past but which provides enough information, ie it signals the levels of activity that have been recorded and what influences they have determined at each cost level. Future decisions can be improved by knowing the past, the disturbing factors and the concrete production possibilities.

The strong point of the RA is that this analysis can also be performed when the values of the fixed and total fixed cost are not known. Starting from the total production cost of the previous year and from the production quantity and considering that next year the activity will remain at least constant, the equation of the total production cost can determined. By establishing the link between the fixed cost, variable cost and quantity, this will help managers in the decision making process, while allowing cost planning and control.

Because often the costs of tracking and calculating the cost elements required for an analysis involve additional costs, and firms are often required to make cost cuts, using this comparative analysis companies can calculate and observe differences in production cost as a result of changes in the quantity produced. Although the analysis has taken into account only the production costs (without general administrative costs), we consider that the analysis is useful because it allows the delimitation of those costs that represent the largest share in the total cost.

References

1. W.M. Baker, *The missing element in cost management: Competitive target costing*, Ind. Manag., **37 (2)**, 29–32 (1995).
2. R. Cooper, R Slagmulder, *Develop profitable new products with target costing*, MIT Sloan Manag. Rev., **40 (4)**, 23–34 (1999).
3. A.D. Jr. Chandler, H. Daems, *Administrative coordination, allocation and monitoring: a*

comparative analysis of the emergence of accounting and organization in the USA and Europe, Acc. Org. and Soc., **4**, 3-20 (1979).
4. P. Skærbæk, K. Tryggestad, *The role of accounting devices in performing corporate strategy*, Acc., Org. and Soc., **35 (1)**, 108-124 (2010).
5. S.Y. Collin, T. Tagesson, A. Anderson, J. Cato, K. Hansson, *Explaining the choice of accounting standards in municipal corporations. Positive accounting theory and institutional theory as competitive or concurrent theories*, Critic. Persp. on Acc., **20 (2)**, 141-174 (2009).
6. Z. Toth, *The current role of accounting information systems*, Club of Economics in Miskolc' TMP, **8 (1)**, 91-95 (2012).
7. Y.W. Kim, S.C. Kim, *Cost analysis of information technology-assisted quality inspection using activity-based costing,* Constr. Manag. and Ec., 163–172 (2011).
8. H.U. Kupper, *Investment-based cost accounting as a fundamental basis of decision-oriented management accounting*, Abacus, **45 (2)**, 249-274 (2009).
9. J. Innes, F. Mitchell, D. Sinclair, *Activity-based costing in the U.K.'s largest companies: a comparison of 1994 and 1999 survey results*, Manag. Acc. Res., **11 (3)**, 349–362 (2000).
10. D.J. Stockton, J.E. Middle, *An aproach to improving cost estimating*, Int. J. Prod. Res., **20 (6)**, 741-751 (1982).
11. C.D.A. Wanderley, J. Cullen, *A case of management accounting change: the political and social dynamics*, Rev. Contab. & Fin., **23 (60)**, 161-172 (2012).
12. C. Horngren, S.M. Datar, M. Rajan, *Cost accounting, a managerial emphasis*, Prentice Hall, 14-th edition, ISBN: 9780132109178, (2006).
13. L.O. Obokoh, G. Goldman, *Infrastructure deficiency and the performance of small- and medium-sized enterprises in Nigeria's Liberalised Economy*, Acta Commercii, **16 (1)**, 339 (2016).
14. B. Nixon, J. Burns, *The paradox of strategic management accounting*, Manag. Acc. Res., **23 (4)**, 229–244 (2012).
15. R.H. Chenhall, F. Moers, *The role of innovation in the evolution of management accounting and its integration into management control*, Acc. Org. and Soc., **47**, 1–13 (2015).
16. B.L. Miller, A.G. Buckman, *Cost allocation and opportunity costs*, Manag. Sc., **33 (5)**, 626-639 (1987).
17. C. Sheu, L.J. Krajewski, *A decision model for corrective maintenance management*, Int. J. Prod. Res., **32**, 1365–1382 (1993).
18. Z. Lin, D. Chang, *Cost–tolerance analysis model based on a neural networks method*, Int. J. Prod. Res., **40**, 1429–1452 (2002).
19. B. Finch, S. Gavirneni, *Confidence intervals for optimal selection among alternatives with stochastic variable costs*, Int. J. of Prod. Res., **44 (20)**, 4329–4342 (2006).
20. P.H. Hsu, H.M. Wee, *Discounting decision for enterprises with high fixed cost and low variable cost*, Intl. Trans. in Op. Res., **13**, 111–124 (2006).
21. C. Pong, M. Falconer, *Full costing versus variable costing: Does the choice still matter? An empirical exploration of UK manufacturing companies 1988–2002*, The British Acc. Rev., **38**, 131–148 (2006).
22. R. Cooper, R. Kaplan, *How cost accounting distorts product costs*, Manag. Acc., **69 (4)**, 20–28 (1988).
23. C. Drury, B.P. Osborne, M. Tayles, *A survey of management accounting practices in U.K. manufacturing companies*, London, U.K.: ACCA Publications (1993).
24. E. Shim, E.F. Sudit, *How manufacturers price products*, Manag. Acc., **76 (2)**, 37–39 (1995).

Education and training in Romania on occupational safety and health in the offshore industry

Mariana Ratiu[1,*], *Corina* Suteu[2], and *Ioan-Florin* Oarga[3]

[1]University of Oradea, Department of Mechanical Engineering and Automotive, 410087 Oradea, Romania
[2]University of Oradea, Department of Psycho-Neurosciences and Rehabilitation, 410073 Oradea, Romania
[3]Technical University of Cluj-Napoca, Faculty of Machine Building, 400641 Cluj-Napoca, Romania

Abstract. This paper is the result of a study on the current situation in Romania regarding education and training on the occupational safety and health in the offshore industry. This study was conducted within a European project, which, among other things, has as intellectual outputs, identifying the main requirements related to learning needs and competencies and identifying the similar safety and health training courses for offshore industry at European level. Romania has transposed the European Union legislation in the area of minimum requirements for the safety and health of workers in the offshore industry and has an adequate institutional framework aligned to the Community requirements in the field. In Romania, there are no special occupations and explicit requirements for occupational safety and health specialists in the offshore industry. But, there are some training centers, which offer theoretical and practical courses regarding safety and health for those graduates or professionals who wish to pursue a professional career in the offshore industry. Also, some universities provide diplomas, courses or training related to the offshore industry.

1 Introduction

Against the backdrop of increasing energy consumption and the reduction of underground natural resources, the offshore industry has been growing in the last time. However, some serious accidents that were happened in this field and that led to environmental disasters, economic damages, and loss of life, illustrated the need for comprehensive measures in this field.

The "offshore industry" means "off the coast industry" and comprises the offshore oil and gas industry and the offshore renewables and marine energy industry (offshore wind, tidal and wave).

In the offshore industry, there is a wide range of challenging occupations, but which requires many hours of work in tough conditions. The occupations in offshore industry are at all levels, from entry-level to experienced professionals, in different fields: drilling, piping, welding, diving, production, rigging, surveying, as well as in quality, environment, safety and health, and other support staffing. Under the laws and regulations in force, all companies from offshore industry have a great responsibility for their employees and the environment. So all specific requirements regarding safety, health and environment have to be integrated into all offshore activities, including in education and training courses of the employees.

2 Romanian strategy and legislation related to occupational safety and health

By now, Romania did have not a strategy on safety and health at national level, adopted by a normative act. During this period, it is in consultation a proposal for the Romanian Occupational Safety and Health Strategy for the period 2017-2020. This strategy represents an opportunity to achieve a safe and healthy work environment and a framework instrument that ensures the correlation with the strategic directions of the European Union on occupational health and safety [1].

In Romania, the field of the occupational safety and health is regulated through three major levels of legislation: primary, secondary and tertiary legislation. The primary legislation is the Law no. 319 of 14 July 2006 on safety and health at work that is mandatory to be applied to all sectors of activity, both public and private, and to employers, workers, and representatives of the workers. According to it, the employers shall ensure such conditions that each worker receives sufficient and adequate safety and health training [2].

The secondary legislation includes preventive measures and rules of application specified in methodological norms for the enforcement of the law no. 319 on safety and health at work [3]; government decisions transposing EU directives; standards (standards on safety, standards of the products).

In the tertiary legislation are detailed preventive measures in different regulations or instructions of the companies, in the completion and implementation of the regulation in the field according to the specific conditions of their activity.

* Corresponding author: mratiu@uoradea.ro

© The Authors, published by EDP Sciences. This is an open access article distributed under the terms of the Creative Commons Attribution License 4.0 (http://creativecommons.org/licenses/by/4.0/).

3 Relevant authorities and institutions with attributions in the field

The Ministry of Labor and Social Justice is the relevant authority in the field of occupational safety and health in Romania, having the main attributions: to elaborate, in cooperation with the Ministry of Health and by consultations with other institutions with attributions in the field, the policy and the national strategy on occupational safety and health; to authorize the persons who perform the services in the field of the protection and prevention; to coordinate, in collaboration with the Ministry of Research and Innovation, the elaboration of the national research programs in the field of occupational safety and health; to organize, together with the Ministry of National Education, the activity of general and/or specific training in the field of occupational safety and health for educational institutions; to approve diverse information and training materials, for a correct information [4].

The Ministry of Health is the responsible central public authority in the field of workers' health at the workplace [5].

The Labor Inspection represents the relevant authority concerning the control of the application of the legislation on safety and health at work [6].

The National Research and Development Institute of Occupational Safety (INCDPM) - "Alexandru Darabont" is a research and development institute accredited by the national authority for scientific research (ANCS) as a Romanian national legal entity of public interest, coordinated by the Ministry of National Education [7].

The National Qualifications Authority (NCA) regulates and coordinates the framework of continuing education and training for adults, develops and manages the National Register of Qualifications and National Register of training providers, coordinates and controls the licensing and authorization process for adult training providers and centres of professional competence assessment and certification of professional competence assessors [8].

4. Legislation and competencies regarding the safety of the offshore operations

To help prevent offshore accidents from happening and to ensure the proper response if they do, the EU adopted the Directive 2013/30/EU on Safety of Offshore Oil and Gas Operations. According to with this directive, "all operators have to take all necessary measures to prevent any major accidents and to have available sufficient physical, human and financial resources to limit the consequences when they do occur." They must obtain a license from an independent authority and must to prepare a proper emergency response plan, to respond to any major hazard. Also, they have to inform the public, before the exploration, about the possible effects of any planned offshore operations [9].

According to this directive, Romania has adopted the Law no. 165/2016 on offshore oil operations safety, which establishes the minimum requirements for the prevention of the major accidents and limitation of the consequences of such accidents involving oil operations in the Black Sea areas under Romanian jurisdiction [10].

The authority responsible for all issues relating to the prevention and response to the major accidents in offshore operations is the European Union Offshore Oil and Gas Authorities Group (EUOAG). It is a forum for the exchange of information and expertise between National Authorities, Third Countries, Industrial Associations, the European Commission and other stakeholders. The EUOAG has the role to promote and facilitate the application of the best practices and highest safety standards in offshore oil and gas operations worldwide [11].

In Romania, the authority responsible for all these is the Regulatory Competent Authority for offshore oil operation at the Black Sea (ACROPO) [12].

5. The minimum requirements for the safety and health of workers in the offshore industry

The Directive 92/91/EEC, which is the eleventh individual Directive within the meaning of Article 16 (1) of Directive 89/391/EEC, amended by Directive 2007/30/EC, establishes the minimum requirements for the protection of the worker's safety and health in the mineral-extracting industries through drilling [13].

The Romanian legislative act according to this directive is Decision no. 1050/2006 on minimum requirements for the safety and health of workers in the drilling industry. This act contains common minimum requirements applicable to the ground and offshore drilling sectors; minimum requirements applicable to the ground drilling sector; special minimum requirements applicable to offshore drilling sector. Under this act, workers must be provided with the necessary information, training and professional development necessary to ensure their safety and health. The employer must ensure that workers receive instructions in their terms so that they do not compromise their safety and health or that of other workers [14].

If the health and safety document so provides, a work permit system should be introduced for the execution of hazardous works and for the execution of works that are not normally dangerous but which, in interaction with other activities, may generate serious hazards. Work permits must be issued before the work begins by the person designated for that purpose and must contain the conditions to be met and the measures to be taken before, during and after work.

For each workplace, written instructions should be developed, including rules to be followed to ensure the safety and health of workers and the safe use of the machinery. These instructions must include information on the use of intervention equipment, as well as on the measures to be taken at or near the workplace in case of emergency.

6. The main requirements related to learning needs and competencies

According to the Law on safety and health at work no. 319/2006, the employers shall designate one or more workers to carry out the activities related to the protection and prevention of the occupational risks, named designated workers, or they shall enlist external services. The training levels required are at average level and high-level, certified by a diploma of studies and a certificates for the graduation of the courses. The representatives of workers with specific responsibilities in the field of the safety and health must follow a basic training program in the field of health and safety at work of at least 40 hours.

The training courses and programs in the field of safety and health at work are performed by authorized vocational training providers, according to the provisions of the Government Ordinance no. 129/2000 regarding the vocational training of adults, republished, with subsequent amendments and completions [15]. The contents of the OSH training courses corresponding to each training level are specified within Annex 6 of the Methodological Norms for the enforcement of the law no. 319/2006.

Through the Decision no 567/2015, the Romanian Government updated the national qualifications framework with eight levels based on labor market requirements and offered training, in accordance between it and the European benchmark, thus linking national qualifications systems in Europe [16].

According to International Standard Classification of Occupation (ISCO 08), Classification of Occupations in Romania (COR) contain the occupations specific to the field of occupational safety and health: at average level: 325721 - OSH Technician; 325723 - OSH Inspector; 325724 - OSH Coordinator; at high level: 226303 - OSH Coordinator; 242304 – OSH Expert [17].

In Romania, there are no special occupations and explicit requirements for OSH specialists in the offshore industry.

7. Romanian Higher Education relevant to offshore industry

In Romania, some universities, located in the Black Sea area or areas with important oil and gas fields, offer programs of studies relevant to the offshore industry.

CONSTANTA MARITIME UNIVERSITY is a public institution of higher education and research that offers the next specializations, degrees or certificates, related to the offshore industry:

- Bachelor of Science Degree in Navigation and Naval Transport; Navigation and Waterborne Transport; Naval Electromechanics;
- Master Degree in Offshore Oil and Gas Technology and Management; Advanced Engineering in the Offshore Oil and Gas Industry;
- International Maritime Organization (IMO) Certificates of competency for aspiring marine officers (cadets) [18].

"MIRCEA CEL BATRAN" NAVAL ACADEMY is a polytechnic higher education establishment, which provides education and training programs both for military cadets/midshipmen/officers of the Romanian Navy/Coast Guard, but also for civilian engineering specialists for the Merchant Marine sector and provides:

- Bachelor Degree in Naval Electromechanics; Electrical Engineering; Navigation and Maritime and River Transports; Naval and Port Engineering and Management;
- Master Degree in Naval Electromechanical Systems; Nautical Sciences; Naval and Port Engineering and Management; Management of Logistic Systems.

It has, in its structure: Damage Control Training Center (firefighting, waterflood simulation); Seamanship and Sailing Training Center [19].

OVIDIUS UNIVERSITY OF CONSTANTA is a public higher education institution that offers the next specialization related to offshore industry:

- Bachelor Degree in Naval equipment and installation; Port and marine installation and equipment;
- Master degree in Engineering of advanced naval systems and equipment [20].

PETROLEUM - GAS UNIVERSITY OF PLOIEŞTI, through the Faculty of Oil and Gas Engineering, is the only faculty in Romania that prepares specialists in the field of oil and gas exploration. It offers the next specializations, related to the offshore industry:

- Bachelor of Science Degree in Oil & Gas Engineering; Transportation, storage and distribution of hydrocarbons;
- Master Degree in Drilling of wells; Oil extraction; Deposit engineering; Technology of transport, storage and distribution of hydrocarbons; Management in the oil industry; Petroleum engineering;
- Doctoral studies in Mine, oil and gas [21].

8. Safety and health training courses for the workers from the offshore industry

Offshore Petroleum Industry Training Organization, commonly known as OPITO, is responsible for the standards in the oil and gas industry, particularly in emergency response and training. The standards set by OPITO are produced and agreed by related groups, and those that passed such standards are recognized worldwide. There are more than 205 approved training providers across the globe, of which two in Romania: Romanian Maritime Training Centre (CERONAV) and Regional Air Services Tuzla [22].

International Minimum Industry Safety Training (IMIST) is an OPITO standard that supports the global oil and gas industry to meet the safety targets. The IMIST standard ensures that the workers have the necessary safety awareness and training to reduce the risks and the number of incidents [23].

CERONAV, with an experience of more than 40 years in vocational education and training, obtained accreditation from OPITO in October 2012 for three training courses: BOSIET - Basic Offshore Safety Induction and Emergency Training; HUET - Helicopter Underwater Escape Training; FOET - Further Offshore Emergency Training, in 2016 gained accreditation for

H2S and Escape Chute courses and in April 2018, for the Digital Course – BOSIET [24].

SEA SURVIVAL SCHOOL, Tuzla International Airport, is the first training provider from Eastern Europe, in response to the emergency situations, in the offshore industry and the air transportation. The training centre was opened for operations in 2010, and currently ensures training courses BOSIET; HUET; FOET; CA-EBS Initial Deployment Training (CA-EBS); Basic H2S Training; International Minimum Industry Safety Training (IMIST), an online course offered on behalf of ATLAS highly recommended for all inexperienced employees in the oil and gas industry [25].

GSP TRAINING CENTER offers the next offshore training courses: BOSIET and HUET (in partnership with CERONAV); Start-Up Your Career in Offshore; Basic Offshore Safety Induction; IADC (International Association of Drilling Contractors) Rig Pass; Advanced Safety Management Training; HSE (health, safety and environment) on-site; H2S Awareness Including Breathing Apparatus; Working at Height – Rescue and Self-Rescue Training; Confined Space – Working and Rescue Training [26].

9. Conclusion

The Romanian Occupational Safety and Health system might be characterized by a comprehensive legislative framework that fully transposed the EU legislation in the area and an adequate institutional framework aligned to the Community requirements in the field. According to the Romanian law in force, there are the minimum training requirements in the field of occupational safety and health corresponding to the average level and higher level OSH training. According to International Standard Classification of Occupation (ISCO 08), Classification of Occupations in Romania (COR) contain the occupations specific to the field of occupational safety and health, necessary to carry out prevention and protection activities. In Romania, there are no special occupations and explicit requirements for OSH specialists in the offshore industry. Authorized vocational training providers perform the training courses and programs in the field of the occupational safety and health. In Romania, there are some training centers, which offer theoretical and practical courses for those graduates or professionals who wish to pursue a professional career in the offshore industry, with specialization on H&S offshore, like BOSIET, HUET, FOET, H2S, ESCAPE CHUTE, IMIST, and others. Also, there are some Higher Education Institutions in Romania, which provides diplomas (Bachelor Degree, Master Degree, Ph.D. Degree), courses or training related to the offshore industry.

This paper was supported by the project Erasmus+ 2017-1-CY01-KA202-026782 – "Development of a VET Training Program to upgrade the Skills on Health and Safety Risks for Offshore Construction Workers".

References

1. sgg.gov.ro - General Secretariat of the Romanian Government
2. Law no 319 of 14 July 2006 on Safety and Health of Workers at Work, Romanian Official Gazette, Part I no. 646 of 26.07.2006
3. Government Decision no. 1425 of 11 October 2006, Methodological Norms for the enforcement of the Law no. 319 on Safety and Health at Work, Romanian Official Gazette, Part I no. 882 of 30.10.2006
4. www.mmuncii.ro - Ministry of Labor and Social Justice
5. www.ms.ro – Ministry of Health
6. https://www.inspectiamuncii.ro – Labor Inspection
7. http://www.inpm.ro - National Research and Development Institute of Occupational Safety
8. http://site.anc.edu.ro - National Qualifications Authority
9. Directive 2013/30/EU of the European Parliament and of the Council of 12 June 2013 on safety of offshore oil and gas operations and amending Directive 2004/35/EC, OJ L 178 of 28.06.2013
10. Law no. 165/2016 on offshore oil operations safety, Romanian Official Gazette, Part I no. 572 of 28.07.2016
11. https://euoag.jrc.ec.europa.eu/ - European Union Offshore Oil and Gas Authorities Group (EUOAG)
12. http://acropo.gov.ro/web/ - Regulatory Competent Authority for offshore oil operation at the Black Sea (ACROPO)
13. Directive 92/91/EEC, which is the eleventh individual Directive within the meaning of Article 16 (1) of Directive 89/391/EEC, OJ L 348, 28.11.1992, p. 9
14. Government Decision no. 1050/2006 on minimum requirements for the safety and health of workers in the drilling industry, Romanian Official Gazette, Part I no. 737 of 29.08.2006
15. Government Ordinance no. 129/2000 regarding the vocational training of adults, republished, Romanian Official Gazette, Part I no. 110 of 13.02.2014
16. Government Decision no. 567/2015 on the national qualifications framework, Romanian Official Gazette, Part I no. 572 of 30.07.2015
17. http://data.gov.ro/dataset/clasificarea-ocupatiilor-din-romania - Classification of Occupations in Romania
18. https://cmu-edu.eu/en - Constanta Maritime University
19. https://www.anmb.ro - "Mircea cel Bătrân" Naval Academy
20. www.univ-ovidius.ro - Ovidius University of Constanta
21. https://www.upg-ploiesti.ro - Petroleum - Gas University of Ploiesti
22. https://www.opito.com - Offshore Petroleum Industry Training Organization
23. https://www.imist-online.com - International Minimum Industry Safety Training
24. https://www.ceronav.ro – CERONAV, Romanian Maritime Training Centre
25. http://www.sea-survival.ro - Sea Survival School
26. http://gsptrainingcenter.com - GSP Training Center

MATEC Web of Conferences **184**, 04006 (2018) https://doi.org/10.1051/matecconf/201818404006
Annual Session of Scientific Papers IMT ORADEA 2018

A multi-criteria decision making approach for public lighting system selection

Maria Cristea[1,*], *Radu Adrian* Tîrnovan[2], *Ciprian* Cristea[3], *Constantin Sorin* Pică[4] and *Cristian* Făgărăşan[5]

[1]Technical University of Cluj-Napoca, Faculty of Electrical Engineering, fagarasan_maria@yahoo.com, România
[2]Technical University of Cluj-Napoca, Faculty of Electrical Engineering, radu.tirnovan@enm.utcluj.ro, România
[3]Technical University of Cluj-Napoca, Faculty of Electrical Engineering, ciprian.cristea@emd.utcluj.ro, România
[4]Technical University of Cluj-Napoca, Faculty of Electrical Engineering, sorin.pica@enm.utcluj.ro, România
[5]Technical University of Cluj-Napoca, Faculty of Electrical Engineering, office.cristif@gmail.com, România

Abstract. There are numerous and different ways to design an energy efficient street lighting system, but not always the best technical solution is the optimal economic solution as well. In this paper, we used two distinct software tools in order to obtain the optimal technical and economic solution for selecting street lighting technology. The first software used was DIALux, which is frequently applied in this field, assisting designers in choosing the most appropriate electrical installations and evaluating the technical performance of the chosen solutions. This tool determined the street lighting system variants for which there are predetermined configurations, so that the lighting parameters obtained are in accordance with the normative requirements of the lighting classes in which the roadways are included. Different ways of locating the luminaires and other parameters were changed to reach the best version that meets the qualitative requirements of the lighting technology. The second software used was the one developed by the authors, based on ELECTRE III method, which evaluated the technical solutions returned by DIALux against several criteria, taking into account the designer's preferences and constraints. The optimal technical and economic solution was obtained after ranking the variants by their performance at each criterion. The simulation experiment and the results returned suggested that the combination of the two software tools is comprehensive and effective and it can be used in this type of decisions at large scale.

1 Introduction

In recent years, public lighting systems are designed pursuing energy efficiency, not only for environmental reasons but also to decrease the energy consumption as well [1, 2]. Energy efficient technologies should always be in accordance with the lighting standards, ensure an adequate photometrical performance, providing visual comfort and safety. These are the main conditions regarding the design of a street lighting installation, which are detailed in [3]. There are several computer programs that can be used to design a street lighting installation, but none of these software tools does not include energy efficiency as an analysis parameter [4].

Different methods were studied and several algorithms were developed in previous research papers in order to obtain energy efficient installations for street lighting systems. Using Sustainable Process Index methodology, a comparative study of different lighting technologies is developed in [5]. The available technologies, such as light emitting diode, compact fluorescent and high pressure sodium were analysed in order to obtain the more suitable green solutions. Another comparative study, between the performances of the light emitting diode and high pressure sodium technologies, is presented in [6], and in [7] are described several schemes using Solid-State Lighting (SSL) in order to increase the energy efficiency of street lighting technology. A wireless sensor network system is presented in [8], which can be used to control and monitor electrical variables. Energy efficiency is one of the requirements and it is demonstrated that the algorithm developed fulfils the requirements. Issues related to smart and advanced street lighting systems are presented in [2] and [9], where the control of the street technology is the main objective.

Besides all the technical design conditions, the economic aspect should also be taken into account when designing a street lighting installation. There are only a few studies presented in the related papers that considered both technical and economic perspectives when designing a street lighting system. Several of these papers are mentioned below. Thus, in [10] is proposed a decision making tool to solve a quadratic integer programming problem in order to reduce energy consumption and to ensure that the optimal allocation solution is selected, regarding energy retrofit interventions. In [11] and [12] the same multi-criteria decision method is used, ELECTRE III, in order to develop an Energy Action Plan and help local authorities to design more sustainable energy systems, and in the second paper to select the most

* Corresponding author: fagarasan_maria@yahoo.com

© The Authors, published by EDP Sciences. This is an open access article distributed under the terms of the Creative Commons Attribution License 4.0 (http://creativecommons.org/licenses/by/4.0/).

appropriate street lighting installation with the best investment.

The rest of the paper is organized as follows: Section 2 describes the methodology of ELECTRE III method and the application of DIALux software; Section 3 presents the case study; in Section 4 the results are discussed and, finally, the conclusions are drawn.

2 Methodology

The software application DIALux was used to design the public lighting system. All the parameters (type of road, lighting class, the number of lanes, maintenance factor, reference value, maintenance cycle, road width and length, variants of the location of luminaires) were introduced and the program returned the most appropriate technical solutions, which represent the alternatives. These solutions were then analysed with the software developed based on ELECTRE III method in order to obtain the optimal economic solution.

The algorithm of the ELECTRE III method consists of the following phases [13-16]:

a) Establishing a set of data – in this phase are defined the following data: the alternatives and the criteria, which will compose the performance matrix, three types of thresholds: indifference (q_i), preference (p_i), and veto (v_i), used to determine outranking relations and the weight indexes (w_i), used to express the importance of each criterion (i).

b) *Construction of the outranking relations* – this phase includes three separate tests: concordance, discordance and credibility test. In each stage, there are calculate indexes for each pair of alternatives, based on separate equations.

The concordance index is calculated based on the statement "alternative (A) outranks alternative (B)", using the following equation (1):

$$c_i(A,B) = \begin{cases} 1, if\ q_i(f_i(A)) \geq f_i(B) - f_i(A) \\ \frac{p_i(f_i(A))+f_i(A)-f_i(B)}{p_i(f_i(A))-q_i(f_i(A))}, if\ p_i(f_i(A)) > f_i(B) - f_i(A) > q_i(f_i(A)) \\ 0, if\ p_i(f_i(A)) \leq f_i(B) - f_i(A) \end{cases} \quad (1)$$

where:
$f_i(A)$- performance of alternative A based on criterion i;
$f_i(B)$- performance of alternative B based on criterion i;
q_i – indifference threshold for criterion i;
p_i – preference threshold for criterion i.

Having the concordance indexes for all pairs, the global indicators, which form the concordance matrix, can be calculated using equation (2).

$$C(A,B) = \frac{1}{W} \cdot \sum_{i=1}^{n} w_i \cdot c_i(A,B) \quad (2)$$

where:
W – sum of all weights;
w – weight of criterion i;
n – number of criteria;
$c_i(A,B)$ – concordance index with respect to criterion i.

The discordance indexes are calculated based on the statement "alternative (A) outranks alternative (B)", using formula (3).

$$D_i(A,B) = \begin{cases} 0, if\ p_i(f_i(A)) \geq f_i(B) - f_i(A) \\ \frac{f_i(B)-f_i(A)-p_i(f_i(A))}{v_i(f_i(A))-p(f_i(A))}, if\ v_i(f_i(A)) > f_i(B) - f_i(A) > p_i\big(f_i(A)\big) \\ 1, if\ v_i(f_i(A)) \leq f_i(B) - f_i(A) \end{cases} \quad (3)$$

where:
v_i – veto threshold for criterion i.

If the decision maker does not assign a veto threshold, the discordance indexes are null and the concordance matrix is the same as the credibility matrix.

In the final test the credibility matrix is developed, based on indexes calculated with equation (4).

$$S(A,B) = \begin{cases} C(A,B), if\ C(A,B) \geq D_i(A,B) \forall i, \\ C(A,B) \cdot \prod_{i \in J(A,B)} \frac{(1-D_i(A,B))}{(1-C(A,B))} \end{cases} \quad (4)$$

where:
$J(A,B)$- the criteria for which $D_i(A,B) > C(A,B)$.

c) *Exploitation of the outranking relations* – this phase includes two distillation procedures, called *descending distillation* and *ascending distillation*. These operations rank the alternatives depending on their performances in two pre-orders, which combined provide the final result.

3 Case study

The methodology presented above was applied to a public road, which has the following features:

a) Urban one-way road traffic with ME2 lighting class, four lanes, maintenance factor=0.57, reference value: external equipment, maintenance cycle: 3 years;
b) Road width: l_0= 14 m, road length: L=1000 m;
c) Variants of the location of luminaires: bilateral face to face, bilateral alternative.

To determine the technical solutions for the public road described above, there have been selected three types of luminaires: SELENIUM, IRIDIUM and MALAGA, combined with different lamps: high pressure mercury vapour discharge lamps (HPL-N), high pressure sodium lamps (SON-T) and light-emitting diode lamps (LED), with different wattage. In order to perform a technical optimization, there are three parameters that can be varied: the distance between pillars (considered between 30–47 m), pillar height (considered 8m, 9m, 10m) and the luminaire inclination. Thus, the data is input in DIALux and there are performed several simulations, which returned all the technical solutions that can be used to illuminate the public road. The first eighteen solutions, alternative A1 to alternative A18, where analysed and are presented in Table 1.

Table 1. The set of alternatives.

Alternatives	Luminaire type	Variants of the location of luminaires	Pillar height [m]	Distance between pillars [m]	Number of pillars [pcs]
A1	PHILIPS SGP340 FG 1xSON-TPP400W TP P1	bilateral alternative	10	47	21
A2	PHILIPS SGS454 C FG 1xSON-TPP400W SGR CP P2_220	bilateral alternative	10	44	22
A3	PHILIPS SGS453 C GB 1xSON-TPP250W SGR CR P1_220	bilateral alternative	10	44	22
A4	PHILIPS SGS453 C GB 1xSON-TPP250W SGR CR P1_220	bilateral face to face	10	41	24
A5	PHILIPS SGP340 FG 1xSON-TPP400W TP P1	bilateral face to face	10	41	24
A6	PHILIPS SGP340 FG 1xSON-TPP250W TP P1	bilateral face to face	10	37	27
A7	PHILIPS SGP340 FG 1xSON-TPP400W TP P1	bilateral alternative	9	42	23
A8	PHILIPS SGS454 C FG 1xSON-TPP400W SGR CP P2_220	bilateral alternative	9	39	25
A9	PHILIPS SGS453 C GB 1xSON-TPP250W SGR CR P1_220	bilateral alternative	9	39	25
A10	PHILIPS SGS453 C GB 1xSON-TPP250W SGR CR P1_220	bilateral face to face	9	36	27
A11	PHILIPS SGP340 FG 1xSON-TPP250W TP P1	bilateral face to face	9	34	29
A12	PHILIPS SGP340 C FG 1xSON-TPP400W TP P1	bilateral face to face	9	32	31
A13	PHILIPS SGP340 FG 1xSON-TPP400W TP P1	bilateral alternative	8	36	27
A14	PHILIPS SGP340 C FG 1xSON-TPP250W TP P1	bilateral alternative	8	31	32
A15	PHILIPS SGS453 C GB 1xSON-TPP150W SGR CR P1_220	bilateral alternative	8	30	33
A16	PHILIPS SGP340 FG 1xSON-TPP250W TP P1	bilateral face to face	8	31	32
A17	PHILIPS SGP340 C FG 1xSON-TPP250W TP P1	bilateral face to face	8	30	33
A18	PHILIPS SGS454 C FG 1xSON-TPP400W SGR CP P2_220	bilateral face to face	8	30	33

For the study developed using the software based on ELECTRE III method, eight criteria had been considered. The first one, *energy consumption* [kWh/1000 h] is determined for each variant, taking in consideration the number of luminaires needed to provide the minimum illuminance. Another criterion is *luminous efficiency* [lm/W], which is inversely proportional with the first one and is calculated as division between the lamp luminous flux and the power consumed by the lamp.

The next criterion, *IK protection* [points], is specified by the manufacturer of the luminaires. The degrees of protection indicate the level of protection to external mechanical impact, and it differs for each type of luminaire.

The *total price of the lighting system* [RON], *installation price* [RON] and *maintenance price* [RON] are usually determined based on an auction, being a public project, where different companies offer a price to complete this work. The bidder who offers the lowest price wins the contract. Thus, these prices should be very advantageous. For this study, the prices were considered from a local company, which often participates at auctions for public lighting projects.

The *lamp lifetime* [h] and *lamp warranty* [months] differs depending on the lamp type, and the values for these two criteria are specified in the data sheet of the product.

The weights and the values for each threshold are given in Table 2, and the performance matrix, which consists of all the alternatives and criteria are presented in Table 3.

Table 2. Thresholds and criteria weights.

Criteria / Weights Thresholds	Energy consumption [kWh/1000h]	Luminous efficiency [lm/W]	IK [points]	Lighting system price [RON]	Installation price [RON]	Maintenance price [RON]	Lamp lifetime [h]	Lamp warranty [months]
Indifference (q)	0.1	0.15	0.1	0.05	0.05	0.05	0.1	0.1
Preference (p)	0.2	0.25	0.2	0.15	0.15	0.15	0.2	0.25
Veto (v)	0	0	0	0	0	0	0	0
Weight (w)	0.2	0.15	0.05	0.25	0.1	0.15	0.05	0.05

Table 3. Performance matrix.

Criteria / Alternatives	Energy consumption [kWh/1000h]	Luminous efficiency [lm/W]	IK [points]	Lighting system price [RON]	Installation price [RON]	Maintenance price [RON]	Lamp lifetime [h]	Lamp warranty [months]
A1	8400	101.7	10	56175	8820	2100	28000	60
A2	8800	114.41	8	62700	9240	2200	28000	48
A3	8800	112.88	8	56980	9240	2200	32000	48
A4	6000	112.88	8	62160	10080	2400	32000	48
A5	9600	101.7	10	64200	10080	2400	28000	60
A6	6750	82.88	10	65205	11340	2700	32000	60
A7	9200	101.7	10	55315	9660	2300	28000	60
A8	10000	114.41	8	64500	10500	2500	28000	48
A9	10000	112.88	8	58000	10500	2500	32000	48
A10	6750	112.88	8	62640	11340	2700	32000	48
A11	7250	82.88	10	62205	12180	2900	32000	60
A12	12400	101.7	10	74555	13020	3100	28000	60
A13	10800	101.7	10	59805	11340	2700	28000	60
A14	8000	82.88	10	62560	13440	3200	32000	60
A15	4950	93.33	8	66495	13860	3300	34000	48
A16	8000	82.88	10	62560	13440	3200	32000	60
A17	8250	82.88	10	64185	13860	3300	32000	60
A18	13200	114.41	8	78870	13860	3300	28000	48

4 Results

In order to obtain the optimal economic solution for street lighting system it was used the software developed by the authors based on the ELECTRE III method algorithm.

After the criteria and alternatives were defined and the direction of the preference for each criterion was identified, the indifference, preference and veto threshold for each criterion were introduced.

The software returns complementary results: three types of matrices (the concordance matrix, the discordance matrix and the credibility matrix) that are determined using the methodology described at Section 2. Figure 1 shows the credibility matrix calculated by the application, which is identical to the concordance matrix because the veto threshold was not defined and the discordance indexes are null.

* Corresponding author: fagarasan_maria@yahoo.com

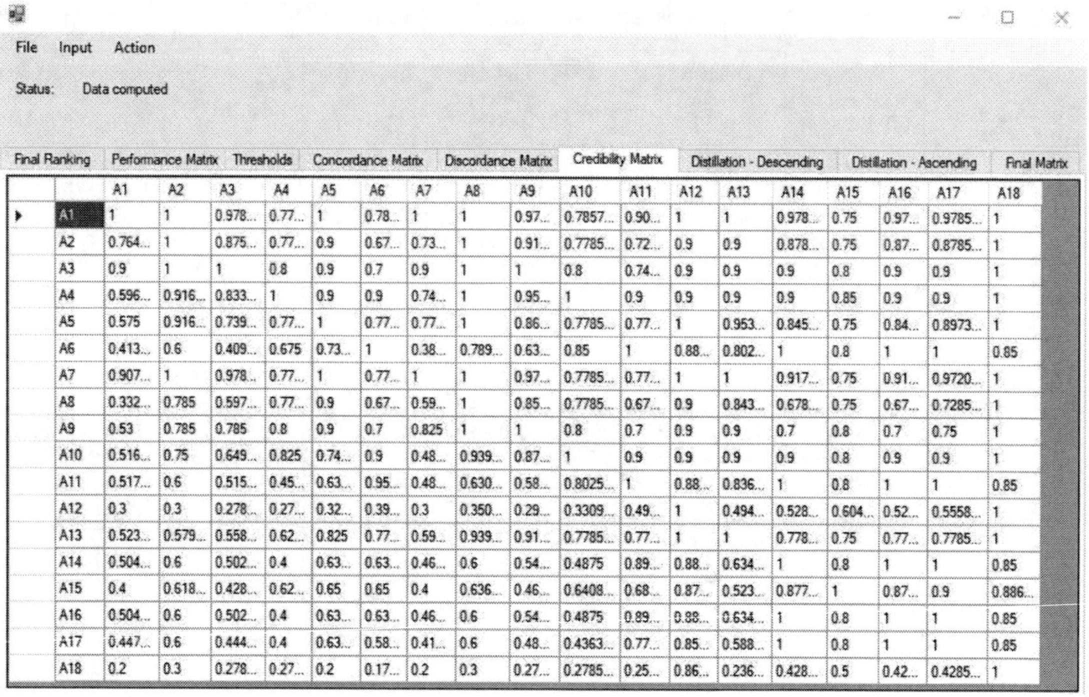

	A1	A2	A3	A4	A5	A6	A7	A8	A9	A10	A11	A12	A13	A14	A15	A16	A17	A18
A1	1	1	0.978...	0.77...	1	0.78...	1	1	0.97...	0.7857...	0.90...	1	1	0.978...	0.75	0.97...	0.9785...	1
A2	0.764...	1	0.875...	0.77...	0.9	0.67...	0.73...	1	0.91...	0.7785...	0.72...	0.9	0.9	0.878...	0.75	0.87...	0.8785...	1
A3	0.9	1	1	0.8	0.9	0.7	0.9	1	1	0.8	0.74...	0.9	0.9	0.9	0.8	0.9	0.9	1
A4	0.596...	0.916...	0.833...	1	0.9	0.9	0.74...	1	0.95...	1	0.9	0.9	0.9	0.9	0.85	0.9	0.9	1
A5	0.575	0.916...	0.739...	0.77...	1	0.77...	0.77...	1	0.86...	0.7785...	0.77...	1	0.953...	0.845...	0.75	0.84...	0.8973...	1
A6	0.413...	0.6	0.409...	0.675	0.73...	1	0.38...	0.789...	0.63...	0.85	1	0.88...	0.802...	1	0.8	1	1	0.85
A7	0.907...	1	0.978...	0.77...	1	0.77...	1	1	0.97...	0.7785...	0.77...	1	1	0.917...	0.75	0.91...	0.9720...	1
A8	0.332...	0.785	0.597...	0.77...	0.9	0.67...	0.59...	1	0.85...	0.7785...	0.67...	0.9	0.843...	0.678...	0.75	0.67...	0.7285...	1
A9	0.53	0.785	0.785	0.8	0.9	0.7	0.825	1	1	0.8	0.7	0.9	0.9	0.7	0.8	0.7	0.75	1
A10	0.516...	0.75	0.649...	0.825	0.74...	0.9	0.48...	0.939...	0.87...	1	0.9	0.9	0.9	0.9	0.8	0.9	0.9	1
A11	0.517...	0.6	0.515...	0.45...	0.63...	0.95...	0.48...	0.630...	0.58...	0.8025...	1	0.88...	0.836...	1	0.8	1	1	0.85
A12	0.3	0.3	0.278...	0.27...	0.32...	0.39...	0.3	0.350...	0.29...	0.3309...	0.49...	1	0.494...	0.528...	0.604...	0.52...	0.5558...	1
A13	0.523...	0.579...	0.558...	0.62...	0.825	0.77...	0.59...	0.939...	0.91...	0.7785...	0.77...	1	1	0.778...	0.75	0.77...	0.7785...	1
A14	0.504...	0.6	0.502...	0.4	0.63...	0.63...	0.46...	0.6	0.54...	0.4875	0.89...	0.88...	0.634...	1	0.8	1	1	0.85
A15	0.4	0.618...	0.428...	0.62...	0.65	0.65	0.4	0.636...	0.46...	0.6408...	0.68...	0.87...	0.523...	0.877...	1	0.87...	0.9	0.886...
A16	0.504...	0.6	0.502...	0.4	0.63...	0.63...	0.46...	0.6	0.54...	0.4875	0.89...	0.88...	0.634...	1	0.8	1	1	0.85
A17	0.447...	0.6	0.444...	0.4	0.63...	0.58...	0.41...	0.6	0.48...	0.4363...	0.77...	0.85...	0.588...	1	0.8	1	1	0.85
A18	0.2	0.3	0.278...	0.27...	0.2	0.17...	0.2	0.3	0.27...	0.2785...	0.25...	0.86...	0.236...	0.428...	0.5	0.42...	0.4285...	1

Fig. 1. Credibility matrix

The next step is to perform the two distillation procedures, based on the credibility matrix. The software ranks the alternatives in two pre-classifications, descending and ascending, which are presented in figures 2 and 3. The descending distillation is computed by selecting first the best ranked alternatives and ending with the worst. The ascending distillation is the reverse operation, successive removal of the worst alternative and ending with the best.

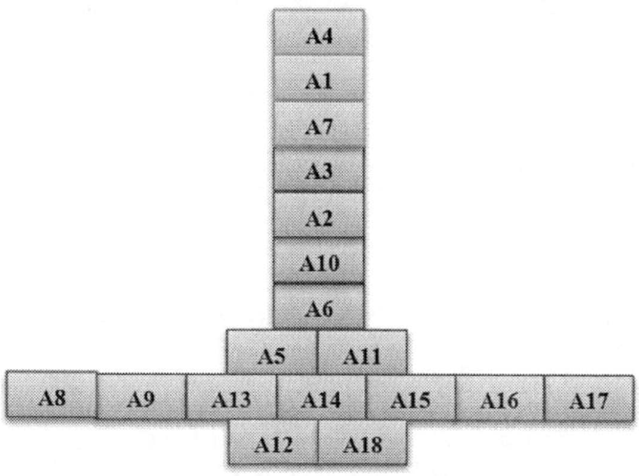

Fig. 2. Descending distillation.

After the descending distillation, A4 is ranked as the best alternative, followed by A1 and A7. At the other extreme, A12 and A18 are at the same place, being ranked as the worst alternatives.

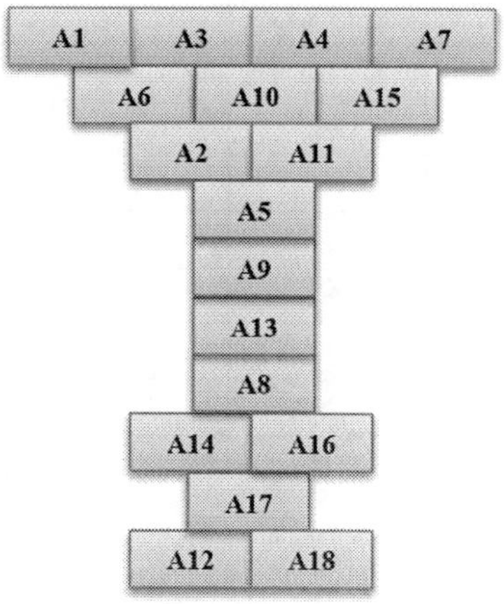

Fig. 3. Ascending distillation.

The results after the ascending distillation show that four alternatives are ranked as the best, the software not being able to distinguish between their performances. These alternatives are A1, A3, A4 and A7. Regarding the worst alternatives, the results are the same as the ones from the descending distillation: A12 and A18.

* Corresponding author: fagarasan_maria@yahoo.com

The two pre-orders obtained after the distillation procedures, are combined and used to determine the final ranking, as presented in figure 4.

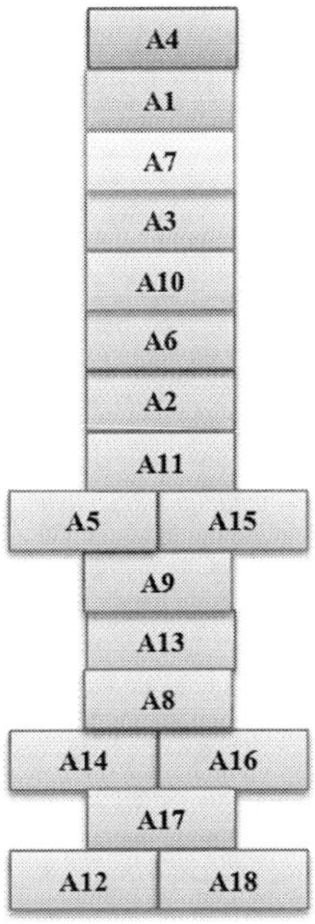

Fig. 4. Final ranking

The final ranking reveals the optimal technical and economic solution for the street lighting system designed in this study, which is alternative A4. Thus, using PHILIPS SGS453 C GB 1xSON-TPP250W SGR CR P1_220 luminaires on pillars which have 10 m height and the distance between them of 41 m is the best solution from all alternatives taken into consideration in this analysis.

The next best ranked alternatives are A1 and A7, meaning using the PHILIPS SGP340 FG 1xSON-TPP400W TP P1 and PHILIPS SGP340 FG 1xSON-TPP400W TP P1 luminaires on pillars which have 10 m and 9 m height, and the distance between them being of 47 m, respectively 42 m.

5 Conclusions

In this paper, the combination of two separate software applications was proposed in order to design a high energy efficient street lighting system. DIALux and the software developed by the authors based on ELECTRE III method were used to select the optimal technical and economic solution from all the alternatives included in the analysis. Eighteen luminaires combined with different types of lamps have been identified as technical solutions, which were obtained by varying different parameters, such as: the location of the luminaires, the distance between the pillars and the pillar height. The technical solutions returned by DIALux were then assessed from economic point of view, by using eight criteria: energy consumption, luminous efficiency, the degree of protection to external mechanical impact, lighting system, installation and maintenance price and lamp lifetime and warranty. After running the software, the final ranking revealed the best ranked alternatives.

The results obtained using the combination between DIALux, to select the technical solutions, and the software based on ELECTRE III method, to point the best economic solution, underline that this concept is comprehensive and effective and it can be used in the lighting technology design field at large scale.

References

1. R. Carli, M. Dotoli, R. Pellegrino, Comput. Oper. Res. **1** (2017)
2. I. Wojnicki, S. Ernst, L. Kotulski, A. Sedziwy, Expert Syst Appl. **41**, 999 (2014)
3. M. Kostic, L. Djokic, Energy **34**, 1565 (2009)
4. O. Rabaza, D. Gomez-Lorente, F. Perez-Ocon, A. Pena-Garcia, Energy **107**, 831 (2016)
5. K. Shahzad, L. Cucek, M. Sagir, N. Ali, M.I. Rashid, R. Nazir, A.S. Nizami, H.A. Al-Turaif, I.M. Ibrahim Ismail, J. Cleaner Prod. **175**, 683 (2018)
6. S. Yoomak, C. Jettanasen, A. Ngaopitakkul, S. Bunjongjit, M. Leelajindakrairerk, Energy Build. **159**, 542 (2018)
7. M. Fontoynont, C.R. Physique
8. R. Pantoni, D. Brandao, Comput. Electr. Eng. **37**, 1147 (2011)
9. D. Jin, C. Hannon, Z. Li, P. Cortes, S. Ramaraju, P. Burgess, N. Buch, M. Shahidehpour, Electr. J. **29**, 28 (2016)
10. R. Carli, M. Dotoli, E. Cianci, IFAC PapersOnLine **50-1**, 14460 (2017)
11. D. Neves, P. Baptista, M. Simoes, C.A. Silva, J.R. Figueira, J. Cleaner Prod. **176**, 251 (2018)
12. J.A. Lobao, T. Devezas, J.P.S. Catalao, Energy Rep. **1**, 110 (2015)
13. G. Fancello, M. Carta, P. Fadda, Transp. Res. Procedia **3**, 175 (2014)
14. C. Cristea, M.Cristea, COSME'16 **4**, 9 (2016)
15. J. Zak, M. Kruszynski, Transp. Res. Procedia **10**, 820 (2015)
16. J. Zak, Transp. Res. Procedia **27**, 43 (2017)

A multi-criteria decision making approach in lighting technology selection for a production hall

Maria Cristea[1,*], *Radu Adrian* Tîrnovan[2], *Ciprian* Cristea[3], and *Constantin Sorin* Pică[4]

[1]Technical University of Cluj-Napoca, Faculty of Electrical Engineering, fagarasan_maria@yahoo.com, România
[2]Technical University of Cluj-Napoca, Faculty of Electrical Engineering, radu.tirnovan@enm.utcluj.ro, România
[3]Technical University of Cluj-Napoca, Faculty of Electrical Engineering, ciprian.cristea@emd.utcluj.ro, România
[4]Technical University of Cluj-Napoca, Faculty of Electrical Engineering, sorin.pica@enm.utcluj.ro, România

Abstract. Nowadays, lighting industry can contribute to a more sustainable future, by reducing the total consumption of electrical energy used to illuminate indoor and outdoor spaces. Thus, lighting system design is one of the important components when thinking about building sustainability. Taking into account the final purpose of the lighting installation (the type of human activity that will take place in the indoor space for which lighting is to be provided), the most suitable lighting technologies are selected considering their characteristics, dimensions, and technical specification. The paper presents how a lighting technology selection for a production hall using a multi-criteria decision making approach. There are considered the most common luminaires and lamps which can assure the illuminance necessary for the hall, which represent the variants, evaluated against several criteria, considering the decision maker's preferences and constraints. The variants are ranked according to their performance for each criterion. The optimal lighting solution obtained after the simulation experiment suggests that ELECTRE III method is a practical decision-making model that can be considered for this type of decisions.

1 Introduction

Lighting system design is one of the most important elements when taking into account building sustainability. In most of the countries with a developed industry, indoor lighting is an important contributor in rising energy cost, because it is used approximately 18% of the total energy consumption [1, 2]. There can be considered three main elements that affect the final energy consumption of a lighting system: the lamp (which is the light source and includes controls and ballasts), the luminaire and the indoor environmental quality (the specifications of the room) [3, 4]. There are different lighting technologies used in indoor spaces, and are compared by three objectives: photometrical performance, energy efficiency and installation performance. Recently, the photometrical performance is frequently considered less important than energy efficiency, which determines poor visual comfort, affecting not only the well-being and satisfaction of the occupants, but also the productivity, and leads to actions with the goal of restoring or preserving the comfort, that can contribute to high energy consumption [2, 5].

European standard EN 12665 defines visual comfort as "a subjective condition of visual well-being induced by the visual environment" and includes the following physical factors: "the amount of light", "the uniformity of light", "the quality of light in rendering colours", and "the risk of glare for occupants" [6, 7]. These factors can describe the lighting requirements that humans need in different indoor spaces [3]. There are multiple lighting technologies that can satisfy the conditions for visual comfort in indoor environments, regardless the kind of human activities that take place there: each type of luminaire can be combined with different lamps, conventional ones (e.g. incandescent and fluorescent lamps) or light emitting diodes (LED's). Each type of lamp has its own features (power consumption, energy efficiency, degree of protection against different environmental factors, the cost and many other) that have to be taken in consideration when choosing the right luminaire-lamp combination for indoor lighting [2, 8, 9].

Hence, this type of decision implies multiple criteria, and there are several methods that can be used to solve this problem, the most common being: Analytical Hierarchy Process (AHP), Preference Ranking Organisation Method for Enrichment Analysis (PROMETHEE), Elimination and Choice Expressing Reality (ELECTRE), Techniques for Order Performance by Similarity to Ideal Solution (TOPSIS), and Simple Multi Attribute Rating Technique (SMART) [10]. To select the optimal lighting technology for a production hall, the method ELECTRE III was chosen, using the software created by the authors in [11].

The rest of this paper is organized as follows: Section 2 describes the methodology of the multi-criteria decision-making method; Section 3 presents the case study and are described the alternatives (the luminaire-lamp combinations that define the technically acceptable

* Corresponding author: fagarasan_maria@yahoo.com

© The Authors, published by EDP Sciences. This is an open access article distributed under the terms of the Creative Commons Attribution License 4.0 (http://creativecommons.org/licenses/by/4.0/).

solutions), the criteria used to rank the alternatives, the decision maker's preference and the existing constrains, and, finally, in Section 4 the results are discussed; in the last section the final conclusions are drawn.

2 Methodology

The different versions of ELECTRE methods (ELECTRE I, II, III, IV and TRI) were conceived to solve complex problems, based on various forms of data and inaccurate information, and which have high unpredictability and contradictory objective. All these methods have the same basic principles, but are different from two points of view: operationally and based on the type of the decision problem discussed [12, 13].

The selection of the optimal lighting technology for the production hall is a complex decision problem and there are various criteria that have to be taken into account. Thus, ELECTRE III was chosen for this type of decision because it works with inaccurate, imprecise and uncertain data and takes into account both qualitative and quantitative criteria, then ranks the alternatives from the best to the worst, considering the decision maker's preferences and constraints [14].

The algorithm of the ELECTRE III method consists of the following phases [11-15]:

a) *Setting up a set of data* – the alternatives and the criteria, used to compare the performance for each alternative, are defined; having these two important elements, the performance matrix can be developed. In order to determine outranking relations with a tolerance for imprecise and uncertain data, for each criterion (i) are associated three thresholds: indifference (q_i), preference (p_i), and veto (v_i). At last, the weight indexes (w_i) are defined, which express the importance that the decision maker assigns to each criterion.

b) *Establishing the outranking relations* - to determine outranking relations between the variants with respect to each criterion there are formed successive pair wise comparisons of two alternatives. The computation procedure includes three tests: concordance, discordance and credibility test.

In the first test are estimated the concordance index for each pair of alternatives by comparing the performance of both variants for all criteria, based on the statement "variant (a) outranks variant (b)"; the value of concordance index is between 0 (the statement is false) and 1 (the statement is true) and in the process are used both indifference and preference thresholds. The global indicators, which form the concordance matrix, are obtain as the weighted sum of all concordance indexes for the considered criteria.

The second test quantifies the strength of the evidence against the statement "variant (a) outranks variant (b)", obtaining the discordance indexes. In order to estimate this outranking relation, the veto threshold needs to be used. If there is not a veto threshold assigned to each criterion by the decision maker, the discordance indexes are null, so there is no need to perform the next test, because the concordance matrix is the same as the credibility matrix.

In the final test the degrees of credibility are calculated, taking into account the concordance indicators and the discordance indexes estimated in the previous tests. The values of the credibility degrees are between 0 and 1, and they compose the credibility matrix, which is used in the next phase.

c) *Implementation of the outranking relations* – after the credibility matrix is obtained, the distillation procedures can be performed. Thus, the alternatives are ranked depending on their performances in two pre-orders: *descending distillation* is constructed by selecting the best ranked variant, called the *first distillate* (D_1), and finishing with the worst, and *ascending distillation* is acquired by selecting the worst rated variants and finishing with the best one.

The results obtained after these two procedures (*descending distillation* and *ascending distillation*) are combined into a final ranking, which will return the final result in a ranking matrix or in an outranking graph. The optimal solution is the best rated alternative, considering all the criteria used in the analysis.

3 Case study

In order to achieve a correct interior lighting design, an initial set of data is required for the room in which the lighting installation is designed: dimensions: width (L_1), length (L_2), height (H) and height of working plane (h_u); room reflectance: walls reflectance (ρ_p), ceiling reflectance (ρ_t) and floor reflectance (ρ_s); average illuminance, minimum (E_{med}).

For this study, a production hall with the following features has been considered:

a) Dimensions: L_1=10 m, L_2= 20 m, H=7 m, h_u=0,8 m;

b) Room reflectance: ρ_p=0,5 , ρ_t=0,7 , ρ_s=0,3;

c) E_{med} = 300 lx.

In order to determine the possible technical solutions that satisfy the lighting conditions for the production hall, there have been selected four luminaires, which were combined with different types of lamps, fluorescent lamps and LEDs, and each one with different wattage. At this point there are hundreds of possible solutions that can be considered. To obtain only the acceptable ones, the utilisation factor (u) method was applied. This factor depends on the room index and room reflectance. The room index (i) is determined as the division between room area and the product of the height of luminaire above working plane (h) and the length and width of the room. After the utilisation factors are determined, the acceptable technical solutions, which represent the alternatives, can be defined. These are presented in Table 1.

There are eight luminaire–lamp combinations that can provide the lighting conditions for the production hall. The first four possible solutions are using the MATIS – FIRA 03 luminaire, which is manufactured by Elba and has an output of 64%, in combination with four types of lamps: MASTER TL-D Reflex 2x36W/840, MASTER TL-D Reflex 2x58W/840, MASTER LEDtube Universal

Table 1. The set of alternatives.

Calculated data			Alternatives	
h [m]	*i*	*u*	**Luminaire type**	**Lamp type**
6.2	1	0.5	MATIS – FIRA 03	MASTER TL-D Reflex 2x36W/840
		0.5	MATIS – FIRA 03	MASTER TL-D Reflex 2x58W/840
		0.5	MATIS – FIRA 03	MASTER LEDtube Universal T8 2x24W/840
		0.5	MATIS – FIRA 03	MASTER LEDtube Universal T8 2x16W/840
		0.51	TCS165	MASTER TL5 HE 2x28W/840
		0.46	TCS165	MASTER TL5 HE 2x35W/840
		0.36	TCW060	MASTER TL-D Reflex 2x36W/840
		0.34	TCW060	MASTER TL-D Reflex 2x58W/840

T8 2x24W/840 and MASTER LEDtube Universal T8 2x16W/840.

MASTER TL-D Reflex, is a conventional source, a fluorescent tube, that directs most of the light were is necessary, thanks to an internal reflector and the result is the same level of illumination, while half of the number of lamps in the luminaire are used; other benefits include the fact that the internal reflector is not affected by dust accumulation (compared to normal TL-D sources) and the efficiency is relatively high, both initially and during its lifetime, with a luminous flux constant. The wattages used for this lamp are 36W and 58W.

MASTER LEDtube Universal T8 is a LED tube and is uniquely designed to allow the direct fitting of the lamp into luminaires, which operate either on electromagnetic ballasts or on high-frequency electronic ballasts or with a direct connection to the electrical network; other benefits include low operating costs, due to lower energy consumption, and lower maintenance costs, thanks to longer lifetime than conventional lamps. The wattages used for this type of lamp are 24W and 16W.

The next two possible solutions use the TCS165 luminaire, which is manufactured by Philips and has an output of 69% for the lamp with 28W and 62% for the lamp with 35W. The type of lamp used is MASTER TL5 HE, which is a conventional source, a fluorescent tube. It has a high luminous efficiency, which results in low energy consumption, while ensuring excellent light output and good colour rendering. Other benefits include easy fastening in ceiling module systems and the high efficiency of the lamp and the luminaire to which it attaches. The wattages selected for this type of lamp are 28W and 35W.

The last two possible solutions use the TCW060 luminaire, which is also manufactured by Philips, and has the lowest output of all luminaires used in this study: 59% for the lamp with 58W and 63% for the one with 36W. The type of lamp used in combination with this luminaire is MASTER TL-D Reflex (fluorescent tube), which is the same as the one used in combination with the first luminaire, MATIS – FIRA 03.

For this case study, eight criteria were proposed in order to perform a complete evaluation of the luminaire-lamp combinations selected using the utilisation factor method.

The first criterion is the *energy consumption* (q_e - [kWh/1000 h]) of the whole lighting installation required to provide the minimum, average illuminance. For this, there has to be determined the number of luminaires (N_c), calculated with the following relation:

$$N_c = \frac{L_1 \cdot L_2 \cdot E_{med}}{u \cdot N_{lc} \cdot \Phi_l} \tag{1}$$

where:
N_c – number of luminaires required [pcs];
L_1 – width of the production hall [m];
L_2 – length of the production hall [m];
E_{med} – minimum, average illuminance [lx];
u - utilisation factor;
N_{lc} – number of lamps in luminaire [pcs];
Φ_l – initial lamp output [lm].

To determine the total energy consumption, the following formula is used :

$$q_e = N_{lc} \cdot q_{el} \cdot N_c \tag{2}$$

where:
q_e – total energy consumption of all luminaires [kWh/1000 h];
N_{lc} – number of lamps in luminaire [pcs];
q_{el} – energy consumption of the lamp [kWh/1000 h];
N_c – number of luminaires required [pcs].

Another criterion used to evaluate the selected alternatives is *luminous efficiency (E_c)*, which is directly related to energy consumption: the higher efficiency the lower energy usage will be, and is determined as division between the lamp luminous flux and the power consumed by the lamp. The unit used for the efficiency is lumens per watt ([lm/W]).

The luminaires require protection against different external factors, therefor the manufacturer specifies the following degrees of protection for each type of luminaire: "Ingress Protection" *(IP)* system indicates the level of protection to ingress of foreign solid objects, such as dust, and moisture; the degree of external mechanical impact protection (*IK*) reflects the property of the luminaire to maintain integrity after being subjected to mechanical shock without affecting the operation of the electrical circuit and without compromising its security. This two criteria, *IP* and *IK protection* ([points]) should be assessed carefully in order to keep the maintenance cost at a reasonable level.

The *total price of the lighting system* (Pr_S - [RON]) may slightly vary, depending on the supplier's pricing policy in relation to its partners. The formula used to determine the total price is:

$$Pr_S = N_c \cdot (Pr_{cil} + N_{lc} \cdot Pr_l) \quad (3)$$

where:
Pr_S – total price of the lighting system [RON];
N_c - number of luminaires required [pcs];
Pr_{cil} – price of the luminaire [RON];
N_{lc} – number of lamps in luminaire [pcs];
Pr_l – price of the lamp [RON].

Another criterion is the *installation price* (Pr_i – [RON]) and is determined as the product between the number of the luminaires required and the unit cost of installation. In this study, the unit cost of installation has been considered to be 4.43 RON, on the assumption that the installation of a luminaire lasts about 30 minutes and the person responsible for this activity has a remuneration of 1450 RON per month.

The last two criteria, *lamp lifetime* ([h]) and *lamp warranty* ([months]), differs depending on the lamp type, if it is an incandescent or fluorescent lamp, or light emitting diode (LED). The lamp lifetime reflects the hours of operation under normal conditions, without taking into account the negative effects that may occur in the network and which may contribute to more rapid degradation of the source. The lamp warranty granted on purchase protects the customers from possible defective products. The values for these two indicators are specified by the manufacturer in the data sheet of the product.

The decision matrix, which represents the performance of all alternatives to be evaluated with respect to the identified criteria, is presented in Table 2. The weights and the values for each threshold are given in Table 3.

Table 2. Performance matrix.

Criteria / Alternatives	Energy consumption [kWh/1000h]	Luminous efficiency [lm/W]	IP [points]	IK [points]	Lighting system price [RON]	Installation price [RON]	Lamp lifetime [h]	Lamp warranty [months]
FIRA03_FL_36W	2352	93.06	40	7	10296.16	126.88	20000	24
FIRA03_FL_58W	2448	89.57	40	7	8306.64	81.56	20000	24
FIRA03_LED_24W	1248	154.17	40	7	11688.04	117.81	50000	60
FIRA03_LED_16W	1216	156.25	40	7	13663.66	172.19	50000	60
TCS165_FL_28W	1860	103.94	20	2	16600.2	135.94	24000	24
TCS165_FL_35W	2204	102.82	20	2	16647.16	131.41	24000	24
TCW060_FL_36W	3360	93.06	65	8	12388.8	181.25	20000	24
TCW060_FL_58W	3944	89.57	65	8	11375.54	131.41	20000	24

Table 3. Thresholds and criteria weights.

Criteria / Weights Thresholds	Energy consumption [kWh/1000h]	Luminous efficiency [lm/W]	IP [points]	IK [points]	Lighting system price [RON]	Installation price [RON]	Lamp lifetime [h]	Lamp warranty [months]
Indifference (q)	0.05	0.1	0.2	0.1	0.05	0.1	0.1	0.15
Preference (p)	0.15	0.25	0.3	0.2	0.1	0.25	0.3	0.3
Veto (v)	0	0	0	0	0	0	0	0
Weight (w)	0.2	0.15	0.05	0.05	0.25	0.1	0.15	0.05

4 Results

The computational experiments for this study were performed using the software developed by the authors based on the ELECTRE III method algorithm. The software is a Windows Forms application, built using the tools Microsoft .NET Framework and Visual Studio 2015.

The language used to create the application is C#, version 4.5, which was developed by Microsoft and is a multi-paradigm programming language encompassing powerful typing, declarative, functional, imperative and object-oriented programming disciplines. The user interface uses Windows Forms, which is a class library included as part of Microsoft .NET Framework, and offer a lot of flexibility and scalability in gathering and displaying data.

MATEC Web of Conferences **184**, 04007 (2018) https://doi.org/10.1051/matecconf/201818404007
Annual Session of Scientific Papers IMT ORADEA 2018

After the developed software is opened, the criteria and alternatives can be defined. For every criterion, the direction of the preference has to be identified. The direction of the preference indicates if the criterion is increasing, like for example the lamp lifetime (the score of the criterion has to be maximized), or diminishing, like for example the energy consumption (the score of the criterion has to be minimized). The directions of preference for the criteria specified in this case study are presented in figure 1.

Fig. 1. Directions of preference for the criteria.

Once all the alternatives and criteria have been introduced, the user has to define the indifference, preference and veto threshold for each criterion. Table 2 and 3 summarize the data that has been loaded in the application. Complementary results are returned by the software. First, three types of complementary matrices are returned (the concordance matrix, the discordance matrix and the credibility matrix). Figure 2 shows the concordance matrix calculated by the application, which is identical to the credibility matrix because there is no veto threshold and the discordance indexes are zero.

Thus, the ascending distillation illustrated that FIRA03_FL_58W and FIRA03_LED_24W were ranked as the best options, followed closely by FIRA03_FL_36W and FIRA03_LED_16W, on the same place, and on the third place by TCS165_FL_28W. Regarding the alternatives, TCW060_FL_36W and TCW060_FL_58W there was no preference between them. The last ranked combination is TCS165_FL_35W, being on the last place.

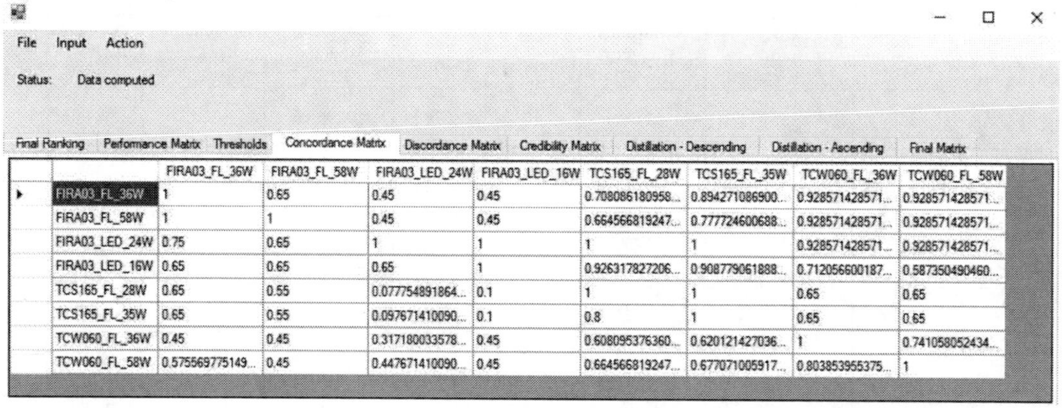

	FIRA03_FL_36W	FIRA03_FL_58W	FIRA03_LED_24W	FIRA03_LED_16W	TCS165_FL_28W	TCS165_FL_35W	TCW060_FL_36W	TCW060_FL_58W
FIRA03_FL_36W	1	0.65	0.45	0.45	0.708086180958...	0.894271086900...	0.928571428571...	0.928571428571...
FIRA03_FL_58W	1	1	0.45	0.45	0.664566819247...	0.777724600688...	0.928571428571...	0.928571428571...
FIRA03_LED_24W	0.75	0.65	1	1	1	1	0.928571428571...	0.928571428571...
FIRA03_LED_16W	0.65	0.65	0.65	1	0.926317827206...	0.908779061888...	0.712056600187...	0.587350490460...
TCS165_FL_28W	0.65	0.55	0.077754891864...	0.1	1	1	0.65	0.65
TCS165_FL_35W	0.65	0.55	0.097671410090...	0.1	0.8	1	0.65	0.65
TCW060_FL_36W	0.45	0.45	0.317180033578...	0.45	0.608095376360...	0.620121427036...	1	0.741058052434...
TCW060_FL_58W	0.575569775149...	0.45	0.447671410090...	0.45	0.664566819247...	0.677071005917...	0.803853955375...	1

Fig. 2. Concordance matrix.

Based on the credibility matrix, the two distillation procedures can be performed. In order to rank the alternatives, the software is exploiting the outranking degrees by the descending and ascending distillation procedures, leading to the first complete pre-classifications.

The first distillation, descending one, consists of selecting the best ranked alternatives initially and ending with the worst. The best alternative (FIRA03_LED_24W) is taken out from the credibility matrix, and then the procedure is gradually repeated for the remaining alternatives, until all of them are ranked. The result of descending distillation showed preference for FIRA03_LED_24W, followed by FIRA03_FL_58W, FIRA03_FL_36W, FIRA03_LED_16W and TCS165_FL_28W. Regarding the last three combinations TCS165_FL_35W, TCW060_FL_36W and TCW060_FL_58W, there was no preference between them. The descending distillation is presented in figure 3 (a).

The ascending distillation, showed in figure 3 (b), uses the same procedure of ranking as the descending one, but with successive removal of the worst alternative.

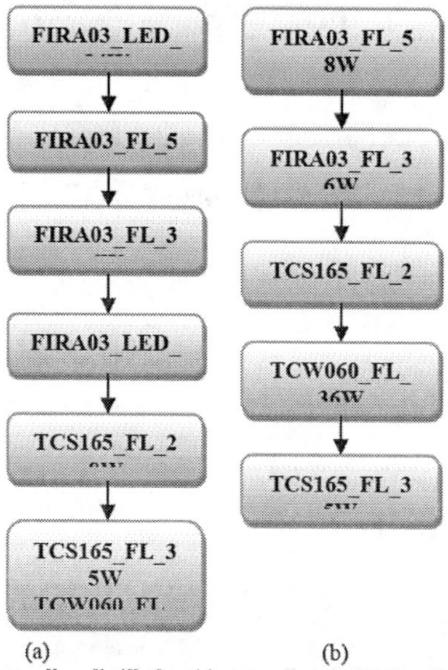

Fig. 3. Descending distillation (a), ascending distillation (b)

Having the results of the two distillation procedures, the final ranking can be identified by calculating the intersection of the two pre-classifications, as presented in figure 4.

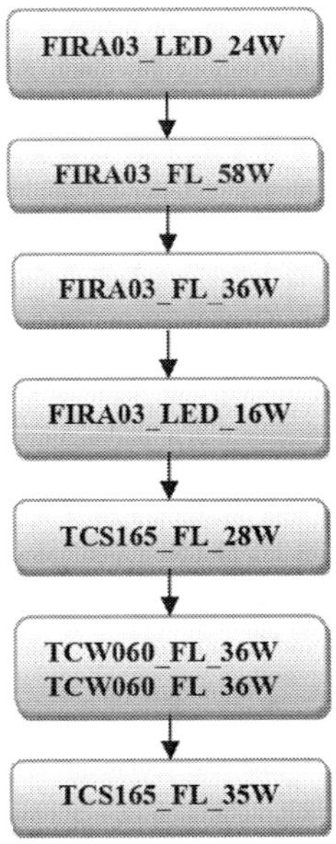

Fig. 4. Final ranking

It can be noticed that FIRA03_LED_24W is the best rated alternative, so it is the optimal lighting solution for the production hall considered in this study. At the other extreme, at the bottom of the ranking, is positioned TCS165_FL_35W, which means that it represents the least preferred luminaire-lamp combination.

5 Conclusions

There are different lighting technologies used in indoor spaces, which should be compared by three objectives: photometrical performance, energy efficiency and performance of the installation. For each of these objectives, great consideration should be given, so that either energy consumption or visual comfort would not be affected.

In this paper, a practical application of the ELECTRE III method is proposed to deal both with the quantitative and qualitative criteria and to select the optimal lighting solution suitable for a production hall. Eight luminaire-lamp combinations have been identified and assessed according to eight criteria that meet the expectations of the beneficiary of the electrical installation project. By running the developed software, the final ranking of the alternatives was obtained. The results reveal that ELECTRE III method may be used in the electrical installations field.

References

1. A.A. Baloch, P.H. Shaikh, F. Shaikh, Z.H. Leghari, N.H. Mirjat, M.A. Uqaili, Renew. Sust. Energ. Rev., **82**, 3007, (2018)
2. F.G. Montoya, A. Peña-Garcia, A. Juaidi, F. Manzano-Agugliaro, Energy Build., **140**, 50, (2017)
3. S. Carlucci, F. Causone, F. De Rosa, L. Pagliano, Renew. Sust. Energ. Rev. **47,** 1016, (2015)
4. B. Stankovic, A. Kostic, M. Jovanovic Popovic, ESD, **19**, 56, (2014)
5. A.M. Atzeri, F. Cappelletti, A. Tzempelikos, A. Gasparella, Energy Build., **127**, 411, (2016)
6. N. Makaremi, S. Schiavoni, A.L. Pisello, F. asdrubali, F. Cotana, *Energy Procedia,* **134**, 306, (Energy Build., Chania, Crete, Greece, 2017)
7. Q. Wang, H. Xu, F. Zhang, Z. Wang, Optik, **129**, 21, (2017)
8. P. Morgan Pattison, M. Hansen, J.Y. Tsao, C.R. Physique, (2017)
9. I. Kiyak, B. Oral, V. Topuz, Energy Build., **148**, 342, (2017)
10. D. Sabaei, J. Erkoyuncu, R. Roy, Procedia CIRP, **37**, 30, (2015)
11. M. Cristea, C. Cristea, *Proceedings of the Annual Sessions of Scientific Papers "IMT Oradea"*, **XIV**, 6, (2015)
12. M.M. Marzouk, Automat. Constr., **20**, 596, (2011)
13. C. Cristea, M. Cristea, COSME'16, **4**, 9, (2016)
14. D. Neves, P. Baptista, M. Simoes, C.A. Silva, J. Rui Figueira, J. Clean. Prod., **176**, 251, (2018)
15. C. Cristea, M. Cristea, COSME'16, **4**, 9, (2016)

The influence of dividend policy on the volatility of shares in the Romanian equity capital market

Ciprian Cristea[1,*], *Maria* Cristea[1]

[1]Technical University of Cluj-Napoca, Faculty of Electrical Engineering, Romania

Abstract. Dividend policy is one of the most debated topics in corporate finance and, at the same time, it represents an important issue from the perspective of both the managers of the firm and the investors. To the best of our knowledge, the literature regarding establishing the relationship between dividend policy and share price volatility of non-financial companies listed on Romanian stock market remains inexistent. This study attempts to identify how the share price volatility of non-financial companies listed on the Bucharest Stock Exchange for a period of sixteen years from 2002 and 2017 is influenced by corporate dividend policy. Results from a cross sectional multiple regression analysis revealed a negative effect of the two components of the dividend policy on the share price volatility. The results support the idea that the lower the dividend yield, the higher the risk to be faced by the shareholder. Growth in assets and share price were negatively related. Positive relationship between firm size and debt ratio to price volatility were identified. However, there was no significant relationship found between earning volatility and price volatility in the Romanian stock market.

1 Introduction

Dividend policy is one of the most debated and challenging topics in modern financial economics both in emerging and developed economies [1]. Dividend policy is a set of guidelines used by a company to establish how much of its earnings should be distributed to the shareholders as dividends and how much to be reinvested in new opportunities. Corporate finance research has investigated why corporations pay dividends, referred to as the "Dividend Puzzle" by [2]. Notwithstanding the comprehensive research on dividend policy over the past decades, no universally accepted explanation is acquired [3]. Hence, more efforts should be implemented to uncover the puzzle.

Dividend policy represents an important decision, usually made by top managers, which demands the final agreement of the board of directors. Managers have to take into account which dividend policy will lead to maximization of shareholder's wealth, how much of company's earnings are needed for investment, and last, but not least they have to analyze the impact of their decision on stock's price [4].

In general, the theoretical framework for explaining the relationship between dividend policy and stock price volatility over the last decades has been dominated by dividend irrelevance theory, bird-in-the-hand theory, signaling theory, agency theory and clientele effects of dividend theories.

Dividend policy proposed by [5] state that dividend payment have no effect on company value and no dividend policy is superior to another under perfect capital market assumptions. The shareholders wealth is not influenced by dividend policy. The company's earnings, coming from firm's investment policy, determine the value of the company. If a company distributes its profits as dividends, then the stock price will be decreased by the amount of a dividend per share on the ex-dividend date. In a perfect market, without taxes and brokerage costs, dividend policy does not alter the shareholder's return.

Bird-in-the-hand theory asserts that in a world of uncertainty and information asymmetry, investors prefer dividends over the capital gains, even if the promised dividend is larger, due to uncertainty of future cash flows [6]. This perspective is consistent with the view that "a bird in hand is worth more than two in the bush" and supports the results of [7-8].

The dividend paid to shareholders encompasses information about the prediction of the company [9]. Even though [5] assumed that investors and management have perfect knowledge about a company, it is unrealistic to believe that they have the same amount of data and, because of this, firms pay dividend to send a signal to their investors that their company is financially stable and remain profitable [10]. Increasing the dividend can be seen as a good sign, notifying the owners that managers are working in their best interest, and thus, can affect the share price positively [6].

Agency theory establishes the role of dividends in controlling agency relationships between managers and shareholders and, by paying dividend, it compels the company to raise additional funds from the capital market, thus decreasing the scope for agency conflicts and exposing the firm to the discipline of the market [11]. Agency cost represents the cost of the conflict of interest between management and shareholders and management,

* Corresponding author: ciprian.cristea@emd.utcluj.ro

© The Authors, published by EDP Sciences. This is an open access article distributed under the terms of the Creative Commons Attribution License 4.0 (http://creativecommons.org/licenses/by/4.0/).

namely when management team, instead of acting on behalf of the owners of the company (shareholders), they act on their behalf [12].

The clientele effect is a theory that explains that due to the fact that investors are taxed in different individual brackets, the difference between dividend taxes and capital gains taxes would lead to clientele preference [13]. Thus, investors who pay low taxes on dividends would be concerned in purchasing high dividend yielding stocks, while investors who pay high taxes on dividends would prefer to buy low dividend yielding stocks [13].

The dividend policy literature is vast and contains numerous theories and hypotheses. Notwithstanding voluminous theorizing and empirical research, no consensus exists among researchers on the impact of dividend policy on share price.

Furthermore, many researchers investigated the association between share price volatility and dividend policy. The common stock's volatility is a measure used to establish risk, indicating the rate of change in the stock's price over a determined period [14]. The smaller the volatility, the smaller possibility of gain or loss in short-term is. Volatility deals with the variance of a stock's price. The price of less volatile stock would not differ notably over time and it is not difficult to predict the future price of this security [4]. Normally, investors select less risky investments [4].

There are several studies examined the relationship between dividend policy and share price volatility. In Malaysia [4] studied the effect of the dividend policy on the stock price volatility from 2005 and 2010. The price volatility is associated negatively with both proxies of the dividend policy. In the UK stock market [14] found a significant and negative relationship between the payout ratio and dividend yield with the stock price volatility. In the US equity capital market dividend policy and stock price volatility is analyzed by ordinary least square regression [10] and, the stock with high dividend is less risky. In Brazil, [15] examined the relationship between the dividend policy and the risk of stocks, in the period 2000-2012, and it was found that the dividend policy is significant in determining the volatility, establishing that the lower the dividend yield, the higher the risk to be faced by the shareholder. [16], on the other hand, found that payout is not related to stock price volatility. In Karachi stock exchange, [17] found the relationship between stock price volatility and dividend yield to be positive and significant.

In this paper, we investigate the relationship between dividend policy and share price volatility of non-financial companies listed on the Bucharest Stock Exchange Market for a period of sixteen years from 2002 to 2017. In this study, multiple regression analysis is applied to explore the association between share price changes and both dividend yield and dividend payout ratio of the Romanian firms mentioned above. This research has important implications for both theory and practice. No study, to the best of our knowledge, has examined the relationship between dividend policy and share price volatility on the Romanian market. Our findings have implications for corporate finance literature, academics, investors, and policymakers in emerging markets.

The rest of the paper is organized as follows: in the next paragraph the methodology is described, moreover, the empirical results are presented in the third section and, finally, the last section summarizes the conclusions.

2 Methodology

The goal of this paper is to study the relationship between dividend policy and share price volatility using multiple least square regressions. Primarily, we conducted a basic test of the relationship between price volatility and the two most important measures of dividend policy: dividend yield and dividend payout ratio.

The model used in this paper is based on the previous literatures commenced by [16]. Following [4, 10, 14, 16], several control variables were included to account for certain factors that affect both stock price volatility and dividend policy, namely, firm size, growth in assets, debt ratio and earnings volatility.

The financial data were collected from the annual reports and financial statements of 175 Romanian non-financial companies listed on the Bucharest Stock Exchange Market over the period 2002-2017. The firms included in the sample had at least one cash dividend during period 2002 to 2017. The data regarding share prices were obtained from Romanian equity capital market and transformed into an appropriate form so as to be used for the study.

The dependent variable is share price volatility. Firstly, to obtain a crude test of the relationship between price volatility and dividend policy, dependent variable is regressed against the most important independent variables, namely, dividend yield and payout ratio utilizing multiple least square regression based on following equation:

$$PV_i = \alpha + \beta_1 \cdot DY_i + \beta_2 \cdot PR_i + e_i \qquad (1)$$

where:
α – constant;
β – regression coefficient;
PV_i – share price volatility for company i;
DY_i – dividend yield for company i;
PR_i – payout ratio for company i;
ε – error term.

The price volatility (PV) is utilized to capture the movement in share price changes and to measure the risk of the stock. For each year, it is computed by taking the difference between highest and lowest share prices and dividing by the average of highest and lowest prices and then by taking the square of it. Following [16], the formula for computing the annually share price volatility is as follows:

$$PV_i = \sqrt{\frac{HP_i - LP_i}{\left(\frac{HP_i + LP_i}{2}\right)^2}} \qquad (2)$$

where:
HP_i – highest share price for company i;
HP_i – lowest share price for company i.

Dividend yield (*DY*) is calculated by taking all cash dividends paid to common stock holders in a year and then dividing them by share price. Payout ratio (*PR*) is computed as the ratio of dividend per share to earning per share.

Dividend yield and dividend payout ratio are closely related to each other. Thus, a multicollinearity problem is likely to arise. To limit this problem, the control variables specified earlier were included. Thereby, the estimation model is modified as follows:

$$PV_i = \alpha + \beta_1 \cdot DY_i + \beta_2 \cdot PR_i + \beta_3 \cdot SZ_i + \beta_4 \cdot DR_i + \beta_5 \cdot EV_i + \beta_6 \cdot GROW_i + \varepsilon_i \quad (3)$$

where:
SZ_i – firm size for company *i*;
$GROW_i$ – growth in assets for company *i*;
DR_i – debt ratio for company *i*;
EV_i – earnings volatility for company *i*.

Size (*SZ*) is one of the control variable measured by using the natural logarithm of total assets [18]. Growth in assets (GROWTH) is determined as the ratio of the change in total assets at the end of the year to the level of total assets at the beginning of the year [14]. Debt ratio (*DR*) is the ratio of total debt to total assets. Earnings volatility (*EV*) is the standard deviation for the ratio of operating profit to total assets.

3 Results

Table 1 shows a broad description of the summary statistics of the variables used in the study. It shows the statistical means and standard deviation. Following [19], if it is assumed that share prices follow a normal distribution and the effect of company's going ex-dividend is not took into consideration, the standard deviation of stock market returns correspond to the determined volatility of this study [14]. It can be computed using [20] formula. Thus, the mean of price volatility 0.8093, is multiplied by the constant 0.6008 which return a result of 48.62%. This is higher that of UK (17.66%) companies [14] and Australian (29.42%) companies [19], but lower that of Pakistani (48.85%) companies [21].

Table 1. Descriptive statistics of variables.

	Mean	Std. Dev.
PV	0.809382	0.793112
DY	0.044468	0.085295
PR	0.399657	0.812162
SZ	17.53487	1.808350
GROW	0.165239	0.704577
DR	0.276865	0.239220
EV	0.019598	0.029877

Table 2 shows the Pearson correlation coefficients for the variables used in this paper. According to our expectation and to the results obtained in other studies [4, 14, 15] we found a statistically negative correlation coefficient between price volatility and dividend yield, but it differs from that of [19] which was positive. Likewise, the correlation between price volatility and dividend payout is statistically negative as expected and is similar to [14, 15].

Table 2. Correlation matrix.

	PV	DY	PR	SZ	GROW	DR	EV
PV	1						
DY	-0.176**	1					
PR	-0.128**	0.422**	1				
SZ	0.057*	-0.011	0.07**	1			
GROW	-0.071**	0.009	-0.047	-0.014	1		
DR	0.082**	-0.143**	-0.112	0.049	0.02	1	
EV	0.01	0.054*	-0.07	-0.23**	-0.021	0.178**	1

Notes: (*) implies significance at 5%, (**) implies significance at 1%

Table 3 shows the results of regression based on equation (1). The results of this regression show that price volatility and the two components of dividend policy have significant negative association.

Table 3. Results of regression based on equation (1).

Variable	Coefficient	t-Statistic	P-values
C	0.896723	36.45901	0
DY	-1.388043	-5.091325	0
PR	-0.064099	-2.238692	0.0253

Consistent with previous findings, the study reports a significant negative relationship between price volatility and dividend yield [4, 14].

In the next phase, the four control variables were added to notice if there would be any change in the coefficients of dividend policy components. Table 4 presents the results of regression based on equation (3). It can be noticed that the significant negative association between share price volatility and dividend yield remains. This is in line with our expectation and to the results obtained in other studies [4, 10, 14, 15]. Thus, the lower the dividend yield, the higher the risk to be faced by the shareholder. Also, the negative association between share price volatility and dividend payout remains and it is significant. The result is consistent with previous findings [4, 14, 15, 21], except that those findings are not statistically significant.

Table 4. Results of regression based on equation (3).

Variable	Coefficient	t-Statistic	P-values
C	0.383919	1.791055	0.0735
DY	-1.297521	-4.719696	0
PR	-0.069068	-2.399844	0.0165
SZ	0.026788	2.230776	0.0259
DR	0.165321	1.820868	0.0688
GROW	-0.082076	-2.749157	0.0061
EV	0.450626	0.607971	0.5433

As table 4 shows, there is a significant positive association between price volatility and firm size. The result is not in line to the results obtained in other studies [10, 14, 15]. The study reveals a negative relationship between price volatility and growth in assets. The result is similar with prior findings [4].

A positive association between price volatility and earnings volatility is found, implying that firms which have more volatility in their earnings have more volatility in their stock price, but the association is not significant.

4 Conclusions

This paper examines the relationship between dividend policy (dividend yield and dividend payout) and the volatility of stock price by applying multiple regression. For this purpose, a sample of 175 nonfinancial companies listed on the Bucharest Stock Exchange market between 2002 and 2017 were selected. Also, the study examined the relationship between stock price volatility and the control variables, namely, firm size, growth in assets, debt ratio and earnings volatility.

The empirical findings report that there is a significant negative relationship both between the payout ratio of a company and the volatility of its share price, and between dividend yield and the volatility of stock price. The findings suggest that the less the payout ratio the higher volatile a share price would be. The study noticed that share price of firms with high dividend yield can be associated with nearby cash, which decrease the uncertainty regarding the cash flows of the company leading to less price volatility. Among the control variables, debt ratio indicates a positive relationship with price volatility implying that the less leveraged a company is, the less volatile would be the share price.

The results of this study reveal that the dividend policy has an impact on the price volatility, and that the managers of the firms have the capacity to affect their company's stock price by adjusting dividend policy that is convenient for their target investors. The results of this paper are useful for academics, analysts and investors in emerging markets.

References

1. G. M. Frankfurter, B. G. Wood Jr., Int. Rev. Financ. Anal., **11**, 111-138 (2002)
2. F. Black, JPM, **2**, 5–8 (1976)
3. H. K. Baker, J. C. Singleton, E. T. Veit, *Survey Research in Corporate Finance: Bridging the Gap between Theory and Practice* (Oxford University Press, Oxford, 2011)
4. M. Hashemijoo, A. Mahdavi Ardekani, N. Younesi, JBSQ, **4**(1), 111-129 (2012)
5. M. Miller and F. Modigliani, J BUS, **34**, 411–433, (1961)
6. H. N. Al-Malkawi, JAAR, **23**, 44-70 (2007)
7. M. J. Gordon, J. Fin., **18**(2), 264-272 (1963)
8. J. Lintner, Rev. Econ. Stat., **44**(3), 249-263 (1962)
9. R. R. Petit, J. Fin., **27**, 993-1007 (1972)
10. K. A. Profilet and F. W. Bacon, BS, **20** (1), 219-231 (2013)
11. R. Manos, V. Murinde, C. J. Green, IREF, **21**, 42–56 (2012)
12. S. A. Ross, R. W. Westerfield, J. Jaffe, B. D. Jordan, *Modern Financial Management* (8th ed., McGraw-Hill International, 2008)
13. J. L. Procianoy and R. S. Verdi, RAE, **8**(1) (2009)
14. K. Hussainey, C. O. Mgbame, A. M. Chijoke-Mgbame, J. R. FINANC., **12**(1), 57-68 (2011)
15. C. Silva and M. A. V. Machado, RC&C, **7**(3), 8-20 (2015)
16. J. Baskin, JPM, **15**(3), 19-25 (1989)
17. M. Asghar, S. Z. A. Shah, K. Hamid, M. T. Suleman, FJSB, **4**(1), 45-52 (2011)
18. I. Jabbouri, RES INT BUS FINANC, **37**, 283–298 (2016)
19. D. E. Allen and V. S. Rachim, J. Appl Econ, **6**, 175-188 (1996)
20. M. Parkinson, J Bus., **53**, 61-65 (1980)
21. Y. Habib, Z. I. Kiani, M A. Khan, GJMBR, **12**(5) (2012)

Cash conversion cycle and corporate performance: evidence from Romania

Ciprian Cristea[1,*], *Maria* Cristea[1]

[1]Technical University of Cluj-Napoca, Faculty of Electrical Engineering, Romania

Abstract. Cash conversion cycle is considered one of the most important measures of management effectiveness, especially the cash flow and liquidity management. This study examines the relationship between cash conversion cycle and corporate profitability for the non-financial companies, from several industries, listed on the Bucharest Stock Exchange for a period of fifteen years from 2002 to 2016. The findings from a cross sectional multiple regression analysis pointed out a negative relationship between cash conversion cycle and the performance of firms. Based on the results from this paper it has been concluded that managers can improve the profitability of their firms by decreasing the number of days in cash conversion cycle.

1 INTRODUCTION

Cash conversion cycle is an important metric, developed by [1], and used to assess the effectiveness of the company's overall financial health. The cash conversion cycle measures the number of days between the expenditure of the company's cash for the acquisition of raw materials to manufacture the products, and the collection of cash obtained from the sale of the finished goods [2]. Cash conversion reveals how the enterprises are performing and, at the same time, helps to dig out the areas where further improvement is demanded [3].

Cash flow management is a vital element of a company's operational strategies [4]. Cash conversion cycle is a measure of working capital efficiency. Enterprises manage working capital using three approaches: conservative, aggressive and moderate [5]. The conservative approach is where a company tends to use short-term sources only in urgent circumstances, while in other cases uses mainly long-term sources of finance for its operations. The aggressive approach utilizes higher levels of current liabilities and less long-term debt or a low level of current assets as percentage of total assets, which may lead to lower levels of liquidity [6]. In the last approach, permanent current assets are financed by long-term sources and short-term sources of finance are used to finance fluctuating current assets [5].

The cash conversion cycle has three elements: inventory conversion period, average receivables period and average payables period. Firms can improve their profitability by reducing their length of cash conversion cycle through decreasing the receivables collection period, prolonging the credit payment period or lessening the inventory selling period [7].

Many studies examined the relationship between the length of cash conversion cycle and corporate profitability. The link between the cash conversion cycle and the company profitability has been studied by [8], using a large sample of Belgian firms, and showed that for a better performance the time duration for collection of receivable should be kept short. [9] analyzed the relationship between cash conversion cycle and profitability on a sample of industrial companies and reported a negative association between the stock turnover days and the return on investment. Thus, low level of inventories was associated with high profitability. [10] found the same results on a large sample of Greek firms from food, textiles and chemical sectors. [11] conducted a study on Spanish small and medium size firms and confirmed a negative association between the profitability and the number of days accounts receivable and inventory days. Thus, companies should delay in making the payments for efficient performance. [12] investigated the relationship between working capital and profitability on a sample of 110 manufacturing companies listed on Borsa Istanbul, over the period of 2005–2014. The results pointed out that working capital negatively impacts on corporate profitability, so that firms could decrease profitability by prolonging cash conversion periods. [13] analyzed the relation between the enterprise's cash conversion cycle and its profitability using dynamic panel data analysis for a sample of Japanese firms during the period 1990-2004. The analysis is applied at the levels of the full sample and divisions of the sample by industry and by size. Results reported a negative relation between the length of the firm's cash conversion cycle and its profitability. [14] examined the relationships between cash conversion cycle and invested capital, liquidity, and performance of small U.S. manufacturing and retail firms. They found that cash conversion cycle is a significant factor in small firm capital needs, liquidity, and performance, and owners may

* Corresponding author: ciprian.cristea@emd.utcluj.ro

© The Authors, published by EDP Sciences. This is an open access article distributed under the terms of the Creative Commons Attribution License 4.0 (http://creativecommons.org/licenses/by/4.0/).

not be proactive in managing cash conversion cycle. [15] tackled the impact of the cash conversion cycle on the profitability of a sample of Italian manufacturing firms. The results indicated that lesser the cash conversion cycle, greater the profitability.

In this paper, we examine the relationship between cash conversion cycle and profitability of non-financial companies listed on the Bucharest Stock Exchange Market for a period of fifteen years from 2002 to 2016. Multiple regression analysis is applied to explore the association between firm's cash conversion cycle and its profitability. This research has important implications because, no study, to the best of our knowledge, has examined the relationship between cash conversion cycle and corporate profitability at the level of full sample and divisions of sample by industry on the Romanian equity capital market. The rest of the paper has the following structure: in the next paragraph the methodology is described, followed by the results of this study and, the final part summarizes the conclusions.

2 Methodology

We selected a pooled ordinary least squares regression model to investigate the relationship between the cash conversion cycle corporate performance by adopting enterprises from different industries as samples.

The financial data were collected from the annual reports and financial statements of 330 Romanian companies listed on the Bucharest Stock Exchange Market, from 12 industries, namely wholesale/retail, construction, pharmaceuticals, manufacturing, plastics, machinery and equipment, metallurgy, chemicals, basic resources, transportation and storage, tourism, utilities, over the period 2002-2016.

The study takes return on assets as measure of profitability to represent dependent variables. The dependent variable is regressed against independent variables, namely, cash conversion cycle, firm size, debt ratio, current ratio and fixed assets ratio based on following equation:

$$ROA_i = \alpha + \beta_1 \cdot CCC_i + \beta_2 \cdot DR_i + \beta_3 \cdot CR_i + \beta_4 \cdot FA_i + \beta_5 \cdot SZ_i + \varepsilon_i \quad (1)$$

where:
α – constant;
β – regression coefficient;
ROA_i – return on assets for company i;
CCC_i – cash conversion cycle for company i;
DR_i – debt ratio for company i;
CR_i – current ratio for company i;
FA_i – fixed assets ratio for company i;
SZ_i – firm size for company i;
ε – error term.

Return on assets indicates how profitable a firm is relative to its assets and is measured as the ratio of net income to total assets. Cash Conversion Cycle is calculated as the number of days of inventory plus the number of days of receivables less the number of days of payables. The formula for computing cash conversion cycle is as follows:

$$CCC_i = DIO_i + DSO_i - DPO_i \quad (2)$$

where:
DIO_i – days inventory outstanding for company i;
DSO_i – days sales outstanding for company i;
DPO_i – days payable outstanding for company i.

The days inventory outstanding ratio is calculated as follows:

$$DIO_i = \frac{Inv_i \cdot 365}{COGS_i} \quad (3)$$

where:
Inv_i – inventory for company i;
$COGS_i$ – cost of goods sold for company i.

The days sales outstanding ratio is determined as follows:

$$DSO_i = \frac{AR_i \cdot 365}{Sal_i} \quad (4)$$

where:
AR_i – account receivable for company i;
Sal_i – sales for company i.

The days payable outstanding ratio is computed as follows:

$$DPO_i = \frac{AP_i \cdot 365}{COGS_i} \quad (5)$$

where:
AP_i – account payable for company i.

Debt ratio is defined as the ratio of total debt to total assets. Current ratio is determined as the ratio of current assets to current liabilities. Fixed assets ratio is obtained as the ratio of fixed assets to total assets. Size of the firm is measured by using the natural logarithm of total assets.

3 Results

Table 1 presents the descriptive statistics for the variables used in this study. This table shows the average of the variables from the full sample and industries between the years 2002 and 2016. Manufacturing is the industry with the highest firm observations, accounting for 33.64% of the total sample size, while, pharmaceutics, chemicals and utilities exhibit the lowest firm observations, accounting for 1.52%. The average cash conversion cycle for Romanian firms was 182.64 days, which is considerably higher that of Italian manufacturing firms (79.34 days) [15] and US manufacturing companies (54.18 days) [14], but lower than the average cash conversion cycle of listed companies on the Athens Stock Exchange (189 days) [16]. We could see that the lowest average cash conversion cycle is found in utilities industry, with an average of 46.89 day, and the highest average value of the cash conversion cycle is found in the machinery and equipment industry, with an average of 251.77 days.

Table 1. Descriptive statistics (mean values of variables).

Industry	No. of companies	Percentage	ROA	CCC	CR	SZ	FA	DR
Wholesale/retail	62	18.79%	0.050423	195.8492	4.967184	16.19128	0.62271	0.270808
Construction	43	13.03%	0.013462	234.6195	3.052749	16.70483	0.587786	0.372402
Pharmaceuticals	5	1.52%	0.054837	166.135	2.780126	18.37409	0.363298	0.367251
Plastics	6	1.82%	0.054777	67.43166	1.163833	16.99983	0.636039	0.397582
Machinery and equipment	24	7.27%	-0.0004	251.7786	2.966687	17.15166	0.55625	0.408223
Metallurgy	9	2.73%	-0.0326	105.4802	1.262787	19.32892	0.566317	0.710661
Manufacturing	111	33.64%	0.016524	173.1453	5.104269	16.88651	0.534869	0.340303
Chemicals	5	1.52%	0.027331	111.3905	1.662096	18.47774	0.668075	0.42066
Basic resources	15	4.55%	0.059306	117.5791	3.514736	18.32981	0.643547	0.37898
Transportation and storage	14	4.24%	0.027475	112.573	4.000161	16.77677	0.675511	0.221699
Tourism	31	9.39%	-0.00725	196.7439	9.656167	16.30307	0.795546	0.187083
Utilities	5	1.52%	0.047511	46.8978	4.277155	21.7455	0.697744	0.24413
Full sample	330	100%	0.022272	182.647	4.713359	16.92038	0.598068	0.328367

It may be concluded that on an average, basic resources companies are facing the highest profitability (0.059), while metallurgy companies are facing the lowest profitability (-0.032), as represented by return on assets.

Figure 1 shows the evolution of cash conversion cycle for the Romanian companies between the years 2002 and 2016.

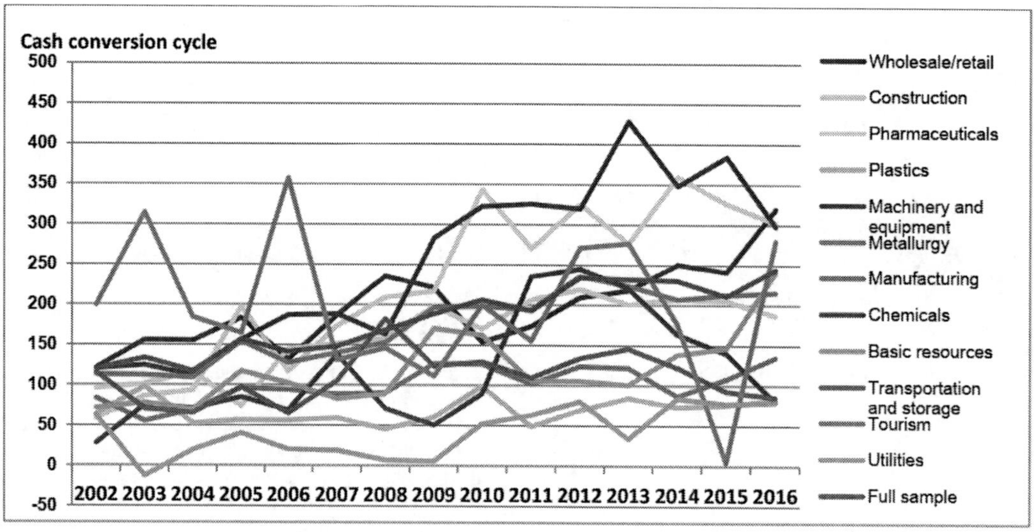

Fig.1 The evolution of cash conversion cycle for the Romanian companies from 2002 to 2016

It can be noticed an upward trend in cash conversion cycle values for the full sample of firms and at the same time utilities companies exhibit the lowest value in most years. Tourism companies are the most volatile in terms of cash conversion cycle values during the analyzed period.

The estimated coefficients based on equation (1) are presented in table 2. The results indicate that the length of the cash conversion cycle had a negative impact on the corporate performance, measured by return on assets for the full sample and for all the industries, except for metallurgy and utilities companies. This indicates that larger the cash conversion cycle is, smaller would be the firm profitability. These results are consistent with the results of the existing cash conversion cycle literature such as [8,12,13,14].

Table 2. Regression results.

* Corresponding author: ciprian.cristea@emd.utcluj.ro

Industry	Constant	CCC	CR	SZ	FA	DR	Adjusted R-squared	F-statistic
Wholesale/retail (Prob.)	0.207931 (0)	-0.00031 (0)	0.000333 (0.2618)	-0.006947 (0)	-0.066744 (0)	-0.01289 (0.0087)	0.209318	46.95729
Construction (Prob.)	0.134451 (0)	-0.00037 (0)	-0.000453 (0.6583)	7.99E-05 (0.9583)	-0.11739 (0)	-0.094129 (0)	0.204751	31.02083
Pharmaceuticals (Prob.)	-0.087492 (0.2303)	-0.00082 (0.1859)	0.011081 (0.1108)	0.00717 (0.0481)	0.025068 (0.6703)	-0.041546 (0.172)	0.225839	5.084099
Plastics (Prob.)	-0.074353 (0.6255)	-0.00336 (0.227)	0.038412 (0.2343)	-0.001904 (0.6441)	0.10558 (0.3144)	0.159371 (0.1332)	0.062080	1.93988
Machinery and equipment (Prob.)	0.001495 (0.9698)	-0.0073 (0)	-0.000836 (0.1532)	0.010107 (0)	-0.152555 (0)	-0.157801 (0)	0.266502	24.76174
Metallurgy (Prob.)	-0.182421 (0.3669)	0.00076 (0.7549)	0.00321 (0.6081)	0.008165 (0.286)	0.151732 (0.1906)	-0.121098 (0.0046)	0.118174	3.653409
Manufacturing (Prob.)	0.051867 (0.0019)	-0.00062 (0)	-0.000187 (0.0099)	0.005749 (0)	-0.141411 (0)	-0.119386 (0)	0.224692	91.88463
Chemicals (Prob.)	-0.229106 (0.2273)	-0.0019 (0.048)	0.017922 (0.2183)	0.021742 (0.0833)	-0.232924 (0.0524)	-0.100966 (0.6096)	0.085768	2.163299
Basic resources (Prob.)	0.21352 (0.0002)	-0.00097 (0.0445)	-0.001306 (0)	-0.001684 (0.3086)	-0.09007 (0.0009)	-0.147457 (0.0004)	0.176609	9.450929
Transportation and storage (Prob.)	-0.117851 (0.0722)	-0.00054 (0.1219)	-0.000975 (0.4917)	0.017414 (0)	-0.166038 (0)	-0.126479 (0.0024)	0.369132	22.06427
Tourism (Prob.)	0.056362 (0.2238)	-0.00187 (0.0897)	-0.00026 (0.0024)	0.006603 (0.0071)	-0.141226 (0)	-0.210717 (0)	0.23233	25.81674
Utilities (Prob.)	0.132428 (0.05)	0.00084 (0.0343)	-0.002831 (0.0158)	0.002497 (0.4753)	-0.15672 (0.0001)	-0.10336 (0.0001)	0.344105	7.190672
Full sample (Prob.)	0.074456 (0)	-0.00035 (0)	-0.000183 (0.0006)	0.002962 (0)	-0.108511 (0)	-0.093308 (0)	0.16975	185.2158

Results of the debt ratio show that the debt level of a company has a significant negative effect on the company's performance for the full sample and for all the industries of the study except plastics companies, where the findings are not statistically significant. The results reported show a negative relationship between the ability of a company to cover its short-term liabilities with its current assets and profitability for the full sample and for all the industries of the paper except wholesale/retail pharmaceuticals, plastics, metallurgies and chemical companies, where the findings are not statistically significant.

4 Conclusions

The cash conversion cycle is an important performance metric for analyzing how well a company is managing its working capital. In this paper, a negative relationship was found between the cash conversion cycle and corporate performance for the full sample of companies, and for all industries except of metallurgy and utilities. The results also show that the debt ratio and current ratio have a negative effect on the enterprise's performance for the full sample. These findings recommend that managers of the Romanian companies can enhance the profitability of their firms by lessening the cash conversion cycle.

References

1. V. D. Richards and E. J. Laughlin, Financ. Manage., **9**(1), 32–38 (1980)
2. C. R. Sathyamoorthi, L. B. Wally-Dima, J. Manage. Res., **7**(5), 7–24 (2008)
3. P. D. Hutchison, M. T. Farris II, S. B. Anders, CPA J., **77**(8), 42-47 (2007)
4. J. R. Kroes, and A. S. Manikas, Int. J. Prod. Econ., **148**, 37–50 (2014)
5. Ukaegbu B., Res. Int. Bus. Financ., **31**, 1–16 (2014)
6. J. C. Van Horne, J. M. Wachowicz, Fundamental of Financial Management (12th ed., Prentice Hall, 2004)
7. R. Anser and Q. A. Malik, JBM, **8**(2), 83-87 (2013)
8. M. Deloof, J. Bus. Finan. Account., **30**, 573-587 (2003)
9. K. Padachi, Int. Rev Bus. Res. Pap., **2**(2), 45–58 (2006)
10. D. P. Koumanakos, Int. J. Prod. Perform. Manag., **57**(5), 335–369 (2008)
11. P. J. G. Teruel and P. M. Solano, Int. J. Manag. Financ., **3**(2), 164-177 (2007)
12. M. Öner, Res. J. Polit. Econ. Manag., **4**, 63–79 (2016)
13. H. Nobanee, M. Abdullatif, M. AlHajjar, Asian Rev. Account., **19**(2), 147-156 (2011)
14. J.J.Ebben, A.C.Johnson, J.Small Bus Entrep, **24**(3), 380-396 (2011)
15. M. Muscettola, Int. J. Bus. Manag., **9**, 25–35 (2014)
16. I. Lazaridis, D. Tryfonidis, J. Financ. Manag. Anal., **19** (1), 26–35 (2006)

MATEC Web of Conferences **184**, 04010 (2018)
Annual Session of Scientific Papers IMT ORADEA 2018
https://doi.org/10.1051/matecconf/201818404010

The requirement elicitation process of designing a collaborative environment – The CRE@TIVE.BIZ case

Lidia Alexa [1,*], and *Silvia* Avasilcai[1]

[1]"Gheorghe Asachi" Technical University of Iasi, Department of Engineering and Management, Romania

Abstract. The software development industry is one that is under a constant pressure to deliver products that are innovative, fast to market and better suited to the costumers' needs and expectations. In this context, the traditional approaches to product development can become obsolete, as the market and the end-users are dynamic and rather unpredictable and the latter's active participation can sometimes make the difference between success and failure. The Agile approach, on the other hand considers extensive collaboration with all stakeholders paramount in all the stages of the software development process and for this, a variety of requirements elicitations techniques are used in order to obtain extensive input that leads to designing and developing enhanced products and applications. This paper presents the requirement elicitation process and techniques used in the development process of an open collaboration and innovation platform – CRE@TIVE.BIZ which is an experimental model dedicated to create a community where all relevant stakeholders from three specific creative industries, namely design, fashion design and crafts, can meet, interact, gain access and share information, knowledge and resources.

1 Requirements elicitation process. The Agile framework

The process of requirements elicitation (RE) is generally accepted as one of the critical activities in software development and it refers to the specific manner used in developing, analyzing and documenting all the system's requirements from stakeholders. Getting the right requirements is considered to be a vital, but difficult part of software development projects [1], as it deals with the significant problem of designing the right software for the customer [2].

Requirements elicitation is all about learning and understanding the needs of users and project stakeholders with the ultimate aim of communicating these needs to the system developers. A substantial part of elicitation is dedicated to uncovering, extracting and surfacing the wants of the potential stakeholders [3].

The conventional methods to the requirements elicitation process used to focus on gathering all the system requirements and preparing the requirements specification documents up front before proceeding to the design phase, but in practice, this approach proves to be detrimental to the development process, as it leaves no room to accommodate any kind of change, which is inevitable due to the business dynamics and the ongoing modifications of costumers' needs and expectations.

In this context, the agile approach to software development and requirements engineering [4] proves to be a better fit, as it welcomes changing requirements even late in the development cycle to accommodate the stakeholders needs.

Agile Methods (AMs) are a family of software development processes designed to deliver products on time, on budget, and with high quality and customer satisfaction, that have become popular during the last few years [5-7]. The aim of these methodologies is to deliver higher quality products faster and better fitting to the costumers needs and this is done through extensive collaboration with the project's stakeholders. This element is one of the pivotal points of the Agile Manifesto which states that the main priority should be given to "individuals and interaction over processes and tools, working software over comprehensive documentation, customer collaboration over contract negotiation, and responding to changes over following a plan" [6].

Agile Methods enforce very little upfront requirements elicitation but instead advocate incremental and iterative discovery throughout and integrated with the software development lifecycle [8], by prioritizing and delivering the most important functionalities first.

In most cases the process of requirements elicitation is performed incrementally over multiple sessions, iteratively to increasing levels of detail, and at least partially in parallel with other system development activities. Typically, the result of this process is a detailed set of requirements in natural language text and simple diagrammatic representations with additional information including descriptions of the sources, priorities, and rationales [3].

Plus, due to the complexity of most software systems, the technical and non-technical requirements can come from a variety of sources: problem owners, project documentation, stakeholders, and other existing systems

* Corresponding author: lidia.alexa@tuiasi.ro

© The Authors, published by EDP Sciences. This is an open access article distributed under the terms of the Creative Commons Attribution License 4.0 (http://creativecommons.org/licenses/by/4.0/).

and so, many of the effective techniques used in the requirements elicitation process do not originate from the traditional areas of software engineering or computer science research but from social sciences, organizational theory, group dynamics, knowledge engineering, and very often from practical experience [3] and are being embedded in iteration-based agile methods [9].

The right combination of elicitation technique and clear proposed objective produce quality software whereby clear proposed objective after doing requirements elicitation is essential to the success of software development projects [10].

2 Short description of CRE@TIVE.BIZ project

CRE@TIVE.BIZ is a complex and innovative platform developed through a research project implemented by the Engineering and Management Department of "Gheorghe Asachi" Technical University of Iasi.

The goal of the project is to develop an experimental model dedicated to creating a community where all relevant stakeholders from specific creative industries can meet, interact, gain access and share information, knowledge and resources.

The platform has three main components:

- **A crowdsourcing component** aimed at engaging relevant stakeholders from creative industries for a common goal, in terms of knowledge discovery & management, distributed human intelligence tasking, *peer-vetted creative production and broadcast search. Through this platform section, users can* address questions and find answers, share creative ideas in order to receive feedback, find solutions to different problems etc.
- **A marketplace component** which is a brokerage section that has the objective to pair offers from support services and raw materials providers with companies from creative industries. *Through this platform section, users can find* relevant providers of products or services needed and can buy and sale products and services related to creative industries.
- **An e-shop component** that offers the opportunity for each company or individual that has an account on the platform to create its own virtual shop in order to present and sell the products to end-users.

3 The requirement elicitation process for CRE@TIVE.BIZ

According to the literature, the typical activities of the requirements elicitation process can be divided into five fundamental types [3]: understanding the application domain, identifying the sources of requirements, analyzing the stakeholders, selecting the techniques, approaches, and tools to use and eliciting the requirements from stakeholders and other sources.

Understanding the application domain – this phase refers to the detailed investigation and examination of the situation or "real world" in which the system will ultimately reside (sometimes called the application domain) [11].

In our case, the domain analysis, focusing on identifying missing online solutions and instruments that entrepreneurs and freelancers from creative industries can benefit from was conducted in 2017 as part of the research process to apply for UEFISCDI financing.

The outcomes of the understanding the application domain stage consisted in two types of documentations: an academic oriented documentation and a business oriented documentation.

The academic oriented documentation focused on studies and analysis conducted by the research team regarding business ecosystems, open innovation and crowdsourcing, as the domain knowledge in the form of detailed descriptions and examples plays an important part in the process of requirements elicitation [3].

The business oriented documentation consisting in: the mission statement of the project, the main goals of the project and implicitly of the platform (experimental model), a list of the main components of the platform, a list of basic functionalities of the platform, a list and general description of the target groups and the main stakeholders.

Another relevant aspect considered in this phase was the project's team experience in designing and developing innovative online instruments for business development, as it proved instrumental in every stage of the requirements elicitation process.

Identifying the sources of requirements - due to the complexity of software development projects requirements may be spread across many sources and exist in a variety of formats [12].

The process of investigating all the relevant sources of requirements for CRE@TIVE.BIZ platform development provided a variety of sources and documents to be used in the requirements elicitation process by the project team. The main sources identified were:

- platform stakeholders – they were further involved in the requirements elicitation process through requirements workshops, namely focus-groups, questionnaires and prototype testing.
- users and subject matter experts: in this case PhD students researching creative industries ecosystems, web designers and web programmers. The PhD students performed extensive research on their expertise subject and delivered reports. Web designers and web programmers participated in requirements workshops to identify, analyze and prioritize the most important features and tasks to be performed in each sprint and later on in the process in prototype testing. At the same time, their input was relevant regarding the aspects of the non functional requirements of the platform, such as security, performance, maintainability etc.
- existing systems and platforms – the academic project team performed a benchmark analysis on different platforms covering the three main components of the CRE@TIVE.BIZ project: crowdsourcing, marketplace and e-shop, to identify common functional and non-functional features.
- project and legal limitations were also taken into account as there are relevant to the implementation of different

features of the platform and they are listed in the financing contract signed between the University and UEFISCDI.

Analyzing the stakeholders - the *customer involvement and interaction* were declared the primary reasons for project success and limited failure [13] and the identification and description of relevant stakeholders that can appropriately define, clarify and prioritize requirements is paramount [14]. Stakeholders are people who have an interest in the system or are affected in some way by the development and implementation of the system and hence must be consulted during the requirements elicitation process [3].

Identifying and analyzing the stakeholders was an extensive and rather difficult process, due to project complexity, but in the end the stakeholders list included:
- companies and individuals active in the three target creative industries, namely: design, fashion design and crafts.
- companies and individuals that can provide support services, raw materials or products for creative industries.
- companies and individuals interested in purchasing products from creative industries companies.
- others, namely individuals interested in participating in crowdsourcing competitions and contests.

All stakeholders were considered to be users as for interactive systems, users play a central role in the elicitation process, because usability can only be defined in terms of the target user population [15].

The stakeholders' needs were analyzed and evaluated for a primary identification of the main requirements and then they were further involved in the requirements elicitation process through requirements workshops – focus groups, by completing questionnaires regarding the platform's functionalities and design and in the prototype testing phase.

Selecting the techniques, approaches, and tools to use - in practice it is generally accepted that most projects imply more than an individual requirements elicitation technique or approach. However, the variety of techniques to be employed depends on the specific context of the project and the organization developing it and is often a critical factor in the success of the elicitation process [15].

The techniques and tools employed throughout the requirements elicitation process were chosen taking into account the project team previous experience in developing online applications and the specifics of the stakeholders involved and consisted in:
- domain analysis
- brainstorming
- requirements workshops
- task analysis
- questionnaires
- prototype testing.

Eliciting the requirements from stakeholders and other sources - once all the sources of requirements and the specific stakeholders have been identified, the actual elicitation of the core requirements begun by using each of the techniques selected in the previous stage.

Domain analysis – in the first stage of the requirements elicitation process which took place in 2017 when the project team started working on the research project consisted in an extensive literature review meant to clarify the research problem and to identify the needs of entrepreneurs from creative industries and the existing online instruments/platforms companies can use for business development.

The domain analysis was followed by several **brainstorming** sessions which had the goal to develop the preliminary mission statement for the project and target system and to establish who are the platform's users, what are the main characteristics of the platform and what problem(s) will solve for the users.

At this stage, the project team has designed the main components and functionalities of the platform, based on the benchmarking analysis conducted.

Once the project received financing, the project team extended the use of requirements elicitation techniques to involve more stakeholders in the process through requirements workshops, task analysis, questionnaires and prototype testing.

The requirements workshops, namely **focus groups** were organized by the project team taking into account two main types of stakeholders/users: one of the focus groups with companies and individual activating in the three creative industries nominated in the project and another with web designers and web programmers. Each focus group had a different motivation and emphasis, but the common goal was to improve platform usability and utility. The focus group guide included questions divided into several subgroups covering: technical aspects, graphical design, contents, structure and navigation.

Each focus group was conducted by a member of the project team who had the facilitator role and apart from asking the questions included in the focus group guide and recording the answers, they also guided and assisted the participants in addressing the most relevant issues in order to receive accurate and complete requirement information. The results of the focus groups were analyzed and transformed in a detailed list of processes which were further decomposed.

After the focus group results examination, the project team conducted a **task analysis** which employed a top-down approach where high-level tasks were decomposed into subtasks. The main objectives in this stage was to construct the hierarchy of the tasks performed by the users in the platform and to determine the necessary knowledge required for users to carry them out. This was a relevant assignment, as the platform is an innovative instrument and the target groups may not fully understand how to use it in order to gain most advantages. Plus, the details resulted from the task analysis stage, were further included in the questionnaires.

The **questionnaires** were designed following QDF (quality function deployment) methodology which is used to translate the customer wants and needs into product or service design characteristics utilizing a relationship matrix. The questionnaires were sent to all the types of stakeholders identified.

Once the platform **prototype** has been designed it was tested with all types of stakeholders in order to receive relevant feedback regarding the functionalities developed and the overall platform usability.

4 Conclusions

The requirements elicitation process is starting to be recognized as a critical activity in any software development process, due to the constant demand for better, faster and more usable applications.

In this context, the project team analyzed and selected the elicitation techniques that better suit the project specific needs, the stakeholders involved and the team members expertize. The goal was to make sure that the requirements elicitation process is properly conducted and the results obtained in the activity will thoroughly cover all the relevant functional and non functional requirements of the platform.

In order to ensure this, the research team opted for a combination between domain analysis, brainstorming, requirements workshops – focus groups, task analysis, questionnaires and prototype testing and an ongoing active involvement with all relevant stakeholders.

At the same time, the requirements elicitation in agile approaches is an ongoing iterative process as any change in business dynamics can influence and produce adjustments in the development of the platform.

Acknowledgment

This work was supported by a grant of the Romanian National Authority for Scientific Research and Innovation, CNCS/CCCDI – UEFISCDI, project number PN-III-P2-2.1-PED-2016-0689, within PNCDI III.

References

1. C. Jones, *Applied Software Measurement: Assuring Productivity and Quality*, McGraw-Hill: New York, (1996).
2. P. Jalote, *An integrated approach to software engineering*. Springer (1997).
3. D. Zowghi, C. Coulin, Requirements Elicitation: A Survey of Techniques, Approaches, and Tools in BT *- Engineering and Managing Software Requirements*, pp. 19-46, (2005).
4. S. Ambler, *Agile Requirements Modeling*, available at: http://www.agilemodeling.com/essays/ agileRequirements.htm, (2012) accessed 4th of May 2018.
5. P. Abrahamsson, O. Salo, J. Ronkainen, J. Warsta *Agile software development methods: Review and analysis*. EPSOO 2002, VTT Publications 478, (2002).
6. K. Beck, M. Beedle, A. Bennekum, A. Cockburn, W. Cunningham, M. Fowler, J. Grenning, J. Highsmith, A. Hunt, R. Jeffries, J. Kern, B. Marick, RC. Martin, S. Mellor, K. Schwaber, J. Sutherland, D. Thomas *Manifesto for Agile software Development,* (2001), available at ttp://www.agilemanifesto.org/. Accessed on 12th of April 2018.
7. D. Cohen, M. Lindvall, P. Costa, Agile software development. DACS State-of-the-Art Report, Data and Analysis Center for Software (2003).
8. R.C. Martin, *Agile Software Development: Principles, Patterns, and Practices*, Prentice Hall: Upper Saddle River, (2003).
9. Y. Zhu, *Requirements engineering in an agile environment*, Uppsala University (2009).
10. K.R. de Oliveira, M. de Mesquita Spínola. POREI: patterns-oriented requirements elicitation integrated--proposal of a metamodel patterns-oriented for integration of the requirement elicitation process. 2007: ACM.
11. P. Zave, M. Jackson, Four dark corners of requirements engineering. *ACM Transactions on Software Engineering and Methodology*, 6 (1), pp. 1-30, (1997).
12. P. Loucopoulos, V. Karakostas, *Systems Requirements Engineering*, McGraw- Hill: London, (1995).
13. A. Eberlein, J. Cesar, S. do Prado Leite, Agile requirements definition: A view from requirements engineering. In *Proceedings of the international workshop on time constrained requirements engineering,* (2002).
14. M. Daneva, E. van der Veen C. Amrit, S. Ghaisas, K. Sikkel, R. Kumar, *et al.,* Agile requirements prioritization in large-scale outsourced system projects: An empirical study, *Journal of Systems and Software*, 86 (5) (2013), pp. 1333-1353.
15. B. Nuseibeh, S. Easterbrook, Requirements Engineering: A Roadmap, *Proceedings of the Conference on The Future of Software Engineering*, pp. 35-46, (2000).

Value co – creation through crowdsourcing: The case of Squadhelp competition platform

Elena Galateanu (Avram)[1*] and *Silvia* Avasilcai[1]

[1]"Gheorghe Asachi" Technical University of Iasi, Department of Engineering and Management, Blvd D. Mangeron 29, 700050, Iasi, Romania

Abstract. The increasing attention for open innovation processes within organizational development context is materialized through new product and/or services development. The main approach into this direction is to enhance the crowdsourcing features within internal innovation processes, especially value co–creation. The emerging industries, such as creative ones, are using this approach in order to innovate and to enhance collaborative power of creative co–producers. By shifting the co–creation from customers only to multiple stakeholders' engagement there can be observed the actors' evolution within the innovation ecosystems. Thus, in this paper there will be assessed the key features of crowdsourcing processes in terms of the most relevant transitions in value co–creation theory. The use of case study approach will reveal relevant insights from the Squadhelp competition platform in terms of innovation, crowdsourcing activities and creative co–producers.

1 Introduction

Innovative products/ services development requires the use of different approaches from the companies' point of view. By participating within a specific business ecosystem, the economic entities acknowledged the major relevance of collaboration enhancement and the use of open innovation and crowds as the engines for boosting the creative potential and innovation processes.

As the concept of business ecosystem evolved into the innovative one, there can be traced various innovation techniques which were used by the leaders within different industries. However, there still remains the necessity to understand how companies co–create value with their partners, especially in case of innovation ecosystems. The most relevant stakeholders within an innovation ecosystem are often seen as resource providers [1]. Within this context the crowds are taken for granted, as an integrative part of innovation ecosystem [1]. From this point of view, this paper aims to present the most relevant key features of innovation processes, to underline the major importance of open innovation within an innovation ecosystem, to identify the specific crowdsourcing activities within an emergent creative ecosystem and to provide an illustrative example by using the case study approach on the Squadhelp competition platform.

2 Innovation ecosystems and the creative potential of the emergent industries

Innovation ecosystems emerged as result from business ecosystems to more specific type of ecosystems, whose main concern is innovation [1]. According to Adner an innovation ecosystem has a customer-centric structure [2], where all involved actors are working collectively in order to create and deliver appropriate value [3]. Such value can be emphasized through the enhancement of creative potential and technology development as it was stated by E. Galateanu (Avram) and S. Avasilcai [4]. Adner and Kapoor emphasized the major importance of technological challenges that needs to be overcome in order for end users to benefit from its development and implementation [5]. Oh, Philips, Park and Lee suggested that the use of innovation ecosystem can be supported by several reasons, such as:

- Different value created by different actors,
- The performance of this system depends on continuous development
- Establishing relationships between actors and emergence of innovative strategies through continuous and voluntary involvement of human resource [6].

The emergent industries can be perceived as a potential source of creativity, the most relevant feature lying in innovation creation. From this point of view Lazzeretti stated that the creative potential can be achieved by a group of individuals by quantifying their feelings and perceptions which bestow on the creative processes a leadership position [4], [7]. By following this statement there can be traced an important transition from customer co–creation to multiple stakeholders' involvement into value creation process in order to unlock the creative potential.

2.1 Co – creation transitions: The key features and motivations

* Corresponding author: egalateanu@tuiasi.ro

© The Authors, published by EDP Sciences. This is an open access article distributed under the terms of the Creative Commons Attribution License 4.0 (http://creativecommons.org/licenses/by/4.0/).

Pera, Occhiocupo and Clarke explored the importance of value co–creation transitions within business ecosystems context. The authors identified two main development trends [8]:

- From consumers to multiple stakeholders' involvement
- From various stakeholders to ecosystems stakeholders.

According to their findings, the relations between various stakeholders can be considered as the main trigger for value creation, especially value co–creation.
The first transition emphasizes the shift from consumer to multiple stakeholders' engagement into value creation process, presented in table 1.

Table 1. The first transition: from consumers to multiple stakeholders

Key Aspects	Description
Prerequisites	• Potential and existing consumers as the main target
Engaging Actors	• Consumers and suppliers
Process	• Knowledge sharing • Collaborative actions • Direct links between actors
Results	• Problem formulation • Negotiation • Resources allocation
Roles	– Co–designers, co–creators, co-producers, co-implementors
Landscape	– Through entire network of business networks

This approach is more likely to characterize the transition from supply chain to value chain, where consumers represent the source of innovation [9].

The second transition implies an ecosystem's point of view. The engagement of multiple stakeholders into value creation process underlines their contributions within a business ecosystem context. From this point of view the key aspects are adaptation and flexibility of the actors [8], presented in table 2.

Table 2. The second transition: from multiple stakeholders to business ecosystem [8]

Key Aspects	Description
Prerequisites	• Stakeholders relationships
Engaging Actors	• A system of stakeholders from different industries
Process	• Adaptation and evolution • Value exchange • Direct links between actors from different networks
Results	• The success of the business ecosystem • Resources integration
Roles	– Co–creators of value
Landscape	– Through entire business ecosystem

From this point of view the value creation process became more complex and is described by the engagement of actors from various business networks. According to Merz, Zarantonello and Grappi the perceived value can be created only by common effort of multiple actors [10]. Customers play an important role within a co–creation process as they are a potential source of design, requirements and customized new products [11].
However, stakeholders' engagement into value co–creation is relevant in terms of expected benefits and Perra, Occhiocupo and Clarke identified a few motives for stakeholders' engagement as follows:

- Exploiting the creative potential of relevant customers
- The use of resources from other stakeholders
- "the constellation of services, goods and design" [8, pp 4035].

Finnerty, Kucherbaev, Tranquillini and Convertino stated that the quality of crowdsourcing tasks or projects relies on the workers motivation [12]. From this point of view the authors introduced a reward – based schema and identified a few types of rewards, as presented in table 3.

Table 3. The second transition: from multiple stakeholders to business ecosystem [12]

Reward Type	Description
No reward	• Without extra reward for project's completion
Courteousness	• Use the word "please" • Asking the participants to finish the tasks in terms of time and accuracy
Fixed	• Gaining benefits despite their poor performance, if it the case
Dynamic	• Different rewards • Calculated based on participants performance • Quicker tasks completion means a higher reward

2.2 The use of crowdsourcing in value co – creation process

The use crowdsourcing mechanisms provides valuable insights into the value co–creation process and the exploitation of creative potential. Dai, Wang, Jin and Ma suggested that crowdsourcing is especially important in terms of performed tasks [13]. The authors stated that there should be identified three main components for crowdsourcing activities: specific tasks, participants and the platform, which facilitates and links the participants to the tasks [13].

Alexa and Avasilcai identified two types of participants in crowdsourcing activities [14]:

- Seekers – those who propose the tasks as specially designed projects
- Solvers – those who provide the creative potential, it can be anybody with no restriction.

The most relevant aspect, as it was pointed by Dai, Wang, Jin and Ma [13], is the decision-making process. Accordingly, Zhang, Shangguan and Yuan linked this strategy to the crowdsourcing platform and stated that the decision–making process is time limited and it should provide fast results to the specific tasks [15]. From this point of view in crowdsourcing is preferred the approach where individual capabilities are matching the specific provided tasks [16].

Crowdsourcing can be seen also as an innovative type of recruitment of potential creative workforce. According to Behrend, Sharek, Meade and Wiebe the crowdsourcing comprises three main features [17]:

- The remuneration of the solvers
- Online recruitment, not restricted by geographical criteria
- The solvers are recruited only for specific tasks.

From this point of view crowdsourcing is narrowing the attention to the target groups [18] and is presenting interesting insights in comparison to open innovation, as follows in table 4.

Table 4 The main differences between open innovation and crowdsourcing [18].

Key Aspects	Open Innovation	Crowdsourcing
Main Target	• Customers' integration • Innovation process within the companies	• Different target groups • A broader coverage of processes
Strategy Actors	• Interaction with different types of stakeholders	• Interaction between the company and the crowd
Type of call	• Predefined by a contract	• Open call launched to the crowd
Openness	• Open type • Defines the base for product development	• Closed type • Intellectual property rights
Purpose	– Collective improvement of the product – Granting access to specific elements of the product	– Creating items individually and / or collaboratively

3 Value co – creation within the Squadhelp community

Squadhelp represents a community of freelancers, namely experts, whose main concern is to innovate and provide a full service and support for brands construction for companies, start-ups, and other actors [19]. The main objective of this community is to provide innovation of traditional agency business model [19]. Established in 2011, Squadhelp nowadays numbers a community of over 70 thousand creatives who contribute to creation, validation and growing new brands [19].

According to their vision there are several aspects which differentiate the community, such as:

- Provides agency experience without specific price tag – through naming contest the clients benefit from support of branding consultants without the implication of special agencies.
- Branding support – through specially designed services Squadhelp allows their clients to benefit from targeting their potential customers, test their choices, avoid legal risks and secure their trademark
- Ensures the highest quality – ensures quality of generated ideas through specially designed algorithm, which provides a ranking for potential creatives

3.1 Squadhelp provided services

As it was stated before the Squadhelp competition platform offers the most quality and comprehensive package of related and support services. The wide range of specially designed services represents an essential difference between competition platforms and competition marketplaces. From this point of view the community includes services as it is shown in table 5, which are related to co – ideation of names or logos, validation of the submitted ideas, online verification of ideas availability, legal advising services, validation of the chosen name and legal support service for trademark or name registration [20] – [26].

3.2 Co – creation within Squadhelp community

The current theories emphasize the relevance of co–creators within a value co–creation process. From this point of view there should be identified the most important actors who participate in the co–designing process within the Squadhelp community.

According to their vision in the value co–creation process there are involved:

- A community of designers – especially for logo designing and naming services
- A community of branding consultants – are relevant for branding services, as well as for platinum customers
- Legal advisors – provide legal assessment of trademarks, assistance in trademark registration process
- Customers – Squadhelp addresses their services to a wide range of actors: companies, small enterprises, start–ups who are engaging into branding activities or are new in the market.

Each of these actors are involved differently into the value co–creation process according to the provided service.

3.2.1 Value co – creation process within the Squadhelp community

According to the crowdsourcing theory value co – creation can embrace the form of competition/contest. From this point of view the Squadhelp defines the co–creation process through several essential stages [27]:

- Start the competition – represents the stage

Table 5 Squadhelp provided services [20] – [26]

Targeting services	Specific actions
Naming	• Contest activities • Interactive observation of active contests • Defining filters criteria • Addresses to different industries
Logo design	• Creation of logo through contests • Offers a logo template • Limited timeline of the contest (not exceed 7 days) • Interactive feedback
Taglines	• Maximum 5 days contest timeline • Engage creative potential of the community • Offers tagline template • Checking the trademark for any conflicts
Business names for sale	• Complementary trademark evaluation action • Data base of name for sale • Narrowing filters by: category, style, length, keywords • Fixed and low pricing package • Offers names for business and brands, and names for premium domain
Audience testing	– Identifies the targeted clients – Fast and reliable feedback from the targeted audience – Resonates the chosen name with clients' perception – Use specific instruments for data collection – Provides an extensive demographic research – Duration of the research up to 3 days
Trademark research & filing	– Legal assessment of the trademark and chosen name – The use of online or mobile channels of communication – Provides support services for trademark registration (U. S. Patent and Trademark Office USPTO) and efficient use
Managed agency services	– Platinum type of competitions – Provides the advising services throughout the entire project – Assistance in contest\s requirements or project – Granting unlimited access to the client – Support and legal services applied

where the customer provides a brief description of the future project. For this stage Squadhelp implemented a specially designed template which comprises the information about basic requirements and which will be distributed to the community

- Ideas generation – at this stage the customer will receive dozens of ideas according to the specified requirements. At the same time those ideas (names or logos) will be verified online for their availability
- "collaborate and communicate" – describe the interactive communication between the customer and the designers. At this stage the customer or project holder should provide. feedback for designers, comment the ideas privately, rank the submissions
- Validation stage – at this stage the customer benefit from assessment process of the chosen name, domain, legal risk, trademark, linguistic evaluation and target groups testing
- Choose the winner – this stage provides information about the winning idea. The customer should choose the best idea and start the registration process of the name.

3.2.2 Contest content

Each started contest comprises basic information such as the brief of the contest, to whom it is addressing, provided reward and the type of reward (guaranteed reward) [28]. Along with the contest briefing there can be traced a short description of the contest holder and the contest timeline.

Also, there can be found statistics of submitted, withdrawn and active ideas and number of rated entries. As the co–creation process is an interactive one and the customer should provide feedback all along the entire submission process, there can be find the statistics of customers'' feedback. From this point of view the contest holder should respond with qualificatives as: "Love it", "Like it", "On right track" and "No, thank you".

3.2.3 Contest pricing packages

At this point each potential contest holder or seeker have the opportunity to choose the pricing package for the future contest. From this point of view Squadhelp provides four different packages [29]:

- Bronze – provided services do not exceed specified budget. According to this package the contest holder benefit from a low budget pricing, at least 100 ideas submitted and online matching availability of the name and domain
- Gold – includes up to 300 expected submissions, along with verification of domain

availability the seeker will benefit also from assistance for trademark assessment, strict privacy terms and promotion on Squadhelp homepage

- Platinum – according to this package the contest holder should expect around 400 submitted ideas. It also includes audience assessment and testing of chosen names, a legal trademark assessment and validation. This package includes almost all services provided by Squadhelp which are also: linguistic assessment, refund options and the use of collaborative tools;
- Managed – this exclusive package promotes one–on–one coaching with branding experts. Basically, this approach is preferred for the seekers who chose to acquire an agency experience.

Based on the provided pricing packages the customers can chose the services they want to benefit from the competition platform.

3.2.4 Additional crowdsourcing instruments and resources

Although the Squadhelp provides an interactive contests platform, additionally it also comprises a few instruments and resources which are relevant for potential or already existing customers. From this point of view in order to choose a name the seeker can chose from names database and receive a suggestion from the marketplace.

Also, there can be found the information for contest holders and for participants. In first case the Squadhelp provides tips for contest holders in terms of ranking choices, how to provide fast and reliable feedback, information regarding domain registration and how to reward the contestants based on the quality of their work [30]. In this case the Squadhelp provides a unique pointing system which can be seen as a quality filter. The contestants there are provided with a few guidelines such as [31]:

- Read and research if the submission idea respects seeker's requirements
- Do not submit low quality ideas
- Comment and ask if there is the case
- Use the provided resources to come up with a new name
- Use the activity feed frequently in order to provide high quality ideas and to adjust the entries.

Along with these general guidelines there can be found also a discussion forum where there can be traced all relevant and frequently asked questions [32] and a blog which comprises all tips and novelties in terms of naming, branding and logo creation for creatives [33].

4 Conclusions

Innovation ecosystems are relevant in the current research as they provide valuable insights in terms of innovation diffusion across and within an organisation. Nowadays, innovation became an essential key feature which describes the most relevant internal or external processes. From this point of view value co–creation represents a development process which promotes innovation. However, there should be pointed that the shift from a single customer to multiple stakeholders' involvement into co–creation process had a major impact on organisational activity. The use of crowdsourcing within value co–creation emphasised the birth of new types of customers naming seekers (contest holders) and solvers (contestants or participants). Based on contest type of value co–creation there can be enhanced the creative potential of young designers or branding experts as it is illustrated by the Squadhelp competition platform. The key feature which differentiate it from a crowdsourcing marketplace is the wide range of provided services. The Squadhelp provides integrative, all in one, branding services starting from choosing the business or brand's name until research, assessment and its validation. Additionally, they offer legal risk assessment and assistance in trademark registration at Trademark Office. Different pricing packages influences contest holders' benefits.

This work was supported by a grant of the Romanian National Authority for Scientific Research and Innovation, CNCS/CCCDI – UEFISCDI, project number PN-III-P2-2.1-PED-2016-0689, within PNCDI III

References

1. L. Scaringella, A. Radziwon, Innovation, entrepreneurial, knowledge, and business ecosystems: Old wine in new bottles? *Technological Forecasting & Social Change* (2017) retrieved from: /10.1016/j.techfore.2017.09.023, to be published
2. R. Adner, Match your innovation strategy to your innovation ecosystem, *Harvard Business Review*, **84 (4),** pp 98 – 107, (2006)
3. B. Walrave, M. Talmar, K. Podoynitsyna, G. Romme, G. Verbong, A multi – level perspective on innovation ecosystems for path – breaking innovation, *Technological Forecasting and Social Change,* (2017) retrieved from: 10.1016/j.techfore.2017.04.011, to be published
4. E. Galateanu (Avram) and S. Avasilcai, Emerging creative ecosystems: platform development process, *Annals of the Oradea University, Fascicle of Management and Technological Engineering*, **XXVI (XVI), 3**, pp 5 – 10 (2017)
5. R. Adner and R. Kapoor, Innovation ecosystems and the pace of substitution: Re – examining technology S – curves, *Strategic Management Journal*, **37 (4),** pp 625 – 648, (2016)
6. D. S. Oh, F. Phillips, S. Park and E. Lee, Innovation ecosystems: A critical analysis, *Technovation* **54**, pp 1 – 6, (2016)
7. L. Lazzeretti, "Cultural and Creative Industries" in *Creative Industries and Innovation in Europe: Concepts, Measures and Comparative Case Studies,*

L. Lazzeretti, Ed. Routledge, London and New York, pp. 1 – 20 (2013)

8. R. Perra, N. Occhiocupo, J. Clarke, Motives and resources for value co – creation in a multi – stakeholder ecosystem: A managerial perspective, *Journal of Business Research* **69 (10),** pp 4033 – 4041 (2016)
9. E. Galateanu (Avram), S. Avasilcai, Value co – creation process in business ecosystem, *Annals of the Oradea University, Fascicle of Management and Technological Engineering* **XXIII (XIII), 3,** pp 169 – 174 (2014)
10. M. Merz, L. Zarantonello, S. Grappi, how valuable are your customers in the brand value co – creation process? The development of a customer co – creation value (CCCV) scale, *Journal of Business Research* **82**, pp 79 – 89 (2018)
11. E. Galateanu (Avram), S. Avasilcai, Symbiosis process in business ecosystem, *Advanced Materials Research* **1036**, pp 1066 – 1071 (2014)
12. A. Finnerty, P. Kucherbaev, S. Tranquillini, G. Convertino, Keep it simple: Reward and task design in crowdsourcing, Proceedings of
13. W. Dai, Y. Wang, Q. Jin, J. Ma, An integrated incentive framework for mobile crowdsourced sensing, *Tsinghua Science and Technology* **21 (2),** pp 146 – 156, (2016)
14. L. Alexa, S. Avasilcai, A. Bujor, An open innovation marketplace: The case of Innocentive, *Annals of the Oradea University, Fascicle of Management and Technological Engineering*, **XXVI (XVI), 3**, pp 1 – 4 (2017)
15. X. Zhang, L. Shangguan, Y. Yuan, A crowd wisdom management framework for crowdsourcing systems, *IEEE Access* **4**, pp 9764 – 9774, (2016)
16. M. Z. Tunio, H. Luo, W. Cong, A. R. Gilal, A. Abro, S. Wenhua, Impact of personality on task selection in crowdsourcing software development: A sorting approach, *IEEE Access* **5**, pp 18287 – 18294, (2017)
17. T. Behrend, D. Sharek, A. Meade, E. Wiebe, The viability of crowdsourcing for survey research, *Behaviour Research Methods* **43**, pp 800 – 813 (2011)
18. Y. Zhao, Q. Zhu, Evaluation on crowdsourcing research: Current status and future direction, *Information System Frontiers* **16**, pp 417 – 434 (2014)
19. Squadhelp, about us, retrieved from: https://www.squadhelp.com/AboutUs, accessed at: 19.04.2018 (2018)
20. Squadhelp, naming service, retrieved from: https://www.squadhelp.com/branding-marketing-naming-contests#contest_type_filter=1, accessed at 19.04.2018 (2018)
21. Squadhelp, Logo design, retrieved from: https://www.squadhelp.com/LogoDesignContests, accessed at: 19.04.2018 (2018)
22. Squadhelp, taglines, retrieved from: https://www.squadhelp.com/AdvertisingSlogansTaglines, accessed at 20.04.2018 (2018)
23. Squadhelp, names for sale, retrieved from: https://www.squadhelp.com/premium-domains-for-sale?q=&hPP=10&idx=Marketplace&p=0, accessed at 20.04.2018 (2018)
24. Squadhelp, audience testing, retrieved from: https://www.squadhelp.com/brand-name-testing, accessed at: 20.04.2018 (2018)
25. Squadhelp, trademark research & filing, retrieved from: http://helpdesk.squadhelp.com/squadhelp-services/trademark-filing-package, accessed at 21.04.2018 (2018)
26. Squadhelp, managed contest service, retrieved from: http://helpdesk.squadhelp.com/squadhelp-services/squadhelp-managed-contest-service, accessed at 21.04.2018 (2018)
27. Squadhelp, how it works? retrieved from: https://www.squadhelp.com/how-it-works, accessed at 22.04.2018 (2018)
28. Squadhelp, contest content, retrieved from: https://www.squadhelp.com/branding-contest/16095/Catchy-name-for-a-consumer-credit/-finance-Website, accessed at 22.04.2018 (2018)
29. Squadhelp, pricing, retrieved from: https://www.squadhelp.com/squadhelp-pricing, accessed at 22.04.2018 (2018)
30. Squadhelp, tips for contest holders, retrieved from: https://www.squadhelp.com/tips-for-contest-holders, accessed at 22.04.2018 (2018)
31. Squadhelp, tips for contestants, retrieved from: https://www.squadhelp.com/tips-for-creatives, accessed at 22.04.2018 (2018)
32. Squadhelp, discussion forum, retrieved from: http://discussion.squadhelp.com/, accessed at 22.04.2018 (2018)
33. Squadhelp, blog, retrieved from: https://www.squadhelp.com/blog/, accessed at 22.04.2018 (2018)

MATEC Web of Conferences **184**, 04012 (2018) https://doi.org/10.1051/matecconf/201818404012
Annual Session of Scientific Papers IMT ORADEA 2018

Occupational health and safety risk management

Ioan-Florin Oarga[1,*], *Mariana* Rațiu[2], and *Ioan-Tudor* Oarga[3]

[1]Technical University of Cluj-Napoca, Faculty of Machine Building, 400641 Cluj-Napoca, Romania
[2]University of Oradea, Department of Mechanical Engineering and Automotive, 410087 Oradea, Romania
[3]Technical University of Cluj-Napoca, Faculty of Automation and Computer Science, 400027 Cluj-Napoca, Romania

Abstract. The ongoing industrial revolution also includes the health and safety domain, which assists this development process. The participants and beneficiaries of this revolution don't have to endure a lowering of the protection levels from the beginning of it. Consequently, the health and safety domain needs to play its role in this historical stage. The risks of injury and illness are managed from the beginning, when the technological processes are being developed, on each production stage, until the end of every technological process. In order to achieve this, the professional risks must be identified, quantified, ranked and evaluated such that, the elaborated measures will eliminate or decrease the risks from the source. This control process of occupational health and safety risks of injury or illness uses documents which are applicable to any organization. These documents are legally required in each state, being instruments of the organization management, working like a system along with all other documents.

1 Risk

We continue to search for the root of the word "risk". A very good synthesis of searches is found on the free encyclopedia Wikipedia [1], which refers to the word "risicu", dating back to Homer's times (11th century). "Risicu" was used to describe threats faced by characters. In the 13th-14th century, the term of commercial risk was used in Italy, after which appeared in France, in the maritime vocabulary (17th century). Oxford English Dictionary cites the oldest use of the word "risqué" in 1621.

1.1 Risk as a concept

1.1.1 Risk or the possibility of loss

The risk is an uncertain event which leads to the possibility of loss, in other words, the occurrence of negative circumstances leading to damage [2]. This future event is used in many areas, starting with commercial, financial, medical, insurance, environment protection, informational, business and last but not least, occupational health and safety.

1.1.2 Risk measurement

The quantitative analysis of risk. There are many formal methods provided to measure the risk. This is a necessity to decide the importance of the risk and the costs required to avoid or decrease it.

The probability of an event or a negative act to occur is estimated through the frequency of the last similar events. Often, the risk is measured as an expected value for an unwanted result, described very well with the formula:

$$R = \sum_{\text{All the cases}} (\text{probability of the event}) \times (\text{expected loss in the case of the event happening}) \quad (1)$$

Statistically , the function of risk is defined as the expected value of a certain loss depending on the used decision rule to decide in an uncertain situation [3]. This risk value takes in account all the loss possibilities.

1.2 Occupational health and safety risk

1.2.1 Health, safety and environment

Health, safety and environment are three different domains, but in practice are often tied together. The strongest bond between them is that a single event can impact all three. Over time, different forms of risk analysis were developed to help understand the events and activities leading to human and environment health risks.

1.2.2 Occupational health and safety

From occupational health and safety point of view, the term "risk" can be defined as the most probable consequence of a hazard, combined with the risk or occurrence probability. As such, the risk is a function of danger and exposure [4].

* Corresponding author: florin.oarga@gmail.com

© The Authors, published by EDP Sciences. This is an open access article distributed under the terms of the Creative Commons Attribution License 4.0 (http://creativecommons.org/licenses/by/4.0/).

2 Risk management in occupational health and safety

2.1 Occupational health and safety in organizations

From occupational health and safety point of view, the "risk" is present in all organizations, regardless of their field of activity. This "risk" can be quantified using several methods, with the purpose of measuring it in relation with its occurrence probability and its maximum possible severity. It is true that the magnitude of the risk is directly influenced by the danger degree of the processes which are unfolding in the organization and the probability of an event occurring. The purpose is to decrease the injury and illness risks for the workers involved in the processes conducted by the organization. This risks mitigation is required in all the stages of the organization's operation: technological processes design, equipment acquisition, upgrading production lines, during conversion or upgrade of manufacturing flows, or even during equipment operation. There were cases when the production processes were radically changed, because the specific legislation became very severe from the health and safety point of view, imposing the risk mitigation. In some situations, whole industries have been abandoned, because the required measures to decrease the professional risks were too costly to continue the industrial activity. A relevant example is the asbestos industry and asbestos derived materials.

2.2 Documents of occupational health and safety system as a part of risk management

The ongoing industrial revolution also includes the health and safety domain, which assists this development process. The participants and beneficiaries of this revolution don't have to endure a lowering of the protection levels from the beginning of it. In the area of health and safety, everything is centred on identifying the professional injury and illness risks and applying measures to diminish the risks until disposal. Risk management uses a document system applicable to a specific organization. These documents are imposed by legal provisions of all states.

Identification, quantification, ranking and evaluation of occupational health and safety injury and illness risks are an important topic in the economic equation. Taking this in consideration, it is very important that the economic interest should not take precedence over safety considerations at work. This control process of occupational health and safety risks of injury or illness uses documents which are applicable to any organization. These documents are legally required in each state, being instruments of the organization management, working like a system along with all other documents.

In Romania, the legislation governing the occupational health and safety area is based on the provisions of Law 319/14.07.2006 [5] published in Official Monitor no. 646, July 26th, 2006. The Law 319/2006 puts into operation the requirements of the Framework Directive 89/391 [6] for occupational health and safety.

To manage the professional injury and illness risks and to apply health and safety legal provisions, the practice of the organizations led to the development of the documents that met the Romanian and EU legal requirements. These documents should start from authorization of the organization regarding occupational health and safety, before beginning the actual activity. For the situations mentioned in the Law no. 359/08.09.2004 [7], the authorization can be obtained using a simplified procedure, at the moment of registration in the Trade Register of individual persons, family association and legal entities. The proof of this authorization is the Certified Statement, issued under Art. 17 (1)(e) of Law no. 359/2004.

The identification of professional risks for each job is done using the Professional risk exposure sheet, which is filled in by the model required from G.D. no. 355/2007 by the employer, using certain information from each job. When the document is drafted, the leader of the workplace, the assigned worker or the health and safety inspector must participate. Often, human resources representatives participate as well. The next document is the one that quantifies the level of professional risk, using a method of evaluation for the injury or illness risk factors, from the health and safety point of view. This quantification is calculated using the chosen method, which takes in account the four elements of the work system [8]: executor, work task, work means and equipment and work environment by jobs. The evaluation method is chosen by the team responsible for the quantification process of the injury or illness risk. Methods used until this moment operate with statistics, mathematical, graphical and scalar methods, with the purpose of quantifying the risk and comparing it with the minimum, medium and maximum accepted level of workload. This comparison is aiming to offer the management real, concrete information about injury or illness risks, allowing to adopt the right measures of fighting and decreasing them. It is very important to know the fact that the evaluation process of professional injury and illness risks is done every time new technologies, techniques, methods or materials are being introduced in the work process. Also, the evaluation study of professional risks is resumed immediately after a work accident. Fighting and decreasing measures of professional risks are filled in the Prevention and protection plan [9], which has a minimum content imposed by legislation: identified professional risks and fighting measures (technical, organizational, sanitary or other) must be listed [10]. These Prevention and protection plans must keep up with the evolution of processes, technologies and working techniques which take place in an organization. Also, every time something new appears in the production process or a work accident happens, the employer is legally required to adjust the Prevention and protection plans, from the health and safety point of view, and their stipulations need to be introduced in training topics of the staff which participates in the production process. To put into operation the measures stipulated in the Prevention and protection plan, occupational health and safety Own Instructions are needed, having the role to complete the legal provisions in the area, taking into account the particularities of activity and enterprise, as well as job particularities and sharing them in the company only after they were approved by the employer [11]. These Own Instructions are included in job specific training topics of each workplace [12]. The topics should contain a training program of each participant to the production processes and testing programs. The participants training is marked in forms, called individual training sheets and have a minimal imposed content [13]. Another way to write down the trainings is the Verbal report of collective training, informing the workers having group activities [14]. These trainings, individual or collective, are completed by occupational health and safety provisions in job descriptions, internal regulations and collective labour agreements, including chapters of occupational health and safety. Their provisions have the role of harmonizing the work tasks, worker's, leader's and employer's health and safety responsibilities, to comply to the provisions of the Prevention and protection plan [15].

Occupational risk management uses measures listed by health and safety documents, the main objective of which is to reduce the risk of injury and illness until elimination. Current practice and legal requirements indicate that the employers should take into account the four elements of the work system, when organizing the activities: executor, work task, work means and equipment and work environment by jobs [16]. To accomplish this, almost always it is necessary to develop a Signalling plan, to complete the informing of the workers and all participants to the production activities [17]. These plans include types of signalling (permanent and occasional) which will be used in the workplace. Most often, they include pictograms (interdiction, warning, compulsory, first aid and emergency situations), acoustic and light signals, trails and markings on the floor, pressured vessels, pipes, code words and gestures. Other present documents in the health and safety area refers to workers equipment, with the purpose of decreasing the risk to which they are exposed [18], the evidence of jobs requiring additional medical exams, or posts with special authorizations. It is imperative to continuously supervise the health status of the workers. Monitoring their status is often the responsibility of specialized medical entities in labour medicine. The management system of occupational risks also includes documents that refer to the technical monitoring of the working equipment, by performing regular inspections and repairs for those that came out of the warranty period or have undergone modifications [19] from their initial state, those that enter the list of work places with serious and/or imminent danger, those entering the I.S.C.I.R. regime, technical equipment books and equipment (in the national language) and the charts of repair/maintenance of the work equipment.

Occupational risk management also includes documents that organize work on health and safety [20] by their own means or by using external services. In this management system, provision should also be made to organize activities during extreme temperatures [21], maternity protection [22], youth protection [23], the organization of mobile and temporary sites [24], the use of hazardous [25], biological [26], carcinogenic or mutagenic substances [27], explosive atmospheres [28].

The documents of the health and safety system as part of the risk management in the health and safety area, are aimed to regulate the activity of this domain in every organization. These documents should comply to the legal requirements and integrate in all documents, to have a systemic effect, with the main goal of decrease to zero the occupational risks.

3 Conclusion

Occupational risk management can be realized on every phase of investment, starting with designing of the work process, choosing the materials and working methods. This risk management has as a unique goal the mitigation to zero of the injury or illness risks. To deeply analyze occupational risks, good practice, together with current legal requirements focus on the four elements of the work system: executor, work task, work means and equipment and work environment by jobs. The analysis of occupational risks of injury or illness contains evaluation methods, which are using statistical, mathematical, graphical and scalar methods. Thus, managerial decisions can be oriented more towards one or more elements of the work system and can keep an equilibrum with the desired profit.

Occupational risks management is done using a set of well-defined documents, which are aimed to comply with the legal requirements, working along with all the other documents in an organization. The identification of professional risks for each job is done using the Professional risk exposure sheet. The next document is the one that quantifies the level of professional risk, using a method of evaluation for the injury or illness risk factors, from the health and safety point of view. Fighting and decreasing measures of professional risks are filled in the Prevention and protection plan. These Prevention and protection plans must keep up with the evolution of processes, technologies and working techniques which take place in an organization. To put into operation the measures stipulated in the Prevention and protection plan, occupational health and safety Own Instructions are needed, having the role to complete the legal provisions in the area. These Own Instructions are included in job specific training topics of each workplace. The topics should contain a training program of each participant to the production processes and testing programs. The participants training is marked in forms, called individual training sheets and have a minimal imposed content. The management system of occupational risks also includes documents that refer to the signalling plans, the evidence of jobs requiring additional medical exams, or posts with special authorizations, technical monitoring of the working equipment, by performing regular inspections and repairs for those that came out of the warranty period or have undergone modifications.

These documents are legally required in each state, being instruments of the organization management, working like a system along with all other documents. The main goal of decrease to zero the occupational risks.

References

1. https://ro.wikipedia.org/wiki/Risc#Etimologie
2. https://en.wikipedia.org/wiki/Oxford_English_Dictionary
3. https://ro.wikipedia.org/wiki/Risc#cite_note-2
4. https://ro.wikipedia.org/wiki/Risc#S%C4%83n%C4%83tate,_siguran%C8%9B%C4%83_%C8%99i_mediu
5. Law 319/14.07.2006 published in Official Monitor no. 646, July 26th, 2006 modified by including the modifications and completions of Law no. 51 of March 19th, 2012 and Decision no. 513 of July 4th, 2017
6. Framework Directive 89/391 of June 12th, 1989, on promotion and improvement measures of occupational health and safety
7. Law no. 359/8 Sept 2004 regarding simplification of registration formalities in the Trade Register of individual persons, family association and legal entities, as well as authorization of legal entities, with additional modifications and completions
8. G.D. 1425/2006 with subsequent amendments and completions. Methodological Norm for the Application of Law 319/2006. Art.15 (1)
9. G.D. 1425/2006 with subsequent amendments and completions. Methodological Norm for the Application of Law 319/2006.Secțiunea 7
10. G.D. 1425/2006 with subsequent amendments and completions. Methodological Norm for the Application of Law 319/2006.Art.46 (2)
11. G.D. 1425/2006 with subsequent amendments and completions. Methodological Norm for the Application of Law 319/2006.Art. 15(1)3
12. G.D. with subsequent amendments and completions. Methodological Norm for the Application of Law 319/2006. Art.15 (1)7
13. G.D. 1425/2006 with subsequent amendments and completions. Methodological Norm for the Application of Law 319/2006. Anexa 11
14. G.D. 1425/2006 with subsequent amendments and completions. Methodological Norm for the Application of Law 319/2006. Anexa 12
15. G.D. 1425/2006 with subsequent amendments and completions. Methodological Norm for the Application of Law 319/2006. Art.15 (1)5
16. G.D. 1425/2006 with subsequent amendments and completions. Methodological Norm for the Application of Law 319/2006.Art.15(1)
17. G.D. 971/2006 On minimum requirements for safety and / or health signaling at work. Art.1,5
18. G.D. 1048/2006 on minimum safety and health requirements at work for workers to use personal protective equipment at work
19. G.D. 1146/2006 regarding minimum health and safety requirements for using working equipments by workers
20. G.D. 1425/2006 with additional modifications and completions. Methodological Norm for the Application of Law 319/2006. Chapter III
21. G.E.O. 99/2000 on measures that can be applied in periods of extreme temperatures for the protection of employed persons
22. G.E.O. 96/2003 on the protection of maternity at work
23. O. no. 753/2006 on the protection of young people at work
24. G.D. 300/2006 on minimum safety and health requirements for temporary or mobile construction sites
25. G.D. 1408/2008 on the classification, packaging and labeling of dangerous substances
26. G.D. 1092/2006 on the protection of workers from the risks related to exposure to biological agents at work
27. G.D. 1093/2006 laying down minimum safety and health requirements for the protection of workers from the risks related to exposure to carcinogens or mutagens at work
28. G.D. 1058/2006 on minimum requirements for improving the safety and health protection of workers potentially at risk from explosive atmospheres

Challenges in Romanian technical higher education from the perspective of the european requirements for its modernization

Felicia Veronica Banciu[1,*], *Eugen* Pămînţaş[1]

[1] Politehnica University Timisoara, Faculty of Mechanics, Materials and Manufacturing Engineering Dep., Bd. Mihai Viteazu No 1, 300222, Timişoara, Romania

Abstract. In today's economically changing environment, higher education institutions faces many challenges regarding permanent adaptations to industry needs, inappropriate financing arrangements, the knowledge triangle of teaching, research and innovation, tailored approaches to learning for students, students centred education, reforms in the education system etc. The problem mentioned in European agenda for higher education modernization linked to the inconsistency between the knowledge and skills graduates gain from the educational process and those required by the economy and society remains one major problem, at least at this moment. In regard to this, the paper aim is to analyses the actual situation regarding the consequences of an admission without exam where the baccalaureate grade is taken into account, and as possible solutions to this major challenge are emphasised possible training methods appropriate to the presented situation, methods with a great emphasis on the practical side and the permanent emphasis of the theoretical background used. In this context, good anticipation and thoroughly prepared restructuring is needed, attention and unified efforts from specialists in education field, industry and whole higher education system in order to face actual situation and be prepared for challenges that this problems will rise in the future.

1 Introduction

In November 2011, European Union (EU) member ministers, worried about the growing gap with the advanced Asian states but especially with the United States, analysed the current situation, identified major shortcomings and established the Agenda for Higher Education Systems Modernization in Europe.

Among the objectives, priority was "to improve the quality and relevance of higher education, so that the curriculum corresponds to the needs of the students, employers and careers of the future, as well as the increase in the number of graduates" [1], one of the main objectives of the Europe 2020 strategy being that by 2020, 40% of young people should graduate a higher education institution or equivalent studies.

The European Commission, aware of the Agenda objectives, in order not to remain only a noticed activity, adopts a support strategy that will be granted to the modernization of higher education in the Member States and according to it, immediately in September 2012, the Commissioner for Education, Culture, Multilingualism and European union youth at that time, Androulla Vassilioiu, constitutes a high-level group for the modernization of higher education, chaired by former Irish President Mary McAleese. In the first report of the aforementioned working group, published on 18.06.2013, 16 recommendations are formulated, of which, on the subject of this article, we mention:

- introduction of compulsory certified vocational training for teachers and other teachers in higher education;
- focusing activities on student support in developing entrepreneurial and innovative skills;
- the creation of a European Academy for teaching and learning. In order to analyse what has been done and what needs to be done to achieve the Education&Training 2020 objectives in terms of higher education, a new agenda for the modernization of higher education systems in Europe [2] was drafted on 30.05.2017.

2 Findings and recommendations of public consultation at European level on the modernization of higher education

This agenda is based on a series of preliminary public consultation of which can be extracted several aspects / specific recommendations which are related to the topic of this paper [3]:

- globalization and digitization directly affects higher education institutions and the economic environment, society (30% of today's occupations did not exist 20 years ago). Routine jobs can be replaced by advanced technologies.
- the problem of the inconsistency between the knowledge and skills graduates gain from the educational process and those required by the economy and society;
- graduates work on positions where they cannot fully use their high-level skills;

* Corresponding author: felicia.banciu@upt.ro

© The Authors, published by EDP Sciences. This is an open access article distributed under the terms of the Creative Commons Attribution License 4.0 (http://creativecommons.org/licenses/by/4.0/).

- people designing higher education courses are insufficiently aware of potential skills needs in the labor market;
- broadening access to higher education - through tailored approaches to learning and better student support and guidance mechanisms;
- that technology was not yet used effectively to improve teaching and learning (the way blended learning is used);
- the challenge that the training of researchers of the future raises;
- inappropriate financing arrangements - especially for countries in central and eastern Europe;
-the growing importance of a range of "transversal", skills, such as critical thinking, problem-solving or communication;
- skills gaps encompass the broad fields of Science Technology, Engineering and Math, where many national authorities and stakeholders see a need to strengthen high-level skills provision;

Attempts to help students acquire the right knowledge and skills:
-importance of increasing the diversity (and quality) of higher education programmes, to be able to respond to diverse student needs and diverse demand for skills;
-support teachers and reward good teaching in higher education; training available to teaching staff in higher educations should be improved;
-increasing the links between higher education and other economic actors, including through involvement in the design and delivery of programs and the provision of work-based learning opportunities;
- *Internationalisation and mobility* help students acquire skills needed for an internationalised world;
- a broader conception of innovation that should include both an entrepreneurial mind-set (as well as a creative mind-set with a long-term focus on seeking sustainable solutions across the disciplinary boundaries;
-strengthening the knowledge triangle of teaching, research and innovation was a key priority for fostering the innovative potential of students and higher education institutions;
- high quality doctoral training to ensure high quality academic research and to support high quality teaching and effective cooperation activities in higher education and deliver the top-level experts needed in other parts of the economy and society outside academia.

Romania, despite the fact that it is making progress on the Europe 2020 [4,5] targets on employment rates, national greenhouse gas emissions, renewable energy, energy efficiency and tertiary education mentions limited progress quality mass education, research and development intensity, and also early school leaving remain some distance away from their respective targets, labour and skills supply are not keeping up with the fast-changing needs of the economy and maybe one of the most important is linked to the weak performance of the education system limits growth prospects in the long run.

Each country through its government and ministries is responsible for organizing and delivering higher education while EU actions provide a general framework and guidelines for studies, teaching and research activities, or policy-making in the field, all together seen in the dimension of the European international space. Reforms in education are medium and long term, with impact and results that take time to be observable, analysed and eventually made changes to them to meet society's demands.

Given the numerous attempts to reform the education system, and especially considering the numerous top-level changes at the level of the ministry that governs education in Romania, one can ask whether the recommendations of the European Commission (EC) Working Group have also taken into account the Romanian realities and are they suitable for this case.

This paper seeks to analyse the state of the Romanian higher education so that it can respond appropriately and reasonably to the above question.

3 Romanian Technical Education in the face to of European requirements

According to a study based on the European Union Labor Force Survey, Romania has a progressive increase with regard to the population aged between 30-34 years and graduating a higher education institution (International Standard Classification of Education ISCED 5-8) (in 2001-2005 period), from 13.9% in 2007 to 26.3% in 2017, thus being near its target of 26.7% (Fig.1). [6]

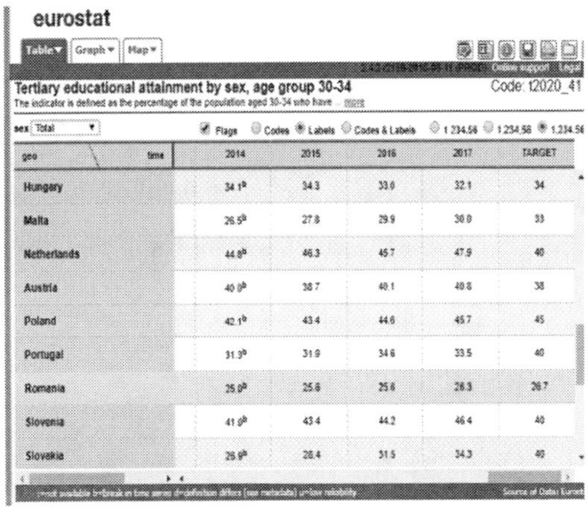
eurostat

Tertiary educational attainment by sex, age group 30-34 — Code: t2020_41

geo \ time	2014	2015	2016	2017	TARGET
Hungary	34.1^{b}	34.3	33.9	32.1	34
Malta	26.5^{b}	27.8	29.9	30.8	33
Netherlands	44.8^{b}	46.3	45.7	47.9	40
Austria	40.0^{b}	38.7	40.1	40.8	38
Poland	42.1^{b}	43.4	44.6	45.7	45
Portugal	31.3^{b}	31.9	34.6	33.5	40
Romania	25.0^{b}	25.6	25.6	26.3	26.7
Slovenia	41.0^{b}	43.4	44.2	46.4	40
Slovakia	26.9^{b}	28.4	31.5	34.3	40

Fig 1. .Tertiary educational attainment by sex, age, group 30-34 [6]

Although Romania is close to the 2020 target for higher education graduates [8], the number of high school baccalaureate graduates is decreasing and the rate of advancement is below 70% (66.7 in 2016, 68.1% in 2017) [7] (Figure 2). It should not be forgotten that in 2004, the baccalaureate graduation was higher than 80% after the first session.

At the same time, compared to the level of students training in the years 2001-2005, the training level of those who accede in higher education in the last years is lower, at least at some specializations. For example, in the school year 2017/2018, at industrial engineering admitted students in year I, 62% have baccalaureate averages lower than 7 (in the case where the minimum grade is 6), a percentage of 24% have averages between 7 and 8 and

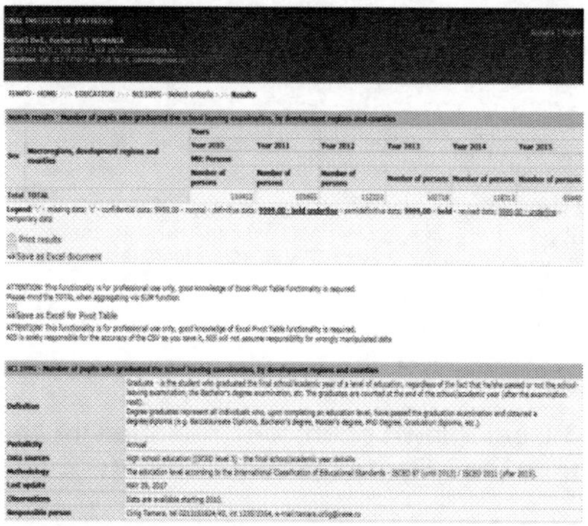

14% have averages over 8.

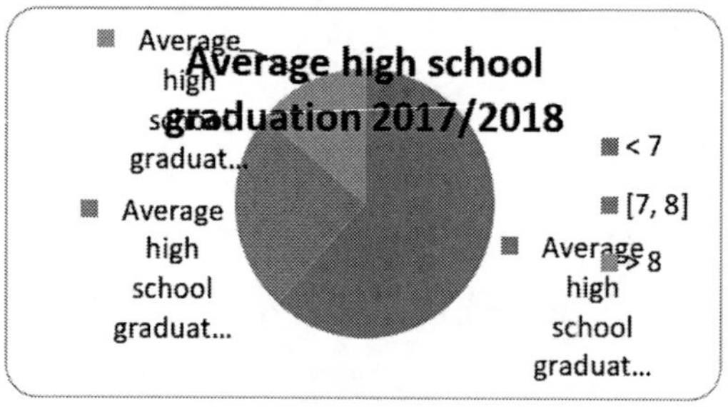

Fig. 2. Number of high school baccalaureate graduates attainment [7]

Fig. 3. Average high school graduation 2017/2018; UPT-Ind. Eng. Domain

4 Coordinates of the Romanian technical education

4.1 Entering in the higher education system

A part of the higher technical education organizes an admission exam, but at some faculties / specializations, for the students who have passed the baccalaureate admission is done without an exam based on evaluation of knowledge and abilities.

The fact that an admission exam is not organized, and thus high school baccalaureate graduates with poor trainees in basic mathematics and physics become students is a real challenge for both them and teachers.

The teachers, in addition to any other soft skills, must provide students with the minimal knowledge of acquiring the necessary skills to the employment of any engineer, skills that are built day by day through sustained work and on a minimal basis that is supposed to be existent. This can raise a series of questions including:

-what skills do these high school students have gained in high school?

- how can the provision of high-level technical skills be strengthened as a basis using a poor background?

- if and under what conditions these students will finish their studies

- how these students will meet the needs of the labor market.

All skills required by students as critical thinking, problem solving, analysing, comparing, extracting information, problem thinking in a certain context, high level skills in the STEM field (Science Technology, Engineering and Math) [3, 8] require the ability to see through principles, require the existence of some skills to learn, the existence of a minimal background in fundamental knowledge - in our case - in physics / mathematics / chemistry.

Action and understanding depend on mastered skills and memory. Also, searching, finding, evaluating and communicating information, building knowledge on a particular technical subject matter involves the existence of both cognitive and technical skills and abilities.

How can those who do not have a solid background, do not have learning skills acquired, will learn constructively by building on existing knowledge / information?

4.2 Challenges in the system

The learning process involves a constant change / adaptation to the understanding of what surrounds us and learning itself imply that you can see other points of view, understand the differences, etc. When students with a baccalaureate became higher education system students, without the skills to learn, to think of certain problems in a context, learning new things, what can the teachers do? What will be and how is the performance of such students measured? How will those high level knowledge / skills be acquired?

The answer is somewhat simple; teachers have to cope with all the challenges mentioned above. How will they deal? An attempt to respond is "student-centred", "content tailored to students' needs", "new teaching / learning / evaluation methods" and other such well-known texts.

However, good quality teaching cannot fill any gaps that students have. In fact, focusing on a student in the above-mentioned case, the poor students background means also lowering the complexity and volume of the taught content, and the evaluation obviously also undergoes changes. Also, related to this subject, there are at least three problems concerning the teachers:

-the course / lab / seminar should contain an atmosphere favourable to professional development, completion of teacher training;

- the university must prepare its staff for teaching - research and future doctoral candidates;

- the poor results of the students have negative repercussions on the teaching staff teaching / evaluation process of the and on the whole image of the university;

Not often, firms employs students in the 4th year bachelor degree, 1st and 2nd years at masters, which is somewhat positive because the attendance at courses / laboratories / seminars suffers and these future engineers will most often be specialized in a very narrow field.

At the speed of technological advancement and the emergence of new technologies, the world will change rapidly, and what they know to do well-that domain - may disappear. That is why it is very important to develop general knowledge in the field, the ability to absorb and process various knowledge combined with adaptive power to help them in a changing environment.

On the other hand, there is a need to think creatively to stimulate new ways of thinking about teaching / learning / evaluation in order to address some of the problems mentioned above and at the same time to maintain a level of appropriate teaching / learning / assessment goals and mission of higher education.

4.3. Possible solutions to challenges

First of all, there is a continuous shift from the terms of education and learning to the training-vocational training. As we know, education is also done outside the school system: family, friends, mass communication, own reading, more recently - the Internet, with deeper effects than those produced within the educational institution.

To improve the learning process, we do not have to look for solutions too much. The learning products to whom are devoted these institutions need only be known to and understood by those who conceive and exploit them.

They are not only well-known but also accepted by the broader view, being integrated into contemporary culture since Seneca's time: "It is a long way of learning, but it is short and effective by exemplifying (to read, practice).

Achieving learning / teaching / evaluation through the use of Information and Communication Technology, open courses are of real importance, but the amount of time needed to prepare them is immense for teachers.

At the same time, the amount of information provided is very high, so processing a large amount of information, learning it for evaluation raises serious difficulties for students. The use of learning materials relies heavily on the visualization aspect, but for the proper assimilation of materials students must still read, think, analyse, compare, and write.

To learn and the learning process itself requires a lot of effort and dedication, that's why it's hard and not everyone can do it successfully.

As possible training methods appropriate to the presented situation - lack of basic knowledge – the training methods, with a great emphasis on the practical side and the permanent emphasis of the theory's presence in practice could be adequate:

-case studies based on real engineering problems [8];-integrated laboratories to show the relation between engineering disciplines (e.g. coordinate measurements, assembly, and tolerance allocation)

- using debate as a way of improving the skills of learning, skills of cooperating, and verbal skills, the formation of the technical vocabulary, their expressions and presentation in technical terms.

-oral evaluations

Also practice stages are very important. Law no. 258/2007 [9] regarding the practice of school and university students and the Order 3955/09.05.2008 [10] define the persons involved in planning, organizing and supervising the deployment of practice- teaching practice supervisor and tutor – person designated by practice partner that will provide/ensure compliance with their training and acquire professional skills planned for internship by practitioner.

Teaching practice supervisor has to establish an internship program individually tailored to the needs of companies, and consistent with the preparation, personal development opportunities and desires of student/students- the student centred internships.

Because practical training is not a process that looks exclusive the student but one in which he relates to others, is integrated into a team, is subject to rules of the institution where he performs internships, is subject of regulations that exist in its higher education institution and so besides the skills, attitudes and skills acquires improves its skills of teamwork, collaboration, communication so necessary for continuous adaptations to labor market changes and permanent challenges.

The idea of not abandoning classical education - the source of so many discoveries / innovations - and also to not abandon a classical education and a knowledge-based education for a student centred one is more and more common among specialists in education, including our country and even more vocal, from abroad: „Classical education then, in this sense, is a life-long process of applying the "tools of learning" [11, 12]- tools that are skills entailed in basic sciences, engineering basic sciences, and specific of engineering, which travels with the student through her/his career as a professional or as an academic."

In other words, basic skills in learning can be used to learn almost anything at any time and. anywhere. It is worth mentioning that few institutions are as conservative as those of education, especially higher education. The tenacity with which they have defended their "identity" and "autonomy" in society is, however, a merit that cannot be neglected.

But traditionalism has weakened the ability to adapt to new realities and to innovate. The choices (options) are easy and we must take into account both the benefits and the disadvantages.

Emphasizing the economic nature brings knowledge to the firm ground of their utility, but the private economy is often under the sign of immediate benefit.

The acceptance of its character implies the increase of the state's responsibility, but also the danger of bureaucratization, of the statist formalism. Is it possible to find a middle way? Yes, it is but only if we accept a training system such as the public domain, with an emphasis on applications, on professional relevance and on flexible and innovative economic criteria.

5 The teacher and the European requirements

5.1 What kind of knowledge he has to teach

Teacher's problem exists and deepens year after year, especially under the contents teached and the quality of the teaching act itself. The causes are many: systemic, selection mode, salary level, conditions for building a career, the speed with which you have to face the many changes imposed by s the initiated and unfinished reforms, etc.

„Ars longa, vita brevis" - says a very famous Latin proverb. Under the conditions listed, when do you have time to "grow" teachers and valuable researchers? This is the hardest problem and the worst repercussions for the future of technical education. Sociologically, the phenomenon of professional mobility is becoming increasingly widespread, both in the workplace and in the profession itself. The same timeframe may include alternate successive stages of learning, with the actual work steps.

In turn, the need for multidisciplinary strongly asserted in the second half of the last century and the disciplines weakened the rigid separatism that has dominated the world for a long time. Research teams have become multidisciplinary.

The disciplines are established through tradition and law, and the freedom to address the subjects of new areas is reduced, if not excluded. The gap between scientific research and the subjects taught in school has deepened. It is neglected that engineering solves problems. Almost all research issues are multidisciplinary, and researchers need to be trained in the spirit of multidisciplinary. The student is thus relieved of the full deepening of a discipline, of which only a part is useful for his purposes. At least one third of the subject matter of overlapping subjects is saved by using modules. Thus, the modular concept becomes a fundamental element for the interdisciplinary of the sciences and for the continuous process of learning. Apart from the benefits related to efficiency, effort savings, simplification and coherence, one more result emerges: the possibility of individual curriculum. It is a perspective that empowers the individual as the source and the main subject of the training. You cannot go to such a different system overnight, but important steps can be done experimentally in any legal and recognized system. This transition can be accomplished as a slow process. It depends on the maturity of universities and on the resistance of traditionalist disciplines.

5.2 For whom and with what results are taught

In the evoked context, at the level of the European Union, as in all developed countries, a decline in the interest of young generations in engineering studies is observed. If demand varies from one country to another, at European level it exceeds the offer.

When the statement "I would like to become a scientist" was analysed, the weight of respondents in the developed countries was very low and the girls had a much higher share than the boys. The answers to the statement "I would like to work in the technical field" were similar. Differences can only be made with regard to the gender of person who responded to this question.

Going forward, it has been argued that not just the interest of young people in the technical field has to be considered, but rather how this topic is treated in school and the perception that young people form in "building" their own identity. To ease this work, teachers should use specially designed materials. It should put more emphasis on technology related to training, practical experience and problem solving in a creative way. At the same time, the examples applied to life events should be more relevant to daily living and much more attractive to boys and girls. Above all, technical training should not be limited to explaining a function.

The need to attract younger generations to technical professions (in particular, engineer) is today, first and foremost, a national problem of the EU Member States, but also a problem of the European Union in its entirety [13]. Technology must, however, be oriented to respond to future problems. For these reasons there is an urgent need for initiatives aimed at ensuring that all those interested in technology also benefit from solid training in the chosen field of activity and those who work in the technical domain to do its best to acquire the knowledge and skills needed to meet the new market demands. The EC consultation has highlighted the importance conferred by industry to Competitiveness competencies. The Expert Group on "New Skills for New Jobs" has clearly demonstrated the importance of coordinating education with industry training to better anticipate future skills demand. Competencies are an essential factor for growth, employment and competitiveness: they are the basis for productivity and innovation. However, Europe faces major challenges related to the supply of skills. Already, the inconsistencies between supply and demand - and in some Member States even lack of skills - are already impeding growth and employment. Investment in training is also investment in innovation, as much of the technical evolution is a consequence of the progressive innovations made by skilled workers and engineers in production units. However, a prerequisite for investing in new skills and training is the need to further develop tools to monitor and anticipate skills needs and inconsistencies. At European level, several instruments are currently being developed for this purpose [14].

6 Conclusions

An uninterrupted reform is needed. The need to adapt education becomes so important in the social, economic and scientific environment which changes very fast so that the reform cannot be effective, with discontinuous changes and abrupt rhytm.

Historical experience has shown that the only good reform in education is the one built into the permanent pressure of the system, which takes place in doses required under the sign of continuity. The school must be flexible, highly adaptable. Appropriate legislation, transcribed in succinct and sufficiently flexible regulations for adaptation to different ages and continuous changes, is, in this context, a major desideratum.

The authors are not yet able to indicate the path that should be followed, but with certainty, even if only because of the arguments presented in this paper, can say that the current pathway of the Romanian higher education, especially the technical one and within it, especially in the field of industrial engineering, is not a good one.

Therefore, we need to stop, suspend the so-called "reforms" in progress and appeal to the entire community, both university and economic-industrial, for a sincere and detailed analysis of the current situation. Let us look back at the moment when the results were still satisfactory, remember the mistakes that followed to learn from them, show the successes in order to promote them and analyse our needs to design the future paths of action.

At the same time, we affirm with certainty that the recommendations presented in the reports of the European Commission high-level group supporting the modernization of the European higher education match as a "glove" to Romanian technical higher education and to achieve their aim "High quality teaching and learning depends on persons and dedicated institutions supported by policies on teaching and learning. Teachers in higher education must receive the training and support they need to perform their role in excellent conditions "[1, 2].

The emphasis placed in Romania on the results of research in higher education is more related to the classification criteria and less to the quality of the educational process. So the finding that many higher education institutions in the EU place too little emphasis on teaching compared to research, even if both are their main missions, is true, as well as the solution proposed by the high European Force by his chairman's statement that this situation needs to be rebalanced and that the role of teaching in defining academic merit needs a stronger accent and recognition.

Last but not least, the importance of knowing the future needs of the economic and industrial society is emphasized. Anticipating employment and skills needs is recognized as having a strategic importance for the success of any enterprise; it influences the competitiveness and long-term prospects of enterprises, irrespective of the industrial sectors in which they are active. In this context, good anticipation and thoroughly prepared restructuring not only can have a positive impact on the sustainable competitiveness of enterprises, but can also generate positive side effects in other markets and sectors, while also improving the overall employability of the workforce and facilitating the transition to new jobs.

References

1. Communication from the Commission to the European Parliament, the Council, The European Economic And Social Committee and The Committee of The Regions on a renewed EU Agenda for Higher Education, Brussels, 20.09.2011, Available at http://eur-lex.europa.eu/legal-content/EN/TXT/PDF/?uri=CELEX:52017DC0247&from=RO, (2011)
2. Communication from the Commission to the European Parliament, the Council, the European Economic and Social Committee and the Committee of the Regions Supporting Growth and Jobs – an Agenda for the Modernisation of Europe's Higher Education Systems, Brussels, 30.5.2017 Available at http://eur-lex.europa.eu/legal content/EN/TXT/PDF/?uri=CELEX:52011DC0567&rid=13, (2017)
3. Key Challenges for Higher Education in Europe - Annex II: Results of the Public Consultation on the EU's Modernisation Agenda for Higher Education, May (2016)
4. 2018 European Semester: Assessment of progress on structural reforms, prevention and correction of macroeconomic imbalances, and results of in-depth reviews under Regulation (EU) No 1176/2011, (2017)
5. COUNCIL RECOMMENDATION on the 2017 National Reform Programme of Romania and delivering a Council opinion on the 2017 Convergence Programme of Romania
6. Tertiary educational attainment by sex, age group Indicator [t2020_41] ; EU Labour Force Survey
7. National Institute of statistics, http://statistici.insse.ro/shop/, last visited April 2018
8. Engineering Education. Research and Development in Curriculum and Instruction, John Heywood, Published by John Wiley & Sons, Inc., Hoboken, New Jersey, ISBN-13978-0-471-74111-4, p. 354, (2005)
9. Law no. 258/19.07.2007 – regarding students practice in Romanian (in Romanian Legea 258/19.07.2007 privind practica elevilor şi studenţilor), MO no. 493/24 July (2007)
10. Order No. 3955/9.05.2008 approving the general framework for internships in undergraduate and master's degree programs organization and the framework agreement on the internship conduct of the in the undergraduate or master's degree programs (in Romanian Ordinul 3955/9.05.2008 privind aprobarea cadrului general de organizare a stagiilor de practică în cadrul programelor de studii universitare de licenţă şi de masterat şi a convenţiei-cadru privind efectuarea stagiului de practică în cadrul programelor de studii universitare de licenţă sau masterat), MO. no. 440 /12/06/2008
11. C. da R. Brito, M. M. Ciampi, et.al. "Engineering Adventure for Youth Generations," in 2016 ASEE Annual Conference & Exposition Proceedings, (2016)
12. C. R. Brito, M. M. Ciampi, R, et. al Rethinking Engineering Education, 2017 IEEE Frontiers in Education Conference (FIE) Indianapolis, USA, ISBN: 978-1-5090-4920-2; Available at http://doi.ieeecomputersociety.org/10.1109/FIE.2017.8190506, (2017)
13. G. Mihăiţă, Actual issues in Engineering (I) (in Romanian probleme actuale in inginerie (I)), Univers Ingineresc, Year XXIV, Nr. 11 (537) 1-15 June 2013, p. 1, (2013)
14. New Orientations in Community Policies, (in Romanian Noi orientări in politici comunitare) Univers Ingineresc, Year XXIV, Nr. 12 (538) 15-30 June 2013, p. 5, (2013)

Absorption of management knowledge in Romanian small and medium-sized enterprises

Gina-Maria Moraru[1,*], and *Daniela* Popa[1]

[1]Lucian Blaga University of Sibiu, Engineering Faculty, Industrial Engineering and Management Department, 4 Emil Cioran Street, 550025 Sibiu, Romania

Abstract. One aim of this paper is to prove that the absorption of management knowledge in Romanian small and medium-sized enterprises (SMEs) has many gaps. Another aim is to find the main causes of this. Paper begins with a short bibliographical study about the most important weakness of the Romanian SMEs: the low availability of skilled staff or experienced managers. In order to discover which phase of knowledge management in these firms is the most problematic, we have made an exploratory research based on the use of Fishbein-Rosenberg model for measuring managers' opinions on this subject. The results have proved that the management knowledge absorption is done to a small extent in Romanian SMEs. In order to find the main causes of the poor management knowledge absorption, we have next developed a research based on questionnaires. We have discovered that the studied problem has its roots both in the business environment and in the academic environment.

1 Introduction

The role of knowledge in achieving organizational performances is essential both in the development of small businesses and of whole economy [1]. In terms of intangible resources of the organization, knowledge and knowledge management are among the most important [2]. Knowledge management helps organization to get strategic competitive advantages and to increase their efficiency in the era of globalization and information explosion [2].

There are several phases of knowledge management. We have selected from literature [3] the most important and numbered them as follows:

- Phase 1: Assessing the knowledge which is available and accessible to the organization;
- Phase 2: Forecasting needs and sources of knowledge;
- Phase 3: Obtaining the necessary knowledge for the company (from inside or outside of it), phase that – according to other specialists [4] – we can called it "the knowledge absorption";
- Phase 4: Administrating, sharing, using, and developing knowledge;
- Phase 5: Harnessing of knowledge.

A firm that recognizes the value of new, external information, assimilates it, and applies it to commercial ends is constantly increasing its innovative capabilities [4]. In order to cope with global competition, small and medium-sized enterprises (SMEs) should do the same. We have to mention that the SME sector has delivered 57.6% of the gross value added generated by the private, non-financial European economy during 2012 [5].

Nowadays, the first two of the most pressing problems faced by European SMEs are finding customers and availability of skilled staff or experienced managers [6].

So, finding customer is the most important danger in the external organizational environment, and low availability of skilled staff or experienced managers is the most important weakness in the SMEs' internal organizational environment. We believe that this weakness can be analyzed according to the knowledge management stages mentioned above [3]. In order to ensure a long-term development of the businesses they run, managers of Romanian SMEs have to discover which stages of knowledge management have to be improved.

2 Exploratory research

The study presented in this paper has had two main objectives: to find out which stage of knowledge management in Romanian SMEs is the most problematic and what are the main causes of this.

In order to discover which stage of knowledge management in Romanian SMEs is the most problematic, we have firstly developed a brainstorming and braiwriting session with 10 managers working in SMEs from different fields of activity in Sibiu County. They have set the key criteria for assessing the problems that arise in the different stages of knowledge management (presented in Table 1): severity of the problem, its repeatability, average duration of a problem, difficulty of the problem solving, and cost of solving. We have codified these criteria as shown in Table 1.

With the intention to use the Fishbein-Rosenberg model [7] for discover the most problematic stage of the

* Corresponding author: gina.moraru@ulbsibiu.ro

© The Authors, published by EDP Sciences. This is an open access article distributed under the terms of the Creative Commons Attribution License 4.0 (http://creativecommons.org/licenses/by/4.0/).

knowledge management, we have asked the participants to give marks on these criteria, using a scale with integer values from 0 to 10. So, we have calculated the average of the scores for each criterion and its importance/weight (Table 1).

Table 1. Criteria used to assess the nonconformities/problems founded in knowledge management phases.

Criterion name	Criterion code	Average score [points]	Weight
Severity	Se	9.2	0.226
Repeatability	Re	7.8	0.192
Duration	Du	7.4	0.182
Difficulty of solving	Di	8.2	0.201
Cost of solving	Co	8.1	0.199
SUM	-	**40.7**	**1**

Then the participants have assessed the knowledge management phases in terms of each criterion, giving them marks from 0 to 10. We have calculated in Table 2 the averages of the scores and divided them with 10, in such a way that all the results to be in the range 0, ..., 1. Using the criteria weights from Table 1, we have then calculated – in the last row of Table 2 – the weighted sum of the average scores obtained by a knowledge management phase for selected criteria.

Table 2. Assessment of knowledge management phases from the perspective of problematic situations.

Criterion code	**Phase 1 Assess**	**Phase 2 Forecast**	**Phase 3 Absorb**	**Phase 4 Use**	**Phase 5 Harness**
(Table 1)	**of knowledge** [points]				
Se	0.67	0.73	0.82	0.63	0.71
Re	0.72	0.55	0.75	0.83	0.69
Du	0.45	0.35	0.88	0.76	0.74
Di	0.38	0.51	0.93	0.57	0.80
Co	0.42	0.61	0.70	0.78	0.65
Weighted sum	**0.532**	**0.558**	**0.816**	**0.710**	**0.718**

According to the results, the most problematic phase is the knowledge absorption, with a weighted sum of 0.816 points, followed by harnessing knowledge, with 0.718 points, and by the phase of administrating, sharing, using, and developing knowledge, with 0.710 points.

According to specialists, an organization's absorptive capacity depends on the absorptive capacities of its members [4]. From this perspective, taking into account the number of attitudes and opinions related to any subject, we would be tempted to believe that knowledge absorption is easier in small enterprises than in big ones. But the results of the brainstorming and braiwriting session show the opposite. An explanation is the fact that absorptive capacity of an organization consists not only in the assimilation of information, but also in the organization's ability to exploit it [4]. So, on one hand, there is a strong connection between knowledge absorption and harnessing of knowledge (third and fifth phases of the knowledge management). On the other hand, absorptive capacity of a firm depends on transfers of knowledge across and within organizational subunits [4], meaning it is partially depended on fourth phase: administrating, sharing, using, and developing knowledge.

Discussions of the session continued then with the impact that a correct absorption of management knowledge could have on each SME in particular and on the economy of the country in general. The ideas of the participants led to the scheme in Figure 1. The effect of the correct absorption of knowledge is gradually expanding from the level of each SME as an entity to that of the SMEs' sector, then to the national economy, and finally to the whole society, leading to its sustainable development (Fig. 1).

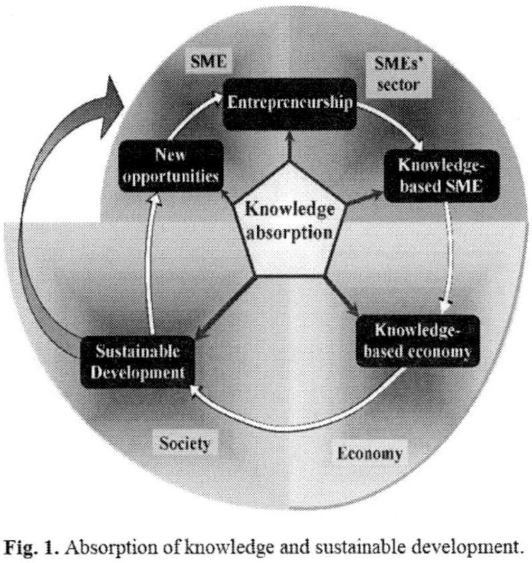

Fig. 1. Absorption of knowledge and sustainable development.

Finally, participants have determined the most important causes of the poor management knowledge absorption in Romanian SMEs. These causes are represented in the Ishikawa diagram of Figure 2, divided into 6 categories: people (managers and employees), methods, equipments, money, academia and other environmental issues.

Participants also have ranked these causes, according 1 point to them with low impact to the studied problem, 3 points to them with medium impact, and 9 points to them with high impact (Table 3). They have selected six causes with high impact: manager's lack of training, poor digitization in SMEs, difficult access to funding, poor partnerships universities - SMEs, fierce knowledge competition, and finding new experienced managers.

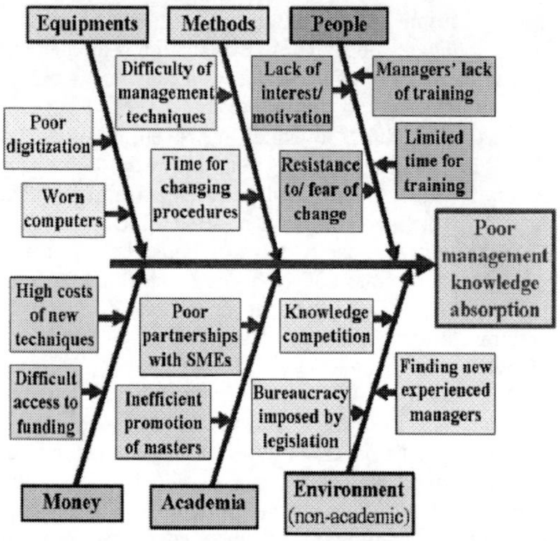

Fig. 2. Ishikawa diagram for the poor management knowledge absorption in Romanian SMEs.

Table 3. Causes of poor management knowledge absorption in Romanian SMEs and their impact on the firms.

Cause	Category	Code	Impact
Managers' lack of training	People	P1	9
Resistance to change/ fear of change	People	P2	3
Lack of interest/motivation	People	P3	1
Limited time for training (other priorities)	People	P4	3
Time for changing procedures	Methods	Me1	3
Difficulty of many management techniques	Methods	Me2	1
Poor digitization	Equipments	Eq1	3
Worn computers	Equipments	Eq2	3
Difficult access to finance	Money	M1	9
High costs of new managerial techniques	Money	M2	1
Poor partnerships universities - SMEs	Academia	A1	9
Inefficient promotion of master programs	Academia	A2	3
Knowledge competition	Environment (non-academic)	E1	9
Finding new experienced managers	Environment (non-academic)	E2	3
Bureaucracy imposed by legislation	Environment (non-academic)	E3	3

We have then used these values (Table 3) in the research based on questionnaires.

3 Research based on questionnaires

The questionnaires were distributed in printed format to 120 managers of SMEs in different fields of activity in Sibiu County. We have presented to respondents a list of the potential causes for the poor management knowledge absorption in Romanian SMEs (according to Table 3) and we asked them to mention which of these causes have generated problems in the management knowledge absorption in their own firm. Only 44 of the managers have agreed to complete the questionnaire, the response rate being 36.67%. The results are presented in Table 4.

Table 4. Scores obtained by the causes of poor management knowledge absorption in SMEs.

Cause	Frequency	Score	Percentage	Cumulative percentage
Managers' lack of training	42	378	23.28%	23.28%
Poor partnerships universities - SMEs	41	369	22.72%	46.00%
Difficult access to finance	39	351	21.61%	67.61%
Knowledge competition	21	189	11.64%	79.25%
Time for changing procedures	15	45	2.77%	82.02%
Finding new experienced managers	14	42	2.59%	84.61%
Inefficient promotion of master programs	14	42	2.59%	87.19%
Poor digitization	13	39	2.40%	89.59%
Resistance to change/ fear of change	11	33	2.03%	91.63%
Lack of interest/ motivation	10	30	1.85%	93.47%
Limited time for training	9	27	1.66%	95.14%
Difficulty of many management techniques	22	22	1.35%	96.49%
High costs of new managerial techniques	21	21	1.29%	97.78%
Worn computers	7	21	1.29%	99.08%
Bureaucracy imposed by legislation	5	15	0.92%	100.00%

We mention that the score given to a cause (third column in Table 4) was obtained by multiplying the frequency of its selection by the respondents (second column in Table 4) with the impact it has on the analyzed problem (last column in Table 3).

The other 76 managers out of the 120 have refused to complete the questionnaire, invoking the following reasons:

- They did not have time or any interest in the studied problem;
- They did not face the problem of absorbing knowledge in their firm;
- They did not want to share their professional problems with us.

We have represented the respondents' opinions in a Pareto chart (Fig. 3), using the coding of the causes of the studied problem in Table 3. According to the results (Table 4 and Fig. 3), four of the fifteen causes mentioned in questionnaire have had representative weights in the management knowledge absorption, cumulating 79.25%: managers' lack of training, poor partnerships universities – SMEs, difficult access to finance, and knowledge competition.

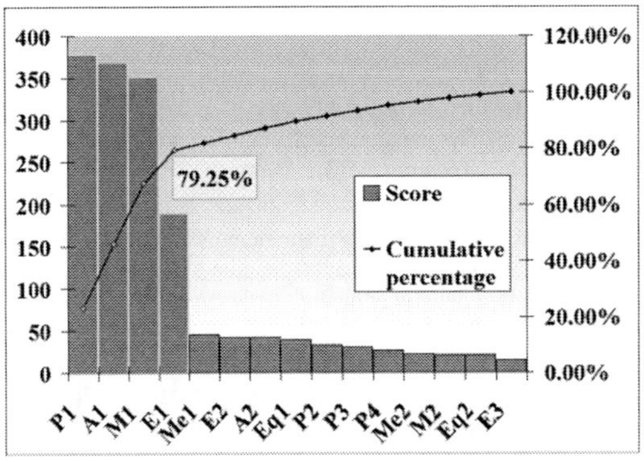

Fig. 3. Pareto chart with main causes of poor management knowledge absorption in SMEs in Sibiu County, Romania.

Finally, the respondents have offered several solutions to the studied problem. We mention only the most important of them:

- Building a national platform for developing SMEs – universities partnerships;
- Creating a government funding scheme for the development of SMEs' managers;
- Stimulating by state management of benchmarking processes between SMEs in the same industry or in different industries;
- Developing of post-graduate courses by universities, focused on the ways in which SMEs can face on global competition in general, and on global knowledge competition in particular.

4 Conclusions

In order to cope with global competition, Romanian SMEs should improve their knowledge management system. The most problematic phase in their knowledge management is the knowledge absorption, in other words, the phase of getting the necessary knowledge from outside and inside the firm.

The exploratory research and the research based on questionnaires presented in this paper were completed with the establishment of four main causes of the studied problem: managers' lack of training, poor partnerships universities – SMEs, difficult access to finance, and knowledge competition. Because the problem has its roots both in the business environment and in the academic environment, the solutions offered by respondents have focused mainly on the junction between these two environments.

References

1. D. Popa, *Increasing performance in SMEs through improving strategic planning in the context of knowledge based economy (Cresterea performantei in IMM-uri prin imbunatatirea planificarii strategice in contextul economiei bazate pe cunostinte)* ("Lucian Blaga" Univ. Pub. House, Sibiu, 2015)
2. S.I. Ceptureanu, Rev. Int. Comparative Management **16/1**, 1 (2015)
3. O. Nicolescu, I. Verboncu (coordinators of Book 1), *Modern approaches in the organizational management and business (Abordari moderne in managementul si economia organizatiei)*, Book 1, Ch. 1, 29-80 (Economic Pub., Bucharest, 2003)
4. W.M. Cohen, D.A. Levinthal, Adm. Sci. Quarterly, **35/1**, 128-152 (1990)
5. European Commission, *Growth prospects for SMEs are encouraging* (Brussels, 26/11/2013), http://ec.europa.eu/growth/content/growth-prospects-smes-are-encouraging-0_en
6. P. Muller, S. Devnani, J. Julius, D. Galiardi, C. Marzocchi, *Annual Report on European SMEs 2015/2016. SME recovery continues* (European Commission, 2016)
7. M.J. Rosenberg, J. Abnorm. Soc. Psychol., **53**, 367-372 (1956)

Strategic management problems in Romanian small and medium-sized enterprises

Gina-Maria Moraru[1,*], and *Daniela* Popa[1]

[1]Lucian Blaga University of Sibiu, Engineering Faculty, Industrial Engineering and Management Department, 4 Emil Cioran Street, 550025 Sibiu, Romania

Abstract. The aim of the paper is to discover several of the strategic management problems that affect the most of Romanian small and medium-sized enterprises. The study presented here is only a part from a larger research developed by us during the last five years, research oriented towards sustainable development opportunities of these enterprises. Technological revolution, knowledge management and global expansion of big companies are just a few challenges to which small businesses have to cope. For this, their strategic management needs to take into account the turbulent business environment, to anticipate and to carry out correctly and in time the necessary changes in the organization. Unfortunately, our study has found gaps in the use of key strategic management tools and in the balanced use of resources in small Romanian firms.

1 Introduction

Strategy is the road on which organization goes after matching its internal resources and skills with the opportunities and risks from the external environment [1]. But, for elaboration of a good strategy, management should take into account that knowledge is nowadays the most important resource of an organization. That is why we propose to apply in the sector of small and medium enterprises (SMEs) the "Configuring for Knowledge" framework used in the institutional domain [2]. This closed loop framework has four essential steps [2, 3]:
- Identifying the strategic business drivers;
- Establishing the knowledge core, that includes both tangible and intangible assets (values, culture, people, technological intangible assets, business capabilities), and the knowledge interrelationships, for constructing a "knowledge value chain";
- Applying just-enough-discipline by focusing on knowledge culture and on speed and precision with which knowledge is disseminated;
- Monitoring and rebalancing the entire process.

Strategic management is the process of managing both the organization's mission and the relationship of the organization to its environment [4]. More and more firms, especially those of small and medium size, feel the need to develop formalized enterprising strategies [5]. This is the firs premise of our study.

The second one is the fact that European SMEs play a decisive role in the labour market [6]. By example, 0.2% of enterprises in the EU-28 non-financial business sector in 2016 were SMEs and employed 93 million people, respectively 67% of total employment in that sector [6]. Moreover, those firms generated 57% of value added in the EU-28 non-financial business sector [6].

And the third premise of our study is the fact that the European SMEs are facing many challenges of the business environment [6]. Technological revolution, knowledge management and global expansion of big companies are only a few of them. To this we can add the growing efforts that organizations need to make to assure their financial resources.

2 Study for determine gaps in strategic management of SMEs

The study presented here is only a part from a larger research developed by us during the last five years, research oriented towards sustainable development opportunities of SMEs. It aim was to discover several of the strategic management problems that affect the most of Romanian SMEs. The methodology of our study has combined the questionnaire technique with the interview technique. We have collected opinions of 40 managers working in SMEs from various fields, in Sibiu County, Romania (Fig. 1).

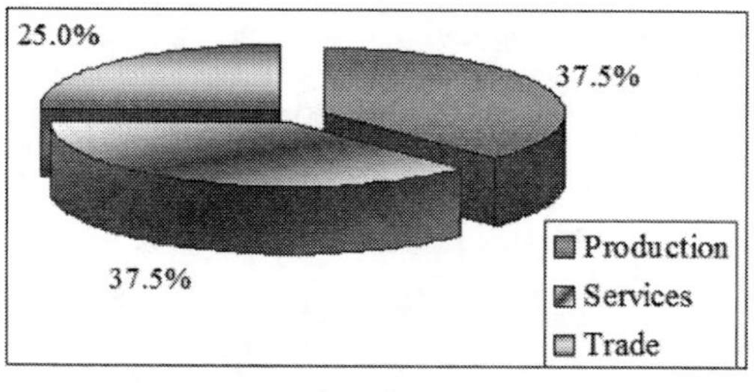

Fig. 1. Respondents' general fields of activity.

* Corresponding author: gina.moraru@ulbsibiu.ro

© The Authors, published by EDP Sciences. This is an open access article distributed under the terms of the Creative Commons Attribution License 4.0 (http://creativecommons.org/licenses/by/4.0/).

We have firstly analyzed their opinion on the influence of environmental factors on their own strategies. Respondents' opinions on the current influence of factors in the general external environment on the business they lead are presented in Figure 2. Unfortunately, they do not foresee for the future any significant evolution of these influences on the studied organizations (Fig. 3).

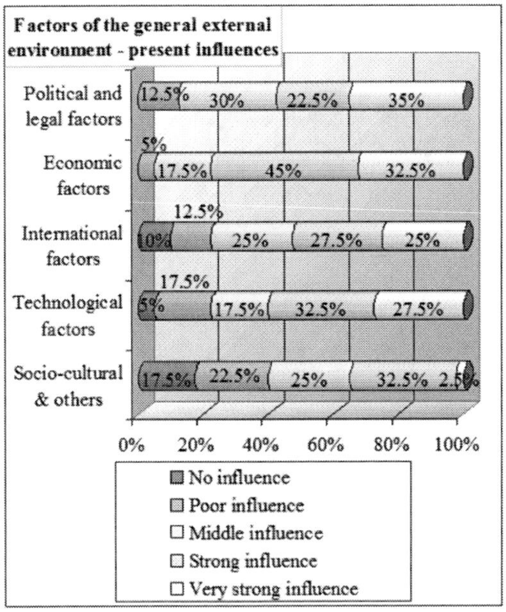

Fig. 2. Current influence of the general external environment on the studied SMEs and on their strategy.

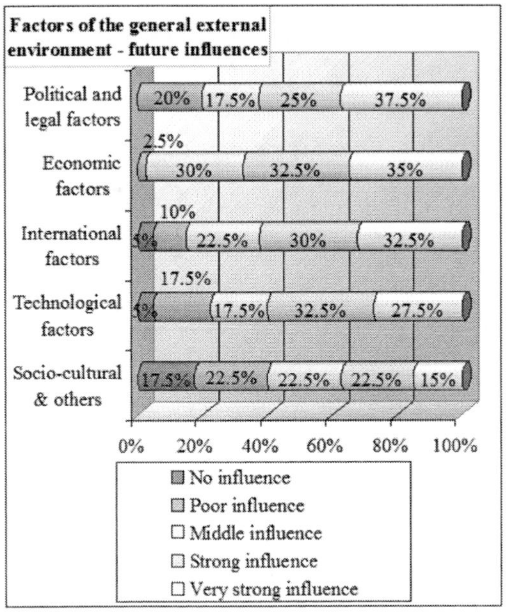

Fig. 3. Future influence of the general external environment on the studied SMEs and on their strategy.

These results (Fig. 2 and Fig. 3) determine us to affirm that SMEs' managers do not consistently monitor the factors of the external environment and do not perceive their increasingly intense turbulence. Moreover, we can conclude that SMEs' managers will build long-term strategies especially on the basis of legislative, political and economic influences, partly neglecting the others (Fig. 2 and Fig. 3).

Respondents' opinions on the current and futures influences of specific external environment on the SMEs are presented in Figure 4, respectively Figure 5.

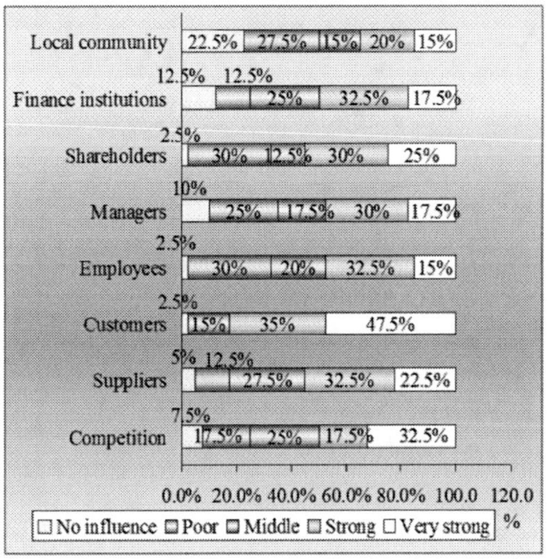

Fig. 4. Current influence of the specific external environment on the studied SMEs and on their strategy.

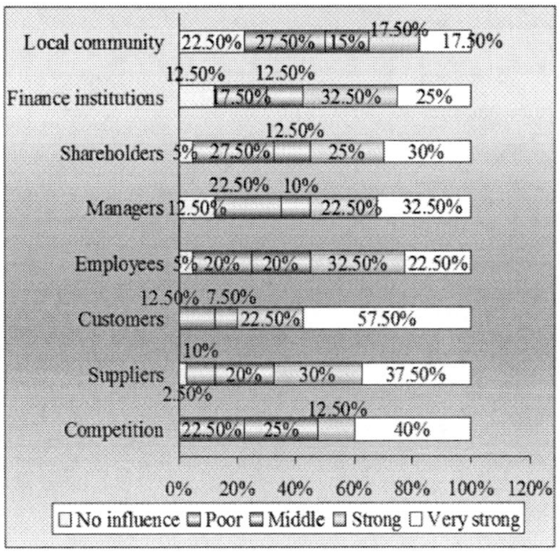

Fig. 5. Future influence of the specific external environment on the studied SMEs and on their strategy.

As we can see, respondents expect to increase in the future the impact of all factors in the specific external environment on SMEs (Fig. 5 versus Fig. 4). So, they are more interested in monitoring the specific external environment than in monitoring the overall external environment.

Regarding the impact of the classic resources (Fig. 6) and of the modern intangible ones (Fig. 7), we observe that our respondents do not perceive their crucial importance in developing the strategy. For developing and enhancing the knowledge in their organization, they will need to put more emphasis on intangible resources (Fig. 7) and to use them all in a more balanced way.

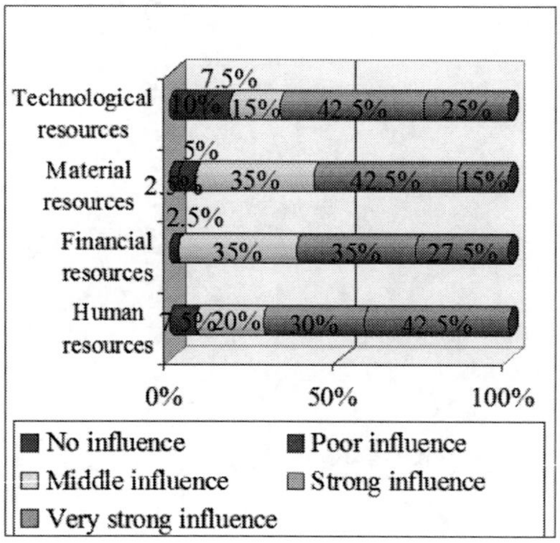

Fig. 6. The impact of the classic resources on the SMEs' strategies.

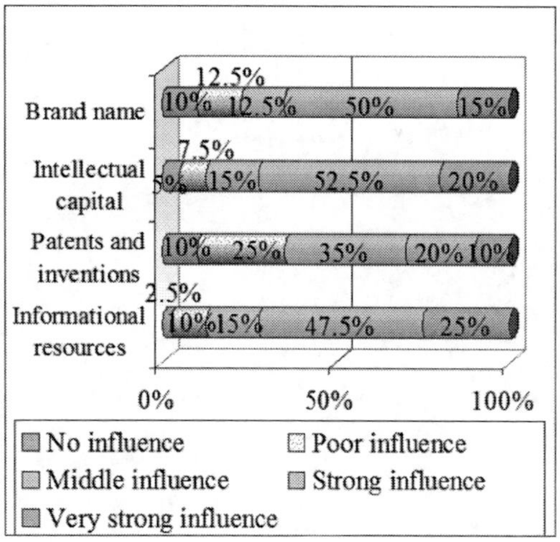

Fig. 7. The impact of the modern intangible resources on the SMEs' strategies.

In the second part of our study, we have questioned the respondents about the types of strategies they have used. Regarding the classification of competitive strategies, the situation of the responses by field of activity is presented in Figure 8.

Studying strategies according to the companies' development intentions, we find that 75% of them have adopted growth strategies (Fig. 9). We must say that most of them are in service and production.

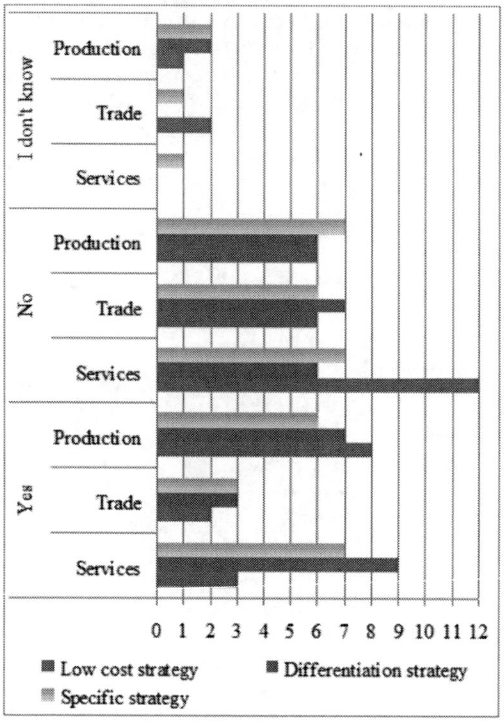

Fig. 8. Competitive strategies used by respondents in SMEs.

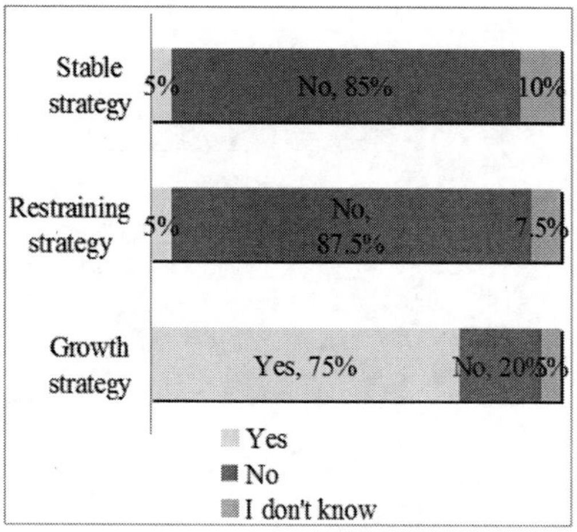

Fig. 9. Strategies adopted in SMEs regarding their development.

The largest share of businesses that have not adopted restraining strategies (Fig. 9) is found in the services sector, followed closely by the productive sector.

Responses to strategies adopted on the basis of the environment are centralised in Figure 10. We note that prospective and analyzing strategies are predominant in

SMEs. 42.5% of the companies turned to defense strategies, and 7.5% said they also adopted destructive strategies (Fig. 10).

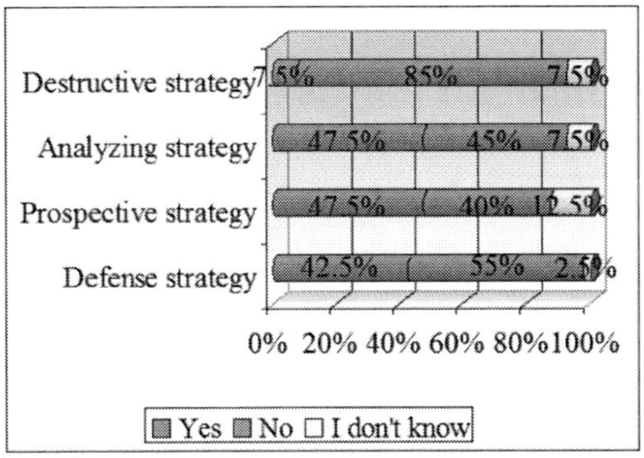

Fig. 10. Strategies adopted in SMEs on the basis of business environment.

Remaining in this classification of strategies, we have to mention that 10 firms in services field have adopt the analyzing strategy (Fig. 11).

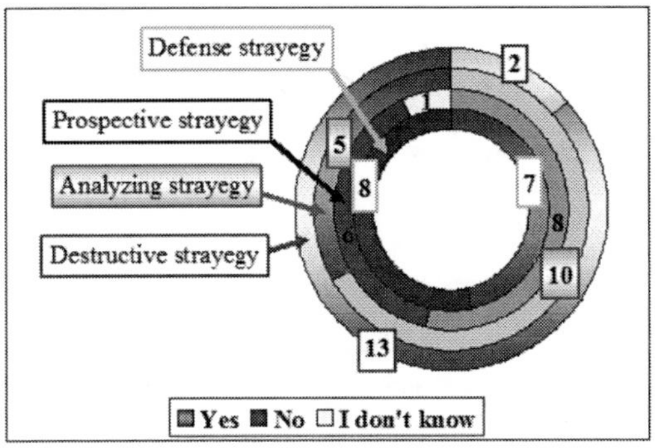

Fig. 11. Strategies adopted in SMEs in services field on the basis of business environment.

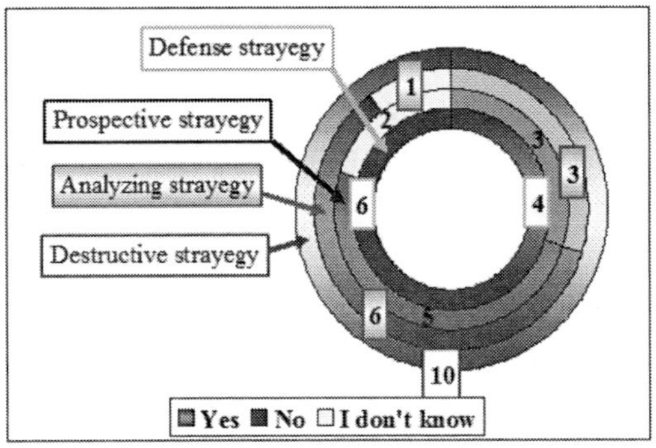

Fig. 12. Strategies adopted in SMEs in trade sector on the basis of business environment.

Only 3 firms in trade sector (Fig. 12) and 6 firms in production field (Fig. 13) have adopted the analyzing strategy.

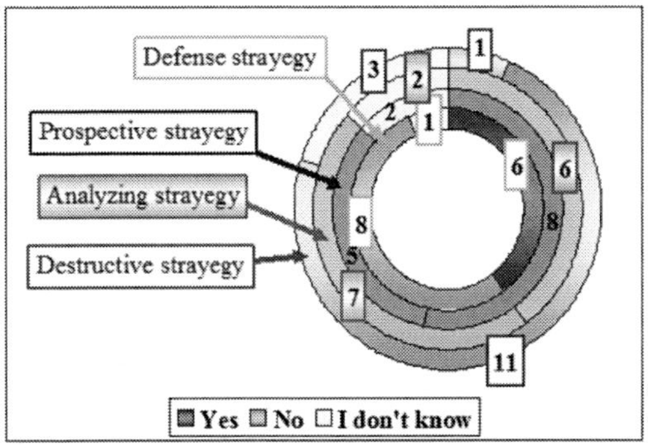

Fig. 13. Strategies adopted in SMEs in production field on the basis of business environment.

So, we can say that managers of services firms are better strategists than those in other areas.

3 Conclusions

Technological revolution, knowledge management and global expansion of big companies are just a few challenges to which SMEs have to cope. For this, their strategic management needs to take into account the turbulent business environment, to anticipate and to carry out correctly and in time the necessary changes in the organization. Unfortunately, our study has found gaps in the use of key strategic management tools and in the balanced use of resources in small Romanian firms.

References

1. R.M. Grant, California Mng. Rev. **33, 3**, 114-135 (1991)
2. A.B. Graham, V.G. Pizzo, European Manag. J. **14, 4**, 338-346 (1996)
3. L.A. Halawi, J.E. Aronson, R.V. McCarthy, Electronic J. Knowledge Manag. **3, 2**, 75-86 (Academic Conferences Ltd, 2005)
4. J.M. Higgins, J.W. Vincze, *Strategic Management: Texts and Cases* (Dryden Press, New York, 1989)
5. O. Nicolescu, C. Nicolescu, *Entrepreneurship and the management of small and medium enterprises (Intraprenoriatul si managementul intreprinderilor mici si mijlocii)* (Economic Pub., Bucharest, 2008)
6. P. Muller, J. Julius, D. Herr, L. Koch, V. Peycheva, S. McKiernan, *Annual Report on European SMEs 2016/2017. Focus on self-employment* (European Commission, 2017)

Blockchain technology – a new approach in business environment

Olga Bucovetchi[1,*], *Dorel* Badea[2], and *Radu D.* Stanciu[1]

[1]University "Politehnica of Bucharest", Economic Engineering Department, 313, Spl. Independentei, 060042, Bucharest, Romania
[2]"Nicolae Bălcescu" Land Forces Academy, Department of Management, 3-5, Revolutiei St., 550170, Sibiu, Romania

Abstract. Given the newest trends in business and the growing interest towards Internet of Things, it became natural that economic environment switched the paradigm of classical business model. Blockchain technology is a very timely topic of discussion both in academic environment and outside it. This element was proven by authors when analyzing the topic towards the number of publication releases and the search engine queries. Furthermore, the authors shaped the framework of blockchains' development highlighting the main advantages and disadvantages for Industry 4.0.

1 Introduction

Specialists discuss that the world is at the start of a new industrial revolution, which is considered the fourth revolution and is called "Industry 4.0" [1-5]. The manner of doing business in any company changes due to several factors, such as: globalization, high level competition, information becoming a key resource or digitalization (both of work space and business environment). Connecting many products to the internet, the presence of highly sensitive sensors, the expansion of wireless communications, developing intelligent machine and real-time data analysis have the potential to turn the way the production is done. Given this, the business environment tends to adapt to this shift of paradigm. As a consequence, businesses turn to cloud or at least try embed more and more features related to Internet of things, not only in terms of marketing strategy , but also by integrating innovative elements such as cryptology or cryptography..

In 2008, Nakamoto [6] introduced the concept of blockchain without using the exact term yet explaining the technology by presenting Bitcoin as virtual currency. Bitcoin is known as the first virtual currency. Yet, in only 10 years , the number of known virtual currencies grew to more then 1590 [7], the most traded elements being the following:

Table 1. Top 5 most traded virtual currencies

Currency	Market capitalisation [$]
Bitcoin	141,568,525,343
Ethereum	70,160,750,450
Ripple	26,661,640,812
Bitcoin Cash	20,283,402,474
EOS	11,347,212,707

Until recently there was no official definition of "blockchain". Still, in March 2018, blockchain was defined as "a digital database containing information (such as records of financial transactions) that can be simultaneously used and shared within a large decentralized, publicly accessible network" and also as "the technology used to create such a database" [8].

Expert Aaron Montemayor Walker explains the technology in few easy steps [9]:
- One party requests a transaction
- Requested transactions are funneled into a P2P (peer-to-peer) network and broadcast to each individual computer (node)
- Individual nodes receive the request and validate the transaction using an algorithm
- Approved transactions are represented as blocks and added to a public ledger
- Once the block is added to an existing chain, transactions are complete and permanent.

A blockchain is characterized by censorship resistance, immutability and global usability, and has a global network of validators called miners, who maintain it through block rewards, named crypto tokens [10,11].

The paper presents, in great, a literature research towards the concept of "blockchain technology". The literature research shall represent the starting point for a deeper analysis in order to help practitioners reshape their businesses, in order to become more globally competitive.

2 Research methodology

The research methodology had the following steps:
- Enquiring Web if Science database towards the number of publications that had "blockchain" as topic and assessing the categories regarding those results
- Investigating "blockchain" as "search topic" on Google Trends.

* Corresponding author: olga.bucovetchi@upb.ro

© The Authors, published by EDP Sciences. This is an open access article distributed under the terms of the Creative Commons Attribution License 4.0 (http://creativecommons.org/licenses/by/4.0/).

- After the comprehensive trend analyses, a wide-ranging literature review was conducted and summarized in order to shape the advantages, disadvantages and possible applications of blockchain technology as the IT impact on the company is not only felt from the outside environment, but also from within the company

Web of Science is an online subscription-based scientific citation indexing service that provides a comprehensive citation search.

It gives access to multiple databases that reference cross-disciplinary research, which allows for in-depth exploration of specialized sub-fields within an academic or scientific discipline. [12]

Google Trends is a public web facility of Google Inc., based on Google Search, that shows how often a particular search-term is entered relative to the total search-volume across various regions of the world, and in various languages.

The horizontal axis of the main graph represents time (starting from 2004), and the vertical is how often a term is searched for relative to the total number of searches, globally. [13]

3 Results

The level of interest towards "blockchain" embedded two layers: the academic environment and the non-specialists' environment. The authors assumed that the more attractive the topic is, the more many publications are released.

This research was conducted on the 1st of May 2018. We collected a number of 594 of publications within the selected period, namely 2013-2017.

As it can easily be observed, the evolution is positive, having an exponential increasing trend. This trend is presumed to be a consequence of the novelty of the topic.

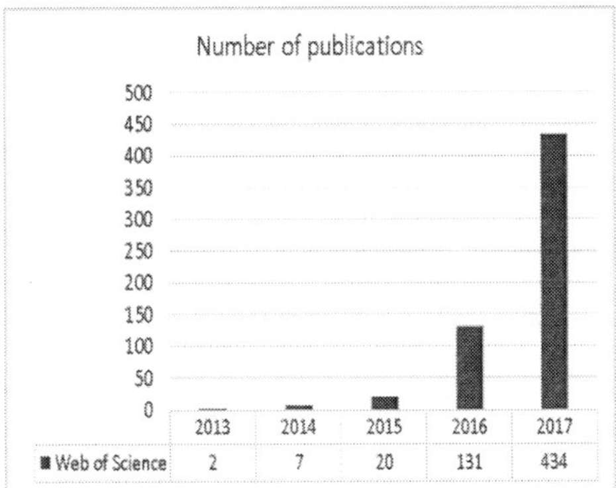

Fig. 1. Evolution of number of publications on "blockchain"

In terms of "Web of Science" domain, most papers deal more with the technology as engineering process, only recently, scientists shifting the paradigm to business features (Table 2).

Table 2. Top 10 papers categories on "blockchain"

Web of Science Categories	No. of papers
computer science information systems	180
engineering electrical electronic	165
computer science theory methods	162
telecommunications	95
computer science interdisciplinary applications	81
computer science software engineering	63
business finance	49
computer science artificial intelligence	46
computer science hardware architecture	45
business	28

As for the non-specialists' interest towards blockchain, Google trends revealed that the interest reached a pick at the end of 2017, nowadays, the trend being negative. This is considered to be also an effect of the negative marketing campaigns conducted by the conventionall financial organizations. Furthermore, many governments try to establish an restictive regulatory framework for "smart contracts".

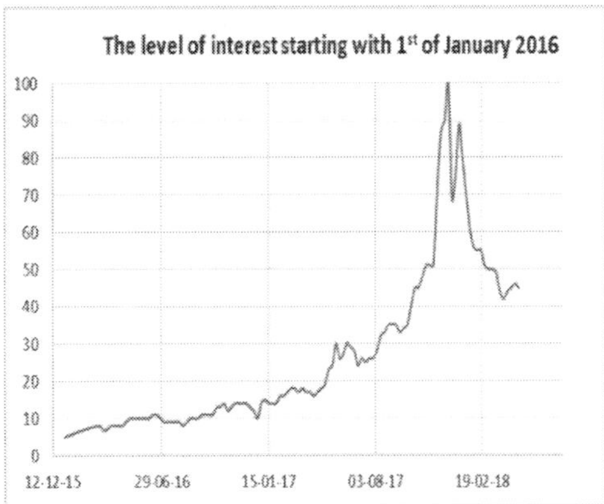

Fig. 2. People's interest towards "blockchain"

The numbers associated in a certain period with interest (as seen in Fig. 2) represent the search interest associated with the highest point in the chart for the specified region and time. Value 100 is related with the maximum popularity of the term. We selected the period after 1st of January 2016 as, starting that date, Google trends implemented a more performant tool in collecting and analyzing these particular type of data.

One may conclude that the trend shifted to negative as Bitcoin value started to decrease. As it can be observed in following tables, the search of "blockchain technology" on Google is correlated mostly with business environment (Bitcoin, wallet, investment, ...)

Table 3. Top 5 search terms related to blockchain

Searched topic	Scoring
Blockchain	100
Bitcoin	14
Wallet	6
Ethereum	3
Blockchain.info	3

Scoring is done on a relative scale, where the most frequently searched topic has 100 points.

Table 4. Related entities.

Term of search	Results
Ethereum	massive growth
Cryptocurrency	massive growth
Technology	massive growth
Investment	massive growth
Ripple	massive growth

According to Google trends "The results marked "Massive Growth" have shown an increase almost impossible to measure, most likely because the topics are new and previously there were few searches for them (or not at all)".

Table 5. Top 5 related queries to "blockchain"

No.	Related query
1.	blockchain bitcoin
2.	bitcoin
3.	blockchain wallet
4.	blockchain technology
5.	what is blockchain

Another layer of the analysis took into consideration the interest of academia in blockchain technology vs. the interest of business sector. It was revealed that the scientific interest is not followed by a putting in practice of the theoretical findings.

Table 6. Top 5 most productive countries

Country	No. of papers
USA	162
China	107
England	64
Germany	49
Italy	39

Despite the results retrieved from Web of Science (see Table 6), L. Lewis states that top 5 countries leading in blockchain industry is [14]:

- Switzerland
- United Kingdom
- Estonia
- Singapore
- Cyprus

Even if most people refer to "blockchain" only in terms of crypto-currencies, this technology has multiple applications: [15]

- As a system of record:
  - For digital identity
  - For authentication in certain context (tokenization)
  - For governments
  - For financial institutions
  - For audit trails
- As a platform:
  - For smart contracting
  - For automated governance
  - For markets
  - For streamlining of clearing and settlement
  - For automating regulatory compliance

There should be mentioned several blockchain platforms that embed those elements:

- Bitcoin [16] is the first and best known blockchain network, mainly oriented on crypto-currency transactions
- Ethereum [17] was launched as an open-source blockchain platform in 2015, defines smart contracts and is designed for a large variety of decentralized applications (DApps)
- Hyperledger [18] is a Linux Foundation Project and incubates and promotes a range of business blockchain technologies, including distributed ledger frameworks, smart contract engines, client libraries, graphical interfaces, utility libraries and sample applications.

The advantages of the blockchain technology are the following [10,19]:

- self-sovereignty - users identify themselves and maintain control over the storage and management of personal data;
- trust - the technical infrastructure offers secure operations (payments or issue of certificates);
- transparency and provenance - to perform transactions in knowledge that each party has the capacity to enter into that transaction;
- immutability - records are written and stored permanently, without the possibility of modification;
- disintermediation - no need for a central controlling authority to manage transactions or keep records;
- collaboration - ability of parties to transact directly with each other without the need for third-parties
- durability – the decentralized networks offer no central point of failure

One the other hand, we shall mention also the disadvantages, as for example:

- large energy consumption
- uncertain regulatory status
- the novelty of the technology
- performance – it is slower than a centralized database
- costs

4 Conclusions

As Holotescu [10] mentioned: "blockchain has surged from a technology with narrow applications related to digital currencies, to one with important applications in many domains, drawing attention to policy makers at different levels too.

In spite of the evidences that blockchain could solve some of the most pressing problems in business and society today, there is a gap in turning many use cases into reality, particularly from a strategic viewpoint.

In October 2017, Pannetta mentioned blockchain as a strategic technology trend it "holds the promise to change industries, and although the conversation often surrounds financial opportunities, blockchain has many potential applications in government, healthcare, content distribution, supply chain and more." [20]

Furthermore, being such new technology, there are many risks that might occur as there was not enough time of testing it therefore "black swan" events are highly probable.

Nevertheless, we shall mention that specialists predict that [9, 21, 22]:

- major institutions will offer pilot programs to the public
- blockchain will reshape the future
- financial institutions will invest more and more in blockchain technology
- blockchain – IoT (Internet of Things) will boom

References

1. H. Lasi, H.G. Kemper, P. Fettke, T. Feld, M. Hoffman, Industry 4.0, Business & Information Systems Engineering 6 4 (2014) doi: 10.1007/s12599-014-0334-4
2. SY Wang, JF Wan, DQ Zhang, D. Li, Ch. Zhang Towards smart factory for industry 4.0: a self-organized multi-agent system with big data based feedback and coordination, Computer Networks, **101** (2016) doi: 10.1016/j.comnet.2015.12.017
3. D. Ivanov, A. Dolgui, B. Sokolov, F. Werner, M. Ivanova, A dynamic model and an algorithm for short-term supply chain scheduling in the smart factory industry 4.0, Int. J. Prod Research **54** (2) (2016) DOI: 10.1080/00207543.2014.999958
4. M. Stonebraker, S. Madden, D. J. Abadi, S. Harizopoulos, N. Hachem, P. Helland, The end of an architectural era (it's time for a complete rewrite), Proceedings of the 33rd International Conference on Very Large Data Bases (VLDB), Vienna, Austria, pp. 1150–1160 (2007)
5. S. Singh, New Mega Trends: Implications for our Future Lives, (Palgrave Macmillan, 2012)
6. S. Nakamoto, Bitcoin: A Peer-to-Peer Electronic Cash System. (2008), available online at: https://bitcoin.org/bitcoin.pdf
7. https://coinmarketcap.com/all/views/all/
8. https://www.merriam-webster.com/dictionary/blockchain
9. https://blog.g2crowd.com/blog/trends/cybersecurity/2018-cs/blockchain/
10. C. Holotescu, Understanding Blockchain Opportunities and Challenges, The 14th International Scientific Conference eLearning and Software for Education Bucharest, April 19-20, 2018, DOI: 10.12753/2066-026X-18-253
11. R. Shulman, How Millionaire Jeremy Gardner And Jinglan Wang Built The Largest Blockchain Education Network (Forbes, 2018)
12. https://en.wikipedia.org/wiki/Web_of_Science
13. https://en.wikipedia.org/wiki/Google_Trends
14. L. Lewis, Countries Leading the Blockchain Innovation Movement, 2018, available online at: https://tokentarget.com/countries-leading-the-blockchain-innovation-movement/
15. https://www.coindesk.com/information/applications-use-cases-blockchains/
16. www.bitcoin.org
17. Ethereum.org
18. https://www.hyperledger.org/projects
19. https://www.e-spincorp.com/2017/11/24/pros-and-cons-of-blockchain-technology/
20. K. Pannetta, Gartner Top 10 strategic trends for 2018 (2017) available online at: https://www.gartner.com/smarterwithgartner/gartner-top-10-strategic-technology-trends-for-2018/
21. https://www.quora.com/What-is-the-future-of-blockchain
22. https://www.digitaltrends.com/computing/beyond-bitcoin-how-blockchain-will-reshape-the-future/

MATEC Web of Conferences **184**, 04017 (2018) https://doi.org/10.1051/matecconf/201818404017
Annual Session of Scientific Papers IMT ORADEA 2018

Study on the degree of use and knowledge of digital marketing elements in Romanian small and medium enterprises

Răzvan Mihai Dobrescu[1,*], *Cristina Petronela* Simion[1], *Iuliana* Grecu[1], and *Cristian Aurelian* Popescu[1]

[1] Politehnica University of Bucharest (UPB); Faculty of Entrepreneurship, Business Engineering and Management

Abstract. This paper aims to identify digital marketing elements, to analyze and to study the industry's degree of knowledge regarding digital marketing, as well as the degree of use of digital technology in the marketing mix of small and medium enterprises(SME's) in Romania. First of all, the basic concepts of digital marketing will be defined and analyzed: e-bussines, e-commerce and e-marketing, and then, in order to identify the degree of knowledge of these elements as well as the degree of use in the promotion process, a statistical research was carried out which had the following objectives: to determine the degree to which SMEs in Romania know the concepts of digital marketing; to identify whether these organizations have developed digital marketing strategies; to identify whether there are people or groups of people who are responsible for digital marketing activities at the organization level.

1. Introduction

We are witnessing today, in a world based on real-time communications, the emergence of new challenges that benefit some and disadvantage others. Thus, the information society, one of the most important factors of the business environment, has made it possible, through digital marketing, for real mutations to appear in everyday life that we would not even have thought about a few years ago.

In a dynamic world marked by rapid technological and economic change, SME's have to adapt and seek new ways to improve their business [1]. While customer requests are on the rise, companies are forced to do more with the same or even fewer resources.

Micro, small and medium-sized enterprises (SME) play an important role in today's society and economy, since they represent 99% of all enterprises in the European Union (EU) [2].

They have always been a central pillar of economic and social life of any country, underpinning the development of the economy and making a major contribution to growth and job creation. The innovative potential, dynamism, flexibility, adaptability and mobility are some of the characteristics that SMEs have and which are considered essential for the development of the economy of any country. Starting from these things, and wishing to build a strong economy, the European Union aims to strengthen the entrepreneurial spirit and create conditions for the development of innovative practices that ultimately increase the performance of SMEs [3].

Marketing and marketing specialists, today need to be open minded and to interlink their efforts with those of their IT colleagues to create profitable synergys that help improve the implementation of marketing strategies and to reward efforts in this regard by obtaining the necesary feedback in real time in order to establish future directions [4].

The elements behind this study stem from the companie's confusion about the differences between concepts of : digital marketing, e-marketing, e-business and e-commerce. Often, those terms overlap, but marketing specialists share the oppinion that the term digital marketing is more comprehensive and its scope of application includes e-business elements (IT infrastructure through which the organization's business strategy is implemented), e-commerce (purchasing and selling) and e-marketing (efforts to promote and sell services on the Internet).

Based on the literature there are clear research gaps in the field of E-Marketing especially in the areas of E-Marketing performance as well as E-Marketing adoption in Small Business enterprises (SBEs). To fill such gaps , there is a need for conducting more research to investigate the relationship between implementing E-Marketing and SMEs performance [5].

In the paper of [6] results suggest that e-business use emerges from technological and internal organizational resources rather than from external pressure. In addition, results show that e-business use contributes positively to firm performance through organizational innovation.

SME has bright future when they can take an advantage of information technology [7]. Beside cost reducing, they can reach wider market share. E-commerce is one of the solution to promote SME products. It was not enough to implement ecommerce without consider the success factor [8].

Starting from this information, the present paper aims to identify the current level of penetration and the degree of knowledge of the digital marketing elements as well as

* Corresponding author: author@e-mail.org

© The Authors, published by EDP Sciences. This is an open access article distributed under the terms of the Creative Commons Attribution License 4.0 (http://creativecommons.org/licenses/by/4.0/).

the degree of use of digital technology in the promotion process carried out in SMEs in Romania, as we consider that these elements can lead to increasing organizational performance .

2. Methodology

The main research method used was the survey, and the main research tool, the questionnaire. Several types of closed questions, as well as appreciation scales, were used to design the questionnaire.

The most costly and most prone to error is the information gathering stage. In the case of investigations, there are four types of problems that we certainly took into consideration, namely: they do not answer honestly, they are hard to get by, they are dishonest and they do not respond at all.

To obtain the raw data, research based on the questionnaire was used. The tool used was the online questionnaire, considering it to be more appropriate to the subject of the paper, it is easy to manage and complete, cost management is reduced compared to other survey tools, the advantage of overcoming spatial and temporal areas [9]. In addition to these advantages, the on-line questionnaire also has some limitations: it is a self-administered questionnaire because additional explanations may be needed on how to fill in the questions, managing the online questionnaire can not be controlled and supervised by an interviewer and due to the lack of the interviewer, abandonment of the questionnaire completion can be recorded.

The questionnaire was designed and designed to meet the proposed objectives. For the final form of the questionnaire, a small group of specialists, both in digital marketing and in statistical surveys, was consulted resulting in several improvement loops, until the final form was sent to the respondents. This final form of the questionnaire contains a total of 32 questions, but in this paper only the answers from some of them are used, as the rest will be the subject of another research.

In order to draw the appropriate conclusions, in the information analysis step, the collected data will be analysed using the IBM-SPSS program, dedicated to statistical study.

3. Results

As stated before, the questionnaire was sent directly by the team members to small and medium size romanian companies and the rate of response was above expectations. Small companies with less than 9 employees and a turnover of less than 2 million Euro account for 48.6% of all the completed questionnaires. The other half of the responses is divided among small enterprises with no more than 49 employees that account for 22.9% of answers, medium enterptises with no more than 249 employees that account for 25.7% of all respondents and non-proffit organizations that represent only 2.9% off our sample.

When analysing the business sectors of the respondent companies we can see that the best represented sector is IT&C accounting for 20% of the companies while construction and commerce account for 11.4% of respondents each. All other sectors represent less than 10% of the sample and are not especially mentioned. The fact that more than 30% of the respondent companies come from IT&C and commerce that are two fields that should heavily rely on digital marketing as the main channel for promoting their products should be considered when analysing the overall results.

A very important metric for our study consists of the age of the company, as we expect that younger companies will use digital marketing to a more extended degree than the well established older competitors. This is due to the fact that younger companies ten to be entrepreneurial, dynamic and have smaller marketing budgets. On the other hand, older companies tend to have larger marketing budgets and this allows them to have a strong online presence and also a larger number of years of presence in the market should allow them the time to consolidate their online footprint. In our sample 34.3% of the respondent companies were formed between 2000 and 2009 and have been on the market for more than 10 years. Only 20% of them were formed in the recent years, starting with 2010 and the other categories are made up by companies that were set up between 1990 and 1999 and account for 21.9% of the sample while companies with more than 30 years in the market account for 23.9% of respondents.

Then we asked the respondents to asses their online presence on a scale from 1 to 5. As expected we found a relatively strong online presence with more than 50% of the companies evaluating their online presence with 4 and 5. 19.4% of all companies said that they have a medium presence online while the other 30.5% a low one.

Investigating further we found that out of the 50% of companies that assessed their online presence as high, the larger part were older companies, established before 1990. This category accounted for 38.9% of the respondents while only 5.6% of them were companies established after 2010 as shown in Fig.1.

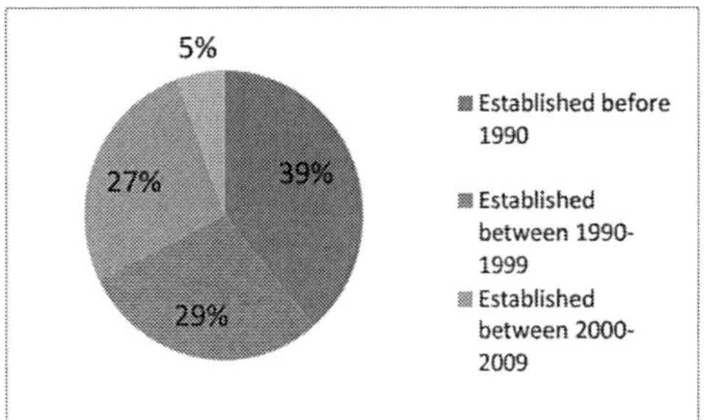

Fig.1. Companies with strong online presence

Even if it was expected than younger companies would be more inclined to using modern digital marketing tools, it is clear that in the case of well established companies the extra time and the availability of founding had a larger influence on the use of digital marketing.

On the other side, among companies that reported a low online presence, 44.4% were companies established after 2010 and 33.1% were companies established between 2000 and 2009. Of the companies that reported a low online presence only 10% were established prior to 1990.

In Fig.2. we can see that companies established before 200 have reported a better online presence than the younger competitors that reported only a medium and weak position.

When asking about the existence of a digital marketing strategy in the company, 40% of the respondents answered that they don`t have a digital marketing strategy implemented and among them 53% said that such a strategy will be developed and implemented. Out of the companies that reported not having a digital marketing strategy, 57.3% are young companies established after the year 2000.

Most of the copanies in the sample, 71.4%, reported to have one or more employees that are responsable for digital marketing activities. The rest of 21.4% use outsourcing in order to coordinate the digital marketing activity and 7.1% do not have anyone responsable for this type of activity. Out of all the companies 35.7% reported to employ a person responsable for all types of activities (e-business, e-commerce and e-marketing), 28.6% only have a person responsable for e-marketing activities and 7.1% of them have a person responsable for e-commerce activities.

When asked how familiar the marketing manager is with the terms of ebusiness, e-commerce and e-marketing we found that most of them knew what e-commerce and e-marketing mean. In the case of e-marketing, 57.3% of the respondents said they are familiar or very familiar with the term while only 17.2% admitting to having limited knowledge of the meaning of the term. In the case of e-commerce 53.6% of managers reported a good and very good understanding of the term meaning and 21.3% reported being less familiar with the term. Regarding e-business, the level of knowledge reported by managers is lower than for the other two terms. Only 46.4% of them reported to be familiar or very familiar with the meaning of e-business, while 3.6% of them admitted to have never heard of the term.

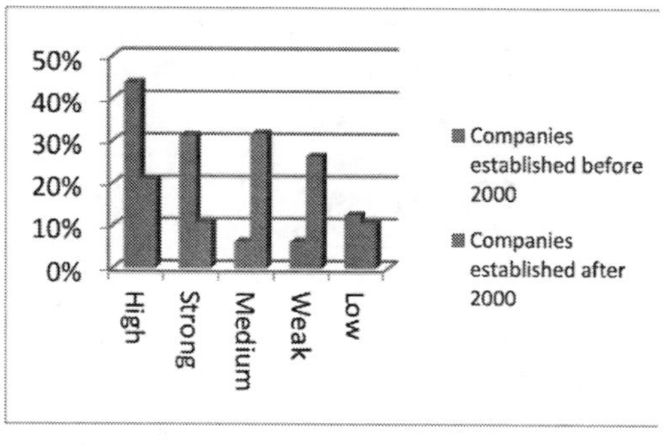

Fig.2. Distribution of companies by online presence

Out of the managers that reported a strong knowledge of the three terms of e-business, e-commerce and e-marketing, 21% reported a medium online presence and not having a digital marketing strategy in place but willing to implement one in the next year. In all those cases the answers came from young companies established after 2010 and we can assume that the lack of time or a tight marketing budget are the factors affecting their moderate usage of modern digital marketing techniques.

When asked if there is any difference between the meaning of the terms ebusiness, e-commerce and e-marketing, 75% of the managers reported they consider that there are differences between the three terms and 25% of them admitted to not know if there are differences. All of the 25% of the managers that did not know wether there are differences among the three termms, also reported to be less familiar or have medium knowledege about each of the terms.

In order to asses the degree of penetration of new digital marketing techniques within the companies we asked the managers witch of the digital marketing components (PPC campaigns; Social Media campaigns; Display Advertising; SEO; Content Marketing; E-Mail Marketing; Website; Mobile...) ranks the first when ordered by budget. Not surprisingly, almost 43% of the respondents said that website maintenance and improvement have the largest budget out of all digital marketing components. Other three components came at closed results. PPC campaigns have the largest budget in the case of 14,3% of the respondents and for 13,9% and 13,4% of them, Social Media campaigns and Display Advertising campaigns rank the first in terms of money spent by the company for digital marketing.

4. Conclusions

Our study's main goal was to to identify the main digital marketing elements and to study and analyze the industry's degree of knowledge regarding digital marketing, as well as the degree of use of digital technology in the marketing mix of SMEs in Romania. For this purpose we used a questionnaire that was sent to a large number of companies and the results were analysed by the team in order to draw conclusions on the main factors that influence the use and adoption of digital marketing techniques.

We aimed at profiling the companies that regularly use digital marketing elements in promoting their products and at understanding what are the key characteristics of those firms.

We found out that digital marketing is a fairly new concept for the Romanian managers and that only around 53% of them are familiarized with terms from digital marketing. Surprisingly, 21% of the managers that reported a strong knowledge of the three terms of e-business, e-commerce and e-marketing and understood the importance of on-line presence, also reported a medium online presence and not having a digital marketing strategy in place in their companies. When asked if there is any difference between the meaning of the terms ebusiness, e-commerce and e-marketing, 25% of the respondents admitted to not know if there are differences.

We found out that more than 50% of the companies consider to have a relatively strong online presence and out of them the larger part were older companies, established before 1990. The older companies made up for around 39 % of the high online presence respondents. Those findings tell us that even if it was expected that younger companies would be the category that used modern digital marketing tools the most, it is clear that in the case of well established companies the extra time and the availability of founding had a larger influence on the use of digital marketing techniques.

In order to strengthen our conclusion on the limited penetration of digital marketing techniques in Romanian SMEs, the main focus in terms of allocation of funds in those companies is the maintenance and improvement of website and not the newest and more sophisticated categories of mobile marketing or content marketing.

Acknowledgement: This work has been funded by University Politehnica of Bucharest, through the "Excellence Research Grants" Program, UPB – GEX 2017. Identifier: UPB-GEX2017, Ctr. No. 94/25.09.2017 (IMkDiP_IMM).

References

1. M. Robu,, *Use of social media components in bussines-SME`s point of view(Utilizarea componentelor Social Media în mediul de afaceri – perspectiva IMMurilo)r*, Munich Personal RePEc Archive, MPRA Paper No. 44585, posted 7, (2013)
2. P. Kuller, M. Vogt, D. Hertweck, M. Grabowski, *A Domain Specific Approach Journal of Innovation Management in Small & Medium Enterprises*. IT Service Management for Small and Medium Sized Enterprises, (2012)
3. L.D. Anghel, E. Matei, Marketing for SME (*Marketingul întreprinderilor mici şi mijlocii)*, Editura ASE, (2005)
4. C.D. Gostian, *Research on marketing of professional training services in the online environment (Cercetare privind marketingul serviciilor de formare profesională în mediul online)*, Cluj Napoca, (2015)
5. Hatem El-Gohary, *E-Marketing - A literature Review from a Small Businesses perspective*, International Journal of Business and Social Science, Editor, Centre for Promoting Ideas, USA, Volume 1, Issue 1, (2010)
6. P. Soto-Acosta, S. Popa & D. Palacios-Marqués, *E-business, organizational innovation and firm performance in manufacturing SMEs: an empirical study in Spain,* Technological and Economic Development of Economy, Volume 22, 2016 - Issue 6,(2016),https://doi.org/10.3846/20294913.2015.1074126
7. R. Pulfer, O. Bucovetchi, *Understanding the Organization's Readiness for Change*, Central and Eastern European Lumen Conference on New Approaches in Social and Humanistic Sciences, Chisinau, Moldova, politica de finanţare.453-458, ed. Medimond, (2016), WOS: 00039152150008
8. F. Fahruzzaman, A. Pribadi Subriadi, *An Analysis of E-Commerce Adoption for Small and Medium Enterprises in Indonesia*, IPTEK Journal of Proceedings Series, Volume 2, No. 1, (2015), DOI: http://dx.doi.org/10.12962/j23546026.y2015i1.1046
9. G. Grossek, *Email for marketing research*, Revista de Informatică Economică, (2004

MATEC Web of Conferences **184**, 04018 (2018)
Annual Session of Scientific Papers IMT ORADEA 2018
https://doi.org/10.1051/matecconf/201818404018

Examination of fiscal expenditures on education and research as human capital investments

Judit T. Kiss[1,*]

[1]University of Debrecen, Department of Engineering Management and Enterprise, 4028 Debrecen, Hungary

Abstract. The economic role of knowledge, education and research as investments in individuals' human capital has become a real focus of attention among researchers and policy makers in Europe over the last few decades. Individuals' high level skills and the versatile knowledge acquired in a formal or informal educational context have become extremely important factors in evolving knowledge-based economies. In this article we primarily analyze fiscal expenditures on education, and on research and development among OECD countries, and examine the change over time of state expenditures on education.

1 Introduction

Research and development, investments in knowledge have an increasingly important role in enabling adaptability to economic change, and in contributing to economic growth at both a national and a regional level.

The economic and social impact of expenditures on health, R&D (Research and Development), education and training as investments into individuals' human capital have been examined over several decades G. S. Becker [1], A. Case, D. Lubotsky, and C. Paxson [2], M. Grossman [3], W. Groot, and H. Maassen van den Brink [4], S. M. Lynch [5], J. Mincer [6], [7], G. Psacharopoulos, and H. A. Patrinos [8], [9], T. W. Shultz [10], [11]. Many researchers have analyzed the positive impact of human capital investments on economic growth, and the positive externalities of expenditures on health, education and training for the economy and society at large G. S Becker, K. M. Murphy, and R. Tamura [12], R. E. Lucas [13], R. R. Nelson, and E. S. Phelps [14], T. W. Shultz [15], S. Rebelo [16], P. M. Romer [17], [18].

The European Commission developed a strategy for smart, sustainable and inclusive growth in 2010. The proposed objectives include, among others, support to increase the proportion of the population aged 30-34 who have completed tertiary education (from 31% to at least 40%), to achieve a target of investing 3% of GDP (Gross Domestic Product) in R&D, and to reduce the proportion of early school leavers to 10% [19].

In what follows, we examine the change over time in public expenditures on education and research and development. We try to analyze whether or not the state spent more on education between 2005 and 2013, and research and development between 2006 and 2012, in OECD countries.

2 The development of public expenditures on education

Usually, the level of public expenditure on education exceeds that of private educational expenditures. This relationship is one of the reasons why private rates of return to education are greater than fiscal rates of return and social rates of return to education. Investment in human capital can be partially characterized by public expenditure on education as a percentage of total public expenditure.

The OECD and EU22 averages for public educational expenditures as a percentage of total public expenditures were 11.2% and 9.9% in 2013. The lowest value for the indicator examined occurred in Hungary (6.8%) in 2013 [Fig.1].

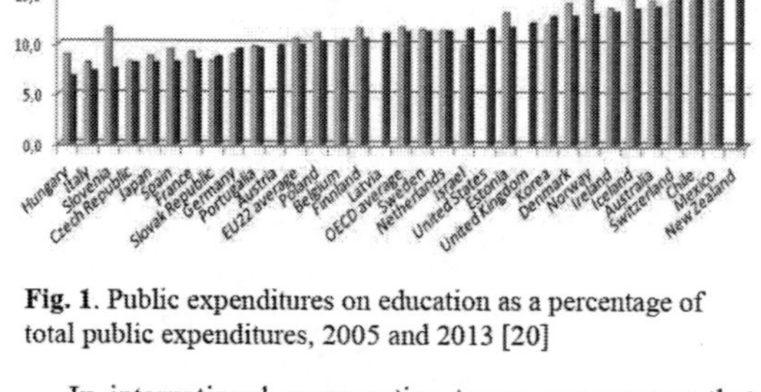

Fig. 1. Public expenditures on education as a percentage of total public expenditures, 2005 and 2013 [20]

In international comparative terms, we can see that between 2005 and 2013 the level of this indicator decreased in most countries. The largest decrease occurred in Slovenia, Mexico, Hungary, and Iceland,

---

[*] Corresponding author: tkiss@eng.unideb.hu

© The Authors, published by EDP Sciences. This is an open access article distributed under the terms of the Creative Commons Attribution License 4.0 (http://creativecommons.org/licenses/by/4.0/).

falling by 4; 3.1; 2.1 and 2.1 percentage points, respectively [Figs. 1, 2].

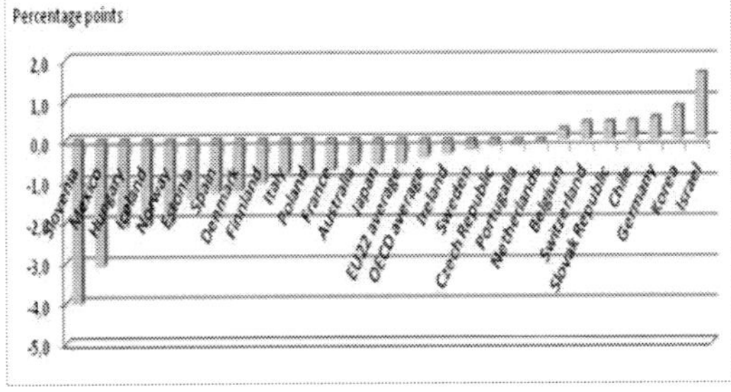

Fig. 2. The change in public expenditures on education as a percentage of total public expenditures, between 2005 and 2013 [20]

It should be noted that public spending on education as a percentage of total public expenditures may decrease if the State provides fewer resources for education compared to the other items of public expenditure, or the amount of total public resources decreases. If the level of State revenues decreases, the State should decrease resources allocated for different areas. This increase in total public expenditures as a percentage of GDP could be observed in most OECD countries for which data were available between 2005 and 2013 [Fig. 3]. However, the indicator decreased moderately (by 0.2 percentage points) in Hungary in the investigated period [Fig. 3]. The greatest increase could be found in Greece (16.73 percentage points); however, interpreting this change can be misleading without a knowledge of GDP data. The decrease in GDP was greater than the increase in the total public expenditure to GDP ratio in Greece during the examined period. This means that total public expenditures decreased in Greece.

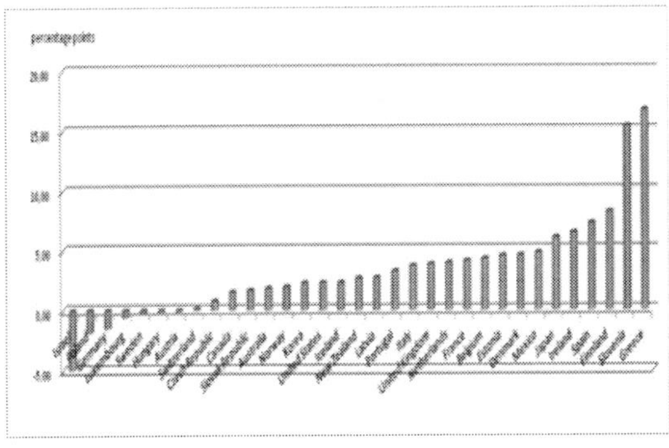

Fig. 3. The change in total public expenditure as a percentage of GDP in OECD countries, between 2005 and 2013 [20]

Despite the increase in the total public spending to GDP ratio, total public expenditures decreased, similarly to Greece, due to the larger decrease in GDP in several countries, including the Mediterranean member states (in Italy, Portugal, Spain) and Ireland.

It is also notable that while the most significant decrease in total public expenditures between in 2005 and 2013 occurred in Israel (by 4.94 percentage points) [Fig. 3], public expenditures on education increased to the greatest extent in Israel (by 1.6 percentage points) [Fig. 2] during the same period. In the case of Israel, it can be established that the state paid particular attention to education as investment in human capital; however, this statement does not necessarily hold for Hungary. Despite the moderate increase in total public expenditures, public spending on education decreased in Hungary between 2005 and 2013. Public expenditures on education between 2005 and 2013 decreased in most countries covered by the investigation.

Expenditures on education, training and research and development can be appreciably different for countries with different levels of development. In the following section, we examine public expenditures on education as a proportion of GDP.

3 Public expenditures on education as a percentage of GDP, between 2005 and 2013

The level of expenditures on education as an investment in human capital can be affected by the economic development of a country. In this section, we examine how public expenditures on education differ across countries with different levels of economic development.

In 2013, public education expenditures as a percentage of GDP ranged from 3.3% to 7.3% in countries involved in the investigation. The OECD and EU22 averages in 2013 were 4.8% and 4.7%, respectively [Fig.4].

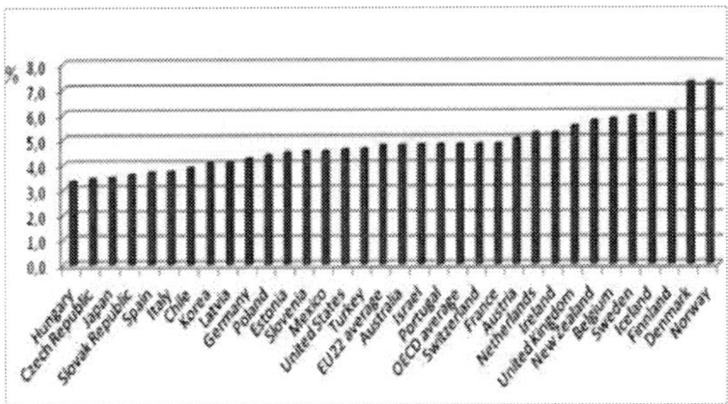

Fig. 4. Public expenditures on education as a percentage of GDP, in 2013 [20]

The highest level of public educational expenditures in relation to GDP occurred in Norway (7.3%), Denmark (7.2%), Finland (6%), and Iceland (6%). However we found the smallest value of this indicator for Hungary (3.3%), where it was 1.5 percentage points lower than the OECD average (and 1.2 percentage points lower than the EU22 average).

Between 2005 and 2013 public education spending as a percentage of GDP increased in most countries for which analyzed data were available. The value of the indicator decreased in 8 (out of 26) countries between 2005 and 2013 [Table 1]. The largest decreases in the indicator occurred in Norway and Hungary (by 1.3 and 1.1 percentage points, respectively) [Table 1].

Table 1. The change in public expenditures on education as a percentage of GDP, between 2005 and 2013 [20]

Country	Change (2005-2013) (percentage points)	Country	Change (2005-2013) (percentage points)
Norway	-1.3	OECD average	0.1
Hungary	-1.1	Germany	0.1
Slovenia	-0.6	Israel	0.2
Poland	-0.6	Portugal	0.2
Iceland	-0.5	Mexico	0.2
Sweden	-0.2	Australia	0.3
Switzerland	-0.1	Japan	0.3
Italy	-0.1	Slovak Republic	0.3
France	0.0	Finland	0.3
Czech Republic	0.0	Netherland	0.4
Estonia	0.0	Korea	0.5
Denmark	0.0	Belgium	0.6
Spain	0.1	Chile	0.6
EU22 average	0.1	Ireland	0.7

On the basis of the two indicators examined, most countries did not take any visible initiatives to respond to the challenges of the knowledge economy.

We continue our study by examining public expenditures on research and development. Given the absence of reliable data for the member states of the OECD for the period 2005 and 2013, the analysis is restricted to the period between 2006 and 2012.

4 Public expenditures on research and development as a percentage of GDP, between 2006 and 2012

In an economy, the public sector generally spends less on research and development compared to other macroeconomic sectors. Public expenditures on research and development in proportion to GDP in 2012 were the highest in Sweden (1.99%), Finland (1.99%), Denmark (1.91%) and Germany (0.96%) [Fig. 5]. The lowest values for this indicator occurred in Romania (0.3%), Hungary (0.43%), and Greece (0.45%). The EU28 and OECD averages in 2012 were 0.74% and 0.71, respectively [Fig. 5].

As we can see in Fig. 6, between 2006 and 2012 public expenditures on research and development as a percentage of GDP increased in all countries, with the exception of four.

Estonia, Korea, and Portugal where it increased by 0.35, 0.3, 0.26; 0.25 percentage points, respectively [Fig. 5; 6]. It can be observed that there was a positive movement in relation to research and development; however, there is room for improvement in Romania, Hungary and Greece. Between 2006 and 2012, public expenditures on research and development as a percentage of GDP decreased in Israel, Hungary, Canada and the United Kingdom by 0.14, 0.07, 0.04 and 0.02 percentage points, respectively. The largest increase can be found in the Czech Republic, It should be noted that the decrease in the indicator examined occurred from a higher level in Israel, Canada and the United Kingdom. However, in Hungary the ratio of expenditure on research and development to GDP (0.5%) was significantly lower than the EU28 (0.65%) and OECD (0.64%) averages in 2006. It cannot be stated that there was a positive change in relation to public spending on research and development in Hungary.

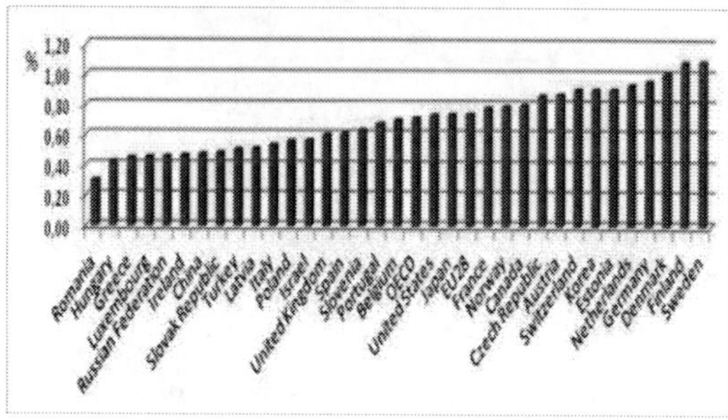

Fig. 5. Public expenditures on research and development

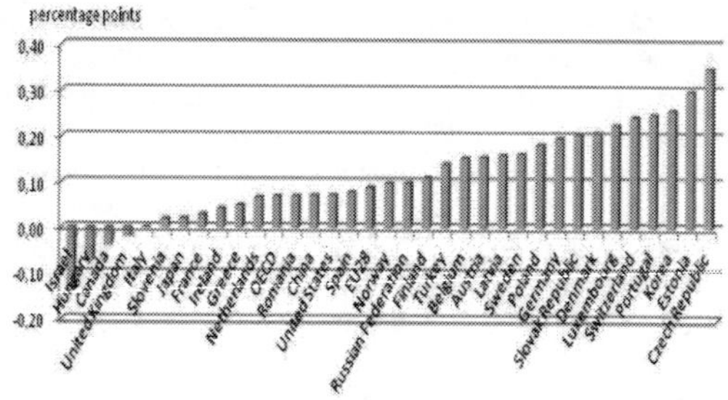

as a percentage of total GDP, in 2012 [21]

Fig. 6. Change in public expenditures on research and development as a percentage of total GDP, in 2012 [21]

5 Conclusion

In this study, we sought to examine the reactions of the public sector to the changing economic environment among OECD countries between 2005 and 2013, a task which requires increasing and versatile knowledge. Among other aspects, we analyzed public expenditures on education as a percentage of total public expenditures, and the change over time in the level of public expenditures on education as a percentage of GDP.

Between 2005 and 2013, public expenditures on education as a percentage of total public expenditures decreased in most OECD countries for which data were available. The indicator decreased to the greatest extent in Slovenia, Mexico, Hungary, and Iceland, falling by 4, 3.1, 2.1 and 2.1 percentage points, respectively. The decrease in the indicator examined in Hungary is noteworthy because in 2013 the lowest value of all was also recorded in Hungary.

The decreases in public expenditure on education in the period examined occurred against a background of rising total public expenditure. This indicates that the State provided fewer resources for education compared to the other areas of public expenditure in several countries among the OECD member states.

We analyzed the development of public education expenditures as a percentage of GDP. This indicator ranged from 3.3% to 7.3% in 2013. In the majority of countries for which analyzed data were available public expenditures on education were lower than the OECD average (4.8%). Similarly to the ratio of public expenditure on education to total public expenditure, in 2013 the lowest value for the ratio of public spending on education to GDP could be found in Hungary (3.3%), a value which was 1.5 percentage points lower than the OECD average. In Hungary, public spending on education was significantly below the EU22 and OECD averages, and the State spent less on education in terms of its economic development and compared to other countries during the period under investigation.

Beside the indicators discussed above, we analyzed public expenditure on research and development. In 2012 public expenditure on research and development as a percentage of GDP was above 1% in Sweden (1.99%), Finland (1.99%), and Denmark (1.91%). In view of the change in this indicator, we can state that there is a positive movement in relation to public spending on research and development, because between 2006 and 2012 the indicator increased in most countries for which analyzed data were available. The largest increase occurred in the Czech Republic, Estonia, Korea, and Portugal, which recorded increases of 0.35, 0.3, 0.26 and 0.25 percentage points, respectively. The public expenditure on research and development to GDP ratio was the lowest in Romania (0.3%), Hungary (0.43%) and Greece (0.45%). However, we found positive changes in relation to this indicator in Romania and Greece, because there was a moderate increase in the ratio, of 0.07 and 0.05 percentage points, respectively.

References

1. G. S. Becker, The Journal of Political Economy. Vol. **70**, Issue 5. Part 2: Investment in Human Beings. pp. 9-49. Oct, (1962)
2. A. Case, D. Lubotsky, and C. Paxson, The American Economic Review, Vol. **92**, No. 5, pp. 1308-1334. (2002)
3. M. Grossman, In. N. E. Terleckyj (ed.), *Household production and consumption*, New York: National Bureau of Economic Research, pp. 147-224. (1976)
4. W. Groot, W. and H. Maassen van den Brink, Economics of Education Review, Vol. **26**, pp. 186-200. (2007)
5. S. M. Lynch, Journal of Health and Social Behaviour, Vol. **47**, No. 4. pp. 324-338. (2006)
6. J. Mincer, The Journal of Political Economy. Vol. **66,** Issue 4, pp. 281-302. (1958)
7. J. Mincer, Journal of Economic Literature, Vol. **8**, No. 1, pp. 1-26. (1970)
8. G. Psacharopoulos, and H. A. Patrinos, World Bank Policy Research Working Paper **2881**. (2002)
9. G. Psacharopoulos, and H. A. Patrinos, Education Economics, vol. **12**(2), pages 111-134. (2004)
10. T. W. Schultz,. The Journal of Political Economy.Vol. **68**, No. 6, pp. 571-583. (1960)
11. T. W. Schultz American Economic Review. Vol. **51**, No. 1, pp. 1-17. (1961)
12. G. S. Becker, K. M. Murphy, and R. Tamura, The Journal of Political Economy, Vol. **98**, No. 5, Part 2: The Problem of Development: A Conference of the Institute for the Study of Free Enterprises Systems, S12-S37. (1990)
13. R. E. Lucas, Journal of Monetary Economics, Vol. **22,** 3-42. (1988)
14. R. R. Nelson and Phelps, E. S, The American Economic Review, Vol. **56**, No. 1/2. 69-75. (1966)
15. T. W. Schultz, Education and Economic Growth. In Social Forces Influencing American Education, ed. N. B. Henry, *National Society for the Study of Education* pp. 46-88. Chicago University of Chicago Press. (1961)
16. S. Rebelo, The Journal of Political Economy, Vol. **99**, No. 3, (Jun., 1991), pp. 500-521. (1991)
17. P. M. RomerNBER Working Paper No. 3173. (1989)
18. P. M. Romer, The Journal of Political Economy. Vol. **98**, No. 5, Part 2: The Problem of Development: A Conference of the Institute for the Study of Free Enterprise Systems. S71-S102. (1990)
19. European Commission, *EUROPE 2020 A strategy for smart, sustainable and inclusive growth*, http://ec.europa.eu/eu2020/pdf/COMPLET%20EN %20BARROSO%20%20%20007%20- %20Europe%202020%20-%20EN%20version.pdf. 2010. Accessed on 03.2016.
20. OECD, *Education at a Glance, 2016. OECD Indicators*. ISBN: 978-92-64-25979-9. (2016)
21. OECD, *Main Science and Technology Indicators Database*, June 2014, Science, technology and Industry Outlook, 2014, 978-92-64-25979-9. http://stats.oecd.org//Index.aspx?QueryId=57863. Accessed on 03.2017.

Measuring the Innovation of Economy throught Global and European Tools

Delia Marcela Pop[1], *Mircea Teodor* Pop[2]

[1]University of Oradea, Industrial Engineering Department, Oradea, Romania
[2] University of Oradea, Engineering and Management Department, Oradea, Romania

Abstract. Global and European innovation assessment and measurement represents a means by which relevant analyses of the innovative process can be carried out, as well as long-term strategies that aim at sustainable economic development. Measurement of innovation is carried out by specialized bodies through specialized studies, using relevant indicators whose number and definition have been modified over time as needed. This paper proposes a comparative analysis of the assessments of a country performances in innovation by means of a global measurement tool (Global Innovation Index) and of a European one (European Innovation Scoreboard).

1 Introduction

From life in caves to living in intelligent buildings, from wagon to today's cars and from the early 20th century industry up to the current industrial platforms, evolution is undeniable. The essence of this evolution is man, his intelligence, his creativity, his knowledge, his perseverance or his inventiveness, characteristics that are associated with him. Throughout this evolutionary process each moment has represented an innovation of the existing means or processes, and the creation of some totally new, based on totally unknown directions until that time.

In the current stage that mankind is going through, these moments of generating the novelty are defined, conceptualized, researched, monitored, even the process of innovation itself being continually subject to innovation. As a result, it is strictly timely and the management and measurement of innovation processes in companies, i.e. in the structures that assure the development of national economies and the global economy.

2 Defining the concepts of innovation

Innovation is the essence of the innovation process. Over time it has been defined in various forms that have also had specific evolution stages. Innovation is considered now as a challenge for industry, services, agriculture etc., but also for the economy in its whole, because innovation is seen as an essential element of competitiveness, as the essence of competitive development. From this perspective, it is increasingly accepted that only the achievement of results in research and technological development has become insufficient to increase competitiveness as they do not create a competitive economic outcome. It is necessary to transform the research results into innovative products and services that will generate a competitive advantage on the market. This is also the current strategic orientation of the European Union that aims at transforming Europe into an Innovation Union which, by stimulating competitiveness, targets the economic development and the quality of life of its citizens.

The conceptualization of innovation starts from defining it as a product or process that targets a marketable application of a novelty, an improvement or an invention. The innovation-related scientific research has deepened the basic knowledge and broadened the perspectives of addressing them. A consequence of this development is the formalization of science called the innovation management that has generated new innovation issues related to the following:
- the scientific foundation of the innovation process and its formal definition.
- the defining of business models by which the results of research and technological development are implemented through innovation in production.
- the emergence of so-called open innovation, i.e. the involvement of external collaborations in a company's innovation (R&D - Research & Development consortia, universities or research institutes, other companies in specific fields or specialists and communities involved in innovation by crowdsourcing) that have an essential role in generating new ideas or technologies, in the more efficient use of company assets or in finding and capitalizing of market-access opportunities even during the innovation cycle.
- improvement in the way innovation is managed.
- development in the methods and tools of measuring innovation and its results.

Correspondent Author: deliampop@gmail.com

© The Authors, published by EDP Sciences. This is an open access article distributed under the terms of the Creative Commons Attribution License 4.0 (http://creativecommons.org/licenses/by/4.0/).

- modelling the innovation stimulation that addresses the process which the innovator has to follow in order to generate and stimulate innovation starting from a research result.
- the generating and developing of specific funding for innovation; in large firms that have their own innovatory processes the funding can be on projects, but in the case of SME's, funding opportunities extend to: grants, microcredits, business angels, venture capital firms, etc.

3 The European Union's strategy for stimulating the innovation

The European Union is a strong participant in the global process of research, technological development and innovation, being the main "knowledge factory" in the world. Thus, in June 2017, the World Intellectual Property Organization (WIPO), a specialized agency of the UN, published the top of the world's most innovative economies included in the Global Innovation Index[1]. The top three positions in the ranking list are held by European countries, of which the second (Sweden) and the third (Netherlands) are European Union countries. Also, the fifth place (UK) and the sixth one (Denmark) are held by EU countries. At the same time, in the top 25 of the most innovative countries in the world, 13 countries belong to the European Union. It has been determined that the community space supplies almost one-third of the scientific and technological production in the world.

But even being in this leadership position, the European Union is subject to strong international competitive pressures, especially from the emerging economies. At the same time, the aging phenomenon of the European population has a major effect on the economic growth of the area. In July 2017, EUROSTAT published a study on the impact of demographic aging in the European Union[2], according to which in 2050 the population aged over 65 will be of 28.5% compared to 2016 when this percentage was 19.2%. As a result, the EU's economic strategy is geared towards the economic growth that can be achieved by stimulating research, technological development and innovation in particular. Innovation is the one that generates jobs and stimulates market competitiveness, elements that have a strong macro-economic growth effect.

In this direction, the European Union, “A Union of Innovation” - the EU's strategy of promoting this process, provides the framework for supporting an environment in which member countries encourage innovation in products, services and business models, including the public sector. This initiative is part of the European Union's growth strategy "Europe 2020" which predicts cumulative investment in research and innovation of 3% of gross domestic product (GDP) by 2020[3]. One of the arguments in this guideline is linked to the fact that investments in research and innovation have a strong multiplier effect in the economy (every euro invested by the EU generates about 13 euros in terms of added value for businesses, meaning a significant multiplier of 13 times). As a result, Orizont 2020 is thought to be the European Union's largest research and innovation-oriented program that has a budget of € 80 billion over the run-up period (2014-2020).

The aim of this major program to stimulate research and innovation is to support areas that are considered as priorities by the leaders of the European Union, which is reflected in seven challenges where specific investment in research and innovation can have a real impact for the benefit of European citizens. These societal challenges are: Health, demographic change and well-being, Food security, sustainable agriculture and forestry, bio-economy and marine and maritime research, including inland waterways, Safe, environmentally friendly and efficient sources of energy, Intelligent, clean and integrated transport, Action to combat climate change, the environment, the efficient use of resources and raw materials, Europe in a changing world, Safe societies.

4 Analyzing and measuring innovation

Innovation is a very complex process whose analysis and macroeconomic management reveals the need to measure it. In order to be quantified, it has been used a number of instruments that group indicators specific to the fields of research - development and innovation. All these tools, at their impact on the real environment of research and innovation, have highlighted, following feed-backs, deficiencies that have been gradually corrected by successive improvements of the indicators used, the ways of collecting data, their processing, resulting reports etc.

Worldwide, one of the most widely used and accepted tools for measuring innovation is the Global Innovation Index. This is an annual report prepared and published by the World Intellectual Property Organization (WIPO), Cornell University (USA) and INSEAD (France)[1]. The analysis of innovation through the Global Innovation Index for 2017 was carried out within 127 functional economies and took into account 81 indicators that characterize global innovation, including the political environment, education, infrastructure, business sophistication, and so on. 2017 was the year when the study was entitled *Innovation Feeding the World* and was geared towards innovation in agriculture and food systems. This orientation has been given by the scientifically determined perspective according to which, at the global level, the earth's population will be of 9.7 billion in 2050, which will lead to a major increase in global food demand and significant pressure on the existing agricultural and food resources. From this perspective, the Index was oriented towards innovation as a means of supporting additional food demand but also as a solution for the sustainable finding and exploitation of food systems.

At the European Union level, the most significant tools for measuring innovation are:
- 1.The European Innovation Scoreboard (EIS) - is annually conceived by the European Commission, providing a comparative analysis of innovation performance in European Union countries, in other European countries as well as in neighbouring countries; with the help of this it assesses, on the basis of uniform criteria, the strengths and weaknesses of the national

innovation systems; the latest European Innovation Scoreboard is the one corresponding to 2016 (published in June 2017) [4] and contains procedural changes that provide an easier way of tracking the performance of innovation over time, and through the new indicators it identifies the level of investment in skills, digital literacy, entrepreneurship and public-private partnerships; The pilot version of this assessment was published in 2000 and since 2001 complete analyses have been prepared and published under this name; In 2017, the methodology for analysis and measurement of performance in innovation contained 29 innovation indicators, divided into three main groups: a) Innovation drivers, b) Business activities and c) Results, and for a summary assessment of the national innovation performance, it is determined the Summary Innovation Index (SII), obtained as a single figure, by aggregating the indicators of the Table, for each country; the SII value varies between 0 (the lowest performance) and 1 (the highest performance); depending on each country's SII, there is an average performance index for innovation in EU countries, an average benchmark for each country's performance; according to this comparison, each country is framed within the following innovation performance groups: a) Leaders in innovation - whose performance exceeds the EU average by at least 20%; b) Innovation followers - with performances below 10% of the average and up to 20% above the EU average; c) Moderate innovators - whose performances are below 10% of the European average but not more than 50%) Modest innovators - with a performance index below 50% of the EU average.

- 2.The Community Innovation Survey (CIS) - is a statistical analysis based on the Oslo Manual methodology; this analysis is coordinated by the statistical office of the European Commission, EUROSTAT; the use of this analysis began in 1993 with CIS 1 Light; the evolution of this analysis is given in Table 1.

Table 1. Community Innovation Survey (CIS).

Indicators	CIS 1 Light	CIS 2	CIS 3	CIS 4	CIS 2006	CIS 2008	CIS 2010	CIS 2012	CIS 2014	CIS 2016
Analyzed period	1993	1997-1998	2000-2001	2002-2004	2004-2006	2006-2008	2008-2010	2010-2012	2012-2014	2014-2016
Total number of participant countries of which	18	18*	18**	29	30	30	31	31	34	***
UE member countries	15	15	15	25	25	26	26	28	28	***

* except for Greece and Ireland that launched the survey in 1999
** except for Norway, Iceland, Luxembourg and Greece that launched the survey in 2002
*** at the time of writing, the final results have not yet been published by EUROSTAT

- 3.The European Trend Chart on Innovation in Europe - is a tool that analyzes innovation in Europe, introduced by the European Commission in 1998 to monitor changes in Member States' innovation policies; under this name was used until 2002.

- 4.Inno-Barometer - is an annual European-level research carried out by the European Commission, which focuses on the activities and attitudes of European countries towards innovation, also establishing trends in firms; starting with 2015, the research has been conducted by means of a sample survey based on a questionnaire mainly observing the way in which companies manage their innovation activities, plan their investments for modernization, pursue the results of innovation, and determine the barriers that limit the marketing of innovative products and services.

5 Comparative use of global and European innovation measurement tools

For a comparative analysis of innovation performance of a country in the European Union at global and European level, two studies were carried out annually, namely: the Global Innovation Index (for Global Analysis) and the Scoreboard European (for country analysis at European level), only the analysis for the countries of the European Union. Romania was used as a country, and the evolution in innovation for 2010-2017 was pursued.

Romania analyzed in terms of the Global Innovation Index in 2017 ranks 42th out of the 127 analyzed economies. The performance of the evaluated economies was determined by a set of 81 indicators distributed into the following groups: institutions, business and political environment, regulatory framework, education, R & D, human capital and infrastructure. The evolution of Romania's position in this ranking since 2010 can be found in Table 2.

Analyzing the data above, it can be noticed that Romania holds, in the analyzed period, the positions between 40 and 55, which means that it is in the first half of the ranking.

But, however, Romania's position is inferior to many countries in the region such as the Czech Republic, Poland, Bulgaria, Slovenia etc. Romania's position was followed, most of the years, by Greece, Turkey, Russia etc., countries considered to have an economic power superior to Romania.

At the same time, over the entire period, Romania's performance is more than 55% of that of the leader, being, in 2010, at almost two-thirds of this.

By conducting an analysis of innovation in Romania through the European Innovation Scoreboard (EIS) for the period similar to the previous analysis but within the European Union, a centralization of the data in Table 3.

Table 2 – The evolution of Romania's position in the Global Innovation Index between 2010 and 2017.

Indicators	2010	2011	2012	2013	2014	2015	2016	2017
Number of analyzed economies	132	125	141	142	143	141	128	127
Number of indicators used	60	60	84	84	81	80	82	81
Leader of classification	Iceland	Switzerland	Switzerland	Switzerland	Switzerland	Switzerland	Switzerland	Switzerland
Leader's score	4,86	63,82	68,20	66,65	64,78	68,30	66,28	67,69
Position of Romania	52	50	52	40	55	54	48	42
Score of România	3,22	36,83	37,80	37,84	38,08	38,20	37,90	39,16
Average score of Romania	66	58	55	57	59	56	57	58

Table 3 - Evolution of Romania's position in the European Scoreboard between 2010 and 2017.

Indicators	2010	2011	2012	2013	2014	2015	2016	2017
Number of analyzed economies	27	27	27	27	28	28	28	28
Number of indicators used	25	25	24	24	25	25	25	27
Leader of classification	Sweden	Sweden	Sweden	Sweden	Sweden	Sweden	Sweden	Sweden
Position of Romania	24	24	25	26	26	28	28	28
Type of innovator	Modest	Modest	Modest	Modest	Modest	Modest	Modest	Modest
Number of countries below the average of performance in EU	16	16	16	16	17	17	17	17

The graph shows that Sweden is still the European leader in innovation, being followed by Denmark and Finland. Romania maintains the last position, being considered a modest Innovator throughout the analyzed period. Since 2016, in the European Union there are only two countries with a modest Innovation state, most of the European countries being moderate innovators.

Conclusions

By the two studies: the Global Innovation Scoreboard and the European Innovation Scoreboard, it results the fact that the performances in innovation is an indicator that needs to be constantly pursued and refined. The global innovation measurement system is based on a larger number of indicators and manages a significantly larger number of countries participating in the research. Both systems have a special complexity, as well as different but relevant indicators and methodologies. By linking these studies and the results of the other innovation measurement tools described in the paper, Europe is still a pole of scientific research and technological development but it is exceeded by the United States, Singapore and Korea in terms of performance in marketing products and services resulting from innovation.

Another issue arising from the overlapping of the two surveys is that all EU countries are ranked in the Global Index to position 44, which is the first third of the ranking. The EU leader, Sweden, is ranked second in the Global Index, the first being a European country but not in the EU. This demonstrates, in both systems of assessing the innovation, the superiority of the European Union from the perspective of innovative processes.

As far as Romania is concerned, it is worth mentioning that for the year 2017 in the global ranking Romania is ranked 42th but it is the penultimate country in the European Union, while Greece is ranked 44th. This positioning reflects the fact that Romania as a member of the European Union has confirmed its position as modest Innovator and the Global Index of Innovation, but it is in the first third of this world ranking, because the European performances in innovation at the global level are superior.

References

1. *The Global Innovation Index 2017. Innovation Feeding the World,* Cornell University, INSEAD, and WIPO, **XVIII –XIX**, (2017)
2. http://ec.europa.eu/eurostat/statistics-explained/index.php/Population_structure_and_ageing
3. https://ec.europa.eu/programmes/horizon2020/what-horizon-2020

Correspondent Author: deliampop@gmail.com

4. http://ec.europa.eu/growth/industry/innovation/facts -figures/scoreboards_ro

Analysis of supporting indicators for innovation in Romania compared to neighboring EU countries

Bungau Constantin[1,*], *Gherghea Ion Cosmin*[1]

[1]University of Oradea, Engineering and Management Department, Romania

Abstract. The paper presents a comparative study of indicators which targeting the facilities and support for the innovation process. The study targets indicators regarding human resources, support structure / infrastructure and legislative provisions. The analysis is carried out over the last 5 years, comparing the indicators in Romania and the neighboring countries (Bulgaria, Hungary, Austria, Slovakia and Poland) and highly developed countries (Germany and France). Based on this study, will be drawing conclusions for improvement of these indicators. The indicators that will be taken into consideration are: Availability of scientists and engineers, Researchers, Quality of the education system, Researchers in Research and Development, Research and development expenditure, Firm-level technology absorption, University-industry collaboration in Research & Development, Quality of scientific research institutions, Gross domestic spending on Research and Development, Property rights, Intellectual property protection.

1 Introduction

There is a significant distinction between invention and innovation. The invention is the process of creating a new idea or a new concept, while innovation is transforming the new concept into a commercial success. In conclusion, the applied invention is an innovation. The innovation process starts from identifying an existing problem, looking for new solutions that then implement it contributes to increasing profitability and competitiveness. [1]

Innovation has an important role in the development of a society, as Tim Stock et all considers that innovation is the "growth engine" of society. In the same paper Tim Stock defines innovation as a multi-stage process, and creativity, motivation and knowledge are the main factors leading to the first phase of innovation. [2].

The literature presents various studies that measure the innovation performance in Romania. [3- 6].

2 Research methodology

In order to carry out this study, there were analysed 11 indicators that aim and support the innovation process, which were structured in 3 groups: human resources, structure support / infrastructure and legislative provisions (according to Table 1). Statistical data was accessed from the World Bank website [7]. The analysis of the indicators has been carried out over the last 5 years, comparing the indicators of Romania with the neighboring countries (Bulgaria, Hungary, Austria, Slovakia and Poland) and the highly developed countries (Germany and France). As a result of data processing, have been generated line graphs that show the upward or downward trend in the last 5 years of each indicator [7].

Table 1. Structuring the analyzed indicators on three groups [7].

Groups	Indicators	Acronym
Human resources	Availability of scientists and engineers	ASE
	Researchers	R
	Quality of the education system	QES
	Researchers in R&D	RRD
Support structure / Infrastructure	Research and development expenditure (% of GDP)	RDE
	Firm-level technology absorption	FLTA
	University-industry collaboration in Research & Development	UICRD
	Quality of scientific research institutions	QSRI
	Gross domestic spending on R&D	GDSRD
Legislative provisions	Property rights	PR
	Intellectual property protection	IPP

3 Data analysis

3.1 Group 1 – Human resources

3.1.1 Availability of scientists and engineers (ASE)

In 2013, Romania ranks last in terms of the ASE indicator, registering a score of 3.64 on a scale of 1 to 7, but it can

* Corresponding author: bungau@uoradea.ro

© The Authors, published by EDP Sciences. This is an open access article distributed under the terms of the Creative Commons Attribution License 4.0 (http://creativecommons.org/licenses/by/4.0/).

see a slight increase by 2015, when it scored 4.13, after which it drops slightly by 2017, recording a score of 3.78, but keeping it higher than in 2013. Although Romania has declined by the past two years, there is an increase in the hierarchical level. From the last ranked in 2013, in 2017 it climbed to 5th place in the 8 countries analyzed. In 2017 it is worth noting the high level of Germany which registered a significant increase, surpassing Romania by 36.24%.

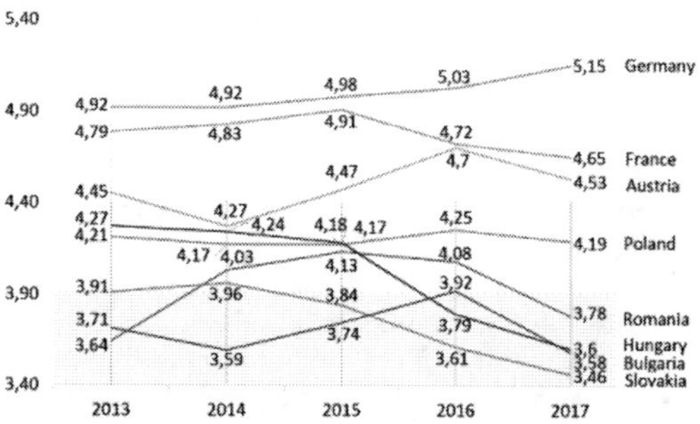

Fig. 1. Availability of scientists and engineers (ASE) [7]

3.1.2 Researchers (R)

Regarding R indicator, we analyzed the available data onto the last 5 years, namely the period 2011-2015. Data onto Bulgaria is missing from this study. This indicator is measured taking into account the number of researchers per 1000 employees. It is also important to note that the data are incomplete in some analyzed countries, according to the chart below. Taking these issues into account, the situation in 2013, the year in which the values of each analyzed country (with the exception of Bulgaria) are identified. In 2013, Romania ranks 6th, registering a score of 27.6. Germany ranks first with 549.28 points, with 49.95% above second place (France) and 1890.14% more than Romania, which is a big difference between the two countries.

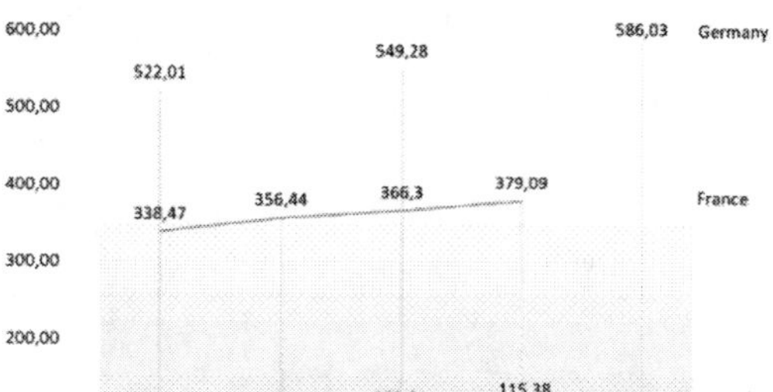

Fig. 2. Researchers (R) [7]

3.1.3 Quality of the education system (QES)

In 2013, Romania ranks 7th in terms of QES indicator, at a close distance from Poland, Hungary and Bulgaria, recording a score of 3.25 on a scale of 1 to 7, and in 2014 it is ranked 4th, then decreasing easy, going down to 7th in 2017.

In terms of QES, in 2017 Germany leads the standings ranking, being at the top of the ranking with a score of 5.37, followed by France and Austria. Compared with Romania, Germany is 91.78% above Romania.

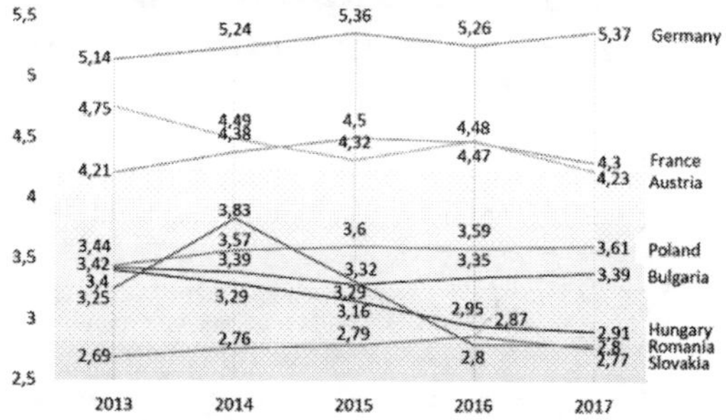

Fig. 3. Quality of the education system (QES) [7]

3.1.4 Researchers in R&D (RRD)

Starting from 2010 until 2014, Romania ranks last in terms of RRD indicator. In 2014, register a scores of 0.92 on a scale of 1 to 7, with 98.91% below Bulgaria which is the next ranked. The top 3 positions are Austria with a score of 4.81, followed by Germany with 4.38 and France with 4.2.

Compared with Romania, the first ranked Austria is above Romania with 422.82%.

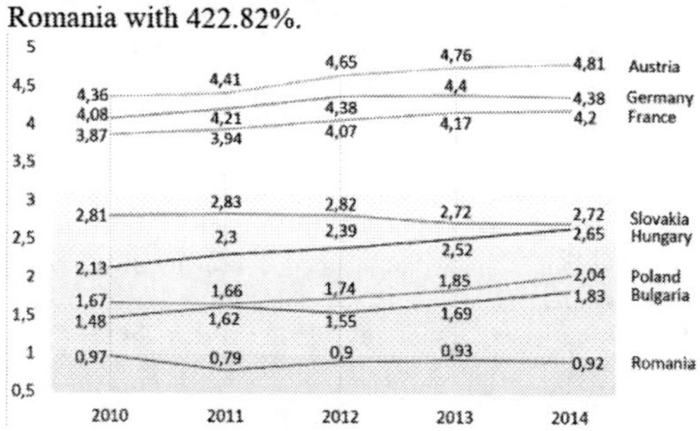

Fig. 4. Researchers in R&D (RRD) [7]

3.2 Group 2 – Support structure/ Infrastructure

3.2.1 Research and development expenditure (RDE)

Starting from 2010 to 2014, Romania ranks last in terms of RDE indicator. In 2014, registering a scores of 0.38 on a scale of 1 to 7, with 110.52% below Bulgaria which is the next ranked. The first 3 positions are Austria with 2.99, followed by Germany with 2.87, and 3rd place France with 2.26.

Compared with Romania, the first ranked Austria is above Romania with 686.84%.

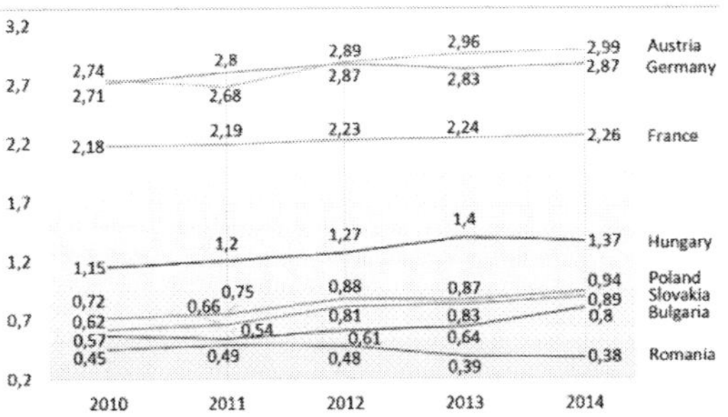

Fig. 5. Research and development expenditure (% of GDP) (RDE) [7]

3.2.2 Firm-level technology absorption (FLTA)

In 2012, Romania ranks 7th from 8 countries in terms of FLTA indicator, recording a score of 4.05 on a scale of 1 to 7, but it can see a slight increase until 2015 when recording a score of 4.44, maintaining the same level in 2016 (the last year analyzed), surpassing Poland and Bulgaria, thus rising to the 6th place.

On the first 3 positions are Germany with 5.74, followed by Austria with 5.68, and 3rd place is France with a score of 5.45.

Compared with Romania, the first ranked Germany is 29.27% above Romania.

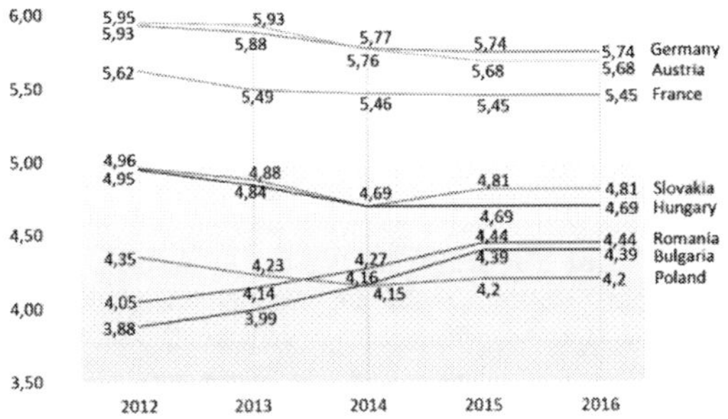

Fig. 6. Firm-level technology absorption (FLTA) [7]

3.2.3 University-industry collaboration in Research & Development (UICRD)

In 2013, Romania ranks 6th out of 8 countries in the UICRD indicator, registering a score of 3.33 on a scale of 1 to 7, then in 2014 and 2015 has increased to 3.59, and then dropped on the last place in 2017, reaching a score of 3.14. Compared to the neighboring countries it is at a relatively small distance.

On the first 3 positions are Germany with 5.37, followed by Austria with 4.81, and third place is France with a score of 4.22.

Compared to Romania, the first ranked, Germany, is with 71.01% above Romania.

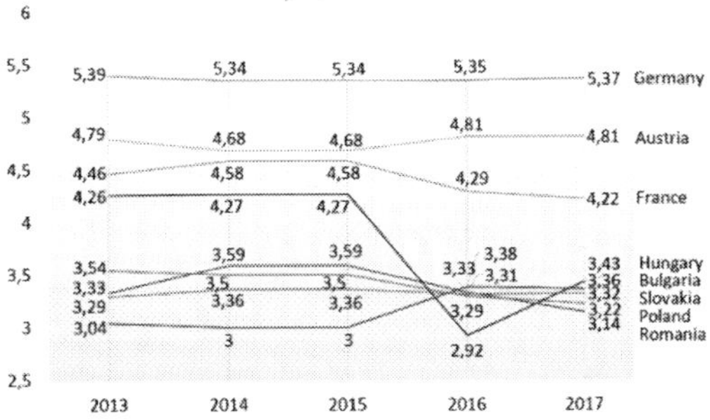

Fig. 7. University-industry collaboration in Research & Development (UICRD) [7]

A detailed study of university-industry collaboration in Research & Development is presented in publications 8 and 9 [8, 9].

3.2.4 Quality of scientific research institutions (QSRI)

In 2013, Romania ranks 6th in terms of QSRI indicator, registering a score of 3.74 on a scale of 1 to 7, and the next year rose slightly overtaking Poland, then dropped to 7th. In 2017 it is again on the 6th place with a score of 3.98, surpassing Bulgaria and Slovakia.

On the first three positions are France with a score of 5.78, followed by Germany with 5.67, and Austria with a score of 5.36.

Compared with Romania, the first ranked France is with 45.22% above Romania.

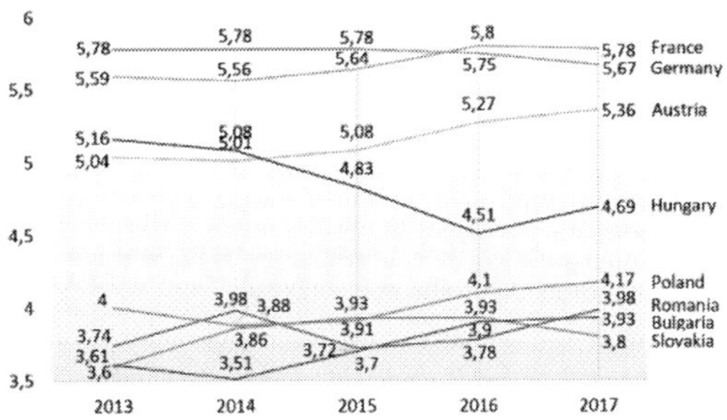

Fig. 8. Quality of scientific research institutions (QSRI) [7]

3.2.5 Gross domestic spending on R&D (GDSRD)

Regarding the GDSDR indicator, it was analyzed the available datas onto the last 5 years (2011-2015). This indicator is measured in millions of USD and as a percentage of GDP. It is also important to note that the data onto Bulgaria are missing from this indicator.

In the first analysis, in 2011, Romania ranked 6th out of the 7 analyzed countries, registering a score of 1.67 over Slovakia.

In 2015 Germany leads the ranking with a score of 101.68, over the place occupied by France with 86.56%, and compared to Romania is over 5195.83%, representing a very large difference between the two countries.

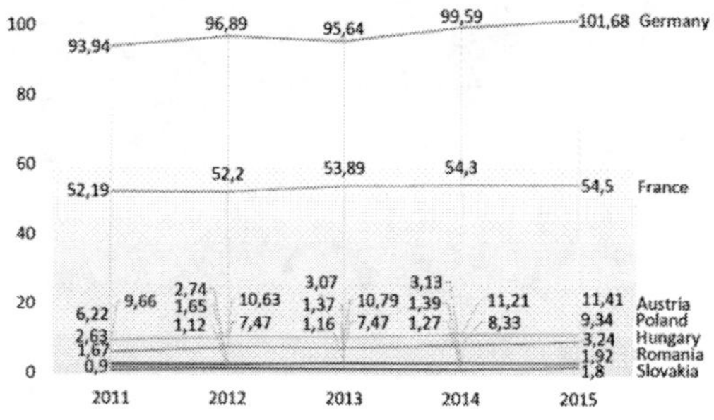

Fig. 9. Gross domestic spending on R&D (GDSRD) [7]

3.3 Group 3 – Legislative provisions

3.3.1 Property rights (PR)

In 2013, Romania ranks 5th in terms of PR indicator, registering a score of 3.94 on a scale of 1 to 7, at the same level as Slovakia.

In the following year, 2014, registered a slight increase exceeding Slovakia, then dropped in 2015, then rose slightly until 2017, registering a score of 4.38 points when it surpassed Slovakia and Poland, ranked 4th and also topped the reported ranking to the neighboring countries.

On the first three positions are Austria with a score of 5.86, followed by Germany with 5.56, and 3rd place is France with a score of 5.37.

Compared with Romania, the first ranked Austria is with 33.78% above Romania.

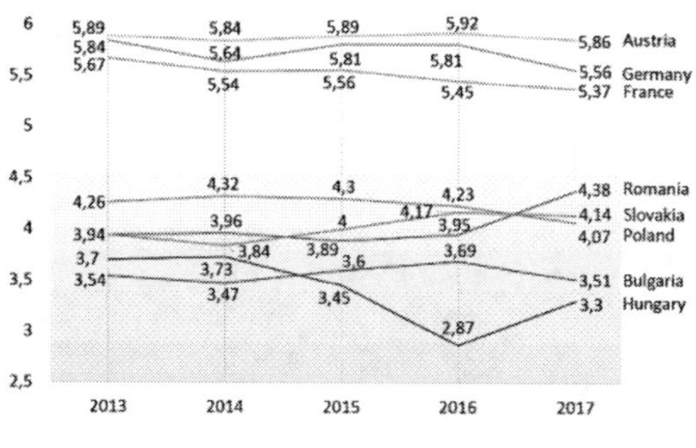

Fig. 10. Property rights (PR) [7]

3.3.2 Intellectual property protection (IPP)

In 2013, in terms of IPP indicator, Romania ranks last from the eight countries analyzed, registering a score of 2.9 on a scale from 1 to 7.

In the following year, 2014 grew slightly beyond Bulgaria, after which it climbed steadily until 2017 when it surpassed all of neighboring countries, being ranked among 4th and also topping the ranking of neighboring countries, recording a score of 4.59.

On the top three positions are Austria with a score of 5.91, followed by France with 5.82, and Germany with 5.74.

In comparison with Romania, the first ranked Austria, is with 28.75% above Romania.

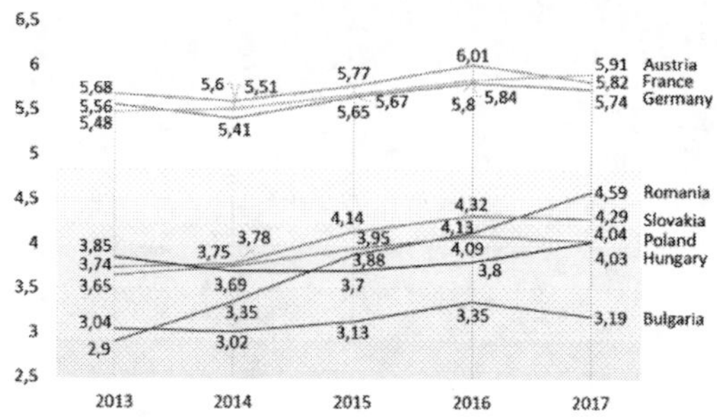

Fig. 11. Intellectual property protection (IPP) [7]

4 Conclusions

The objective of this study was to analyze the 11 indicators that aim and support the innovation process, which were structured in 3 groups, namely human resources, structure support / infrastructure and legislative provisions.

Conclusion Group 1 – Human resources

In terms of the ESA indicator, Romania is ranked 5th in the eight countries analyzed. Compare with the first ranked, Germany, Romania is over 36.24%. As for the comparison of Romania with the neighboring countries, Romania is located above Slovakia, Bulgaria and Hungary, but under Poland with 10.84%.

From the point of view of the R indicator, from the 8 countries analyzed, Romania ranks 6th. Compare with the first ranked, Germany, Romania is surpassed by 1890.14%. Regarding the comparison of Romania with the neighboring countries, Romania is situated above Slovakia, being overtaken by the other neighboring countries.

In terms of the QES indicator, out of the 8 countries analyzed, Romania ranks 7th. Compare with the first ranked, Germany, Romania is surpassed by 91.78%. Regarding the comparison of Romania with the neighboring countries, Romania is located above Slovakia, but under Hungary, Bulgaria and Poland. As for Poland, the top of the ranking among the neighboring countries Romania is at a distance of 28.92%.

From the point of view of the RRD indicator, out of the 8 analyzed countries, Romania ranks last. Compare with the first-ranked, Austria, Romania is surpassed with 422.82%. Regarding the comparison of Romania with the neighboring countries, Romania is located under all of the neighboring countries. Slovakia, the first ranked among the neighboring countries, Romania is at a distance of 195.65%.

From the point of view of human resources, Romania is best positioned to the ASE indicator, ranking 5th out of the 8 analyzed countries, and the weakest indicators is RRD.

The biggest difference between Romania and the first ranked country in the 1st group is the Resource (R) indicator, where Germany records a rate of 1890.14% above Romania.

Conclusion Group 2 – Support structure/Infrastructura

From the point of view of the RDE indicator, out of the eight analyzed countries, Romania ranks last. Compare with the first-ranked, Austria, Romania is surpassed with 686.84%. Compare with the neighboring countries, Romania is situated under all of the neighboring countries. Compared with Hungary which is on top of the ranking among the neighboring countries, Romania is at a distance of 260.52%.

In terms of the FLTA indicator, out of the 8 countries analyzed, Romania ranks 6th. Compare with the first ranked, Germany, Romania is surpassed with 29.27%. Regarding the comparison of Romania with the neighboring countries, Romania is located above Bulgaria and Poland, but under Hungary and Slovakia. Slovakia, at the top of the ranking among the neighboring countries, compared to Romania is at a distance of 8.33%

In terms of the UICRD indicator, out of the 8 analyzed countries, Romania ranks last. Compare with the first-ranked, Germany, Romania is surpassed with 71.01%. Compared with the neighboring countries, Romania is situated under all of the neighboring countries, regarding Hungary, which is on top of the ranking among the neighboring countries, Romania is at a distance of 9.23%.

Romanian researchers have a very high potential (many new methods and applications are published in relevant specialist journals, but the major problem is the low financial capacity of universities to promote inventions, to be taken over and then applied on a large scale in production, industry, agriculture, etc.) [10, 11, 12].

In terms of the QSRI indicator, out of the 8 countries analyzed, Romania ranks 6th. Compare with the first ranked, France, Romania is surpassed with 45.22%. Compared with the neighboring countries, Romania is located above Slovakia and Bulgaria, but under Hungary and Poland, Regarding Hunagry which is on the top of the ranking among the neighboring countries, Romania is at a distance of 17.83%.

In terms of the GDSRD indicator, out of the 7 analyzed countries (because the data for Bulgaria are missing), Romania is ranked 6th. Compare with the first ranked, Germany, Romania is surpassed with 5195.83% (wich is a big difference). As for the comparison of Romania with the neighboring countries, Romania is situated above Slovakia, but under the other neighboring countries, and compared to Poland which is on the top of the ranking among the neighboring countries, Romania is at a distance of 386.45%.

In terms of Support structure, Romania is best placed to the FLTA and QSRI indicators, ranking 6th out of the 8 analyzed countries, and the worst indicators are RDE and UICRD.

The biggest difference between Romania and the first ranked country within Group 2 is indicator 7, Gross domestic expenditure on R & D (GDSRD), where Germany records a 5195.83% over Romania.

Conclusion Group 3 – Legislative provisions

From the point of view of the PR indicator, out of the 8 countries analyzed, Romania ranks 4th. Compare with the first ranked, Austria, Romania is surpassed with 33.78%. Compare with the neighboring countries, Romania is located above all of neighboring countries, surpassing the next neighboring country, Slovakia with 5.79%.

From the point of view of the IPP indicator, out of the 8 countries analyzed, Romania ranks 4th. Compare with the first ranked, Austria, Romania is surpassed with 28.75%. Regarding the comparison of Romania with the neighboring countries, Romania is located above all the neighboring countries, surpassing the next neighboring country, Slovakia with 6.99%.

From the point of view of legislative provisions, Romania is best placed on the IPP indicator, being on the 4th place, surpassing Slovakia by 6.99%, and the weakest place on the PR indicator, being also on the 4th place, surpassing Slovakia with 5.79%.

The biggest difference between Romania and the first ranked country within Group 3 is the indicator 1, property rights (PR), where Austria records a percentage of 33.78% over Romania.

In conclusion, out of the three groups of indicators analyzed, Romania is best placed in terms of Group 3 - Legislative provisions, averaging 31.26% compared to the first rank, followed by group 1 - human resources with an average of 610.24%, and the worst ranking is Group 2 - support structure / infrastructure, averaging 1205.63% compared with the first ranked countries.

The field of innovation and ethical criteria in research should not be neglected [13-25].

References

1. Zeynep Tuğçe Şimşit , Özalp Vayvay, Özgen Öztürkc, An outline of innovation management process: building a framework for managers to implement innovation, Proceedings of the conference: Procedia - Social and Behavioral Sciences **150**, pages 690 – 699 (2014).
2. Tim Stock, Michael Obenaus, Amara Slaymaker, Günther Seliger, A model for the development of sustainable innovations for the early phase of the innovation process, Proceedings of the conference: Procedia Manufacturing **8** pages 215 – 222 (2017).
3. C. I. Ivanov, S. Avasilcai, Performance measurement models: an analysis for measuring innovation processes performance, Proceedings of the conference: Procedia Social and Behavioral Sciences, Challenges and innovations in management and leadership 12th international symposium in management, Volume: **124,** Pages: 397-404 (2014).
4. S. Avasilcai, Performance evaluation in Romanian industrial organisations, Proceedings of the conference: Intelligent Manufacturing & Automation, Annals of daaam for 2007 & proceedings of the **18th** International Daaam Symposium: Intelligent Manufacturing & Automation: focus on creativity, responsibility, and ethics of engineers, Pages: 27-28 (2007).
5. V. Baesu, C.T. Albulescu, Z.B. Farkas, A. Draghici, Determinants of the high-tech sector innovation performance in the European Union: a review, Proceedings of the conference: Procedia Technology, Volume: **19**, Pages: 371-378 (2015).
6. L. Bacali, R.C. Cordos, Intelligent Manufacturing & Automation, Study on the innovation activity in the

Romanian firms, Proceedings of the conference: Proceedings of the Romanian Annals Of Daaam For 2007 & Proceedings Of The 18th International Daaam Symposium: Intelligent Manufacturing & Automation: Focus On Creativity, Responsibility, And Ethics Of Engineers, Pages: 41-42 (2007).

7. https://tcdata360.worldbank.org/topics/inn?country=ROU, accessed in April, 2018.
8. A. Draghici, C.F. Baban, M.L. Gogan, L.V. Ivascu, A Knowledge Management Approach for The University-Industry Collaboration in Open Innovation, Proceedings of the conference: Procedia Economics and Finance, Volume: **23**, Pages: 23-32 (2015).
9. N. Barkoczi, L. Bacali, M. Lobontiu, Current status of the interferences between technology diffusin and marketing management, Proceedings of the conference: Management between profit and social responsability, Pages 324-331 (2014).
10. S. Bungau, D. M. Tit, C. Iovan, L. Copolovici, R. Boscencu, G. Cioca, D. Copolovici, New method for Simultaneous Determination of Ascorbic and Acetylsalicylic Acids in effervescent Tablets, Rev Chim. – Bucharest, **68**, nr. 11, pages 2495-2503 (2017).
11. M. Badea, C. Chiperea, M. Balan, L. Floroian, P. Restani, J.L. Marty, C. Iovan, D. M. Tit, S. Bungau, N. Taus, New Approaches for Electrochimical Detection of Ascobic Acid, Farmacia, **66**, nr. 1, pages 83-90, (2018).
12. D. Mosteanu, G. Barsan, P. Otrisal, L. Giurgiu, R. Oancea, Obtaining the Volatile Oils from Wormwood and Tarragon Plants by a New Microwave Hydrodistilation Method, Revista de Chimie, **68** (11), pages 2499-2502, (2017).
13. F. Blaga, C. Bungau, I. Stanasel, T. Vesseleniy, R. Pancu, D. Craciun, Researches regarding rotary-tilting table precision evaluation method using the ball-bar system, Academy Series a-Mathematics Physics Technical Sciences Information Science, Volume: **16** Issue: 4, Pages: 513-521 (2015).
14. C. Bungau, R. Pancu, T. Buidos, Contributions to the Design of the Load Station Concept used at the Flexible Manufacturing Cell, Proceedings of the conference: Engineering Solutions and Technologies in Manufacturing, Volume: **657**, Pages: 818-822 (2014).
15. C. Bungau, M. Ganea, R. Pancu, New Tendency of Machine Tool Study Based on Virtual Machining Software Applied at University of Oradea, Industrial and Service Robotics, Volume: **613,** Pages: 396-401 (2014).
16. C. Bungau, F. Blaga, C. Gherghea, Kaizen implementation for cost reduction in manufacturing process product "driver control board, Proceedings of the conference: International Conference on Production Research - Regional Conference Africa, Europe and the Middle East, Pages: 55-58 (2014).
17. C. Bungau, I. C. Gherghea, M. Prichici, Value Stream Mapping Analysis, Efficiency Methods of Operational Management, Proceedings of the conference: Twenty Years after: How Management Theory Works, Pages: 188-198 (2010).
18. L. Ivascu, B. Cirjaliu, A. Draghici, , *Business model for the university-industry collaboration in open innovation*, Proceedings of the conference: Procedia Economics and Finance, Volume: **39**, Pages: 674-678 (2016).
19. B.L. Blaga, L. Bacali, R.C. Cordos, Ethics - a priority within the 21st century's organizations, Proceedings of the conference: 2nd Review of Management and Economic Engineering International Management Conference Pages: 384-389 (2011).
20. G. R. E. Maries, D. Chira, C. Bungau, T. Costea, L. Moldovan, Determining the Influence of the Processing Temperature by Injection and of the Subsequent Pressure on the Surface's Hardness and Indentation Modulus of the Products Made of HDPE, PMMA, PC plus ABS through Nanoindentation - G-Series Basic Hardness Modulus at a Depth Method, Materiale Plastice, Volume: **54** Issue: 2 Pages: 214-220 (2017).
21. L. S. Csokmai, R. C. Tarca, C. Bungau, G. Husi, , A Comprehensive Approach to Off-line Advanced Error Troubleshooting in Intelligent Manufacturing Systems, International Journal of Computers Communications & Control, Volume: **10,** Issue: 1, Pages: 30-37 (2015).
22. G. R. E. Maries, D. Chira, C. Bungau, The Influence of Processing Temperatures of (HDPE), (PMMA), (PC plus ABS) on Some Mechanical Properties of Items Obtained Through Injection, Materiale Plastice, Volume: **52,** Issue: 4, Pages: 452-456 (2015).
23. D. Chira, G. R. E. Maries, C. Bungau, The Influence of Subsequent Pressure of (HDPE), (PMMA), (PC plus ABS) on Some Mechanical Properties of Items Obtained Through Injection, Materiale Plastice, Volume: **52,** Issue: 4, Pages: 572-577 (2015).
24. M. Baban, C. F. Baban, C. Bungau, G. Dragomir, R. M. Pancu, Estimation of the Technical State of Automotive Disc Brakes Using Fuzzy Logic, International Journal of Computers Communications & Control, Volume: **9,** Issue: 5, Pages: 531-538 (2014).
25. T. Vesselenyi, C. Bungau, G. Husi, Research equipment and computation techniques for Intelligent Building applications, Proceedings of the conference: International Symposium on System Integration, Pages: 228-233 (2014).

MATEC Web of Conferences 184, 04021 (2018) https://doi.org/10.1051/matecconf/201818404021
Annual Session of Scientific Papers IMT ORADEA 2018

Study on improving silicone losses in the process of mixing components of soldering machines

Mihai Victor Zerbes[1,*], and *Liliana Georgeta* Popescu[1]

[1]Lucian Blaga University of Sibiu, Industrial Engineering and Management Department, 550025, Victoriei Bld. no.10, Sibiu, Romania

Abstract. The paper presents a summary of the steps of the 6 Sigma methodology applied in the gluing section within an automotive company form Sibiu, Romania. We describe the synthesis of each step of the DMAIC model, a well-known model in the application of the 6 Sigma methodology. In the last chapter, we outline some general conclusions regarding the use of the 6 Sigma methodology presenting the benefits both within the automotive organization from Sibiu and its general use by industrial organizations.

1 Introduction

This study is conducted at an automotive company form Sibiu in Gluing Division. The approach is based on production processes. The purpose of this case study is to present a problem-solving project based on the 6 sigma methodology. This is a real problem faced by the organization, linked to the involvement of employees in research-innovation projects.

This company is part of a bigger multinational corporation. Its global products include: steering wheels, airbag systems, airbag fabrics, pyrotechnic capsules, safety belts, child safety chairs, electronic components and interior car sensors. The subsidiary of Sibiu has, as its main field, the manufacture of airbags for motor vehicles.

The airbag consists of the following parts: textile, sensor, electronic control unit and pyrotechnic capsule. Textile parts of the airbag are different depending on the type of airbag and model. Thus, it is distinguished: the driver's/steering wheel, passenger, knee, side and curtain airbags (Fig. 1), which in turn differ according to the model of the vehicle for which are intended.

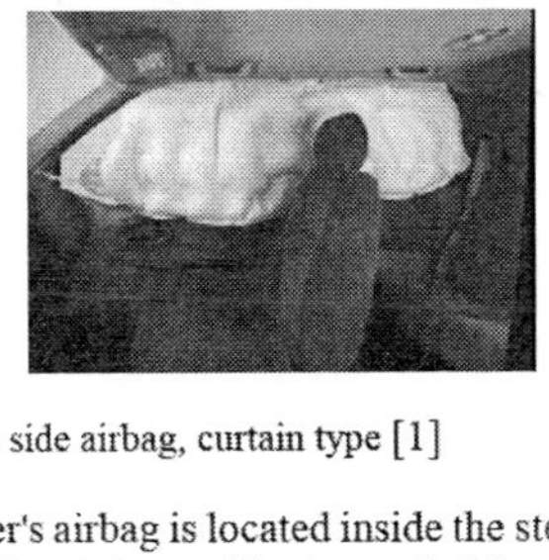

Fig. 1. Head side airbag, curtain type [1]

The driver's airbag is located inside the steering wheel. Head side airbag is located in the roof of the car. They can have different tubular shapes, or mounted in the door depending on application needs.

During the production of these products there were identified delays due to the silicon losses occurred during the mixing of components in the Gluing section.

2 Methodology

The Six Sigma methodology - a methodology initiated in the 1980's at Motorola as a challenge to achieve a reduction in the number of defective products - was used to identify the causes and to solve the problems [2]. In order to achieve this effect, a thorough analysis of the causes and possibilities of correction was needed. Sigma is a term taken from statistics, used to measure the variation of a process to specifications or to another reference such as consumer requirements. The higher sigma level, the better the customer's requirements. Thus, it can be said: the probability of events being in the range [3]:

- $[-\sigma; \sigma]$ is of 68%;
- $[-2\sigma; 2\sigma]$ is of 95%;
- $[-3\sigma; 3\sigma]$ is of 99,73%;
- $[-6\sigma; 6\sigma]$ is of 99,9997%

so, the probability that the event is not in the range $[-6\sigma; 6\sigma]$ is of $3{,}4\mathrm{x}10^{-6}$%:

A six sigma improvement process comprises a series of steps, grouped by type of activity as follows [4]:

Step 1. *Definition* - the deficiencies to be addressed will be clearly specified and the estimated improvement defined in measurable terms. It establishes a team for the project and allocates the resources and time it takes for the project to succeed.

Step 2. *Measurement / Analysis* - At this stage the team discovers the true causes of the deficiency.

Step 3. *Improve* - Establish improvements for identified causes.

Step 4. *Control* - the improvement team develops and implements control elements.

Step 5. *Multiply the results* - once the improvement team has positive results, there are two more responsibilities: to support other employees in organization with similar problems to apply what the team has learned from the improvement project, and to

* Corresponding author: mihai.zerbes@ulbsibiu.ro

© The Authors, published by EDP Sciences. This is an open access article distributed under the terms of the Creative Commons Attribution License 4.0 (http://creativecommons.org/licenses/by/4.0/).

nominate other projects for solving in order to correct a deficiency.

3 Case study - applying the sigma methodology in the case of silicon losses occurring in the components mixing process

This case study was conducted in the Gluing section, starting from the production processes [5].

3.1 Defining

The project is applied in the gluing section for soldering machines. A 5W1H questionnaire (Table 1) applied to 25 people in the Gluing section was used as a tool for formulating the problem.

Table 1. The 5W1H questionnaire

5W1H	Questions	Answers
Which	Which is the opportunity / insatisfaction?	The difference between component A and B of 40491 €/ 8 months
How	How do we know about this possibility?	Following inventory made periodically
When	When does this happen?	When mixing the silicone components A and B
Where	Where is it going?	On soldering machines (M3)
Who	Who is involved?	Operators / exchange, engineers
Why	Why is this aspect of the business a problem?	This affects us financially by throwing out the silicone component that expires with no pair

Following the analysis of the results of the 5W1H questionnaires, the problem to be solved was formulated as follows: "Silicone losses when mixing the components".The project team consisted of project manager (Quality Engineer), Quality Team Chief, Prototype Engineer, Process Enhancement Engineer, Logistics Team Coordinator, Maintenance Engineer, Chief Laboratory Team.

3.2 Measuring range

The structure of the measurement stage consists of 7 steps for which specific actions have been taken. The key milestones for the masonry phase are: customer voice, setting measurement aspects, validating measurement systems, detecting influence factors, measuring planning, visualizing graphical measurements and performance of the process.

3.2.1 Customer voice

In the measurement phase, the expectations, requirements and specifications of the internal clients are summarized in Table 2.

Table 2. Customer voice

Customer	Expectations	Requirements	C T Q	Specifications	Weight
Gluing	Observance of ration of 1:1	Decrease in proportion between component A and B from 2.5% to 1.6%	Big difference of 1: 5 between component A and B (Race)	according to the 1: 1 ratio	25%
	Technological losses	Not to exceed more than 10%	Incorporation into normal consumption	maximum 10%	55%
Laboratory	Fitted parts according to the standard	The proportion of component A and B is 1: 1	Resistance to silicon at break> 175N	>175N	10%
Logistics	Reduce silicone scrapping due to expiration	Proportional use of components A and B	Compliance with the date of validity of components A and B	Using FIFO of silicone barrels	10%

3.2.2 Establishing aspects of measurement

In order to better understand the change process, the measurement factors have been prioritized (Table 3)

Table 3. Prioritization of measurement factors Y

No.	Measurement factor Y	Weight
1	Ration	25%
2	Resistance to tearing >175N	10%
3	Costs	10%

3.2.3 Validation of measurement systems

Validation of measuring systems was done with a specialized "Measurement System Analysis" system shown in Fig. 2.

3.2.4 Detection of influence factors

In order to detect the possible causes of the problem, the Ishikawa diagram was used. The main cause was "The Stock Difference Between Components A and B Generated by Ratio Variation".

3.2.5 Planning the measurements

The mixing process is influenced by the following parameters: D1-speed on the route; D2-starting speed; D3-speed of absorption.

Leakage velocity may be influenced by viscosity (different viscosity between the two components). The influence of viscosity on the mixing process can be expressed by the difference between rations due to the slower or faster leakage of components A and B through the pipe. The monitoring of the mixing process is presented in Tables 4 and 5.

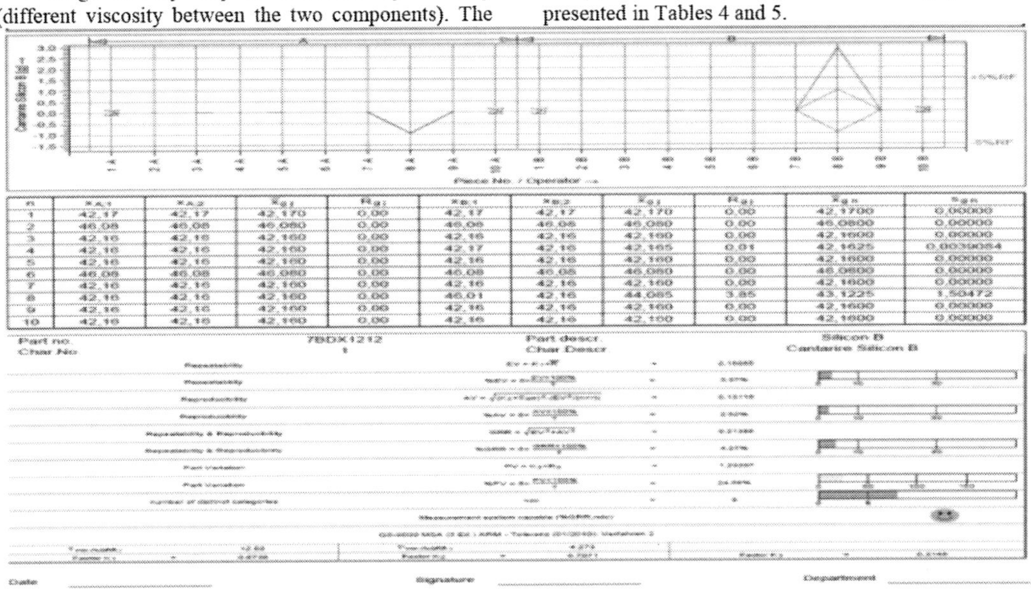

Fig. 2. Analysis of measurement systems using electronic balance

3.2.5 Planning the measurements

The mixing process is influenced by the following parameters: D1-speed on the route; D2-starting speed; D3-speed of absorption.

Leakage velocity may be influenced by viscosity (different viscosity between the two components). The influence of viscosity on the mixing process can be expressed by the difference between rations due to the slower or faster leakage of components A and B through the pipe. The monitoring of the mixing process is presented in Tables 4 and 5.

Table 4. Monitoring of the mixing process during January 24-February 03 2017 on first exchange.

Data	Silicone A Viscosity	Silicone A Temperature	Silicone B Viscosity	Silicone B Temperature
24.01	341,4	29,3	371,5	28,9
25.01	332,2	28,7	362,3	28,2
26.01	326,6	26,7	386,2	26,6
27.01	360	27,3	341,5	28,3
28.01	300,9	28	322,8	28,3
30.01	372	23	338,1	23,4
31.01	378,2	25,6	333,3	25,7
01.02	378,3	25,9	318,3	25,8
02.02	321,9	26,2	274,8	26
03.02	392	25,3	356,6	25,2

Table 5. Monitoring of the mixing process during January 24 – February 06 2017 on second exchange

Data	Silicone A Viscosity	Silicone A Temperature	Silicone B Viscosity	Silicone B Temperature
24.01	350	28,4	356,3	29,4
25.01	352,2	28,3	332,2	28,7
26.01	336,6	28,7	348,9	29,1
27.01	355,3	28,7	316,5	29,2
28.01	355,4	27,6	279	28,1
30.01	315	27,7	361	27,4
31.01	315,2	27,8	351,3	26,8
01.02	320	28	299,1	27,7
02.02	201,2	27,8	351,3	27
03.02	387	26,1	297,7	25,4
06.02	311,6	27,3	291	26,7

A range of values for the three machine parameters for silicone type A and type B has been established and their influence on the ratio by two methods has been monitored. Ratio by method 1 - the classical method used in the line is allowed to flow 40 seconds silicone type A and then type B, after which weigh.

Fig.3. The weighing process of the type A component [1]

https://doi.org/10.1051/matecconf/201818404021

Ratio by Method 2 consists of depositing A and B type silicon on the panel, according to a route corresponding to each model (BMW F10).

Fig. 4. Silicone component type A deposition process [1]

The values obtained by changing the parameters on the ratio through the two test methods for components A and B are presented in Table 6. Following the modification of the parameters, the values are obtained which are highlighted in the graphs of Fig. 5 and Fig. 6, taking the and base ratio in percent (difference between components A and B in grams).

Table 6. Data collection sheet

rosu A2			alb B2			A	B	M3 TEST 1	A	B	M3 TEST 2	A	B	M3 TEST 3	Medie	A	B	M1 TEST 1	A	B	M1TEST 2	A	B	M1 TEST 3	Medie
D21	D22	D23	D31	D32	D33																				
800	800	-450	842	698	-450	44.3	49.9	-5.98	46.6	49.9	-3.44	46.5	49.23	-2.83	-4.082	58.2	63.6	-4.44	58.2	62.5	-3.56	58.3	62.6	-3.57	-3.8584
850	800	-450	842	698	-450	48.5	53	-4.45	48.78	53.3	-4.38	48.8	53.03	-4.19	-4.34218	60.9	66.2	-4.24	66.2	61	4.08	61.1	66.1	-3.98	-1.37995
900	800	-450	842	698	-450	49.6	50.2	-0.59	49.9	50.2	-0.30	49.9	50.38	-0.44	-0.44302	62.7	62.8	-0.07	66.9	62.9	3.01	62.7	63	-0.29	0.884811
847	750	-450	842	698	-450	47.9	51.4	-3.53	48.29	52	-3.69	48.3	51.84	-3.49	-3.56982	59.9	64.1	-3.45	60	64.3	-3.53	60.2	64.3	-3.31	-3.43187
847	800	-450	842	698	-450																				
847	850	-450	842	698	-450	48	51.1	-3.10	47.97	51.4	-3.47	48.2	51.74	-3.51	-3.35972	59.6	65.1	-4.42	59.8	65.2	-4.31	59.9	61.7	-1.48	-3.40541
847	800	-400	842	698	-450	48	50.5	-2.51	48.48	51	-2.56	48.6	50.97	-2.37	-2.47979	59.6	65.1	-4.38	59.8	65.1	-4.27	60	64.9	-4.00	-4.21278
847	800	-450	842	698	-450																				
847	800	-500	842	698	-450	49	51.9	-2.87	49.69	52.2	-2.42	49.7	49.74	0.00	-1.7604	61.9	65.8	-3.06	62.3	64.4	-1.67	62.4	65.7	-2.54	-2.42458
847	800	-450	800	698	-450	49.8	50.2	-0.38	50.19	50.7	-0.50	50	50.95	-0.90	-0.59216	62.5	64.8	-1.80	62.6	65	-1.92	62.7	65.2	-1.96	-1.89399
847	800	-450	850	698	-450																				
847	800	-450	900	698	-450	50.1	55.6	-5.19	49.98	55.8	-5.54	50.2	55.93	-5.37	-5.36615	62.6	69.1	-4.98	62.5	69.5	-5.29	62.6	69.7	-5.36	-5.20889
847	800	-450	842	650	-450	50.2	53.6	-3.25	50.29	53.5	-3.13	50.1	53.68	-3.42	-3.26601	62.5	67.1	-3.53	62.6	67.2	-3.51	62.7	67.3	-3.49	-3.50834
847	800	-450	842	700	-450																				
847	800	-450	842	750	-450	46.5	49.4	-3.01	48.88	50.9	-1.99	48.9	50.98	-2.07	-2.35648	59.4	64.3	-3.93	60.2	64.5	-3.50	60.3	63.9	-2.94	-3.45482
847	800	-450	842	698	-400	48.4	49.9	-1.57	48.79	50.4	-1.62	48.8	50.32	-1.55	-1.58115	60.4	61.4	-0.85	60.4	62.5	-1.74	58.5	62.7	-3.50	-2.02883
847	800	-450	842	698	-450																				
847	800	-450	842	698	-500	47.9	49.8	-1.90	47.97	50	-2.06	48	46.63	1.45	-0.8395	59.2	61.6	-2.02	59.2	61.7	-2.03	59.5	61.7	-1.89	-1.98095

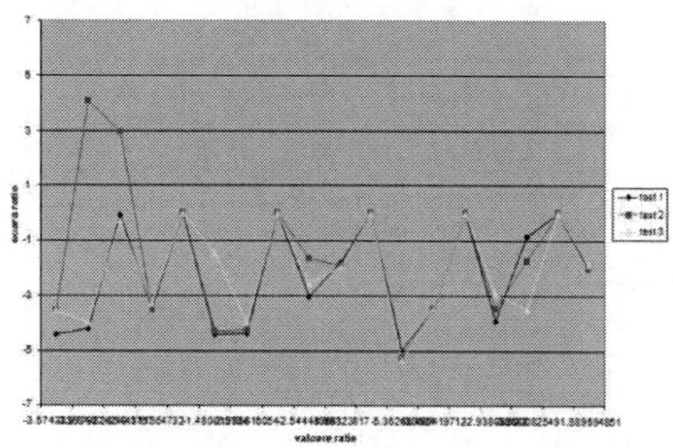

Fig. 5. Ratio by method 1 following tests 1, 2, 3

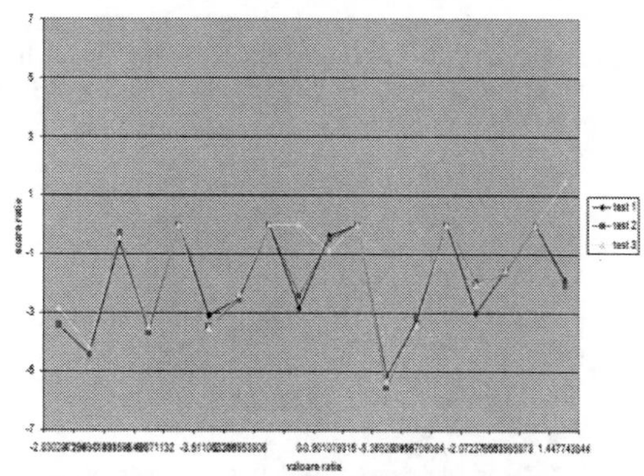

Fig. 6. Ratio by method 2 following tests 1, 2, 3

3.3 Analyze

3.3.1 Find and prioritize the main causes that impact on performance

For identification of the root cause the following questionnaire was made (Table 8).

Due to the fact that the ratio is higher than the maximum admissible 5%, the consumption of the two components A and B is different.

Table 9 shows the difference in Kg between January and May.

Following the calculations, the difference between consumption (formula 1) and ration (formula 2) was double.

$$Consumption = A / (B - 1) \qquad (1)$$

$$Ration = [(A - B) / (A + B)] * 100 \qquad (2)$$

The different perception of consumption and ration has made for values of rationality considered good (difference to 2.5% versus maximum specification of 5%), in fact to be too high for the proposed objective, because the calculation formulas for consumption and ration give totally different results. Consumption 1% different from ration 1%.

Table 8. The 5 WHY questionnaire

No.	Causes	Why	Why	Why	Why	Why
1	The stock gap generated by misinterpretation between ration and consumption	Because the stock gap is calculated according to the consumer formula logistics department, and in the production line the difference between components A and B is calculated using the different ration formulas.	The calculation, accounting and technical formulas have not been correlated.	The technical objective did not take into account the inventory, but only the technical performance of the product (the strength of the silicone cord at break)	The specified technical range of ration allowed increased consumption without taking into account the implications in the inventory of components.	The specified range for the allowed ration is up to 5% and the ration optimization has not been taken into consideration in order to obtain the technical parameter (breaking strength) with a more balanced consumption.
2	Different parameters for the same machine	Because there are 2 heads on the machine that work individually	Because each head is powered by 2 independent motors	the two motors operate after a set of independent parameters.	Because that was the way the machine was built.	-
3	The obvious lack of possible losses	Because there is a record only for ventilation	Because it was not considered necessary to record the other losses	It was considered that the influence is insignificant in relation to the large difference between the two components	-	-

Table 9. Comparative analysis between the two methods

Month	Machine M3P5		Machine M3P5	
	A	B	A	B
January	1448,62	1477,16	1365,97	1410,19
February	1685,53	1657,45	1630,63	1653,23
March	1381,38	1395,26	1311,45	1321,44
Aprilie	1270,46	1341,6	1259,91	1315,31
May	1264,77	1321,84	1135,74	1198,62
Amount	7050,76	7193,31	6703,7	6898,79
Consumption	-1,98		-2,83	
Ration	-1,00		-1,43	

3.3.2 Updating project benefits (based on root cause analysis)

Following the implementation of the optimal parameters on head 5, the average ration of 0.28% was obtained, comparing it with the average of the ration in April (2.71%) there was a decrease of the ration by 10.33%, but comparing with the change in machine parameters is a drop of only 6%. By making an analogy for head 6 (which rate in April was 2.17%) to bring it to a value of 0.22%, (which means a 8.11% reduction in the rate) should be we modify machine parameters not by 8%, but by 4%.

3.4 Improving

3.4.1 Generating ideas for solutions

The evaluation criteria for the alternatives previously developed are: total cost, impact on the problem, cost / benefit relationship, resilience / impact on change, implementation time, uncertainty about efficacy.

The main solutions that were found following a brainstorming session (Table 10):

Table 10. Developing alternatives for improvement

Cause	Alternative 1	Alternative 2
a. Not respecting the ration	**a.1.** Reducing the specified range from a maximum of 5% to 2.5%	**a.2.** Adjustment of rations in process to 2.5% and their validation by silicon breaking strength > 175N
b. Deregulation of parameters on machine 3 head 6	**b.1.** Adjusting the parameters of the machine 3 the head 6 according to the optimal parameters set for the head 5	**b.2.** Following the ratios obtained from the alignments of the head 6 with optimal parameters from head 5
c. Lossless monitoring	**c.1.** Drafting Loss Sheets and communicating these losses to the logistics department	

3.4.2 Prioritize solutions

In order to evaluate the improvement alternatives against these criteria, the team used as a quality tool the alternative selection matrix (Table 11) where it used the following notations: 3 - Very favorable impact, 2 - Favorable medium impact, 1 - Poor impact .

Table 11. Matrix for selecting the alternatives

Cause	The alternative of improvement	Selection criteria						Total
		a	b	c	d	e	f	
a. Not respecting ration	a.1 Reducing the specified range from a maximum of 5% to 2.5%	1	1	3	1	2	2	10
	a.2 Adjustment of rations in process to 2.5% and their validation by silicon breaking strength> 175N	2	3	1	3	3	2	14

Cause	The alternative of improvement	Selection criteria						Total
		a	b	c	d	e	f	
b. Deregulation of parameters on machine 3 head 6	b.1 Adjusting the parameters of the machine 3 the head 6 according to the optimal parameters set for the head 5	2	2	2	3	2	2	13
	b.2 Follow the ratios obtained from the headline 6 alignments with optimal parameters from head 5	1	2	1	3	2	2	11
c. Lossless monitoring	c.1 Drafting Loss Sheets and communicating these losses to the logistics department	1	3	2	1	2	2	11

Adjustment of process ratios to 2.5% and their validation by silicon breaking strength> 175N - prioritization of ideas was made on the basis of the large difference between the result obtained by the calculation formula of consumption and ration

Adjusting the machine parameters 3 the head 6 according to the optimal parameters set for the head 5 - on the assumption that it is the same machine and the same type of motors it was considered that optimal parameters obtained for M3 C5 can be applied to C6.

3.4.3 Designing solutions

Due to the change in machine parameters, there were little differences between settings 3 and 10 where the track speed was higher for silicone A and large differences between settings 1 and 12 where track speed was higher for silicon B. For machine 3 head 5 the parameters giving the best result for the 0.22 ration were the ones corresponding to the setting 3. The next step was to modify the current parameters (setting 2) with the parameters at setting 3 on both the head 5 and the head 6. From the top data it can see that setting 3 is the optimal one because a lower ration is obtained by modifying parameter D21.

3.4.4 Validation of solutions

After changing the machine parameters to setting 3 to validate the 0.22% rate on both the head 5 and 6, the ration was made and the silicone resistance to breakage was tested by 2 methods. The ration on the head 5 confirms the initial result, on the head 6 the ratio exceeds the limit of 5% admitted. Results are confirmed by the silicone tear strength test by: Method 1 - Rapid drying of the parts (in the oven 30 minutes at 80° C) and Method 2 (classic) after 24 hours under normal temperature conditions.

3.4.5 Measuring the Impact of Improvement

As a result of the improvement of parameters D21, D31, only a reduction of the ration from 8,33% to 7,69% during the period January-May 2015 was obtained on the machine 3 head 5, compared to the target value (3,5%) representing improvement of 13.25%.

3.5 Control

Based on the results of the DOE parameter analysis, the improvement was achieved with the modification of parameters D21 and D31 only (parameters D21 and D31 have a major influence on ration, D22, D32 have little influence, and D23, D33 insignificant infinity). So extending to all silicone machines can be applied in a simpler way just by doing 4 attempts.

4 Conclusions

As a result of the improvement of the parameters only on the machine 3, the head 5 achieved an improvement of 13.25% which represents 573 kg of silicon B, and 1 kg of silicon costs 12.54 €. Financial benefits 7185 € / 5months

Which means a benefit of 1437 € / 1 month.

If the action applied to the machine 3 the head 5 would extend with the same results on M1C1 machines (machine 1 head 1), M1C2 (machine 1 head 2), M2C3 (machine 2 head 2), M2C4 (machine 2 head 4), and M3C6 (machine 3, head 6) would have a financial benefit of 7185 eur.

Six Sigma is a tool that, if used correctly, can identify the key areas of a business with the advantage that all of the improvements measured and obtained through this technique are directly transformed with financial results. 6 sigma is a philosophy that relies on good resource-saving and good behavior practices that are geared to the continuous improvement of current work and life so that everything we do and everything we live is more enjoyable, better, more beautiful, cheaper and simpler.

References

1. Automotive company from Sibiu, *Company Technical Documentation*, Sibiu (2017)
2. M. V. Zerbes, *Advanced planning and design of processes and products in the automotive industry, (Planificarea si proiectarea avansata a proceselor si produselor în industria auto)*, Editura Techno Media, Sibiu (2017)
3. C.V. Kifor, C. Oprean, *Quality Engineering. Improving methodology - 6 Sigma, (Ingineria Calitatii. Imbunatatirea 6 Sigma)*, Editura Universitatii "Lucian Blaga", Sibiu (2006)
4. M. V. Zerbes, C. V. Kifor, G. Budau, *Six sigma methodhology in wood industry organizations*, Proceedings of the 7th International Conference on Wood Science and Engineering in the Third Millennium, "Transilvania" University of Brasov, June 4 - 6, (2009)
5. Automotive company, *QMS documentation*, Sibiu (2017)
6. I. Masaaki, *Gemba Kaizen: A Commonsense Low-cost Approach to Management, (Gemba Kaizen, O abordare practică cu costuri reduse, a managementului,)* Editura Finmedia, Bucharest (1997)

MATEC Web of Conferences **184**, 04022 (2018) https://doi.org/10.1051/matecconf/201818404022
Annual Session of Scientific Papers IMT ORADEA 2018

The asymmetry of income distribution between stakeholders within the framework of industrial park structures

Andrey Plakhin[1*], *Igor* Semenets[2], and *Ekaterina* Ogorodnikova[1]

[1]Ural State University of Economics, Department of Management, 8th March str. 62, Ekaterinburg, Russian Federation
[2] Slavonic University of the Republic of Moldova, Department of Economics, Florilor str. 28/1, Chisinau, Moldova

Abstract. The purpose of this article is to prove the author's hypothesis about the formation of an asymmetry in the distribution of income of stakeholders in the industrial park structure, due to the increased impact of economies of scale on the activities of the residents of the park. In this article, the authors propose a model of income distribution, on the basis of which the prevailing role of the state as the beneficiary of the results of activity within the framework of the park structure was proved. It also indicates a clear asymmetry of income distribution, formed due to economies of scale between other stakeholders. The revealed characteristics can be claimed for the justification of concepts and other documents of a program nature in the formation of industrial park structures.

1 Introduction

The modern vector of the development of the Russian economy is largely based on the formation of mesoeconomic objects that have a common infrastructural base. The objects of the meso-level include new park structures for the Russian economy. The works of M. Mădălina, P.A. Constantin [1], M. Mani, S.M. Hosseini, T. Ramayah [2], I. Siskos and L.N. Van Wassenhove [3], A.E. Plakhin, E.S. Ogorodnikova [4], [5] present a detailed description of the park structures. Clusters, new industrial areas, special economic zones and other territorial localization of industrial production, which can be called the unifying category "industrial park structure" [5].

The effectiveness of the use of industrial parks structures in the world is confirmed quantitatively. There are more than 400 industrial parks in the USA, 200 parks in Germany, 200 parks in Germany, 262 in Turkey, 140 in the Czech Republic, over 70 in Japan, more than 60 in Poland, 34 in Slovakia, 14 in Costa Rica , in Romania - 9. Vietnam has created about 200 parks, which account for 25% of GDP and 40% of attracted investments [6]. About 400 industrial parks have been created in China over the past 40-50 years.

The experience of creating parks in different countries has shown that this is the most understandable, profitable and therefore popular with the investor product. 444 industrial parks have been created in the regions of Russia [7], which is comparable to countries such as China and the United States, but their effectiveness is still far from the industrial parks of these countries.

But features of the park structure make it ineffective to apply a wide range of methods of administrative management. In the authors' opinion, the processes of managing the development of mesoeconomy objects largely depend on indirect regulatory instruments, the integrated application of which relates to the stakeholder approach to management. This methodology makes it possible to take into account the specific features of the interests of groups of persons participating in activities and using their own criteria for assessing the functioning of the park structure from the point of view of its own interests. Several authors have proposed a step-by-step mechanism for analyzing relationships with stakeholders [8], [9], [10], [11].

R.S. Gondarev and N.V. Raznov believe that the most advanced way to interact with stakeholders today is bridging, which implies strategic partnership and pooling of resources [12]. However, these same authors point to the difficulties in organizing interaction between the parties.

R.W. Nuzhdin and A.N. Polozov see the main goal of stakeholder management in achieving positive synergy. They believe that if synergy is expressed in increasing the added value and profitability of the parties involved, it means that the cost is added for integration when the possibilities of economic potentials are used as efficiently as possible [13].

A.M. Hein, M. Jankovic, W. Feng and others adapt the stakeholder value network approach from the strategic management literature to the industrial symbiosis context as a means to provide insights into the power of stakeholders of an industrial symbiosis [14].

Some authors emphasize the digital accompaniment of the stakeholder approach to managing complex systems. In particular M. Neubauer and C. Stary present several case studies highlighting the latest findings in Industry 4.0 projects utilizing S-BPM features. They provides a realistic portrait of the status quo based on current

* Corresponding author: apla@usue.ru

© The Authors, published by EDP Sciences. This is an open access article distributed under the terms of the Creative Commons Attribution License 4.0 (http://creativecommons.org/licenses/by/4.0/).

findings, and outlines the future activities to be pursued in order to establish stakeholder-centred digital production systems [15].

S. Ismulyati and G. Ginting offer conceptual modeling to clarify more deeply in how connection formed by building relationship with primary stakeholder can support open innovation through co-innovating and co-creating value [16].

A review of the literature sources showed that the issues of stakeholder participation in the implementation of joint projects, as well as the distribution of revenues from their implementation to all stakeholders, are still relatively insufficiently studied.

2 Methods

In several works mentioned above, one can note the manifestation of the category of "added value", which combines the use of the stakeholder approach and industrial park structures, the creation of which is aimed at obtaining external effects, including synergies.

This conclusion allows us to talk about the effectiveness of economic methods of managing the park structure, which can include mechanisms for the distribution of income or additional "added value", formed due to external effects. The distribution tools help to stimulate certain areas, primarily the development of human capital and regional infrastructure. Such tools will correspond to the adaptation principles of the stakeholder approach, which considers development through the prism of satisfying the multidirectional interests of multiple groups of influence.

To understand the subject, research authors proposed the concept of "asymmetry" reflecting the uneven distribution of income. Objects of control can be both alignment of "asymmetry" and its artificial formation. In the first case, it is possible to use the methods of institutional regulation, by creating transfer funds that redistribute income between stakeholders. In the second case, the "asymmetry" of income distribution can be an economic lever to stimulate the creation of individual infrastructure facilities, or to implement the necessary functions within the framework of the park structure.

To solve this scientific problem, the authors formed a model of income distribution among the stakeholders of the park structure (Fig. 1).

In this model, the main stakeholders of the park structure and the types of income they receive are listed. It should be noted that some of the influence groups receive income not directly, but with the mediation of direct participants in the operational process. For example, the state that regulates the distribution of tax revenues forms many aspects of the activity of the park structure.

Using this model, the initiators of the creation of an industrial park structure can justify the effectiveness of the project, attract investors and receive the appropriate support of the state as the most interested participant and the largest beneficiary.

Individual participants in turn can also see the full picture of effects taking into account the asymmetry of income distribution from joint activity.

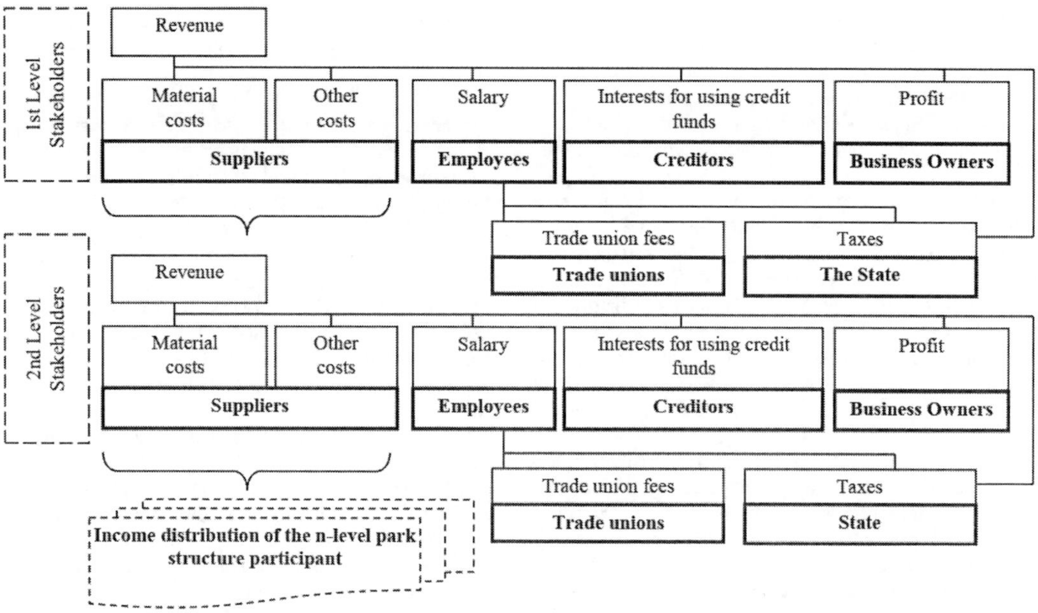

Fig. 1. Model of income distribution among stakeholders of the park structure

3 Results

As an object of research, this article selects the industrial park "Grabtsevo", the development of which is carried out in the Kaluga region of Russian Federation within the project "Development of the infrastructure of industrial parks: the formation of a cluster of production of cars and auto components".

The project also includes industrial parks "Rosva" and "Kaluga-Yug". It is in these parks that a large number of auto cluster enterprises are concentrated. At the same time, large pharmaceutical manufacturers also preferred the site in "Grabtsevo".

As of the beginning of 2016, 12 industrial parks operate in the Kaluga region, the total area of which is 7,556.5 hectares. In territory of industrial parks in aggregate carry out the activity of 94 companies-residents.

It should be noted that in recent years, the Kaluga region among the subjects of the Russian Federation demonstrates a stable growth dynamics in all important economic indicators. Large-scale modernization of the regional economy, a significant increase in labor productivity, the use of advanced production technologies and, finally, the active attraction of investments - such, according to experts, are its components of successful social and economic development. An effective system of working with investors has been created in the region. Competitive advantages among Russian regions are not only favorable geographical position and proximity to the huge capital market, but also low risks of investment, provision of tax incentives and subsidies, competitive tariffs for natural monopolies, administrative support from the state authorities of the region [17].

The analysis of the distribution of tax and customs revenues according to levels of the budgetary system indicates their disproportionality. The regional level receives only a small part of its huge investments in the creation of parks in the form of tax revenues. On average, from one ruble of costs incurred, the regional budget has returned 1.35 rubles, that is, after almost a decade, the region has reached the self-sustainability of these projects. As for the overall economic efficiency, more optimistic results are obtained here. On average, one ruble of expenses had 12.5 rubles of profit. The greatest value has a multiplier effect in the park "Grabtsevo" - 24.76 rubles. The lowest figure is observed in the park "Rosva" - 4.45 rubles [18].

Table 1 shows the dynamics of commissioning of production facilities located in the park.

As can be seen from the Table 1 on the territory of the park structure manufacturing enterprises of related activities are located, therefore, the principles of the park structure formation are observed and the object corresponds to the research objectives.

Table 1. Commissioning of the main production capacities of the industrial park "Grabtsevo"

Name of resident	Commissioning year	Activity type
Volkswagen Group Rus LLC	2015	manufacture of internal combustion engines for cars
	2007	cars production
Gestamp-Severstal-Kaluga LLC	2010	manufacture of stamped parts for car bodies
Benteleer Automotive LLC	2010	manufacture of car suspension parts
Raidel Automotive Rus LLC	2010	manufacture of carpets for car doors
Yapp Rus Automotive Systems LLC	2011	production of plastic fuel tanks
FUIAO Glass Rus LLC	2013	manufacture of automobile glasses
Branch of JSC "Magna Automotive Rus"	2010	production and painting of car bumpers
	2016	construction of the second stage of production facilities and infrastructure of the plant for the production of auto parts

It is important to note that the project of the Volkswagen concern in the industrial park Grabtsevo in Kaluga has become one of the largest not only in the Kaluga region, but also in Russia. For the German company, other major automakers with a worldwide name came to the region - Volvo Trucks (a plant was built in the industrial park Kaluga-Yug), PSA Peugeot Citroen Mitsubishi (the assembly plant is located in the industrial park Rosva).

In the course of the study, information was collected and summarized on the distribution of income among the main groups of stakeholders in the industrial park "Grabtsevo" (Fig. 2) [19].

As can be seen in the figure, the largest beneficiary of the project is the state, since a little less than half of the generated income goes to different levels of the budget. The collected funds go to the public sector and are channelled into the formation of the social environment and infrastructure of the region.

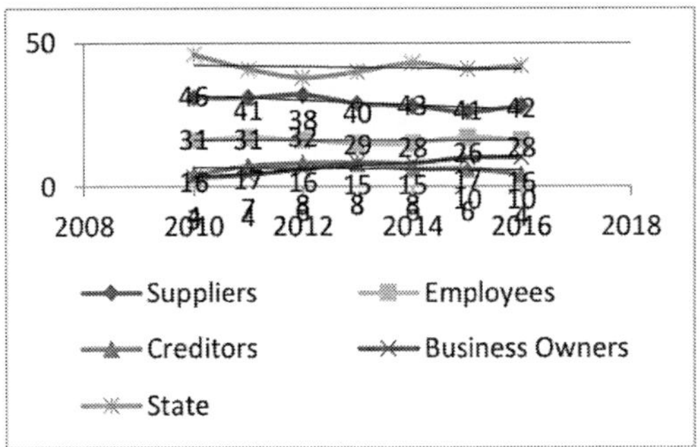

Fig. 2. Asymmetry distribution of income of stakeholders in the industrial park "Grabtsevo" 2010-2016, %

Discussions and Conclusions

Asymmetric dynamics is observed in the distribution of the additional effect when new park capacities are introduced. There is a decrease in the share of material costs, which is due to the effects of cooperation among participants in the industrial park and the share of creditors, showing a decrease in dependence on borrowed capital. If we consider the income of employees, then no significant changes in the structure of income distribution are observed. This fact is an example of the asymmetry of income distribution, the correction of which is possible by creating equalizing or incentive funds in the park. In particular, it is possible to propose the creation of educational centres, for example, a corporate university for the automotive industry.

Such a project will allow to direct additional "added value" to improve the quality of human capital of industrial park enterprises and ultimately significantly improve the efficiency of its work.

References

1. M. Mădălina, P.A. Constantin, Industrial parks and business incubators as clustering incipient forms, Quality - Access to Success **18**, 304-307 (2017).
2. M. Mani, S.M. Hosseini, T. Ramayah, Parks as business opportunities and development strategies, Business Strategy Series **13** (2), 96-101 (2012).
3. I. Siskos, L.N. Van Wassenhove, Synergy Management Services Companies: A New Business Model for Industrial Park Operators, Journal of Industrial Ecology **21** (4), 802-814 (2017).
4. A.E. Plakhin, E.S. Ogorodnikova, I.G. Generalov, Institutional conditions for the creation and methodological aspects of assessing the effectiveness of clusters: Russian and foreign experience, Azimuth of scientific research: economics and management **2** (19), vol. 6, 61-65 (2017).
5. A.E. Plakhin, Strategic management for the development of industrial park structures of cluster type: monograph, M: INFRA-M, 218 (2017).
6. O.N. Belenov, Yu.V. Shurchkova, T.Yu. Smolyaninova, Industrial parks in Russia and abroad: the experience of creation and development, Modern economics: problems and solutions **7,** 78-92 (2014).
7. Association of Industrial Parks (official site), industry review "Industrial parks of Russia", issue 4, (2016), available at: http://www.indparks.ru/ (accessed 04/06/2018).
8. V.O. Morozov, Determination of the strategic positions of the stakeholder-organization, Territory of new opportunities. Bulletin of the Vladivostok State University of Economics and Service **1** (19), 138-14 (2013).
9. E.K. Timofeeva, Corporate Social Responsibility: Sustainable Development and Interaction with Stakeholders, Ethnosociology and Interethnic Culture **4** (82), 44-49 (2015).
10. I.N. Tkachenko, Application of the methodology of corporate governance for the analysis of stakeholders in public-private partnership projects, Administrative Sciences in Modern Russia **1**, vol. 1, 40-44 (2014).
11. I.N. Tkachenko, K.K. Sivokoz, Usage of AGILE and SCRUM flexible technologies to manage project stakeholders, Manager **4**, 85-95 (2017).
12. R.S. Gondarev, N.V. Raznova, Building a Business Development Model Based on Integration of the Organizational Changes Concept and Stakeholder Management Tools, Economics and Management in the 21st Century: Development Trends **14**, 7-15 (2014).
13. R.V. Nuzhdin, A.N. Polozova, Business analysis of the opportunities of stakeholder-management in achieving positive synergy, Bulletin of the Voronezh State University of Engineering Technologies **4** (70) , 437-445 (2016).
14. A.M. Hein, M. Jankovic, W. Feng and others, Stakeholder power in industrial symbioses: A stakeholder value network approach, Journal of Cleaner Production **148**, 923-933 (2017).
15. M. Neubauer, C. Stary, S-BPM in the Production Industry: A Stakeholder Approach, 1-232 (2016).
16. S. Ismulyati, G. Ginting, Encouraging open innovation with open sources development for small medium industries (smis) productivity: Stakeholder approach, International Journal of Applied Business and Economic Research **13** (7), 5535-5548 (2015).
17. A.D. Artamonov, Investments and innovations - the basis of the future of the Kaluga region, Economics and Management **12**, 99-103 (2010).
18. T.A. Burtseva, A.A. Saveliev, Performance of creation of industrial parks in the regions of Russia, Regional economy and management: electronic scientific journal **1-2** (49), 234-244 (2017).
19. The system of professional analysis of markets and companies "Spark-Interfax". Available at: http://www.spark-interfax.ru/ (accessed 04/06/2018).

New directions in the development of neuromarketing and behavioral economics

Andrey Plakhin[1*], *Igor* Semenets[2], *Ekaterina* Ogorodnikova[1] and *Maria* Khudanina[1]

[1] Ural State University of Economics, Department of Management, 8th March str. 62, Ekaterinburg, Russian Federation
[2] Slavonic University of the Republic of Moldova, Department of Economics, Florilor str. 28/1, Chisinau, Moldova

Abstract. The article substantiates the relevance of neuromarketing studies based on an understanding of the basic functional mechanisms of the consumer's nervous system. Such studies include high-precision hardware testing of unconscious human responses to audio-visual or any other stimulus. Neuromarketing is considered as a symbiosis of separate disciplines: psychology and psychophysiology, traditional marketing and economics, neurobiology, econometrics, etc. In the work comparison with traditional marketing researches on such parameters as: objectivity of an estimation of results received during researches is made; also effectiveness of application in practices was investigated; sample for the study is selected; cost of research and interpretation of results are determined. The foreign experience of using neuromarketing technologies and behavioural economics on the example of such organizations as Singapore Airlines, Nike, Starbucks, automotive companies, universities, etc. was studied. Comparison with the Russian level of development of innovative marketing is made. Recommendations are given on the introduction of neuromarketing in the Russian automotive industry using the example of AvtoVAZ. The efficiency and value of carrying out neuromarketing research in the enterprise and the possible economic effect are calculated. The main tasks of modern higher education are marked and innovations in the learning process are proposed.

1 Introduction

In modern microeconomics, presumably, there is no marketer who does not set a goal to correlate the product with the buyer and give people what they really want most. Today it is not enough to understand and manage price policy: it is absolutely necessary to know the nature of a person, including his emotions. This is important, as the consumer becomes more demanding and sophisticated, complicating behaviour patterns. He understands that the product is not only a collection of consumer properties and services, but it is both emotions and feelings that appear when buying and using. It is the emotional processes and thoughts associated with branding that are the predominant reactions to product quality [1]. To understand such reactions, neuromarketing research is carried out, which today no longer has a creative solution, but a rational approach [2], [3].
In this article, the authors tried to characterize neuromarketing, its goals and objectives, as well as involvement in the behavioural economy. The advantages and disadvantages of technologies that help to compare neurovisualization with the tools of traditional marketing are considered. Further, the importance of categories such as emotions, motivational structure and cognitive distortions of a person is indicated. In the first part, you can find out what categories they are and why it is so important to consider them when forming a commercial offer. Then the analysis of other channels of influence on the consumer is presented: various forms of visual representation of a product or service, their tangible features, musical and aromatic support.
In the second part, foreign experience of introducing neuromarketing technologies, features and structure of innovations in European companies are considered. Comparative statistics are presented, which shows a relatively low level of development of marketing, including neuromarketing, innovations in Russia [4].

At the end of the article, a conclusion is drawn that neuroimaging in Russian organizations can be applied regardless of the scope and type of activity. In particular, in the automotive industry or education.

2 Theoretical aspects of neuromarketing

2.1 Content and objectives of neuromarketing research

Neuromarketing is a new direction in marketing, which uses brain research - neuroimaging - to study the behaviour of the buyer. In fact, this is one of the tools of the customer relationship management (CRM), built on the basis of understanding the fundamental functional mechanisms of the nervous system. It includes high-precision hardware testing of unconscious human responses to audio-visual or any other stimulus. Such irritant can be advertising, label or packaging, particularly

* Corresponding author: apla@usue.ru

© The Authors, published by EDP Sciences. This is an open access article distributed under the terms of the Creative Commons Attribution License 4.0 (http://creativecommons.org/licenses/by/4.0/).

a product or service. Knowledge of how irritants affect the main 4 channels of communication: hearing, sight, touch and smell through consciousness, make it possible to analyse the behaviour of consumers, change their thoughts even before understanding and making a choice [5].

At the same time, it should be noted that, according to the Russian Federal Law on Advertising No. 38-FZ of 13 March 2006, Article 9 [6] states: "Radio, television, video, audio and film products or other products shall not be used and distributed hidden advertising, that is, advertising, which has an impact on their consciousness not realized by consumers of advertising...". Therefore, neuromarketers should not pursue the goal of deceiving or imposing an unnecessary product on the buyer, but they can investigate their attention and reaction to the advertised product for the most effective promotion [7].

Thus, neuromarketing can be described as an interdisciplinary science, since the development of events uses already accumulated or acquired knowledge from psychology and psychophysiology, neurobiology, traditional marketing, neuroeconomics, economics and econometrics and other branches of knowledge.

The combination of all these sciences provides an opportunity for a fuller and deeper understanding and classification of the buyers value system, the structure of their motivation, correcting socially directed and life-affirming factors of self-perception and self-esteem.

All these factors can be expressed in the exact price positioning, or in the development of precisely those products that will be more in demand by the potential consumer (both because of the presence of specific physical characteristics in this product, and because of some emotional attractiveness not reducible to operationalization) or in the ability to communicate accurately with their customers, or in the optimal layout of retail space, or in maximizing product life cycle and brands themselves.

Therefore, we can identify the following main tasks of neuromarketing:

1. In the paper "Neuromarketing: Myths, Real Goals and Traps" [8] Oleg Klepnikov and Maria Chernova compare neuromarketers with artists: they combine a systematic approach and an uncontrollable subconscious process. In the case of neuromarketing, it is necessary to combine all theoretical and empirical knowledge of traditional economic sciences with applied sciences and brain research and consumer behaviour.
2. Neuromarketers in the course of experimental and scientific activity determine the reasons and sequence of making decisions about the purchase or refusal of a particular product.
3. Nomination of the most logical ways, or in some cases illogical, for promotion of goods in accordance with the true needs of the potential buyer, identified in the course of research. The development of such strategies for places of sale, which would fall into the focus of selective (selective) customer perception.

In this case, in our opinion, it is not necessary to resort to "ultrasound" brain activity every time: it is enough to interpret the buyer's reaction with the help of a number of the above-named disciplines and experiments that confirm this interpretation.

2.2 Advantages and disadvantages of neuromarketing: comparison with traditional methods

Analysis of literature sources allows to formulate the following main advantages of neuromarketing in comparison with traditional marketing research, and some disadvantages of neuromarketing technologies. Let us also consider which of the shortcomings are more or less justified.

The most important privilege is an objective evaluation of the subjective reaction of a person to a product or service, or their component. Possession of such information can give an advantage in the form of influence on purchasing behavior so that the costs of conducting neurovisualization studies will be justified and paid off by improving the product and increasing sales.

It has been proven that neuromarketing technologies provide marketers with information that can not be obtained through conventional marketing methods. According to experts, 95% of all thoughts occur in our subconscious, and it is these data that can not be fully obtained through surveys or focus groups, and some of them can be misinterpreted.

Let's look at an example. Baileys liqueur and Red Bull drink did not like the focus groups at the initial stages, and according to preliminary tests it was revealed that the subjects did not even agree to try the drinks more at all. However, the products were brought to the market, and now it is sold superbly throughout the world [9].

No less important is that the results of many neuromarketing research are now available to a wide range of people. This means the applicability of such data in building a development strategy and planning sales not only in large companies, but also in small and medium-sized businesses. It is the society that makes information open: we have moved to the digital level and every day we are creating a huge amount of archival data.

On the other hand, there arises the problem of protecting the privacy of clients. This, as well as the desire to keep secret the interpretation of the data, does not allow some companies to publish the results of the conducted studies. Nevertheless, although some of the information remains open, the tools for its analysis remain closed [10].

Another disadvantage is the inadequate sample in the studies. Back in 1979, the work of Amos Tversky and Daniel Kahneman "Prospect Theory: An Analysis of Decision under Risk " was published [11], in which it was demonstrated how strongly economic theories differ from people's behavior in real life. Today, a whole science has emerged about the description of the behavior of buyers - the behavioral economy - which is scientifically supported by neuromarketing. Although it was previously thought that the error factor in decision-making can not be predicted, according to the work of Tversky and Kahneman, economic rationality is systematically violated, but errors in decision making are not only widespread, but also predictable [12].

This is partly written about by Dan Ariely in his book "Predictable Irrationality" [12]. Therefore, it can be stated that this is one of the reasons why a large sample should not become a decisive factor when choosing a method of marketing research.
The second reason speaks for itself: the Nielsen company conducted an experiment that showed that the neurosearch for a sample of 100 people gives the same result as for a sample of 20 people [13].
Further, according to the generally accepted opinion, and this is the strongest "anti-stimulus", neuromarketing technologies are one of the most expensive methods of research.

Thus, the scientific direction in marketing analysis gives an answer to the question what the consumer really needs, and how to sell him a "useful" product. Neuromarketing can be characterized as an accessible method of research and promotion of goods, and effective and wearing an objective shade.

3 Russian and foreign experience of using neuromarketing technologies and behavioural economy

To date, there are some more than 250 research centres around the world dealing with neuromarketing. There is also an opportunity to pass special educational courses on neuromarketing in such universities as "Erasmus University of Rotterdam" (Netherlands), "INSEAD" (France), "Zeppelin" (Germany) and "Stanford" (USA) [14].
In the international arena, companies of various industries apply the behavioural economy in practice, including neuroimaging. Let's give some examples.
Sensual branding is used by "Singapore Airlines". The company enjoys a special colour scheme and the smell of "Stefan Floridian Waters", which is used by flight attendants. This smell is also impregnated with towels and other attributes on board.
"Starbucks" is asking for a high price not only for coffee, but also for a unique combination of interior design, exotic smells, background music from around the world.
Constant neuromarketing research is conducted by "Nike". The company is well-oriented in the market due to the behavioural economy. More than 10 applications that track the overall physical condition of a person, allow consumers to add all the necessary information to a well-known brand.
From the examples above, it can be seen that innovations can be introduced in various ways: with the use of special equipment, impact on sensorics or the development of applications; in all kinds of spheres and branches.
Most European companies already take into account behavioural patterns in their pricing policies and advertising, and traditional methods are inherent in organizations that have long held a stable position in the market due to the old technology of doing business or without growth tendencies.
Moreover, Russian companies are lagging behind the Western ones on the introduction of innovative technologies for about 2 to 3 years [15].

According to experts, in Russia people are more exposed to neuromarketing than Europeans. For example, from the use of sound, you can expect sales to increase by 10% in the West, while at the same time in the Russian company sales in this way are increased by a third [16].
According to studies of the European market of neuromarketing among countries in terms of the cost of neuromarketing activities confident leader is Germany, with a total amount of over 100 million USD per year. In Russia, the amount is about 3 times less (Fig. 1) [17].

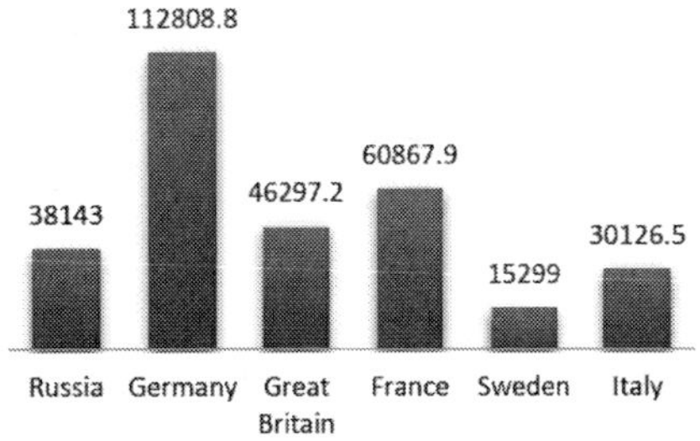

Fig. 1. Domestic expenditure on research and development, million US dollars

It is rather indicative also the analysis allowing to estimate a share of the companies on the countries which in one form or another apply marketing innovation tools in the activity (Fig. 2) [17].

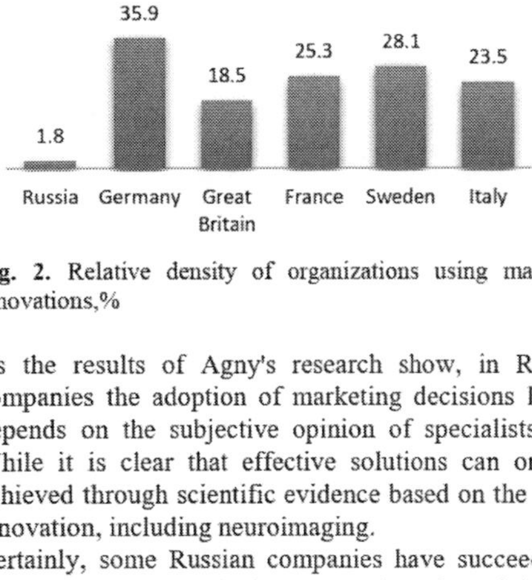

Fig. 2. Relative density of organizations using marketing innovations,%

As the results of Agny's research show, in Russian companies the adoption of marketing decisions largely depends on the subjective opinion of specialists [17]. While it is clear that effective solutions can only be achieved through scientific evidence based on the use of innovation, including neuroimaging.
Certainly, some Russian companies have succeeded in conducting neuromarketing research and applying the behavioural economy in practice. The brain laboratory, NeuroTrend, OLYMPIA and other research centres offer to conduct neuroimaging and find out the true consumer preferences. In the process, many professional skills and knowledge are integrated. For example, in St. Petersburg, the leading doctors - psychotherapists, candidates of philological sciences, astronomers and even mentalists united in the promotion company. All this contributes to the effective and reasonable development of the brand.
As for the direct application of the behavioural economy in practice, we can give an example of some companies that do not hide this information: Makfa, Uvelka, Borjomi, Cherkizovo, Faberlic, Sela, Dixi, leading Russian mobile operators, etc. As you can see, the largest companies preferred innovative technologies.

Organizations that are not ready to widely modify and promote business, realize the need for neuroeconomics in B2B. For example, using an outsourcing call centre.

Obviously, in Russia it is necessary to develop neuromarketing and to identify new directions, where the use of the behavioural economy is necessary. Since in the global science the share of the Russian Federation is no more than 2%, then efforts should focus on the introduction of innovations.

4 New directions of neuromarketing development and behavioural economy in Russia

4.1 Car production

The automotive industry in Russia is constantly changing and is facing a lot of different problems. One of the reasons for the backwardness of the industry is that enterprises pay insufficient attention to research and development. Let's consider in dynamics specific density of the organizations which are carrying out marketing innovations in the process of production of vehicles and equipment in their total number for 2009 - 2015 (Fig. 3) [8].

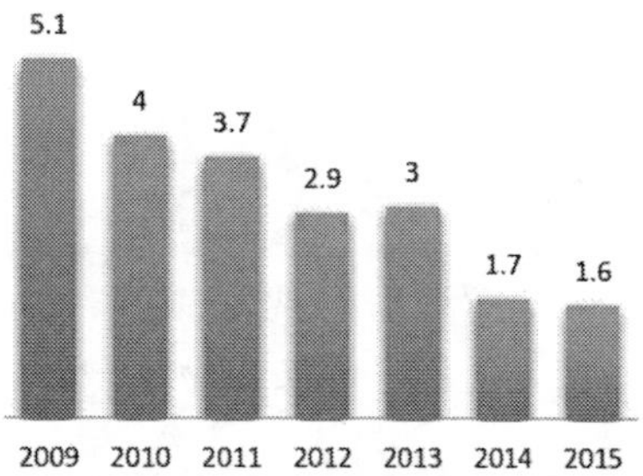

Fig. 3. Specific weight of organizations engaged in marketing innovations in the production of vehicles and equipment in the total number of organizations surveyed for 2009-2015,%

As can be seen from the graph, every year the number of organizations implementing innovative marketing solutions is decreasing.

At the same time, a number of European automakers, systematically conducting neuromarketing research, can observe the following dynamics of profit growth (Fig. 4).

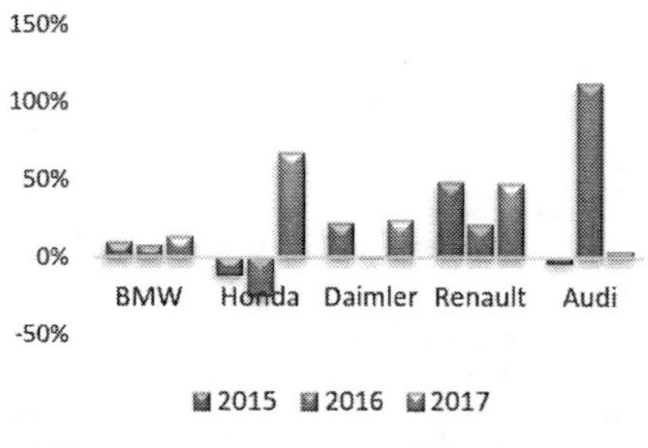

Fig. 4. The growth rate of net profit in 2015-2017 for companies with the practice of introducing neuromarketing

BMW uses the slogan created by neuromarketers: "The Ultimate Driving Machine", which focuses on the complex of pleasant experiences that the consumer will experience while driving. BMW, Honda and Audi also have their own fragrance, which is also an instrument of sensory influence on a person's consciousness. Daimler conducts research on the impact of the product on the consumer and develops special advertising campaigns; thanks to one of these advertising sales grew by 12%, etc.

In order to build a strategy for the development of production and sales, let us turn to the leading Eurasian automotive companies that have already introduced neuromarketing technologies, and advertising campaigns are built according to the canons of the behavioural economy.

Let's take AvtoVAZ as an example. AvtoVAZ is a Russian automotive company, the largest car manufacturer, with a market share of 19.5%. In 2017, 45% of cars were sold through state support programs. 17 new LADA auto centres were opened, and retail sales grew by 31% [18].

In 2016, the expenses of the auto concern for advertising increased 4-fold, and in 2017 they decreased 2-fold. Let's consider the marketing strategy of AvtoVAZ in more detail.

First, the Russian manufacturer is preparing in the next ten years the release of 12 completely new SUVs built on the platforms of the most popular cars.

Looking at this situation from the other side, it is appropriate to use the Zaltman metamorphism method (ZMET), widely used in neuromarketing and in the automotive industry, by the transnational automotive concern Daimler.

At the first stage of the target audience, it is suggested to select pictures reflecting the emotional state of each individual person. Pictures should encourage the creation of associative links with respect to any statements without direct binding. For example, pictures associated with investing or proper nutrition. Then the researcher conducts interviews with each participant within 1-2 hours.

In the process of research it is possible to reveal a latent reaction to a car, to uncover thoughts, feelings and desires about a product. These are deep processes of consciousness contained within the human mindset of people, and reactions to auditory, visual, kinaesthetic information or another stimulus. The result can be used when changing the shape, design, type, colour, quality of the upholstery material, etc.

The basis of the model is a symbiosis of neurobiology, psychoanalysis, psychology, art and theory to identify mental models that determine human behaviour and other sciences.

Thus, it is not necessary to change and adjust the autoconstruction completely to the European model, but to create one.

It is also appropriate to use such devices as EEG, eye-tracking, polygraph and others. For example, the South Korean automotive company Hyundai used the EEG to test its prototypes. The activity of the brain was measured for various design features and it was studied what kind of stimulation is likely to lead to a purchase. The results

of this study influenced the decision to change the exterior design of the cars themselves.
Secondly, an aggressive advertising campaign. Advertising war with AvtoVAZ is carried out by the largest car dealers Citroen, Audi, Volkswagen, Ford, Hyundai, who have long considered the behavioural economy.
Nevertheless, AvtoVAZ is also trying to affect emotion emotionally the consumer's consciousness using consonant associations in the advertising company to the Vesta model.
Since there are two ways of forming loyalty and "falling in love" with the consumer towards the manufacturer, it is probably worth paying attention not only to the emotional component, but also to cognitive distortions.
The idea is not to create and use distortions, but to develop critical thinking in favour of the Russian automotive industry.
For example, the binding heuristic (anchor effect). Consciously there is a binding to the European standards of automobile designs. People are ready to overpay for the brand and the framing it created. This is what we call "brand equity".
There are several ways to deal with a distorted consciousness: develop critical thinking and encourage the buyer to reassess values and views. The choice of a car can be caused by a "trap of incomplete information", when decision-making is based on a few unpleasant facts, while there are hundreds of them, and the consumer does not need to know anything more about them.
The easiest way to point people to their own mistakes is with the help of accurate measuring equipment: a lie detector, an MRI, etc.
Third, the development of e-commerce. With the advent of the new function of buying cars online, new prospects opened on the site of AvtoVAZ. Neuromarketing tools can be used in advertising texts and graphics, NLP; They can affect conversion and total sales.
We calculate the overall economic efficiency of the proposed innovations in AvtoVAZ.
First of all, we will list the activities and their cost characteristics for a sample of 200 to 1500 people, depending on the duration and quality of the event (Table 1).

Table 1. Proposed list of neuromarketing activities and their cost characteristics

Neuromarketing event	Cost characteristics
Conducting ZMET using additional equipment: traditional marketing tools and neuromarketing, including: EEG; MRI; polygraph; eye-tracker	From 60,940 thousand rubles to 110,750 thousand rubles
Attraction of highly qualified personnel for holding events	
Interpretation of resultsв	
Formation of a new heuristic binding using an advertising campaign	The same costs as with a traditional advertising campaign (from 1 to 2 billion rubles.)
Development of commercial trade	

Further, we will estimate the level of quality of neuromarketing research (Table 2).

Table 2. The system of indicators used in forecasting the correspondence of the results of the study with the objectives of the study

Index	Value and range of evaluation of the study quality
Deadlines	6 months
The value of the results (Table 3)	Average level. Opens prospects of the Russian car industry development, increase in sales volume, etc.
Level of results novelty	Getting new results
Actuality and relevance of the study	The high significance of the research, both from the point of view of the automotive industry, and in general of the economy in Russia
Possible scale of results implementation	Branch
Technical and methodological levels of the study	Full compliance of the technical means and techniques used with modern achievements of science
Volumes of results implementation	Partial implementation of the results, depending on the professionalism of the researchers interpreting the findings

Table 3. Evaluation of the effectiveness of information (by peer review).

$K = f(i; c; d) = i * c * d = 0{,}45$	
The average level of information value	
I	0,5
C	0,9
D	1

Based on the experience of foreign automotive companies and concerns, the introduction of neuromarketing research can increase profit growth up to 24% (Daimler). However, taking into account the average level of the value of information and the high cost of marketing research, experts predicted growth of the company until 2020, the average increase in profits of foreign companies, which arose through neuroimaging, the additional profits can be expected to increase to 10%.

Given the above, it seems appropriate to consider the use of neuromarketing technologies in AvtoVAZ, as many European companies have already successfully implemented these innovations and were able to increase sales.

4.2 Education and training of specialists

Even today in Russia you can take courses in neuromarketing in major universities, namely the Moscow State University. Lomonosov, MGIMO, REU and some others.
At the same time, it is quite obvious that it is necessary to modernize the education system in the sphere of marketing.
The main tasks of modernization include:
- the flexibility and effectiveness of vocational training programs;

- conformity of technologies used in training, the needs of modern economy;
- training of specialists who will later be engaged in research work.

Flexibility and effectiveness of training programs can be provided only through the introduction of modern disciplines. It is quite obvious that classical marketing does not allow us to examine the state of the market from all points of view. In this regard, it is advisable to allow students to assess the current status of enterprises with the help of tools of the behavioural economy, incl. neuroimaging.

Therefore, it is necessary to introduce neuromarketing tools into the learning process: methods, work with equipment, methods of interpretation of data; To improve skills and disseminate knowledge about innovations among students and listeners. Influence emotions and motivational structure.

Thus, it is possible to train more researchers and specialists in non-traditional marketing, to develop more capacious and effective training programs. This will increase the economy of the state as a whole, conduct effective sales and conduct successful advertising campaigns, reducing the budget for expenses to obtain unreliable information.

Neuromarketing technologies have the potential to conquer the Russian market. Neuroimaging makes it possible to deeply understand such questions as the motivational structure of the consumer, his values and gives opportunities for segmentation at a qualitatively new level. It has a wide scope for application both in manufacturing, retail trade, and education.

References

1. P. Bardeen, Hacking Marketing. 97-110 (2014).
2. R. Dooly, Neuromarketing. How to influence the subconsciousness of the consumer. 59-69 (2018).
3. D. Lewis, Neuromarketing in action. How to penetrate the buyer's brain. 40-51, 113-115 (2015).
4. E.V. Avdeychikova, S.P. Trubitsyna, V.V. Burgat, Innovations in trading activities: Russian and foreign experience. Moscow. 140-148 (2016).
5. E. Sternberg, Neurologics: What explains the strange things that we do unexpectedly for ourselves 302-313, 315-320 (2017).
6. On advertising: Federal Law No. 38-FZ of March 13, 2006 (as amended on December 31, 2017).
7. M. Akulich, Neuromarketing. *Publishing Solutions*. 30-31, 46-48, 143-150 (2018).
8. M.A. Chernova, O.E. Klepikov, Neuromarketing: myths, real goals and traps, available at: http://www.marketing.spb.ru/lib-around/science/neuromarketing.htm (accessed 07/04/2018).
9. RBC Magazine: research and market news. available at: http://marketing.rbc.ru/ (accessed 07/04/2018).
10. Lada.online, available at: http: //lada.online.ru/ (accessed 07/04/2018).
11. D. Kahneman, A. Tversky. Prospect Theory: An Analysis of Decision under Risk. Econometrica (1979).
12. D. Ariely, Predictable irrationality. 166-170, 172-180 (2008).
13. R. Thaler, New behavioral economics. 66-69 (2017).
14. W. Davis, The industry of happiness. 245-261 (2017).
15. Adview: Using neuromarketing to increase sales, available at: http: //adview.ru/ (accessed 07/04/2018).
16. Rosstat: state statistics, available at: http://gks.ru/ (accessed 07/04/2018).
17. Agny: the center of neuromarketing technologies, available at: http://agny.ru/ (accessed 07/04/2018).
18. LessWrong: a series of articles by Eliezer Yudkovsky on rational thinking, available at: http://lesswrong.ru/ (accessed 07/04/2018).

MATEC Web of Conferences **184**, 04024 (2018) https://doi.org/10.1051/matecconf/201818404024
Annual Session of Scientific Papers IMT ORADEA 2018

Solving the problem of logistics center location based on the AHP method

Sabina Kauf[1], *AgnieszkaTłuczak*[2,*]

[1]Opole University, Faculty of Economics, Ozimska str. 46a, 45-058 Opole
[2]Opole University, Faculty of Economics, Ozimska str. 46a, 45-058 Opole

Abstract. The aim of this paper is to present the method of locating logistic centres in a regular hierarchical logistics network. The problem is to determine the optimal location of logistic centres with given selection criteria. Among many methods, the authors will decide on the use of the AHP method, which is an effective tool for selecting a location.

1 Introduction

Logistics centers are an important part of the infrastructure of each country. They are an important factor of economic development, they influence on the flow of goods and they contribute to the increase of efficiency of supply chains. Global forecasts indicate that logistics centers, both industry, local, regional and international, determine global supply networks. This role is attributed to distribution centers, warehousing and distribution centers, trans-shipment terminals, central warehouses, warehouse bases or logistics platforms. Their aim is to create added value in the flow of goods, to alleviate or eliminate the nuisance of environmental logistic activities, as well as to support and initiate desired directions of business development, mainly in the area of logistics [1, 2].
Although the first centers were created more than thirty years ago, commonly accepted definition has not been developed to this day. This is due to the lack of absence of standard methodologies or decision criteria [3]. However, without going into a wider definition discourse (overview of ways of defining the logistics center are presented by Higgins, Ferguson and Kanaroglou [4]), we appeal only to the recognition of logistic centers in the infrastructural context, i.e. as transhipment points and intermodal terminals with the highest degree of complexity in which they combine streams of cargoes flowing through global supply chains. In this way, centers define, among others Tsamboulas and Dimitropoulos (1999) and Winkler and Seebacher (2011): "logistic interconnection points in a local and long-distance goods transport" [5, 6, 7, 8].
The construction of a logistics center is a strategic decision, mainly due to the irreversible effects resulting from the choice of location. This indicates the particular importance of the localization problem [9, 10]. Decisions in this respect are made in relation to a specific economic and spatial system, which creates specific consequences for actions aimed at the optimal deployment of newly planned facilities. Depending on the nature of the localized center and the type of logistics network to be part of, there is a different degree of complexity of the location problem [11, 12, 13]. As a consequence, various mathematical methods and formulas can be used to solve the problem. Bearing in mind the importance of the problem of the center's location and the multiplicity of possible methods in this study, the AHP method was presented to determine the location of the logistics center [14, 15, 16, 17].

2 Literature review

In the literature, a lot of space is devoted to the methods of location of logistic objects in a logistics network (including Owen, Daskin, 1998; Özcan, Celebi, Esnaf 2011; Van Thai, Grewal, 2005; Drezner, Hamacher, 2002; Church, Murray, 2009 Farahani, Hekmatfar 2009, Farahani, SteadiSeifi MS, Asgari 2010 [2, 18, 19, 20, 21, 22, 23, 24]). The classification of localization problems according to the criterion function, decision variables and system parameters is also considered. When solving the problem of object location in the network, various aspects should be taken into account, including its structure and forms of object connections. In this case, the problem is searching for answers, among others for questions such as: is each node assigned to one or many objects, is the connection node - node with the omission of the logistic object acceptable and whether the connection object - object are full or partial?
Choosing the right location of a logistics center is not easy [25]. It is usually taken after taking into account many criteria, such as: (1) technical (usable) - including road density indicator, efficient telecommunications network, number of potential contractors, storage infrastructure, intellectual resources; (2) economic - capital expenditures, annual operating costs, cost measures (eg. the cost of moving a unit of cargo) indicator of

[*] Corresponding author: atluczak@uni.opole.pl

© The Authors, published by EDP Sciences. This is an open access article distributed under the terms of the Creative Commons Attribution License 4.0 (http://creativecommons.org/licenses/by/4.0/).

development trends, share of international turnover, (3) hard to measure, including reliability, flexibility, expandability [26]. Brenke, based on research carried out in 2000, established the following hierarchy of location factors: (1) proximity to markets, (2) labor costs, (3) supply of qualified labor, (4) business support measures, (5) proximity to the motorway, (6) energy costs, (7) proximity to suppliers [27]. Mentzer (2008) points to three basic economic factors determining location decisions: land, labor and capital [28], and Mindur as criteria for the location of logistics centers, taking into account the possibility of adapting existing logistics infrastructure, proposes, among others [29] (1) the possibility and cost of buying land for the investment, (2) the cost of land treatment, (3) connection to the existing railway network, (4) the cost of building additional warehouses or office facilities, (5) the cost of building parking and parking yards, (6) the cost of supplying new facilities in the media, (7) the cost of preparing technical documentation [29]. The presented criteria show that searching for the location of a logistic center is usually reduced to determining the lowest costs, taking into account many places of obtaining raw materials and distribution places for finished products [26, 28, 29]. And that means the primacy of economic factors [1]. This fact was already cited in 1929 by Weber (1929) [26], writing: "[...] *uses and simple set-kilometers to optimize the location and set of customers*".

Location problems can be solved using various methods that can be classified according to [20, 30]:

• approaches to formulating the problem - mathematics rankings and programming, optimization problems; number of assessment criteria - simple, multi-criteria;

• approaches to risk management and its division - mathematical, expert estimates;

• topographic nature of the network - continuous and discrete location problems;

• number of objects to be considered for the location - one or many objects.

Farahani, SteadiSeifi and Asgari (2010) provide an overview of localization problems formulated as problems with multi-criteria decisions [20]. Klose and Drexl, on the other hand, focus on optimization methods, using mathematical programming [30, 31]. Daskin (1995) presented the method of distance minimization, also called the gravitational method [18]. To evaluate the location methods are used descriptive and analytical and descriptive, not requiring knowledge of the sophisticated mathematical methods. An interesting approach to solving the problem of the location of logistic facilities is to look for a variant of the number and location of these objects, for which the sum of transport and storage costs is minimal [25]. Noteworthy are also methods of linear optimization, static one-criterion methods with analytical and numerical implementation [27]. The multi-criteria optimization methods, e.g. AHP, TOPSIS, are also used to solve the problem of warehouse locations.

3 AHP method

Considering several potential locations of the warehouse, the decision maker is the most convenient to choose localization. During the analysis of individual variants, it may be guided by criteria depending on its individual preferences. In such situations, it is advisable to use multi-criteria analysis methods such as AHP. The AHP method has gained wide recognition in decision-making processes. Literature analysis points to several hundred articles using AHP / ANP methods in various fields and is used in many countries, e.g. in the United States, Turkey, China, Japan, Brazil, the Czech Republic and Poland, where it gains more and more supporters. In the area of transport, this method was used to assess transport systems [32, 33, 34], evaluation of the railway corridor in Memphis [35], analysis of multi-criteria problems in road transport [36], telematics technologies in the management of a fleet of hazardous materials [37, 38] and reverse logistics [39, 40]. Evaluation of large investment projects, evaluation of alternative fuels, selection of optimal transport policy reducing impact on climate change. These are just a few of the many uses of this method. The authors cite the following example of the use of AHP from the article by M. Berrittell, A. Certa, M. Enea and P. Zito "An Analytic Hierarchy process for the evaluation of transport policies to reduce climate change impacts" [41]. Analytic Hierarchy Process (AHP) is a commonly accepted and used tool for making complex decisions based on a large number of criteria. The basis of AHP is the hierarchical decomposition of the assessment criteria. The hierarchy of validity in the AHP method has a predetermined structure. First, the purpose of the decision-making process is chosen, then the evaluation criteria and the solution variants are set [41, 42, 43, 44]. In AHP, the decision maker has an impact on the entire process - AHP is a method that allows you to choose the best solution from among many variants. The decision-maker assesses possible variants of choice in terms of specific criteria for the possibility of assessing the value of these criteria in terms of the objective according to their knowledge and experience-based recognition [5, 42, 45].

The use of the AHP method follows the following steps:

1. Construction of a pairwise comparison matrix for all criteria in pairs and matrices enabling the comparison of potential locations in pairs in relation to all criteria. Assessment criteria by pair comparison is made by the decision maker according to the scheme [5, 45, 46]:

1 - equal importance (both factors contribute equally to achieving the goal);

3 - a small advantage (judgments and experience slightly prefer one factor over another);

5 - strong advantage (judgment and strong experience put one factor above the other);

7 - very strong advantage (one factor is strongly put above the other, practice confirms this);

9 - absolute advantage (the advantage of one factor over the other is absolute and confirmed in the highest degree).

Intermediate values 2, 4, 6 and 8 are assigned in the case of difficulty in the classification of the result and when we think that it lies between the characteristics specified in the classification [47, 48, 49].

This step completes the formation of a matrix A= $[a_{ij}]_{nxn}$, there is a $\frac{n(n-1)}{2}$ in possibility of comparisons. Characteristic for these matrices is that a_{ii} = 1 and [17, 18, 19, 28, 29, 33, 50, 51]:

$$a_{ji} = \frac{1}{a_{ij}} \quad i\neq j,\ i,j = 1,2,\ldots,n \qquad (1),$$

The next step of the AHP method is the normalization of matrix elements A = $[a_{ij}]$ according to the formula [5, 40, 41, 45, 46, 47]:

$$\hat{a}_{ij} = \frac{a_{ij}}{\sum_{i=1}^{n} a_{ij}},\ j=1, 2, \ldots, n \qquad (2)$$

Based on the normalized matrix elements the individual preferences indexes are determined referred (forming a column vector C=$[c_i]$ $(i = 1,2,\ldots,n)$ according to the formula [41, 43, 51]:

$$c_i = \frac{\sum_{j=1}^{n} \widehat{a_{ij}}}{n} \ (i = 1,2,\ldots,n) \qquad (3)$$

which indicate the place of the ith variant in the ranking - the higher the value, the higher the ranking. The individual ranking for criteria is determined in the same way.

The final step of the presented method is the determination of vector C=$[c_i]$ (i=1, 2, ...,m), multicriteria preference indexes, whose coordinates are calculated according to the formula [41, 42, 51]:

$$c_i = \sum_{k=1}^{n} c_k c_i \qquad (4)$$

The c_i values determine the position of i-th variant in ranking (the higher the value, the higher the rank).

When many pairwise comparisons are performed, some inconsistencies may typically arise.

Analysis of matrix consistency is made through appropriate transformations of the matrix's own vectors [43, 48, 49, 52], thus calculating the Consistency Ratio (CR) according to formula (2):

$$CR = \frac{\lambda_{max} - n}{r(n-1)} \qquad (5)$$

where: λ_{max} - maximum value of the own vector of the preference matrix of n, n - number of compared characteristics. It is assumed that the coefficient of consistency is accepted (comparison of characteristics is consistent) when the CR value is less than or equal to 10%. Then rankings of particular characteristics are calculated for the first and subsequent criteria respectively. Subsequently, rankings for variants (sub-criteria) are calculated.

4 Application the AHP method in solving the problem of the location of a logistics center

The first step, regardless of the chosen method, is the analysis of various conditions related to a given investment process and participating in it entities, aimed at indicating appropriate variants and decision-making factors. Catalog of example criteria and sub-criteria, determining the implementation of the localization of warehouse, has been presented in figure 1. It should be noted that these factors are only of a nature exemplary, developed on the basis of selected thematic publications. In a real situation, the choice of individual factors should be justified by the results of relevant analyzes (including applicable legal regulations) made by an expert group. It is worth noting that the accuracy of the work analytical results translate into the reliability of results obtained by the AHP method.

Due to the theoretical nature of the problem decision, only selected elements of the AHP algorithm are indicated. The choice of location should start with the construction of a hierarchical decision model (Fig.1).

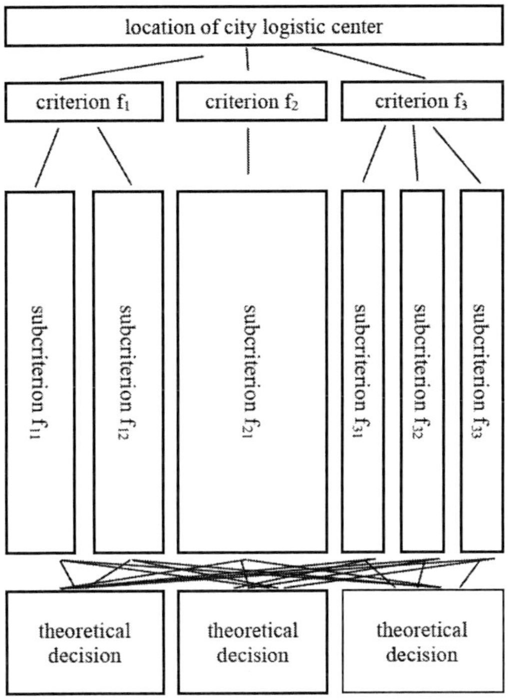

Fig. 1. Hierarchical model (decomposition of the decision problem)

Suppose we are considering a discrete problem of multi-criteria decision making, in which six decision-making variants are assessed on the basis of three criteria. The first two criteria are unmeasurable, while the third, maximized criterion is known to the decision-maker as real numbers (tab. 1). At the beginning the pairwise comparison matrix should be determined (tab. 2). Using the Saaty method, the c scale vector is as follow (0.076; 0.657; 0.267)

Table 1. Value of criterion f_3

	a^1	a^2	a^3	a^4	a^5	a^6
$f_3(a^i)$	19	35	24	29	24	28

Table 2. The pairwise comparison matrix

	f_1	f_2	f_3
f_1	1,000	0,143	0,333
f_2	5,000	1,000	7,000
f_3	5,000	0,333	1,000
s_j	11,00	1,48	8,33

In the end of first step of the AHP method the "own value" or "characteristic value" of matrix should be found, in this research it is λ_{max}=3,690. The value of the modified CR consistency measure will be calculated according to formula 5, and it is equal CR=0,595, which indicates the existence of sufficient consistency of assessments.
Next the decisions-maker has to compare of individuals decision-making variants with respect to each criterion f_1 and f_2. Using the Saaty method, standardized matrices of variants were created (tab. 3, tab. 5), relative to the first and second criteria f_1i f_2 on the basis of which the c_1 and a c_2 scale vectors were found (tab. 4).

Table 3. Normalized matrix for criterion f_1

	a^1	a^2	a^3	a^4	a^5	a^6
a^1	0,548	0,459	0,404	0,329	0,283	0,250
a^2	0,078	0,153	0,173	0,164	0,189	0,286
a^3	0,061	0,031	0,058	0,066	0,047	0,179
a^4	0,183	0,306	0,288	0,329	0,425	0,143
a^5	0,069	0,031	0,058	0,066	0,047	0,107
a^6	0,061	0,022	0,019	0,047	0,009	0,036

The values of the modified consistency measure CR for the comparison of options with respect to the first criterion f_1 is 0.003, while for the second criterion f_2 it is 0.228. This indicates that there is sufficient consistency of assessments. Due to the fact that the value of the third criterion f_3 are real numbers, the scale vector c_3 was determined as the normalized values of the third criterion in table 6.
In next step C matrix was determined, columns in this matrix are scale vectors for all criteria. The C matrix is presented in table 7.
The product of the matrix C and the vector of the scale c (formula 4) forms the final vector of the scale c on the basis of which the ranking of decision variants was created. The ranking of variants is as follow: a_3=0,063; a_5=0,076; a_2=0,134; a_4=0,172; a_1=0,260; a_6=0,295.

Table 4. Scale vectors c_1, c_2

c_1	0,379	0,174	0,073	0,279	0,063	0,032
c_2	0,304	0,094	0,025	0,156	0,047	0,373

Table 5. Normalized matrix for criterion f_2

	a^1	a^2	a^3	a^4	a^5	a^6
a^1	0,168	0,271	0,300	0,517	0,356	0,210
a^2	0,056	0,054	0,100	0,034	0,238	0,084
a^3	0,024	0,011	0,033	0,015	0,010	0,060
a^4	0,056	0,271	0,167	0,103	0,198	0,140
a^5	0,024	0,014	0,100	0,021	0,040	0,084
a^6	0,672	0,379	0,300	0,310	0,158	0,421

Table 6. Normalized value of criterion f_3, scale vector c_3

	$f_3(a^i)$	c_3
a^1	19	0,119
a^2	35	0,220
a^3	24	0,151
a^4	29	0,182
a^5	24	0,151
a^6	28	0,176

.**Table 7.** C matrix – scale vectors for all criteria

	c_1	c_2	c_3
a^1	0,379	0,304	0,119
a^2	0,174	0,094	0,220
a^3	0,073	0,025	0,151
a^4	0,279	0,156	0,182
a^5	0,063	0,047	0,151
a^6	0,032	0,373	0,176

5 Conclusion

The construction of logistic centers is a big financial challenge and creates a lot of problems, like most business-investments construction. However, correctly located on the basis of preliminary research and simulations, it can certainly be a source of income not only in financial terms but also in terms of environmental protection.
Activities related to the location of logistic centers and their very construction should be implemented at a priority pace. Summing up, the location, of such a huge undertaking as, the logistics center should be preceded by a series of detailed studies and simulations that would clearly indicate and justify the construction of such a facility.

Among many methods that can help in choosing the right location, the AHP method should be indicated. Based on the research presented in this article, it can be concluded that AHP method is proper to choose location for logistics center. The analyzes carried out were aimed at selecting, from the point of view of commercial activities, the location of a new logistics center. The proposed procedure of selecting a location allows to meet the criterion of transport accessibility and equal attractiveness for as many potential clients as possible.

References

1. F. Chen, Z. Drezner, J. K. Ryan, Simchi-Levi D., *Quantifying the Bullwhip Effect in a Simple Supply Chain: The Impact of Rorecasting, Lead Times, and Information*. "Management Science", Vol. 46, No 3, (2000)
2. R.L. Church, A.T. Murray, *Business Site Selection, Location Analysis and GIS*, John Wiley and Sons, New York (2009)
3. D. A. Tsamboulas, I. Dimitropoulos, *Appraisal of investments in European nodal centres for goods – freight villages: A comparative analysis*, Transportation 26(4):381-398 (1999)
4. C. D. Higgins, M. Ferguson P. S. Kanaroglou, *Varieties of logistics centres: developing a standardized typology and hierarchy*, The Transportation Research Board Volume 2288 (2012)
5. S. Kauf, A. Tłuczak, *Optimization of logistics decisions (Optymalizacja decyzji logistycznych)*, Warszawa, Difin, 30-50 (2016)
6. H. Winkler, G. Seebacher, *Management of freight villages: findings from an exploratory study in Germany*, International Journal of Logistics: Research and Applications, Vol. 14, No. 4, pp. 271-283 (2011)
7. C. Rao, M. Goh, Y. Zhao, J. Zheng, *Location selection of city logistics centers under sustainability*, Transportation Research Part D, Vol.36, p.29(16) (2015)
8. Y. Kayikci, *A conceptual model for intermodal freight logistics centre location decisions*, Procedia - Social and Behavioral Sciences, Vol.2(3), pp.6297-6311 (2010)
9. L. Yang, X. Ji, Z. Gao, K. Li, *Logistics distribution centers location problem and algorithm under fuzzy environment*, Journal of Computational and Applied Mathematics, Vol.208(2), pp.303-315 (2007)
10. O. Tkach, I. Voloshchuk, *Regional logistics center: formation and functioning specific*, Aktualnì Problemi Rozvitku Ekonomìki Regìonu, Vol.2(13), pp.127-134 (2013)
11. G. Peng, Z. Yang, Z. Feng, H. Liman, M. Chenhao, *Design and implementation of urban logistics distribution center location optimizati on system*, Journal of Hebei University of Science and Technology, Vol.38(1), pp.19-2501 (2017)
12. V. Klapita, Z. Švecová, *Logistics centers location*, Transport, Vol.21(1), p.48-52 (2006)
13. H. Liu, L. Liu, *Multi-Objective Location Model Research and Application in the City Emergency Logistics Based on Different Product Materials*, Applied Mechanics and Materials, Vol.63-64, p.277 (2011)
14. L. Li, J. Zhang, *Logistics Distribution Center Location Optimization Model An Example Study*, MATEC Web of Conferences 100, 02026 (2017)
15. H. Xiang, H. Xiao, Y. Wuwei, *Research optimization on logistics distribution center location based on adaptive particle swarm algorithm*, Optik - International Journal for Light and Electron Optics, Vol.127(20), pp.8443-8450 (2016)
16. B. Wang, H. Xiong, Ch. Jiang, *A Multicriteria Decision Making Approach Based on Fuzzy Theory and Credibility Mechanism for Logistics Center Location Selection*, The Scientific World Journal, Vol.2014 (2014)
17. M. Făgărăhan, C. Ciprian, *Logistics center location - selection using multicriteria decision making*, Annals of the Oradea University: Fascicle Management and Technological Engineering (2015)
18. M. S. Daskin, *Network and discrete location: models, algorithms, and applications*, John Wiley & Sons, New York (1995)
19. Z. Drezner, H. Hamacher, *Facility Location: Applications and Theory*, Springer, Berlin (2002)
20. R. Farahani, M. S. Steadi Seifi, N. Asgari, *Multiple criteria facility location problems: A survey*. "Applied Mathematical Modelling", 34, 1689-1709 (2010)
21. R. Farahani, M. Hekmatfar, (Eds.) *Facility Location: Concepts, Models, Algorithms and Case Studies*, Heidelberg, Physica-Verlag (2009)
22. S. Owen, M. Daskin, *Strategic facility location: A review*, European Journal of Operational Research, 111, 423 – 447 (1998)
23. T. Özcan, N. Celebi, S. Esnaf, *Comparative analysis of multi-criteria decision making methodologies and implementation of a warehouse location selection problem*, Expert Systems with Applications, Vol. 38, pp. 9773 – 9779 (2011)
24. V. Van Thai, D. Grewal, *Selecting the Location of Distribution Centre in Logistics Operations: A Conceptual Framework and Case Study*, Asia Pacific Journal of Marketing and Logistics, Vol. 17, No. 3, pp. 3-24 (2005)
25. A. Rushton, P. Croucher, P. Baker, *The Handbook of Logistics and Distribution Management*. Third ed., Kogan Page, London, (2006)
26. A. Weber, *Theory of the location of industries*. University of Chicago Press, pp. 41-42 (1929)
27. K. Brenke, *How East German Companies Estimate Location Conditions in Their Region - Survey Results (Wie ostdeutsche Unternehmen die Standortbedingungen in ihrer Region einschätzen – Ergebnisse einer Umfrage)* Deutsches Institut für Wirtschaftsforschung, Berlin (2003)
28. J. T. Mentzer, 7 keys to facility location, "Supply Chain Management Review" Vol. 12, Iss. 5, pp. 25-31 (2008)
29. L. Mindur, Methodology of locating and shaping logistic centers in Poland (*Metodyka lokalizacji i kształtowania centrów logistycznych w Polsce)*. Kolejowa Oficyna Wydawnicza, Warszawa (2000

30. A. Klose, A. Drexl, *Facility location models for distribution system design,* European Journal of Operational Research, 162, 1, 4-29 (2005)
31. R. Ballou, *Business Logistics: Supply Chain Management*, 5th edition. Pearson Education, p. 551 (2004)
32. I. A. Kaysi, M. A. Abdul-Malak, *Decision Structuring and Robustness Analysis in Selecting Urban Transit Alternatives*; Journal of Public Transportation 4(1), 73-102 (2001)
33. T.L. Saaty, *Fundamentals of the Analytic Network Process. Multiple Networks with Benefits, Costs, Opportunities and Risks*, Journal of Systems Science and Systems Engineering, published at Tsinghua University, Beijing, (to be published) 2004, Vol. 13, No. 3, September (2004)
34. W. Chmielorz, M. Zborowski, *The use of the AHP / ANP method in the confrontational method of model design of information systems,* (Wykorzystanie metody AHP/ANP w konfrontacyjnej metodzie projektowania wzorcowego systemów informatycznych), *Economic IT* (Informatyka Ekonomiczna), 2014, Issue 31, pp.195-211 (2014)
35. R. Banai, *Fuzziness in Geographical Information Systems: contributions from the Analytic Hierarchy Process*, International Journal of Geographical Information Systems. 7, 4, pp. 315- 329 (1993)
36. J. Żak, Multi-criteria decision support in road transport (*Wielokryterialne wspomaganie decyzji w transporcie drogowym).* Poznań: Wydawnictwo Politechniki Poznańskiej (2005)
37. K. G. Zografos, *Analytic Hierarchy Process (AHP): It's application in FTS business model assessment*, Athens University of Economics and Business Transportation Systems and Logistics Laboratory. Conference Proceedings, Cremona (2005)
38. K. G. Zografos, K.N Androutsopoulos, *A Decision Support System for Integrated Hazardous Materials Routing and Emergency Response Decisions*, Transportation Research Part C, Vol. 16(6), pp. 684-703. (2008)
39. M. Bouzon, K. Govindan, R. C. Taboada, L. M. S. Campos, *Identification and analysis of reverse logistics barriers using fuzzy Delphi method and AHP*, Resources, Conservation & Recycling, March-April 2016, Vol.108, pp.182-197 (2016)
40. Ch. Prakash, M. K. Paura, *Integration of AHP-TOPSIS method for prioritizing the solutions of reverse logistics adoption to overcome its barriers under fuzzy environment,* Journal of Manufacturing Systems, October 2015, Vol.37, pp.599-615 (2015)
41. M. Berrittella, A. Certa, M. Enea, P. Zito, *An Analytic hierarchy process for the evaluation of transport policies to reduce climate change impacts* http://www.feem.it/Feem/Pub /Publications/WPapers/default.htm, january 2007.
42. M. Brunnelli, *Introduction to the Analytic Hierarchy Proces,* Springer (2015)
43. L. G. Vargas, *An overview of the Analytic Hierarchy Process and its applications*, European Journal of Operational Research, Vol. 48, 2–8 (1990)
44. T. L. Saaty, *The Analytic Hierarchy Process*, McGraw-Hill, New York (1980)
45. W. J. Hurley, *The analytic hierarchy process: a note on an approach to sensitivity which preserves rank order*, Computers & Operations Research, Vol. 28, 185–188 (2001)
46. Y. Li, X. Liu, Y. Chen, *Selection of logistics center location using Axiomatic Fuzzy Set and TOPSIS methodology in logistics management,* Expert Systems With Applications, June, Vol.38(6), pp.7901-7908 (2011)
47. J. Perez, *Some Comments on Saaty's AHP*, Management Science, 1995, Vol.41(6), p.1091-1095 (1995)
48. T. L. Saaty, *Decision making for leaders; The Analytic Hierarchy Process for decisions in a complex world*, RWS Publications, Pittsburgh, (2001)
49. T. L. Saaty, *The Analytic Hierarchy Process*, RWS Publications, Pittsburgh, (1998)
50. T. L. Saaty, *Decision making with Dependence and Feedback: The Analtic Network Process*, RWS Publications, Pittsburgh, P.A. (2001)
51. M. Andriejova, Z. Kimakova, D. Marasova, *Using ahp method at the determination of the optimal selection criteria of conveyor belts,* Annals of the Faculty of Engineering Hunedoara, 2013, Vol.11(2), pp.155-160 (2013)
52. D. Božanić, D. Pamučar, S. Karović, *Use of the fuzzy AHP - MABAC hybrid model in ranking potential locations for preparing laying-up positions* , Vojnotehnički Glasnik, 01 July 2016, Vol.64(3), pp.705-729 (2016)

Allocation of logistic risk-investment in public-private-partnership – use of fuzzy TOPSIS method

Sabina Kauf[1], *Agnieszka* Tłuczak[2,*]

[1]Opole University, Faculty of Economics, Ozimska str. 46a, 45-058 Opole
[2]Opole University, Faculty of Economics, Ozimska str. 46a, 45-058 Opole

Abstract. Appropriate risk assessment related to the project, and then its distribution between partners is the core of cooperation in the framework of public-private partnership (PPP). Risk analysis for investment project in the PPP formula includes: preliminary identification of risks associated with the investment, preliminary assessment and forecast of probability of occurrence of risks in the recommended option of investment implementation, initial analysis of the distribution of risks between a public and private entity depending on the legal and organizational model adopted, including the determination of the appropriate separation between the parties to the contract. The article presents the possibility of using multicriteria methods, in particular fuzzy TOPSIS, for risk assessment and selection of the best contractor for a given project under PPP.

1 Introduction

In modern world, participation of infrastructure, financial investment in private-public partnership is still growing. More and more often investments for building logistic centers are realized in that way. As an experiences from Germany and Italy show, this model of investment realization was, strong support of the government and budget money is one of main features of success of their building. The most often in the basis of a decision about letting private capital to public. Money was an idea that this solution would positively influence not only for time of investment realization but also on savings of budget money and keeping deadlines with work. It should be mentioned that act about PPP in Poland doesn't put a duty of estimation and analysis of risk in enterprises [1]. Necessary condition to join common investment realization is only analysis of its effectiveness and potential dangers. According to art. 11 act. 1 from act about analysis of investment danger it should make it possible to identify risks, describe their optimal allocation and influence on level of public debt and public financial deficit. Optimal risk allocation in the context of projects under public-private partnership should be based on the transfer of particular types of risk. This risk should be transferred for this participant who has the greatest opportunities and opportunities to effectively reduce a given threat [2]. However, it should be remembered, that this is quite abstract concept. In practice, many types of risks are strongly correlated with each other. It is therefore difficult to separate them from each other and achieve a theoretical optimum by allocating to different participants. Nevertheless risk and rules of its share should be an every time point of a discussion [2, 3].

Creating logistic Centre is an expensive investment for many years with lots of risk, which can dishearten potential investors to participation in the project. Those risks are mostly visible in case when initiators want to realism an event in traditional way. Public Sector is worried about difficulties with gaining seers for created logistic infrastructure that contains many point elements. Private sector is afraid of big costs of investment, which should realism not only commercial goals but also social ones, that are not in a circle of private units' interest. Although banks are worried that public sector would not gain enough own money, which could make a guarantee of enterprises being real. In this context private-public-partnership looks to be perfect solution. Creating a partnership is equivalent to effort of allocation of financial benefits from enterprises and risk of its realization [4]. Although all risk is connected to social acceptance of investment realization through external units, is in public sector side. As a law says, responsibility for completing a task is on her. This make asymmetric share of risk. It's similar in case of curve effect experience: external contractor can use it without any constraints, public paymaster only on certain area and in reference to certain tasks [3]. Finding proper proportion between risk and expected benefits is one of key elements of success of public-private partnership. Having following under account in this article we will indicate to method of estimation of infrastructure investment risk realized for public-private-partnership with use of multi criteria method.

Taking the above into consideration in this study, the method of assessing the risk of infrastructural investments realized within public-private-partnership using multi-criteria methods will be indicated [5].

* Corresponding author: atluczak@uni.opole.pl

© The Authors, published by EDP Sciences. This is an open access article distributed under the terms of the Creative Commons Attribution License 4.0 (http://creativecommons.org/licenses/by/4.0/).

2 Literature review – defining infra-structure risk

The extent of risk issues has exceeded all limits. The risk is connected with the most fundamental aspects of psychology, mathematics, statistics and history. Dice and a roulette wheel, as well as the stock market or the bond market are natural laboratories for risk research, because they are very easily quantifiable and their language is the language of numbers.

The word "risk" comes from the old Italian *risicare*, which means "dare". In this sense, the risk is a choice rather than an inevitable destiny. Risk is present in all areas of human activity when it is unable to control or accurately predict the future. The terms risk and uncertainty often co-exist or are even identified, but they do not mean the same [6, 7, 8].

Certainty is no doubt. One of the meanings of this term is "state no doubt ", which may be a sufficient definition. Antonym of certainty there is uncertainty, which is defined as "doubting the ability to predict the effects of current activities". It is obvious that this term describes the state mind. We are dealing with uncertainty when we see that the effects of those present actions cannot be known with absolute certainty. Risk is the potential variability of events [9]. It is present almost in everything that people do. If this is the case, the result of the action cannot be precisely predicted and uncertainty is felt. Man is exposed to risk wherever activity generates potential benefits or losses that cannot be predicted. Risk is an objective concept and therefore it can be measured.

The first concept of economic risk theory was published by A. H. Willett in 1901. A. H. Willett noticed the diversity of the meaning of the term commonly used in everyday life and made an attempt to unify it. According to him, the risk is the state of the environment, which should be related to the degree of uncertainty of occurrence, and not to the likelihood of materialization [10]. According to this theory, the risk understood as the state of the environment is objective and correlates with subjective uncertainty, and the impression whether the illusion of randomness is only an effect imperfections of human knowledge about objective laws that govern processes in the outside world [8, 11]. The second concept was presented by F. H. Knight in 1921. The main the purpose of F. Knight's work was to quantify the uncertainty. In this concept, the risk is measurable uncertainty. Uncertainty which cannot be measured, it is the uncertainty *sensu stricto*, which Knight he called environment unmeasurable uncertainty. F. H. Knight "explicitly stated, that uncertainty was confused with risk, but only risk it is measurable, while uncertainty is unmeasurable [12].

The third basic concept was developed by the Commission for US Insurance Matters in 1966. According to her the risk is uncertainty about a specific event in the conditions two or more possibilities. In this sense, it is a measurable uncertainty, whether the intended objective of the action will be achieved [7, 13].

The extent of understanding risk and uncertainty means that they are not they are perceived in a homogeneous and unambiguous manner. Their definitions evolving with the development of science are still used interchangeably. In order to distinguish the two concepts, one should pay attention to the objective one and the measurable nature of risk as well as subjective and irrational nature uncertainty [10]. Dependencies and differences between these two concepts illustrate table 1.

Table 1. Relationships and differences between the economic understanding of risk and uncertainty.

Risk	Uncertainty
Realized by business entities part of the uncertainty.	Accidentality resulting from the event from unforeseen behaviors or course of phenomena.
Rational, objective, empirical character	Immeasurable, subjective nature, subconscious.
A posteriori - it refers to cognition, resulting from experience, reasoning about something by way of induction	A priori - it refers to cognition pure, undistorted knowledge from other similar experiences.
An uncertain event, but it's known probability of occurrence	Unpredictable or unlikely happening.
An uncertain event whose effects materialization can be previously estimated.	An accidental event effects of materialization are not possible for an earlier estimate.
Measurable uncertainty of achievement assumed goals	Unpredictability of failure to achieve assumed goals.
Connects with the bonus for taking.	No bonus for acting in a state of uncertainty.

Idea of 'risk' in PPP context never gain one definition and its measure still gives some problems. In the literature concept of 'risk' has many meanings. Its connected to realization of plans to keep existing status quo. It's a consequence of ma king decision because every single one of them is connected to necessity of forecasting future midsection of surroundings .Because exact prediction of the future isn't possible, those action may not bring desired effects [11, 14]]. Oxford English Dictionary defines risk as "*chance or possibility of danger, loss, injury, etc.*". Although more of definitions concentrates on uncertainty and possibility of an event happening. In this context we can define risk as uncertainty connected to future midsections of surrounding or effects of decision. Moreover many people says that uncertainty and risk aren't identical, saying that risk is objectively co-dependent from uncertainty at it is subjective. In this point of view risk is objectified uncertainty of unwanted event happening [11]. In this context risk is defined by OECD: „risk *as the probability that the actual outcome (for example, sales, costs, and profits) will deviate from the expected outcome*" [15]. According to Furnell risk is "*the chance of an event occurring which would cause actual project circumstances to differ from those assumed when forecasting project benefit and costs*"[16]. *According to*

 https://doi.org/10.1051/matecconf/201818404025

UK's Orange Book, risk is "*uncertainty of outcome, whether positive opportunity or negative threat, of actions and events*"[17]. This concept includes possibility of an event with its consequences. That's why "*The risk has to be assessed in respect of the combination of the likelihood of something happening, and the impact which arises if it does actually happen. Risk management includes identifying and assessing risks and then responding to them*" [17]. The definitions quoted above indicate that the risk in relation to investment projects is most often understood as the possibility or the danger of achieving real results that differ from the planned ones. And although these deviations can be both beneficial and unfavourable, in practice the term risk is used in a negative sense.

Noteworthy is the fact that the term uncertainty often appears when defining risk. In the literature on the subject, one can find a position that these concepts are identical [18, 19]. Most authors emphasize that despite the close relationship between these terms, they are not synonymous [6, 13]. Uncertainty it is a broader concept, but the risk is a derivative of uncertainty and is measurable. One of the first researchers in this issue was F.H. Knight, who announced the theory of uncertainty, measurable and unmeasurable.

The one we cannot measure is uncertainty of *sense stricte* [20], and basic criteria that differs uncertainty and risk is possibility to use theory of probability. Uncertainty we can meet when possibility of ginning wanted effects is unknown. Risk occurs when effect of decision we can describe according to mathematical, statistical or estimative possibility. In situation of uncertainty of we cannot predict the decision or describe it with help of any measures. In both cases there are deviations from predicted results [5]. Not to starting wide terminological discourse over risk and uncertainty we can accept that risk is connected to not completing expectations about certain event and it's described by a probability of an event by this formula:

$$\sum_{j=1}^{n} p_j = 1$$

It should be remembered that necessary condition of risk is existence of uncertainty, which means that every single risk situation is uncertain, although there can occur uncertainty without risk. From long-term investment realization projects point of view a prior meaning has correct management of risk. That's why government should concentrate on those risks that can bring big losses. From public sector point of view, about investment project success is connected to those features: social results, loss of reputation or image but also risk of breaking the partnership with private investor. Moreover there is a risk every time with lateness of realization of the project and unpredictable increase of costs, there than planned. That's why good allocation of risk and its control of level of constructing conditions of agreement with private investor is important. Concentration of public sector mainly on those risk that determine share of roles and obligations of partners is necessary. We should we conscious that in All time of project investors (private and public) are coping with uncertainty, which makes a catalogue of risk actually unlimited and even the most detailed descriptions may not be able to include them. Nevertheless thank to long duration of agreement and its big content, risk should be underestimated [2, 21]. The risk assessment is the key, being a process used to identify and determine the probability of destructive factors and their consequences. Kaplan and Garrick (1981) presented the risk assessment as a set of three questions to be answered [22]:

(1) What can go wrong ?,
(2) What is the likelihood?
(3) What are the consequences?

Usually in estimation of infrastructure risk investment we can use numerical and quality methods [23, 7]. Among most used numerical methods we should pay attention of ranking method, that uses probability of occur of event and its consequences in a scale from 1 to 5 [1, 4], sensitivity analysis and Monte Carlo simulation [24, 25]. Usually is estimation of risk in projects realized for PPP, its method TOPSIS finds its use , mainly because projects realized because of multi-level partnerships and every single one of them has their own risk features and sources of success. TOPSIS method allows to overall and synthetic analysis of estimation of risk from perspective of all life of the project.

3 TOPSIS method

In recent years, it can be find a wide range of applications of the TOPSIS method. TOPSIS was extended by Chen in 2000 to fuzzy environments, which used a fuzzy linguistic value as a substitute for the directly given crisp value in the grade assessment. This method is used, among others, for the choice of supplier [16, 26, 27], selection of projects and risk assessment [28], evaluation of websites [28], selection of the object [29], etc. It is also possible to find hybrid approaches. In [30], it's proposed an integration of the Fuzzy TOPSIS method of linear programming for credit risk assessment. Sun and others [31] propose creating decision models and rankings based on integration of Fuzzy AHP and Fuzzy TOPSIS methods.

The TOPSIS method (Technique for Order Preference using Similarity to Ideal Solution) is a useful tool that is used to rank variants (alternatives, criteria) during the decision making process. The factor that distinguishes this method is the use of a measure of relative distance to the best solution, which is a model (ideal) and the worst solution, which is anti-ideal [29].

In decision-making analysis it could be found two problems:

1. imprecise defining the decision issue
2. imprecise defining the decision-maker preference.

The reason of this situation can be the lack of precise determination of criteria's level of realization by analyzed variants and uncertainty of consequence of some variants of choice [24].

The main difference between classical TOPSIS and Fuzzy TOPSIS is concerned to definition of decision matrix. In fuzzy TOPSIS method this matrix is defined by triangular

 https://doi.org/10.1051/matecconf/201818404025

fuzzy numbers. The fuzzy decision matrix choose the appropriate linguistic variables for the alternatives with respect to criteria. Decision matrix is made by fuzzy assessment of decision variants due to further criteria's. These assessments are given by formula [28]:

$$\tilde{f}_k(a^j) = (l_{jk}; m_{jk}; u_{jk}) \quad (1)$$

and this are triangular positive fuzzy numbers where: l_{jk} - pessimistic assessment of j-th variant based on the k-th criterion; m_{jk} – most expected assessment of j-th variant based on the k-th criterion; u_{jk} – optimistic assessment of j-th variant based on the k-th criterion; $l_{jk}\geq 0$; $m_{jk}\geq 0$; $u_{jk}\geq 0$.

In the process of risk assessment of logistic investments, the following stages of calculation can be distinguished on the Fuzzy TOPSIS method [26, 28, 29, 31]:

normalization of the fuzzy decision matrix $\tilde{N} = [\tilde{z}_{jk}]$ where:

$$\tilde{z}_{jk} = \left(\frac{l_{jk}}{maxu_{ik}}; \frac{m_{jk}}{maxu_{ik}}; \frac{u_{jk}}{maxu_{ik}}\right) \quad (2)$$

- benefit criteria;

$$\tilde{z}_{jk} = \left(\frac{minu_{ik}}{u_{jk}}; \frac{minu_{ik}}{m_{jk}}; \frac{minu_{ik}}{l_{jk}}\right) \quad (3)$$

– cost criteria.

1. Calculate the weighted normalized fuzzy decision matrix:

$$\tilde{V} = [\tilde{r}_{jk}] \quad (4)$$

where:

$$\tilde{r}_{jk} = w_j\tilde{z}_{ij} \quad (5)$$

for i=1,…,m and j=1,…,m.

2. Identify the fuzzy positive ideal solution (FPIS)

$$(f_k(\tilde{a}^+) = \tilde{v}_k^+ = max\tilde{r}_{ik}) \quad (6)$$

and fuzzy negative ideal solution (FNIS)

$$(f_k(\tilde{a}^-) = \tilde{v}_k^- = min\tilde{r}_{ik}) \quad (7)$$

for i=1,…,m and k=1,…,n.

3. Calculate the distance of each alternative from ideal and negative ideal:

$$d_j^+ = \sum_{k=1}^{n} d(\tilde{r}_{jk}, \tilde{v}_k^+) \quad (8)$$

and

$$d_j^- = \sum_{k=1}^{n} d(\tilde{r}_{jk}, \tilde{v}_k^-) \quad (9)$$

for j=1,…,m.

4. The closeness coefficient S_j represents the distances to fuzzy positive ideal solution and the fuzzy negative ideal solution:

$$S_j = \frac{d_j^-}{d_j^+ + d_j^-} \quad (10)$$

for j=1,…,m.

5. The alternative with highest closeness coefficient represents the best alternative and is closest to the Fuzzy Positive Ideal Solution and farthest from Fuzzy Negative Ideal Solution.

4 Application of fuzzy TOPSIS method to assessment of risk of logistics investment in PPP projects

We are considering a discrete problem of multi-criteria decision making. The decision problem is: how to choose a contractor who will be realize logistic project PPP. We consider three criteria which represents the possibility of risk: f_1-the risk of delay in completing construction works, f_2- the risk associated with a change in the technology of providing services, f_3-the risk of inability to deliver the contracted number of services. Six of contractor will be assessed due to these criteria.

Table 2. Fuzzy decision matrix

			Offer of a logistics operator					
			a_1	a_2	a_3	a_4	a_5	a_6
criteria	f_1	l	1	3	8	1	1	5
		m	5	5	9	2	2	9
		u	7	6	12	7	5	10
	f_2	l	3	2	1	4	3	3
		m	8	3	2	8	5	6
		u	10	5	4	9	5	6
	f_3	l	5	4	4	3	7	1
		m	7	7	5	5	9	3
		u	9	10	12	10	12	4

The profitable criteria will be considered, and each of them will be defined by a triangular fuzzy number. Assessment of decision variants against the criteria forming the decision making matrix is provided by the table 2.

The next stage involves steps as outlined in Section 3. Construct a normalized fuzzy decision matrix as shown in Table 3. The step of data normalization is necessary to overcome differences between the units. Normalization also enables valuation measure in the same range of values which is usually between zero and one. In the range system, 1 represents the highest value in upward movement where 0 represents the lowest value.

Table 3. Fuzzy normalized decision matrix for the selection

			Offer of a logistics operator					
			a_1	a_2	a_3	a_4	a_5	a_6
criteria	f_1	l	0,08	0,25	0,67	0,08	0,08	0,42
		m	0,42	0,42	0,75	0,17	0,17	0,75
		u	0,58	0,5	1	0,58	0,42	0,83
	f_2	l	0,3	0,2	0,1	0,4	0,3	0,3
		m	0,8	0,3	0,2	0,8	0,5	0,6
		u	1	0,5	0,4	0,9	0,5	0,6
	f_3	l	0,42	0,33	0,33	0,25	0,58	0,08
		m	0,58	0,58	0,42	0,42	0,75	0,25
		u	0,75	0,83	1	0,83	1	0,33

The next step is constructing a weighted normalized fuzzy decision matrix. For this purpose, the weight vector was as follow (0,35; 0,25; 0,4). To get multi criteria index, data from each of the criteria need to be aggregated. A lot of various methods can be implemented to do this. It this paper weighted mean was used, which can be calculated in two ways: by using arithmetic and geometric mean. Index based on arithmetic mean is generally more popular because of easily understood and implemented.

The next step is to get the fuzzy positive ideal solutions (FPIS) and fuzzy negative ideal solutions (FNIS). After getting the ideal solutions, the next step is to calculate the distance of the alternatives from (FPIS) and (FNIS) using equation (8) and (9), respectively (tab. 6). The last step is to find the closeness coefficient of each alternative, which is calculated according to formula (10): a_1=0,57; a_2=0,42; a_3=0,63; a_4=0,47; a_5=0,49; a_6=0,40.

Based on table 6, it can be seen that the coefficients of the third alternative is the highest value followed by first and the fifth. Based on the coefficients, an alternative to selecting the firms listed on third should be the first choice, followed by selecting the firms listed first and the last one is to select a firm on the sixth.

Table 4. Weighted normalized fuzzy decision matrix

			Offer of a logistics operator					
			a_1	a_2	a_3	a_4	a_5	a_6
criteria	f_1	l	0,03	0,09	0,23	0,03	0,03	0,15
		m	0,15	0,15	0,26	0,06	0,06	0,26
		u	0,2	0,18	0,35	0,2	0,15	0,29
	f_2	l	0,08	0,05	0,03	0,1	0,08	0,08
		m	0,2	0,08	0,05	0,2	0,13	0,15
		u	0,25	0,13	0,1	0,23	0,13	0,15
	f_3	l	0,17	0,13	0,13	0,1	0,23	0,03
		m	0,23	0,23	0,17	0,17	0,3	0,1
		u	0,3	0,33	0,4	0,33	0,4	0,13

In essence, the greater the value of the coefficient indicates the priorities of the decision to be made. This method not only allows the decision maker to provide the rank of each alternative, but also shows the degree of likelihood of alternative selection.

Table 5. Fuzzy positive ideal solutions (FPIS) and fuzzy negative ideal solutions (FNIS)

	Criteria								
	f_1			f_2			f_3		
	l	m	u	l	m	u	l	m	u
FPS	0,23	0,26	0,35	0,10	0,20	0,25	0,23	0,30	0,40
FNS	0,03	0,06	0,15	0,03	0,05	0,10	0,03	0,10	0,13

5 Conclusion

Ensuring the highest quality of logistic services provided is an important element of establishing cooperation in the field of private-public-partnership. Choosing the best logistics service provider can be a choice in the field of multicriteria methods. Many decisions are burdened with risk; the aim is to minimize it.

The article presents a risk assessment procedure for the implementation of PPP projects. The subject of the analysis was potential offers of logistic operators evaluated in terms of the occurrence of risk in the implementation of projects. These offers were evaluated on the basis of three criteria: the risk of the completion of contracted works, the risk of a delay in the delivery of the services. The paper presents the application of Fuzzy TOPSIS method. The main advantage of this method is the ability to quantify the hierarchy of decision variants and the quantitative risk assessment

References

1. A. Akintoye, T. Craig, F. Eamon, *Risk analysis and management of Private Finance Initiative Projects*, "Engineering, construction and Management", 5/1, 9-21 (1998)

2. J. F. Al-Bahar, *Risk Management in Construction Projects: a systematic Analytical Approach for Contractors*, PhD Thesis, University of California, Berkeley, 6-27 (1989)
3. C. Andersen, *Competition and Cooperation in Public Service Production, [w:] Public Management - A New Generation in Business and Practice (Wettbewerb und Kooperation in der öffentlichen Dienstleistungsproduktion,* [w:] *Publik management – Eine neue Generation in Wirtschaft und Praxis),* Hrsg. K. Birkholz, C. Mass, P. Siebert, Potsdam, 311 (2006)
4. D. Asenova, M. Beck, *A financial perspective on risk management in public-private partnership,* [in:] Public-Private Partnerships: Managing Risks and Opportunities. Ezekiel Ch., Alasdair F. (ed.) Blackwell Science (2006)
5. S. Kauf, A. Tłuczak, Optimization of logistics decisions (*Optymalziacja decyzji logistycznych),* Warszawa, Difin, 30-50, (2016)
6. C. A. Williams Jr., R. M. *Heines, Risk Management and Insurance*, ed. 3. McGraw-Hill Publishing Co., New York, 67 (1989)
7. M. Power, *The risk management of everything,* The Journal of Risk Finance, Vol. 5 Issue: 3, 58-65 (2004)
8. A. H. Willett, *The Economic Theory of Risk and Insurant,* Columbia University Studies in Political Science, Nr. 2 (1901)
9. P. L. Bernstein, *Against Gods. Unusual risk history (Przeciw Bogom. Niezwykłe dzieje ryzyka),* WIG-Press, Warszawa, 19–20 (1990)
10. T. Vasari, *Risk, risk perception, risk management – a review of the literature,* Public Finance Quarterly, vol. 1, (2015)
11. A. H. Willet, *The Economic Theory of Risk Insurance*, The University of Pennsylvania Press, Philadelphia, 6 (1951)
12. F. H. Knight, Risk, *Uncertainty and Profit*, University of Bosto Press, Boston (1921)
13. D. Dziawgo, Credit-rating. *Risk and bonds on the international financial market* (*Credit-rating. Ryzyko i obligacje na międzynarodowym rynku finansowym),* PWN, Warszawa, 13 (1998)
14. *Risk Allocation and Contractual Issues,* http://www.dtf.vic.gov.au/ Publications/Infrastructure-Delivery-publicati, (14.02.2018)
15. OECD, *Public-Private Partnerships: In the Pursuit of Risk-Sharing and Value for Money*, 13 (2008)
16. C. H. Furnell, *Risk Identification and Risk Allocation in Project Finance Transactions*, paper presented at the Faculty of Law, The University of Melbourne, cited in Partnerships Victoria (2001)
17. *The Orange Book Management of Risk - Principles and Concepts*, London 2004
18. G. J. Alexander, W.F. Sharpe, J.V. Bailey, *Fundamentals of Investment,* Prentice-Hall, Englewood Cliffs (1993)
19. W. F. Samuelson, S.G. Marks, *Managerial economics (Ekonomia menedżerska),* Warszawa (2009)
20. E. H. Knight, *Risk, Uncertainty and profit*, Cosimo Classics, New York, 233 (2005)
21. PMBOK, A *Guide to the Project Management Body of Knowledge,* Project Management Institute, PMBOK, US, 111-122 (2004)
22. G.J. Alexander, W.F. Sharpe, J.V. Bailey, *Fundamentals of Investment,* Prentice-Hall, Englewood Cliffs (1993)
23. Ch. Ezekiel, F. Alasdair, *A construction perspective on risk management in public-private partnership,* [in:] *Public-Private Partnerships: Managing Risks and Opportunities*. Ezekiel Ch., Alasdair F. (ed.) Blackwell Science (2003)
24. D. Grimsey, M.K. Lewis, *Evaluating the risks of public private partnerships for infrastructure projects,* "International Journal of Project Management", no 20, 107-118 (2000)
25. K. P. Dey, O.S. Ogunlana, *Selection and application of risk management tools and techniques for build-operate-transfer projects*, Industrial Management & Data System, no. 4, 334-346 (2004)
26. S. Gupta, A. Gupta, *Fuzzy multi criteria decision making approach for vendor evaluation in a supply chain,* Interscience Management Review (IMR), Vol. 2, Issue 3, 10–16 (2012)
27. F. E. Boran, S. Genc, M. Kurt, D. Akay, *A multi-criteria intuitionistic fuzzy group decision making for supplier selection with TOPSIS method*, Expert Systems with Applications: An International Journal archive, Vol. 36 Issue 8, 11363-11368 (2008)
28. G. Kabir, M. A. Hasin, *Comparative analysis of topsis and fuzzy topsis for the evaluation of travel website service quality*, International Journal for Quality, Vol. 6, No. 3, 169–185 (2013)
29. I. Ertuğrul, N. Karakaşoğlu, *Comparison of fuzzy AHP and fuzzy TOPSIS methods for facility location selection,* Int J Adv Manuf Technol, 39, 783–795 (2008)
30. Y. Tansel İç, *Development of a credit limit allocation model for banks using an integrated Fuzzy TOPSIS and linear programming,* Expert Systems with Applications: An International Journal, Vol. 39, Issue 5, 5309–5316 (2012)
31. C. C. Sun, *A performance evaluation model by integrating fuzzy AHP and fuzzy TOPSIS methods,* Expert Systems with Applications, vol. 37, no. 12, 7745–7754 (2010)

AUTHOR INDEX

AUTHOR INDEX

EDP Sciences – Web of Conferences
17, Avenue du Hoggar
Parc d'Activité de Courtabouf
BP 112
F-91944 Les Ulis Cedex A
France

ISBN 978-1-5108-6858-8

9 781510 868588

CURRAN ASSOCIATES INC.
proceedings
.com

Asia Conference on Mechanical Engineering and Aerospace Engineering (MEAE 2018)

MATEC Web of Conferences Volume 198 (2018)

Wuhan, China
23 - 25 June 2018

edp sciences